D0742499

A *Flora* OF THE *Marshes* OF CALIFORNIA

HERBERT L. MASON

A *Flora*

OF THE

Marshes

OF CALIFORNIA

UNIVERSITY OF CALIFORNIA PRESS

BERKELEY AND LOS ANGELES

UNIVERSITY OF CALIFORNIA PRESS, BERKELEY AND LOS ANGELES

UNIVERSITY OF CALIFORNIA PRESS, LTD., LONDON, ENGLAND

© 1957, BY

THE REGENTS OF THE UNIVERSITY OF CALIFORNIA

LIBRARY OF CONGRESS CATALOG CARD NUMBER: 57–7960

MANUFACTURED IN THE UNITED STATES OF AMERICA

DESIGNED BY JOHN B. GOETZ

4 5 6 7 8 9 0

Preface

The flora of California, in the past one hundred and fifty years of exploration and study, has become reasonably well known. There are, however, certain areas of the state which, because of difficulties of access or for other reasons, have not been fully investigated botanically, and thus their floras are still not completely known. One such flora is that found in the marshlands scattered over the state.

From the current manuals the erroneous impression might easily be gained that floras of the marshes and ponds are best developed in the southern part of California, since many, if not most, of the marshland species are recorded only for that region. Actually, however, these floras occur widely over the state, and, with few exceptions, the species populations are also widespread. The degree of differentiation in their floristics seems to be more the result of edaphic conditions and conditions associated with altitude than of climatic differences associated with geographic position. For instance, the salt marshes and the fresh-water marshes have their characteristic floras, as do also the lowland and the upland marshes, and these floras vary only slightly throughout the length of California.

The present state of incomplete knowledge concerning the marsh floras has made difficult the routine identification of plants by the nonbotanist and by those working in the applied fields of botany such as game management. Work-

ers naturally have assumed that published floras were based on relatively complete knowledge. Consequently, to find unrecorded plants or those growing far
out of their published range has been a disconcerting experience.

The drainage of many of the marshlands and their conversion to agriculture
have led inevitably to problems of game management. In addition, the depre-
dations of waterfowl on agricultural crops have brought the interests of the
sportsman into conflict with those of the agriculturalist. Demands have been
made that something be done to alleviate this situation. These problems have
required that the game manager assess his area with a view to establishing facts
about the natural food supply for waterfowl, and to creating conditions that
would increase and improve this food supply.

To provide a basis for the study of this problem, a practical manual for the
identification of the marsh plants has been needed, and the present work is
designed to meet that need. The work was sponsored under the joint auspices
of the California Department of Fish and Game, the Federal Fish and Wildlife
Service of the United States Department of the Interior, and the Department
of Botany of the University of California. Funds were provided for a field survey through a Pittman-Robertson Grant for five years. This study was made
possible with funds of federal aid in Wildlife Restoration Project W20-R. Delay
in its approval until late in the first season reduced the period of effectiveness of
the grant to four years of field work. To all who have had a part in the ful-
fillment of the grant I here express my gratitude.

The stated aim of this sponsored project was to make a floristic survey of the
feeding and resting areas along the major flyways of aquatic birds, namely the
coastal area, the Central Valley, the northeastern plateau, the valleys east of
the crest of the Sierra Nevada, and the areas to the south through the Imperial
Valley. It became necessary, however, to do much more extensive field work
than this, in order that the taxonomic problems raised in the survey might be
solved. These problems pertained chiefly to determining the morphological
limits of species and to seeking data with respect to their geographic range. As
a result, field work was done in irrigation ditches, rice fields, streams, ponds,
and lakes, as well as in swamps, marshes, and bogs over much of the area of
California, with the exception of the high montane regions. However, when
the species known from these montane habitats were necessary for identification
or helpful in facilitating it, they were included in the taxonomic treatment.

Since the work is designed chiefly to be a source for identification of the
materials encountered in studies of the food habits of waterfowl, we allowed
ourselves some latitude with respect to the inclusion of plants other than marsh
plants. Many waterfowl graze over the floodplains adjacent to waterways.
Because the seed of plants on these floodplains finds its way into resting and
feeding ponds formed by the fall and winter rains, these plants comprise a conspicuous element in waterfowl food. Many of the common members of this
marginal flora are therefore included.

The plan of the work called for: first, bringing the taxonomic material abreast of published research on each group; second, evaluating the taxonomic problems evident in the local plant materials on which no monograph is available, or on which monographs are available but are at variance with the evidence from the local flora; third, giving the geographic and ecological range of the species; and fourth, illustrating the more important species.

The field work was executed by sampling; in no locality was it possible to make an exhaustive survey. This limitation was dictated by the law of diminishing returns and the availability of time. This method has, I think, permitted the accrual of an adequate knowledge of the more common species, as well as of many less common species, to make the finished work very useful throughout the state.

It is impossible to anticipate the fluctuating recurrence of a species that is locally common in an area for a year or two and then disappears there, perhaps to reappear at a later time—a phenomenon observed for several species in the course of this study. Hence, further range extensions are bound to be reported in the future. Also, some plants not observed in this survey are likely to be discovered, and still others have yet to become established in California.

The taxonomic presentation is documented by herbarium vouchers, the first set of which is deposited in the Herbarium of the University of California, at Berkeley. The illustrations were drawn mainly from fresh material, although some, such as those of the grasses and sedges, had to be done from dried material during winter months. Most of the illustrations are by Mary Barnas Pomeroy and may be identified by the initials "CMB"; a few are by Patricia Verret Reinholtz, Robert Miller, and Emily Patterson Reid. In addition, Mrs. Reid is responsible for those illustrations employed in conjunction with the keys and glossary. Since the excellence of the drawings will provide a conspicuous element of the utility of the work, I wish to express my appreciation to these artists.

The field assistants included Dr. Verne Grant, Dr. Irving Schneider, Mr. Malcolm Nobs, Mr. S. Galen Smith, and Mr. G. Thomas Robbins. In addition, Dr. Grant aided in the preliminary draft of the Cyperaceae, Mr. Nobs aided in the preliminary draft of the grasses and the genus *Polygonum,* Mr. Smith helped to make the preliminary draft of the Lemnaceae, and Mr. Robbins contributed to the final draft of several groups, as well as to the solution of many taxonomic, bibliographic, and nomenclatural problems. Many changes were made in the final draft, and for these I take full responsibility, but I wish to express appreciation for the groundwork laid by these assistants.

Finally, I take this opportunity to express my great appreciation to my colleagues Miss Annetta Carter, Miss Francia Chisaki, Dr. Helen Sharsmith, and Miss Isabelle Tavares, who have assisted both with botanical problems and with the manuscript in so many ways that it is impossible even to mention them all. Miss Carter, Miss Chisaki, Miss Tavares, and Dr. Sharsmith edited the manuscript in its last draft before it was submitted to the University of California Press, and in so doing were able to make significant contributions by discovering deficiencies and errors and by rounding out the botanical treatment. Miss Chisaki, in addition, prepared the treatment of the Boraginaceae and organized the preliminary manuscript for several other groups. Dr. Sharsmith organized the preliminary manuscript in many groups, often so efficiently that very little further attention was required. She also critically reviewed the manuscript for the glossary and made many helpful suggestions by way of added information, judgment, and opinions. Dr. John Ingram, Mr. Howard Arnott, and Mr. S. Galen Smith also gave much time and helpful effort during the editorial phase of the work. For their continuing loyalty and sympathetic understanding I am truly grateful. As I review their part in this work I am filled with deep humility when I inscribe my name as author.

HERBERT L. MASON

Contents

INTRODUCTION   1

TAXONOMY   15

    GENERAL KEY TO THE FAMILIES   15

    TAXONOMIC TREATMENT   27

GLOSSARY   839

INDEX   859

Introduction

"A flora of the marshes of California," the subject of this work, carries with it an assumption that there is a unity embracing floristics and ecology which expresses itself in communities of plants in such a manner that one can easily circumscribe his research problem and know precisely what to include and what to exclude. As research proceeds, however, it soon becomes clear that this is true only to a very limited extent and pertains to only a few facets of the problem. This is not peculiar to the California problem, for one need only turn to the literature to realize that all research in this field has artificial boundaries. To set seemingly natural ecological boundaries is often to cut artificially across the problem and exclude much of it. For instance, if we include only the wholly aquatic species, as was done by Muenscher,[1] we eliminate the very important amphibious species. If we include also the amphibious species, we are drawn immediately up on the shore, where the naturalness of communities and the overlapping of their species lead us farther and farther from water.

I have therefore arbitrarily included all the species known by me to occur typically or commonly on wet lands or in water. A few others are included when they are deemed useful to the problems of identification. In addition, because the work was financed to aid in the management of waterfowl, I have

---

[1] W. C. Muenscher, *Aquatic Plants of the United States,* Comstock Publishing Company, Inc., Ithaca, New York, 1944.

1

included plants marginal to these habitats that are known to enter significantly into the food habits of waterfowl. It should be clear that the main body of the species included, namely those growing on wet lands or in water, are representative of a rather wide range of habitats and habitat conditions.

In dealing with plants from such diverse habitats, no single term suffices to express the relationship between floristics and environment. Yet this work is intended as a unit, and there is need for a term which refers to all the plants characteristic of habitats with an abundance of water, as contrasted with the remainder of the plants in the California flora. For this purpose I shall use the terms "marsh plant" and "marsh flora." Of all the terms available, that referring to marshes is perhaps broadest in scope, both ecologically and geographically. It will, I hope, be clear from the context when I am using the term in this broad sense and when I am referring to marshes in particular. In the latter case I shall usually employ such terms as "salt marsh," "alkali marsh," or "freshwater marsh." For other special habitats, such as "vernal pool," "lake," "riparian habitat," and so on, the usual term will be employed.

THE HABITATS REPRESENTED

I shall break with precedent and define my problems in terms of habitat rather than plant community. If in any given aquatic or marsh habitat in a given particular locality a definitive community exists, it does not follow that in some other locality having what may appear to be the same habitat a similar community will be found. Indeed, in our problem variation in floristics is so great, and the consequent lack of regularity in the recurrence of similar communities of plants is so pronounced, that the concept of the community as an indicator of a given set of environmental conditions does not hold. Variation in the floristic composition of a community is not so much successional, in the sense of moving toward a climax, as it is fluctuant with respect to the presence and absence of conspicuous species populations.

In addition, a classification of communities which reflects the relation of organisms to their environment demands that every included individual contribute to the characterization or community concept through characters common to all individuals. Since these common characters, if such exist, are not morphological in nature, they must be physiological—hence, experiment will be required to discover them. Any other basis of classification would result either in a specialized system (for example, a system which is based upon some dominant feature that is useful to the classifier, but which in its empirical basis would ignore most of the other kinds of associated plants), or it would result in a system that would subordinate all other aspects of the community to those features which convey a sense of homogeneity to the classifier. Whatever is subordinated contributes no more to the meaning of any statement relative to the community than if it were ignored. Classifications like these "blanket in" all associated individuals along with those significant to the classification. Beyond their association, such individuals contribute nothing to the meaning of "the community" in discourse. Such classifications, although they may have other important uses, are of very limited value to ecology.

Since habitat is the locus of the selective operation that prevails between the plant and its environment, the kinds of habitats are meaningful to the phenomenon of natural selection. Like most complex natural phenomena, habitats vary and intergrade, but there are certain prevailing basic features that will serve the purposes of classification. At least we have characters to employ that are common to the phenomena being classified and are thus meaningful to the objectives of the classification, for the features of the habitat lend themselves to a descriptive nomenclature referable to common experience.

A nomenclature founded on habitat type permits us to speak from any base we may choose. We may speak in broad terms such as "aquatic habitat," or we may wish to speak of "ponds," "lakes," "streams," or "vernal pools." We may speak broadly of marshes, or we may qualify by specifying salt marshes, fresh-water marshes, or the artificial marshes such as the cultivated rice field. Each such term refers to a habitat type. These may be aggregated into organizational levels, so that any such level may be spoken of unitarily as a "habitat type."

To one familiar with such habitats it will be clear that it is not always possible to refer to any particular situation by a single unitary term. This results from intergradation between habitats, and nomenclature will then be facilitated by use of combined terms such as "marshy lake" or "boggy swamp." Lakes merge with marshes; riparian situations merge with lacustrine sites; bogs, marshes, and swamps merge with one another, making it impossible to classify intergrading forms unitarily. Even habitat extremes such as salt marshes and fresh-water marshes intergrade both geographically and seasonally in the lowlands about San Francisco Bay. The Suisun marsh is filled mostly with salt water in the summer and fall when local streams are dry, and mostly with fresh water in the winter and spring when there is heavy runoff from all streams.

A similar situation prevails with respect to both the vegetation and the flora. In these marsh habitats we have plant species with members that range from submersed or floating aquatics to amphibious, riparian, and strand plants rooted in wet mud; or we may have species composed of plants that begin their life as aquatics and end it as extremely spiny xerophytes. Examples of the former are *Elatine californica* and *Polygonum amphibium;* examples of the latter are *Navarretia Jepsonii* and *Eryngium Vaseyi.* In general, however, the members of a given species adhere rather closely to a given growth form, although some may vary widely in response to the special environmental condition in which they may occur. Members of some species grow under a single set of environmental conditions, whereas the members of other species will occur under the selective regime of different sets of environmental conditions. The common cat-tail (*Typha latifolia*) occurs along salt and alkaline marshes near the coast and in the Central Valley, and in fresh water from sea level to middle altitudes in the Sierra Nevada. Modern ecogenetic researches suggest that many, if not most, of such widely ranging species are made up of genetically distinct ecological races with respect to their tolerances of the various local combinations of environmental phenomena. A given broad climatic complex will select certain individuals that are genetically adapted to this range of conditions. Within this population there will be further ecological selection in accordance with other environmental phenomena. Where these phenomena occur in distinctive geographic patterns the selected individuals will be accordingly distributed. Thus it is now apparent that both ecologically and morphologically the species is often a highly complicated variable aggregate. The broader the geographic and ecological range, the more is this likely to be true.

As we seek common denominators between the various habitats that marsh plants occupy, it is obvious that the many aspects displayed by water in the habitat are highly characteristic, and that in some instances the condition of water may be diagnostic and thus may serve as a variable in the classification of habitats. Whether the water is standing or flowing; whether it is permanent or intermittent; fresh, alkaline, or salt—all these considerations provide convenient bases of classification, since these are the features that tend to characterize the major kinds of habitats. The following outline provides a practical, though rough, sort of classification that will comprehend all the major characteristics of wet-land habitats, although some readers may find it convenient and desirable to subdivide further or to fit other situations into the sequence.

I. Water Standing or Essentially So
   Presence of water permanent and level fairly persistent
     Open water surface the most conspicuous feature
       Fresh water
         Lakes
         Ponds
       Salt water
         Salt lakes
         Bays and oceans
         Estuaries
     Vegetation more conspicuous than water surface
       Vegetation dominantly herbaceous
         Marshes
           Alkaline marshes
           Salt marshes
           Brackish marshes
           Fresh-water marshes
         Bogs
           Quaking bogs
           Floating bogs
       Vegetation dominated by trees or shrubs
         Swamps

Presence of water intermittent or at least the level widely fluctuating
Intermittence seasonal
Vernal pools
Vernal marshes
Intermittence tidal
Salt-water marshes
Seasonally salt and fresh-water marshes
Fresh-water marshes subject to tidal influence

II. Water Flowing
Live streams
Intermittent streams
Irrigation ditches
Hillside bogs
Streamside marshes

III. Wet Soil Adjacent to Habitats with Standing or Flowing Water
Strand areas
Riparian lands
Lacustrine lands
Seasonally wet floodlands

It will be noted that I have not employed climate directly in any role in this classification, however significant it may be to the existence of any of the included habitats. This does not imply that climate is not significant as a variable in marsh habits. On the contrary, there are many species that are restricted to distinctive climatic conditions, and all marsh species occur geographically within their characteristic climatic ranges. However, since moisture is usually at an optimum in most of the habitats under consideration, the climatic variables that are significant are temperature and length of season. These variables express themselves geographically as broad gradients, and are, therefore, less useful in any classification system than those distinctive features which tend to emphasize boundaries.

In their ecogenetic evolution, marsh plants have met the climatic situation through their seasonal adaptations. With few exceptions they belong to one of the last groups to break seasonal dormancy and come into flower. They are chiefly summer bloomers, whether in Del Norte County or Imperial County, at sea level or in the high mountains. They mature and ripen their seed in the early fall. Within a given widely ranging species this process requires less time as we go northward and as we move into high altitudes. This is because the effective season is shorter and has had a selective effect on the biota. It is becoming increasingly clear that many native species have a genetically based clinal differentiation affecting geographic distribution of the individuals in response to length of season. Such "response phenomena" are similar to the phenomena of "earliness" and "lateness" in farm crops which we manipulate genetically in breeding programs to produce plants adapted to special conditions.

Because of such variables it is necessary to understand the full ecological scope of a species before employing it in any indicator capacity. Likewise, it is such variables that distort the homogeneity of communities.

LOCATION AND EXTENT OF MARSHLANDS

Marshlands occur throughout California wherever there is adequate water and the drainage pattern is such as to permit the saturation of the soil, or to permit water to stand on the soil. They are extensively developed at the mouths of rivers, along embayments and estuaries, and along lake margins. They tend to develop along streams wherever the gradient is low and the channel about filled to capacity, or where there is an overflow onto low ground. Thus all the major rivers have marshes along their lower courses and at their mouths. In addition, local topographic features that block water movement and develop lakes and ponds have their aquatic floras, and as the lakes and ponds become silted in, marshes often develop.

Along the coast, large marshes occur around lagoons, bays, and estuaries. Notable are those along the Humboldt County coast, especially around Humboldt Bay. Extensive marshes occur around San Francisco Bay and along the estuaries and lowlands of coastal southern California. Flanking the lower reaches of the Sacramento and San Joaquin rivers and especially on their delta, extensive fresh-water marshes occur. Farther up the valleys of these rivers their floodplains are characterized by vernally wet alkaline marshes. Marshes are common along the Pit River in Modoc, Lassen, and Shasta counties and in the intermontane basins of northeastern California. There are also marshes in the Imperial Valley where water seeps to the surface, and along the Colorado River. Many small habitats for marsh plants have been created artificially by local features of irrigation systems.

Ponds and lakes are common in the mountains; in the valleys the ponds and lakes usually reflect the meandering courses which the rivers had before they were brought under control by diking. Minor topographic features contribute to the formation of vernal pools on mesas and in valleys. Swamps in California are chiefly found in association with rivers.

Marsh habitats as they now exist, however, represent only fragments of these habitats as they originally prevailed in California, especially in the Central Valley and around the bays. In the Central Valley, diking has claimed for agriculture much of the former marshland, and along the bays filling has claimed them for industrial sites. The records of the early explorers mention difficulties in getting their horses across the Central Valley because of the extensive marshlands. Thomas Coulter, an Irish physician and botanist, records that he had fairly authentic information from trappers that the so-called "Tule Lakes" of the Central Valley were nowhere nearly so large as had been originally reported

—"neither was over 100 miles long"! Incidentally, he depicts them on his map as draining directly into the south end of San Francisco Bay, near the place where San Jose now stands. He visited California in 1831. As late as 1912 *Sagittaria Sanfordii* was reported as occurring in a pure stand of nearly one hundred acres at Banta, near Tracy. This area is today all under cultivation. The delta land south and west of Stockton has been reclaimed from marshes. Where drainage permits the runoff of water, many of the alkaline marshes have been converted into rice fields by adding fresh water and running it through the area. Although these are still marshes in a technical sense, to a large extent their flora is agriculturally controlled. The hog wallows and their vernal pools are being filled by leveling, and their floras are rapidly disappearing. Altogether, draining, diking, and leveling have added many thousands of fertile acres to the cultivated lands of California.

In spite of the drastically reduced area of marshlands resulting from the activities of man, the floras of the marshes (except for those of the vernal pools) have continued to increase in numbers of species by the introduction of weeds, especially through rice culture. Rice fields provide enormous temporary areas of suitable habitat for many aquatic weeds that originated in Asia and in the Valley of the Nile. Many of these species find their way into the natural marshlands and become a part of our flora.

FLORISTICS AND VEGETATION

Generalizations regarding the floristic organization of the marsh and wet-land habitats are difficult, because such organization centers around the intergrading environmental variables that not only account for different combinations of habitat conditions, but, through natural selection, permit a high degree of overlapping of species between habitats. Communities of plants therefore are rarely definitive in relation to what may appear to be a distinctive habitat. The three most important sets of environmental variables are:

1. The relative permanence of water, or the character of the intermittence of water in the habitat
2. The relative salinity and the hydrogen-ion concentration of the soil solution
3. The habitat variables related to seasonal temperatures and length of the growing period

Some aspects of each of these three sets of variables are evident in every marsh or wet-land habitat. They combine in various ways to produce exceedingly complex habitat diversity. In addition, other local features often may produce important local effects, such as turbidity of the water and the presence of special mineral ions in the soil solution.

The first set of variables produces such diverse conditions as standing or flowing water of marshes, ponds and lakes, bogs and streams. This set is also concerned with tidal overflow and the seasonal intermittence of water in the vernal marshes and pools in regions of winter rain and high summer evaporation. Some of the most significant aspects of floristic diversity relate to these phenomena.

The second set of variables produces such conditions as salt lakes and marshes and the gradient to fresh-water lakes and marshes. Like the first, this second set is also concerned with the range of hydrogen-ion concentration, from the alkaline marshes of the regions of high evaporation rate, on the one hand, to the acid bogs and marshes where organic acids tend to accumulate, on the other. Certain aspects of relative salinity and hydrogen-ion concentration exhibit some seasonal fluctuation due to local rainfall. The Chenopodiaceae and *Cordylanthus* in the Scrophulariaceae, to the extent that they occur in our marshes, seem to owe their speciation to selection by these variables.

The third set of variables is related to the gradient of seasonal temperatures as these occur over topography and latitude, especially as they become manifest in the length of the growing season. This set affects the areal pattern of each species, because (1) each individual of the species is characterized by a range of tolerance for temperature and length of season, and (2) the species displays variation in these tolerances among its individuals, both locally and geographically, and thereby complicates its ecological and geographical pattern. Again, certain groups of plants are found toward one extreme, while others occur toward the other extreme.

Superimposed over the geographic mosaic constituted of different combinations of these three sets of variables is a fourth, the biota, the members of which are adapted to different combinations of the environmental conditions that prevail locally in the habitat. This fourth variable is especially manifest in the genetic variation in the tolerances of the different aspects of the environment. This variation may be between the individuals of different populations of a given species, or it may be within each species population. In this respect it may be said that the species has become elaborated over the habitats and that no two species populations are physiologically alike. Nor are any two individuals of the species within such populations apt to be exactly alike. Any particular species may itself represent a gradient in its physiological variation over its geographic area in response to any phenomenon, such as temperature or length of season; or it may be divisible into local geographic races in response to the local occurrence of some variant in the environment; or it may be made up of races representing many complicated combinations of these in their adaptations to environmental gradients, or to local conditions of presence or absence of any environmental phenomenon.

The net result of this variation is of very great importance to the ecologist and to the range manager concerned with populations of wild plants. All wild species exhibit the genetic variation described above, but there are differences in the degree of complexity involved. It is not often possible to transfer members of wild species from one habitat into another and have them do equally well, or even grow. Furthermore, the association of plants is not an adequate indicator of the detailed conditions that prevail. The different members of any given plant community are in the community for different environmental reasons, and, where variation in the species exists, the members of what may be accepted as the same community may, in the different areas in which they occur, be representative of very different environmental complexes.

GEOGRAPHY

The marsh floras of California are divisible into two distinctive geographic patterns with respect to relationships pertinent to habitat type. There are the geographically widespread species that occur chiefly in the permanently wet habitats, especially where there is an abundance of permanent water. Most of the species in these habitats are widespread, and there are very few local endemics. The second group comprises the typically western species and the local endemic species that seem to be mostly confined to those habitats where water is seasonally intermittent, such as the vernal pools and the vernal marshes.

For the most part, species of widespread occurrence display little morphological variability in California and therefore do not usually merit taxonomic segregation from the species as a whole. Most species of *Potamogeton* and many of the common species of *Scirpus* and *Typha* are in this group. However, these plants probably reflect considerable physiological diversity from their eastern relatives, since their geographic range includes areas of very different length of day and very different length of growing season. Most of the species extending to the Atlantic coast are in permanently wet habitats.

On the other hand, the typically western groups display considerable morphological variability, and most of the taxonomic problems, except those involving species names based on European types, were encountered among these plants. The vernal pools and the shore lines of seasonally fluctuating lakes provide the habitats for the greatest number of local endemic species, and for the most part these are in genera and families that are either confined to the West or are most highly developed here. *Downingia* and *Navarretia* are examples. The vernally wet alkaline marshes likewise have been the locus of much speciation for such plants as *Atriplex* and *Astragalus,* and are also the chief habitat of the endemic *Hibiscus californicus*. Marshes of these alkaline types, being rather common over the Southwest and West, have much in common in their floras, but few of their species extend very far eastward. Since these alkaline habitats are the product of arid climates, this relationship is to be expected.

FLORISTIC CHANGES

Although the floristic features of the marshes and other wet-land habitats may often appear to be long-enduring, it is a readily observable matter that some marshes, and certain aspects of all marshes, undergo rather rapid change, even where the physical features of the habitat seem rather constant. Some of these changes are no doubt successional, but there are many species of wet habitats that behave as though they were transient: they are abundant one season and gone the next, and may return again a few years later. The term "transient" as applied to such plants perhaps gives a wrong impression of what happens. There is very little doubt that seeds are present, but they may fail to germinate every year or the plant population may be reduced to a few inconspicuous individuals. Examples of such phenomena may be cited. In 1945, *Wolffia columbiana* was collected at Dune Lake in San Luis Obispo County. This was the first record of this genus in California. In June the plants formed a cover over the lake that gave the appearance of a painted surface. The following June, after very diligent search, only two or three individuals of this minute plant were located. However, in nearby Oso Flaco Lake, where none had been reported the previous year, *Wolffia columbiana* was abundant. In the Sacramento Valley during the first two years of the survey, *Sagittaria Greggii* was so rare that only three or four stations were located. In 1950 it was without question the commonest species of its genus in this valley. It is perennial, but may behave as an annual. There is some evidence that this increase may have resulted from unclean rice seed, for the *Sagittaria* was especially abundant in rice fields, but this was by no means the only place where it occurred in profusion.

Sometimes disappearance may be related to the activity of birds and animals that feed on the plant, temporarily exterminating it locally. I doubt that this would explain such an occurrence as that reported for *Wolffia*. All parts of some species of *Sagittaria,* however, are eaten by birds. Early in the migratory season the leaves are stripped by waterfowl, especially sprig, and later the tubers are sought out and dug by geese and bottom-feeding ducks and hence only such seed as are overlooked carry the species over in these habitats. The same holds for certain species of *Potamogeton* and for *Ruppia*. The plants are literally torn to pieces by birds in their search for seeds and tubers.

Successional changes are also rapid, and a condition of stability is soon reached if it is possible. If an area becomes flooded it is not long before marsh species make their appearance, one of the first being the cat-tail. *Typha* is soon followed by species of *Scirpus*. I suspect that the sequence here is as much the result of the omnipresence of wind-blown cat-tail seed as of any other factor. Areas were observed, however, in which *Scirpus* appeared before the cat-tail. Both *Typha* and *Scirpus* species begin in shallow water and, through clonal growth, may migrate out into deeper water. We have seen no evidence of the germination and establishment of these seedlings in deep water. Stout rhizomes grow through or upon the surface layers of soil and become interlaced. Because of the close stands of the erect parts of the *Typha* and *Scirpus* plants, conditions are favorable for the catching and holding of debris and the rapid building up of the margin of the marsh or lake in which growth was initiated. Where such activity occurs along irrigation ditches, the area for flow of water soon becomes reduced and this choking of the waterway represents a factor of economic loss in ditch maintenance. So rapid are changes of this nature that relatively stable conditions in fresh-water marshes seem to prevail only where the deeper water impinges upon the marsh. Under salt-water or alkaline or brackish conditions, these changes appear to operate much more slowly. The more rapid changes found in fresh-water marshes may possibly be the result of the much larger number of species available to and capable of entering into a fresh-water succession, so that very slight changes are taken advantage of. In contrast, there are relatively few species of flowering plants available that can tolerate the conditions found in salt and brackish water habitats.

ECONOMIC PROBLEMS RELATING TO BOTANY

In the field investigations on which this publication is based, certain observations of a botanical nature were made relative to crop damage by migratory birds. In weedy rice fields, various species of *Sagittaria* are often conspicuous and occupy, in all, a fairly large area. About the time the first birds come in from the north, the leaves of these plants are voraciously sought and eaten. A patch will be eliminated in two or three days, leaving open water in the rice field. It is this open water that attracts large flocks of birds too numerous for the area, and the result is that the rice is badly trampled and cannot be harvested. Obviously, clean agriculture would eliminate much of the crop damage traceable to these causes. When this situation was brought to the attention of one grower, he replied that the weed seeds were a marketable crop for poultry feed. This fact would make clean agriculture very costly. Obviously, the expense of freeing the rice field from weeds would have to be weighed against the anticipated amount of damage by wildfowl. The trouble is that such damage, being a "hit-or-miss" phenomenon, is never anticipated, so preventive measures are not taken.

There are certain facts relative to the botany of marshlands that are important to the game manager. There are three main problems, namely, resting sites, nesting sites, and food. The resting sites must have water for the birds to rest upon. The other two problems are botanical and agricultural. Where there are no adequate nesting sites and the food supply is inadequate, it is often possible, by planting, to correct these situations. Much money has been poured into such projects, often without significant results. One important source of the trouble may be properly designated by the forester's term "provenience," that is, obtaining seed that is genetically adapted to the habitat in which it is to be grown. Most seed that has been grown by duck clubs in California has been purchased from eastern and southern sources, and little permanent improvement has resulted. The seed most likely to yield successful results is that obtained from plants growing successfully in the immediate vicinity. To put such seed on the market requires that someone harvest it. It must be remembered that wild plants are often very diverse genetically, and it is not to be expected that they can be grown as successfully as agricultural crops. The farmers' crops have been highly selected over many generations and represent relatively pure stocks genetically. In dealing with wild plants one must rely upon natural selection to select, out of the diverse seed planted by the sower, those which will function in the available environment. Often this may seem to be a highly wasteful procedure. However, this is all that there is to work with, until someone can breed or select strains more locally suitable.

Usually, however, where conditions are adequate, the habitat is producing plants to capacity naturally. In discussions of these problems with gun club managers, it was observed that concepts of available food supply are generally predicated on the number of birds available for the hunter. If there are few birds, the hunter is apt to reason that there is inadequate food. However, inadequate food is an indication of too many birds, and not too few. Investigation has usually disclosed ample food. The birds were absent for some other reason. The big problem is the number and size of the marshes and ponds. Accommodation of the number of migratory birds that regularly pass through California demands the restoration of some of the marshlands that have been reclaimed or else provision of other marshes developed on sites not suitable to agriculture. Some of the alkali land that lies too low to be successfully irrigated might well be flooded for this purpose. Here again, however, is another point of conflict with agriculture, for these lands are now being grazed by cattle or sheep. An appraisal with respect to maximum return is required, and these facts can then function in the competition between diverse interests. If the area is worth more to the sportsman, he should be able to put up a price that the cattleman cannot afford to turn down. If it is worth more for the production of cattle, that fact will dictate the course of the competition.

Conservation, however, is a concern outside the competitive sphere of the cattleman and the sportsman. The fact is that, in general, many more birds survive each year than the sportsmen take. This residue alone crowds the feeding and resting sites and makes it imperative that increased sites be developed for the benefit of the birds.

The task of developing marshlands presents no serious problems. Given adequate water of proper depth, a marsh is inevitable. A satisfactory balance between open water and vegetation is best initiated by properly surfacing the soil area before flooding, so that there will be both deep and shallow water, with some areas of saturated soil at or slightly above the surface. This will provide resting, nesting, and feeding sites and will also give some areas over which birds can graze and islands on which they can rest out of reach of predators.

FLORISTIC TREATMENT

The following floristic treatment, as is characteristic of most such works, will be found to be varied in its detailed taxonomic handling, especially in respect to revisionary problems of a taxonomic nature. To have dealt completely with each group would have required several lifetimes, or a large group of collaborators. When a problem has especially intrigued me, I have gone into it in greater detail. Much of the other work has been compiled from the researches of others so as to bring the groups into line with current taxonomic thought. In a few places where it was obvious that the authors were frustrated by their problems, I have attempted some changes. Some of these, I am sure, will be found to be an expression of my own frustration rather than that of the monographer! No human being is adequate to the task of avoiding this in an avowedly floristic treatment. No local situation ever provides the means of solving all the problems that it presents, and time has not permitted seeking elsewhere.

Some may question my choice of certain names. In general, I have attempted to follow current versions of the International Rules of Botanical Nomenclature. However, since in problems peculiar to taxonomy the expression of differences in taxonomic judgment and of differing objectives and conclusions in defining relationships demands a flexible nomenclature, nomenclature can only be stabilized within the framework of an understandable synonymy. Therefore the name employed, so long as it is within the framework of an understandable synonymy, is much less important than the adequacy of the description or the relations of the species to others. For the purposes of floristic writing, if a name has had its origin in a minor divergence of taxonomic judgment, I feel no pangs of professional conscience when I employ one that has enjoyed long local usage. However, I have attempted the clarification of nomenclature in all situations in which the origin stems from violations of rules pertaining to purely bibliographic matters. It is intended that the descriptions of families and genera shall pertain primarily to such plants as enter our area and problem, although no effort has been made to limit them to such usage.

Taxonomy

General Key to the Families

I. Plants not producing flowers or seeds; reproducing by spores produced in sporangia.

PTERIDOPHYTA

Spores produced in sporangia borne in the axils of quill-like leaves (plants
appearing onion-like at base) .... ISOETACEAE (p. 33)
Spores produced in sporangia borne otherwise than in axils of leaves.
Stems tubular and jointed (rushlike), with scale-like leaves; sporangia borne
on a terminal conelike spike .... EQUISETACEAE (p. 35)
Stems not tubular or jointed; leaves foliaceous.
Plants large, terrestrial, sporangia borne on underside of leaves; spores
of one kind .... POLYPODIACEAE (p. 27)
Plants small, aquatic or subaquatic; sporangia borne in special sporo-
carps; spores of two kinds.
Plants floating in water; leaves minute, loosely imbricate, entire or
2-lobed .... SALVINIACEAE (p. 31)
Plants rooting in mud, leaves 4-foliate or filiform, petioled

MARSILIACEAE (p. 29)

II. Plants producing true flowers and seeds.
SPERMATOPHYTA

A. Plants with leaves usually parallel-veined; vascular bundles of stem distinct and scattered; cotyledon 1; flowers usually with parts in multiples of three. MONOCOTYLEDONEAE

Ovary inferior.
   Leaves alternate or basal; plants rooted, erect.
      Perianth irregular, segments free or variously united
ORCHIDACEAE (p. 389)
      Perianth regular, segments united at least at base; leaves equitant
IRIDACEAE (p. 387)
   Leaves opposite or whorled; plants wholly submersed; perianth regular, segments united below, forming a long, filamentous tube from the top of the sessile ovary . . . . HYDROCHARITACEAE (p. 119)
Ovary superior.
   Perianth present, petal-like or of 4–6 papery blades or appearing as though attached to stamen filaments.
      Pistil 1.
         Inflorescence without a spathe.
            Anthers subsessile on a bractiform, perianth-like appendage
JUNCAGINACEAE (p. 95)
            Anthers with filaments well developed.
               Perianth segments petal-like . . . . LILIACEAE (p. 375)
               Perianth brown and papery . . . . JUNCACEAE (p. 347)
         Inflorescence with a spathe.
            Flowers crowded as though embedded in the spadix; plants rooted in soil . . . . ARACEAE (p. 324)
            Flowers not appearing as though embedded in a spadix; plants submersed or free-floating . . . . PONTEDERIACEAE (p. 343)
      Pistils 3 to many.
         Perianth of petal-like and sepal-like segments.
            Flowers in a raceme; ovaries 3 . . . . . . SCHEUCHZERIACEAE (p. 95)
            Flowers in a panicle; ovaries many . . . . ALISMACEAE (p. 103)
         Perianth segments alike, free from one another but appearing as though attached to stamen filament; flowers in spikes or heads
POTAMOGETONACEAE (p. 49)
   Perianth absent, or reduced to bristles or scales, sometimes the flowers in a perianth-like involucre.
      Leafless, minute floating plants . . . . LEMNACEAE (p. 327)

Leafy, erect or submersed plants.
　Submersed or floating aquatics.
　　Flowers clustered in the leaf axils.
　　　Leaves entire; pistils 4 . . . . ZANNICHELLIACEAE (p. 89)
　　　Leaves serrate or toothed; pistils solitary and naked
　　　　　　　　　　　　　　　　　　　　NAJADACEAE (p. 83)
　　Flowers in spikes; umbels, heads, or headlike clusters.
　　　Flowers in a few-rayed umbel . . . . RUPPIACEAE (p. 81)
　　　Flowers in spikes or heads.
　　　　Flowers in flattened spikes; plants in tide pools or saline
　　　　　estuaries . . . . ZOSTERACEAE (p. 91)
　　　　Flowers in heads or headlike clusters . . SPARGANIACEAE (p. 43)
　Erect, rooted plants with at least the tops emersed.
　　Flowers not in axils of dry, chaffy bracts.
　　　Flowers in heads or headlike whorls . . . SPARGANIACEAE (p. 43)
　　　Flowers in spikes or spikelike racemes or sometimes some solitary
　　　　in basal leaf axils.
　　　　Spikes dense; pistillate flowers below, staminate above
　　　　　　　　　　　　　　　　　　　　TYPHACEAE (p. 36)
　　　　Spikes racemose.
　　　　　Flowers all in racemose spikes; bractiform, perianth-like
　　　　　　appendages usually 6 . . . . . . JUNCAGINACEAE (p. 95)
　　　　　Some flowers solitary in the axils of leaf bases (pistillate)
　　　　　　and others (staminate and perfect) forming a weak,
　　　　　　erect or ascending, spikelike inflorescence; bractiform,
　　　　　　perianth-like appendage 1 . . . . LILAEACEAE (p. 101)
　　Flowers in the axils of dry, chaffy bracts; grass or sedgelike plants.
　　　Leaves solitary at nodes, in 2 rows; stems hollow except at nodes;
　　　　leaf sheaths split; bracts 2 (lemma and palea) to each flower
　　　　　　　　　　　　　　　　　　　　GRAMINEAE (p. 123)
　　　Leaves basal or in 3 rows; stems mostly solid; leaf sheaths not
　　　　split; bracts 1 to each flower; perianth often represented by
　　　　bristles . . . . CYPERACEAE (p. 203)

　B. Plants with leaves usually pinnately or pal-
　　mately veined; vascular bundles of stem usu-
　　ally in a ring, flowers with parts usually in
　　multiples of 2, 5, or many; cotyledons 2.
　　　　　　　　DICOTYLEDONEAE

　　　　1. Plants woody at base

Flowers monoecious or dioecious, in catkins, spikes, headlike clusters, or
　crowded panicles.

KEY TO MONOCOTYLEDONS AND DICOTYLEDONS

FLOWERS WITH PARTS USUALLY IN MULTIPLES OF 3 (RARELY MORE OR LESS)

LEAVES USUALLY PARALLEL VEINED (RARELY NET VEINED)

VASCULAR BUNDLES DISTINCT AND SCATTERED

COTYLEDONS 1

MONOCOTYLEDONEAE

SOME FLOWER TYPES

SOME LEAF TYPES

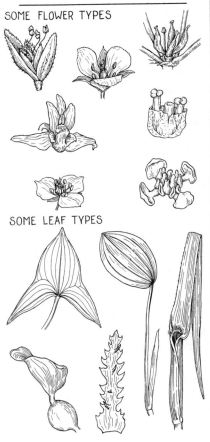

FLOWERS WITH PARTS USUALLY IN MULTIPLES OF 2, 5 OR MANY (RARELY 3)

LEAVES PINNATELY OR PALMATELY VEINED (RARELY PARALLEL VEINED OR RIBBED)

VASCULAR BUNDLES OF STEMS USUALLY IN A RING (SCATTERED IN A FEW AQUATICS)

COTYLEDONS 2 (RARELY REDUCED TO 1 IN A VERY FEW AQUATICS)

DICOTYLEDONEAE

SOME FLOWER TYPES

SOME LEAF TYPES

Fig. 1. Illustrated key to monocotyledons and dicotyledons.

Leaves alternate and simple.
  Flowers in catkins.
    Calyx none; flowers 1 to a scale.
      Fruit a capsule; seeds with long hairs; flowers dioecious
        SALICACEAE (p. 397)
      Fruit a waxy-coated, berry-like seed, glabrous; flowers monoecious
        MYRICACEAE (p. 408)
    Calyx present; flowers 2 or 3 to a scale, staminate flowers in catkins,
      pistillate flowers in short spikes . . . . BETULACEAE (p. 409)
    Flowers in headlike clusters . . . . PLATANACEAE (p. 542)
  Leaves opposite or sometimes fascicled.
    Flowers in erect catkins (maritime shrubs) . . . . . . BATIDACEAE (p. 473)
    Flowers not in catkins.
      Leaves simple and palmately lobed . . . . ACERACEAE (p. 568)
      Leaves simple and oblong or pinnately compound
        OLEACEAE (p. 649)
Flowers mostly perfect.
  Perianth segments not clearly differentiated into sepals and petals, that is,
    either all petal-like or all sepal-like (see also Vitaceae and Cornaceae);
    all free from one another.
    Flowers usually solitary, with large, dark red perianth segments; leaves
      opposite, deciduous . . . . CALYCANTHACEAE (p 519)
    Flowers in cymose clusters, small, perianth yellowish green; leaves alter-
      nate, evergreen . . . . LAURACEAE (p. 519)
  Perianth segments clearly differentiated into sepals (small in *Vitis*) and
    petals, sometimes the sepal lobes absent or reduced.
    Plant a woody vine . . . . VITACEAE (p. 571)
    Plant erect, not climbing.
      Petals free from one another.
        Ovary superior.
          Pistils more than 1 . . . . ROSACEAE (p. 542)
          Pistil 1 . . . . TAMARICACEAE (p. 591)
        Ovary inferior . . . . CORNACEAE (p. 634)
      Petals united into a tube.
        Ovary superior . . . . ERICACEAE (p. 636)
        Ovary inferior.
          Stamens free, not united into a tube.
            Flowers in involucrate pairs, the corolla tube gibbous below
              CAPRIFOLIACEAE (p. 742)
            Flowers in heads, these cymosely arranged
              RUBIACEAE (p. 741)
          Stamens united into a tube around the style; flowers in heads
            surrounded by bracts . . . . COMPOSITAE (p. 769)

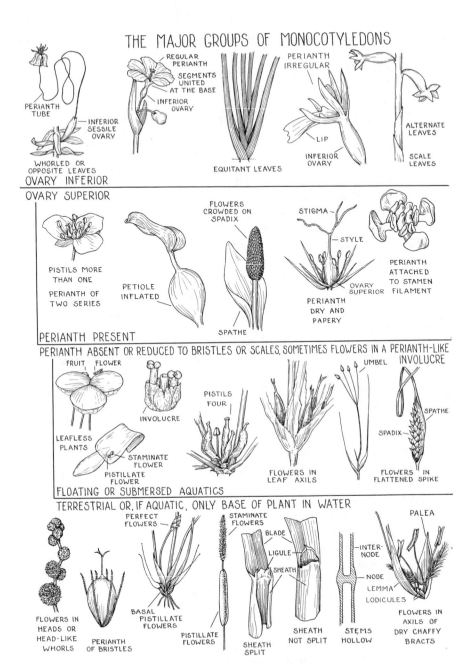

Fig. 2. The major groups of monocotyledons.

2. Plants wholly herbaceous

Corolla absent; calyx present or absent.

Flowers in a stout spike subtended by petaloid white bracts

SAURURACEAE (p. 397)

Flowers usually not in a spike, but if in a spike, the spike not subtended by white petaloid bracts.

Ovary superior.

Corolla absent, calyx present.

Pistil 1.

Ovary 1-celled.

Herbage with stinging hairs; stamens 4 . . URTICACEAE (p. 411)

Herbage without stinging hairs.

Fruit indehiscent.

Leaves without stipules; sepals herbaceous; plants often scurfy . . . . CHENOPODIACEAE (p. 449)

Leaves with stipules; fruit of 3-angled achenes

POLYGONACEAE (p. 413)

Fruit dehiscent.

Sepals 4 or 5, distinct or nearly so.

Capsule many-seeded, the valves separating longitudinally . . . . CARYOPHYLLACEAE (p. 479)

Capsule 1-seeded, circumscissile

AMARANTHACEAE (p. 472)

Sepals united, calyx 5-lobed.

Paired leaves unequal in size, the larger one with a long petiole . . . . AIZOACEAE (p. 475)

(*Cypselea*)

Leaves sessile, the lower leaves opposite, the upper ones usually alternate . . . . PRIMULACEAE (p. 641)

(*Glaux*)

Ovary 2- to 5-celled . . . . AIZOACEAE (p. 475)

Pistils more than 1 . . . . RANUNCULACEAE (p. 493)

Calyx and corolla absent.

Leaves dissected; ovary 1-celled . . . CERATOPHYLLACEAE (p. 493)

Leaves entire; ovary 4-celled . . . . CALLITRICHACEAE (p. 555)

Ovary inferior.

Leaves alternate, divided; flowers dioecious, the staminate flowers with 8–12 stamens, the pistillate flowers sometimes with as many as 4 stamens . . . . DATISCACEAE (p. 593)

Leaves opposite, simple or submersed leaves finely dissected; flowers perfect; stamens 1–4.

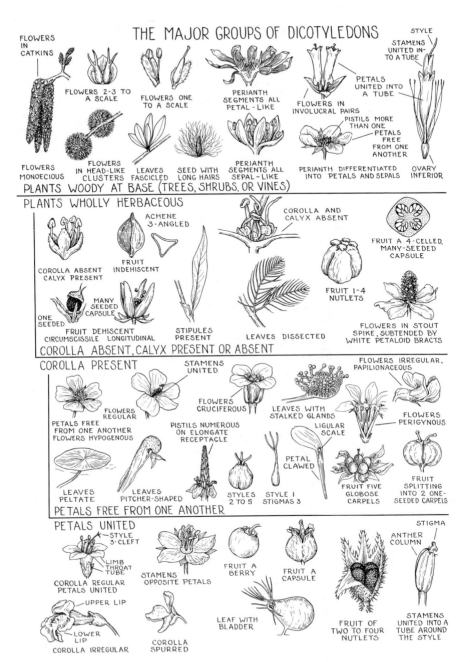

Fig. 3. The major groups of dicotyledons.

Fruit a 4-celled, many-seeded capsule . . . . . ONAGRACEAE (p. 602)
(*Ludwigia*)
Fruit of 1–4 nutlets . . . . HALORAGACEAE (p. 613)
Corolla present.
  Petals free from one another.
    Ovary superior.
      Flowers hypogynous.
        Stamens more than 10.
          Pistils more than 1.
            Leaves not peltate . . . . RANUNCULACEAE (p. 493)
            Leaves peltate . . . . CABOMBACEAE (p. 491)
          Pistil 1.
            Leaves alternate or basal.
              Stamens distinct.
                Leaves pitcher-shaped, erect
                        SARRACENIACEAE (p. 531)
                Leaves deeply cordate, floating; sepals petal-like, yellow
                        NYMPHAEACEAE (p. 489)
              Stamens united . . . . MALVACEAE (p. 571)
            Leaves opposite . . . . HYPERICACEAE (p. 577)
        Stamens less than 10.
          Pistils more than 1.
            Pistils numerous, on an elongate receptacle
                    RANUNCULACEAE (p. 493)
            Pistils 4 or 5.
              Leaves opposite, simple; pistil forming a follicle
                    CRASSULACEAE (p. 535)
              Leaves alternate, usually divided; fruit of 5 globose carpels
                    LIMNANTHACEAE (p. 561)
          Pistil 1.
            Flowers cruciferous (4 sepals, 4 petals, 6 stamens)
                  CRUCIFERAE (p. 521)
            Flowers not cruciferous.
              Ovary 1-celled.
                Flowers regular.
                  Sepals 2 . . . . PORTULACACEAE (p. 477)
                  Sepals more than 2.
                    Leaves opposite.
                      Placenta central . . . CARYOPHYLLACEAE (p. 479)
                    Placenta basal-parietal
                      FRANKENIACEAE (p. 589)

Leaves alternate or basal.
Leaves plane; fruit an achene
PLUMBAGINACEAE (p. 647)
Leaves with stalked glands; fruit a capsule
DROSERACEAE (p. 533)
Flowers irregular.
Leaves compound, flowers papilionaceous
LEGUMINOSAE (p. 547)
Leaves simple . . . . VIOLACEAE (p. 592)
Ovary 3- to 5-celled; stipules present; leaves entire, opposite
ELATINACEAE (p. 577)
Flowers perigynous; stamens on the calyx tube.
Stipules none; leaves simple.
Stigmas 3 or 4.
Inflorescence a 1-flowered scape . . . PARNASSIACEAE (p. 541)
Inflorescence various, not a 1-flowered scape
SAXIFRAGACEAE (p. 535)
Stigma and style 1 . . . . LYTHRACEAE (p. 595)
Stipules present; leaves simple or compound . . ROSACEAE (p. 542)
Ovary inferior.
Flowers in umbels.
Fruit splitting to form 1-seeded carpels . . UMBELLIFERAE (p. 619)
Fruit berry-like . . . . ARALIACEAE (p. 617)
Flowers not in umbels.
Style 1; sepals and petals 4 . . . . ONAGRACEAE (p. 602)
Style none; stigmas 4; leaves in whorls . . HALORAGACEAE (p. 613)
Petals united.
Ovary superior.
Corolla regular.
Stamens opposite petals . . . . PRIMULACEAE (p. 641)
Stamens alternate with petals.
Fruit a capsule (sometimes a berry in Solanaceae).
Style 3-cleft; ovary 1- to 3-celled . . POLEMONIACEAE (p. 658)
Style not 3-cleft.
Twining or trailing plants . . . . CONVOLVULACEAE (p. 655)
Erect plants.
Ovary 2-celled . . . . SOLANACEAE (p. 695)
Ovary 1-celled.
Leaves opposite . . . . GENTIANACEAE (p. 650)
Basal leaves alternate . . . MENYANTHACEAE (p. 654)
Fruit of 4 nutlets . . . . BORAGINACEAE (p. 666)

Corolla irregular.
　Fruit a capsule.
　　Ovary 1-celled; corolla spurred; leaves with bladders
　　　　　　　　　　　　　　LENTIBULARIACEAE (p. 731)
　　Ovary 2-celled; leaves not bearing bladders
　　　　　　　　　　　　　　SCROPHULARIACEAE (p. 701)
　Fruit of 2–4 nutlets.
　　Style entire; herbage in ours not aromatic
　　　　　　　　　　　　　　VERBENACEAE (p. 675)
　　Style split at apex; herbage usually aromatic . . LABIATAE (p. 679)
Ovary inferior.
　Stamens distinct.
　　Leaves alternate .... CAMPANULACEAE (p. 745)
　　Leaves opposite or whorled .... RUBIACEAE (p. 741)
　Stamens united in a tube around the style.
　　Flowers not in heads; fruit a many-seeded capsule
　　　　　　　　　　　　　　LOBELIACEAE (p. 745)
　　Flowers in heads; fruit a 1-seeded achene . . COMPOSITAE (p. 769)

Taxonomic Treatment

POLYPODIACEAE. FERN FAMILY

Ours ferns with erect pinnate or pinnatifid leaves from stout rhizomes. Leaves coiled in bud. Sporangia borne in sori in clusters on the lower surface of leaves or along their margins, commonly covered by a membrane or indusium.

Sori short, curved oblique to midvein .... *Athyrium*
Sori elongated in chainlike rows parallel to the midvein .... *Woodwardia*

Although several other ferns at times occur in habitats such as those covered by this work, they are not consistently found in bogs, marshes, or along the margins of lakes or streams and hence are omitted here. Among these are *Adiantum pedatum,* five-finger fern; *Lomaria Spicant,* deer fern; *Aspidium nevadensis,* Sierran water fern; and *Cystopteris fragilis,* fragile fern.

ATHYRIUM

1. Athyrium filix-femina (L.) Roth in Roem., Arch. Bot. 2^1:106. 1799. Lady fern.

Large, erect fern of marshy or boggy ground; leaves 6–12 dm. long, 2–3 dm. wide, bi- or tripinnately dissected, the pinnae oblong, acute to lanceolate, the lobes of pinnules toothed to apex; sori oval to oblong, oblique along midvein; indusium attached to the vein along the upper margin, toothed or ciliate along free margin.

Bogs and marshes along the coast: from Santa Barbara County northward, North Coast Ranges, Sierra Nevada, and mountains of southern California; cosmopolitan.

According to Mr. Anson Blake of Berkeley, a former student of E. L. Greene and former resident of Stockton, this species was once abundant in the delta islands around Stockton. It became rare here after the drainage of these islands and their conversion to agriculture.

At high altitudes this species is replaced by *Athyrium americanum* (Butters) Maxon, the alpine lady fern, which differs from it in having skeleton-like blades and in having no indusium.

WOODWARDIA

1. Woodwardia Chamissoi Brack., U. S. Expl. Exped. (Wilkes) 16:138. 1854. Giant chain fern.

Large ferns from stout, woody, oblique rhizomes beset with persistent leaf bases and chaffy scales; leaves 1–3 m. long, pinnate, the pinnae deeply pin-

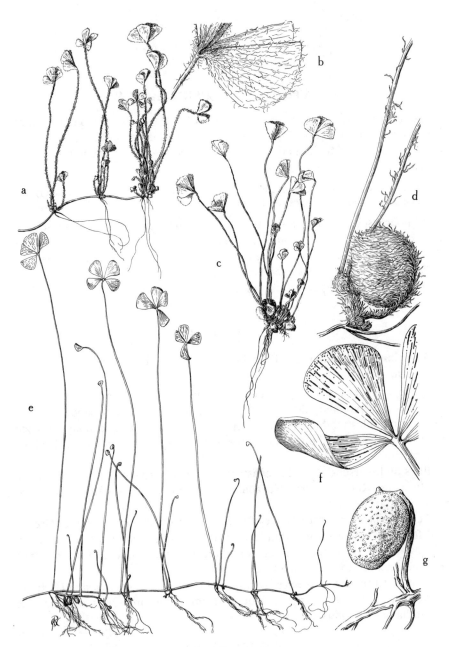

Fig. 4. *Marsilea mucronata: a,* habit, terrestrial plant, with densely pubescent leaves and petioles, arising from slender rhizome, × 1⅕; *b,* leaf detail, terrestrial plant, × 2½; *c,* habit, terrestrial plant showing sporocarps, × ⅔; *d,* sporocarp, terrestrial plant showing dense pubescence, × 4; *e,* habit, aquatic plant with elongate, slender, completely submersed petioles, their glabrous leaf blades floating, × ⅖; *f,* leaf detail, aquatic plant, × 2½; *g,* sporocarp, aquatic form, showing blunt teeth near junction with stalk, × 4.

natifid, the lobes somewhat falcate and spinulose-serrate or again pinnatifid; sori oblong-linear, in chainlike rows on either side of midvein; indusium fixed to a veinlet along its outer margin, the inner margin free.

In bogs or along streams at low and middle altitudes: throughout California, but especially in the Coast Ranges and at middle altitudes in the Sierra Nevada; cosmopolitan.

MARSILIACEAE. MARSILEA FAMILY

Perennials from slender, filiform rhizomes. Leaves filiform and bladeless or with a 4-parted blade. Sporocarps on short peduncles, arising from near the base of the leaf stalks, 2- to 4-valved, the valves thick, upon germination emitting an organized or disorganized gelatinous substance to which the megasporangia and microsporangia are attached; the spores of two kinds, megaspores and microspores.

Leaves with distinct blade and petiole, the 4-lobed blade of submersed plants floating on surface .... *Marsilea*
Leaves linear-filiform, without distinct blade, the tips uncoiling as the plants mature
Pilularia

MARSILEA

1. Marsilea mucronata A. Br., Am. Jour. Sci. II. 3:55. 1847. Fig. 4.

Marsilea vestita of American authors, not Hook. & Grev.

Amphibious plant from slender rhizomes often in water of seasonally fluctuating depth; leaves of submersed plants glabrous on elongate, slender petioles, the 4-parted blade floating on the surface, the leaves of terrestrial plants pubescent, on short, erect, wiry, filiform petioles; sporocarps flattened, 3–5 mm. broad, pedunculate with 2 conspicuous teeth above the point of attachment, the teeth rounded and blunt or obsolete, the sporocarps of terrestrial plants densely hairy, the aquatic form, when present, glabrous, the sporocarps opening by 2 valves from which a band of gelatinous tissue is emitted, bearing sori containing megasporangia and microsporangia.

Common in ponds and ditches and along their margins: throughout the valleys and foothills where summer temperatures are high; east to Atlantic.

Marsilea oligospora Gooding (Bot. Gaz. 33:66, 1902) has long been listed as occurring in Tulare County, California. Detailed and careful work by Margaret Stason (Bull. Torrey Club 53:473–478, 1926) on the range of variation in these plants has clearly shown that *M. oligospora* cannot be separated from *M. mucronata* (*M. vestita* of authors). The most striking characters of *M. oligospora* are to be regarded as chance-coördinated variables of little or no genetic significance.

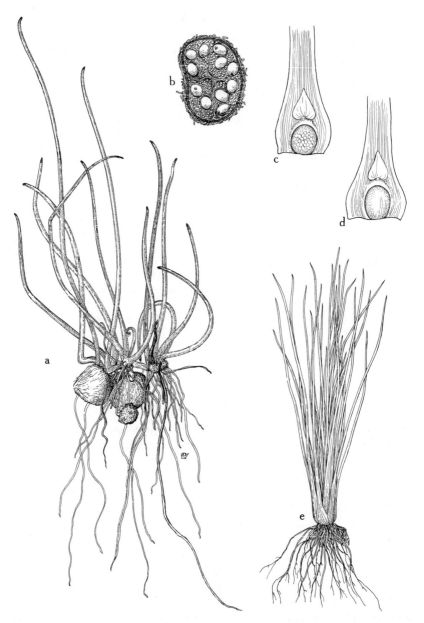

Fig. 5. *a* and *b, Pilularia americana: a,* habit, showing the filiform, bladeless leaves, the young ones coiled, and the stalked sporocarps, × 6; *b,* sporocarp (cross section), showing the sporangia, × 12. *c–e, Isoetes Bolanderi: c,* megasporangium on adaxial side of leaf base, the upper part partially covered by the velum, the ligule free, × 3; *d,* microsporangium on adaxial side of leaf base, the upper portion partially covered by the velum, the ligule free, × 3; *e,* habit, × ⅘.

PILULARIA

1. Pilularia americana A. Br., Monatsb. Kön. Acad. Wiss. Berlin 1863:435. 1863. Fig. 5.

Inconspicuous plants, technically like *Marsilea* but appearing strikingly different because they have no leaf blades and they are usually smaller in stature; sporocarps globose, 2 mm. broad.

Margins of ponds and vernal pools: valleys and foothills throughout the state where summer temperatures are high; Arkansas. Rarely collected, but undoubtedly frequently overlooked.

SALVINIACEAE

Free-floating, "mosslike" plants often completely covering the surface of ponds and ditches and coloring it a velvety green or, in late summer and fall, red. Stems 1–3 cm. long, pinnately branched, rooting from below, the roots hanging in the water. Leaves 2-lobed, imbricately disposed over upper surface of plant, usually concealing stems, the surface minutely papillose. Sporocarps usually in pairs on lower side of stem; massulae within microsporangia beset with septate or nonseptate, terminally barbed hairs or glochidia; microspores 6–100; megasporocarp rounded below, conical above, often described as "acorn-shaped," containing a single megaspore.

AZOLLA

All too often, the characters by which the species of *Azolla* are identified are not very satisfactory; in particular, size of plant, character of branching, overlapping of leaves, and number of microsporangia are so variable as not to be convincing criteria for identification. The characters used in the following key, however, are reasonably stable.

Svenson, H. K., The New World species of *Azolla*. Am. Fern. Jour. 34:69–84. 1944.

Glochidia of microsporangic massulae nonseptate; plants 2–6 cm. across
1. *A. filiculoides*
Glochidia of microsporangic massulae septate; plants 1–3 cm. across ... 2. *A. mexicana*

1. Azolla filiculoides Lam., Encycl. 1:343. 1783. Fig. 6.

Floating plants with pendent roots, 2–6 cm. across, pinnately branched; leaves minutely papillose, 2-lobed, imbricate; glochidia of microsporic massulae usually nonseptate; megaspore 1, the wall pitted.

Common in fresh-water ponds and ditches almost throughout California; western states, South America.

Adequate material for critical study is necessary for a satisfactory treatment of this genus.

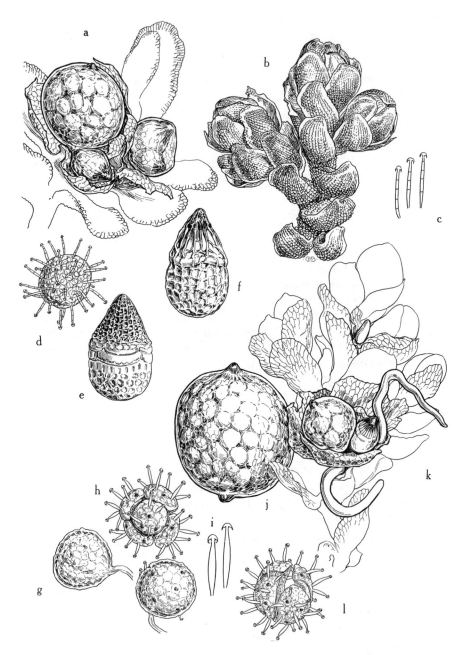

Fig. 6. (For explanation, see facing page.)

2. Azolla mexicana Presl, Abh. Böhm. Gesell. Wiss. V. 3:150. 1845. Fig. 6.

Similar in general appearance to *Azolla filiculoides,* but the plants usually smaller and the glochidia of the microsporangic massulae clearly septate (under a microscope); megaspore 1, the wall pitted.

Common in fresh water, in valleys and adjacent foothills: Central Valley, Santa Clara County, Modoc County, southern California; southwestern states, Mexico.

ISOETACEAE. QUILLWORT FAMILY

Aquatic or terrestrial perennial herbs with a short 2- or 3-lobed stem or corm, the roots from the sinus of the lobes of the corm. Leaves all caespitose, linear and often quill-like, expanding at base to enclose sporangium. Sporangium completely or partially covered by a membranous velum and subtended above by a triangular or rounded ligule, bearing either powdery microspores or tetrahedral megaspores, the megaspores ridged on the angles, smooth or variously tubercled on the faces.

ISOETES

Isoetes presents an inordinately difficult problem in the delimitation and identification of its species. This results from the variability of the characters that have been used as criteria for defining the species. Often, even those characters deemed important enough to separate sections of the genus will vary so much on an individual plant that confusion in identification results. Such variable characters include the lobing of corms, the markings on spores, the extent of the velum, and the shape of the ligule. For our purposes we follow the treatment employed by Pfeiffer, at least as concerns the species recognized.

Pfeiffer, Norma E., Monograph of Isoetaceae. Ann. Mo. Bot. Gard. 9:79–232. 1922.

Plants terrestrial or amphibious, from the margins of ponds or streams or growing on wet
    soil; stomata abundant.
  Corm usually 3-lobed; velum usually completely covering sporangium or nearly so.
    Leaves more than 8 cm. long .... 1. *I. Nuttallii*
    Leaves less than 8 cm. long .... 2. *I. Orcuttii*
  Corm 2-lobed; velum narrow to broad but not completely covering the sporangium
                                                  3. *I. Howellii*

Fig. 6. *Azolla. a–e, A. mexicana:* a, part of a fertile plant: (left) globose microsporocarp with megasporocarp at its base, (right) pair of megasporocarps enclosed in one indusium (uncommon), × 20; b, habit, top view, × 12; c, septate glochidia of microsporic massulae, × 100; d, microsporic massula, × 40; e, megaspore covered by tip of indusium, × 40. *f–l, A. filiculoides:* f, megaspore covered by tip of indusium, × 40; g, young stalked microsporangia, showing a few glochidia of the massulae protruding from ruptured wall, × 40; h, separating massulae, × 40; i, nonseptate glochidia, × 100; j, microsporocarp containing a large number of microsporangia (from same plant as k), × 20; k, part of fertile plant viewed from below and showing the roots and a small microsporocarp with a megasporocarp at its base, × 20; l, separating massulae, × 40.

Plants wholly and normally submersed, growing on the bottoms of lakes or ponds; stomata few or lacking; corms usually 2-lobed.
Stomata few; markings on megaspore wrinkled, tuberculate or spiny.
Ligule cordate; tubercles of megaspore often confluent into minute wrinkles
.... 4. *I. Bolanderi*
Ligule deltoid; megaspores spiny, tuberculate .... 5. *I. Braunii*
Stomata none; ligule short-deltoid; tubercles of megaspore confluent into minute cristate
ridges .... 6. *I. occidentalis*

1. Isoetes Nuttallii A. Br. in Engelm., Am. Nat. 8:215. 1874.

Corm usually somewhat 3-lobed; leaves 13–60, slender, 5–20 cm. long, somewhat 3-angled, erect, the membranous margin of expanded base extending well above sporangium; stomata numerous; ligule small, triangular; sporangium oblong, completely covered by velum; megaspores white, smooth or covered with minute papillae.

Amphibious or terrestrial on margins of pools or along streams: North Coast Ranges, Santa Cruz Mountains, Sierra Nevada; north to British Columbia.

2. Isoetes Orcuttii A. A. Eat., Fern Bull. 8:13. 1900.

Similar to *Isoetes Nuttallii*, but differing in the smaller size of its leaves and spores. It is with misgivings that it is here retained as distinct.

Replacing *I. Nuttallii* in similar habitats in southern California and ranging north to central California; south to Baja California, Mexico.

3. Isoetes Howellii Engelm., Trans. Acad. Sci. St. Louis 4:385. 1882.

Corm usually 2-lobed; leaves often many, 5–25 cm. long, slender in amphibious or terrestrial form, often quill-like in submersed forms, the membrane of lower leaf margin extending well above the sporangium; stomata many; velum partially covering the sporangium; ligule elongate-triangular; sporangium oblong; megaspores white, more or less obscurely marked with minute tubercles.

Partially or completely submersed in shallow water of ponds, streams, or vernal pools, from sea level to an elevation of 9,000 feet, almost throughout California; north to British Columbia and the northern Rocky Mountains, south to Baja California, Mexico.

A small form that is almost coextensive with the species has been described as *Isoetes Howellii* var. *minima* (A. A. Eat.) Pfeiffer (*op. cit.,* p. 142).

4. Isoetes Bolanderi Engelm., Am. Nat. 8:214. 1874. Fig. 5.

Plants submersed; corm usually conspicuously 2-lobed; leaves 6–15 cm. long, conspicuously quill-like; stomata very few; ligule small, cordate; velum not covering sporangium; sporangium small, orbicular to oblong; megaspores white to bluish, the tubercles sometimes aggregated into wrinkles.

Bottoms of lakes and ponds in shallow to deep water, the common species of high montane lakes: Sierra Nevada; British Columbia to Mexico.

Plants small in all characters have been recognized as *Isoetes Bolanderi* var.

pygmaea (Engelm.) Clute (Fern Allies 228, 258, 1905). California, Nevada, and Arizona.

5. Isoetes Braunii Dur., Bull. Soc. Bot. France 11:101. 1864.

Plants from bottoms of ponds and lakes; corm usually 2-lobed; leaves 10–30 cm. long, membranous border at base; stomata few; ligule deltoid; velum partially covering sporangium; sporangium oblong; megaspores white, beset with minute spines, these sometimes confluent into ridges.

Reported from lakes about the headwaters of the Sacramento and Trinity rivers, rare in California. A common species in the eastern United States.

6. Isoetes occidentalis Henderson, Bull. Torrey Club 27:358. 1900.

Submersed plants; corms usually 2-lobed; leaves 5–20 cm. long, quill-like, broadly membrane-margined at base, the membrane extending upward 2–3 times the length of the sporangium; ligule short-triangular; velum narrow, not covering sporangium; sporangium orbicular; megaspores cream-colored, the tubercles confluent into irregular cristate ridges.

Ponds and lakes: northern Sierra Nevada, and mountains around the head of the Sacramento River.

EQUISETACEAE. HORSETAIL FAMILY

Rushlike plants from perennial rhizomes, wide-creeping, freely branched, rooting at the nodes. Aerial stems usually erect, cylindrical, fluted, simple, jointed with whorled branches at the solid sheathed nodes, the internodes usually hollow, the surfaces often silicated, ridged, or grooved. Leaves nodal, minute, united lengthwise to form cylindrical sheaths, their tips (teeth) free or connivent, persistent or deciduous. Fertile stems with conelike terminal spikes (strobili) composed of many closely appressed, stalked, shield-shaped scales set in at right angles to the cone axis, these bearing a few sporangia beneath. Spores uniform, green, each provided with 4 hygroscopic elaters or bands.

Many species of *Equisetum,* such as *E. arvense, E. Funstonii, E. kansanum,* and *E. laevigatum,* often invade moist alluvial situations, but are not characteristic marsh plants; they also grow in shaded dry areas.

EQUISETUM

Spikes tipped with a rigid point; aerial stems evergreen, tall and rigid, usually without
    whorls of branches .... 1. *E. hiemale*
Spikes blunt or rounded at the tip.
    Aerial stems all alike, the fertile branches with numerous whorls of branches
                                                      2. *E. palustre*
    Aerial stems of two kinds, the fertile ones whitish or brownish, the sterile ones green
        and much-branched .... 3. *E. telmateia*

1. Equisetum hiemale var. **californicum** Milde, Nova Acta Acad. Leop.-Carol. 32²:517. 1867.

Aerial stems persisting through several seasons, rigidly erect, stout, 0.5–2.5 mm. tall, 25- to 40-angled, rough, the ridges with 2 rows of silica tubercles; sheaths a little longer than broad, cylindrical, at maturity ashy with black bands at top and base, the constituent leaves with a deep central groove, obscurely 4-carinate, the teeth long and flexuous, brown, late-deciduous; spikes 1–3 cm. long, the apex sharp-pointed.

Beside streams and in marshy places in the hills and mountains: southern California to Humboldt County; north to Alaska, east to Nevada, Arizona, and New Mexico.

2. Equisetum palustre L., Sp. Pl. 1061. 1753.

Annual; stems slender, 25–90 cm. tall, much-branched above, 5- to 10-grooved, the grooves separated by roughish, narrow ridges, the central canal of stem very small; sheaths loose and somewhat widened upward, greenish, the teeth black but whitish-margined; branches simple, few in whorls; spikes 1–2.5 cm. long, terminating the stem, smaller spikes sometimes terminating the branch tips.

Wet places: San Mateo County and Lake County; widespread in northern part of Northern Hemisphere.

3. Equisetum telmateia Ehrh., Hannov. Mag. 1783:287. 1783. Giant horsetail.

Equisetum telmateia var. *brauni* Milde, Monogr. Equis. 246. 1865.
Equisetum maximum of California authors.

Plant with tuber-bearing rhizomes; fertile stems erect, 25–60 cm. tall, usually in a cluster, whitish or brownish, smooth, succulent, soon withering; sheaths loose, the lower part whitish, the upper part brownish, 2–5 cm. long, the lower ones longer than the internodes, the teeth 20–30, attenuate, coherent in groups of 2 or 3; spikes stout, 4–8 cm. long, 1–2.5 cm. thick; sterile stem erect, greenish, 0.5–3 m. long, 20- to 40-furrowed, the ridges smooth, the sheaths cylindrical with 20–40 brownish teeth, the branches spreading in whorls, numerous, solid, simple, 4- to 6-angled, the keel rough with crossbands of silica.

Swamps and borders of streams: southern California to British Columbia.

TYPHACEAE. CAT-TAIL FAMILY

Rhizomatous perennials with distichously arranged leaves. Leaves linear, glabrous, sheathing at the base, the sheath cylindrical, open, tapering or auricled at the shoulder, the auricle often scarious or membranous. Flowers monoecious, in spikes, the staminate spikes above the pistillate ones. Pistillate flowers on compound pedicels interspersed with bracts and sterile flowers; ovary on a stipe beset with long hairs; style linear; stigma linear-lanceolate. Sterile floret on an elongate, hairy stipe terminating in a swollen, aborted ovary, much longer than the functional ovary. Staminate flowers ephemeral, densely packed in the spike, and interspersed with simple or forked hairs or with linear or linear-lanceolate bracts; stamens on branched filaments.

TYPHA

In our treatment of the genus *Typha* we were confronted by the usual complex of variation that is to be noted wherever two or more species of this genus occur together. Our first impression was that specific lines were impossible to establish, but continued field observation led to an approach to the problem that was both rational and practicable. Beginning with the treatment in current manuals, we were led to assume that there were two species in the California flora. If we accepted this, the natural assumption was that intermediates between these two could only be interpreted as having arisen through hybridization. A very common apparent intermediate possessed much lighter-colored pistillate spikes. It was not long before we discovered that in many places this type was growing entirely by itself and not in association with any other type, and that here the population was much more homogeneous than populations found where two or more types were growing together. We concluded that we had a third species, and we soon identified it as *Typha domingensis*. A similar pattern of occurrence was observed for the other two species. This led us to conclude that we were dealing with three entities which apparently hybridize with one another, and that the hybrids are capable of establishing themselves locally through backcrossing without the loss of identity of the parent stock. *Typha glauca* is not known except in association with *T. latifolia* and *T. angustifolia,* and even under these circumstances it is rare. Its intermediate character led us to regard it as a hybrid between these two species.

Hotchkiss, N., and H. Dozier, Taxonomy and distribution of North American cattails. Am. Midl. Nat. 41:237–254. 1949.

Leaf sheaths usually open at throat, margins free, tapering to the blade, occasionally truncate but rarely auricled; leaves light green to yellow-green, 6–15 mm. wide; pith of stem base white.
  Pistillate spike buff-colored, usually 6–10 times as long as thick; stigmas linear; leaves convex on the back; pollen 1-celled, golden yellow; staminate bracts linear, often laciniate, brown . . . . 1. *T. domingensis*
  Pistillate spike dark brown, usually almost 6 times as long as thick; stigmas ligulate to lanceolate; leaves usually flat on back; pollen 4-celled, orange or yellow; staminate bracts simple, hairlike, white . . . . 2. *T. latifolia*
Leaf sheaths usually closed at throat, margins free but parallel, auricled above, rarely tapering at throat to the blade; leaves dark green to glaucous green; pollen 1-celled.

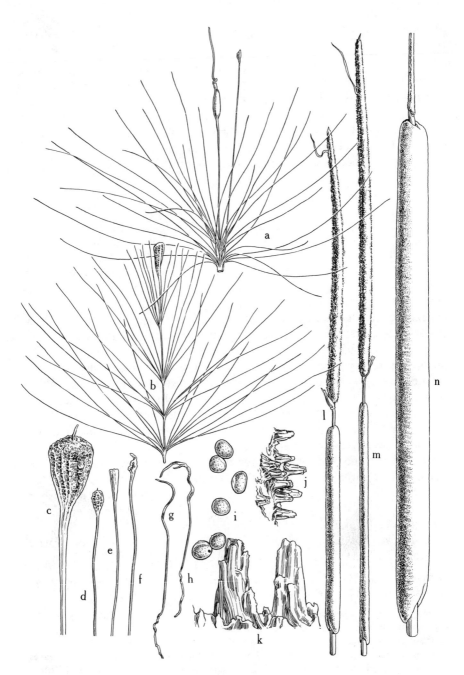

Fig. 7. (For explanation, see facing page.)

Leaves 5–6 mm. wide, dark green; pith of stem base white; sterile ovary flat-topped; stigmas linear; pollen lemon yellow .... 3. *T. angustifolia*

Leaves 6–12 mm. wide, glaucous green; pith of stem base buff; sterile ovary tapered above or rounded; stigmas lanceolate to ligulate; pollen golden yellow

4. × *T. glauca*

1. Typha domingensis Pers., Syn. Pl. 2:532. 1807. Fig. 7.

Plants 1–2 m. tall; sterile stems with 6–9 leaves and white pith; leaves 6–12 mm. wide, light yellowish green; sheaths tapering at throat to the blade, scarious-margined above; spikes as tall as or somewhat shorter than leaves, the staminate and pistillate parts usually separated by a distance as great as the diameter of the pistillate spike or greater, the distance often varying from 0.5 cm. to 6 cm.; pistillate spike light brown, becoming buff or gray, usually 10 times or more as long as thick, 15–25 cm. long, 1.5–2.5 cm. thick; flowers arranged on stout compound pedicels which when stripped of appendages are rough and openly spaced; bracts with light brown, translucent, obovate or apiculate blades on slender stalks; fertile flowers with a light brown, linear, deciduous stigma; sterile flowers with obovoid abortive ovaries, much longer than the functional ovaries; stamens on branched filaments, interspersed with linear, cuneate, laciniate brown bracts; pollen golden yellow, 1-celled or occasionally in pairs.

Coastal and valley marshes at low altitudes: throughout California; east to Atlantic and south through tropical areas.

2. Typha latifolia L., Sp. Pl. 971. 1753. Common cat-tail. Fig. 8.

Plant usually coarse and stout; pith of the stem base white; leaves 12–16, 8–20 mm. broad, nearly flat, light green; sheaths cylindrical but open to base, the scarious upper margin tapering to blade, rarely truncate or slightly auricled; spike-bearing stems subequal to or longer than leaves; pistillate and staminate spikes usually contiguous, rarely separated; pistillate spike dark greenish brown to reddish brown, in age becoming blotched with white, usually about 6 times as long as thick, 10–18 cm. long, 1.8–3 cm. thick; flowers without bracts or the bracts hairlike, on slender, often hairlike, compound pedicels; stigma medium brown to dark brown, lanceolate-ovate, conspicuously fleshy, persistent; sterile flowers with an ellipsoid aborted ovary, tipped by a rudimentary style and much longer than the functional ovary; stamens on branched filaments often 2 or 3 to a cluster; pollen 4-celled, elsewhere reported as orange for the species, but ours yellow.

Fig. 7. *Typha domingensis: a,* fertile pistillate flower, showing mature ovary and the surrounding hairs originating at base of stipe, bract attached, × 8; *b,* sterile pistillate flower terminating in a swollen, aborted ovary, hairs surrounding stipe in whorls, × 8; *c,* aborted obovoid ovary tipped by rudimentary style, × 20; *d–f,* typical bracts, showing variations in the swollen tips, × 12; *g* and *h,* bracts of staminate flowers, slender, simple or laciniate, with dark brown, shiny tips, × 12; *i,* 1-celled pollen grains, occasionally in pairs; *j* and *k,* compound pedicels of pistillate spike, *j,* × 9, *k,* × 40; *l* and *m,* spike, showing area of separation between the staminate part (above) and the pistillate part (below), × ¼; *n,* pistillate spike, × ⅖.

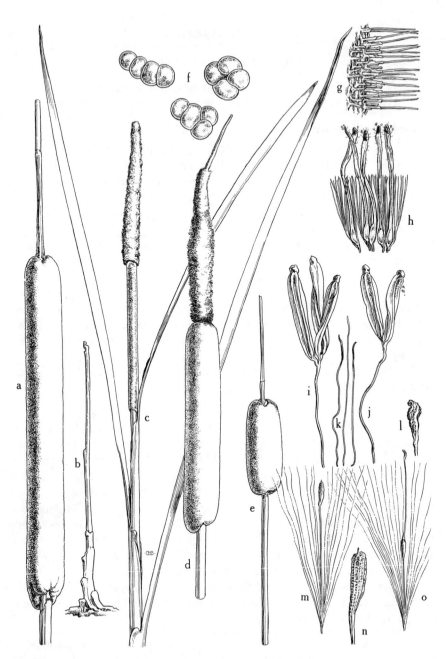

Fig. 8. (For explanation, see facing page.)

Commonly occurring with the other species in coastal and valley marshes, and to the exclusion of the other species in the mountains at elevations above 3,000 feet; widely distributed in the Northern Hemisphere.

3. Typha angustifolia L., Sp. Pl. 971. 1753. Narrow-leaved cat-tail. Fig. 9.

Slender perennial 0.5–1.5 m. tall; pith of stem white; stems about ⅔ as long as the leaves; leaves narrow, plano-concave or plano-convex or strongly convex on the back, 5–6 mm. wide, dark green; sheaths appearing cylindrical below but actually open to base, usually conspicuously auriculate above, rarely some sheaths tapering to the blade, the auricles scarious-margined; pistillate and staminate spikes usually separated by a distance twice as great as the diameter of the pistillate spike or greater, rarely less than 0.5 cm. or more than 12 cm. apart; pistillate spike dark brown to reddish brown or in age becoming greenish brown or mottled, usually 6–10 times as long as broad, 8–20 cm. long, 1.8–2.5 cm. thick; pistillate flowers arranged on compound pedicels which when stripped of appendages appear smooth; bracts spatulate, truncate, their blades dark brown, opaque, and firm, slender-stalked; fertile flowers pediceled, the stipe densely long-hairy, the style long, slender, bearing a dark brown, linear stigma; sterile flowers long-stipitate with a broad, flat-topped, inflated, terminal, aborted ovary with a rudimentary style, much longer than the functional ovary; stamens on branched filaments, or sometimes sessile, often 2 or 3 to a cluster, the anthers opening by longitudinal slits, the connective clavate, often swollen and truncate above; pollen 1-celled, lemon yellow; bracts hairlike to linear, simple or forked, brown.

Coastal and valley marshes at low elevations throughout California; widely distributed in Northern Hemisphere.

4. × Typha glauca Godron, Fl. Lorraine 3:20. 1844.

Typha latifolia var. elongata Dudley, Bull. Cornell Univ., Sci. 2:102. 1886.
Typha angustifolia × latifolia Kronfeld, Verhandl. K. K. Zool.-Bot. Gesell. Wien 39:167. 1889.
Typha elongata (Dudley) Kronfeld, Verhandl. K. K. Zool.-Bot. Gesell. Wien 39:176. 1889.
Typha latifolia × angustifolia Figert, Deutsch. Bot. Monatsschr. 8:55–57. 1890.
Typha angustifolia var. longispicata Peck, N. Y. State Mus. Rep. 47:162. 1894.
Typha angustifolia var. elongata (Dudley) Wieg., Rhodora 26:1. 1924.

Fig. 8. *Typha latifolia: a*, pistillate spike, × ⅖; *b*, single, compound pedicel of pistillate spike, × 20; *c*, upper part of plant, showing distichously arranged leaves and young contiguous spike with staminate flowers (above) and pistillate flowers (below), × ⅕; *d*, somewhat older spike, × ⅖; *e*, variation in spike size, × ⅖; *f*, 4-celled pollen grains; *g*, group of compound pedicels of pistillate spike, × 4; *h*, young pistillate flowers, the pedicel not yet elongated, and fascicled, hairlike bracts, × 12; *i* and *j*, stamens on branched filaments, × 6; *k*, staminate bracts, commonly white or brown-tipped, × 6; *l*, oblanceolate, fleshy stigma, × 12; *m*, sterile pistillate flower with ellipsoid, aborted ovary tipped by rudimentary style, the surrounding hairs, like those of fertile flower, originating at base, × 4; *n*, sterile ovary, light brown, × 12; *o*, pistillate flower with mature, functional ovary, × 4.

Fig. 9. (For explanation, see facing page.)

Under the name *Typha glauca,* and under those names listed in the above synonymy, various authors have included what seem to me to be various hybrids and hybrid derivatives emanating from natural crossing (and back-crossing with either parent) of *T. latifolia* and *T. angustifolia.* The one out-standing character of × *T. glauca* is the presence of yellow pith in the stem immediately above the rhizome. In all other respects it is intermediate be-tween the two putative parent species, and it is not known to us except in association with both parents.

SPARGANIACEAE. BUR-REED FAMILY

Herbaceous perennials with fibrous roots and creeping rootstocks, erect or floating, simple or branched stems, and linear alternate leaves, sheathing at the base. Flowers monoecious, in globose heads on the upper part of the stem and branches, the staminate heads uppermost, sessile or peduncled, the lower heads pistillate, with leaflike bracts. Perianth of 3–6 linear-subulate scales. Stamens commonly 5, their filaments distinct; anthers oblong or cuneate. Ovary sessile, 1-celled (rarely 2-celled). Fruit 1- or 2-celled and 1- or 2-seeded, nutlike, the head burlike from the prominent beaks of the fruits.

SPARGANIUM

Sparganium is seriously in need of a revision based upon extensive field study. Many of the technical characters which have been used in separating the species, such as the presence or absence of supra-axillary heads, the cross section of the leaf, the position of the sepals, and the precise shape of the fruits, appear to be quite meaningless. Knowledge concerning the effect of the depth of the water, and of the length of time that it is present, on width and shape of leaves, and on whether the leaves are floating or emersed, is essential

Fig. 9. *Typha angustifolia: a,* swollen, aborted ovary with rudimentary style, × 20; *b,* sterile, long-stipitate flower with terminal, aborted ovary, the hairs on stipe in whorls, terminating in club-shaped or ligulate tips, × 8; *c,* young spike, showing area of separa-tion between staminate spikes (above) and pistillate spikes (below), × ⅖; *d,* single com-pound pedicel of pistillate spike, × 40; *e,* group of compound pedicels, appearing smooth, × 8; *f,* cluster of spatulate, truncate bracts, with transitional forms resembling abortive ovaries, occurring frequently among sterile flowers, × 8; *g* and *h,* upper part of plant, showing distichous leaf arrangement and young flowering spikes, × ⅛; *i,* cluster of young anthers surrounded by bracts, filament not yet elongated, × 6; *j–l,* mature stamens, 2–6 anthers in a cluster sessile on a single filament, × 6; *m–o,* staminate bracts—linear, simple, and forked types, × 6; *p,* 1-celled pollen grains; *q,* group of young fertile and sterile pis-tillate flowers, the pedicels not yet elongated, × 12; *r,* swollen tip of pistillate bract, × 40; *s,* pistillate bracts, × 8; *t,* auricle of sheath, × ⅘; *u* and *v,* mature pistillate flowers with functional ovaries, long styles, and linear stigmas, the pedicels of varying length and sur-rounded by basal hairs, × 8.

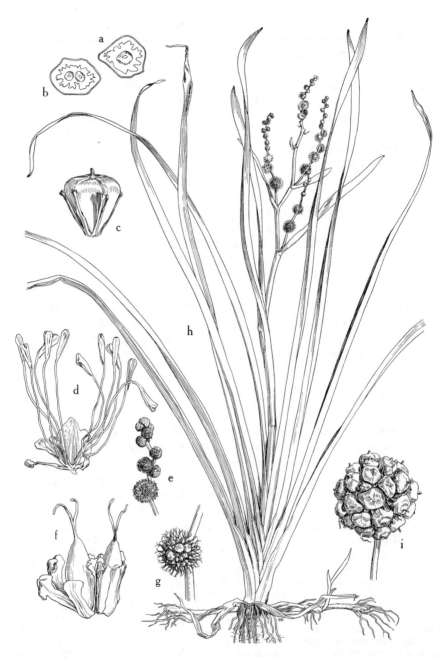

Fig. 10. *Sparganium eurycarpum: a* and *b,* 1-seeded and 2-seeded fruits (cross sections), × 1½; *c,* mature fruit, × 1½; *d,* paired staminate flowers, usually with 1 broad perianth scale and several long-clawed scales expanding into a spatulate apex, the anthers elliptic-clavate, × 6; *e,* staminate inflorescence showing globose heads, × ⅖; *f,* young, sessile pistillate flowers, showing the perianth scales with spatulate apex, the scales broader than those of the staminate flowers, × 4; *g,* young fruiting bur, showing the long, 2-lobed style branches, × ⅔; *h,* habit of plant, × ⅛; *i,* mature fruiting head, the styles broken off, × ⅖.

to the understanding of the large group of species centering around *S. simplex, S. americanum, S. chlorocarpum, S. multipedunculatum,* and *S. fluctuans.* Such information is especially pertinent in a group that is essentially amphibious and variable. In other words, are some of the variables the product of habitat, or are they all genetic?

Stigmas predominantly 2-lobed, rarely simple (occasionally a simple stigma among many 2-lobed stigmas); plants coarse, 1–2 m. tall; inflorescence branched.
  Fruits truncate or depressed at summit, 6–10 mm. long, 6–8 mm. wide at summit
                                                                        1. *S. eurycarpum*
  Fruits rounded or tapering above, 4–7 mm. long, 3–5 mm. wide at summit
                                                        1a. *S. eurycarpum* var. *Greenei*
Stigmas predominantly simple, rarely 2-lobed (occasionally a 2-lobed stigma among many simple stigmas); plants slender, erect or floating (sometimes stout in *S. angustifolia*), 1–8 dm. tall; inflorescence simple.
  Fruiting burs 15–25 mm. broad; staminate heads usually more than 1.
    Leaves 5–12 mm. broad, flat or rounded above to conspicuously keeled below, usually erect and emergent, rarely floating; fruiting burs 20–25 mm. broad
                                                                        2. *S. simplex*
    Leaves 1–5 mm. broad, rounded on back, usually floating, ours rarely erect and emergent; fruiting burs 10–20 mm. broad .... 3. *S. angustifolium*
  Fruiting burs 8–12 mm. broad; staminate heads usually solitary .... 4. *S. minimum*

1. Sparganium eurycarpum Engelm. in Gray, Man., 2d ed., 430. 1856. Broad-fruited bur reed. Fig. 10.

Sparganium californicum Greene, Bull. Calif. Acad. Sci. 1:11. 1884.

Stem stout, erect, branching, 5–18 dm. tall; leaves 5–10 dm. long, 7–17 mm. wide, flat, somewhat keeled below, as long as or slightly shorter than inflorescence; pistillate heads 2–6 on the main stem or on branches, sessile or usually peduncled, 2–2.5 cm. in diameter in fruit; staminate heads 8–12; perianth scales long-clawed, expanding into a spatulate apex, irregularly shallowly lobed and hyaline-margined at apex, ⅔ to ¾ as long as the fruits; anthers 1–1.5 mm. long, elliptic-clavate; style branches commonly 2, occasionally 1, filiform, about 2 mm. long; fruits sessile, hard and thick at maturity, cuneate-obpyramidal, irregularly and obtusely 3- to 5-angled, 6–10 mm. long and 6–8 mm. wide at apex, the top truncate, depressed or very shallowly rounded, the beak 2–3 mm. long.

Fresh-water or brackish marshes: coastal southern California, near the coast and in the Coast Ranges from San Luis Obispo County to Del Norte County, Central Valley, also in Plumas and Modoc counties; Pacific states, across southern Canada, eastern states.

Sparganium californicum Greene appears to have been based on exceptionally large specimens of *S. eurycarpum.*

Fig. 11. *Sparganium eurycarpum* var. *Greenei: a,* group of staminate flowers, showing irregular perianth scales which are clawed and spatulate at apex, × 6; *b,* upper part of stem, with flowering branches, × ⅓; *c,* 1-, 2-, and 3-seeded fruits (cross sections), × 1½; *d,* inflorescence with staminate heads and maturing fruiting burs, × ⅓; *e,* young pistillate flowers with branched styles, × 4; *f,* mature fruit, × 1½; *g,* mature fruiting head, the fruits smaller and considerably more numerous than those of *Sparganium eurycarpum* and with tapering apex, × ⅘.

1a. Sparganium eurycarpum var. **Greenei** (Morong) Graebner in Engler, Pflanzenr. 4¹⁰:13. 1900. Fig. 11.

Sparganium Greenei Morong, Bull. Torrey Club 15:77, pl. 79. 1888.

Similar to the species in many of its characters, but differing mainly in having shorter fruits (5–8 mm. long) which are only slightly more than ½ as wide (3–5 mm.) at the apex and which are rounded or usually tapering to a stout beak; pistillate heads averaging fewer (2 or 3) on a branch and at maturity varying from 1.5 cm. to 2 cm. in diameter; styles with 2 branches most frequent, but flowers with single, unbranched styles often occurring intermixed.

Scattered throughout the Pacific coast range of the species.

The lack of any distinct geographical pattern separating this variety from the species, as well as its close morphological similarities to the species, are the chief reasons for retaining it as a variety.

2. Sparganium simplex Huds., Fl. Angl., 2d ed., 401. 1778. Simple-stemmed bur reed. Fig. 12.

Stem rather stout, but sometimes slender, 3–10 dm. tall; leaves 4–10 dm. long, 4–8 (or –10) mm. wide (or sometimes as much as 15 mm. wide), flat or triangular-keeled toward base, well overtopping the inflorescence; bracts flat or slightly keeled; the leaves and bracts often conspicuously margined near the base; inflorescence usually simple, the pistillate heads 2–5, the lowest ones peduncled, the upper ones sessile, at least some of them supra-axillary; staminate heads 3–8, congested or confluent; perianth scales oblanceolate, erose at broadened apex; anthers 1–1.5 mm. long, elliptic-clavate; fruiting heads 15–18 mm. in diameter; fruits brown or greenish brown, stipitate, 4–6 mm. long, fusiform, often constricted at the middle; stigma linear, about 1 mm. long.

Mucky bottoms of shallow ponds, along streams and sloughs: San Bernardino Mountains, Sierra Nevada from Tulare County north, east of the Sierra Nevada crest from Mono County to Modoc County, Siskiyou Mountains, near the coast in North Coast Ranges; cooler parts of Northern Hemisphere.

3. Sparganium angustifolium Michx., Fl. Bor. Am. 2:189. 1803.

Sparganium affine Schnizl., Typh. 27. 1845.

Slender, usually submersed aquatic, the leaves and stems floating or below the surface; leaves 1–1.5 mm. wide, often very long, curved on the back; inflorescence usually simple, the bracts dilated at the base, often floating, boatlike on the surface and each containing a staminate or a pistillate head, or sometimes the inflorescence erect and emersed; sepals of pistillate flowers borne at base of the constricted lower part of the ovary or at base of the stipe; fruits sessile or stipitate, often both kinds in same head; staminate heads usually more than 1, often 2 or more, somewhat confluent.

Usually in shallow or deep water at elevations of 4,000–12,000 feet, the common floating form of high montane lakes: San Bernardino Mountains,

Fig. 12. *Sparganium simplex: a,* mature fruit, × 4; *b,* mature fruiting head, × ⅘; *c,* young pistillate flowers, stipitate, × 4; *d,* group of staminate flowers with irregular perianth scales, × 6; *e,* habit, showing the triangular-keeled leaves and bracts and the fruiting burs, × ⅖; *f,* young plant with leaves overtopping the staminate inflorescence, × ⅖; *g,* 1-seeded fruit (cross section), × 4.

throughout the Sierra Nevada and higher North Coast Ranges, east of the Sierra Nevada crest from Mono County to Modoc plateau; cooler parts of North America.

4. Sparganium minimum (Hartm.) Fries, Summa Veg. Scand. 68:560. 1849.

Slender submersed or suberect plants, 1–8 dm. long; leaves 2–8 mm. wide, flat, without evident keel; inflorescence simple, the fruiting heads 1–4, all in axils of bracts, 8–12 mm. in diameter; staminate head solitary; fruits sessile or stipitate, beaked.

In California known only from the Lake Tahoe region (*Nobs & Smith 1509*); Oregon to Alaska, east across the continent, south to New England.

Readily identified by the very small fruiting heads. The species is very close to *Sparganium hyperboreum* Laestad, differing chiefly in its beaked fruit and the always simple inflorescence.

POTAMOGETONACEAE. PONDWEED FAMILY

Perennial aquatic herbs from rhizomes or detached winter buds, submersed or floating in quiet fresh or brackish water. Leaves alternate or opposite, of one or two kinds, the floating leaves broad, the submersed ones narrower, often threadlike or linear. Stipules present, free or adnate to leaf base, sometimes sheathing the stem, or united by their margins in front of leaf to form a ligule which may be inserted at the junction of the petiole and stem or at the base of the blade. Flowers in spikes on axillary peduncles, the bud enclosed by a stipular sheath. Stamens 4, the connective of each produced into a broad, sepaloid structure resembling a perianth segment and here called a sepaloid connective. Pistils 4.

POTAMOGETON

Because of intergradation, possibly due to hybridization, identification by means of existing keys is often difficult and unsatisfactory. The entities herein accepted are presumed to represent the main centers around which variation occurs. A few cases of obvious hybridization are known. The hybrids are usually characterized both by a high degree of sterility of the pollen and by imperfectly formed fruits.

Two interesting morphological problems present themselves in this genus. One is the character of the so-called sepaloid connective, which behaves as a perianth segment; the other is the character of the stipules. The question in respect to the sepaloid connective is whether it truly arises from the connective or whether it is a true perianth segment to which the stamen is adnate. Similar structures occur in *Lilaea* and *Triglochin*. The question concerning the stipules

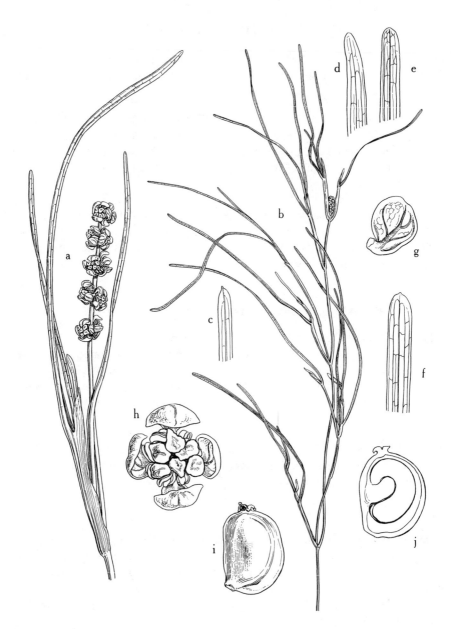

Fig. 13. *Potamogeton filiformis: a,* apical part of plant, showing the fleshy, linear leaves with adnate stipules sheathing the young leaf blades and the 5 regularly spaced flower whorls, × 2; *b,* habit, × ⅘; *c–f,* leaf tips, showing variation from blunt to submucronate, × 4; *g,* sepaloid connective, showing pronounced veins, × 8; *h,* flower, × 8; *i,* achene, showing rounded back and nearly central, wartlike beak, × 8; *j,* achene (longitudinal section), × 8.

is, does the situation in *Potamogeton* give evidence of the stipular origin of ligules generally?

Fernald, M. L., The linear-leaved North American species of *Potamogeton,* section *Axillares.* Mem. Am. Acad. Arts & Sci. II. 17:1–183. 1932.

Stipules adnate to the leaf base forming a sheath enfolding stem; leaf appearing as though arising from top of sheath.

  Plants with submersed leaves only; fruiting spikes slender, if capitate see no. 5 below.

    Leaves all linear-filiform, sometimes setaceous, less than 1 mm. broad; free part of stipules not united to form a ligule.

      Stigma disc-shaped, sessile or short-stalked .... 1. *P. filiformis*

      Stigma not discoid, the stigmatic tip prolonged and with little evident swelling

                                                      2. *P. pectinatus*

    Leaves more than 1 mm. broad; free part of stipules united in front of leaf to form a ligule.

      Leaf margins entire; lobes of ligule entire; leaf blades not auricled at junction with sheath; nutlets usually without keel .... 3. *P. latifolius*

      Leaf margins minutely serrulate; lobes of ligule often lacerate; leaf blades often auricled at junction with sheath; nutlets prominently 3-keeled

                                                  4. *P. Robbinsii*

  Plants with long-petioled, floating leaves and sessile, linear, submersed leaves, if floating leaves absent the fruiting spikes of the submersed parts capitate and sessile or on very short peduncles .... 5. *P. diversifolius*

Stipules axillary and free from the leaf base, either free from one another or united to form a cylinder around the stem, or the margins enfolding the stem not united.

  Leaves all submersed and essentially alike; petioles short or absent.

    Leaves lanceolate, oblong, or ovate.

      Leaves coraate at base, sessile, or partially clasping the stem.

        Stems straight, usually not white; stipules not conspicuous; leaf base completely clasping the stem or nearly so .... 6. *P. Richardsonii*

        Stems zigzag, usually white; stipules conspicuous; leaf base clasping not more than ½ of the stem .... 7. *P. praelongus*

      Leaves tapering to the base, sessile or petioled.

        Leaves broadly lanceolate-attenuate, large, serrulate only at tip .. 8. *P. illinoensis*

        Leaves oblong and crisped, serrulate throughout, rounded at tip .... 9. *P. crispus*

    Leaves linear-filiform or setaceous.

      Leaves broadly linear and grasslike, 2–5 mm. broad .... 10. *P. zosteriformis*

      Leaves filiform or setaceous, less than 2 mm. broad.

        Stipules connate into a cylinder, clasping the stem at least below.

          Stipules with conspicuous veins, persisting as fibers on disintegration of stipules

                                                  11. *P. fibrillosus*

          Stipule fibers not conspicuous, disappearing with stipules upon their disintegration.

            Keel of fruit conspicuously dentate-winged; spikes capitate .. 12. *P. foliosus*

            Keel of fruit not dentate; spikes short-cylindric .... 13. *P. pusillus*

        Stipules ligulate, open to base on the side of stem opposite leaf

                                            14. *P. Berchtoldii*

  Leaves of two kinds, submersed and floating, the floating leaves with broad blades and long petioles.

    Submersed leaves sessile or petioled, lanceolate, oblong or ovate, at least some well over 5 mm. wide.

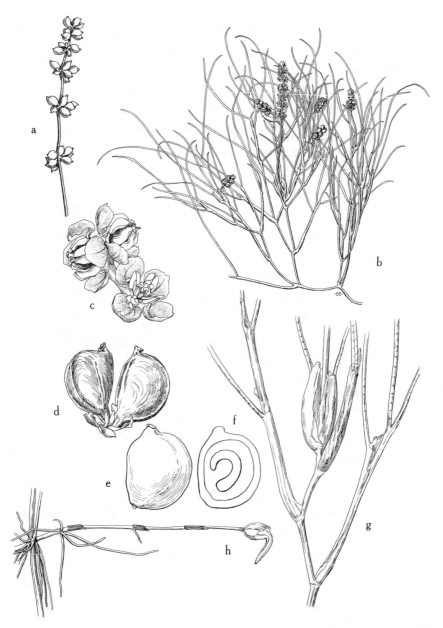

Fig. 14. *Potamogeton pectinatus: a,* moniliform spike with mature achenes, × 1⅕;
b, habit, showing slender, branching stems and linear-filiform, submersed leaves, arising
from rhizome, × ⅔; *c,* upper flowers of spike, × 4; *d* and *e,* variation in achenes (usually
obliquely ovoid, with a short, wartlike beak), *d,* × 5½, *e,* × 5; *f,* achene (longitudinal
section), × 5; *g,* stipules sheathing stem or loosely investing it and somewhat inflated,
the linear leaf appearing to originate at the top of the sheath, × 3; *h,* rhizome with winter
corm, × 1⅕.

Floating leaves, when present, delicate and thin, no sharp distinction between blade
and petiole, often absent, the submersed leaves not arcuately folded; stems
and peduncles usually red .... 15. *P. alpinus*
Floating leaves coriaceous, with sharp distinction between blade and petiole.
Submersed leaves ovate-lanceolate, arcuately folded or falcate in outline, sessile
or on short petioles .... 16. *P. amplifolius*
Submersed leaves lanceolate, flat, on long petioles .... 17. *P. nodosus*
Submersed leaves linear or filiform, usually less than 5 mm. wide, or, if broader, then
very unequal in size and shape.
Submersed leaves linear, usually bladeless and filiform, 0.8–2 mm. wide, blade
when present linear-lanceolate and on a very long petiole; floating leaves broad,
many-veined, base of blade subcordate .... 18. *P. natans*
Submersed leaves linear to lanceolate or oblanceolate, 1–15 mm. wide, sessile.
Submersed leaves ribbon-like, that is, with essentially parallel margins, 6–20 cm.
long, in California to 5 mm. wide .... 19. *P. epihydrus*
Submersed leaves linear to linear-lanceolate or linear-obovate, often very unequal
in size, usually tapering to tip and base, 3–12 cm. long, to 15 mm. wide
20. *P. gramineus*

1. Potamogeton filiformis Pers., Syn. Pl. 1:152. 1805. Fig. 13.

Potamogeton interior Rydb., Fl. Colo. 13. 1906.

Slender, much-branched, wholly submersed plant of brackish waters, with
horizontal stolons bearing white tubers 1–2 cm. long; stipules adnate to leaf
and sheathing the stem, the sheaths 0.4–2.2 cm. long, connate below, the tips
free, scarious, 1–5 mm. long; leaves setaceous, to 12 cm. long, 0.2–0.5 mm.
wide, blunt; peduncles filiform, flexuous, to 1 dm. long; spike moniliform,
1.5–5 cm. long, with 2–5 whorls, the upper whorls 3–12 mm. apart, the lower
ones 0.7–2.5 cm. apart; connectives 0.5–1 mm. long; styles almost wanting;
nutlets sessile, 2–2.7 mm. long, 1.5–2 mm. wide, rounded on back, the beak
short, wartlike, nearly central.

Ponds, slow streams, and ditches, widely distributed but not common: central Sierra Nevada, San Francisco Bay region, Modoc and Lassen counties;
across northern part of North America, Eurasia.

Differs from *Potamogeton pectinatus* primarily in its sessile stigma, nearly
beakless fruit, and obtuse leaves.

2. Potamogeton pectinatus L., Sp. Pl. 127. 1753. Sego pondweed. Fig. 14.

Potamogeton vaginatus Turcz., Bull. Soc. Imp. Nat. Moscow 27^2:66. 1854.

Wholly submersed plant from thickly matted rhizomes, perennial by fleshy
winter corms, often growing in very large masses; stems very slender, to 2.5 m.
long; leaf blades linear-filiform, 5–35 cm. long, appearing to originate at top
of the sheath, the stipules adnate to the leaf and forming a closely or loosely
investing sheath 2–3 cm. long around the stem, those surrounding the inflorescence becoming somewhat inflated, the leaf tips acute to attenuate or sometimes obtuse to apiculate, in older plants several types sometimes occurring on
the same plant; peduncles slender, 5–25 cm. long; spikes conspicuously inter-

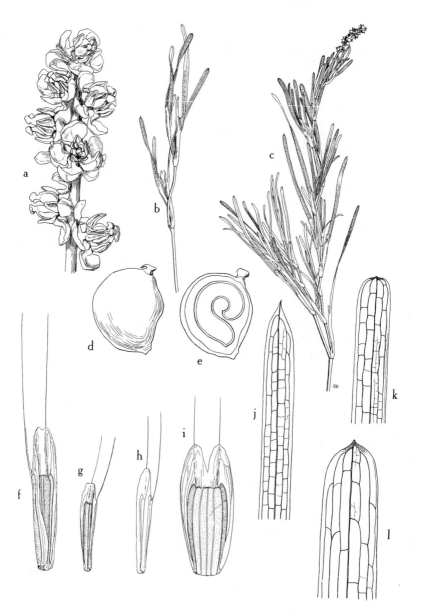

Fig. 15. *Potamogeton latifolius: a,* interrupted flowering spike, showing the reflexed sepaloid connectives, × 2; *b,* young branch, × ⅖; *c,* habit, × ⅖; *d,* achene, × 6; *e,* achene (longitudinal section), showing tip of curved embryo directed toward base of seed, × 6; *f–i,* sheaths and ligules, showing variation in ligule apices, × 1½; *j–l,* leaf tips, showing variation in apices and venation, × 4.

rupted by 2–6 unequally remote whorls, the axis of spike becoming lax at maturity; nutlets plump, 2.5–5 mm. long, obliquely ovoid, with a short beak or the beak reduced to a wartlike process.

Floating below the surface in fresh or brackish water; the most common species, often occurring in great masses: throughout California; cosmopolitan.

Fruits mature in California plants about the middle of July. In resting ponds along the main flyways, these plants are so eagerly sought by the ducks that they almost completely disappear by the end of August.

3. Potamogeton latifolius (Robbins) Morong, Mem. Torrey Club 3:52. 1893. Western pondweed. Fig. 15.

Potamogeton pectinatus var. latifolius Robbins in Wats. in King, Geol. Expl. 40th Par. 5:338. 1871.

Coarse, wholly submersed plant from creeping rhizomes, rooting at the nodes; stems slender to stout, whitish, 2–6 dm. tall, much-branched, the ultimate branchlets short, at least in age, wholly submersed, usually very leafy; stipules adnate to the leaf base, clasping the stem as a sheath, sometimes cylindric below and confluent across the top of the sheath to form a conspicuous hyaline ligule, the sheath 7-nerved, 5–15 mm. long, the nerves arched and confluent at the top as they enter the blade, the outer nerves supplying the stipules and the ligule with a few somewhat obscure branches, the ligule 3–7 mm. long, truncate or rounded across the top or retuse, emarginate, or even 2-lobed, becoming irregularly lacerate in age; leaves linear, entire, 2–10 cm. long, 2–5 mm. wide, the blade somewhat thick, 3- to 5- or to 7-nerved, the lateral nerves often merging with the margin, the midvein entering the rounded or acute tip, the others inarching and anastomosing just below the tip, cross-nerves abundant and forming a pattern of somewhat quadrate but unequal and irregular lacunae; peduncles and spikes 2–8 cm. long, slender, the spike cylindric, interrupted, 8- to 26-flowered; part of filament above the sepaloid connective elongating prior to anthesis, that below the sepaloid connective elongating during anthesis; blade of connective concave, elliptic, entire, 2 mm. long, 2–3 mm. wide, the stalk becoming 1–2 mm. long; anthers ellipsoid; pistils narrowly ovoid; style stout, about ½ as long as ovary, somewhat oblique, persistent; stigma capitate, flattened; nutlets 3–4 mm. long, 2–3 mm. wide, somewhat quadrate in outline, the median keel obscure, the lateral keels obsolete, the coil of embryo complete, the tip directed toward base of seed.

Northeastern California, Crane Ranch south of junction of Merced and San Joaquin rivers in Merced County; east in the Great Basin to Utah, Arizona, and Texas.

As seen in the herbarium, Potamogeton latifolius appears to be of two types, namely, a short, narrow, thin-leaved type and a long, broad, thick-leaved type. Two specimens taken on the same day at Amedee, Honey Lake Valley, Lassen County, illustrate these types. When sufficient material is available for study,

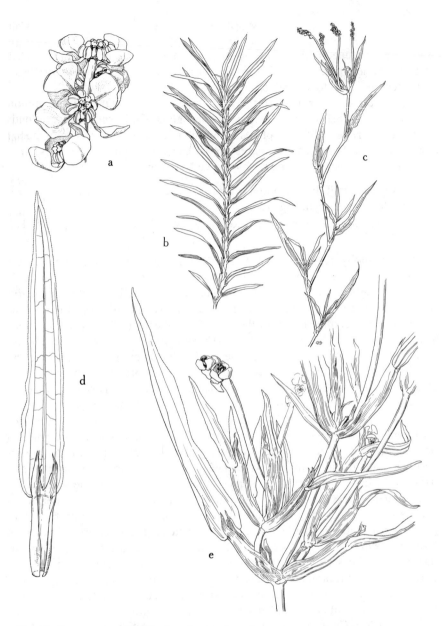

Fig. 16. *Potamogeton Robbinsii: a,* flowers in spike, × 6; *b,* sterile branch, with closely spaced, 2-ranked leaves, × ⅖; *c,* fertile part of plant, showing long internodes and zigzag stem, × ⅖; *d,* leaf, showing conspicuous stipular sheath and free, lacerate stipule, × 2; *e,* fertile stems, showing the few-flowered spikes on clavate peduncles, × 1½.

however, these differences break down completely. Apparently these types only represent growth stages, for no significant differences in other respects have been noted that cannot be found on single plants.

4. Potamogeton Robbinsii Oakes, Mag. Hort. Hovey 7:180. 1841. Fig. 16.

Wholly submersed plant with stems to 2 m. long, from creeping rhizomes, usually clothed with closely spaced, 2-ranked leaves, the internodes increasing in length in the fertile parts of plant; leaves with a conspicuous stipular sheath 5–12 mm. long, culminating in free, lacerate, white, membranous stipules 1–3 cm. long, the blades linear to lanceolate, 3–7 cm. long, auriculate at junction with sheath, finely nerved, the margin entire or minutely serrulate; inflorescences on slender, clavate peduncles, few-flowered, often in pairs at a single node; nutlets said to be 3-keeled on back, the lateral keels rounded.

Northeastern California, Sierra Nevada south to Yosemite region; north to British Columbia, east to Atlantic.

Fruiting material of this species is exceedingly rare. We have seen none. The plant is reported generally as being sterile and bearing much aborted pollen. Yet the species is widely distributed over the northern part of the continent and has very distinctive characters. It is suspected of being a hybrid, yet it would be difficult to account for its distinctive characters as stemming from any other living species. Hagström suggested that this plant was an ancient hybrid, and that its fertile ancestry was to be sought in the western United States. This was probably suggested by the fact that the first fruit ever taken was a single one from Oregon. We can, however, vouch for the fact that the species is as sterile in the western United States as it is elsewhere. Morong suggests that fruiting is dependent on the lowering of water levels.

5. Potamogeton diversifolius Raf., Med. Repos., 2d hexade, 5:354. 1808. Fig. 17.

Potamogeton dimorphus of California authors, not Raf.

Delicate plant, perennial by winter buds; submersed leaves narrowly linear, 1–6 cm. long, the apex acute but not setaceous, the stipules adnate, the free ligule often 1–2 times as long as the sheath; floating leaves 1–4 cm. long, the petiole 1–2 times as long as the blade, the stipules free or barely adnate, lanceolate, 5–10 mm. long, the blade elliptic, 5–30 mm. long, 3–8 mm. wide, the tip acute, the base tapering to petiole; spikes in axils of submersed leaves capitate, 1- to 4-flowered, sometimes cleistogamous, often on short, recurved, clavate peduncles or sometimes subsessile; spikes in axils of floating leaves capitate to cylindric, few- to 50-flowered, subsessile to stout-peduncled, the peduncles rarely longer than spikes; anthers sessile, the sepaloid connectives apparently sessile; nutlets usually green, with a conspicuous, somewhat undulate or denticulate keel, the lateral angles evident, keeled or sometimes rounded, the face flat to slightly depressed, the beak toothlike to almost obsolete, erect or laterally directed, the embryo strongly coiled.

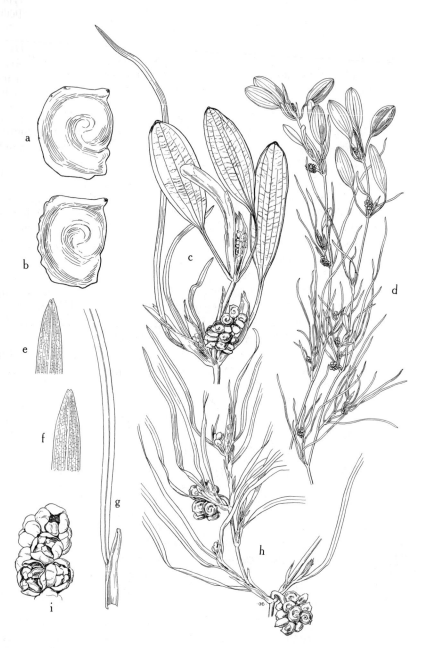

Fig. 17. *Potamogeton diversifolius: a* and *b,* achenes, showing the angular, often denticulate, outline of the dorsal keel and the strongly coiled embryo, × 16; *c,* upper part of
stem, the floating leaves elliptic, the submersed leaves linear, × 2; *d,* habit, showing the
numerous, capitate, subsessile spikes, × ⅔; *e* and *f,* tips of submersed leaves, × 10;
g, linear leaf blade arising from stipule, and the long, free ligule, × 4; *h,* mature capitate
spikes, showing reflexed peduncles in axils of submersed leaves, × 2; *i,* flowers in spike,
× 8.

Fig. 18. *Potamogeton Richardsonii: a,* habit, showing the leaves becoming progressively shorter toward tip of branch, × ⅖; *b,* achene with obscure keel, × 6; *c,* achene, showing variation in shape, × 6; *d,* achene (longitudinal section), × 6; *e,* part of a flowering spike, × 4; *f,* part of stem, showing cordate, clasping leaf bases with the persistent fibers of the deteriorating stipules in their axils, × 1⅕; *g,* leaves, showing venation, undulate margins, and young, lanceolate stipules, × 1⅕.

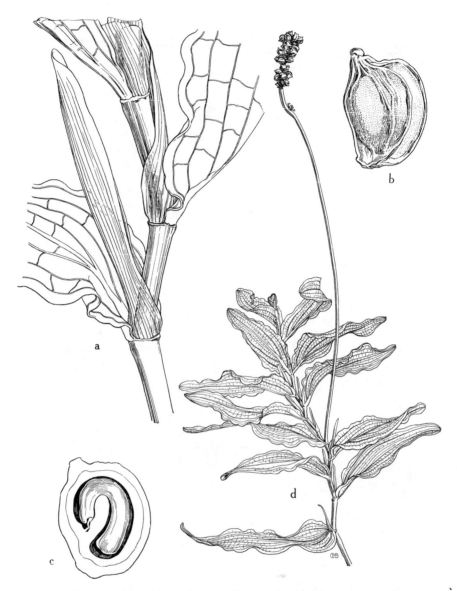

Fig. 19. *Potamogeton praelongus: a,* part of stem, showing zigzag nodes, base of undulate-margined submersed leaves, and the oblong-lanceolate stipule without a keel, × 4;
b, achene, showing sharp keels and stout beak, × 6; *c,* achene (longitudinal section), × 6;
d, habit, × ⅖.

Ponds and irrigation ditches: Central Valley, Boggs Lake in Lake County, Modoc County; east to Atlantic mainly in the southern states, Mexico.

For an account of the identity of and the tangled synonymy between *Potamogeton diversifolius* Raf., *P. dimorphus* Raf., and *P. spirillus* Tuckerm., see Fernald (*op. cit.,* p. 102).

6. Potamogeton Richardsonii (Bennett) Rydb., Bull. Torrey Club 32:599. 1905. Fig. 18.

Potamogeton perfoliatus var. *Richardsonii* Bennett, Jour. Bot. Brit. & For. 27:25. 1889.

Plant usually from a tangled mat of rhizomes in fairly deep water; stems terete, branched, not zigzag; leaves sessile, the blades ovate to broadly lanceolate, cordate-clasping at base, 5–10 cm. long, the margins undulate, the stipules axillary, lanceolate, at length deteriorating into persistent fibers; peduncles of spikes subequal to or somewhat longer than spikes; nutlets about 4 mm. long and 2 mm. wide, obscurely keeled to rounded on back, the lateral angles obscurely developed.

Deep lakes and ponds or canals: Sierra Nevada north to Modoc and Siskiyou counties; widespread across northern part of North America.

The species may be readily distinguished by its undulate clasping leaves and the persistent strands of fibers from the deteriorating stipules. Hagström expresses the opinion that although this is to be regarded as a distinct species, it seems to have arisen through hybridization between *Potamogeton praelongus* and *P. perfoliatus,* since it is quite intermediate in many characters between these two.

7. Potamogeton praelongus Wulf in Roem., Arch. Bot. 3:331. 1805. Fig. 19.

Plant from stout rhizomes; stems simple or branched, white or whitish, often zigzag; leaves all submersed, sessile, ovate to lanceolate, cordate or tapering to base but not wholly cordate-clasping, never clasping more than halfway, 5–20 cm. long, 1–3 cm. broad, sometimes strongly undulate, the tip often hooded, splitting when drying, the stipules white, oblong-lanceolate, persistent, without keel, 3–8 cm. long; peduncles thickened upward, 5–20 cm. long; spikes of 6–12 whorls of flowers; nutlets obovate, sharply keeled and with a stout beak.

High lakes in usually deep, cold water, infrequently collected: Sierra Nevada at Webber Lake in Sierra County and Willow Lake in Plumas County; widespread across northern part of North America, Eurasia.

In herbaria this species is often confused with *Potamogeton alpinus* and with *P. illinoensis.* From the former it may be distinguished by its larger-sized leaves and zigzag white stems; from the latter by these stem characters and by its broadly sessile leaf bases and usually rufous color.

Fig. 20. *Potamogeton illinoensis: a,* habit, showing profusion of crowded leaves, conspicuous stipules, and long, stout peduncle, $\times$ ⅖; *b,* part of flowering spike, $\times$ 4; *c,* achene, showing strong dorsal keel, smooth face, and short beak, $\times$ 6; *d,* achene (longitudinal section), $\times$ 6.

8. Potamogeton illinoensis Morong, Bot. Gaz. 5:50. 1880. Fig. 20.

Potamogeton lucens of American authors, not L.

Plant often profuse from a network of interlaced horizontal rhizomes; leaves all submersed, 6–20 cm. long, lanceolate to elliptic on short petioles spreading horizontally, these often unequal in length, the blade 11- to 15-nerved, the margins entire or minutely serrulate, the stipules 3–8 cm. long, bicarinate, usually membranous and often attenuate; peduncles 5–15 cm. long, stout; spikes 3–8 cm. long, often fruiting densely; nutlets 3–4 mm. long, with a strong keel and the lateral angles often conspicuous, the faces often smooth, the beak nearly erect.

In ponds and canals: Clear Lake in Lake County, San Francisco, valleys and foothills bordering the Sacramento and San Joaquin valleys, south to San Diego County; widespread across southern Canada and the United States.

9. Potamogeton crispus L., Sp. Pl. 126. 1753. Fig. 21.

Plant from a network of slender rhizomes, perennial by detached, fleshy winter buds; stems with very short internodes; leaves thickened, strongly dentate, all submersed, sessile, the blade 4–8 cm. long, 5–6 mm. wide, oblong, rounded at apex, the margin usually serrulate, the stipules axillary, 3–5 mm. long, often uniting around the stem at the base to form a sheath, or ligulate and free on opposite side, lacerate and soon deteriorating; spikes on slender, often recurved, somewhat clavate peduncles, usually few-flowered; nutlets 4–6 mm. long with a stout, erect or somewhat curved beak as long as or longer than the body of nutlet, the back keeled, the keel entire or somewhat denticulate, the lateral angles obscure.

Slow-running streams or canals, at low altitudes: San Joaquin and Sacramento valleys, San Francisco Bay region; eastern states and southeastern Canada, Europe.

A well-marked species, clearly defined by its very common, gall-like winter buds, the well-developed, stout beak of the fruit, the oblong, crisp leaves and the short, often corona-like stipules.

10. Potamogeton zosteriformis Fern., Mem. Am. Acad. Arts & Sci. II. 17:36. 1932. Fig. 22.

Potamogeton compressus of California authors, not L.

Plant from slender rhizomes, perennial by detached winter buds; stems branched, flattened, 0.7–4 mm. wide, somewhat constricted at the nodes; leaves sessile, broadly linear, 5–20 cm. long, 2–5 mm. wide, many-nerved, obtuse or subacute or mucronate at apex, the stipules axillary, green to hyaline, rounded or truncate at apex, 2–6 cm. long, at first clasping the stem, at length opening and deteriorating; peduncles slender, subclavate, 1–2 times as long as the spike, 2–6 cm. long; spikes 2–3 cm. long; nutlets 5 mm. long, 3 mm. wide,

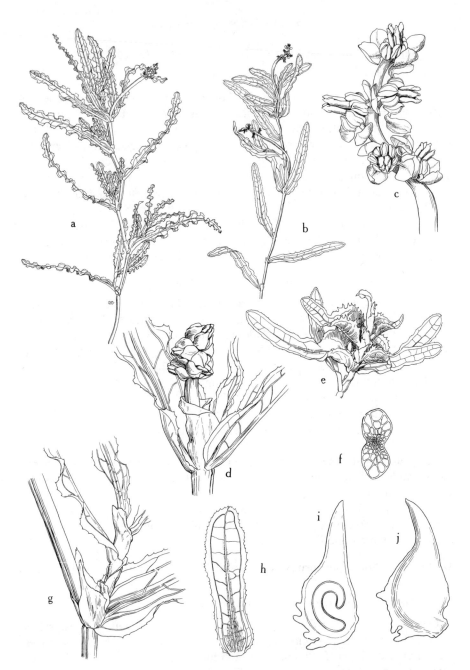

Fig. 21. *Potamogeton crispus: a,* habit, × ⅖; *b,* branch with maturing spikes, × ⅖; *c,* few-flowered spike, × 4; *d,* young flowering spike, emerging from sheathing stipules, × 3; *e,* winter bud, showing fleshy stems, short internodes, and thickened, foliaceous bud scales with strongly dentate, broadened bases, × 1⅕; *f,* stem (cross section), × 6; *g,* ligulate stipule, × 1½; *h,* young leaf, showing venation, × 3; *i,* achene (longitudinal section), × 6; *j,* achene, showing the somewhat curved beak and variation in the denticulate dorsal keel, × 6.

Fig. 22. *Potamogeton zosteriformis: a,* flowers, × 6; *b,* achene, showing the sharp keel, rounded lateral angle, and depressed face, × 6; *c,* stem (cross section), × 6; *d,* winter bud, × ⅖; *e,* rhizome developing from detached winter bud, × ⅖; *f,* habit, showing the long, clasping stipules, × ⅖; *g,* mucronate leaf tip, × 2; *h,* ruptured lower stipule and deteriorating upper stipule, × 2; *i,* achene (longitudinal section), × 6.

Fig. 23. *Potamogeton foliosus: a,* spike, showing short, clavate peduncle, × 8; *b,* achene, showing the thin, undulate-toothed keel, × 10; *c,* habit, × ⅔; *d,* fruiting spike, showing connate and ruptured stipules, × 4; *e,* leaf tip, × 8.

sharply keeled, the lateral angles rounded, somewhat depressed on faces, the beak apical, somewhat reclining, or sometimes erect.

Streams, canals, and ponds: Clear Lake in Lake County, San Joaquin River delta, abundant in northeastern California; across the northern United States and southern Canada.

11. Potamogeton fibrillosus Fern., Mem. Gray Herb. 3:51. 1932.

Plant resembling *Potamogeton foliosus;* stems to 6 dm. long, somewhat compressed; leaves linear-acute, 2–4 cm. long, 1–2 mm. wide, the stipules connate, at length splitting, rigidly fibrous, the fibers becoming conspicuous as the stipules disintegrate; peduncles usually short-clavate, sometimes slender-cylindrical, 4–12 mm. long; spikes capitate to short-cylindric; nutlets obliquely obovoid, compressed, 1.5–2.5 mm. long, rounded on the back or with a narrow keel that may be obscurely toothed.

Ponds, ditches, and slow streams: Crescent City in Del Norte County; north and east through Oregon, Washington, Montana, and Wyoming.

Under this name Fernald described a species based upon a specimen collected on "P" Ranch in Harney Valley in Oregon (*Cusick 2598*). Under *Potamogeton pusillus* var. *typicus* (now *P. Berchtoldii* Fieber) he listed a California plant collected by H. E. Parks near Crescent City, Del Norte County, California. These two specimens have in common the singularly fibrous disintegrating stipules, a character which served to distinguish *P. fibrillosus* from *P. pusillus* and *P. Berchtoldii,* and from which the name *P. fibrillosus* is derived. The California specimen differs in having a somewhat more slender peduncle, approaching some of the shorter-peduncled specimens of *P. pusillus* in this respect. It is my belief that these two specimens are conspecific and that this Del Norte County specimen is the only known record of *P. fibrillosus* in California.

12. Potamogeton foliosus Raf., Med. Repos., 2d hexade, 5:354. 1808. Fig. 23.

Potamogeton foliosus var. *californicus* Morong, Mem. Torrey Club 3:40. 1893.

Plant from a dense mat of slender rhizomes, often rooting at the nodes; stems compressed-filiform, sometimes as much as 1 m. long but often short, much-branched and bushy; leaves narrowly linear, tapering at the base, acute to cuspidate at tip, 1–10 cm. long, 0.3–2.5 mm. wide, the stipules connate into a tubular sheath soon rupturing and deciduous, the veins inconspicuous to conspicuous, those of stipules sometimes persisting as fibers; peduncles short-clavate, 3–10 mm. long; spikes subglobose or subcapitate, the internodes between the few-flowered whorls very short; nutlets obliquely orbicular to obovoid, 1.5–2.5 mm. long, with a thin, undulate, toothed keel, the beak erect, submarginal.

Ponds, slow streams, and irrigation ditches in the valleys and mountains; east to Atlantic, south to the West Indies and Mexico.

Fig. 24. *Potamogeton pusillus: a,* flowering spike, × 6; *b,* achene, obliquely obovoid, smooth, with slightly recurved beak, × 8; *c,* achene (longitudinal section), × 8; *d,* habit, showing narrowly linear submersed leaves, × ⅔; *e,* stem (cross section), × 20; *f,* part of stem, showing young, tubular stipules in upper part and split, disintegrating stipules at base, × 2; *g* and *h,* winter buds, × 1½.

Two subspecies, based upon the size and relative density of branching of the plants, are usually delineated in this species; in California, however, they have no geographic status, because each subspecies is found throughout the range of the other. The differences are related, for the most part, to internode length. The typical variety, *Potamogeton foliosus* var. *foliosus,* has long internodes, with the result that the plants are often as much as a meter in length. *Potamogeton foliosus* var. *macellus* Fern. (*op. cit.,* p. 46) is a dense plant.

Potamogeton foliosus may be readily distinguished from *P. pusillus* by the conspicuously short, clavate peduncles and the toothed, winged keels of the fruit.

13. Potamogeton pusillus L., Sp. Pl. 127. 1753. Fig. 24.

Plants perennial from cormlike winter buds; stems slender, often capillary, somewhat flattened, to 1 m. long; leaves all submersed, narrowly linear, 0.3–3 mm. wide, 1–7 cm. long, 3- to 5-nerved, often with a pair of minute glands at junction with stem, the midvein often prominent beneath; stipules membranous to scarious, 6–15 mm. long, usually slender, tubular with the margins united around the stem at least to above the middle, sometimes those subtending an inflorescence becoming split and opening or expanding, soon becoming truncate or lacerate across the tip and ultimately disintegrating completely or sometimes reduced to fibrous shreds, usually evident only in younger parts of plant; peduncles slender, 1.5–8 cm. long, rarely somewhat enlarged upward; spikes cylindric, 6–12 mm. long, usually interrupted, the whorls 3–5; nutlets 2–3 mm. long, obliquely obovoid, smooth, with a low, broad, somewhat obscure keel, rarely with a slightly winged or inconspicuously toothed keel, the faces sometimes plump but more often conspicuously impressed and flattened, strongly rounded on back, adaxial margin arching, the beak erect above the adaxial margin or slightly recurved.

Common in slow, fresh or brackish or sometimes alkaline streams, irrigation ditches, and ponds, often in large masses: almost throughout California, one of the very common species of the valleys, foothills, and mountains; widespread in Northern Hemisphere.

There are two forms common in California: the typical form has leaves 1–3 mm. wide, and *Potamogeton pusillus* var. *minor* (Bov.) Fern. & Schub. has leaves less than 1 mm. wide.

The name *P. panormitanus* has been used in American literature for this entity, but Dandy and Taylor (Jour. Bot. Brit. & For. 76:89–92, 1938) on studying the Linnaean type of *P. pusillus,* found that the type of *P. panormitanus* was identical with it. Furthermore, they found that *P. pusillus* of American authors, not of Linnaeus, should be referred to *P. Berchtoldii.*

Potamogeton pusillus is one of the very common species along the flyways and might readily be confused with the sego pondweed, *P. pectinatus.* It may be distinguished from this, however, by its much shorter and less distantly

Fig. 25. *Potamogeton Berchtoldii: a,* habit, showing dense, short, somewhat spreading leaves, × ⅖; *b,* young flowering spike surrounded by sheathing stipules, × 5; *c,* sepaloid connective, × 12; *d,* habit of a plant with leaves longer and more linear-setaceous than those of the plant in *a,* × ⅖; *e* and *f,* typical leaf tips, showing venation, × 8; *g,* achene (longitudinal section), × 12; *h,* achene, showing the rounded, obscurely keeled back and the marginal, erect beak, × 12; *i,* winter bud, × 2.

interrupted spikes and much smaller seeds, and by the fact that its stipules are not adnate to the leaf margins and hence the leaves do not appear as though sheathing the stem. From *P. foliosus* it differs in its longer, less clavate peduncles of the spikes, and in lacking the conspicuously winged and toothed keels of the fruits. From *P. Berchtoldii* it may be distinguished by its cylindrical sheathing stipules.

The plants produce an abundance of seed, and in the fall of the year they produce quantities of winter buds, about the size of peas, that are within the reach of bottom-feeding birds.

14. **Potamogeton Berchtoldii** Fieber, Potamogeta Böhmens 40. 1838. Fig. 25.

Potamogeton pusillus of Fern., Mem. Am. Acad. Arts & Sci. II. 17:80. 1932. Not L.

Plant from winter buds and without elongated rootstock; stems capillary, simple or freely branching; leaves linear to linear-setaceous, 1.8 cm. long, 0.5–2.5 mm. wide, often with a pair of translucent glands at the base; stipules flat, or the margins inrolled, not connate into a cylinder, those of the inflorescence usually broad; peduncles slender, scarcely thickened upward, 5–40 mm. long; spikes capitate to subcylindric, becoming somewhat interrupted in fruit; nutlets plump, obovoid to obliquely obovoid, rounded on back with an obscure keel, the beak marginal and erect.

Lakes and ponds at middle and higher altitudes of the Sierra Nevada, apparently rare in California; widespread in Northern Hemisphere.

Under the name *Potamogeton pusillus* var. *tenuissimus,* Fernald cited four California specimens from the Sierra Nevada. It was subsequently shown by Dandy and Taylor (Jour. Bot. Brit. & For. 76:90–92, 1938) that the taxon which had been going under the name of *P. panormitanus* was in reality *P. pusillus* L., and that the one which had been going under the name *P. pusillus* was *P. Berchtoldii. Potamogeton Berchtoldii* differs both from the typical *P. pusillus* L. and from *P. foliosus* in having open ligulate stipules. Our material seems otherwise very close to *P. pusillus* L.

15. **Potamogeton alpinus** Balbis, Misc. Bot. 13. 1804. Fig. 26.

Rhizomatous, slender perennial; stems simple or few-forked, the internodes usually shorter than leaves, the whole plant reddish brown or olivaceous; submersed leaves very thin, sessile, tapering at both ends or rounded apically, lanceolate to elliptic-oblong, 2–8 cm. long and 1–2 cm. wide; floating leaves, when present, gradually differentiated from submersed leaves, lanceolate, tapering to a short petiole; stipules membranous, very inconspicuous, deteriorating rapidly, soon-deciduous; spikes often shorter than terminal leaves; nutlets obliquely obovoid, 2.5 mm. long, the beak nearly central to lateral.

In lakes and ponds, usually at middle and upper elevations: Sierra Nevada to the Modoc plateau; east to Atlantic, eastern Asia.

The common, high montane pondweed in California, *Potamogeton alpinus,*

Fig. 26. *Potamogeton alpinus: a,* upper part of plant, showing young stipules, × 4; *b,* habit, showing lanceolate leaves tapering at both ends, × ⅖; *c,* single flower, × 8; *d,* sepaloid connective, × 8.

Fig. 27. *Potamogeton amplifolius:* a, upper part of stem, showing floating leaves, the stout, upwardly thickened peduncle, and acute stipules, × ⅖; b, habit, showing rhizome, arcuate submersed leaves, broad stipules, and densely whorled flowers, × ⅖; c, sepaloid connective, × 6; d, achene (longitudinal section), × 6; e, achene, showing the flat sides and prominent beak, × 6; f, single flower, × 6.

Fig. 28. *Potamogeton nodosus: a,* submersed leaf, $\times$ ⅖; *b,* rhizome and young shoots, showing stipules and attenuate scales, $\times$ ⅖; *c,* venation in submersed leaf blade, $\times$ 2; *d,* upper part of stem, showing elliptic, long-petioled, floating leaves, $\times$ ⅖; *e,* achene, showing strongly developed dorsal and lateral keels and sculptured surface, $\times$ 8; *f,* spike, $\times$ 4.

is readily distinguished from *P. praelongus* by the absence of white zigzag stems. From *P. amplifolius* it is distinguished by its generally smaller, sessile, submersed leaves and much more slender rhizomes.

16. Potamogeton amplifolius Tuckerm., Am. Jour. Sci. II. 6:225. 1848. Fig. 27.

Plant from stout rhizomes; stems simple or branched near the top, often rufous; submersed leaves variable, from short-lived, lanceolate, short-petioled to persistent, broadly lanceolate and ovate, and folded along the midvein, the blade 8–20 cm. long, 25–75 mm. broad, tapering to petiole 1–6 mm. long, the stipules becoming fibrous and stringy, 3–10 cm. long; floating leaves similar to the upper submersed leaves to ovate or elliptic, round-tipped, rounded or tapering to the base, 5–10 cm. long, 25–50 mm. wide, the stipules usually 2-keeled; peduncles often thickened apically, 5–11 cm. long; spikes with 9–16 whorls of flowers, 4–8 cm. long when mature; nutlets 3–5 mm. long, obovate, rounded on back, cuneate at base, the sides flat, the beak prominent.

Lakes, at middle and lower altitudes: Sierra Nevada, North Coast Ranges; east across southern Canada and the United States to Atlantic.

Readily distinguished from *Potamogeton illinoensis* by its folded, arcuate leaves with strictly entire margins and by the rufous color common to plants of higher altitudes.

17. Potamogeton nodosus Poir. apud Lam., Encycl. Meth. Bot. Suppl. 4:535. 1816. Common American pondweed. Fig. 28.

Potamogeton americanus Cham. & Schl., Linnaea 2:226. 1827.

Plant from a network of rhizomes with triangular to attenuate scales at the nodes; stems 1–3 m. long; submersed leaves on petioles 2–15 cm. long, rarely the lowermost leaves sessile or subsessile and linear, the stipules soon decaying, 3–9 cm. long, the blades linear to elliptic-lanceolate, 1–4 cm. wide, acute at each end; floating leaves long-petioled, 3–12 cm. long, 1.5–4 cm. wide, often many-nerved, the blades oblong to elliptic, rounded at base, the apex rounded or obtuse, the stipules lanceolate, 3–9 cm. long; peduncles stout, 3–15 cm. long; spikes 3–7 cm. long, thick; nutlets ovate, the abaxial keel strongly developed, the lateral keels often muricate and the intervening intervals sometimes when dry appearing as though sculptured, the beak short, erect.

Very common in pools and ditches, often with very long stems: almost throughout California at low and middle altitudes; cosmopolitan.

Potamogeton nodosus is distinctive because of its usually very long-petioled submersed leaves (if sessile, the blades are linear and as much as 1–2 dm. in length). It is the most common species with floating leaves occurring at low and middle altitudes. At middle and higher altitudes *P. nodosus* is replaced by *P. gramineus. Potamogeton nodosus* is the plant that has been referred to as *P. americanus* in most western American botanical literature.

Fig. 29. *Potamogeton natans: a,* achene (longitudinal section), × 6; *b,* flower, × 4; *c,* habit, showing the long, linear, submersed leaves and broadly elliptic floating leaves, and the linear-lanceolate stipules, × ⅖; *d,* achene, showing strong keel on the back, × 6.

18. Potamogeton natans L., Sp. Pl. 126. 1753. Broad-leaved pondweed. Fig. 29.

Stems branching from a horizontal rhizome, otherwise usually simple; submersed leaves without blades, 1–3 dm. long, 0.8–2 mm. wide, rarely with a poorly developed blade, the stipules linear, 6–8 cm. long; floating leaves broadly elliptic to oblong, often subcordate at base, broadly rounded at apex, 25- to 27-nerved, the petiole longer than blade, the stipules 5–12 cm. long, linear-lanceolate, membranous; spikes in the axils of floating leaves, 3–6 cm. long, on stout peduncles, 1½–3 times as long as the spike; nutlets 3–5 mm. long, strongly keeled on the back, the lateral angles scarcely evident, the beak erect.

Marshy ponds and lakes, montane and coastal: San Bernardino Mountains, Sierra Nevada, North Coast Ranges; widespread in northern part of Northern Hemisphere.

This species is readily distinguished from *Potamogeton nodosus* by the broad basal part of the blade and by the petiole-like submersed leaves.

An apparent hybrid of *P. natans* and *P. illinoensis* was found growing in Clear Lake, Lake County, in association with its putative parents. Its characters are somewhat intermediate between those of the two species, and both pollen and seeds are sterile. The submersed leaves are much like those of *P. illinoensis,* although smaller; the floating leaves are more ellipsoid than those of *P. natans,* but they differ, in the hybrid, in tending to merge in character with the submersed leaves.

19. Potamogeton epihydrus Raf., Med. Repos., 2d hexade, 5:354. 1808. Fig. 30.

Rhizomes slender, often matted; stems compressed, simple or somewhat branching, often dichotomous at the tip; submersed leaves sessile, linear, ribbon-like, usually parallel-sided except for the narrowly acute tip, 5–25 cm. long, 2–8 mm. wide, the stipules of submersed leaves ligulate, 4 cm. long; floating leaves usually opposite, the blade oblong to elliptic, 2–8 cm. long, 4–35 mm. wide, rounded at tip, tapering below to a flattened petiole, the stipules ligulate, often attenuate; often many leaves transitional between floating and submersed types; peduncles 1–15 cm. long, stout; spikes 1–4 cm. long; nutlets round-obovoid, 2–4 mm. long, often 3-keeled, depressed or pitted on sides, the beak an obscure tooth.

Ponds and ditches: northeastern plateau, North Coast Ranges, middle elevations of northern and central Sierra Nevada; across the northern United States and southern Canada.

Potamogeton epihydrus is especially characterized by the ribbon-like, sessile submersed leaves, rarely more than 5–6 mm. wide. Variation has led to the recognition of *P. epihydrus* var. *Nuttallii* (Cham. & Schl.) Fern., upon the basis of specimens of a slightly smaller size. It is doubtfully distinct. See also the discussion under *P. gramineus.*

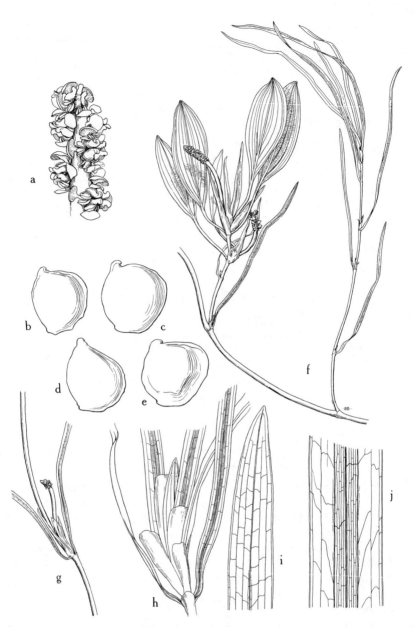

Fig. 30. *Potamogeton epihydrus: a,* upper part of spike, × 4; *b–e,* achenes, all taken from the same spike, showing great variability in outline and surface, × 6; *f,* habit, showing a flowering branch with floating leaf blades and sheathing stipules, and a sterile shoot with linear, submersed leaves, × ⅔; *g,* flowering spike among submersed linear leaves, × ⅔; *h,* ligulate stipules with truncate to emarginate apex, common in submersed leaves, × 1½; *i,* tip of narrowly linear leaf blade, × 4; *j,* venation in broad, linear leaf, × 4.

Fig. 31. *Potamogeton gramineus: a,* tip of compact flowering spike, × 4; *b,* keeled stipules on flowering branch, × 1½; *c,* submersed lower part of stem, showing the sterile branches with leaf variations and the young stipules clasping the stem, × ⅖; *d,* upper part of stem, showing submersed as well as floating leaves, × ⅖; *e,* achene with obscure keels, × 8; *f,* submersed foliage, showing transitional forms, × 1⅓; *g,* young, linear leaf, showing venation and tip, × 3.

Fig. 32. *Ruppia maritima: a* and *b,* variations in habit, the stems sometimes very long and slender or sometimes with short, zigzag nodes, × ⅖; *c,* peduncle bearing 2 young flowers, each consisting of 2 large, bicellular anthers and 4 pistils, × 8; *d,* 2 flowers, after fertilization, × 8; *e,* development of the long-pediceled fruits following fertilization of the 2 flowers (note elongated, coiled peduncle), × 2; *f,* mature nutlet, hard and black, × 8; *g,* 2 stipular sheaths of the alternate, capillary, succulent leaves, × 2; *h,* habit variation, × ⅖; *i,* serrate leaf tip, × 20.

20. Potamogeton gramineus L., Sp. Pl. 127. 1753. Fig. 31.

Potamogeton heterophyllus of Jepson, Man. Fl. Pl. Calif.; Abrams, Illus. Fl. Pacif. States. Not *P. heterophyllus* Schreb.

Plant from a mass of rhizomes; stems slender, occasionally fistulose, 2–15 dm. long; submersed leaves abundant, typically sessile (occasionally petioled), linear to lanceolate or oblanceolate, 3–12 cm. long, 1–15 mm. wide, acute and often with a short-attenuate tip, the stipules persistent; floating leaves on slender petioles, the blades ovate to elliptic, 1.5–7 cm. long, 1–3 cm. broad, usually shorter than petioles, the stipules lanceolate, somewhat keeled, persistent, 5–30 mm. long; peduncles stout, 2–10 cm. long; spikes compact, 1–4 cm. long when mature; nutlets obovate, 1.5–3 mm. long, obscurely keeled, the beak somewhat recurved.

Ponds and marshes: Modoc and Siskiyou counties, south along the Sierra Nevada to the Yosemite region and Mono County; widely distributed in northern part of Northern Hemisphere.

A variety, *Potamogeton gramineus* var. *maximus* (Morong) Fern., is based upon the relative size of the submersed leaves. It differs from the typical form in having at least some of its submersed leaves between 7 and 13 cm. long. Both the typical form and the variety occur in California, the latter being more abundant, the former occurring in middle altitudes of the Sierra Nevada.

Potamogeton gramineus may be easily distinguished from *P. nodosus* by its usually abundant, narrow ("gramineus") submersed leaves of great diversity of size and normally on short petioles. *Potamogeton nodosus* ordinarily has much larger submersed leaves, which are usually on long petioles. From *P. epihydrus, P. gramineus* differs in its larger submersed leaves, which are lanceolate rather than linear.

RUPPIACEAE. DITCH-GRASS FAMILY

Aquatic herbs of brackish or saline ponds or marshes. Leaves alternate or opposite, linear, sheathing at base. Flowers enclosed in a spathelike sheath, at length being raised on a much-elongated peduncle to the surface of the water. Perianth absent. Anthers 2, bilocular. Pistils 4, at length stalked. Stigma sessile, peltate. Fruit a nutlet.

Often included in Potamogetonaceae, from which it differs in the possession of 2 stamens, the absence of petaloid connectives, and the elongation of the stipe of the ovary after anthesis.

RUPPIA

1. Ruppia maritima L., Sp. Pl. 127. 1753. Ditch grass. Fig. 32.

Much-branched aquatic herb; plants 6–10 dm. long; leaves alternate, capillary, 2–10 cm. long, with membranous stipular sheath 6–10 mm. long, the

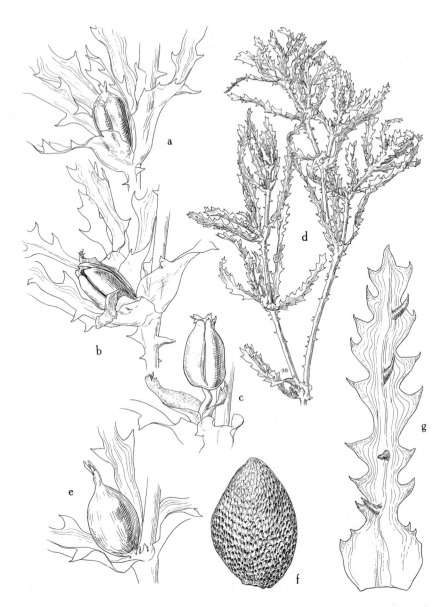

Fig. 33. *Najas marina*: *a–c*, development of anther: *a*, anther enclosed in sessile spathe in leaf axil, × 8; *b*, anther beginning to elongate and rupture spathe, × 8; *c*, mature anther, showing short filament, × 8; *d*, habit, showing the stems beset with prickles, and the spiny-toothed leaves, × 1½; *e*, mature pistillate flower, showing the 3 stigmas and the intra-vaginal scales at base, × 8; *f*, mature seed, × 10; *g*, leaf blade, showing the coarse, spiny-toothed margins, the spines on the outer side along the midrib, and the rounded shoulders of the leaf sheaths, × 6.

free part of the sheath short or wanting; flowers on slender axillary peduncles which at first are short and surrounded by the spathelike base of the enclosing leaf, each peduncle bearing 2 naked flowers each composed of 2 large, bicellular anthers and 4 pistils with sessile peltate stigmas, in anthesis the peduncles remaining short, or elongating and carrying the flowers above the water surface, after pollination the elongated peduncles sometimes coiling and again submersing the floral structures; fruit a pediceled, obliquely pointed nutlet, 2 mm. long, the pedicel 4–10 mm. long.

Alkaline, subsaline, or brackish water: throughout California; cosmopolitan.

An exceedingly variable complex which I treat here as a single species, but not without misgivings. Setchell, in "The Genus Ruppia L." (Proc. Calif. Acad. Sci. 25:469–478, 1946), discusses the problems relative to nomenclatural types. The behavior of the peduncle both before and after anthesis is the source of much confusion in the taxonomy of the group. Those who separate *Ruppia maritima* from *R. spiralis* do so on the basis of the following characters:

Peduncles either short or elongate, not spirally coiled after anthesis . . . . *R. maritima*
Peduncles elongate, becoming spirally coiled after anthesis . . . . *R. spiralis*

Variation is so rampant with respect to peduncle behavior that as yet we have found no basis for its clarification. The form described under the name *Ruppia spiralis* Dum. is depicted in our figure 32.

NAJADACEAE. WATER-NYMPH FAMILY

Slender, branching, submersed, fresh-water annual herbs, the branches forming either open and diffuse or much-branched, condensed plants. Leaves linear, usually spiny-toothed, apparently opposite but each pair consisting of a lower leaf and an upper leaf on opposite sides of the stem, or leaves seemingly whorled or appearing fascicled because many are crowded in the axils. Leaf bases dilated, forming conspicuous sheaths, the shoulders of these truncate, obliquely rounded, or drawn out into auricles of various lengths, a pair of minute, hyaline, cellular scales (intravaginal scales) often present within the sheath. Flowers monoecious or dioecious, usually solitary in the leaf-sheath axils, but sometimes several together. Staminate flowers consisting of a single stamen usually enclosed by a perianth-like envelope (spathe) ending above the anther in 2 thickened lips. The anthers 1- or 4-celled, at first nearly sessile but becoming short-stalked at maturity. Pistillate flowers naked, consisting of a single ovary, tapering into a short style bearing 2–4 linear stigmas. Fruit a nutlet, enclosed in a loose and separable membranous coat.

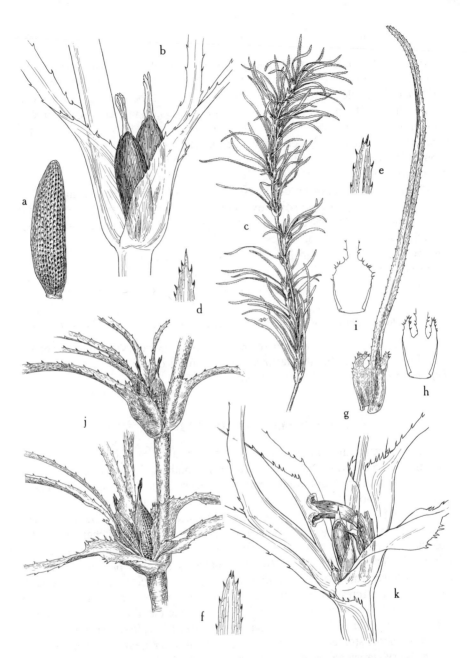

Fig. 34. *Najas graminea: a,* mature seed, with reticulations forming longitudinal rows, × 16; *b,* mature pistillate flowers in axils of leaf sheath, showing slender intravaginal scales at base, × 14; *c,* habit, showing the slender, recurving leaves, × 1⅕; *d–f,* leaf-tip variation, × 14; *g–i,* leaf, showing toothed margin of the blade and variation in the leaf-sheath auricles, × 6½; *j,* pistillate flowers, showing intravaginal scales and shouldered leaf sheaths, × 8; *k,* staminate flowers in various stages of development, including a de-hisced anther on an elongated, stout filament, surrounded by ruptured spathe, × 14.

NAJAS

Rendle, A. B., A systematic revision of the genus *Najas*. Trans. Linn. Soc. London II. 5:379–436, pls. 39–42. 1899.

———, Supplementary notes on the genus *Najas*. Trans. Linn. Soc. London II. 5:437–444. 1900.

Clausen, R. T., Studies in the genus *Najas* in the northern United States. Rhodora 38:333–345. 1936.

Leaves coarsely toothed; internodes and back of leaves often spiny . . . . 1. *N. marina*
Leaves minutely toothed (teeth often visible only under magnification); internodes and back of leaves unarmed.
  Leaf sheaths usually distinctly auriculate . . . . 2. *N. graminea*
  Leaf sheaths truncate or rounded, not auriculate.
    Seeds apparently smooth and shining (but finely reticulate under magnification)
                                                   3. *N. flexilis*
    Seeds dull, with distinct, squarish, pitted reticulations . . . . 4. *N. guadalupensis*

1. Najas marina L., Sp. Pl. 1015. 1753. Holly-leaved water nymph. Fig. 33.

Stems branching near the base, often armed with prickles on the internodes; leaves narrowly linear, 1–4 cm. long, 1–3 mm. wide, with coarse, spiny-toothed margins and often spiny on the abaxial side along the midrib, the teeth broad, many-celled at base, tapering to a strong, sharp-pointed, yellowish brown cell; leaf sheaths with broadly rounded shoulders, entire or with a few minute teeth; 1 or 2 minute intravaginal scales at base of sheath, broad-based and somewhat obliquely tapering, 5–6 mm. long; flowers dioecious; staminate flowers 3–4 mm. long, the anther 4-celled; pistillate flowers 3-4 mm. long; stigmas 3, sometimes one shorter than the others; fruit and seed more or less ovoid, apparently smooth, but finely reticulate at maturity.

Rare and local in fresh-water ponds: Imperial, San Diego, San Luis Obispo, and Lake counties.

The California material has been segregated varietally by Rendle, upon the basis of two southern California collections displaying slender leaves in which the spines along the midveins tend to be confluent at their bases, forming a winglike rib. Variation in this respect is so great that the variants can scarcely be segregated meaningfully.

2. Najas graminea Raffeneau-Delile, Fl. Egypte 282, pl. 50, fig. 3. 1813. Rice-field water nymph. Fig. 34.

Naias graminea var. *Delilei* Magnus, Ber. Deutsch. Bot. Gesell. 1:523. 1883.

Stems branching, usually very slender, grasslike, varying in length from 0.5 dm. to 6 dm.; lateral branches often very short, with leaves tufted and recurved near tips of shoots; leaves 1–2.5 cm. long, 0.5–1 mm. wide, light green, thin and translucent, acute, tipped with 1 or more short spines; leaf sheaths usually with 2 prominent auricles nearly ½ as long as the body of the sheath, these and the blades very minutely spine-tipped, the spines single-celled protrusions of 1 or 2 marginal cells; intravaginal scales minute, slender; flowers monoe-

Fig. 35. *Najas flexilis: a,* mature staminate flower, showing dehisced anther and ruptured spathe, × 8; *b,* young, sessile, staminate flower enveloped by spathe, × 8; *c,* pistillate flowers in axil of leaf sheath, showing variations in stigmas, × 8; *d,* habit, plant completely submersed, showing the fascicled leaves, × 1⅕; *e,* mature seed, shiny, yet finely reticulate under magnification, × 12; *f–h,* leaf blades, showing minute teeth and obliquely sloping, somewhat unequally shouldered leaf sheaths, × 5.

cious, often several at the same node; staminate flowers short-stalked, the anther 4-celled; pistillate flowers 1.5–3 mm. long; stigmas 2 or 3; fruit solitary or 2–4 crowded together at the bases of the dwarf shoots, narrowly oblong or ellipsoidal, tapering at the tip, more or less distinctly reticulate.

In rice fields (introduced from the warmer regions of the Old World): Butte County to Colusa County.

3. Najas flexilis Rostkov. & Schmidt, Fl. Sedim. 382. 1824. Slender water nymph. Fig. 35.

Naias canadensis Michx., Fl. Bor. Am. 2:220. 1803.

Stems branching, slender, to 2 m. long; leaves narrowly linear, 1–2.5 cm. long, 1–2 mm. wide, acuminate or acute, thin and translucent, very minutely toothed, numerous and crowded on the upper parts of the branches, the teeth consisting of protrusions of usually 1 marginal cell; leaf sheaths with obliquely sloping shoulders, the margins bearing several very minute teeth; intravaginal scales filiform, less than 1 mm. long; flowers monoecious; staminate flowers 2.5–3 mm. long, the anther 1-celled; pistillate flowers about 3 mm. long, the stigmas 2–4, usually 3; seed narrowly elliptic or lanceolate-ovoid, apparently smooth and shining but finely reticulate under magnification.

Infrequent in fresh-water pools and streams: southern California at Laguna Beach in Orange County, Sawtelle in Los Angeles County, San Bernardino in San Bernardino County (according to Munz, Man. So. Calif. Bot. 25, 1935). Jepson gives the range as north to San Francisco. No California material has been seen by us. Nearly throughout North America and Europe.

Najas flexilis is an extremely variable species in size and habit of growth. In quiet waters it often assumes a short, bushy habit a few decimeters high, whereas in streams it may develop slender stems 9–18 dm. long. Campbell's morphological work on *Najas* was probably based on *N. guadalupensis* rather than on *N. flexilis,* for his material was collected in the lakes near Stanford University in Santa Clara County. Although early collections in our herbaria were erroneously identified as *N. flexilis,* if this species occurs in California at all it is rare. We have, however, seen typical *N. flexilis* from Oregon and from western Arizona; hence it seems reasonable to expect it in California. *Najas flexilis* is readily identified by its shiny seeds with fine, almost obscure reticulations. In the other California species the seeds are not shiny, and their reticulations are very conspicuous.

4. Najas guadalupensis Morong, Mem. Torrey Club 3:60. 1893. Common water nymph. Fig. 36.

Naias flexilis var. *guadalupensis* A. Br., Jour. Bot. Brit. & For. 2:276. 1864.
Naias microdon var. *guadalupensis* A. Br., *loc. cit.*

Stems branching, 3–6 dm. long, usually slender and threadlike but habit very variable; leaves narrowly linear, flat or somewhat crisped, 1–2.5 cm. long,

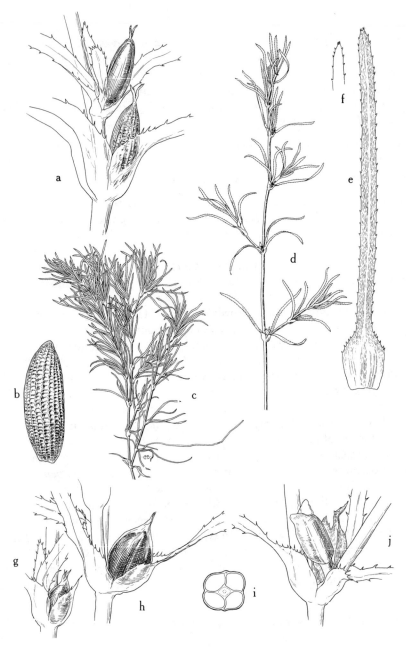

Fig. 36. *Najas guadalupensis: a,* young and mature pistillate flowers, borne singly in leaf-sheath axils, × 8; *b,* mature seed, dull but distinctly reticulate, × 16; *c,* habit, showing plant with threadlike, crowded leaves, × ⅘; *d,* habit, showing plant with less crowded leaves, × 1⅕; *e* and *f,* leaf blade, showing marginal and apical teeth, × 6½; *g* and *h,* young staminate flowers borne singly in leaf-sheath axils, the anthers still enveloped by the spathe, × 8; *i,* anther (cross section), × 12; *j,* mature anther at anthesis, showing ruptured spathe, × 8.

0.5–1 mm. wide, abruptly acute, often tipped with 1–4 short spines, the marginal teeth single-celled, very minute, sometimes wanting; shoulders of leaf sheaths oblique or rounded, few-toothed or sometimes entire, the teeth very minute; flowers monoecious; staminate flowers 2–3 mm. long, the anther 4-celled; pistillate flowers 2–3 mm. long, the ovary with 2 or 3 stigmas and usually with 1 or 2 sterile spiny processes; seed ellipsoid, dull but distinctly reticulate.

Pools, irrigation ditches, and rice fields: in Sacramento and San Joaquin valleys, near the coast from San Luis Obispo County north to San Francisco, and southern California in Orange and San Bernardino counties, our most common and widely distributed species; across the United States and south to Central America.

ZANNICHELLIACEAE

Submersed aquatic herbs from slender rhizomes. Leaves alternate or opposite, sometimes appearing as though whorled at some nodes, linear, entire, sheathing at base by membranous stipules, sometimes those subtending the flowers reduced to sheaths. Flowers monoecious or dioecious, solitary or in cymes, in the leaf axils, the pistillate flowers terminal on a very short shoot, the staminate ones arising from its lower bracteoles. Perianth, when present, of 3 free scales or in pistillate flowers cuplike. Stamens 3–1; anthers mostly 2-celled, sometimes 1-celled, longitudinally dehiscent; pollen either globose or thread-like. Pistils of 4–9 free carpels; style short; stigma asymmetrically peltate to spatulate, rarely 2- to 4-lobed; ovule solitary, pendulous. Fruit sessile or stipitate, often dentate-ridged on back. Seed without endosperm, the cotyledon coiled.

ZANNICHELLIA. GRASS WRACK

1. Zannichellia palustris L., Sp. Pl. 969. 1753. Fig. 37.

Slender, branching aquatic herb with a creeping rhizome; stems submersed, filiform, sparsely branched, 3–10 dm. long; leaves opposite, submersed, filiform, entire, flat, 1-nerved, 2–10 cm. long, acute, with sheathing membranous stipules; flowers sessile, unisexual, in bud surrounded by a spathe; staminate flowers consisting of 1 stamen, the filament slender, bearing a 2-celled anther; pistillate flowers 4 (2–6); style short; stigma peltate; fruit an obliquely oblong, flattened, beaked nutlet, pediceled, 2–4 mm. long, dentate on the back, the flask-shaped beak 0.8–1.5 mm. long; seed orthotropous, the cotyledon coiled.

Common, especially in brackish and subsaline habitats, ponds, streams, and ditches: throughout California; cosmopolitan. For a species so widespread, *Zannichellia palustris* is amazingly uniform.

Fig. 37. (For explanation, see facing page.)

ZOSTERACEAE

Marine aquatics with creeping stems and 2-ranked, long, linear, grasslike leaves. Flowers apetalous, monoecious or dioecious, borne in 2 ranks on one side of a flattened spadix. Spathe thickened and terminated by a long leaflike appendage. Staminate flower a single sessile anther enclosed by hyaline scales. Pollen filamentous. Pistillate flower a single pistil with 2 stigmas. Ovary 1-celled.

Plants dioecious.
Leaves not more than 3–6 mm. broad; ovary sagittate-cordate; tide pools along the ocean
Phyllospadix
Plants monoecious.
Leaves 8–12 mm. broad; ovary ovoid; bays and estuaries .... *Zostera*

PHYLLOSPADIX. SURF GRASS

Submersed marine plants with thickened rootstocks and slender stems. Leaves narrowly linear, elongated, caespitose. Flowers dioecious, borne in 2 rows on the side of a flattened spadix, each flower covered by a thick appendage, the whole inflorescence enclosed by a spathe with membranous edges and produced beyond the spadix as a leaflike appendage; staminate spadices with numerous sessile stamens in 2 rows; pistillate spadices with rudimentary anthers alternating with the pistils. Pistil attenuate into a short style. Fruit a nutlet, beaked by the short, persistent style, sagittate-cordate at base, with 2 downwardly produced horns which often bear deflexed bristles on the inside.

Dudley, W. R., *Phyllospadix*, its systematic characters and distribution. Zoe 4:381–385. 1894.

Inflorescence cauline on elongate peduncles (more than 1 dm. long); spadices several
1. *P. Torreyi*
Inflorescence basal or nearly basal on short peduncles (1–6 cm. long); spadices usually
solitary .... 2. *P. Scouleri*

1. Phyllospadix Torreyi Wats., Proc. Am. Acad. Arts & Sci. 14:303. 1879. Fig. 38.

Stems simple or branched, usually winged in lower part, flat, 3–6 dm. long; leaves flat when young, becoming folded or terete, 2 m. long or less, 0.5–2 mm.

Fig. 37. *Zannichellia palustris: a,* branch with submersed, filiform, 1-nerved leaves, showing stipular sheaths and flowers in lower axil, the staminate flower comprised of a single stamen arising at the base of the short, stout peduncle which bears 4 (usually 2–5) pistils surrounded by a spathe, × 4; *b,* fruit (longitudinal section), × 8; *c,* involucre or spathe with 2 young pistillate flowers and a single staminate flower, all from the same axil, × 16; *d,* habit, showing the long, opposite, filiform, submersed leaves and maturing fruits in the axils, × ⅖; *e,* fruit, showing toothed ridges as revealed by normal deterioration of outer coat in old fruits, × 8; *f* and *g,* mature, undried fruits before deterioration of coat, × 8.

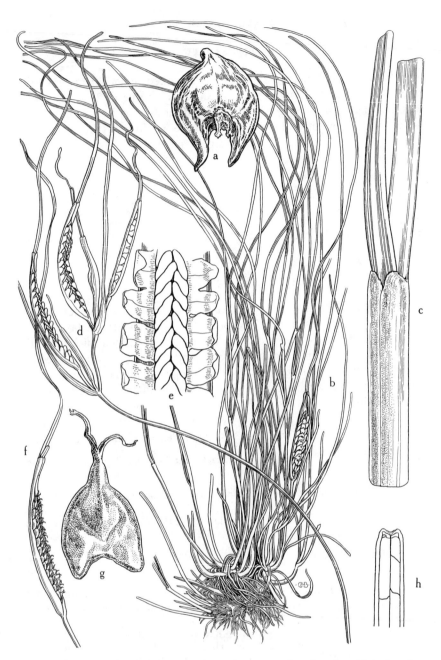

Fig. 38. *Phyllospadix.* *a–c, P. Scouleri: a,* mature winged nutlet, × 4; *b,* submersed habit, showing the short flowering stem bearing a solitary pistillate inflorescence enclosed by the spathe, × ⅖; *c,* leaf sheath, × 8. *d–h, P. Torreyi: d,* flowering branch, showing a group of staminate inflorescences, × ⅔; *e,* part of staminate inflorescence, showing the 2 rows of anthers with their subtending bracts, × 2; *f,* pistillate inflorescence, × ⅔; *g,* winged nutlet, × 6; *h,* leaf tip, × 4.

wide, the primary nerves 1–3 or the leaves sometimes nerveless; flowering stems abundant, elongated, usually 20–30 cm. long; spadices cauline, in 2 or 3 pairs, usually about 5 cm. long, each pair subtended by several bracts; staminate flowers numerous, arranged in 2 rows; pistillate flowers numerous, in 2 rows; fruit flask-shaped, 2–3 mm. long, with 2 projecting and obliquely striate wings, and also sometimes winged on the back.

In surf along ocean shores, but tending to occur in more quiet waters than *Phyllospadix Scouleri,* such as in tide pools and protected coves among rocks: throughout the length of the state; Baja California, Mexico.

2. Phyllospadix Scouleri Hook., Fl. Bor. Am. 2:171. 1839. Fig. 38.

Stem simple or branched, somewhat winged in lower part, 1–4 dm. long; leaves flat, 0.5–2 m. long, 2–4 mm. wide, entire or rarely serrulate, obtuse at the apex, the primary nerves 3; flowering stems not common; peduncles short, 1–6 cm. long; spadices usually solitary or sometimes in pairs, basal or sub-basal; spathe prominently scarious-winged; staminate spadix about 6 cm. long; anthers about 4.5 mm. long; fruit flask-shaped, about 3.5 mm. long, with obliquely striate wings.

On rocks in heavy surf and exposed ocean shores: from Santa Barbara north to Vancouver Island; Japan.

ZOSTERA

1. Zostera marina L., Sp. Pl. 968. 1753. Eel grass, wrack.

Zostera marina var. (?) *latifolia* Morong, Bull. Torrey Club 13:160. 1886.

Marine plant, wholly submersed, with slender rootstock; stems leafy, 1–3 m. long; leaves alternate, linear, obtuse at apex, 3–15 dm. long, 6–12 (sometimes 2–8) mm. wide, with 3–7 principal nerves; flowers monoecious, arranged alternately in 2 rows on a one-sided spadix, enclosed in a spathe formed by the sheathing base of the subtending leaf; staminate flowers escaping from spathe at anthesis and discharging glutinous filamentous pollen; pistillate flowers with stigmas protruding through the spathe at anthesis and dropping off before the anthers of the same spadix open; fruit flasklike, beaked by the persistent style; seeds strongly 20-ribbed, about 3 mm. long, 1 mm. in diameter, truncate at both ends, the ribs showing through the thin pericarp.

Shallow waters in bays and estuaries, usually on muddy bottoms in the intertidal zone: widely distributed along the Pacific and Atlantic coasts.

The leaves of the Pacific coast plant are usually somewhat broader than those of the Atlantic coast plant, and for this reason it has been referred to *Zostera marina* var. *latifolia* Morong. Several detailed studies by Setchell on the morphology and phenology of this species have provided evidence that the differences which separate the two forms are mainly quantitative and seem to be associated with temperature variables throughout the range of the species.

Fig. 39. *Triglochin maritima: a,* inflorescence, × ⅖; *b,* habit, showing racemes raised above the leaves, × ⅛; *c,* young flower, showing bractiform, perianth-like appendages (anthers enclosed) and stigmas of slender papillae, × 8; *d,* flower, showing maturing anthers, each within a perianth-like appendage, and maturing carpels, × 8; *e,* flower, showing the 2 series of perianth-like appendages, each appendage with a dehisced anther, the fruit nearly mature, × 8; *f,* mature fruit, showing the conspicuous appendage scars below, × 4; *g,* fruit (cross section), all carpels fertile, × 6; *h,* separate mature carpel, × 6; *i,* entire ligule, × 4.

Any difference which may exist in the genetic constitution of the two forms should be the subject of further experiment and study.

Setchell, W. A., A preliminary survey of the species of *Zostera*. Proc. Nat. Acad. Sci. Wash. 19:810–817. 1933.

SCHEUCHZERIACEAE

A monotypic family consisting of a single genus *Scheuchzeria* and species *S. palustris*.

SCHEUCHZERIA

1. Scheuchzeria palustris L., Sp. Pl. 338. 1753.

Perennial marsh herb with creeping rhizomes; stems leafy, erect, 1 to several, 7–15 cm. tall; leaves linear, 6–15 cm. long, with broad, sheathing base and a ligule at junction of sheath with the tubular blade; flowers perfect, regular, in bracteate racemes on slender pedicels; perianth segments 6 in 2 slightly dissimilar series, 2–3 mm. long, white, membranous, persistent, distinct; stamens 6, inserted at base of perianth segments; pistils 3, slightly united at base, the stigmas sessile or on a short, stout style, plumose or papillose; follicles 3, coriaceous, inflated, 1- or 2-seeded, 4–8 mm. long, divaricate from base.

Alpine bogs: Sierra Nevada from Tuolumne County to Plumas County, infrequently collected; widely distributed over cooler regions of Northern Hemisphere.

JUNCAGINACEAE. ARROW-WEED FAMILY

Annual or perennial marsh herbs from rhizomes or tubers. Leaves basal, linear, sheathing, ours with blade terete or semiterete. Inflorescence a spike-like raceme borne on a naked scape, the flowers on short, slender or stout pedicels. Flower dioecious or perfect, regular to slightly irregular, bractless. Bractiform perianth-like appendages usually 6, in 2 series, each appendage bearing a stamen attached to its base or on some the stamen absent. Anthers 2-celled, subsessile, opening by slits. Pistil superior, of 6 or 4 (or 3) connate to weakly united carpels, these sometimes separating in fruit; styles short or absent; stigmas often papillate or plumose; ovule 1 per carpel, basal, erect. Fruit of distinct or weakly united, dehiscent or indehiscent carpels, these erect or recurved only at apex, sometimes with hooked spines at base.

TRIGLOCHIN

Herbaceous perennial. Leaves broadly sheathing at base, the sheath culminating above in an entire or 2-lobed ligule, the blade semiterete. Scapes and

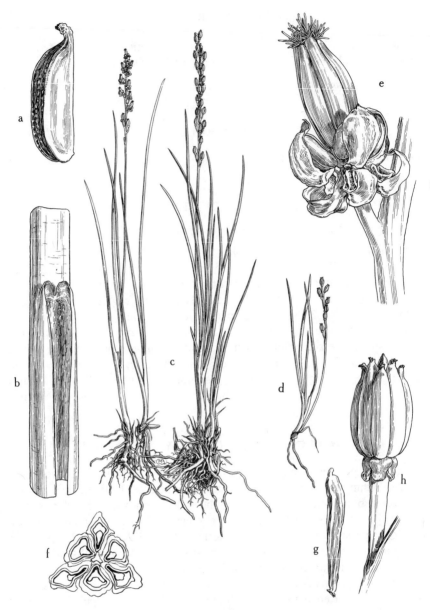

Fig. 40. *Triglochin concinna: a,* mature fruit, × 8; *b,* 2-lobed ligule, × 8; *c,* habit, showing scapes longer than the leaves, × ⅖; *d,* variation of habit, × ⅖; *e,* flower, showing the 2 series of bractiform, perianth-like appendages, each appendage with an attached anther, lower series of anthers past dehiscence, upper series before dehiscence, × 8; *f,* mature fruit (cross section), all carpels fertile and separating from the slender carpophore, × 8; *g,* seed, × 16; *h,* fruit with conspicuous scars at base resembling reflexed perianth parts, × 6.

racemes longer than or shorter than the leaves. Each perianth-like appendage of the flower usually deciduous with its attached stamen and often leaving a conspicuous, enlarged scar which simulates a reflexed perianth part at the base of the fruit. Stamens 6–3 (–1), subsessile, the anthers often broader than high, or ovoid, rarely much longer than broad. Carpels joined to a central carpophore from which only the fertile carpels separate on maturity; stigmas of slender papillae. Seed linear, loosely enclosed in the indehiscent carpel.

A genus of 12 species, with world-wide distribution but conspicuously developed in Australia and South America.

The nature of the bractiform, perianth-like appendages represents an interesting problem in floral morphology. Within a flower there are what at first appear to be two series of bractiform appendages with three appendages to a series; a series of three stamens appears to intervene between the two series of appendages, and still another series of three stamens appears to arise above the upper series. If this were the correct interpretation, then the structure would not be a perianth in the ordinary sense, but would be morphologically homologous to the so-called "sepaloid connective" of *Potamogeton*. The homology with the sepaloid connective is further suggested by the fact that the anthers are sessile on the appendages in a manner that implies that the appendage is related to the filament in its origin and organization. An alternative hypothesis is to regard what appears to be a unit flower as an inflorescence (N. W. Uhl, Studies in the floral morphology and anatomy of certain members of the Helobiae, doctoral diss., Cornell Univ., 1947), the staminate flowers then being each composed of a subtending bract and a single stamen, as in *Lilaea*. Such an interpretation is complicated by the presence in *Triglochin* of united or partially united carpels.

In *Triglochin striata* the flower is reduced and only the outer perianth-like appendages bear stamens, at least in the specimens that I have seen.

Fertile carpels 6; stamens usually 6, sometimes 3.
 Ligule of the leaf sheath entire; the stems often densely tufted from a proliferating caudex or oblique rhizome, 1–10 dm. high .... 1. *T. maritima*
 Ligule of the leaf sheath 2-lobed or at least deeply emarginate; plants usually not tufted but well spaced along slender, naked rhizomes, occasionally arising from a stout, oblique, scaly rhizome, 0.4–5 dm. high .... 2. *T. concinna*
Fertile carpels 3, alternating with 3 sterile carpels; stamens 1 or 2 or rarely 3; ligule entire; plants from slender rhizomes, 0.5–3 dm. high .... 3. *T. striata*

1. Triglochin maritima L., Sp. Pl. 339. 1753. Fig. 39.
 Triglochin elata Nutt., Gen. 1:237. 1818.

Coarse or slender plant with few to many tufted scapes 1–10 dm. tall from a proliferating caudex or stout, short rhizome covered with persistent leaf bases; leaves 1–8 dm. long, 2–5 mm. broad, the ligule entire; scapes longer than the leaves, 1–4 dm. long, terminated by a raceme and densely clothed with many pedicellate flowers, the pedicels somewhat ascending to decurrent,

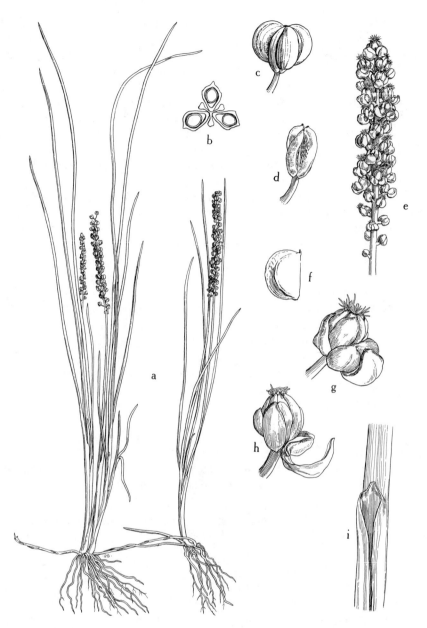

Fig. 41. *Triglochin striata: a,* habit, showing leaves longer than or as long as the scapes, × ⅖; *b,* fruit (cross section), showing 3 fertile carpels separating from the 3 persistent sterile ones, × 8; *c,* mature fruit, with 3 fertile carpels alternating with 3 sterile ones, × 8; *d,* persistent sterile carpels, resembling a contracted winged fruit after dispersal of fertile carpels, × 8; *e,* spikelike raceme, × 4; *f,* separate fertile carpel, × 8; *g,* young flower, showing the one enlarged anther and its enlarged, bractiform, perianth-like appendage, the other anthers hidden by their respective appendages, × 16; *h,* flower, showing anther after dehiscence, × 16; *i,* entire ligule, × 4.

2–6 mm. long; flowers with 6 perianth-like appendages each bearing an attached stamen; pistil of 6 fertile carpels united around the slender carpophore; fruit with carpels united, indehiscent; seeds linear.

Saline and alkaline marshes along the coast and in the interior, sea level to an elevation of 9,000 feet. Widely distributed throughout the Northern Hemisphere.

2. Triglochin concinna Davy, Erythea 3:117. 1895. Slender arrow grass. Fig. 40.

Triglochin maritimum var. debilis of Jepson, Fl. Calif. 3:77. 1912. In part. Not M. E. Jones.

Slender plant usually well spaced and spreading from a slender rhizome, the latter with conspicuous internodes and bracteate nodes; leaves 8–20 cm. long, the sheaths membranous-margined, terminating above in a 2-lobed ligule; scapes longer than the leaves, terminated by a strict raceme, the rachis of which may be either straight or zigzag, the pedicels slender; flower with 6 perianth-like appendages each bearing an attached stamen; carpels 6, united, usually all fertile, the fruiting carpels separating readily from the slender carpophore, indehiscent; seeds slender, needle-like.

Coastal salt marshes: Baja California, Mexico, to British Columbia.

A variety, Triglochin concinna var. debilis (M. E. Jones) J. T. Howell (Leafl. West. Bot. 5:18, 1947), is similar to the species but often coarser and then with many more flowers per spike, the base of stem and rhizome often clothed with coarse fibers of the leaf sheath. A detailed study may indicate the desirability for specific segregation of this plant.

Wet places of the Great Basin and desert regions: Modoc County to Mojave Desert; east to Colorado.

3. Triglochin striata Ruiz & Pav., Fl. Per. 3:72. 1802. Fig. 41.

Usually slender plants 1–2 dm. tall, well spaced along a slender, horizontal or oblique rhizome 2–8 cm. long; leaves 5–20 cm. long, the ligule entire, rounded or somewhat attenuate; scapes shorter than or not longer than the leaves, 5–15 cm. high; raceme spikelike, the pedicels slender, 1–2 mm. long, not elongating in fruit, spreading; perianth-like appendages of flower 3, 4, or 5, greenish yellow, often very unequal in size, only 1 or 2, rarely 3, of these appendages bearing stamens, if stamens 3, then 1 or 2 usually much larger than the third; anthers broader than high, becoming ovate on dehiscence; pistil of 3 fertile carpels alternating with 3 sterile carpels; fruit subglobose, definitely 3-angled, the indehiscent fertile carpels separating from the persistent sterile carpels, the latter appearing as though they were a somewhat contracted winged fruit; seeds linear, about 1 mm. long.

Saline and brackish marshes in warm temperate regions along the coast; North and South America.

Fig. 42. *Lilaea scilloides: a,* scapose spike of perfect flowers, × 4; *b,* perfect flower from spike, showing the single, bractiform, basally gibbous, perianth-like appendage arising from the base of the sepaloid connective of the sessile anther, the papillae of the stigma shorter than those of the basal pistillate flowers, × 20; *c,* habit, aquatic type of plant, × ²⁄₅; *d* and *e,* threadlike styles of basal pistillate flowers, showing capitate stigmas beset with long, ephemeral papillae, × 8; *f,* flattened fruit of perfect flower, showing the abaxial undulate keel and lateral wings (side view), × 4; *g,* flattened fruit of perfect flower (abaxial view), × 4; *h,* emarginate ligule of sheath, × 2; *i,* fruit of pistillate flower, showing hornlike structures at summit and the laterally arising style, × 4.

LILAEACEAE

A monotypic family consisting of the single genus *Lilaea* and species *L. scilloides*.

LILAEA

1. Lilaea scilloides (Poir.) Haum., Publ. Inst. Invest. Geogr. Buenos Aires
10:26. 1925. Flowering quillwort. Fig. 42.

Lilaea subulata Humb. & Bonpl., Pl. Aequin. 1:222, pl. 63. 1808.

Acaulescent annual, terrestrial on wet soil or aquatic, then with the tips of leaves and flowers or at least the stigmas emersed; leaves terete, linear, 3–35 cm. long, their adnate, hyaline stipules sheathing the stem and converging across the top to form a low, somewhat emarginate ligule; flowers of three very distinctive types, each leaf axil giving rise to 2 basal pistillate flowers enclosed in the sheathing leaf base and to a solitary scapose spike of all perfect or both perfect and staminate flowers; pistillate flowers composed of what appears to be a single carpel with an oblong, angled ovary and a threadlike style 2–30 cm. long arising laterally and at right angles to the top of the ovary or sometimes reflexed, but at length ascending and ending in a capitate stigma; perfect and staminate flowers of the spike with a single, bractiform, perianth-like appendage arising from the base of the connective of the sessile anther, the appendage gibbous at the base and caducous with the anther; styles very unequal in length in spicate perfect flowers, 1–20 mm. long; fruits of these perfect flowers bractlike, flattened, with dorsal keel and lateral wings undulate, 4–6 mm. long, 2–3 mm. wide; fruits of the basal pistillate flowers angled and often with hooks or horns at the summit.

Very common in wet soil around ponds, lakes, and slow streams in valleys of central California, less common in the mountains; coastal region from British Columbia to Mexico and South America, eastward sparingly to the Rocky Mountains.

The illustration in K. F. Martius (Fl. Bras. 3³: pl. 118, 1894), depicting the spicate inflorescence with each fruit occurring in the axil of a bract, is in error, inasmuch as the bractiform, perianth-like appendage actually falls off with the anther and no other bract exists. This erroneous illustration apparently served as the basis for an adaptation by Hutchinson (Families Fl. Pl. 2:40, fig. 8, *c,* 1934), which carries this error with it. We have not seen anthers stalked as in Hutchinson's figure (*ibid.,* fig. 8, *f*). Although Jepson (Fl. Calif. 78, 1912) reports that the spikes have pistillate flowers below, perfect flowers in the middle, and staminate flowers above, we have seen no pistillate flowers in the spike. The usual situation is for all the flowers of the spike to be perfect, with occasionally some of the upper flowers or some of the lower ones staminate.

Some controversy exists concerning the nature of the bractiform, perianth-like appendage. Jepson refers to it as a bract, whereas Hutchinson considers it

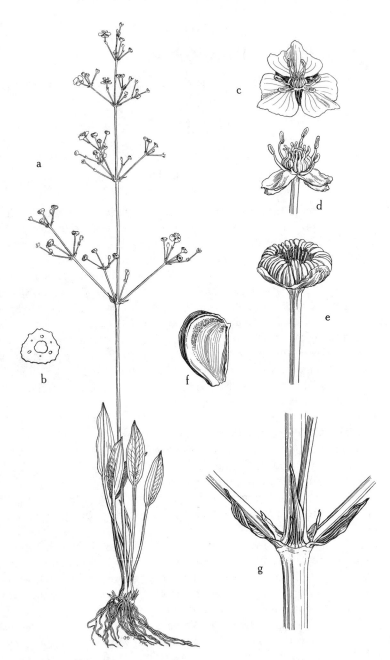

Fig. 43. *Alisma Plantago-aquatica* subsp. *brevipes: a,* habit, showing whorled flowering branches subtended by papery bracts, × ¼; *b,* stem (cross section), × 4; *c,* flower, × 2; *d,* flower with petals removed, showing the gibbous sepals, × 3; *e,* ring of mature achenes, × 4; *f,* mature achene, showing rounded back, × 8; *g,* node, showing bracts subtending flowering branches, × 2.

a "single perianth segment." Whatever its morphological character, its intimate association with the anther suggests, as in the Juncaginaceae, a close morphological agreement with what has been termed a "sepaloid connective" in the Potamogetonaceae.

ALISMACEAE. WATER-PLANTAIN FAMILY

Emersed or rarely submersed lacticiferous aquatics with sheathing, long-petioled basal leaves and scapose stems culminating in verticillate, bracteate racemes and panicles. Flowers regular, perfect, monoecious, dioecious, or polygamous. Sepals 3, green, persistent, rarely growing with the fruit. Petals 3, deciduous. Stamens 6 to many, anthers linear, dehiscing by slits on the margins or on the back. Pistils numerous, distinct, arranged in a ring or irregularly packed over the surface of the receptacle, 1-celled, ours all becoming 1-seeded achenes in fruit.

Achenes arranged in a ring on the receptacle.
    Achenes rounded on the back, the ring of fruits smooth; petals entire or minutely erose
                                                             Alisma
    Achenes conspicuously horned on the back, the ring of fruits in a stellate pattern; petals
        toothed or incised .... *Damasonium*
Achenes densely packed over the surface of the receptacle.
    Flowers all bisexual; fruiting heads simulating a bur; leaves never sagittate, sometimes
        cordate or truncate; stems and petioles often ridged or angled .... *Echinodorus*
    Flowers unisexual or bisexual but at least the upper ones staminate; fruiting heads usually
        smooth; leaves sagittate, hastate, or simple, never truncate or cordate; stems and
        petioles usually smooth and rounded, rarely angled or ridged .... *Sagittaria*

ALISMA. WATER PLANTAIN

Aquatic herbs, perennial or sometimes behaving as annuals, emersed or growing on wet mud, rarely submersed. Leaves basal, erect or rarely floating, with lanceolate or oblong-ovate blades, rarely reduced to ribbon-like phyllodes. Flowers small, 3–10 mm. broad, perfect, numerous, on 3-bracteate pedicels unequal in length. Sepals 3, green, persistent. Petals 3, white or occasionally rose or pink. Stamens 6–9. Pistils separate, arranged in a more or less 3-sided whorl on the receptacle. Fruit an achene, 2- or 3-ribbed or rounded on the back, lateral ribs present or absent.

Samuelsson, G., Die Arten der Gattung *Alisma* L. Arkiv för Bot. 24A(7):1–46. 1932.

Inflorescence 2–6 times as long as the longest leaves, rarely the scape shorter than longest
    leaves; leaf blades 2–10 cm. wide; petals 3–6 mm. long, white or pink to rose; achenes
    distinctly uni- or bisulcate, often both types on same plant
                               1. *A. Plantago-aquatica* subsp. *brevipes*
Inflorescence from much shorter than leaves to slightly longer; leaf blades 1–15 mm. broad;
    petals 1–3 mm. long; achenes indistinctly bisulcate, the median rib broad and rounded
                                        2. *A. Geyeri*

Fig. 44. *Alisma Geyeri: a,* habit, linear-leaved plants, showing blades much longer than the inflorescence, × ⅖; *b–d,* variations in mature achenes, × 8; *e* and *f,* variations in habit, × ⅖

1. Alisma Plantago-aquatica subsp. **brevipes** (Greene) Samuelsson, Arkiv för Bot. 24A(7):19. 1932. Fig. 43.

Alisma brevipes Greene, Pittonia 4:158. 1900.

Erect perennial 1–12 dm. tall, in shallow water or on wet mud; leaves normally with distinct blade and petiole, blades 5–20 cm. long, linear-lanceolate to broadly elliptic or truncate or subcordate at base; inflorescence on an erect scape with several whorls of branches, each with 1 or more whorls of flowers or further compounded into verticillate branches much longer than the leaves, each branch and each pedicel subtended by 2 or 3 lanceolate, papery bracts; flowers hypogynous, perfect; sepals 3, plane or somewhat gibbous, green; petals 3, white or sometimes rose to pink, 3–6 mm. long, rhombic in outline, margins entire or minutely erose; stamens 6, filaments glabrous; pistils numerous, in a single, often obscurely 3-sided whorl; styles 1–1.5 mm. long, as long as or longer than ovary; achenes often 3-ribbed on back, 2–3 mm. long, 1.5–2 mm. wide, the beak on the inner angle, erect or suberect.

Common in fresh water in the valleys and mountains: especially abundant in the Sacramento–San Joaquin delta region and north in Sacramento Valley; east in the northern states and Canada to Atlantic.

Our subspecies varies only slightly from typical *Alisma Plantago-aquatica,* which inhabits the Old World. *Alisma subcordatum* of the eastern United States, characterized by a shorter style and smaller flowers, may be expected in California, but as yet we have not seen it. There is a strong temptation to treat the plants that have pigmented corollas as having subspecific status, since in California the pigment tends to characterize populations and occurs with some geographic regularity. To date, however, we are unable to correlate the pigment character with any other morphological character. We find that the range of variation of the pigmented plants parallels almost exactly that of the white-flowered forms in stature and in leaf shape.

2. Alisma Geyeri Torr. in Nicollet, Rep. Hydrogr. Miss. R. 162. 1843. Fig. 44.

Alisma gramineum var. *Geyeri* Samuelsson, Arkiv för Bot. 24A(7):43. 1932.

Submersed or amphibious perennial herb 5–20 cm. high, erect or ascending, or, when plant submersed, leaves and stems floating; leaves usually erect, with long petioles and linear-lanceolate to lanceolate blades, or blades wholly absent; inflorescence a scapose, verticillate panicle 1–10 cm. long, shorter than leaves, branchlets and pedicels subtended by 2 or 3 lanceolate, papery bracts; flowers 5–7 mm. broad; sepals green, persistent; petals usually white, rhombic, entire or somewhat erose; stamens 6–9; pistils in an obscurely 3-sided whorl; achenes often orbicular to orbicular-cuneate, 2.5 mm. in diameter, the beak on the inner margin, erect, strongly 3-ribbed on back.

Ponds and irrigation ditches: Modoc and Lassen counties; north to Washington, east to Minnesota.

Although we admit close relationship between *Alisma Geyeri* and *A. gra-*

Fig. 45. *Damasonium californicum: a,* habit, showing whorls of flowers and fruitlets, × ⅖; *b,* apex of stem, showing buds, flowers, maturing fruits, and scarious bracts subtending the pedicels, × 1⅕; *c,* stamen, × 12; *d,* achene, showing horn and ribs, × 4.

mineum, we depart from the treatment accorded *A. Geyeri* by Samuelsson and retain it as distinct, because of its generally stocky habit and short inflorescences, its geographic distinctness in North America, and its relation to the linear-leaved phases of the plant in our area. The linear-leaved submersed plants give the impression of constituting a genetic and geographic entity in their own right rather than simply being deeper-water modifications of the species. Plants of this leaf type were collected in deep water as well as in shallow water. We also have found the typical form of *A. Geyeri* from both shallow and deeper water with no evidence of the linear-leaved form in the colony. On the other hand, in some localities, we have seen obvious intergradation between the types. *Alisma gramineum* subsp. *Wahlenbergii* is described by Samuelsson as a deeper-water form of *A. gramineum* and occurring with it. The entire problem is in need of further study, however, before the taxonomic status of these forms will be clear.

DAMASONIUM

1. Damasonium californicum Torr. in Benth., Pl. Hartw. 341. 1857. Star water plantain. Fig. 45.

Machaerocarpus californicus (Torr.) Small, N. Am. Fl. 17:44. 1909.

Emersed aquatic or amphibious perennial; leaves sheathing, basal, erect, spreading or floating, usually with a long, slender petiole and a simple, entire, lanceolate or elliptic blade, or the blade absent; scapes 1 to several, erect or ascending with 1 to several verticels racemosely or paniculately disposed; pedicels very unequal in length owing to a difference in size but eventually becoming subequal, subtended by a thin, broadly ovate-lanceolate, scarious bract; sepals oblong or ovate, 4–5 mm. long, somewhat hooded, green and persistent; petals 8–10 mm. long, white with a yellow patch near base, rarely pink, broadly cuneate, the blade rhombic, irregularly toothed or incised above, deciduous; stamens 6, 2–3 mm. long, 2 in front of each petal, the filaments glabrous, subulate with a broad base; pistils 1-ovuled, 6–10 in a single whorl on the receptacle; styles elongate, stout, erect in anthesis but soon spreading radially; achenes becoming horned by the enlargement of the style, ribbed on the angles, with a prominent angular shoulder and depressed flat faces, spreading radially from the receptacle.

Plants of vernal pools or the margins of intermittent streams or on wet mud: widely but discontinuously scattered in northern California, Modoc and Lassen counties, Sacramento Valley, local in the Sierra Nevada foothills, Sonoma County, and Lake County.

Damasonium californicum differs from Old World species in having a solitary ovule in each carpel. This has led some to regard it as generically distinct and accounts for the name *Machaerocarpus* which appears in some of our literature.

Fig. 46. *Echinodorus cordifolius: a,* mature achene, × 8; *b,* upper part of inflorescence, showing maturing burlike fruits, × ⅘; *c,* whorl of flowers, × ⅘; *d,* flower, showing the arrangement of the 12 stamens, × 1½; *e,* stamen, × 8; *f,* habit of mature plant, × ⅕; *g,* habit of young submersed plant, showing transition stages from early linear to mature cordate leaf blades, × ⅕.

ECHINODORUS

1. Echinodorus cordifolius (L.) Griseb., Abh. Kön. Gesell. Wiss. Göttingen 7:257. 1857. Fig. 46.

Ours amphibious annuals, erect, 2–6 dm. tall; leaves exceedingly diverse from seedling to adult, beginning as submersed, linear-undulate blades, becoming broadly lanceolate-attenuate with broadly winged petioles, at length becoming rigidly erect and emersed on rigid-angled petioles, blade cordate-ovate with coarsely reticulate venation; scapes erect, longer than leaves, strongly ribbed, 1 to several, each bearing a verticillate panicle, the branchlets and pedicels subtended by linear-lanceolate bracts; pedicels often very unequal in length; flowers perfect; sepals broadly triangular, green and persistent; petals greenish white, rhombic-orbicular; stamens about 12, the filaments glabrous; pistils densely packed over the surface of receptacle; style longer than ovary; fruiting heads simulating a bur by virtue of the stout, persistent, beaklike styles.

Very common in shallow water, especially in ponds, irrigation ditches, and rice fields. In deeper water the plants rarely produce normal adult foliage leaves and never flower, but develop large, ribbon-like submersed leaves.

Throughout the valleys at low altitudes; east to the southeastern states, south to tropics.

Echinodorus radicans (Nutt.) Engelm. has been reported in the early literature of California as occurring in the southern part of the state. We have seen nothing that could be referred to this species, and suspect that either it was transient, or it was a plant of *E. cordifolius* which, through some accidental cause related to fluctuation in water level, simulated the prostrate habit of *E. radicans* and was misidentified.

SAGITTARIA

Aquatic herbs, annual and perennial, if perennial bearing rhizomes that terminate in an erect corm or tuber which persists through the winter as the parent plant dies. Leaves usually erect or sometimes on large plants spreading from the base, occasionally floating on the surface of the water, usually with a stout petiole and a sagittate blade which is sometimes described as consisting of a terminal lobe and 2 lateral or basal lobes, sometimes the juvenile leaves bladeless and in some species the adult leaves also bladeless or the blades reduced to lanceolate entire structures. Inflorescence verticillate, simple or branched, the individual flowers subtended by bracts, monoecious or occasionally dioecious or with some flowers perfect, usually the lower flowers pistillate and the upper ones staminate, sometimes the lower flowers perfect and the upper ones staminate or a few perfect flowers between the staminate and pistillate zones, the unisexual flowers often with rudimentary stamens or pistils,

Fig. 47. *Sagittaria calycina: a,* inflorescence, showing apical staminate flowers and basal perfect flowers, pedicels of young flowers ascending and those of older flowers stout and recurved, × ⅔; *b,* whorls of perfect flowers, × ⅔; *c,* habit of young plant, showing onto-genetic leaf variation, × ⅓; *d,* bracts subtending flowers, × 1⅕; *e,* winged achene, show-ing the ascending beak, × 6; *f,* stamen, the filament beset with papillate hairs, × 8; *g,* habit, showing fruiting inflorescence and a widely divergent sagittate leaf blade, × ⅕.

or these lacking. Sepals 3, triangular, usually becoming reflexed and withering-persistent in fruit, occasionally erect and growing with the fruit. Petals 3, white or sometimes with a colored spot at base, blade somewhat quadrate or orbicular and abruptly narrowed to a short claw. Stamens usually numerous, hypogynous, in those plants with perfect and staminate flowers often dimorphic, those of the perfect flowers with short anthers, those of the staminate flowers with long anthers; filaments glabrous or pubescent. Pistils numerous, spirally arranged on the receptacle, which often bears lactiferous cells. Fruit an achene with an erect or divergent, beaklike, persistent style.

Since the preparation of the manuscript for this treatment, a monograph of the genus *Sagittaria* by Clifford Bogin has appeared (Revision of the genus *Sagittaria* [Alismataceae], Mem. N. Y. Bot. Gard. 9:179–233, 1955). Such changes as seemed necessary have been made herein, but some other differences of treatment have not been followed, for reasons indicated in connection with the species concerned.

Fruiting heads maturing on stout, recurved pedicels; sepals not reflexed in fruit; filaments of stamens densely clothed with papillae; flowering indeterminate, scapes 1 to many.
 Plants annual; leaves varying from flat, bladeless phyllodes to terete petioles with elliptic or at length sagittate blades; lower flowers perfect, the upper ones staminate; sepals growing with the fruit, accrescent; beak of achene horizontal ...... 1. *S. calycina*
 Plants perennial by heavy rhizomes; leaves a triquetrous phyllode or sometimes with an elliptic or lanceolate terminal blade; lower flowers pistillate, rarely with a few functional stamens; upper flowers staminate; sepals not growing with the fruit, spreading but not reflexed; beak of achene erect .... 2. *S. Sanfordii*
Fruiting heads maturing on slender spreading or ascending pedicels; sepals reflexed in fruit; flowers monoecious or dioecious, rarely an occasional flower perfect; filaments of stamens glabrous; flowering scapes usually 1 or 2.
 Basal leaf lobes much longer than the terminal lobe (2–3 times as long); beak of achene erect .... 3. *S. Greggii*
 Basal leaf lobes subequal to or shorter than the terminal lobe.
  Terminal leaf lobe much longer than the basal ones, often twice as long or longer; beak of achene short, stout, and erect; plants of middle and high altitudes
   4. *S. cuneata*
  Terminal leaf lobe subequal to or slightly longer than the lateral ones; beak of achene long, horizontal or oblique; plants of valleys and foothills
   5. *S. latifolia*

1. Sagittaria calycina Engelm. in part, in Torr., Mex. Bound. Surv. Bot. 212. 1859. Fig. 47.

Lophotocarpus calycinus (Engelm.) J. G. Smith, Rep. Mo. Bot. Gard. 6:60. 1895.
Lophotocarpus californicus J. G. Smith, Rep. Mo. Bot. Gard. 11:146. 1899.
Sagittaria montevidensis subsp. *calycina* (Engelm.) Bogin, Mem. N. Y. Bot. Gard. 9:197. 1955.

Emersed aquatic, only the early stages completely submersed, erect annual, 1–5 dm. tall; leaves erect or often widely spreading at base, in ours usually with a stout, spongy petiole and a broad, sagittate or hastate blade, rarely,

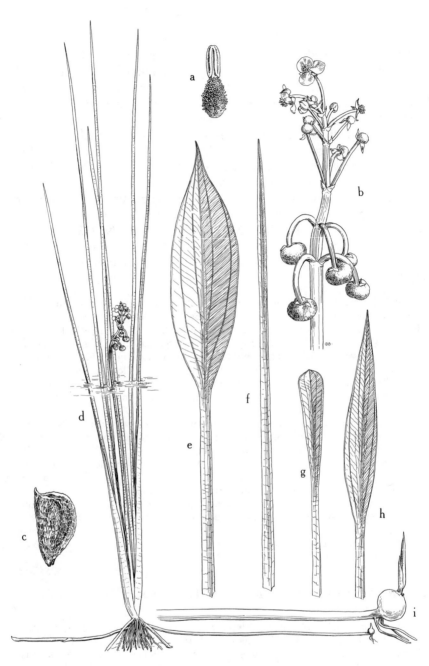

Fig. 48. *Sagittaria Sanfordii: a,* stamen, the stout, ovoid filament densely beset with papillate hairs, × 8; *b,* upper part of inflorescence, showing whorls of staminate and pistillate flowers, and the maturing fruits on recurved pedicels, × ⅔; *c,* mature achene, showing the strong keels and irregular reticulation, × 8; *d,* habit, showing the rhizomes, the naked phyllodes, and the short scape, × ⅛; *e–h,* leaf-blade variation, × ⅖; *i,* globose corm formed at tip of rhizome, × ⅘.

except in juvenile plants, the blades absent or elliptic; lateral lobes of blade as long as or shorter than terminal lobe, usually widely divergent; scapes erect or the base reclining and the tips erect, nearly as long as or somewhat longer than the leaves, usually with several verticels of flowers, each subtended by a triangular or broadly lanceolate bract; pedicels slender and ascending in anthesis, becoming stout and recurved in fruit, 1–3 cm. long; lower flowers perfect, the upper flowers staminate; sepals orbicular, growing with and closely investing the fruiting head; petals white, the blades orbicular, longer than sepals; stamens several in the perfect flowers, anthers orbicular, much shorter than filaments, many in the staminate flowers; filaments stout and closely beset with papillate hairs, somewhat longer than the slender anthers; achene with a horizontal or ascending beak, winged on both margins.

Common throughout the interior valleys of California and in southern California.

The generic status of *Lophotocarpus* is discussed by Mason (Madroño 11:263–270, 1952).

Sagittaria montevidensis Cham. & Schl. (Linnaea 2:156, 1827) has been repeatedly reported in the literature as having escaped cultivation at Stockton, California, and as having been collected by Sanford and sent to J. G. Smith, who illustrated it (Rep. Mo. Bot. Gard. 6: pl. 29, 1895). We have not collected this species, nor have we seen specimens in herbaria. There is some doubt in our minds that the California material was other than a stout specimen of *S. calycina* Engelm. It is reported as having a brownish purple spot at the base of the petals. The plant figured by J. G. Smith (*loc. cit.*) cannot be distinguished from typical *S. calycina*.

2. Sagittaria Sanfordii Greene, Pittonia 2:158. 1890. Fig. 48.

Rhizomatous perennial, emersed aquatic in shallow water; tip of rhizome at length becoming a globose corm 8–15 mm. wide; plants 3–10 dm. high; leaves erect, 2–10 dm. long, triquetrous, typically a naked phyllode with a sheathing base, or less commonly having an elliptic or lanceolate blade 5–15 cm. long; scapes erect in anthesis, shorter than the leaves, at length reclining and floating as the fruits mature; flowers in several whorls, the lowermost pistillate, the upper flowers staminate, each subtended by a triangular bract; sepals of the pistillate flower not accrescent, spreading and soon surpassed in length by the growing fruit; petals white, orbicular with a cuneate claw; stamens sterile, rarely functional; staminate flowers with many anthers and few rudimentary carpels, withering-persistent on scape; filaments stout, ovoid, longer than anthers, densely beset with inflated papillae but when dry often appearing as though glabrous; fruiting heads on stout, recurved pedicels 2–3 cm. long; achenes with an erect beak, strongly keeled or winged on back, the lateral angles ribbed or winged.

Slow-running or standing water: Central Valley.

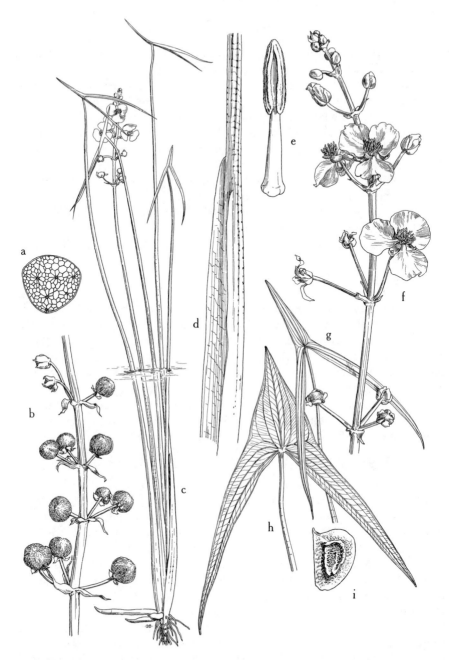

Fig. 49. *Sagittaria Greggii: a,* scape (cross section), $\times$ 1½; *b,* whorls of maturing fruits, showing the reflexed sepals, and long bracts subtending the pedicels, $\times$ ⅔; *c,* habit, showing narrowly sagittate leaf blades, $\times$ ⅕; *d,* leaf-base sheath, $\times$ ⅖; *e,* stamen, showing glabrous filament with dilated base, $\times$ 8; *f,* tip of inflorescence, showing whorls of staminate flowers, and pistillate flowers below beginning to mature, $\times$ ⅔; *g* and *h,* leaf-blade variations, $\times$ ⅖; *i,* mature achene, showing the tubercled, irregularly thickened lateral ribs, $\times$ 8.

Descriptions in the literature all state that the filament of the stamen is glabrous. This is at variance with our observations on living material, in which, we note without exception, the filament is clothed with thin-walled, clavate hairs. These may vary in density from one collection to another but are present on all plants observed. Apparently on drying these hairs collapse and are not evident, for on dried specimens the filaments appear to be glabrous. From these observations it seems clear that studies in the current literature have been based entirely upon dried specimens.

3. Sagittaria Greggii J. G. Smith, Rep. Mo. Bot. Gard. 6:43. 1895. Fig. 49.

Erect aquatic of shallow water, 2–10 dm. tall; tip of ephemeral rhizome at length becoming a globose perennial corm, or plant behaving as an annual; leaves erect, the blades sagittate, 1–2 dm. long, the basal lobes 2–3 times as long as the terminal, linear to linear-lanceolate, sometimes acuminate, the submersed juvenile leaves with blades entire or lacking; inflorescence simple or branched, subequal to or longer than leaves; lower flowers pistillate, upper ones staminate, occasionally a few flowers perfect; pistillate flowers on slender, ascending, often unequal pedicels, the pedicels 1–3 cm. long; sepals becoming reflexed, not growing with fruit; petals white, blades orbicular, claws cuneate; rudimentary stamens in a single whorl, sometimes a few with pollen; staminate flowers withering-persistent, rarely with rudimentary pistils; stamens numerous, the filaments longer than anthers, glabrous, somewhat dilated at base; fruiting heads depressed-globose; achenes obovate, 2–3 mm. long, winged, the lateral ribs irregularly thickened and winged or tubercled, curved to orbicular in outline, the style beak short and erect, occasionally pushed in a lateral direction as the achene matures.

Common in irrigation ditches and rice fields: Central Valley; south to Mexico.

There is some confusion in the literature concerning the type and type locality of this species. The description of the species is based upon a collection by Sanford made in 1893 at Stockton, California. It is "dedicated to Dr. J. Gregg, whose *no. 833,* collected in May 1849, at Zamora, Michoacan, Mexico, seems to be the same as our Californian plant." This statement by J. G. Smith reflects some doubt whether the Mexican specimen actually is the same as the plant described as *Sagittaria Greggii.* The fullness of citation and dedication has led some to regard the Gregg specimen as the type.

Bogin (Mem. N. Y. Bot. Gard. 9:222, 1955) includes this by "monographer's choice" under *Sagittaria longiloba* Engelm. ex Torr. in J. G. Smith, because of an intergradation of "key characters," although he acknowledges a generally larger stature of Californian and Mexican plants upon which the name *S. Greggii* rests. We choose not to alter the status *S. Greggii* from current local usage, pending at least a field study of the problem, since much is lost in herbarium material that might be significant to the case.

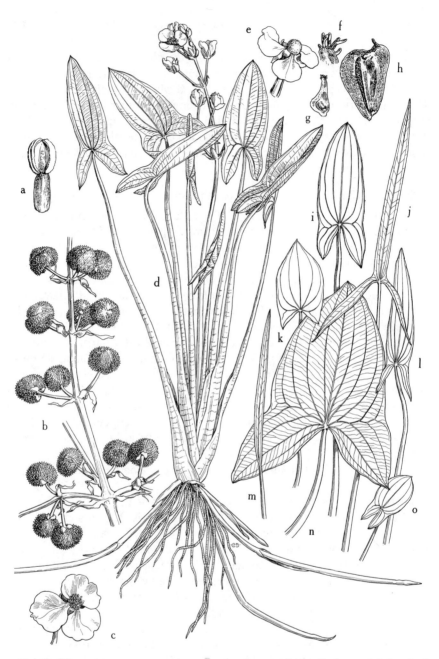

Fig. 50. *Sagittaria cuneata: a,* stamen, showing short, glabrous filament, $\times$ 6; *b,* whorls of maturing fruits, $\times$ ⅖; *c,* staminate flower, $\times$ ⅘; *d,* habit, showing rhizomes, inflorescence, and the somewhat spreading leaves, $\times$ ⅖; *e,* pistillate flower, $\times$ ⅘; *f,* papillate stigma, $\times$ 40; *g,* ovary terminating in stout style with papillate stigma, $\times$ 8; *h,* mature achene, showing wings and the erect, beaklike, persistent style, $\times$ 6; *i–o,* leaf-blade variations (note that the basal lobes are generally shorter than terminal lobe), $\times$ ⅖.

4. Sagittaria cuneata Sheldon, Bull. Torrey Club 20:283. 1893. Fig. 50.

Sagittaria arifolia Nutt. ex J. G. Smith, Rep. Mo. Bot. Gard. 6:32. 1895.
Sagittaria arifolia var. *stricta* J. G. Smith, Rep. Mo. Bot. Gard. 6:34. 1895.

Emersed erect rhizomatous aquatic of shallow water or wet soil, 1–5 dm. tall; leaves erect or spreading, occasionally prostrate, approximately as long as or shorter than inflorescence, the blades sagittate or linear to lanceolate-elliptic in juvenile submersed leaves, the basal lobes shorter than the terminal lobe; inflorescence simple or branching, the lower flowers pistillate, the upper ones staminate; pistillate flowers on slender pedicels of unequal lengths in the axils of lanceolate bracts; sepals not growing with the fruit; petals with orbicular blades and cuneate claws; staminate flowers with numerous stamens; anthers about as long as the oblong, glabrous filaments; fruiting heads depressed-globose; achene obovate, winged dorsally and ventrally, ventral wing somewhat raised, terminating in the erect, beaklike, persistent style.

Middle and high altitudes in the mountains and plateau lands: North Coast Ranges, Sierra Nevada, San Bernardino Mountains, north to Modoc and Siskiyou counties; western states, east across southern Canada.

Bogin (Mem. N. Y. Bot. Gard. 9:227–228, 1955) reports that the range of this species overlaps that of *Sagittaria latifolia* var. *latifolia*. This is not the situation found in our area, however; here, *S. cuneata* ranges from middle to high altitudes, whereas *S. latifolia* is confined to low altitudes.

5. Sagittaria latifolia Willd., Sp. Pl. 41:409. 1805. Wapato. Figs. 51 and 52.

Emersed aquatic of shallow water, erect, 3–20 dm. tall, perennial through the production of corms at the ends of rhizomes; leaves erect, the petioles stout on large plants and more slender on small plants, the blades exceedingly variable, 1–5 dm. long, 2–30 cm. wide, usually sagittate (rarely otherwise in ours), lateral lobes nearly as long as the terminal lobe, the terminal lobe from broadly obtuse, through deltoid, acute and acuminate, to linear, in some changing greatly with the season, the lateral lobes from widely divergent to almost parallel; scapes erect, shorter than to somewhat longer than leaves, usually 1 or 2 maturing, monoecious or dioecious, frequently with a few perfect flowers, ours glabrous throughout, often much-branched; flowers in verticels, usually the lower flowers pistillate, the upper ones staminate, frequently some perfect ones between staminate and pistillate flowers, occasionally all perfect or all staminate or the early inflorescences appearing as dioecious and the later ones as monoecious, each flower subtended by a deltoid or lanceolate bract; sepals triangular, becoming reflexed in fruit; petals white, the blade broadly and abruptly narrowing to the claw; pistillate flowers with or without rudimentary stamens; staminate flowers with or without rudimentary pistils; stamens with a stout filament longer than the anther, glabrous; fruiting heads depressed-globose, on slender, divergent petioles; achenes each with a stout beak extend-

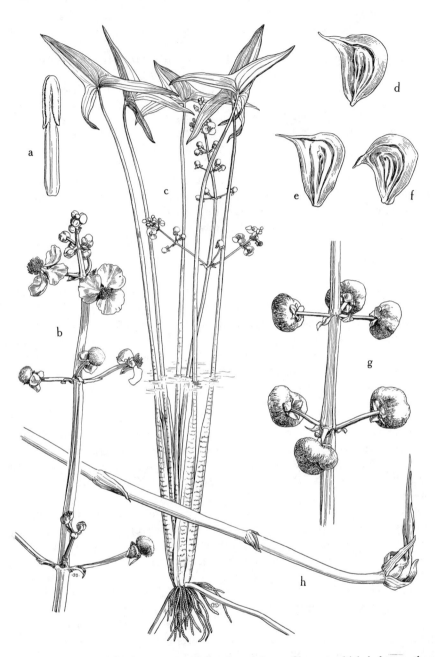

Fig. 51. *Sagittaria latifolia: a,* stamen, showing glabrous filament which is longer than anther, × 8; *b,* inflorescence, showing whorls of staminate flowers and of pistillate flowers, the sepals reflexed, × ⅔; *c,* habit, showing rhizomes and branched inflorescence, × ⅕; *d–f,* mature achenes, the margins with broad, corky, and laterally disposed wings, × 6; *g,* part of inflorescence, showing whorls of mature fruits, × ⅔; *h,* corm at the end of a rhizome, × ⅖.

ing at right angles to the cuneate body or somewhat reflexed, the margins with a broad, corky wing.

This is the plant that has been confused with *Sagittaria sinensis* of Sims. It was presumed by Parish to have been introduced into California by the Chinese, but there is ample evidence of its prior occurrence in the West and of its use by the Indians of the Northwest, where it bore the name wapato. The Chinese, on coming to California, used it for food and may have cultivated it somewhat. In so doing they may have extended its range into the southern part of the state.

5a. Sagittaria latifolia var. latifolia.

Terminal lobe of leaf acute or acuminate with no great seasonal variation. Plants very variable in size.

This is the common, widespread form of the species occurring throughout the major interior valleys: occasional in southern California; rare at elevations above 3,000 feet.

5b. Sagittaria latifolia var. obtusa (Muhl. ex Willd.) Wieg., Rhodora 27:186. 1925. Fig. 52.

Sagittaria obtusa Muhl. ex Willd., Sp. Pl. 41:409. 1805.

Plant usually very large; leaves more or less constant in shape throughout the growing season, the terminal lobe of leaf broadly obtuse.

Occasional along major rivers and irrigation ditches, often occurring in great masses: Central Valley; east to Atlantic.

Bogin (Mem. N. Y. Bot. Gard. 9:219, 1955) cites this, without comment, as synonymous with *Sagittaria latifolia* var. *latifolia* and thereby, in the literature, effectively conceals the problem. We have seen no plants having any intermediate leaf type in our extensive field study of this group.

HYDROCHARITACEAE. FROGBIT FAMILY

Aquatic herbs rooted in mud or floating, ours submersed except for the flowers in anthesis. Leaves basal or cauline, in ours whorled at the nodes. Flowers dioecious or less often perfect, regular, 1 to several, ours in a sessile spathe, the pistillate flowers usually solitary, the staminate ones several, both on slender, pedicel-like perianth tubes or sometimes long-pediceled, reaching the surface of the water. Perianth in 1 or 2 series of 3 each, free, the outer segments valvate, the inner ones imbricate. Stamens 3 to many; anthers 2-celled. Ovary inferior, 1-celled with 3–6 parietal placentae; styles often much-branched. Seeds few to many; embryo straight.

Although not treated in this flora, *Vallisneria,* the commonly cultivated aquarium and garden-pond plant, is to be expected as a transitory escape in shallow aquatic habitats.

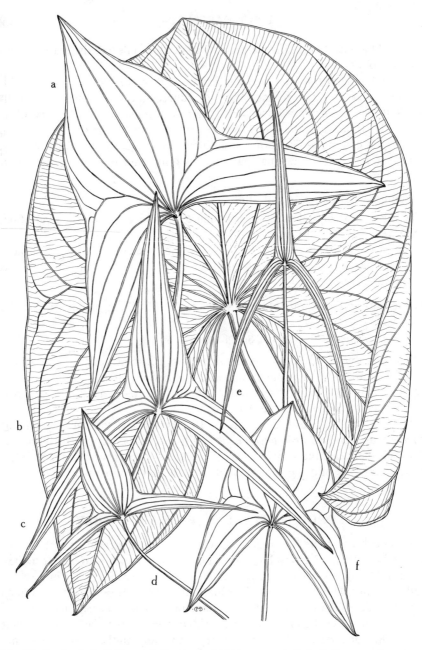

Fig. 52. *Sagittaria latifolia,* variation in leaf blades. *a, c–f, S. latifolia: e,* early seasonal phase, and *f,* late seasonal phase of same plant; *b, S. latifolia* var. *obtusa.* All × ⅖.

ELODEA. WATERWEED

Submersed, elongated, branching, leafy herb. Leaves opposite or whorled, often crowded, 1-nerved, thin, minutely serrate or entire, sessile. Flowers dioecious or polygamous, arising from an ovoid or tubular, 2-cleft spathe. Staminate flowers solitary, rarely in 3's, sessile or nearly so, but long-exserted by a pedicel-like extension of the hypanthium, early separating from the plant and floating on the surface of the water; sepals 3, petals 3, or the latter sometimes wanting; stamens 9, in 2 series, the outer series of 6 and the inner one of 3. Pistillate flowers solitary, sessile, with an elongated hypanthium tube and a 6-parted limb; sepals 3; petals 3; ovary inferior; stamens none or represented by 3 rudimentary filaments. Hermaphrodite flowers when present like the pistillate ones, but with 3–9 stamens. Fruit coriaceous, indehiscent, oblong or spindle-shaped, producing 1–5 spindle-shaped seeds.

The name *Elodea* has been proposed for conservation to the International Committee.

Marie-Victorin, Frère, *L'Anacharis canadensis*. Contr. Lab. Bot. Univ. Montréal 18:1–43. 1931.
Weatherby, C. A., On the nomenclature of *Elodea*. Rhodora 34:114–116. 1932.

Leaves in 3's, sometimes opposite, 5–15 mm. long; flowers solitary in the spathe
1. *E. canadensis*
Leaves in whorls of more than 3, 15–35 mm. long; flowers usually 2 or 3 in a spathe
2. *E. densa*

1. Elodea canadensis Michx., Fl. Bor. Am. 1:20. 1803. Fig. 53.

Elodea Planchonii Casp., Jahrb. Wiss. Bot. 1:408. 1858.
Philotria Planchonii Rydb., Bull. Torrey Club 35:462. 1908.

Stems slender; leaves in 3's or the lower ones opposite, ovate-oblong, 6–12 mm. long, 1.5–4 mm. wide (average 2 mm.), acute, crowded and overlapping near tips; staminate flowers solitary, 4–5 mm. long, remaining attached by a long pedicel, the spathe obovoid-clavate, about 10 mm. long, on a peduncle 5–10 mm. long, the sepals elliptic, the petals wanting, the anthers oblong, 3–4 mm. long, subsessile; pistillate flowers solitary, sessile, the spathe linear or linear-lanceolate, sessile, the perianth tube 3–5 cm. long, very slender, the sepals 2.5–3 mm. long, linear, the stigmas 3, linear; fruit linear or lanceolate-linear.

Frequent in lakes and ponds: scattered localities throughout California; cooler parts of North America.

The name *Elodea Planchonii* is frequent in American literature and is here regarded as synonymous with *E. canadensis*. The difficulty appears to be related to the fact that when Micheaux described *E. canadensis* he had before him a plant with hermaphrodite flowers, apparently a rare condition. Later, when the characteristic dioecious plants were discovered, they were described as new. Hence, a single entity has two names.

Fig. 53. *Elodea. a–d, E. densa:* a, stem with whorls of lanceolate leaves, showing their minutely serrate margins, × 1⅓; *b,* spathe and the flowers, × 1½; *c,* habit, showing long-peduncled staminate flowers, × ⅖; *d,* staminate flower, × 2. *e–g, E. canadensis: e,* obovoid-clavate spathe with staminate flower on long, threadlike peduncle, × 4; *f,* habit, the staminate flower at the surface of the water in anthesis, × ⅔; *g,* mature staminate flower, showing the 9 anthers and 3 rudimentary stigmas, × 5.

2. Elodea densa (Planch.) Casp., Monatsb. Kön. Acad. Wiss. Berlin 1857:49. 1857. Fig. 53.

Stems stout, 2–3 mm. in diameter; leaves numerous, crowded and overlapping; blades lanceolate to linear-lanceolate, 1.5–2 cm. long, 2–5 mm. wide, acuminate, minutely serrate, whorled; spathe of the staminate flowers sessile, narrowly conic, about 1 cm. long, usually 2 or 3 flowers to a spathe, the staminate flowers remaining attached by a long, pedicel-like extension of the hypanthium, the sepals green, oval to ovate, 3–4 mm. long, the petals white, 9–11 mm. long; pistillate flowers not known to us.

Infrequent in lakes, ponds, and irrigation ditches: scattered locations throughout California. Native of South America.

GRAMINEAE. GRASS FAMILY

Perennial or annual herbs (woody in *Arundo*) with hollow or solid stems (culms) closed at the nodes, and 2-ranked, parallel-veined leaves, composed of a sheath enclosing the culm and a blade, with a hairy or membranaceous appendage (ligule) between them on the inside. Flowers perfect or sometimes unisexual, arranged in spikelets. Spikelets consisting of a short axis (rachilla) and 2 to many 2-ranked bracts, the lower 2 (glumes) empty, the succeeding ones (lemmas) each bearing in its axil a single flower and between the flower and the rachilla a 2-nerved bract (palea), the lemma, palea, and included flower constituting the floret. Stamens 1–6 (usually 3), the anthers 2-celled. Pistil 1, with 2 (rarely 1 or 3) styles, and usually plumose stigmas. Spikelets mostly grouped in spikes or panicles (sometimes racemose) at ends of the main culms and branches. Fruit a caryopsis or grain.

Hitchcock, A. S., Manual of the grasses of the United States. U. S. Dept. Agr. Misc. Publ. 200. 1st ed., 1935. 2d ed., rev. by Agnes Chase, 1950.

KEY TO THE TRIBES

Spikelets 1- to many-flowered, usually compressed laterally, the reduced florets (if any) above the perfect florets (except in the Phalarideae); articulation usually above the glumes .... Subfamily FESTUCOIDEAE
Spikelets with 2 (rarely 1) staminate, neuter, or rudimentary florets below the terminal perfect floret .... PHALARIDEAE
(*Anthoxanthum, Phalaris*)
Spikelets without sterile lemmas below the perfect floret.
   Spikelets 1-flowered, borne in fascicles.
      Glumes suppressed or wanting; articulation below the glumes.
         Spikelets unisexual; ovulate spikelet terete, the staminate and ovulate spikelets borne on different parts of the plant or inflorescence .... ZIZANIEAE
(*Zizania*)
         Spikelets perfect, strongly compressed laterally .... ORYZEAE
(*Leersia, Oryza*)

Glumes usually prominent (somewhat reduced in some species of *Muhlenbergia*), borne in open, contracted, or spikelike panicles, but not in true spikes or one-sided racemes .... AGROSTIDEAE
(*Agrostis, Alopecurus, Calamagrostis, Crypsis, Gastridium, Heleochloa, Muhlenbergia, Phleum, Polypogon*)
Spikelets 1- to many-flowered, if 1-flowered then borne in true spikes, fascicles of 2–5 spikelets, or sessile in 2 rows on one side of a continuous rachis.
Spikelets 1- to several-flowered, all but the lowest floret commonly sterile and variously modified, borne in 2 rows on one side of a continuous rachis, forming one-sided spikes or spikelike racemes, these solitary, racemose, or digitate CHLORIDEAE
(*Beckmannia, Cynodon, Leptochloa, Spartina*)
Spikelets 1- to many-flowered on opposite sides of a jointed rachis, forming a spike HORDEAE
(*Elymus, Hordeum, Monerma, Parapholis*)
Spikelets 2- to many-flowered, borne in panicles.
Glumes large, as long as the lowest floret; awns (when present) arising from the back of the lemmas .... AVENEAE
(*Deschampsia*)
Glumes relatively small, shorter than the first floret; awns (when present) arising from the tip or from the bifid apex of the lemmas .... FESTUCEAE
(*Arundo, Cynosurus, Dactylis, Distichlis, Eragrostis, Festuca, Glyceria, Monanthochloë, Neostapfia, Orcuttia, Phragmites, Pleuropogon, Poa, Puccinellia*)
Spikelets with 1 perfect terminal floret and a sterile or staminate floret below, usually represented by a sterile lemma, compressed dorsally, falling entire, singly, or together with parts of the axis .... Subfamily PANICOIDEAE
(*Andropogon, Sorghum*, in ANDROPOGONEAE)

KEY TO THE GENERA

A. Spikelets with 1 perfect terminal floret and a sterile or staminate floret below, usually represented by a sterile lemma only, 1 glume sometimes wanting; the rachilla articulated below the spikelets, the spikelets thus falling entire.

Fertile lemma and palea firmer than the glumes, usually hardened.
Spikelets subtended by 1 or more bristles, these distinct at base .... *Setaria*
Spikelets not subtended by bristles.
Spikelets in open panicles .... *Panicum*
Spikelets short-pediceled or subsessile on one side of the panicle branches.
Second glume mucronate, the sterile lemma mucronate or awned .... *Echinochloa*
Second glume and sterile lemma awnless.
Racemes more or less digitate at the summit of the culms; first glume minute, but evident .... *Digitaria*
Racemes panicled.
First glume and the rachilla joint forming a swollen, ringlike callus below the spikelet; back of fruit turned away from the rachis .... *Eriochloa*
First glume wanting (occasionally present in *Paspalum distichum*); back of the fruit turned toward the rachis .... *Paspalum*

Fertile lemma and palea thin, hyaline, the glumes hardened; sterile lemma like the fertile one.
Spikelets unlike, the sessile one perfect, the pedicellate spikelet usually sterile.
Racemes of several joints, silky-villous .... *Andropogon*
Racemes reduced to 1 or a few joints, these in a compound panicle, not silky-villous
Sorghum
Spikelets all perfect, surrounded by a conspicuous tuft of soft hairs .... *Erianthus*

AA. Spikelets 1- to many-flowered, the reduced florets, if any, above the perfect florets (except in *Antho-xanthum* and *Phalaris*); the rachilla usually articulated above the glumes (except in *Crypsis, Leersia, Oryza, Polypogon, Alopecurus, Spartina,* and *Agrostis semiverticillata*).

Spikelets with 2 sterile or staminate florets below the fertile one.
Lower florets reduced to small, awnless, scale-like lemmas; spikelets much compressed laterally .... *Phalaris*
Lower florets consisting of awned, hairy, sterile lemmas longer than the fertile floret; spikelets subterete .... *Anthoxanthum*
Spikelets with no sterile florets below the 1 or more fertile florets.
Spikelets unisexual, falling entire; pistillate spikelets on the ascending upper branches, the staminate spikelets on the spreading lower branches of the same panicle
Zizania
Spikelets perfect, usually articulate above the glumes.
Glumes wanting (minute in *Oryza*); pedicels articulate below the 1-flowered spikelets.
Annual .... *Oryza*
Perennial with slender, creeping rhizomes .... *Leersia*
Glumes present.

a. Spikelets sessile or subsessile in spikes or spikelike racemes.

Spikelets on opposite sides of the continuous or disarticulating rachis.
Spikelets solitary at each node of the rachis, sunken in hollows in the rachis.
First glume wanting .... *Monerma*
First glume present, the pair of glumes standing in front of the spikelets .. *Parapholis*
Spikelets more than 1 at each node of the rachis.
Spikelets 3 at each node of the rachis, 1-flowered, the lateral ones either with reduced imperfect florets or reduced to awns .... *Hordeum*
Spikelets 2 (sometimes 3) at each node of the rachis, 3- to 6-flowered, the lateral ones not reduced to awns .... *Elymus*
Spikelets subsessile on one side of the continuous rachis; spikes digitate or racemose on a common axis.
Spikelets 1-flowered.
Spikes digitate, rachilla articulate above the glumes .... *Cynodon*
Spikes racemose, erect or nearly so, rachilla articulate below the glumes, the spikelets falling entire.
Glumes equal in size, broad and boat-shaped .... *Beckmannia*
Glumes unequal in size, narrow .... *Spartina*
Spikelets 2- to several-flowered .... *Leptochloa*

aa. Spikelets in open or spikelike panicles.

Spikelets 1-flowered.
Articulation below the glumes, the spikelets falling entire.
Glumes long-awned; panicles open or dense but scarcely spikelike (except in *Polypogon monspeliensis*) .... *Polypogon*
Glumes awnless; panicles narrow, dense, spikelike .... *Alopecurus*
Articulation above the glumes.
Glumes usually longer than the lemma.
Glumes compressed-carinate, ciliate on the keels; panicle dense, spicate .. *Phleum*
Glumes not compressed-carinate or ciliate.
Glumes somewhat inflated at base; lemma long-awned .... *Gastridium*
Glumes not inflated at base; lemma awned or awnless.
Florets bearing a tuft of hairs at the base from the short callus, the hairs at least ½ as long as the lemma .... *Calamagrostis*
Florets without hairs at the base or with short hairs, rarely as much as ½ as long as the lemma .... *Agrostis*
Glumes not longer than the lemma, usually shorter.
Lemma awned from the tip or mucronate .... *Muhlenbergia*
Lemma awnless.
Inflorescence capitate in the axils of broad bracts .... *Crypsis*
Inflorescence with oblong, dense, spikelike panicles .... *Heleochloa*
Spikelets 2- to several-flowered.
Lemma usually shorter than the glumes; the awn arising from the back of the lemma, usually below the middle, and straight or bent .... *Deschampsia*
Lemma usually longer than the glumes; the awn terminal and straight or none.
Lemmas divided at the summit into 5–9 awnlike lobes .... *Orcuttia*
Lemmas awnless, with a single awn, or if with 3 then the lateral awns minute.
Tall, stout reed with large, plumelike panicles.
Lemmas hairy; rachilla naked .... *Arundo*
Lemmas naked; rachilla hairy .... *Phragmites*
Low or rather tall grasses; panicles not plumelike.

b. Plants dioecious; lemmas glabrous; grasses of saline or alkaline soils.

Plants low, creeping; spikelets obscure, scarcely differentiated from the short, crowded, rigid leaves .... *Monanthochloë*
Plants erect from creeping rhizomes; spikelets in narrow, simple, exserted panicles
Distichlis

bb. Plants not dioecious.

Spikelets of two forms, sterile and fertile intermixed; panicle dense, spikelike, somewhat one-sided .... *Cynosurus*
Spikelets all alike in the same inflorescence.
Lemmas 3-nerved, the nerves prominent, often hairy .... *Eragrostis*
Lemmas 5- to many-nerved, the nerves sometimes obscure.
Lemmas fan-shaped; glumes wanting; inflorescence dense, cylindric ..... *Neostapfia*
Lemmas not fan-shaped; glumes present; inflorescence not cylindric.
Spikelets in elongate, loose racemes .... *Pleuropogon*
Spikelets in open or contracted panicles.
Lemmas mostly obtuse, awnless, the nerves parallel, not converging at summit or only converging slightly.
Spikelets strongly compressed, crowded in one-sided clusters ...... *Dactylis*

Spikelets not strongly compressed, not crowded in one-sided clusters.
Nerves faint; usually of saline soil .... *Puccinellia*
Nerves prominent; fresh-water marshes .... *Glyceria*
Lemmas awned or acute at apex, the nerves converging toward summit.
Lemmas mucronate or awned, keeled at least toward apex .... *Festuca*
Lemmas awnless, usually rounded on the back .... *Poa*

AGROSTIS. BENT GRASS

Annual or usually perennial herbs. Culms glabrous. Blades flat or sometimes involute, scabrous. Panicle open or contracted; spikelets small, 1-flowered, disarticulating above the glumes (except in *Agrostis semiverticillata*). Glumes equal in size or nearly so, acute, acuminate, or sometimes awn-pointed, carinate, usually scabrous on the keel and sometimes on the back. Lemma obtuse, usually shorter and thinner than the glumes, awnless or dorsally short-awned, often hairy on the callus. Palea usually shorter than the lemma, usually small and nerveless or obsolete.

In addition to the species treated below, *Agrostis avenacea* Gmel. is sometimes found in marshy areas.

Palea minute or none.
Lemma with a more or less exserted, bent awn .... 1. *A. longiligula*
Lemma awnless.
Panicle very open and diffuse at maturity, the branches spikelet-bearing only near the
ends .... 2. *A. scabra*
Panicle contracted, somewhat dense and spikelike, the branches spikelet-bearing to
the base .... 3. *A. exarata*
Palea ½ to nearly as long as the lemma.
Glumes scabrous on keel only .... 4. *A. alba*
Glumes scabrous on back as well as on keel .... 5. *A. semiverticillata*

1. Agrostis longiligula Hitchc., U. S. Dept. Agr., Bur. Pl. Indus. Bull. 68:54, pl. 36. fig. 3. 1905. Long-tongue bent grass.

Perennial; culms erect, glabrous, 2–7 dm. tall; sheaths smooth; ligules 4–6 mm. long; blades flat or involute, scabrous, 1–3 mm. wide, often somewhat setaceous at base; panicle usually purplish, oblong, open or somewhat dense, 10–20 cm. long, the capillary, very scabrous branches ascending and nearly naked near the base; glumes narrow, acute, scabrous on the keel, the first glume 3–4 mm. long, the second a little shorter; lemma 2.5 mm. long, bearing near the middle a more or less exserted, bent awn; palea minute or none.

Marshes and bogs near the coast: Marin County to Del Norte County; southern Oregon.

The Marin County plants, in which the awns are not exserted and the ligules are as much as 1 cm. in length, have been designated as *Agrostis longiligula* var. *australis* Howell.

Fig. 54. *Agrostis exarata: a,* leaf sheath, ligule, and blade, × 5; *b,* spikelets in lower part of panicle, × 3; *c,* habit, showing the leafy culms and young, close panicle, × ⅖; *d,* upper part of culm, showing panicle, × ⅖; *e,* floret, × 14; *f,* spikelet, the glumes each with a scabrous keel, × 14.

2. Agrostis scabra Willd., Sp. Pl. 1:370. 1797. Tickle grass.

Perennial; culms very slender, tufted, 2–8 dm. tall; sheaths glabrous or scaberulous; ligules 1–3 mm. long; blades mostly basal, flat or loosely involute, often subfiliform, scabrous, 1–2 mm. wide; panicle large, 1–3 dm. long, very diffuse, usually purple, the fine capillary branches scabrous, verticillate (especially the lower ones), spikelet-bearing near the ends; glumes 1.5–2.5 mm. long, subequal in size, acute or acuminate, scabrous on the keels; lemma about ⅔ as long as the glumes, awnless; palea wanting.

Moist meadows: throughout the Sierra Nevada, east of the Sierra Nevada crest from Inyo County to Modoc County, San Jacinto and San Bernardino mountains at elevations above 4,500 feet; occasional throughout North America.

At high altitudes this species passes into *Agrostis scabra* var. *geminata* (Trin.) Swallen, which has shorter culms and panicle branches.

This species was included in *A. hiemalis* in Hitchcock (*op. cit.*, 1935), but differs from it in having broader cauline leaves and slightly longer spikelets, which are longer-pediceled and more loosely arranged at the ends of the branchlets.

3. Agrostis exarata Trin., Gram. Unifl. 207. 1824. Spike bent grass. Fig. 54.

Perennial; culms in small tufts, erect or somewhat decumbent at base, glabrous, rather leafy, 2–10 dm. tall; sheaths glabrous or slightly scabrous; ligules prominent, 2–5 mm. long; blades flat, 1–8 mm. wide, scabrous; panicle narrow, from somewhat open to close and spikelike, 5–25 cm. long; glumes acuminate or tipped with a short awn, nearly equal in size, scabrous on the keel and often scaberulous on the back, 2.5–4 mm. long; lemma about 2 mm. long, awnless, the callus glabrous; palea less than 0.5 mm. long.

Marshes, wet meadows, and along streams: near coast and in Coast Ranges from Ventura County to Del Norte County, Sierra Nevada, east of the Sierra Nevada crest from Mono County to Modoc County, mountains of southern California chiefly at elevations of 4,000–8,000 feet, occasional in Central Valley; western North America from Alaska to Mexico.

4. Agrostis alba L., Sp. Pl. 63. 1753. Redtop, creeping bent grass, spreading bent grass. Fig. 55.

Agrostis stolonifera L., Sp. Pl. 62. 1753.

Perennial; loosely tufted, often with stolons or sometimes rhizomatous; culms erect, ascending, or decumbent, to about 1 m. tall; sheaths smooth; blades flat, 2–10 mm. wide, scabrous; ligules 2–5 mm. long, rounded and often lacerate at apex; panicle from broadly pyramidal with the branches spreading after flowering to narrowly oblong or linear with the branches erect and appressed after flowering, varying from dark purple to straw-colored and whitish or greenish; glumes acute, 2–3.5 mm. long, subequal in size, scabrous

Fig. 55. *Agrostis alba: a,* scabrous branchlets of panicle, × 6; *b,* leaf sheath, ligule, and blade, × 6; *c,* habit, showing the decumbent culms and flat leaf blades, × ⅓; *d,* floret, showing lemma, × 20; *e,* floret, showing the short, emarginate palea, × 20; *f* and *g,* grains (caryopses), × 20; *h,* habit, upper part of culm, showing panicle, × ⅓; *i,* spikelet, showing the glumes, each with scabrous keel, × 16.

on the keel; lemma obtuse, 1.5–2.2 mm. long, usually without an awn; palea ½–⅔ as long as the lemma.

Widely distributed in moist ground throughout California, except at higher elevations. Native of Europe.

The taxonomy, as well as the nomenclature, of this extremely variable species is in a very confused state. The members of the group apparently comprise an intergrading series of forms, some of which have been recognized as distinct species by some authors but regarded by other writers simply as varieties of a single, polymorphic species. The concept here employed is that applied by Hitchcock.

5. Agrostis semiverticillata (Forsk.) C. Chr., Dansk Bot. Ark. 4³:12. 1922. Water bent grass. Fig. 56.

Agrostis stolonifera L., Sp. Pl. 62. 1753. In part.
Agrostis verticillata Vill., Prosp. Pl. Dauph. 16. 1779.

Perennial; culms usually somewhat decumbent at base, sometimes with long, creeping and rooting stolons, 3–9 dm. tall, light green or glaucous green; sheaths glabrous or slightly roughened, occasionally somewhat inflated; ligules truncate, dentate, sometimes scaberulous, 2–6 mm. long; blades slightly scabrous, relatively short, 3–10 cm. long, 2–5 mm. wide; panicle contracted, lobed or verticillate (especially at base), light green or purplish, 3–15 cm. long, the branches ascending, densely flowered from the base; spikelets usually falling entire, 1.75–2 mm. long; glumes obtuse or barely acute, scabrous on back and keel; lemma 1 mm. long, awnless, truncate and toothed at apex; palea nearly as long as the lemma.

Moist ground along streams and irrigation ditches at low to middle elevations throughout California, except in the Sacramento Valley, apparently only occasional in the Sierra Nevada and San Joaquin Valley. Native of Europe.

Except for the absence of awns on the glumes and lemma, this species bears a marked resemblance to *Polypogon lutosus* (Poir.) Hitchc. *Agrostis* is traditionally defined as including plants with awnless glumes and spikelets which disarticulate above the glumes. In *A. semiverticillata*, however, the articulation is below the glumes, and the spikelets therefore fall entire, as in the related genus *Polypogon*.

ALOPECURUS. FOXTAIL

Perennial or annual grasses with flat blades, loose sheaths, and soft, dense, spikelike panicles. Spikelets 1-flowered, disarticulating below the glumes and falling entire, strongly compressed laterally. Glumes equal in size, awnless, usually united at base, the keel ciliate and the lateral nerves usually appressed-pubescent. Lemma slightly shorter than the glumes, obtuse, the infolded margins united adaxially below, bearing from below the middle a slender, abaxial awn, this included or exserted. Palea wanting.

Fig. 56. *Agrostis semiverticillata: a,* habit, showing culms with decumbent base, short, horizontal leaf blades, and panicles, × ⅕; *b,* young floret, showing the truncate lemma, toothed at apex, × 12; *c,* branchlets of panicle, showing inflated base, × 4; *d,* flowering spikelet, the glumes scabrous, × 20; *e* and *f,* leaf sheath, dentate ligule, and scabrous blade, × 4.

Fig. 57. *Alopecurus aequalis: a,* floret, the lemma bearing an awn below the middle, × 12; *b,* habit, showing short basal leaves, the tall culms, and narrow-cylindric panicles, × ⅖; *c,* spikelet, showing the ciliate glumes, the awn of lemma protruding, × 12; *d,* grain, × 12; *e,* floret, variation in the awn of lemma, × 12; *f,* leaf sheath, ligule, and scabrous blade, × 4.

Fig. 58. *Alopecurus geniculatus:* *a,* habit, showing the cylindric panicles, the awns of lemmas conspicuous, × ⅖; *b,* floret, showing the long, curved awn of lemma attached below the middle, × 12; *c,* leaf sheath, ligule, and scabrous blade, × 4; *d,* spikelet, showing the ciliate glumes and the long awn of lemma, × 12.

In addition to the species treated below, *Alopecurus pratensis* is sometimes found in marshy areas.

Awn exserted beyond glumes not more than 1 mm. or included, usually attached slightly
    below the middle .... 1. *A. aequalis*
Awn exserted beyond glumes 1.5 mm. or more, usually attached well below the middle
    and about 1 mm. from the base, often bent.
    Glumes 2.5–3 mm. long; perennial .... 2. *A. geniculatus*
    Glumes 3–5 mm. long; annual .... 3. *A. saccatus*

1. Alopecurus aequalis Sobol., Fl. Petrop. 16. 1799. Short-awn foxtail. Fig. 57.

Perennial; culms erect or somewhat decumbent below and rooting at the nodes, glabrous, 2–6 dm. tall (or taller in some aquatic forms); sheaths glabrous, usually somewhat inflated; ligules 3–5 mm. long; blades slightly scabrous, 1–4 mm. wide, sometimes tufted at base; panicles more or less exserted, narrow-cylindric, 2–7 cm. long, 4–5 mm. wide; glumes 2–2.5 mm. long, ciliate on the keel, appressed-pubescent on the sides, especially below; lemma glabrous, the awn attached at or slightly below the middle, straight or slightly bent, included or exserted about 1 mm.; anthers about 1 mm. long.

Chiefly in water or wet soil: Sierra Nevada north to Modoc and Siskiyou counties, North Coast Ranges in Humboldt and Mendocino counties, mountains of southern California in San Bernardino and San Diego counties; widespread in Northern Hemisphere.

This species shares many of the morphological characteristics of *Alopecurus geniculatus,* and individuals of the two species are sometimes difficult to distinguish. It usually differs from *A. geniculatus,* however, in having less decumbent culms and shorter rooting bases, greenish spikelets (often purplish-tipped in *A. geniculatus*), a shorter, less conspicuous awn (the panicle therefore not having the bristly appearance which is usual in *A. geniculatus*), and a slightly broader spikelet. The glumes, in addition to being appressed-pubescent on the lateral margins, are somewhat more noticeably pubescent on the sides, especially below.

2. Alopecurus geniculatus L., Sp. Pl. 60. 1753. Water foxtail. Fig. 58.

Perennial; culms decumbent or long-decumbent at base, rooting at the lower nodes, glabrous, often bent above (only erect in dwarf forms), 1–6 dm. long above the rooting base; sheaths glabrous, usually somewhat inflated; ligules usually 2–4 mm. long; blades minutely scabrous above, 1–4 mm. wide; panicles 2–7 cm. long, 4–6 mm. wide; glumes 2.5–3 mm. long, the tips often purplish, ciliate on the keel, glabrous or appressed-pubescent on the lateral margins; lemma glabrous, the often purplish awn bent, exserted about the length of the spikelet or farther; anthers about 1.5 mm. long.

In water or wet places at low to middle elevations: Marin County to Humboldt County, San Diego County, Shasta County, east of the Sierra Nevada crest from Inyo County to Modoc County. Native of Europe.

3. Alopecurus saccatus Vasey, Bot. Gaz. 6:290. 1881. Pacific foxtail.

Alopecurus californicus Vasey. Bull. Torrey Club 15:12. 1888.

Annual; culms erect or geniculate below, glabrous, 1.5–5 dm. tall, often dwarf; sheaths glabrous, the upper ones often conspicuously inflated; ligules thin, 3–5 mm. long; blades scabrous, 1–2 (–4) mm. wide; panicles oblong to linear, exserted or, especially in dwarf specimens, included at base, 2–6 cm. long, 4–8 mm. wide; glumes 3–5 (usually 3.5–4) mm. long, ciliate on the keel, appressed-pilose on the lateral nerves, especially below, often darker green or somewhat purplish at tips; lemma glabrous, the awn attached about 1 mm. from the base, bent, exserted, 3–7 mm. (usually 3–4 mm.); anthers about 1 mm. long.

Wet ground: Santa Barbara County to Humboldt County, Central Valley from Kern County to Sacramento County, occasional in coastal southern California; Pacific states.

ANDROPOGON. BLUESTEM

1. Andropogon glomeratus B.S.P., Prelim. Cat. N. Y. 67. 1888. Bushy beard grass.

Andropogon macrourum Michx., Fl. Bor. Am. 1:56. 1803.

Perennial; culms tufted, erect, 5–10 dm. tall, compressed, freely to bushy-branching toward the summit; sheaths scabrous, occasionally sparsely villous, the lower sheaths broad, carinate, overlapping; ligules villous; blades elongate, 3–8 mm. wide; racemes paired, 1–3 cm. long, aggregated into a dense, feathery, fan-shaped to oblong inflorescence 1–3 dm. long, the common peduncle enclosed by a slightly dilated, spathelike sheath about as long as the racemes; the spikelets in pairs at each node of a slender, articulate rachis, one sessile and perfect, the other slender-pedicellate, sterile, reduced to a subulate glume or wanting; rachis and pedicels of sterile spikelet long-villous, the fertile spikelet 3–4 mm. long, its lemma bearing a straight awn 1–1.5 cm. long.

Moist or wet springy slopes and seepage areas: low elevations in coastal southern California, ascending into the foothills and ranging eastward to Death Valley, Inyo County, also near the coast in Ventura, Marin, and Sonoma counties; eastern and southern states, southwestern states, Central America.

A form of *Andropogon virginicus* L., a closely related species occupying similar habitats, may also be present in the California flora. Except for a somewhat less congested inflorescence, *A. virginicus* (at least in its California aspect) so closely resembles *A. glomeratus* that it has not been possible to make a clear separation of the two species. This form of *A. virginicus* has been collected in Sonoma, Yolo, and Butte counties.

ANTHOXANTHUM. VERNAL GRASS

1. Anthoxanthum odoratum L., Sp. Pl. 28. 1753. Sweet vernal grass.

Sweet-smelling perennial; culms tufted, erect, slender, 2–6 dm. tall; blades flat, 2–5 mm. wide; panicle spikelike, long-exserted, greenish yellow or brownish yellow, acute, 2–8 cm. long; spikelets 8–10 mm. long; glumes sparsely pilose or scabrous, the first about ½ as long as the second; spikelets with 1 terminal perfect floret and 2 sterile lemmas, the latter subequal in size, appressed-pilose with golden hairs, the first with a short, included awn arising from below the apex, the second awned from near the base, the awn twisted below, bent, slightly longer than the second glume; fertile lemma about 2 mm. long, brown, smooth and shining; palea 1-nerved, rounded on the back, enclosed in the lemma.

Established in meadows, pastures, waste land, and occasionally in marshes, sometimes cultivated: near coast and in Coast Ranges from Marin County to Del Norte County, also Placer County. Native of Eurasia.

ARUNDO

1. Arundo donax L., Sp. Pl. 81. 1753. Giant reed.

Tall perennial, 2–6 m. tall; culms stout, in large clumps from thick, knotty rhizomes; blades numerous, elongate, 5–7 cm. wide on the main culm, conspicuously 2-ranked, spaced rather evenly along the culm, the margin scabrous, the base cordate and somewhat hairy-tufted; panicle terminal, dense, plumelike, erect, 30–60 cm. long; spikelets several-flowered, 12 mm. long, the florets successively smaller, the summits of all about equal; glumes somewhat unequal in size, membranaceous, slender-pointed; lemmas thin, 3-nerved, densely and softly long-pilose, gradually narrowed at summit, the nerves ending in slender teeth, the middle one becoming an awn.

Established along irrigation ditches and streams (occasional in marshes): central and southern California. Native of the warmer regions of the Old World.

BECKMANNIA

1. Beckmannia Syzigachne (Steud.) Fern., Rhodora 30:27. 1928. American slough grass. Fig. 59.

Annual; culms light green, erect, rather stout, 3–10 dm. tall; blades flat; panicle 10–25 cm. long, narrow, more or less interrupted, the spikes crowded, 1–2 cm. long, appressed or ascending; spikelets 1-flowered, laterally compressed, subcircular, nearly sessile and closely imbricate, in 2 rows along one side of a slender, continuous rachis, disarticulating below the glumes, falling

Fig. 59. *Beckmannia Syzigachne: a,* panicle, showing the ascending spikes, × ⅔;
b, habit, × ⅛; *c,* floret, × 12; *d,* spikelet, laterally compressed, × 8; *e,* grain, × 12;
f, leaf sheath and ligule, × 4.

entire, 3 mm. long; glumes equal in size, inflated, obovate, 3-nerved, transversely wrinkled and with a deep keel; lemma narrow, 5-nerved, acuminate with the apex protruding beyond the glumes; palea nearly as long as the lemma.

Marshy flats and ditches: near the coast from Marin County to Humboldt County, Siskiyou County, east of the Sierra Nevada crest from Mono County to Modoc County; across North America, Asia.

CALAMAGROSTIS. REED GRASS

Perennial, usually moderately tall grasses, mostly with creeping rhizomes. Panicles open or usually narrow, sometimes spikelike. Spikelets small, 1-flowered, the rachilla disarticulating above the glumes, prolonged behind the palea as a short, commonly hairy bristle. Glumes about equal in size, acute or acuminate. Lemma shorter and usually more delicate than the glumes, awned from the back. The callus bearing a tuft of hairs, these often copious and as long as the lemma. Palea shorter than the lemma.

Hairs of callus and rachilla ⅔ to nearly as long as the lemma; awn straight
1. *C. crassiglumis*
Hairs of callus and rachilla ⅓–½ as long as the lemma; awn more or less bent.
Panicle branches naked below, spreading .... 2. *C. Bolanderi*
Panicle branches spikelet-bearing from base, rather stiffly ascending ... 3. *C. nutkaensis*

1. Calamagrostis crassiglumis Thurb. in Wats., Bot. Calif. 2:281. 1880.

Culms rather rigid, 1.5–4 dm. tall, with short rhizomes; lower sheaths overlapping; blades flat or somewhat involute, smooth, firm, 3–5 mm. wide; panicle narrow, dense, spikelike, 3–10 cm. long, dull purple; glumes 3–4 mm. long, ovate, rather abruptly acuminate, purplish, minutely scabrous, firm or almost indurate; lemma about as long as the glumes, broad, obtuse, or abruptly pointed, the awn attached at about the middle, straight, about as long as lemma; callus hairs abundant, about 3 mm. long; rachilla 1 mm. long, the hairs reaching to apex of lemma.

Swales and bogs near the coast: from Marin County to Del Norte County; north to Washington.

This rather well-marked species bears some resemblance to *Calamagrostis inexpansa* Gray of the Sierra Nevada.

2. Calamagrostis Bolanderi Thurb. in Wats., Bot. Calif. 2:280. 1880. Redwood reed grass.

Culms erect, 1–1.5 m. tall, with slender rhizomes; sheaths minutely scabrous; ligules 3–5 mm. long; blades flat, 4–9 mm. wide, scattered; panicle open, 10–20 cm. long, the branches whorled, spreading, naked below, the longer ones 5–10 cm. long; glumes 3–4 mm. long, purple-margined, scabrous, acute; lemma very scabrous, slightly shorter than the glumes, the awn from

Fig. 60. *a–d, Crypsis niliaca: a,* spikelet, the minutely hispid glumes and lemma about equal in length, × 12; *b,* culm, showing leaf blades and inflorescence, × 1½; *c,* grain, × 12; *d,* habit, × ⅕. *e–i, Heleochloa schoenoides: e,* spikelet, the glumes ciliate-scabrous on the keels, × 12; *f,* floret, × 12; *g,* upper part of culm, showing panicle partly enclosed by the inflated leaf sheath, × 1; *h,* seed, × 12; *i,* habit, showing the tufted, branching culms and flat leaf blades, × ⅖.

near the base; awn bent about 2 mm. below its tip, exserted near the bend; callus hairs short; rachilla pilose, 1–2 mm. long.

Bogs and moist ground near the coast: North Coast Ranges from Sonoma County to Humboldt County.

3. Calamagrostis nutkaensis (Presl) Steud., Syn. Pl. Glum. 1:190. 1855. Pacific reed grass.

Culms stout, 1–1.5 m. tall, with short rhizomes; ligules 4–8 mm. long; blades elongate, 5–12 mm. wide, flat, becoming inrolled, gradually narrowed into a long point, scabrous; panicle pale green or purplish, narrow, rather loose, 15–30 cm. long, the branches rather stiffly ascending, the longest 4–7 cm. long; glumes 5–7 mm. long, broadly lanceolate, acuminate, minutely scabrous except on the keel; lemma about 4 mm. long, minutely scabrous, the awn from near the base; awn rather stout, slightly bent, about as long as the lemma or shorter; hairs of callus and rachilla scarcely ½ as long as the lemma.

Bogs and marshes: Coast Ranges near the coast from San Luis Obispo County to Del Norte County; north to Alaska.

CRYPSIS

1. Crypsis niliaca Fig. & De Not., Mem. Acad. Torino II. 14:322. 1854. Prickle grass. Fig. 60.

Crypsis aculeata Ait., Hort. Kew. 1:48. 1789.

Spreading or prostrate annual, forming mats to 3 dm. in diameter, or often depauperate and only 1–2 cm. wide; culms freely branching; blades small, sharp-pointed; inflorescences capitate, 4–5 mm. high, each in the axils of a pair of broad spathes consisting of enlarged sheaths with short, rigid blades; spikelets 1-flowered, disarticulating below the glumes; glumes about equal in size, narrow, acute, 3 mm. long, minutely hispid; lemma broad, thin, about as long as the glumes, scabrous on the keel; palea similar to the lemma and about as long.

Established in vernal pools and on overflowed lands: Central Valley, occasional in Coast Ranges from Santa Clara County to Humboldt County, northeast of the Sierra Nevada crest from Plumas County to Modoc County. Native of Europe.

CYNODON

1. Cynodon dactylon (L.) Pers., Syn. Pl. 1:85. 1805. Bermuda grass.

Perennial with extensively creeping, scaly rhizomes or strong, flattened stolons; flowering culms flattened, usually erect or ascending, 1–4 dm. tall, with numerous sterile shoots; ligule a conspicuous ring of white hairs; blades short, stiffish, flat or involute, often conspicuously 2-ranked; spikes slender, 2.5–5.5 cm. long, 4 or 5 digitately arranged at the summit of the culms; spikelets 1-flowered, imbricate, 2–2.5 mm. long, purplish; glumes strongly keeled, the second glume longer; lemma firm, shining, strongly compressed and keeled, longer than the glumes.

Important pasture and lawn grass in the southeastern states, but usually regarded as a serious weed in California, where it has become established in the warmer parts of the state, especially on irrigated lands. Native of warmer parts of the Old World.

CYNOSURUS. DOGTAIL

1. Cynosurus cristatus L., Sp. Pl. 72. 1753. Crested dogtail.

Perennial; culms slender, tufted or bent at base, 3–8 dm. tall; blades flat, 2–4 mm. wide; panicle spikelike, linear, more or less curved, 3–8 cm. long; spikelets of two kinds, sterile and fertile intermixed, the fertile spikelet sessile, nearly covered by the short-pediceled sterile one, each about 5 mm. long, the pairs imbricate in a one-sided panicle; sterile spikelets consisting of 2 glumes and several narrow, acuminate, 1-nerved, scabrous lemmas on a continuous rachilla; fertile spikelets 2- or 3-flowered, the glumes narrow, the lemmas broader, scabrous, rounded on the back, acuminate or with awned tips about 1 mm. long, the rachilla disarticulating above the glumes.

Known only from a bog near Eureka, Humboldt County, and from Los Angeles (habitat not stated). Native of Europe.

DACTYLIS

1. Dactylis glomerata L., Sp. Pl. 71. 1753. Orchard grass.

Erect perennial; culms tufted, 6–12 dm. tall; blades 2–8 mm. wide, elongate, lax; ligules conspicuous, lacerate; panicle 5–20 cm. long, the branches few, stiff, ascending or appressed; spikelets to 8 mm. long, few-flowered, compressed, finally disarticulating between the florets, nearly sessile in dense, one-sided terminal fascicles; glumes unequal in size, acute, hispid-ciliate on the sharp keel; lemmas 5–8 mm. long, compressed, ciliate on the keel, faintly 5-nerved, awn-tipped; palea hyaline.

Common in irrigated pastures and orchards and occasionally escaped into wet lands and waste places: throughout California at lower altitudes; widespread in North America.

DESCHAMPSIA. HAIR GRASS

Low or moderately tall annuals or usually perennials. Leaves flat or involute. Spikelets in narrow or open panicles, 2-flowered, disarticulating above the glumes and between the florets, the hairy rachilla prolonged beyond the upper floret as a stipe, this sometimes bearing a reduced floret. Glumes nearly equal in size, keeled, acute or acuminate, shiny green or purplish-tinged. Lemmas thin, truncate and toothed or erose at summit, bearded at base, bearing a slender awn from or below the middle, the awn straight, bent, or twisted.

Plants annual; leaves few, scattered and not tufted at base of culm; awn sharply bent
1. *D. danthonioides*
Plants perennial; leaves numerous and mostly tufted at base of culm; awn straight or slightly bent.
Inflorescence open or contracted, the branches ascending or spreading; leaves linear-involute .... 2. *D. caespitosa*
Inflorescence elongate, very slender, the branches closely appressed-ascending; leaves filiform .... 3. *D. elongata*

1. Deschampsia danthonioides (Trin.) Munro ex Benth., Pl. Hartw. 342. 1857. Annual hair grass.

Aira danthonioides Trin., Mém. Acad. St. Pétersb. VI. Math. Phys. Nat. 1:57. 1830.
Deschampsia danthonioides var. *gracilis* Munz, Man. So. Calif. Bot. 45, 597. 1935.

Annual; culms slender, erect, 1.5–6 dm. tall; blades few, short, narrow, involute; panicle narrow or somewhat open, 7–25 cm. long, the capillary branches ascending, naked below, bearing a few short-pediceled spikelets toward the ends; glumes 4–8 (usually 5–7) mm. long, 3-nerved, acuminate, smooth except scabrous on the keel, often purplish-tipped, longer than the florets; lemmas smooth, shining, somewhat indurate, 2–3 mm. long, the awns sharply bent, 4–6 mm. long, arising from below the middle of the lemma, extending well beyond the glumes.

Common and widespread except in the deserts, occurring in vernal pools and moist to wet meadows: throughout California, ascending to an elevation of 6,000 feet or higher in the mountains; western North America from Alaska to Baja California in Mexico, also Chile.

There is considerable variation in this species with respect to the compactness of the panicle and the number and size of spikelets. In *Deschampsia danthonioides* var. *gracilis* from southern California, the spikelets are only 4–5 mm. long, but these plants appear to grade into the usual form with larger spikelets; there appears to be no basis other than size of spikelets for regarding *D. danthonioides* var. *gracilis* as distinct.

2. Deschampsia caespitosa (L.) Beauv., Ess. Agrost. 91, 149, 160, pl. 18, fig. 3. 1812. Tufted hair grass.

Aira caespitosa L., Sp. Pl. 64. 1753.

Perennial; culms densely tufted, erect, 5–15 dm. tall (alpine forms reduced); leaves mostly basal, flat or folded, 1.5–4 mm. wide, short or often elongate; panicle open, nodding (condensed, with short, usually appressed branches in *Deschampsia caespitosa* subsp. *holciformis*), 10–25 cm. long, capillary, the scabrous branches spikelet-bearing toward the ends; spikelets 3.5–7 mm. long, green or purple-tinged, the florets distant, the rachilla joint ½ as long as the lower floret; glumes acute, glabrous or minutely scabrous; lemmas smooth, the awns from near the base, from straight and included to slightly bent and twice as long as the spikelet.

Bogs, meadows, or edges of marshes: San Bernardino Mountains at elevations of 6,500–8,500 feet; Sierra Nevada at elevations of 5,000–12,000 feet, to Siskiyou County; along the coast from Del Norte County south to Monterey County; widely distributed in arctic and temperate regions of North America and Europe.

In an experimental study, Lawrence (Am. Jour. Bot. 32:298–314, 1945) demonstrated the existence of several ecotypes within this species. In California these several ecotypes are treated taxonomically as belonging to three subspecies. Of these, *Deschampsia caespitosa* subsp. *genuina* (Reichenb.) Lawr. is the most common and widespread. The other two subspecies, *D. caespitosa* subsp. *beringensis* (Hultén) Lawr. and *D. caespitosa* subsp. *holciformis* (Presl) Lawr., are restricted to coastal habitats and show various degrees of intergradation with subspecies *genuina*. The following key, adapted from Lawrence's treatment, will serve to distinguish these subspecies:

Panicle open at time of flowering, the branches spreading or drooping.
Glumes 3.5–4.5 mm. long .... *D. caespitosa* subsp. *genuina*
Glumes 5–7 mm. long .... *D. caespitosa* subsp. *beringensis*
Panicle condensed at time of flowering, more densely flowered, erect or ascending; glumes
4–7 mm. long .... *D. caespitosa* subsp. *holciformis*

3. Deschampsia elongata (Hook.) Munro ex Benth., Pl. Hartw. 342. 1857. Slender hair grass.

Aira elongata Hook., Fl. Bor. Am. 2:243, pl. 228. 1840.

Perennial; culms densely tufted, slender, erect, 3–12 dm. tall; blades soft, 1–1.5 mm. wide, flat or folded, those of the basal tuft filiform-involute; panicle very narrow, 15–30 cm. long, the capillary branches strictly appressed; spikelets on short, appressed pedicels; glumes 4–6 mm. long, 3-nerved, as long as or slightly longer than the florets, more or less purplish-tinged; lemmas 2–3 mm. long, smooth and shining, somewhat indurate, the awns straight, to twice as long as the glumes.

Moist or wet soil in meadows, along streams, on open or wooded slopes: mountains of southern California at elevations of 4,500–8,000 feet; Sierra Nevada at elevations of 3,000–10,000 feet, north to Siskiyou and Modoc counties; along the coast and in the Coast Ranges from San Luis Obispo County to Del Norte County; western North America from Alaska to Mexico, also Chile.

DIGITARIA. CRAB GRASS, FINGER GRASS

1. Digitaria sanguinalis (L.) Scop., Fl. Carn., 2d ed., 1:52. 1772. Hairy crab grass.

Syntherisma sanguinale Dulac, Fl. Hautes-Pyr. 77. 1867.

Annual; usually much-branched and spreading at base; culms 3–6 dm. long, rooting at the decumbent base, the flowering culms prostrate or ascending; sheaths more or less papillose-pilose; blades lax, 5–10 mm. wide, pubescent to minutely scabrous; racemes 5–15 cm. long, 3–12, digitately arranged, sometimes 1 or 2 whorls of racemes a short distance below the terminal ones; spikelets in pairs, one subsessile, the other short-pediceled, 3–3.5 mm. long, usually appressed-pubescent between the smooth or scabrous nerves; first glume minute but evident; second glume about ½ as long as the spikelet, narrow, ciliate; sterile lemma strongly nerved.

Common weed in waste and cultivated areas, along irrigation ditches, and in pastures, and very common in lawns in the warmer parts of California. Native of Europe.

Smooth crab grass, *Digitaria ischaemum* (Schreb.) Schreb. ex Muhl., also a weed in waste and cultivated areas, is known in California from a few scattered localities. It differs from *D. sanguinalis* chiefly in its glabrous sheaths and subracemose inflorescence.

DISTICHLIS. SALT GRASS

Plants dioecious perennials. Culms rather rigid, erect from extensively creeping or deeply running scaly rhizomes, forming dense colonies. Leaf blades noticeably, often stiffly, 2-ranked, flat or somewhat involute. Spikelets in open or dense spikes, few- to many-flowered. Glumes unequal in size, broad, acute, keeled, 3- to 7-nerved. Lemmas closely to loosely imbricate, firm, the pistillate lemmas coriaceous, faintly 9- to 11-nerved. Palea as long as the lemma or shorter, the margins dilated near the base, 2-keeled, serrate or ciliate on the keels, the pistillate palea coriaceous.

Beetle, A. A., The North American variations of *Distichlis spicata*. Bull. Torrey Club 70:638–650. 1943.

Reeder, John R., The status of *Distichlis dentata*. Bull. Torrey Club 70:53–57. 1943.

Plants mostly more than 3 dm. tall; blades not conspicuously 2-ranked; panicle more than 10 cm. long .... 1. *D. texana*
Plants mostly less than 3 dm. tall; blades more or less conspicuously 2-ranked; panicle rarely more than 5 cm. long .... 2. *D. spicata*

1. Distichlis texana (Vasey) Scribn., U. S. Dept. Agr., Div. Agrost. Circ. 16:2. 1899.

Culms erect from a decumbent base, 30–60 cm. tall, with extensively creeping rhizomes and long, stout stolons; blades flat, firm, 20–40 cm. long, 2–6

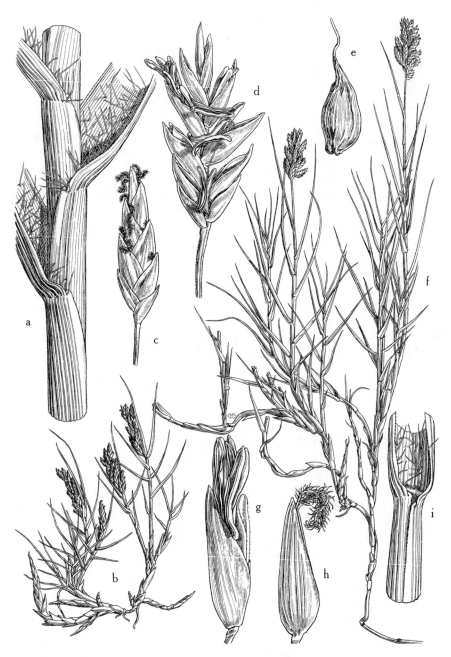

Fig. 61. *Distichlis spicata: a,* culm, leaf sheaths, and ciliate base of leaf blades, × 4;
b, habit, pistillate plant, × ⅖; *c,* pistillate spikelet, × 4; *d,* staminate spikelet; × 4;
e, mature grain, hard and nutlike, × 8; *f,* habit, pistillate plant, × ⅖; *g,* staminate floret,
× 8; *h,* pistillate floret, × 6; *i,* leaf sheath, base of blade, and ciliate ligule, × 6.

mm. wide; panicle narrow, pale, 10–25 cm. long, somewhat interrupted, the branches appressed; spikelets somewhat compressed, 4- to 8-flowered, 1–1.5 cm. long; first glume 2–3 mm. long, the second 3–4 mm. long, acute; lemmas of pistillate spikelets closely imbricate and appressed, about 8 mm. long, with 3 prominent nerves, the margins broad, hyaline; palea of pistillate spikelets shorter than the lemma, strongly dilated below, the keels with narrow, erose or toothed wings; lemmas of staminate spikelets more spreading, about 6 mm. long; palea of staminate spikelet about as long as the lemma, not dilated, the keels minutely scabrous, not winged.

Reported from Salton Sink, Imperial County; Texas and northern Mexico.

2. Distichlis spicata (L.) Greene, Bull. Calif. Acad. Sci. 2:415. 1887. Salt grass. Fig. 61.

Culms 1–4 dm. tall, erect or in coastal plants sometimes prostrate and strongly stoloniferous; blades numerous, spreading or sometimes closely ascending or erect, either as long as or longer than or sometimes shorter than the spikes; spikes green, drying straw brown, or in coastal plants often purplish-tinged, 1–6 cm. long, ovate to oblong; spikelets mostly 1–2 cm. long, the pistillate spikelets often congested and more or less closely imbricate, the staminate ones usually less congested, not as closely imbricate and with the individual spikelets more easily distinguished; the first glume 2–3 mm. long, the second 3–4 mm. long; lemmas 3–6 mm. long, the pistillate lemmas more coriaceous and closely imbricate than the staminate ones, sometimes with a broad hyaline margin; palea 3–5 mm. long, rather soft, narrowly or broadly winged below, often with hyaline margins, the keels minutely serrate or serrate-ciliate to near the base, less frequently dentate, with or without a prominent marginal vein, occasionally with a few long hairs on the back.

Widespread at low to middle elevations throughout California: salt marshes and sandy flats along the coast from San Diego County to Del Norte County, marshes and alkaline flats in the Central Valley and north to Siskiyou County, coastal and foothill southern California east into the deserts, and east of the Sierra Nevada crest from Inyo County to Modoc County; north to British Columbia, east to Atlantic, southern states, Mexico.

The California material of this extremely variable species was treated by Hitchcock (*op. cit.,* 1935) under three separate specific entities. Mrs. Chase, in her revision of Hitchcock's Manual (*op. cit.,* 1950), reduces one of these entities, *Distichlis dentata,* to synonymy under *D. stricta.*

Arguments given by Beetle (*op. cit.*) provide evidence that *D. spicata* and *D. stricta* may not be distinct species. The two species are quite similar morphologically and show much intergradation of characters. The only distinctions which Hitchcock (*op. cit.,* 1935) used to separate them were the following: *D. stricta,* inflorescences less congested, florets more numerous in spikelet, distribution more to the interior; *D. spicata,* inflorescences more congested,

Fig. 62. *Echinochloa colonum: a,* culm, leaf sheath, and ciliate leaf base, × 3; *b,* spikelet, × 12; *c,* floret, adaxial view, showing indurated palea, × 12; *d,* floret, abaxial view, showing indurated lemma, × 12; *e,* habit, showing decumbent stems rooting at the nodes, × ¼.

florets less numerous in spikelet, distribution more toward the coast. Until the problem can be clarified by genetic study, both species are here placed under *D. spicata*.

ECHINOCHLOA. WATER GRASS

Our species annual. Culms more or less stout, often succulent. Sheaths compressed. Ligules none. Blades flat, linear, elongate. Panicles rather compact, composed of short, densely flowered, one-sided racemes. Spikelets 1-flowered, plano-convex, often stiffly hispid, subsessile. First glume about ½ as long as the spikelet, pointed. Second glume and sterile lemma equal in size, pointed, mucronate, or the glume short-awned and the lemma long-awned, sometimes conspicuously so, the sterile lemma enclosing a membranous palea and sometimes a staminate flower. Fertile lemma plano-convex, indurate, smooth and shining, short-acuminate, the margins inrolled below, the apex of the palea not enclosed.

Racemes simple, 1–2 cm. long; awn of sterile lemma reduced to a short point
  1. *E. colonum*
Racemes more or less branched, 2 or more cm. long; awn of sterile lemma 5–10 mm. long.
  Glumes and sterile lemma bearing spines with conspicuous blister-like bases; fertile lemma without a ring of hairs below the tip . . . . 2. *E. pungens*
  Glumes and sterile lemma usually without blister-based spines (these marginal if present); fertile lemma with a ring of minute hairs below the tip . . . . 3. *E. crusgalli*

1. Echinochloa colonum (L.) Link, Hort. Berol. 2:209. 1833. Fig. 62.

Culms erect or procumbent and rooting at the nodes, rather lax, smooth, 3–6 mm. wide; panicle 5–15 cm. long; racemes 5–10, 1–2 cm. long, closely appressed-ascending or slightly spreading, single or occasionally 2 approximate, the lower raceme usually distant as much as 1 cm.; spikelets about 3 mm. long, crowded, nearly sessile; second glume and sterile lemma short-pointed, rather soft, pubescent, faintly nerved, the nerves minutely hispid-scabrous.

Established on moist ground in waste and cultivated lands in southern California, San Joaquin Valley. Native of tropical regions.

2. Echinochloa pungens (Poir.) Rydb., Brittonia 1:81. 1931.

Echinochloa crusgalli var. *frumentacea* of authors, in part.

Tufted annual; culms to 1.5 m. tall, erect or spreading; panicles erect or nodding, to 2 dm. long; spikelets covered with spinelike hairs arising from yellowish blisters; awns 5–10 mm. or more long.

Rice-field and orchard weeds on wet soil: California.

According to R. W. Pohl in his How to Know the Grasses (p. 156, 1954), this may have been introduced from Europe.

Fig. 63. *Echinochloa crusgalli: a*, panicle, × ⅖; *b*, leaf sheath and
ciliate leaf base, × 3; *c*, habit, × ⅖.

Fig. 64. *Elymus triticoides: a,* node, showing group of spikelets, × 4; *b,* floret, the
lemma removed to show the lodicules, × 4; *c,* leaf sheath and ligule, × 4; *d,* floret, × 4;
e, habit, × ⅛; *f,* inflorescence, × ⅖.

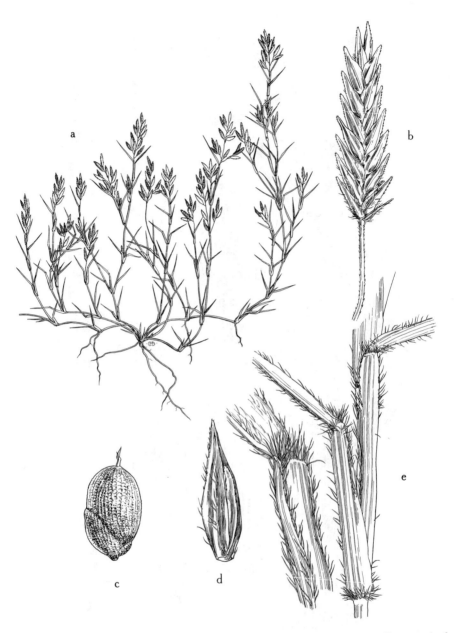

Fig. 65. *Eragrostis hypnoides: a,* habit, showing slender, creeping culms, divergent leaf blades, and elliptic panicles, × ⅖; *b,* spikelet, showing hairs on backs of glumes and lemmas, × 8; *c,* mature seed, × 40; *d,* floret, × 20; *e,* leaf sheaths and blades, × 8.

3. Echinochloa crusgalli (L.) Beauv., Ess. Agrost. 53, 161. 1812. Barnyard grass. Fig. 63.

Panicum crusgalli L., Sp. Pl. 56. 1753.

Culms erect to decumbent, stout, as much as 1 m. or even 1.5 m. tall, often branching at the base; sheaths glabrous; blades elongate, 5–15 mm. wide; panicle erect or nodding, purple-tinged, 1–2 dm. long; racemes spreading, ascending, or appressed, the lower racemes somewhat distant, as much as 1 dm. long, sometimes branched, the upper ones approximate; spikelets crowded, about 3 mm. long, excluding the awns; internerves hispidulous; nerves strongly hispid; 1 to several hispid hairs, 2–4 mm. long, at base of spikelets; awns variable, mostly 5–10 mm. long on at least some of the spikelets, sometimes as much as 5 cm. long.

Widespread on moist or wet soil in cultivated or waste places at low to middle elevations throughout California. One of the principal pests of rice fields.

Two varieties of this species have been included in various California manuals: *Echinochloa crusgalli* var. *mitis* (Pursh) Peterm. and *E. crusgalli* var. *zeyalensis* (H.B.K.) Hitchc. They intergrade so freely that they are distinguished only with difficulty.

ELYMUS. WILD RYE

1. Elymus triticoides Buckl., Proc. Acad. Nat. Sci. Phila. 14:99. 1863. Creeping wild rye. Fig. 64.

Perennial, usually with extensively creeping rhizomes; culms single or in small clusters, 2–3.5 mm. in diameter, glabrous or rarely with a fine pubescence on the upper internode, glaucous especially at the nodes; leaves glabrous or sparsely hairy, rarely pubescent, glaucous or less frequently bright green; blades mostly 3–6 mm. broad, flat or soon involute; inflorescence a loose to dense spike 8–20 cm. long, 1 or 2, occasionally 3, spikelets at a node; spikelets sessile, or rarely on short pedicels, mostly 8–15 mm. long with 3–6 florets; glumes subulate, 5–15 mm. long, as long as or shorter than the first lemma; lemmas 6–10 mm. long, smooth or scabrous, short-awned from an acute or minutely bifid apex.

Widely distributed throughout California, except in the deserts, and ascending in the mountains to an elevation of about 7,500 feet, frequent on the dried or moist edges of meadows and flats, usually in heavy, often alkaline soil, also flourishing as a weed in waste places; western states.

This is an extremely polymorphic species which exhibits a great deal of variation with respect to thickness of culms, width of leaves, number of spikelets at a node, glaucousness of herbage, and cytological pattern (F. W. Gould, Notes on the genus *Elymus*, Madroño 8:42–47, 1945). In *Elymus triticoides* subsp. *multiflorus* Gould, the culms are stouter (3.5–5 mm. in diameter) than in the plants described above, the leaf blades are broader (6–15 mm.), there are often more spikelets at a node (3–6), and the spikelets may be longer (17–25 mm.) and have more numerous florets (6–9). Of somewhat infrequent occurrence is *E. triticoides* var. *pubescens* Hitchc., in which the sheaths and blades are pubescent.

ERAGROSTIS. LOVE GRASS

1. Eragrostis hypnoides (Lam.) B.S.P., Prelim. Cat. N. Y. 69. 1888. Teal grass, creeping love grass. Fig. 65.

Annual; culms slender, creeping, 1–4 dm. long, rooting at the nodes, forming mats; blades 1–3 cm. long, 1–2 mm. wide, divergent, scabrous or pubescent on the upper surface; sheaths loose, smooth or pubescent, the throat and base pilose; panicle elliptic, usually simple, loosely few-flowered, 1–5 cm. long; spikelets lanceolate-oblong, 10- to 30-flowered, 5–10 mm. long, often with a few hairs on backs of glumes and lemmas, especially toward the base; glumes about 1 mm. long, the second glume slightly longer than the first; lemmas 1.5–2 mm. long, the lateral nerves conspicuous, the keel scabrous; palea about ½ as long as the lemma.

Wet ground and sand bars along streams: Sacramento and San Joaquin valleys, Coast Ranges from Lake County to Siskiyou County; temperate and tropical regions of the Western Hemisphere.

Five additional species of *Eragrostis* which are not infrequently found in waste places and irrigated lands throughout California are: *E. cilianensis* (All.) Lutati (stink grass), *E. pilosa* (L.) Beauv. (India love grass), *E. poaeoides* Beauv. ex Roem. & Schult., *E. diffusa* Buckl., and *E. Orcuttiana* Vasey. The first three species are introductions from Europe, and the last two are native to California.

ERIANTHUS. PLUME GRASS

1. Erianthus ravennae (L.) Beauv., Ess. Agrost. 14, 162, 177. 1812. Plume grass, Ravenna grass.

Stout perennial, to 4 m. tall; culms stiff, glabrous; leaves flat, long, long-pointed, more or less scabrous on the nerves, the midrib prominent and lighter-colored; sheaths very long, many-striate, scabrid, with long hairs at throat; panicle plumelike, 6 dm. long, narrow, soft-silky and shiny from the great quantities of hairs in tufts at the base of each spikelet; spikelets 2 at the nodes of a jointed disarticulated rachis, one sessile, 1-pediceled, perfect, mostly 1-flowered; glumes nearly equal in size, the first glume somewhat double-keeled and the second keeled toward the apex; one lemma sterile and awnless, the other fertile, with a short, slender awn.

Known only from the Imperial State Game Refuge, four miles south of Calipatria, Imperial County. Native of southern Europe. Cultivated for ornament in the warmer regions of the United States.

ERIOCHLOA. CUP GRASS

Our species annual. Culms often branching. Panicles terminal, with several to many spreading or appressed rather closely arranged racemes. Spikelets dorsally compressed, more or less pubescent, solitary or sometimes in pairs, short-pediceled or subsessile, in 2 rows on one side of a narrow rachis. Lower rachilla joint thickened, forming a more or less ringlike, usually dark-colored callus below the second glume; first glume reduced to a minute sheath about the callus and adnate to it; second glume and sterile lemma about equal in size; fertile lemma indurate, minutely papillose-rugose, mucronate or awned, the margins slightly inrolled.

Second glume with an awn usually less than 1 mm. long and usually merely awn-tipped.
  Blades glabrous; fruit apiculate . . . . 1. *E. gracilis*
  Blades pubescent; fruit with an awn nearly 1 mm. long . . . . 2. *E. contracta*
Second glume with an awn more than 1 mm. long, often as long as the glume itself
  3. *E. aristata*

1. Eriochloa gracilis (Fourn.) Hitchc., Jour. Wash. Acad. Sci. 23:455. 1933. Southwestern cup grass.

Annual; culms erect or decumbent at base, 4–10 dm. tall; blades flat, glabrous, mostly 5–10 mm. wide; racemes several to numerous, approximate, ascending to slightly spreading, 2–4 cm. long, the axis and rachis softly pubescent, the pedicels short-pilose; spikelets 4–5 mm. long, rather sparsely appressed-pubescent, acuminate, or the glume sometimes tapering into an awn point as much as 1 mm. long; sterile lemma empty; fruit about 3 mm. long, apiculate.

Irrigated lands in Riverside and Fresno counties; southwestern states, Mexico.

Eriochloa gracilis var. *minor* (Vasey) Hitchc. is a somewhat smaller plant with more crowded and less acuminate spikelets. It is probably better regarded as only a minor form of the species.

2. Eriochloa contracta Hitchc., Proc. Biol. Soc. Wash. 41:163. 1928. Prairie cup grass.

Annual; culms erect or sometimes decumbent at base, pubescent, 3–7 dm. tall; blades pubescent, usually not more than 5 mm. wide; panicle usually less than 15 cm. long, contracted, cylindric, the racemes appressed, closely overlapping, 1–2 cm. long, the axis and rachises villous; spikelets 3.5–4 mm. long, excluding the awn tip, appressed-villous; glume awn-tipped; sterile lemma slightly shorter than glume, acuminate, empty; fruit 2–2.5 mm. long, with an awn nearly 1 mm. long, the awn upwardly barbed.

Irrigated lands in Imperial County, also reported from Merced, Merced County (W. W. Robbins, Alien plants growing without cultivation in California, Univ. Calif. Agr. Exp. Sta. Bull. 637, 1940); middle-western and western states.

Similar to *Eriochloa gracilis,* but differing in its pubescent blades, shorter racemes, and the awned fruit.

3. Eriochloa aristata Vasey, Bull. Torrey Club 13:229. 1886. Bearded cup grass.

Annual; culms erect or spreading at base, 5–8 dm. tall; blades flat, mostly 10–12 mm. wide, glabrous; racemes several, ascending, overlapping, 3–4 cm. long, the rachis pilose, the pedicels bearing several long, stiff hairs; spikelets about 5 mm. long; the glume and sterile lemma appressed-villous on the lower ½ or ⅔, with the upper part minutely scabrous, tapering into awns, the awn of the glume about as long as the spikelet; fruit 3.5 mm. long, apiculate.

Alluvial land along the Colorado River at Fort Yuma, Imperial County; Arizona, northern Mexico.

FESTUCA. FESCUE

1. Festuca rubra L., Sp. Pl. 74. 1753. Red fescue.

Plants perennial; culms usually loosely tufted, bent or decumbent at the reddish or purplish base, occasionally closely tufted, erect to ascending, 4–10 dm. tall; lower sheaths brown and fibrillose; blades smooth, soft, usually folded, very narrow, mostly clustered at the base; panicle 3–20 cm. long, usually contracted and narrow, the branches mostly erect or ascending; spikelets 6–12 mm. long, 4- to 7-flowered, pale green or glaucous, often purple-tinged; glumes unequal in length, the first glume ⅓–½ as long as the second; lemmas 5–7 mm. long, smooth or scabrous near the apex, varying from awnless to awned; awn ½ as long as the lemma.

Salt marshes, sandy ocean bluffs, moist or wet meadows: along the coast and in Coast Ranges from Monterey County to Siskiyou County, Sierra Nevada to an elevation of 9,000 feet, north to Modoc County, San Bernardino Mountains at elevations of 6,500–7,500 feet; Rocky Mountains, Eurasia, Africa.

Three additional species, *Festuca californica* Vasey, *F. megalura* Nutt., and *F. myuros* L., have occasionally been found in low swales or recently dried areas bordering marshy sites; and *F. arundinacea* Schreb., an escape from irrigated pastures, occurs in ditches and on marshy lands.

GASTRIDIUM

1. Gastridium ventricosum (Gouan) Schinz & Thell., Mitt. Bot. Mus. Univ. Zürich 65:39. 1913. Nit grass.

Plants annual; culms 1–5 dm. tall; foliage scant, the blades flat, scabrous; panicle 3–8 cm. long (or in robust specimens 10–14 cm. long), dense, shining, spikelike; spikelets 1-flowered, slender, about 5 mm. long; glumes long-acuminate, somewhat swollen at the base, scabrous on the keels, the second glume about ¾ as long as the first; lemmas much shorter than the glumes, hyaline, globular, pubescent, truncate, with a delicate, somewhat bent awn 5 mm. long; palea about as long as the lemma.

Established usually on open, dry ground, but occasionally found in marshy sites along streams or around vernal pools: common in the coastal counties, occasional in the Sacramento and San Joaquin valleys, and common on overgrazed ranges in the central Sierra Nevada foothills. Native of Europe.

GLYCERIA. MANNA GRASS

Usually tall aquatic or marsh perennials with creeping and rooting bases. Culms simple. Sheaths closed or open. Blades flat, soft. Panicles open or contracted. Spikelets few- to many-flowered, linear or ovate to oblong, subterete or slightly compressed, the rachilla disarticulating above the glumes and between the florets. Glumes unequal in size, short, obtuse or acute, usually scarious, 1- to 3-nerved. Lemmas broad, convex on the back, firm, usually obtuse, scarious at the apex, prominently 5- to 9-nerved.

Church, George L., A cytotaxonomic study of *Glyceria* and *Puccinellia*. Am. Jour. Bot. 36:155–165. 1949.

Leaf sheaths free and overlapping; second glume 3-nerved, the nerves sometimes obscure
1. *G. pauciflora*
Leaf sheaths connate; second glume 1-nerved.
  Panicle open, lax; stamens 2.
    First glume not more than 1 mm. long, branches spreading and often drooping
2. *G. elata*
    First glume more than 1 mm. long, usually about 1.5 mm. long, branches nodding
      only at the summit .... 3. *G. grandis*
  Panicle contracted, the branches appressed; stamens 3.
    Lemmas minutely scabrous on the nerves, otherwise glabrous .... 4. *G. borealis*
    Lemmas minutely scabrous or hirsute between the distinctly scabrous nerves.
      Lemmas 2.5–4 mm. long; anthers less than 1 mm. long .... 5. *G. leptostachya*
      Lemmas 4–5.5 mm. long; anthers 1–1.5 mm. long .... 6. *G. occidentalis*

1. Glyceria pauciflora Presl, Rel. Haenk. 1:257. 1830. Weak manna grass. Fig. 66.

Culms 3–12 dm. tall, from a decumbent rooting base; sheaths smooth or minutely scabrous, free and overlapping; blades thin, flat, lax, minutely scabrous, mostly 8–20 cm. long, 5–15 mm. wide; panicle oblong or pyramidal, open or rather dense and spikelike, nodding, 8–20 (or –25) cm. long, the branches ascending or spreading, rather flexuous, naked below, the spikelets crowded on the upper half; spikelets 4- to 7-flowered (usually 5- or 6-flowered), 4–6 mm. long; glumes broadly ovate or oval, purplish-tinged, the first glume

Fig. 66. *Glyceria pauciflora: a,* leaf sheath, blade, and ligule, × 2; *b,* upper part of culm, showing panicle, × ⅕; *c,* branch of panicle, the spikelets crowded on the upper half, × 6; *d,* panicle, × ⅔; *e,* floret, showing palea and rachilla, × 12; *f,* floret, showing lemma, × 12; *g,* grain, showing subbasal, oblong hilum, × 20; *h,* habit, lower part of culm, showing the flat, lax leaf blades, × ⅕.

1–1.5 mm. long, the second glume 1.5–2 mm. long, 3-nerved, the nerves some-times obscure, the margins erose-scarious; lemmas oblong, 2–3 mm. (usually 2.5 mm.) long, with 5 prominent nerves and an outer short, faint pair near the margins, minutely scabrous on the nerves and somewhat so between them, the tip rounded, scarious, somewhat erose, usually with a purplish band below the scarious tip; caryopsis with a subbasal and oblong hilum.

Marshes, shallow water, and wet meadows: along the coast from San Mateo County to Del Norte County, ascending to an elevation of 5,000 feet, thence east to Modoc County and south to Mono County, and south in the Sierra Nevada to Tulare County chiefly at elevations of 4,000–8,000 feet; western North America from Alaska to New Mexico.

At higher altitudes in the Sierra Nevada this species appears to intergrade with *Glyceria erecta* Hitchc., which may be only an alpine subspecies.

In a recent taxonomic study of *Glyceria,* Church (*op. cit.*) has reported sig-nificant data for the erection of a new genus *Torreyochloa,* to which he refers this species and several allied species.

2. Glyceria elata (Nash) Hitchc. in Jepson, Fl. Calif. 1:162. 1912. Tall manna grass.

Culms erect, smooth, succulent, dark green, 1–2 m. tall; sheaths scabrous; blades flat, usually 6–9 mm. or sometimes only 4 mm. wide, scabrous; panicle large and diffuse, becoming oblong, 15–30 cm. long, the branches naked below, the lower ones usually reflexed at maturity; spikelets 3–5 mm. long, oblong or ovate-oblong, usually 6- to 8-flowered; glumes broad, obtuse, much shorter than the lower lemmas, often nerveless, the first glume about 1 mm. long, the second nearly 2 mm. long; lemmas firm, 2–2.5 mm. long, obovoid, obtuse or acutish, prominently 7-nerved, the apex distinctly scarious; stamens 2; palea apex with a narrow slit.

Chiefly montane in wet meadows, swampy woods, or along streams: Sierra Nevada from Fresno County to Plumas County at elevations of 5,000–8,500 feet, North Coast Ranges from Mendocino County to Siskiyou County ascend-ing to an elevation of 5,000 feet, occasional in San Jacinto and San Bernar-dino mountains at 5,000–7,500 feet; western states, British Columbia.

Glyceria elata and *G. pauciflora* are very similar in general appearance, and the two species are sometimes confused. There are a number of distinct char-acters, however, by which *G. elata* may be recognized. In addition to connate leaf sheaths and one-nerved upper glumes, *G. elata* has much more elongate and narrower leaf blades, more nerves (seven) on the lemmas, a distinctly scarious (but not erose) lemma apex, two stamens, stigmas plumose only in the upper half, and a caryopsis with a long, linear hilum.

Glyceria striata (Lam.) Hitchc., another species which closely resembles *G. elata,* has been reported from northern California. We have seen a few spec-imens, chiefly from Humboldt County, which appear to resemble those of

Fig. 67. *Glyceria grandis: a,* floret, showing palea, × 16; *b,* spikelets, solitary on tips of branchlets, × 8; *c,* habit, lower part showing the conspicuous joints of culm and the long, lax leaf blades, × ⅕; *d,* habit, upper part of culm, showing panicle, × ⅕; *e,* leaf sheath, blade, and ruptured ligule, × 4; *f,* floret, showing lemma, the strong nerves papillose, × 16; *g,* young upper leaf, the sheath and ligule enclosing culm, × 4.

G. striata in being more slender, shorter, and lighter green plants with narrower (2–4 mm. wide), often folded blades, and with slightly shorter spikelets (2–3 mm. long) and glumes (about 0.5 mm. long).

3. Glyceria grandis Wats. ex Gray, Man., 6th ed., 667. 1890. American manna grass. Fig. 67.

Culms stout, 1–1.5 m. tall from a perennial base; leaf blades flat, 6–12 mm. wide; panicle large, compound, 20–40 cm. long, somewhat nodding at tip; spikelets 4- to 7-flowered, 5–6 mm. long; glumes 1.5–2 mm. long; lemmas purplish, 2–2.5 mm. long; palea slightly longer than lemma.

Marshes and stream banks: northeastern California; eastern Oregon, Arizona, New Mexico, across the northern United States to Atlantic, and north to Alaska.

4. Glyceria borealis (Nash) Batchelder, Proc. Manchester Inst. 1:74. 1900. Northern manna grass. Fig. 68.

Culms erect or decumbent and rooting at the base, slender, 3–10 dm. tall; sheaths smooth or slightly scabrous, keeled; blades flat or folded, usually 2–6 mm. wide, very narrow; panicle mostly 2–4 dm. long, very narrow, the branches as much as 1 dm. long, bearing several closely appressed spikelets; spikelets mostly 6- to 12-flowered, 1–1.5 cm. long; glumes oblong, scarious, the first glume 1.5–2 mm. long, the second 3–4 mm. long; lemmas rather thin, obtuse, 3–4 mm. long, strongly 7-nerved, broadly scarious at the tip, minutely scabrous on the nerves, otherwise glabrous.

Shallow water in wet meadows or lake margins: Sierra Nevada from Tuolumne County and north to Shasta County, chiefly at elevations of 4,000–7,000 feet; east of the Sierra Nevada crest from Mono County to Modoc County; across North America in cooler areas.

5. Glyceria leptostachya Buckl., Proc. Acad. Nat. Sci. Phila. 14:95. 1863. Slim-head manna grass. Fig. 69.

Culms 1–1.5 m. tall, rather stout or succulent, often decumbent and rooting at the nodes; sheaths slightly rough; blades flat, minutely scabrous on the upper surface, 10–30 cm. long, 4–7 mm. wide, rarely 1 cm. wide; panicle 2–4 dm. long, very narrow, the branches ascending, mostly in 2's or 3's, several-flowered; spikelets 1–2 cm. long, 8- to 14-flowered, often purplish; first glume 1.5–3 mm. long, the second 3 mm. long; lemmas firm, broadly rounded or truncate and scarious toward the apex, 2.5–4 mm. long; 7-nerved, minutely scabrous on and between the nerves; anthers less than 1 mm. long.

In shallow water in marshy places: Sonoma and Marin counties; north to Washington.

This species may be confused with *Glyceria occidentalis,* from which it is distinguished by its somewhat taller, more robust habit, its relatively longer blades (5–15 cm. long in *G. occidentalis*), and its shorter lemmas and anthers.

Fig. 68. *Glyceria borealis: a,* panicle, × ⅖; *b,* habit, showing the slender culms, the leaves, and the panicles, × ⅕; *c,* spikelet, × 6; *d,* grain, × 20; *e,* floret, showing the palea and the broadly scarious tip of lemma, × 12; *f,* leaf sheath, blade, and ligule, × 4.

Fig. 69. *Glyceria leptostachya: a,* leaf sheath, blade, and ligule, × 3; *b,* spikelet, the rachilla zigzag, × 3; *c,* floret, showing palea, × 12; *d,* floret, showing the scabrous lemma, the apex scarious, × 12; *e,* habit, showing the succulent, decumbent culm, rooting at the nodes, and the flat, ascending leaves, × ⅖; *f,* seed, × 12; *g,* panicle, × ⅖.

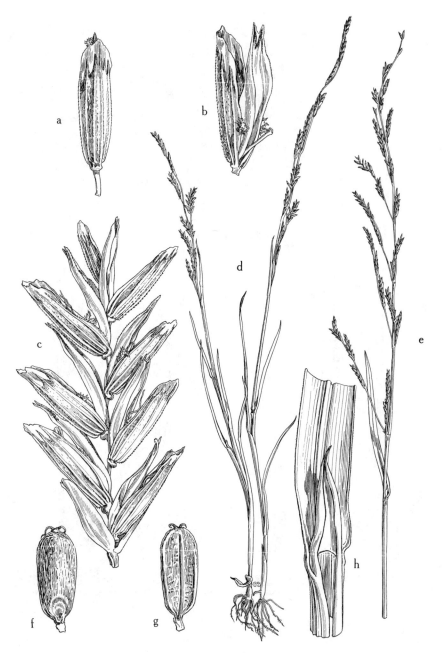

Fig. 70. *Glyceria occidentalis: a* and *b,* floret, showing the scabrous lemma and equally long palea, × 8; *c,* spikelet, each floret with enlarged base, × 6; *d,* habit, showing long, narrow panicles, × ⅖; *e,* panicle, spreading at anthesis, × ⅖; *f* and *g,* seed, × 10; *h,* leaf sheath, blade, and ligule, × 6.

6. Glyceria occidentalis (Piper) J. C. Nels., Torreya 19:224. 1919. Northwestern manna grass. Fig. 70.

Culms soft, ascending from a decumbent rooting base, 6–10 dm. tall; blades 3–12 mm. wide, 5–15 cm. long (or longer), smooth beneath, somewhat scabrous on the upper surface; panicle long and narrow, somewhat spreading at anthesis, 3–5 dm. long; spikelets 1.5–2 cm. long; first glume usually about 2 mm. long, the second one much longer; lemmas sometimes tinged with purple near the tip, 4–6 mm. long, acutish, rather strongly scabrous, 7- to 9-nerved, the nerves more or less prominent; palea about as long as, sometimes slightly longer than, its lemma; anthers 1–1.5 mm. long.

Marshy places near the coast: Marin County to Humboldt County; western states, British Columbia.

The taxonomic position of this and the apparently closely related species *Glyceria fluitans* (L.) R. Br. is somewhat confused. The characters which Hitchcock employs in his Manual (*op. cit.,* 1950) have not proved too satisfactory for the separation of the two entities. In the recent treatment of the genus by Church (*op. cit.*), the following characters have been given for the separation of the two species:

Lemmas 4–5.5 mm. long; anthers 1–1.5 mm. long .... *G. occidentalis*
Lemmas 5.5–7 mm. long; anthers 1.5–2.5 mm. long .... *G. fluitans*

HELEOCHLOA. SWAMP TIMOTHY

1. Heleochloa schoenoides (L.) Host, Icon. Gram. Austr. 1:23, pl. 30. 1801. Fig. 60.

Plants annual; culms tufted, branching, erect to spreading and bent, several-noded, 1–3 dm. long; sheaths glabrous, striate, often inflated, especially below the panicles; blades flat, with involute, slender tips, 2–4 mm. wide, usually less than 5 cm. long; panicle oblong, dense, spikelike, 1–4 cm. long, 8–10 mm. thick, partly enclosed by the inflated leaf sheaths; spikelets 1-flowered, about 3 mm. long, much compressed; glumes unequal to nearly equal in size, ciliate-scabrous on the keels, narrow, acute; lemma broader, thin, 1-nerved, a little longer than the glumes; palea nearly as long as the lemma.

Established in vernal pools or at pool margins: Central Valley from Merced County to Butte County, also in Lake County. Native of Europe.

HORDEUM. BARLEY

Annual or perennial. Blades flat. Spikes dense, bristly. Spikelets 1-flowered, 3 at each node of the jointed rachis, the middle one usually sessile, the lateral ones pediceled. Rachilla disarticulating above the glumes and in the central

Fig. 71. *Leersia oryzoides: a,* panicle, × ⅖; *b,* seed, × 8; *c,* habit, showing the slender, creeping rhizomes and the culms with decumbent bases, × ⅛; *d,* spikelet, laterally compressed, the glumes wanting, × 8.

spikelet prolonged behind the palea as a bristle and sometimes bearing a rudimentary floret. Lateral spikelets usually imperfect, sometimes reduced to bristles. Glumes narrow, often subulate and awned, rigid, standing in front of the spikelet. Lemma rounded on the back, tapering into a usually long awn.

Hordeum Hystrix Roth, a species omitted from the treatment below, is common on inundated lands, especially in alkaline soil.

Covas, Guillermo, Taxonomic observations on the North American species of *Hordeum*. Madroño 10:1–21. 1949.

Glumes and awns 1.8–8 cm. long; florets of lateral spikelets reduced to 1–3 awns
1. *H. jubatum*
Glumes and awns mostly less than 1.5 cm. long; florets of lateral spikelets reduced, but with evident body . . . . 2. *H. brachyantherum*

1. Hordeum jubatum L., Sp. Pl. 85. 1753. Foxtail barley.

Perennial; culms erect or decumbent at base, tufted, 3–6 dm. tall; blades 2–5 mm. wide, scabrous and usually short-pubescent; sheaths smooth or the lower ones sometimes pubescent; spike often nodding, 5–10 cm. long (or longer), often as wide as long; lateral spikelets reduced to 1–3 spreading awns; glumes of perfect spikelet awnlike, 2.5–8 cm. long, spreading; lemma 6–8 mm. long with an awn as long as the glumes, the awn upwardly barbed.

Moist open ground, along ditches, and in waste places, often on alkaline or saline soil. A troublesome weed, especially in irrigated lands. Widely distributed throughout California, but not common.

Covas (*op. cit.*) states that *Hordeum jubatum* var. *caespitosum* (Scribn.) Hitchc. (which differs from *H. jubatum* only in having shorter awns and glumes) appears to be an entity intermediate between *H. jubatum* and *H. brachyantherum* and suggests the probability that it evolved from the hybridization of these species.

2. Hordeum brachyantherum Nevski, Acta Inst. Bot. Acad. U.R.S.S. I. 2:61. 1936. Meadow barley.

Hordeum boreale Scribn. & Smith, U. S. Dept. Agr., Div. Agrost. Bull. 4:24. 1897.

Perennial; culms tufted, erect or sometimes spreading, 1–5 dm. tall; blades soft, usually glabrous, sometimes scabrous or shortly pubescent, 3–9 mm. wide; spike slender, 2–8 cm. long; glumes all setaceous, 8–15 mm. long, those of the central spikelet often scarcely longer than the palea; the rachilla prolonged, usually extending beyond the middle of the palea; lateral spikelets pediceled, the pedicels usually curved, the florets much reduced.

Meadows, flats, marshes and their borders, often in subalkaline or saline soils: widely distributed throughout California except in the deserts, ascending in the mountains to an elevation of 10,000 feet; western states to Alaska, east to Newfoundland.

Hordeum californicum Covas & Stebbins is sometimes found in wet places,

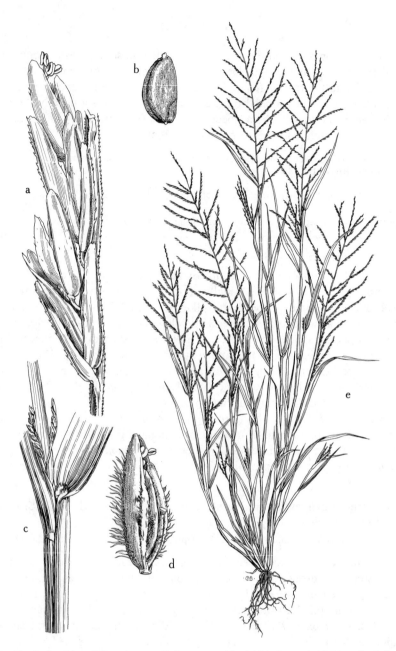

Fig. 72. *Leptochloa filiformis: a,* spikelets on rachis, × 20; *b,* seed, × 20; *c,* leaf sheath and ligule, × 4; *d,* floret, × 20; *e,* habit, showing the long panicles and spreading-ascending spikes, × ⅖.

such as streamsides, but it is more commonly found on open, grassy slopes where there is good drainage and the soil is not too heavy.

LEERSIA

1. Leersia oryzoides (L.) Swartz, Prodr. Veg. Indiam Occ. 21. 1788. Rice cut-grass. Fig. 71.

Plants perennial, with slender, creeping rhizomes; culms slender, often decumbent at base, glabrous, 1–1.5 m. tall, the nodes conspicuously pubescent; sheaths and blades strongly retrorse-scabrous, the blades flat, acuminate at apex, 1–2 dm. long, mostly 4–12 mm. wide; panicle terminal and axillary, open, 1–2 dm. long, the branches flexuous, scabrous, finally wide-spreading, sometimes crowded or whorled, naked on the lower ⅓ to ½; spikelets 1-flowered, strongly compressed laterally, disarticulating from the pedicel, loosely imbricate, oblong-elliptic, 4–5 mm. long, 1.5–2 mm. wide; glumes wanting; lemma papery, broad, oblong to oval, sparsely hispidulous, with a hispid-ciliate keel, usually 5-nerved, the lateral pair of nerves close to the margins, the intermediate nerves sometimes faint; palea as long as the lemma, much narrower, usually 3-nerved.

Marshes, borders of ponds, ditches, and rivers, often forming dense zones: Central Valley from Fresno County to Shasta County, along north coast from Sonoma County to Humboldt County, also Inyo County; warmer parts of North America, Europe.

LEPTOCHLOA. SPRANGLE-TOP

Our species tall annuals. Blades flat or sometimes involute. Panicle of numerous spikes or racemes, scattered along a common axis. Spikelets 2- to several-flowered, subsessile, crowded or somewhat distant along one side of a slender rachis. Glumes unequal or subequal in size, 1-nerved. Lemmas obtuse or acute, sometimes 2-toothed and mucronate or short-awned from between the teeth, 3-nerved, the nerves sometimes pubescent or pilose.

Spikelets 2- or 3-flowered; 1–2 mm. long .... 1. *L. filiformis*
Spikelets 6- to 12-flowered; 5–10 mm. long.
    Lemmas awned .... 2. *L. fascicularis*
    Lemmas awnless or mucronate only .... 3. *L. uninervia*

1. Leptochloa filiformis (Lam.) Beauv., Ess. Agrost. 71, 161, 166. 1812. Red sprangle-top. Fig. 72.

Annual; branched at the somewhat decumbent base, foliage and panicles often reddish or purple; culms 4–10 dm. tall, or often dwarf, erect, glabrous; blades flat, thin, lax, 0.5–3 dm. long, 3–10 mm. wide, glabrous or sparsely

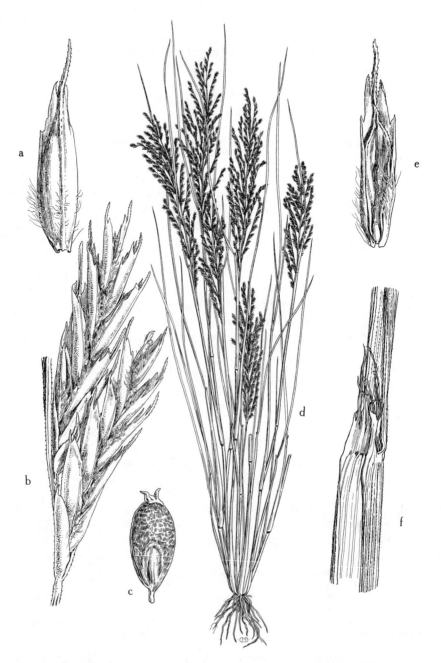

Fig. 73. *Leptochloa fascicularis: a,* floret, showing awned lemma with bifid apex, × 12; *b,* spikelets, × 8; *c,* grain, × 20; *d,* habit, showing the branching culms and the panicles, × ¼; *e,* floret, showing palea, × 12; *f,* leaf sheath and fimbriate ligule, × 4.

papillose and long-pilose near the base, the margins scabrous; sheaths sparsely papillose-pilose, the lower ones often almost glabrous; panicle somewhat viscid, 1–4 dm. long, about ½ the length of the plant; spikes several to numerous, lax, spreading-ascending, 0.5–1.5 dm. long; spikelets 2- or 3-flowered, 1–2 mm. long, rather distant on the rachis, the upper floret scarcely longer than the glumes; glumes acute-acuminate, subequal in size; lemmas 1–1.5 mm. long, obtuse, minutely bilobed, awnless, pubescent on the nerves and sometimes on the internerves toward the base.

Known only from overflow land along the Colorado River in Imperial County. Native of South America.

2. Leptochloa fascicularis (Lam.) Gray, Man. 588. 1848. Bearded sprangle-top. Fig. 73.

Annual, somewhat succulent; culms erect or decumbent-spreading at base, simple or usually freely branching, 3–10 dm. tall, glabrous; blades flat to loosely involute, scabrous, to 5 dm. long, long-attenuate, 1–5 mm. wide, erect; sheaths smooth; scabrous panicle more or less included, mostly 1–3 dm. long, often shorter, occasionally longer, the racemes several to many, as much as 1 dm. long, usually ascending or appressed, or at maturity spreading; spikelets loosely arranged, slightly overlapping, narrow, 8–10 mm. long, 6- to 12-flowered; glumes acuminate, the first glume narrowly lanceolate, 2–3 mm. long, the second broader, 3–4 mm. long; lemmas 3–4 mm. long, acute, narrow, 3-nerved, the midnerve usually produced into an awn about 1–3 mm. long from between a bifid apex, the 3 nerves villous on the lower half.

Along ditches and in wet alkaline places: Central Valley from Kern County to Butte County, also reported as a street weed in San Francisco; temperate and tropical areas in the Western Hemisphere.

3. Leptochloa uninervia (Presl) Hitchc. & Chase, Contr. U. S. Nat. Herb. 18:383. 1917. Mexican sprangle-top. Fig. 74.

Leptochloa imbricata Thurb. in Wats., Bot. Calif. 2:293. 1880.

Annual; culms erect, simple or sparingly branched, 3–10 dm. tall, glabrous; sheaths glabrous or scabrous; blades firm, flat or loosely involute, scabrous, attenuate, 1–4.5 dm. long, 1–4 mm. wide; panicle 1–3 dm. long, oblong; spikes numerous, approximate, usually stiffly ascending but sometimes ascending-spreading, the lower ones 4–9 cm. long, the upper ones gradually shorter and closer together; spikelets lead-colored, 6- to 9-flowered, 5–7 mm. long, appressed; first glume narrow, acute, 1–1.5 mm. long, the second one much broader, abruptly acute or obtuse, mucronate; lemmas 2–3 mm. long, abruptly subacute or obtuse, minutely lobed, apiculate, the lateral nerves more or less excurrent, the margins pubescent near the base.

Ditches and moist alkaline places: coastal and desert southern California, Kern and Merced counties; scattered areas in the Western Hemisphere.

Fig. 74. *Leptochloa uninervia: a,* spikelets, × 8; *b* and *c,* floret, showing palea and the apiculate lemma, the margins basally pubescent, × 16; *d,* habit, × ⅕; *e,* grain, × 16; *f,* leaf sheath and the bilobed ligule, × 4.

This species bears some resemblance to *Leptochloa fascicularis,* but the panicle is denser and somewhat narrower in outline, the glumes are broader and more obtuse, and the lemmas are apiculate but awnless.

MONANTHOCHLOË. SHORE GRASS

1. Monanthochloë littoralis Engelm., Trans. Acad. Sci. St. Louis 1:437, pls. 13 and 14. 1859. Salt cedar.

Plants dioecious; culms tufted, wiry, extensively creeping, the short branches erect, 1–3 dm. tall; blades falcate, subulate, 5–10 mm. long, conspicuously 2-ranked in distant to approximate clusters; spikelets 1 to few, nearly concealed in the leaves, 3- to 5-flowered, the uppermost florets rudimentary; glumes wanting; lemmas rounded on the back, narrowed above, several-nerved, those of the pistillate spikelets like the blades in texture; palea narrow, 2-nerved.

Salt marshes and tidal flats along the coast: San Diego County to Santa Barbara County; southern and southeastern states, Mexico, Cuba.

MONERMA

1. Monerma cylindrica (Willd.) Coss. & Dur., Expl. Sci. Alger. 2:214. 1855. Fig. 75.

Lepturus cylindricus Trin., Fund. Agrost. 123. 1820.

Annual; culms bushy-branched, erect or spreading, 1–5 dm. tall; spike slender, cylindric, 0.5–2 dm. long, curved, narrowed upward; spikelets 1-flowered, embedded in the hard, articulate rachis and falling at maturity attached to the joints; first glume wanting except on the terminal spikelet, the second glume 5–6 mm. (sometimes 4 mm.) long, acuminate, flush with the surface of the rachis, indurate, several-nerved, longer than the joint of the rachis; lemma 4–5 mm. long, pointed, with back to the rachis, hyaline; palea a little shorter than the lemma, hyaline.

Established on mud flats and around salt marshes near the coast: San Diego County to Los Angeles County, Monterey County to Sonoma County, inland to San Joaquin and Stanislaus counties; native of Europe.

MUHLENBERGIA. MUHLY

Our species perennials, tufted or rhizomatous. Culms simple or much-branched, the inflorescence a narrow or open panicle. Spikelets 1-flowered, disarticulating above the glumes. Glumes as long as the lemma or usually shorter, obtuse to acuminate or awned, the first glume sometimes small or

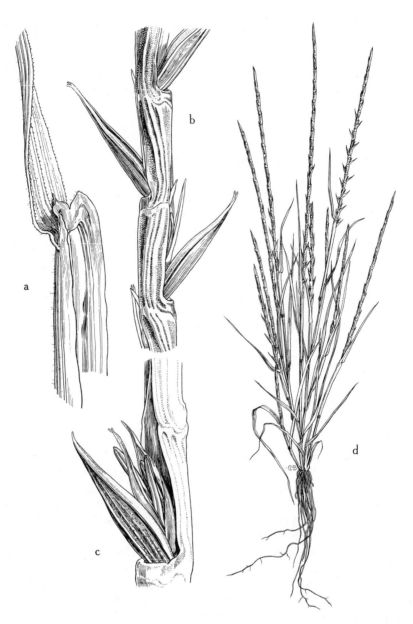

Fig. 75. *Monerma cylindrica: a,* sheath with 3-lobed ligule and scabrous leaf blade, × 6; *b,* part of spike, × 6; *c,* spikelet, embedded in the hard, articulate rachis, × 8; *d,* habit, showing the erect culms and spikes, × ⅖.

rarely obsolete. Lemma membranous, usually rather firm, with a very short, usually minutely pilose callus, the apex acute, extending into a straight or flexuous awn, or sometimes only mucronate. Palea thin, about as long as the lemma.

In the strict sense, none of the species of *Muhlenbergia* are aquatic or marsh plants, but those treated here are most often found in association with other plants that are characteristic marsh plants. Additional species which may occasionally be found in marshy habitats include *M. andina* (Nutt.) Hitchc., *M. squarrosa* (Trin.) Rydb., *M. rigens* (Benth.) Hitchc., and *M. utilis* (Torr.) Hitchc.

Plants with creeping rhizomes.
   Panicle open, diffuse at maturity .... 1. *M. asperifolia*
   Panicle narrow, dense, but more or less interrupted .... 2. *M. californica*
Plants without creeping rhizomes; roots fibrous .... 3. *M. filiformis*

1. Muhlenbergia asperifolia (Nees & Meyen) Parodi, Rev. Agron. Univ. Buenos Aires 6:117. 1928. Alkali muhly.

   Sporobolus asperifolius Nees & Meyen in Nees, Nova Acta Acad. Leop.-Carol. 19, Suppl. 1:141. 1843.

Perennial with slender, scaly rhizomes; culms pale or glaucous green, branching at base, spreading, slender, compressed, glabrous, 1–6 dm. tall, the branches ascending or erect; sheaths compressed-keeled, usually overlapping, glabrous; ligules truncate, erose-toothed, about 0.5 mm. long; blades flat, crowded, scabrous, mostly 2–5 cm. long, 1–2 mm. wide; panicle slender, diffuse, 5–15 cm. long, about as wide as long, the capillary scabrous branches eventually wide-spreading, the panicle at maturity breaking away; spikelets 1.5–2 mm. long, occasionally 2-flowered; glumes acute, glabrous, scabrous on the keel, sometimes minutely and sparsely pilose on the back, from ½ to nearly as long as the spikelet; lemma thin, broad, minutely mucronate from an obtuse apex; palea about as long and as broad as the lemma.

Occasional in marshy, wet, or often alkaline soil, along ditches and stream banks: chiefly in coastal southern California from San Diego County to San Luis Obispo County, San Joaquin Valley (Kern, Tulare, and Merced counties), east of the Sierra Nevada crest from Inyo County to Modoc County and thence west to Siskiyou County; across North America, southern South America.

2. Muhlenbergia californica Vasey, Bull. Torrey Club 13:53. 1886. California muhly.

Perennial, the base more or less creeping and rhizomatous; culms usually several, somewhat woody below, puberulent below the nodes, 3–6 dm. tall; sheaths minutely scabrous, keeled; ligules obtuse or truncate, ciliate, somewhat pilose, about 1 mm. long; blades flat, scabrous, 5–15 cm. but usually less than 10 cm. long, 4–5 mm. wide; panicle narrow, dense, but more or less interrupted, 5–15 cm. long, the axis and branches scabrous; spikelets 3–5 mm. long; glumes slightly shorter than the spikelets, narrow, acuminate, scabrous, awn-tipped; lemma scabrous, about 3 mm. long, acuminate, awn-tipped or with awn 1–2 mm. long, with sparse callus hairs about ½ as long as the lemma; palea acuminate, as long as the body of the lemma.

Along ditches and streams in coastal southern California, particularly in the San Bernardino Valley, ascending in the mountains to an elevation of 7,000 feet.

3. Muhlenbergia filiformis (Thurb.) Rydb., Bull. Torrey Club 32:600. 1905. Pull-up muhly.

Perennial or sometimes apparently annual, with fibrous roots or decumbent, creeping bases; culms tufted, erect or somewhat spreading, glabrous, filiform, usually 0.5–1.5 dm. or sometimes as much as 3 dm. tall; ligules thin, hyaline, 1–2 mm. long; blades flat, glabrous beneath, scabrous-pubescent on the upper surface, 1–3 cm. long, 1 mm. wide; panicles numerous, narrow, interrupted, few-flowered, usually less than 5 cm. long; glumes ovate, about equal in size, obtuse or acutish, awnless, 1 mm. long; lemma lanceolate, acute, 2 mm. long, mucronate, minutely pubescent, minutely scabrous at the tip, 1 mm. long, the callus glabrous.

Chiefly in wet meadows: throughout the Sierra Nevada at elevations of 4,500–10,000 feet, North Coast Ranges from Lake County to Siskiyou Mountains above 5,000 feet, east of the Sierra Nevada crest from Mono County to Modoc County, San Jacinto and San Bernardino mountains at elevations of 5,000–9,000 feet; western and midwestern states and north to British Columbia.

NEOSTAPFIA

1. Neostapfia colusana (Davy) Davy, Erythea 7:43. 1899. Colusa grass.

Anthochloa colusana Scribn., U. S. Dept. Agr., Div. Agrost. Bull. 17:221, fig. 517. 1899.

Viscid annual; culms ascending from a decumbent base, 0.5–3 dm. tall; leaves overlapping, pale green, scarious between the nerves, loosely folded around the culm, not differentiated into sheath and blade, 1–1.2 cm. wide at the middle, tapering to each end, minutely ciliate, with raised glands on the margins and nerves; panicle spikelike, pale green, partially included, 3–7 cm. long, 8–12 mm. wide, the upper part of the axis bearing, instead of spikelets, lanceolate-linear, empty bracts; spikelets subsessile, usually 5-flowered, 6–7 mm. long, imbricate; glumes wanting; lemmas fan-shaped, very broad, many-nerved, 5 mm. long, ciliolate-fringed.

Known only from vernal pools in Colusa, Stanislaus, and Merced counties.

This genus has been confused with the South American genus *Anthochloa* Nees. The differences between *Neostapfia* and *Anthochloa* have been summarized by Hoover (Leafl. West. Bot. 2:273–274, 1940).

ORCUTTIA

Annuals with fibrous roots, pilose throughout to nearly glabrous, at maturity viscid and with a characteristic lemon-like odor. Culms tufted, fragile at the nodes. Blades involute in age. Ligule represented only by a row of hairs. Inflorescence spicate or with spikelike racemes, the spikelets compressed, several-flowered, persistent in fruit and the rachilla not disarticulating. Glumes cleft into 2–5 acute teeth or sometimes the glumes acute and entire. Lemmas several-nerved, with 5–9 acute or awnlike teeth at apex. Palea with 2 green keels, the apex obtuse or somewhat 3-lobed.

Hoover, Robert F., The genus *Orcuttia*. Bull. Torrey Club 68:149–156. 1941.

Lemma apex with 2–4 very short teeth on either side of a prominent apical point
1. *O. Greenei*
Lemma apex parted above the middle into 5 subulate, rather long teeth, the teeth equal or unequal in length, acute to awn-tipped.
Spike 2–5 cm. long; teeth of lemma unequal in length, the middle tooth longest.
Lemmas 3–6 mm. long.
Plant rather sparingly pilose, light green; southern California ... 2. *O. californica*
Plant conspicuously pilose, grayish; San Joaquin Valley
2a. *O. californica* var. *inaequalis*
Lemmas 6–8 mm. long .... 2b. *O. californica* var. *viscida*
Spike 5–10 cm. long; lateral teeth of lemma equal in length to the median tooth.
Upper spikelets densely crowded, all except the lowermost many-flowered
3. *O. pilosa*
Spikelets not crowded, all except the uppermost few-flowered .... 4. *O. tenuis*

1. Orcuttia Greenei Vasey, Bot. Gaz. 16:146. 1891. Beardless orcuttia.

Culms 0.5–3 dm. tall, erect or decumbent at base; nodes hairy, often purplish; leaves more or less hairy throughout or only on the upper surface of the blades; spike 2–9 cm. long, broadened toward apex; spikelets spirally arranged, the lowest few spikelets slightly overlapping, the upper ones congested, 5- to 15-flowered; glumes 3-5 mm. long, toothed or subentire at apex; lemmas 5–7 mm. long, sparingly pilose, somewhat conspicuously so at apex, obliquely truncate and with 2–4 very short teeth on either side of a longer apical point; palea obtusely 3-lobed at apex, hyaline.

Vernal pools: eastern edge of the Central Valley from Tulare County to Tehama County.

2. Orcuttia californica Vasey, Bull. Torrey Club 13:219. 1886.

Culms 0.5–1.5 dm. tall, spreading or ascending, sparingly to moderately pilose; spike 2–5 cm. long; spikelets alternate on opposite sides of the axis, ascending, tending to be all directed to one side, rather few (4–10), the lower spikelets not overlapping and even the upper ones not densely crowded, 4- to 12-flowered; glumes 2–4 mm. long, parted into 3 subulate lobes; lemmas sparsely hairy or nearly glabrous, 4–5 mm. long, parted above the middle into 5 subulate teeth, the middle tooth slightly longer than the lateral ones.

Vernal pools and mud flats: coastal Los Angeles and Riverside counties; Baja California, Mexico.

2a. Orcuttia californica var. **inaequalis** (Hoover) Hoover, Bull. Torrey Club 68:154. 1941.

Orcuttia californica var. *inaequalis,* from vernal pools, eastern margin of the San Joaquin Valley from Tulare County to Stanislaus County, has a wider distribution and is apparently more common than the typical form. It can usually be distinguished from *O. californica* by the densely crowded upper spikelets and by the grayish color resulting from the dense pilosity of the herbage. Most of its other morphologic characters merge with those of the typical form.

2b. Orcuttia californica var. **viscida** Hoover, Bull. Torrey Club 68:155. 1941.

Orcuttia californica var. *viscida,* from vernal pools at the eastern edge of the Sacramento Valley in Sacramento County, is distinguished from the species and from *O. californica* var. *inaequalis* in having longer lemmas with conspicuously awned teeth. In its general habit it resembles the variety *inaequalis*.

3. Orcuttia pilosa Hoover, Bull. Torrey Club 68:155. 1941.

Culms 0.5–2 dm. tall, usually 4–50 in dense tufts, erect but decumbent at base, often with long basal leaves which disappear early, very viscid at maturity, pilose throughout but the backs of the leaves nearly glabrous; spikes 5–10 cm. long, the spikelets restricted to the upper ½ to ⅓ of the taller stems but arising nearly as far down as the base of the shortest; spikelets 8–18, 2-ranked, strictly erect or slightly curved outward, the upper ones densely crowded, about 10- to 40-flowered; glumes about 3 mm. long, irregularly 3- to 5-toothed; lemmas 4–5 mm. long, parted above the middle into 5 equally long teeth, the teeth slightly awn-tipped or merely acute.

Known only from vernal pools on east side of San Joaquin Valley in Stanislaus and Madera counties.

This species bears a close resemblance to *Orcuttia tenuis,* and the two forms may not be specifically distinct. However, these two plants occupy discontinuous areas and in their general habit are quite distinct. In *O. pilosa* the spikelets are more densely crowded and have more numerous florets.

4. Orcuttia tenuis Hitchc., Am. Jour. Bot. 21:131. 1934.

Culms 0.5–1.5 dm. tall, 1–18 in tufts, slender, strictly erect or, when crowded, decumbent at base, under favorable conditions producing long, floating juvenile leaves, becoming very viscid at maturity; leaves thinly pilose throughout or sometimes only on the upper surface of the blades; spike 5–9 cm. long, the lowest spikelets usually below the middle of the stem, often nearly at the base of the stem; spikelets 6–12, not crowded, strictly erect, sparsely pilose or sometimes glabrous, 2- to 10-flowered (all except the uppermost few-flowered), the lower ones distant; glumes 3–5 mm. long, with 2–5 teeth at apex; lemmas 4.5–6 mm. long, parted above the middle into 5 equally long, awn-tipped teeth.

Vernal pools in upper Sacramento Valley east of Sacramento River in Shasta and Tehama counties, also in Goose Valley, northeastern Shasta County.

ORYZA

1. Oryza sativa L., Sp. Pl. 333. 1753. Rice.

Annual; culms erect, 1–2 m. tall; blades elongate, flat; panicle rather dense, drooping, 1.5–4 dm. long; spikelets 1-flowered, laterally compressed, disarticulating below the glumes, 7–10 mm. long, 3–4 mm. wide; glumes 2, much shorter than the lemma, narrow; lemma rigid, keeled, 5-nerved, the apex mucronate to long-awned; lemma and palea papillose-roughened and with scattered, appressed hairs.

Cultivated chiefly on former marshlands in the upper Sacramento Valley; sparingly naturalized in wet places.

PANICUM. WITCH GRASS

Annuals or perennials. Spikelets more or less compressed dorsiventrally, in open or compact panicles, rarely in racemes. Glumes herbaceous, nerved, usually very unequal in size, the first glume often minute, the second one typically as long as the sterile lemma; sterile lemma of the same texture as the glumes, simulating a third glume, and bearing in its axil a membranaceous or hyaline palea and sometimes a staminate flower, the palea rarely wanting. Fertile lemma chartaceous-indurate, typically obtuse, the nerves obsolete, the margins inrolled over an enclosed palea of the same texture.

Two additional species, which are not separately described here, are occasionally found in association with other characteristic marsh plants. These are *Panicum dichotomiflorum* Michx., a sparingly naturalized species of the eastern United States, and *P. thermale* Boland., a species of wet saline soil, which occurs mostly in the vicinity of hot springs, at a few scattered localities throughout California.

Plants annual, spikelets 2–3.5 mm. long, acute to acuminate . . . . 1. *P. capillare*
Plants perennial; spikelets usually less than 2 mm. long (average 1.8 mm.), obtuse
2. *P. occidentale*

1. Panicum capillare L., Sp. Pl. 58. 1753. Common witch grass.

Panicum capillare var. *occidentale* Rydb., Contr. U. S. Nat. Herb. 3:186. 1895.
Panicum barbipulvinatum Nash in Rydb., Mem. N. Y. Bot. Gard. 1:21. 1900.

Annual; culms stout, erect or somewhat spreading at the base, 2–8 dm. tall; sheaths papillose-hispid to nearly glabrous; blades 0.5–2.5 dm. long, 5–15 mm.

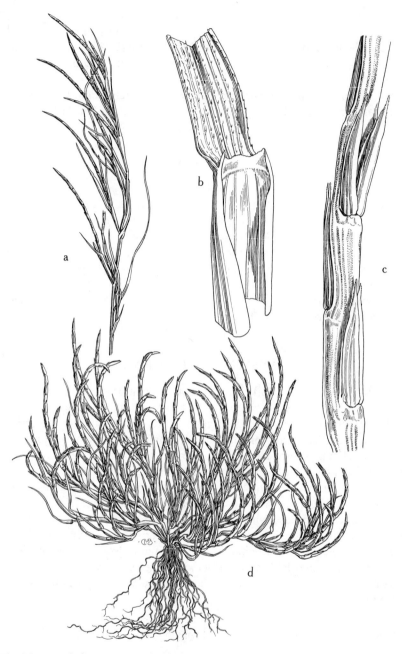

Fig. 76. *Parapholis incurva: a,* habit variation, culm erect, × ⅖; *b,* leaf sheath, ligule, and blade, × 8; *c,* part of spike, showing spikelets embedded in the cylindric, articulate rachis, × 6; *d,* habit, the culms decumbent and the spikes strongly curved, × ⅖.

wide, hispid on both surfaces; panicles densely flowered, very diffuse, often ½ the length of the entire plant, the basal part of the panicle often included in the leaf sheath, the entire panicle breaking out of the leaf sheath at maturity, the branches becoming wide-spreading; spikelets 2–3.5 mm. long, elliptic, acute to acuminate; first glume acute, ½ as long as the spikelets; second glume and sterile lemma acuminate.

Moist soil in waste and cultivated lands, along irrigation ditches, and in sandy places along streams: widespread weed throughout California, but the localities somewhat scattered, absent from the deserts and higher mountains.

Panicum capillare is a common weed in the eastern United States, but it is somewhat less frequent in California. *Panicum capillare* var. *occidentale* has been considered to be a more common western representative of the species and to differ from the species in having a lower average height, shorter, less pubescent blades which are often crowded toward the base, more exserted and divaricate panicles, and longer spikelets that are slightly narrower and somewhat more acuminate.

2. Panicum occidentale Scribn., Rep. Mo. Bot. Gard. 10:48. 1899.

Perennial; culms yellowish green, leafy toward base, 1.5–4 dm. tall, spreading, sparsely pubescent; nodes pubescent; sheaths sparsely papillose-pubescent; ligules ciliate, 3–4 mm. long; blades firm, erect or ascending, 3–8 cm. long, 4–9 mm. wide, nearly glabrous on upper surface, appressed-pubescent beneath; panicle 4–8 cm. long; spikelets 1.5–2 mm. long, pubescent; glumes obtuse, or sometimes tipped with a short mucro.

In meadows and along streams, sometimes in sphagnum bogs: scattered stations in the Sierra Nevada from Mariposa County to Amador County chiefly at elevations of 2,000–4,500 feet, more frequent along the coast and in Coast Ranges from Santa Cruz County to Del Norte County, also in San Diego County; north to British Columbia and Idaho.

Panicum pacificum Hitchc. & Chase, a more common species in California, bears a close resemblance to *P. occidentale,* but differs in having more copious pubescence and more leafy culms. It occurs chiefly in sandy places along streams, in moist rock crevices, or on seepage areas.

PARAPHOLIS

1. Parapholis incurva (L.) C. E. Hubb., Blumea, Suppl. 3. 14. 1946. Fig. 76.

Pholiurus incurvatus Hitchc., U. S. Dept. Agr. Bull. 772:106. 1920.

Low annual; culms tufted, erect or decumbent, spreading and somewhat bushy-branched at base, 1–3.5 dm. tall; blades short, narrow; spike slender, cylindric, curved, 7–10 cm. long; spikelets 1- or 2-flowered, 7 mm. long, pointed, embedded in the cylindric articulate rachis and falling attached to the

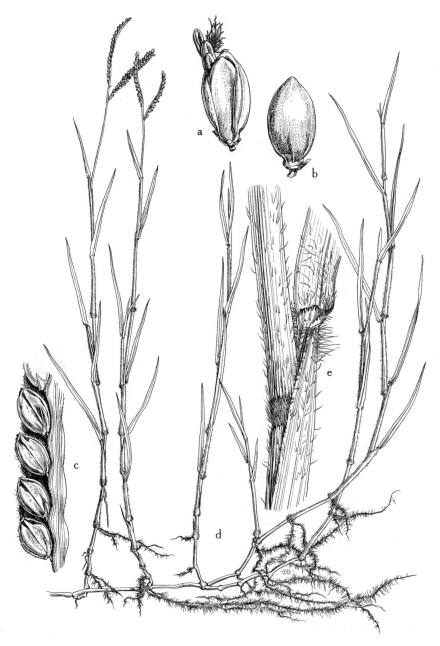

Fig. 77. *Paspalum distichum: a,* floret, showing palea, × 8; *b,* floret, showing lemma, × 8; *c,* rachis, showing 2 rows of spikelets, × 4; *d,* habit, showing the decumbent rooting base, the flat leaf blades, and the paired racemes, × ¼; *e,* leaf sheath, ligule, and node, densely pubescent, × 2½.

joints; glumes 2, placed in front of the floret and enclosing it, coriaceous, 5-nerved, acute, asymmetric, appearing like halves of a single split glume; lemma with its back to the rachis, smaller than the glumes, hyaline, 1-nerved; palea a little shorter than the lemma, hyaline.

Established in salt marshes and tidal flats along the coast from San Diego County to Humboldt County; native of Europe.

The erect plants of this species bear a marked resemblance in habit to *Monerma cylindrica* (Willd.) Coss. & Dur. In the latter species, however, the first glume of the spikelets is wanting, and there is a tendency for the spikelets to spread away from the rachis at maturity, the latter a feature not evident in *Parapholis incurva*.

PASPALUM

Perennial. Racemes digitate or racemose at summit of culm, spikelike, 2 to many. Spikelets 1-flowered, plano-convex, usually obtuse, subsessile, solitary or in pairs, in 2 rows on one side of a narrow or dilated rachis, the back of the fertile lemma toward it. First glume usually wanting; second glume and sterile lemma about equal in size. Fertile lemma usually obtuse, it and the palea indurated, the margins inrolled.

Racemes 2, approximate or nearly so at the summit of the culm, sometimes a third raceme
    below, 2.5–6 cm. long .... 1. *P. distichum*
Racemes commonly 3–8, 5–10 cm. long, spikelets fringed with long, white, silky hairs
                                                                    2. *P. dilatatum*

1. Paspalum distichum L., Syst. Nat., 10th ed., 2:855. 1759. Knotgrass. Fig. 77.

Culms erect from a decumbent rooting base, 3–6 dm. tall, often with extensively creeping stolons, glabrous, or the nodes pubescent; sheaths glabrous or sometimes pubescent; blades flat, glabrous, rarely pubescent, 5–15 cm. long, the upper blades shorter; racemes 2, paired or closely approximate, sometimes a third one below the second, ascending or appressed, commonly incurved, usually 2.5–6 cm. long; spikelets singly disposed, 2.5–3 mm. long, elliptic, abruptly acute, pale green; first glume occasionally present and nearly as long as the spikelet, glabrous; second glume appressed-pubescent; sterile lemma glabrous.

In marshes, around ponds, or along streams: near the coast from San Diego County to Del Norte County, occasional in the Coast Ranges, throughout the Central Valley, and east of the Sierra Nevada crest (Lassen and Inyo counties). A troublesome pest in irrigation ditches. Warmer parts of Western and Eastern hemispheres.

2. Paspalum dilatatum Poir. in Lam., Encycl. 5:35. 1804. Dallis grass. Fig. 78.

Culms tufted, leafy at base, glabrous, mostly 5–15 dm. tall, ascending or erect from a decumbent base; lowest sheaths noticeably pubescent; blades

Fig. 78. *Paspalum dilatatum: a,* floret, showing lemma, × 8; *b,* floret, showing palea, × 8; *c,* rachis, showing the 2 rows of hairy spikelets, × 4; *d,* habit, showing the noticeably pubescent lowest sheaths, the arching leaves, and the spreading racemes, × ⅕; *e,* upper leaf sheath, pubescent only around ligule and on base of blade, × 4.

1–2.5 dm. long, 3–12 mm. wide; racemes usually 3–8, densely flowered, spreading, 5–10 cm. long; spikelets in pairs, ovate, acute, overlapping, 3–3.5 mm. long, fringed with long, white, silky hairs and sparsely silky on the surface.

Established chiefly along streams and irrigation ditches in the Central Valley, also at scattered stations near the coast from Santa Clara County to Sonoma County, and in Los Angeles and Imperial counties. Important forage plant, extensively cultivated in the southeastern states and in irrigated pastures in California. Native of South America.

PHALARIS. CANARY GRASS

Erect annuals or perennials. Blades numerous, flat. Panicles usually dense, spikelike. Spikelets laterally compressed, with 1 terminal perfect floret and 2 reduced sterile lemmas below, the rachilla disarticulating above the glumes, the usually inconspicuous sterile lemmas falling closely appressed to the fertile floret. Glumes equal in size, strongly keeled, sometimes winged on the keel. Sterile lemmas reduced to 2 small, narrow scales; fertile lemma coriaceous, shining in fruit, shorter than the glumes, enclosing the palea.

Three annual species, all of them introduced from Europe, are known to occur as weeds in rice fields. Two of the species, *Phalaris minor* Retz. (Mediterranean Canary grass) and *P. paradoxa* L. (gnawed Canary grass), are rather widely distributed in California, where they are found in rice fields and in other grain fields, on waste land, and along roadsides, especially in heavy soils. The third species, *P. brachystachys* Link, apparently has a more restricted distribution: it grows on waste ground and in rice fields in the northern Sacramento Valley.

Phalaris tuberosa var. *stenoptera* (Hack.) Hitchc., a perennial introduced as a forage grass (Harding grass), has become established near river mouths in Humboldt County and along the edges of salt marshes in Marin County.

Plants annual; glumes somewhat scabrous on the sides, as well as on the prominent nerves
$\qquad$ 1. *P. Lemmonii*
Plants perennial; glumes usually scabrous only on the keel and lateral nerves.
Panicle ovoid or oblong, 2–5 cm. long, dense, not interrupted below ... 2. *P. californica*
Panicle narrow, 5–20 cm. long, usually interrupted below and the branches spreading in anthesis .... 3. *P. arundinacea*

1. Phalaris Lemmonii Vasey, Contr. U. S. Nat. Herb. 3:42. 1892. Lemmon Canary grass.

Annual; culms 3–9 dm. tall; panicle 4–15 cm. long, subcylindric or lobed toward base, often purplish-tinged; glumes 5–6 mm. long, narrow, acuminate, scabrous, not winged on the keel; fertile lemma ovate-lanceolate, acuminate, 3.5–4.5 mm. long, brown at maturity, appressed-pubescent except the acu-

Fig. 79. *Phalaris arundinacea: a,* floret, showing fertile and sterile lemmas, × 8; *b,* spikelet, showing the strongly keeled glumes, fertile lemma, and palea, × 8; *c,* panicle, interrupted below, × ⅔; *d,* leaf sheath, ligule, blade, and node, × ⅖; *e,* habit, showing creeping rhizome, × ⅛; *f,* upper part of culm, showing panicle, × ⅛.

minate tip; sterile lemmas 1 or 2, about 1 mm. long, ⅓ or less as long as the fertile lemma.

Low, wet places in fields, on dried mud flats, or sometimes bordering salt marshes near the coast: San Diego County to Mendocino County, San Francisco Bay region, and Central Valley from Madera County to Sacramento County.

2. Phalaris californica Hook. & Arn., Bot. Beechey Voy. 161. 1841. California Canary grass.

Perennial, culms erect, 7.5–15 dm. tall, often in dense tufts; blades rather lax, 6–15 mm. wide; panicle ovoid or oblong, 2–5 cm. long, 1.5–2.5 cm. thick, often purplish-tinged; glumes 5–7 mm. (mostly 6 mm.) long, narrow, tapering from below the middle to an acute apex, the keel minutely scabrous, sharp but not winged; fertile lemma ovate-lanceolate, 4–5 mm. long, rather sparsely appressed-pubescent, the palea often exposed; sterile lemmas narrow, about ½ as long as the fertile one, villous.

Moist or wet soil in marshes, along streams, or in woods: near the coast and in Coast Ranges from San Luis Obispo County to Humboldt County; southwestern Oregon.

3. Phalaris arundinacea L., Sp. Pl. 55. 1753. Reed Canary grass. Fig. 79.

Perennial with creeping rhizomes, glaucous; culms erect, 6–15 dm. tall, glabrous; panicle 5–20 cm. long, pale green or tinged with purple, narrow and dense or interrupted below, the branches spreading during anthesis, the lower ones as much as 5 cm. long; spikelets 5–6 mm. long; glumes about 5 mm. long, sharply keeled, narrow, acute, longer than the lemmas, the keel scabrous, wingless or very narrowly winged; fertile lemma lanceolate, 3–4 mm. long, shining, with a few appressed hairs in upper part; narrow, scale-like sterile lemmas villous, 1 mm. long.

Sloughs, marshes, and stream banks: in San Joaquin Valley, in lower Sacramento Valley, and along the north coast from Sonoma County to Del Norte County, thence east to Siskiyou County; Northern Hemisphere. Occasionally cultivated as a forage grass in wet places.

PHLEUM. TIMOTHY

Our species perennial. Culms erect. Blades flat. Panicles terminal, dense, spikelike. Spikelets 1-flowered, laterally compressed, disarticulating above the glumes. Glumes equal in size, membranaceous, strongly keeled, abruptly awn-pointed or gradually acute. Lemma shorter than the glumes, hyaline, broadly truncate, 3- to 5-nerved. Palea narrow, nearly as long as the lemma.

Panicle narrowly cylindric, 3–15 cm. long .... 1. *P. pratense*
Panicle broadly cylindric or oblong, 1–5 cm. long .... 2. *P. alpinum*

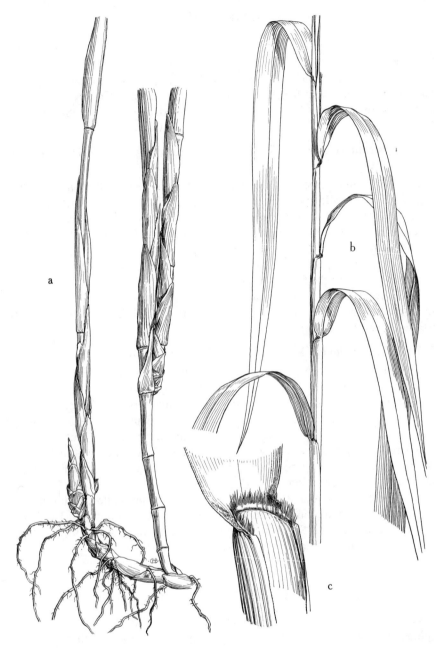

Fig. 80. *Phragmites communis: a,* habit, lower part of culm and rhizome, × ⅖; *b,* habit, culm and leaves, × ⅖; *c,* leaf sheath, base of blade, and ciliate ligule, × 1½.

1. Phleum pratense L., Sp. Pl. 59. 1753. Timothy.

Culms 5–12 dm. tall, forming large clumps from a swollen or bulblike base; blades elongate, mostly 4–8 mm. wide, scabrous above; panicle narrow-cylindric, commonly 3–15 cm. long, 5–7 mm. thick, sometimes longer, the spikelets crowded, spreading; glumes 3.5–4 mm. long, truncate, with a stout awn 0.75–1 mm. long, pectinate-ciliate on the keel.

Commonly cultivated grass which is sparingly escaped in moist meadows and along ditches and streams at low to middle elevations throughout California. Native of Europe.

2. Phleum alpinum L., Sp. Pl. 59. 1753. Alpine timothy.

Culms 2–6 dm. tall, glabrous, from a decumbent, somewhat creeping, densely tufted base; blades mostly less than 15 cm. long, 3–6 mm. wide; panicle 1–5 cm. long or broadly cylindric; glumes 5 mm. (sometimes 7 mm.) long, oblong, hispid-ciliate on the keel, the stoutish awns 2 mm. long, giving the head a bristly appearance.

Common in wet mountain meadows in the San Jacinto and San Bernardino mountains at elevations of 7,000–8,500 feet, throughout the Sierra Nevada and north to Siskiyou County chiefly at elevations of 5,000–12,000 feet, also near the coast in moist meadows and wet places from Marin County to Del Norte County, ascending to an elevation of 5,000 feet in the North Coast Ranges. Native of Europe.

PHRAGMITES. REED

1. Phragmites communis Trin., Fund. Agrost. 134. 1820. Common reed. Figs. 80 and 81.

Perennial; culms robust, erect, 2–4 m. tall, with stout, creeping rhizomes, these sometimes on the surface forming leafy stolons as much as 9 m. long; blades broad, flat, 2–6 dm. long, 1–5 cm. wide; panicle terminal, tawny, 1.5–4 dm. long, the branches ascending, rather densely flowered; spikelets several-flowered, 12–15 mm. long, the rachilla clothed with long, silky hairs as long as or shorter than the florets, disarticulating above the glumes and at the base of each joint between the florets; lowest floret staminate or neuter; glumes lanceolate, acute, unequal in size, the first glume about ½ as long as the upper glume, the second one shorter than the florets; lemmas narrow, long-acuminate, glabrous, 3-nerved, the florets successively smaller; palea much shorter than the lemma.

In fresh-water marshes, on banks of streams, and along irrigation ditches: well established in wet places in the Colorado and Mojave deserts, less common and at scattered localities along the coast from San Diego County to Del Norte County, in the delta region, also east of the Sierra Nevada crest from Mono County to Modoc County; cosmopolitan.

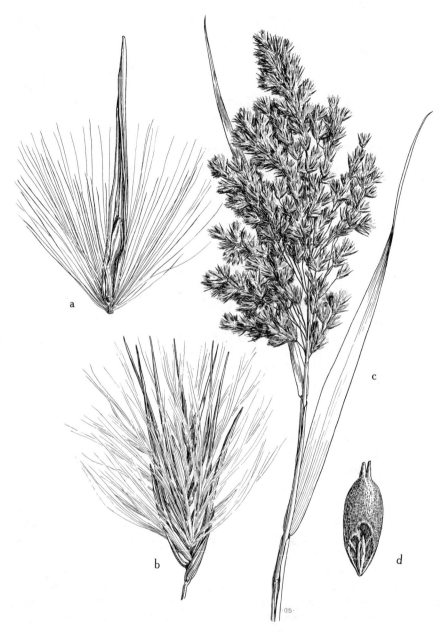

Fig. 81. *Phragmites communis: a,* floret, showing the long-acuminate lemma, the short palea, and the long, silky hairs on rachilla joint, × 6; *b,* spikelet, showing tne glumes and the florets successively smaller, × 4; *c,* habit, upper part of culm and panicle, × ⅖; *d,* grain, × 16.

PLEUROPOGON. SEMAPHORE GRASS

Soft, glabrous perennials, rhizomatous. Culms erect. Inflorescence racemose, loose. Spikelets linear, on a slender, flexuous axis, 5- to 20-flowered, compressed laterally, the rachilla disarticulating above the glumes and between the florets. Glumes unequal to subequal in size, membranous to almost hyaline, scarious at the somewhat lacerate tip. Lemmas thin and flexible to thick and firm, 7-nerved or 7-ribbed, awned or awnless, the apex often thin and scarious, entire or 2-toothed. Palea with both edges folded toward the side away from the rachilla, the folds produced laterally into wings or awns, the apical part of each wing almost always free.

Benson, Lyman, Taxonomic studies. I. A revision of the semaphore grasses; the genus *Pleuropogon*. Am. Jour. Bot. 28:358–360. 1941.

Lemmas 4–7.5 mm. long; spikelets erect or somewhat spreading.
Lemmas with an awn 5–12 mm. long .... 1. *P. californicus*
Lemmas awnless or with a mucro 0.5 mm. long .... 2. *P. Davyi*
Lemmas 8–9 mm. long; spikelets often reflexed or drooping at maturity ... 3. *P. refractus*

1. Pleuropogon californicus (Nees) Benth. in Vasey, U. S. Dept. Agr. Spec. Rep. 63:40. 1883.

Culms tufted, erect or decumbent at base, 3–7 dm. tall; blades linear, flat or folded, abruptly acute, 3–15 cm. long, 2–5 mm. wide; raceme 8–20 cm. long, with 5–10 rather distant, short-pediceled spikelets; spikelets erect or somewhat spreading, 7- to 15-flowered, 1.5–3 cm. long; glumes unequal in size, the first glume 1.5–3 mm. long, the second one 3–6 mm. long, erose; lemmas scabrous, 4–6 mm. long, rather thick and firm, prominently 7-nerved, the tips scarious, erose, bearing from a usually bifid apex an awn 5–12 mm. long; wings of palea cleft, forming a tooth about the middle, the apical free part of the wing 1 mm. or commonly 2–2.5 mm. long.

Low, wet places in fields: San Francisco Bay region and north along the coast to Humboldt County, also in the San Joaquin Valley in Stanislaus County.

2. Pleuropogon Davyi Benson, Am. Jour. Bot. 28:360. 1941.

Culms erect, about 1 m. tall; blades linear but slightly broader below the middle, abruptly acute, 1–3 dm. long, 6–9 mm. broad; raceme 2–3.5 dm. long; spikelets sessile or subsessile, rather distant, erect or ascending, mostly 10- to 18-flowered, 2–5.5 cm. long; glumes unequal in size, the first glume usually slightly shorter, 2–5 mm. long; lemmas 5.5–7.5 mm. long, acute or truncate at apex, awnless or with a mucro 0.5 mm. long, thick and firm, strongly 7-ribbed and scabrous on the ribs, commonly very shallowly 2- or 3-toothed along the margin near the apex; palea oblong, ⅔ to fully as long as the lemma, noticeably winged, the apical free part of each wing about 1 mm. long or rarely shorter.

Wet meadows, slough margins, and creek beds at elevations of 1,000–2,000 feet in Lake and Mendocino counties.

3. Pleuropogon refractus (Gray) Benth. in Vasey, U. S. Dept. Agr. Spec. Rep. 63:40. 1883. Nodding semaphore grass.

Culms tufted, 1–1.5 m. tall; blades linear, attenuate or abruptly acute, often with a sharp mucro, 1 dm. or mostly 1.5–3 dm. long, 5–7 or –12 mm. wide, the uppermost blades nearly obsolete; raceme 2.5–3 dm. long; spikelets finally reflexed or drooping, 6- to 12-flowered, 2–3.5 cm. long; glumes noticeably unequal in size, the first glume 3.5–5 mm. long, the second one 5–7 mm. long, scarious; lemmas 8–9 mm. long, commonly divergent, thin and flexible, 7-nerved (sometimes faintly), the apex erose, usually bifid, bearing an awn 5–20 mm. long from between the apical teeth or from the apex; palea linear, 7–8 mm. long, puberulent, acute, truncate, or emarginate, narrowly winged, the wings free apically for 0.2–1 mm.; rachilla joints 3 mm. long.

Wet meadows and along streamlets: Humboldt and Mendocino counties to an elevation of 3,500 feet; north to Washington.

A form which is known at present only from Mendocino and Marin counties has been named *Pleuropogon refractus* var. *Hooverianus* Benson. It differs from *P. refractus* mainly in having shorter awns (2–3 mm. long) and shorter rachilla joints (2–2.5 mm. long), the base of each joint being glandular-swollen; in *P. refractus* the rachilla joints are not glandular-swollen at the base, or are only very slightly so. J. T. Howell (Leafl. West. Bot. 4:247, 1946) lists additional characters which, in his view, constitute a basis for regarding the variety *Hooverianus* as a distinct species.

POA. BLUEGRASS

1. Poa trivialis L., Sp. Pl. 67. 1753.

Perennial; culms erect from a decumbent base, often rather lax, scabrous below the panicle, 3–10 dm. long; sheaths retrorsely scabrous (or minutely so), at least toward the summit; ligules 4–6 mm. long; blades scabrous, 2–4 mm. wide; panicle oblong, 6–20 cm. long, the lower branches about 5 to a whorl; spikelets usually 2- or 3-flowered, 3–3.5 mm. long, the rachilla disarticulating above the glumes and between the florets, the uppermost floret reduced or rudimentary; lemmas 2.5–3 mm. long, acute, glabrous except for the slightly pubescent keel, prominently 5-nerved, the web of hairs at base of lemma conspicuous.

Established in stream beds and moist places near the coast from Marin County to Humboldt County. Native of Europe.

In addition to *Poa trivialis, P. pratensis* L. and *P. compressa* L. are sometimes found in marshy areas.

POLYPOGON

Annual or perennial usually decumbent herbs. Blades flat, scabrous. Panicles dense, bristly, spikelike, sometimes lobed. Spikelets small, 1-flowered, the pedicel disarticulating a short distance below the glumes, leaving a short-pointed callus attached. Glumes equal in size, entire or 2-lobed, awned from the tip or from between the lobes, the awn slender, straight. Lemma much shorter than the glumes, hyaline, usually bearing a slender, straight awn shorter than the awns of the glumes.

Plants annual.
  Lobes of glumes hispidulous, ⅓ or less the length of the glumes .. 1. *P. monspeliensis*
  Lobes of glumes silky-ciliate, glabrous on sides, about ½ the length of the glumes
                                                  2. *P. maritimus*
Plants perennial.
  Awn of lemma conspicuous, usually exserted beyond the glumes; common
                                                3. *P. interruptus*
  Awn of lemma inconspicuous, included within the glumes, sometimes obsolete or wanting; rare .... 4. *P. elongatus*

1. Polypogon monspeliensis (L.) Desf., Fl. Atlant. 1:67. 1798. Rabbit-foot polypogon.

Annual; culms erect or decumbent at base, somewhat clustered, 1–10 dm. long (sometimes depauperate); sheaths glabrous or minutely roughened, occasionally somewhat inflated or loose; ligules 5–6 mm. long, nerved, scabrous, 4–15 cm. long, 2–7 mm. wide; panicle dense, spikelike, usually oblong (but sometimes narrowly ovate), soft-bristly, 2–8 (or –15) cm. long, 1–2 cm. wide, tawny yellow when mature; glumes hispidulous below, strigose in upper half, about 2 mm. long, the lobes about ⅓ or less the length of the glumes, the awn 6–8 mm. long, rarely longer; lemma smooth and shining, about ½ as long as the glumes, the awn included or slightly exserted beyond the glumes, somewhat bent or curved.

Widely established in moist places at low to middle elevations throughout California; native of Europe.

2. Polypogon maritimus Willd., Gesell. Naturf. Freunde Berlin (n.s.) 3:443. 1801.

Annual; culms erect or ascending, sometimes decumbent at base, 2–3 dm. tall (often depauperate), glabrous; sheaths glabrous; ligules to 6 mm. long, scaberulous, weakly nerved, narrowed to a usually lacerate tip; blades scabrous, usually not more than 5 cm. long, 2–4 mm. wide; panicle dense, spikelike or sometimes interrupted, oblong-cylindric, bristly, usually not more than 2–8 cm. long, 1 cm. wide; glumes lanceolate, 2.5–3 mm. long, hispidulous below, the deep lobes about ½ as long as the glumes, silky-ciliate but glabrous on the sides, the awns 7–10 mm. long.

Locally established in moist places: Humboldt, Lake, and Marin counties,

Fig. 82. *Polypogon elongatus: a,* spikelet, showing the hispidulous, awned glumes, × 8; *b,* leaf sheath, ligule, and blade, × 4; *c* and *d,* upper parts of culms with spikelike, interrupted panicles, × ⅕; *e,* habit, lower part of plant, × ⅕.

Sierra Nevada foothills in Tuolumne and Amador counties, and the upper Sacramento Valley in Butte County; native of Europe.

This species bears a very close resemblance to *Polypogon monspeliensis,* and the two plants may prove, upon further study, not to be specifically distinct. *Polypogon maritimus* can usually be recognized, however, by its more deeply lobed and silky-ciliate glumes and its somewhat narrower and slightly lobed panicle.

3. Polypogon interruptus H.B.K., Nov. Gen. & Sp. 1:134, pl. 44. 1815. Ditch polypogon.

Polypogon lutosus Hitchc., U. S. Dept. Agr. Bull. 772:138. 1920.

Perennial; culms tufted, the decumbent bases sometimes rooting, 3–9 dm. tall; sheaths glabrous, the upper sheaths sometimes inflated; ligules 2–6 mm. long, scaberulous, nerved, laciniate; blades more or less scabrous, 4–15 cm. long, commonly 3–6 mm. wide; panicle oblong-lanceolate, more or less interrupted or lobed, 4–15 cm. long, often purplish; glumes narrow, equal in size, 2–3 mm. long, scabrous on back and keel, the minute teeth acute, the awns 2–3 mm. (or sometimes 5 mm.) long; lemma about 1 mm. long, smooth and shining, minutely toothed at the truncate apex, the awn exserted beyond the glumes; palea nearly as long as the lemma.

Widely established along streams and irrigation ditches at low elevations throughout California except in the Central Valley, occasional in the deserts; western and southeastern states, South America.

This species bears a very close resemblance to *Agrostis semiverticillata,* from which it differs mainly in having awned glumes and lemmas.

4. Polypogon elongatus H.B.K., Nov. Gen. & Sp. 1:134. 1815. Fig. 82.

Perennial; culms erect or often decumbent at base, glabrous, rather stout, as much as 1 m. tall; sheaths glabrous, somewhat nerved, lacerate at the rather broad summit, to 8 mm. long; blades scabrous on the margins, glabrous or somewhat scabrous on the surfaces, to 20 cm. long and 1 cm. wide; panicle erect, in ours rather dense and spikelike but somewhat interrupted in the lower part, 15–30 cm. long, the branches closely flowered to base; glumes hispidulous (especially on keel), 2–3 mm. long, gradually narrowed to an awn 2–3 mm. long; lemma 1.5 mm. long, the awn arising from below the tip, 1–2 mm. long or sometimes obsolete.

Known from salt marshes in Contra Costa and San Luis Obispo counties. Probably introduced from Central or South America.

The type of this species and most of the collections from Central and South America that we have studied have panicles which are somewhat more open and interrupted, especially below, which often nod at the tip. Whether the California plants, with their denser, spikelike panicles, are distinct from the species cannot be determined from the literature or the small number of collections available for our study.

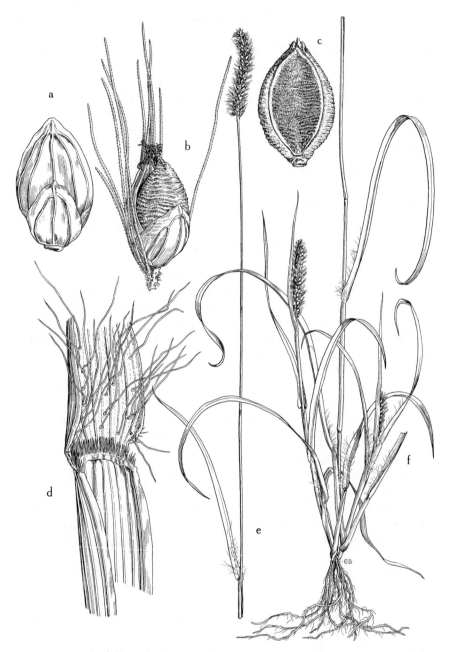

Fig. 83. *Setaria lutescens: a,* spikelet, showing first glume and sterile lemma, × 10; *b,* spikelet, showing fertile lemma and the upwardly barbed, slender bristles on branchlet, × 10; *c,* floret, showing palea, × 10; *d,* leaf sheath and ciliate ligule, × 6; *e,* habit, upper part of culm, showing spikelike panicle, × ⅖; *f,* habit, lower part, showing the leaf blades with villous base above sheath, × ⅖.

PUCCINELLIA. ALKALI GRASS

Glabrous annuals or perennials with narrow, flat or involute blades and narrow or open panicles. Spikelets several-flowered, the rachilla disarticulating above the glumes and between the florets. Glumes obtuse, unequal in size, the first glume 1-nerved, the second 3-nerved. Lemmas firm, rounded on the back, obtuse to acute, scarious and often erose at the tip, glabrous or slightly pubescent at the base, obscurely 5-nerved, the nerves parallel. Palea about as long as the lemma or somewhat shorter.

Swallen, Jason R., The Alaskan species of *Puccinellia*. Jour. Wash. Acad. Sci. 34:16–23. 1944.

Panicle narrow, the branches ascending or appressed; lemmas 3–4 mm. long .. 1. *P. grandis*
Panicle open at maturity, the branches spreading to reflexed.
    Lemmas broadly obtuse or truncate, 1.5–2 mm. long .... 2. *P. distans*
    Lemmas somewhat narrowed above, obtuse to subacute, 2–3 mm. long
                                           3. *P. Nuttalliana*

1. Puccinellia grandis Swallen, Jour. Wash. Acad. Sci. 34:18–19. 1944.

Perennial; culms 3–9 dm. tall, densely tufted, erect or bent at the lower nodes; sheaths glabrous; blades firm, flat or involute, elongate, mostly 2–3.5 mm. wide; panicle 1–3 dm. long, pyramidal, the scabrous branches at first appressed but often at length stiffly spreading, usually naked at the base; spikelets 7–15 mm. long, 5- to 12-flowered, appressed, rather prominently tinged with purple; first glume 2–3 mm. long, obtuse or sometimes subacute; second glume 3–3.5 mm. long, broader than the first glume, obtuse, often minutely toothed; lemmas 3–4 mm. long, rather abruptly narrowed to an obtuse or subacute apex, sparsely pilose at the base; palea obscurely ciliate on the keels; anthers mostly 1.3–1.5 mm., rarely as much as 2 mm. long.

Wet saline soil and salt marshes along the coast from San Mateo County to Humboldt County; north to Alaska.

2. Puccinellia distans (L.) Parl., Fl. Ital. 367. 1848. Weeping alkali grass.

Perennial; culms erect or bent at base, 2–4 dm. tall, sometimes taller; blades flat or becoming involute, mostly 1.5–3 mm. wide; panicle pyramidal, loose, 5–10 cm. long, ovoid, loose, the branches fascicled, rather distant, the lowest ones spreading or finally reflexed, the longer ones naked for ½ of their length or more; spikelets pale or purplish, 4- to 6-flowered, 4–5 mm. long; first glume less than 1 mm. long, second glume 1.5–2 mm. long; lemmas rather thin, obtuse or truncate, 1.5 mm. or usually about 2 mm. long, with a few short hairs at base; anthers about 0.8 mm. long.

Established on moist, usually alkaline soil at widely scattered localities throughout California; native of Eurasia.

3. Puccinellia Nuttalliana (Schult.) Hitchc. in Jepson, Fl. Calif. 1:162. 1912.

Perennial; culms usually erect, slender, often glaucous, rather stiff and firm at base, mostly 3–6 dm., rarely to 1 m. tall; blades 1–3 mm. wide, flat or

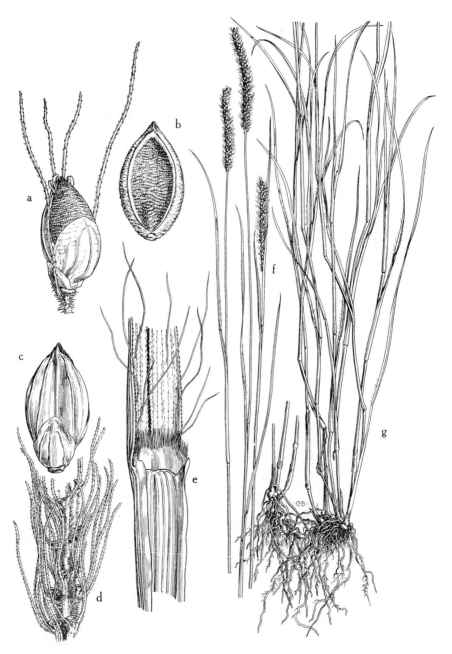

Fig. 84. *Setaria geniculata: a,* spikelet, showing fertile lemma and the few upwardly barbed bristles on branchlet, × 12; *b,* floret, showing palea, × 12; *c,* spikelet, showing first glume and sterile lemma, × 12; *d,* rachilla, the bristles remaining and the spikelets having fallen off from the branchlets, × 6; *e,* leaf sheath, showing long-ciliate ligule and the sparsely set, long hairs at base of the scabrous blade, × 6; *f,* habit, upper part, show-ing the slender linear panicles, × ⅖; *g,* habit, lower part, showing the knotty, branching rhizomes and the erect, ascending leaf blades, × ⅖.

becoming involute; panicle pyramidal, open, mostly 1–2 dm. long, the distant scabrous branches clustered, spreading, naked below, as much as 1 dm. long; spikelets 3- to 6-flowered, 4–7 mm. long, the florets rather distant, the rachilla often exposed; glumes mostly acutish, the first glume about 1 mm. long, the second about 2 mm. long; lemmas 2–3 mm. long, somewhat narrowed into an obtuse, ciliate-fringed apex, slightly pubescent at base; anthers about 0.7 mm. long.

Moist alkaline lands east of the Sierra Nevada crest from Mono County to Modoc County and thence west to Siskiyou County; middle-western and western states to southern Canada.

A related but apparently rarer species in California is *Puccinellia Lemmonii* (Vasey) Scribn. It occupies a similar range and occurs in alkaline soil, but differs from *P. Nuttalliana* in having longer anthers (1.5–2 mm. long), leaves mostly involute and in a tuft at the base of the culm, and lemmas somewhat acute.

SETARIA. BRISTLE GRASS, MILLET

Annuals or perennials. Panicles narrow, terminal, in our species dense and spikelike. Spikelets subtended by 1 to several bristles (sterile branchlets), falling free from the bristles, awnless. First glume broad, usually less than ½ as long as the spikelet, 3- to 5-nerved; second glume and sterile lemma equal in size, or the glume shorter, several-nerved. Fertile lemma thick and rigid, smooth or transversely wrinkled.

Setaria viridis (L.) Beauv., not treated below, is a species introduced from Europe and sparingly established as a weed in California, but rather frequent in the rice fields of the northern part of the state.

Bristles below each spikelet numerous.
    Plants annual; panicle 5–10 mm. broad .... 1. *S. lutescens*
    Plants perennial from short, knotty rhizomes; panicle 3–5 mm. broad .. 2. *S. geniculata*
Bristles below each spikelet 1–3 .... 3. *S. verticillata*

1. Setaria lutescens (Weigel) F. T. Hubb., Rhodora 18:232. 1916. Yellow bristle grass. Fig. 83.

    Panicum glaucum of authors, not L.; *Chaetochloa glauca* of American authors.
    Chaetochloa lutescens Stuntz, U. S. Dept. Agr., Bur. Pl. Indus. Inventory Seeds 31:36, 86. 1914.

Annual, branching at the base; culms erect to prostrate, mostly 5–10 dm. tall, compressed; sheaths keeled; blades as much as 2.5 dm. long and 1 cm. wide, twisted in a loose spiral, villous toward the base on the adaxial surface; panicle dense, evenly cylindric, spikelike, yellow at maturity, mostly 4–10 cm. long, about 5–10 mm. thick, the axis densely pubescent; bristles 5–20 in a cluster, the longer ones 2–3 times as long as the spikelet; spikelets 2 or 3 mm. long.

Fig. 85. *Setaria verticillata: a,* spikelet, showing short first glumes, and the single, down-wardly barbed bristles on branchlets, × 16; *b,* floret, showing lemma, × 16; *c,* auricled leaf sheath and ciliate ligule, × 4; *d,* panicle, × ⅘; *e,* floret, showing palea, × 16; *f,* habit, upper part, showing panicles, × ⅕; *g,* habit, lower part, showing the lax, arching leaf blades and roots at the nodes, × ⅕.

Established as a weed in waste places, in fields, and along ditches at scattered localities usually at low elevations: coastal southern California, Central Valley, and near the coast from Santa Clara County to Humboldt County, also Inyo County; native of Europe.

2. Setaria geniculata (Lam.) Beauv., Ess. Agrost. 51, 169, 178. 1812. Knotroot bristle grass. Fig. 84.

Setaria gracilis H.B.K., Nov. Gen. & Sp. 1:109. 1815.
Chaetochloa geniculata (Lam.) Millsp. & Chase, Field Mus. Nat. Hist., Bot. Ser. 3:37. 1903.

Perennial, with short knotty branching rhizomes to 4 cm. long; culms 3–10 dm. tall, slender, compressed, often bent at base; blades 1–3 dm. long, 3–7 mm. wide; panicle slender, linear, 4–10 cm. long, 3–5 mm. thick; bristles yellowish, 5–7 in a cluster, 1–3 times or even 6 times as long as the spikelet; spikelets 2–2.5 mm. (sometimes 3 mm.) long.

Established as a weed at scattered localities on moist or dry sites in open, waste places at low elevations: coastal southern California, Central Valley, and near the coast as far north as Marin County; Western Hemisphere.

3. Setaria verticillata (L.) Beauv., Ess. Agrost. 51, 178. 1812. Fig. 85.

Chaetochloa verticillata Scribn., U. S. Dept. Agr., Div. Agrost. Bull. 4:39. 1897.

Plants annual; culms to 1 m. long, simple or more often much-branched at base, geniculately spreading and rooting at nodes; leaf blades flat, thin, scabrous and sparingly pilose, 5–10 mm. wide, 10–20 cm. long; panicle erect, not rigid, slightly tapered, sometimes interrupted at base, 5–15 cm. long, 7–15 mm. thick; bristles 1 below each spikelet, retrorsely scabrous and 1–3 times as long as the spikelet; spikelet 2 mm. long; fruit finely rugose.

Along ditch banks and in waste places: scattered localities in California; native of Europe.

SORGHUM

1. Sorghum halepense Pers., Syn. Pl. 1:101. 1805. Johnson grass.

Perennial with extensively creeping, scaly rhizomes; culms erect, stout, glabrous, 5–15 dm. tall; blades flat, 6–20 mm. wide, the midrib prominent, white; panicle open, 1.5–3 dm. long; spikelets usually in pairs; one spikelet sessile and fertile, 4.5–5.5 mm. long, ovate, the glumes appressed-silky, becoming partly glabrate and shiny, the lemma with a bent, readily deciduous awn 1–1.5 cm. long; sterile spikelet pedicellate, well developed, usually staminate, lanceolate, 5–7 mm. long, the pedicels about 3 mm. long, densely pubescent, the glumes membranaceous, nerved, glabrous or sparsely pubescent.

A common forage plant cultivated throughout the southeastern United States but considered a noxious weed in California, where it has become well established along irrigation ditches and on cultivated land. Native of the Mediterranean region.

Sorghum vulgare Pers., the common sorghum of cultivation, is sometimes found as an escape and is cultivated as a food plant for ducks in Imperial County. Another member of the genus, *S. sudanensis,* Sudan grass, occurs in California as a volunteer on wet lands.

SPARTINA. CORD GRASS

1. Spartina foliosa Trin., Mém. Acad. St. Pétersb. VI. Sci. Nat. 4:114. 1840. California cord grass.

Perennial with extensively creeping scaly rhizomes; culms 3–12 dm. tall, stout, as much as 1.5 cm. thick at base, somewhat spongy, usually rooting from the lower nodes; blades 8–12 mm. wide at the flat base, gradually narrowed to a long, involute tip, smooth throughout; inflorescence dense, spikelike, 10–25 cm. long; spikes numerous, approximate, closely appressed, 3–5 cm. long; spikelets 1-flowered, very flat, 9–12 mm. long or occasionally longer, sessile and closely imbricate, disarticulating below the glumes; glumes firm, glabrous or hispid-ciliate on the keel, acute, the first glume narrow, ½–⅔ as long as the second, smooth, the second glume sparingly hispidulous and striate-nerved; lemma hispidulous on sides, mostly smooth on keel, shorter than the second glume; palea thin, longer than the lemma.

In salt marshes and on tidal flats from San Diego County to Del Norte County; Baja California, Mexico.

Spartina gracilis Trin., a species of moist alkaline lands known at present only from Inyo County, has distinct spikelets not aggregated into a dense, spikelike inflorescence as in *S. foliosa.*

ZIZANIA. WILD RICE

1. Zizania aquatica L. var. **angustifolia** Hitchc., Rhodora 8:210. 1906. Northern wild rice.

Tall annual; culms robust, to 1.5 m. long, often long-decumbent at base and rooting at the nodes, spongy, but usually thickened at the nodes; sheaths glabrous, somewhat inflated above; blades flat, 5–12 (or –15) mm. wide, densely pubescent at the base on both surfaces and on the nodes, otherwise minutely scabrous; ligules 5–10 mm. long, ovate, hyaline, acute or somewhat lacerate at the summit; panicles large, 3–5 dm. long, terminal, monoecious, the lower branches ascending or spreading, bearing 1–15 pendulous, reddish staminate spikelets on short, capillary pedicels, the upper branches ascending, at maturity erect, bearing 2–6 appressed pistillate spikelets on short, club-shaped pedicels;

the staminate spikelets early-deciduous and the pistillate spikelets tardily deciduous; spikelets 1-flowered, disarticulating from the pedicel; glumes obsolete, represented by a small, collar-like ridge; pistillate spikelet terete, angled at maturity, 4.5–8 cm. long, bearing a long, bristle-like awn 2.5–6 cm. long (the body of the spikelet 2–3 cm. long); pistillate lemma rather firm and tough, strawlike with a somewhat lustrous, glabrous surface, appressed-scabrous on the margins, at the base and summit, along the awn, and sometimes on the 3 nerves, the lemma closely clasping the palea by a pair of strong lateral nerves; caryopsis narrowly cylindrical, about 1.5 cm. long, pale brown to dark brown.

Known only from a pasture in Indian Valley, Plumas County, where a small colony was found growing in a canal in six to twelve inches of water in association with *Typha latifolia* and *Sparganium eurycarpum*. Probably introduced from eastern North America.

Zizania aquatica L. var. *angustifolia* is a native of the northeastern United States and ranges westward to northern Indiana. The typical form, which has a similar range, differs from the variety chiefly in spikelet characters (N. C. Fassett, A study of the genus *Zizania,* Rhodora 26:153–160, 1924). In the type, pistillate lemmas are rather thin and papery and are minutely scabrous over the entire surface, and the aborted spikelets are very slender and shriveled, without a definite body.

CYPERACEAE. SEDGE FAMILY

Grasslike or rushlike herbs, annuals with fibrous roots, or perennials with rhizomes. Culms (stems) solid, or rarely hollow, terete to variously angled. Leaves mainly basal, alternate, commonly 3-ranked, the blades narrow, and the sheaths closed. Inflorescence commonly subtended by 1 to several involucral leaves. Inflorescence a simple or compound umbel or raceme, or a capitate cluster, of spikelets, or a terminal spike. Spikelets and spikes composed of a series of scales (bracts) subtending individual florets. Scales spirally imbricated or 2-ranked, persistent or deciduous. Florets perfect or unisexual. Perianth represented by several bristles, or by an inner scale, or absent. Stamens 1–3. Pistil 1, the ovary superior, 1-celled, with 1 ovule, and a single bifid or trifid style. Fruit a lenticular or trigonous achene. Embryo minute. Endosperm mealy.

Florets perfect, or perfect and staminate; achene naked.
  Scales of the spikelet all bearing achenes, or empty basal scales not more than 2; florets
    all perfect.
    Scales of the spikelet spirally imbricated.
      Bristles present.
        Bristles much exserted beyond the scales.
          Bristles smooth and white .... *Eriophorum*
          Bristles barbed .... *Scirpus criniger*

Bristles included within the scales.
   Style base persistent as a tubercle on the summit of the achene; involucral leaves wanting .... *Eleocharis*
   Style base deciduous from the summit of the achene, tubercle wanting; 1 or more involucral leaves present (absent in *S. clementis*) .... *Scirpus*
Bristles absent.
   Hyaline scale present between floret and rachis of spikelet .... *Hemicarpha*
   Hyaline scale absent.
      Style base swollen .... *Fimbristylis*
      Style base not swollen .... *Scirpus koilolepis, S. cernuus*
Scales of the spikelet 2-ranked.
   Bristles present; achene beaked .... *Dulichium*
   Bristles absent; achene not beaked .... *Cyperus*
Basal 3 to several scales of the spikelet empty; spikelet consisting of both staminate and perfect florets.
   Bristles present; inflorescence a capitate cluster or close cyme.
      Style base deciduous from the summit of the achene; achene trigonous .. *Schoenus*
      Style base persistent as a tubercle on the achene; achene lenticular
                                      Rynchospora
   Bristles absent; inflorescence a loose compound umbel .... *Cladium*
Florets all unisexual; achene surrounded by a saclike perigynium .... *Carex*

CAREX

Grasslike perennials with rhizomes. Culms (stems) mostly triangular, the flowering culms either with lower leaves bladeless or nearly so or with well-developed blades. Leaves 3-ranked, the upper ones (bracts) subtending the spikes, elongate or short or sometimes wanting. Plants monoecious (or rarely dioecious), the flowers arranged in spikes of many several-ranked scales. Spikes 1 to many, either wholly pistillate, wholly staminate, or partly staminate and partly pistillate. Staminate flowers with 3 (rarely 2) stamens. Pistillate flowers with a single pistil with 2- or 3-cleft style, the pistil surrounded by a saclike organ, the perigynium. Achene hard, lenticular, triangular or plano-convex, completely surrounded by the perigynium.

Mackenzie, K. K., Cyperaceae–Cypereae. N. Am. Fl. 18:1–478. 1931–1935.

Achenes lenticular or plano-convex; stigmas 2.
   Spikes elongate, linear to lanceolate (occasionally oblanceolate), well separated or at least not congested.
   Bracts prominently sheathing.
      Perigynia rounded or truncate at apex, 2.5–3 mm. long .... 1. *C. Hassei*
      Perigynia short-tapering at apex, 2.5–3.75 mm. long .... 2. *C. salinaeformis*
   Bracts sheathless, or the lowest sometimes sheathing.
      Achenes constricted in the middle from one or both sides.
         Lower leaf sheaths strongly filamentose; perigynia shining brown
                                      3. *C. obnupta*
         Lower leaf sheaths little or not at all filamentose; perigynia dull light brown
                                      4. *C. Lyngbyei*

Achenes not constricted in the middle.
  Lower or middle sheaths breaking and becoming filamentose.
    Beak of perigynium bidentate; scales strongly rough-awned .. 5. *C. Barbarae*
    Beak of perigynium entire or merely emarginate; scales not rough-awned.
      Leaf blades 5–12 mm. wide; pistillate spikes 5–20 cm. long .. 6. *C. Schottii*
      Leaf blades 2–6 mm. wide; pistillate spikes 5 cm. long or less .. 7. *C. senta*
  Lower sheaths not breaking and becoming filamentose.
    Perigynia nerveless ventrally or with obscure, impressed nerves
                                             8. *C. sitchensis*
    Perigynia conspicuously nerved or ribbed ventrally, the nerves raised.
      Perigynia coriaceous, the beak bidentate .... 9. *C. nebraskensis*
      Perigynia membranaceous, the beak entire .... 10. *C. Hindsii*
Spikes short, ovoid to oblong, densely congested or at least contiguous.
  Culms arising singly or a few together from long-creeping rhizomes.
    Rhizomes slender, light brown.
      Perigynium tapering into a beak nearly 2 mm. long .... 11. *C. Douglasii*
      Perigynium abruptly narrowed into a very short beak 0.25–0.5 mm. long
                                      12. *C. simulata*
    Rhizomes stout, brownish black to black.
      Scales dark chestnut brown; perigynia 2.5–3.5 mm. long ... 13. *C. praegracilis*
      Scales lighter-colored; perigynia 3–4.5 mm. long .... 14. *C. pansa*
  Culms caespitose or from short rhizomes.
    Perigynia narrowly to broadly wing-margined.
      Leaf sheaths green-striate on adaxial surface, hyaline only at mouth
                                        15. *C. feta*
      Leaf sheaths white-hyaline on adaxial surface.
        Lowest 1 or 2 bracts elongated, usually extending conspicuously beyond the
          head, dilated and many-striate at base .... 16. *C. athrostachya*
        Lowest 1 or 2 bracts shorter than or occasionally extending beyond the head.
          Perigynia nerveless or obscurely nerved ventrally .... 17. *C. sub-bracteata*
          Perigynia prominently nerved ventrally.
            Scales shorter than perigynia, exposing most of perigynia above; spikelets
              congested in a dense, ovate head .... 18. *C. Harfordii*
           Scales about as long as perigynia, concealing perigynia above or nearly
              so; spikelets loosely approximate in a slender, oblong head
                                  19. *C. Tracyi*
    Perigynia sharp-edged or thick-margined.
      Sheaths finely wrinkled adaxially.
        Perigynia lanceolate, the beak longer than or of nearly the same length as the
          body (2–2.5 mm. long) .... 20. *C. stipata*
        Perigynia ovate to ovate-lanceolate, the beak about ½ as long as body (1–2
          mm. long).
          Perigynia 3.5–4.5 mm. long, strongly nerved ventrally ...... 21. *C. densa*
          Perigynia 3–3.5 mm. long, nerveless or nearly so ventrally .. 22. *C. vicaria*
      Sheaths not finely wrinkled adaxially.
        Sheaths red- or purplish-dotted adaxially; perigynia ovate.
          Perigynia white-punctate, greenish-straw-colored or brownish at maturity
                                  23. *C. arcta*
          Perigynia not white-punctate, brown or brownish black at maturity.
            Leaf blades 2–6 mm. wide .... 24. *C. Cusickii*
            Leaf blades 1–2.5 mm. wide .... 25. *C. diandra*

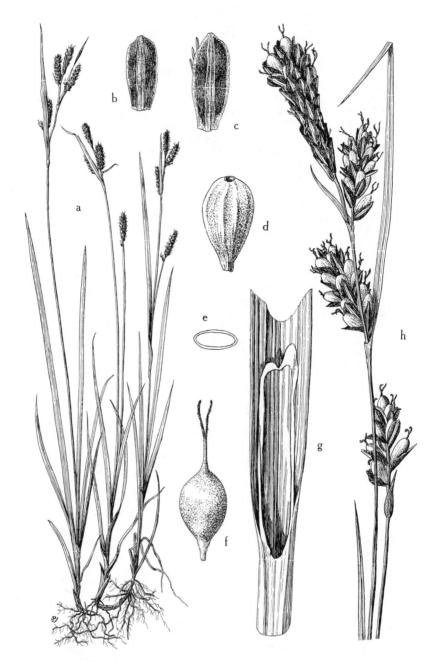

Fig. 86. *Carex Hassei: a,* habit, showing separate pistillate and staminate spikes, the staminate spikes above, × ⅖; *b,* scale of pistillate flower, × 6; *c,* scale of staminate flower, × 6; *d,* perigynium, rounded at apex, × 12; *e,* achene (cross section), × 12; *f,* pistillate flower with perigynium removed, × 12; *g,* ligule, × 8; *h,* inflorescence, the terminal spike staminate at base and pistillate at apex, the other spikes entirely pistillate, × 3.

Sheaths not red- or purplish-dotted adaxially; perigynia ovate-lanceolate, 3.75–4.5 mm. long .... 26. *C. phyllomanica*
Achenes triangular; stigmas 3.
Spike solitary, staminate above .... 27. *C. leptalea*
Spikes more than 1.
  Style jointed with achene, deciduous.
    Perigynia densely pubescent; staminate spikes usually 2, the lower one much shorter .... 28. *C. lanuginosa*
    Perigynia glabrous, or sometimes sparsely pubescent, if pubescent then staminate spike solitary (sometimes with a few perigynia).
      Lowest bract long- or short-sheathing; terminal spike usually staminate.
        Rhizomes with long, creeping stolons.
          Spikes 1–4 cm. long, the perigynia few (5–25) to a spike; leaves 0.5–5 mm. wide.
            Perigynium beakless, but with slightly pointed or rounded apex 29. *C. livida*
            Perigynium strongly beaked, the orifice oblique ...... 30. *C. californica*
          Spikes 3.5–14 cm. long, the perigynia numerous; leaves 8–18 mm. wide 31. *C. amplifolia*
        Rhizomes with short, ascending stolons, or not stoloniferous.
          Perigynium ciliate-serrulate on margin above middle and on beak.
            Scales dark reddish brown .... 32. *C. luzulina*
            Scales hyaline, yellowish-tinged .... 33. *C. sonomensis*
          Perigynium smooth on margin and beak, not serrulate.
            Pistillate spikes 5–10 mm. long, short- to long-peduncled 34. *C. viridula*
            Pistillate spikes 10–35 mm. long, short- to long-peduncled.
              Leaf blades 3–10 mm. wide .... 35. *C. gynodynama*
              Leaf blades 1.75–3.5 mm. wide .... 36. *C. mendocinensis*
      Lowest bract sheathless; terminal spike pistillate above ..... 37. *C. Buxbaumii*
  Style continuous with the achene, persistent as a rigid beak.
    Basal sheaths pubescent.
      Perigynia glabrous .... 38. *C. atherodes*
      Perigynia pubescent .... 39. *C. Sheldonii*
    Basal sheaths glabrous.
      Perigynia nerveless, except for the marginal nerves .... 40. *C. spissa*
      Perigynia several- to many-nerved.
        Lower pistillate spikes more or less nodding on long, slender peduncles.
          Teeth of perigynium beak 0.5–1 mm. long .... 41. *C. hystricina*
          Teeth of perigynium beak 1.25–2 mm. long, recurved or spreading 42. *C. comosa*
        Lower pistillate spikes erect, sessile or short-peduncled, not nodding.
          Perigynium abruptly contracted into beak, 3–8 mm. long.
            Basal sheaths breaking and becoming more or less filamentose; perigynia ascending at maturity .... 43. *C. vesicaria*
            Basal sheaths little, if at all, filamentose; perigynia spreading at maturity 44. *C. rostrata*
          Perigynium tapering into beak, 7–10 mm. long .... 45. *C. exsiccata*

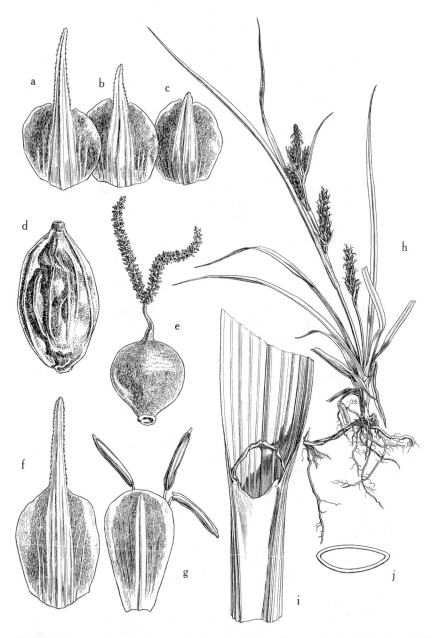

Fig. 87. *Carex salinaeformis: a–c,* scales of pistillate flowers, those at the base of the spike having long, scabrid awn, × 8; *d,* perigynium, × 12; *e,* pistillate flower with perigynium removed, × 12; *f,* scale of staminate flower from lower part of spike, showing the long, scabrid awn, × 8; *g,* staminate flower from upper part of spike, the subtending scale obtuse, × 8; *h,* habit, showing the leaves overtopping the culm, × ⅔; *i,* short, emarginate ligule, × 8; *j,* achene (cross section), × 12.

1. Carex Hassei Bailey, Bot. Gaz. 21:5. 1896. Fig. 86.

Rhizomes slender; culms 1–6 dm. tall, sharply triangular, roughened above, overtopping the leaves; leaf blades flat, channeled, 2–4 mm. wide; staminate spike terminal, solitary, peduncled, 6–15 mm. long, often pistillate at apex; pistillate spikes 2–4, linear-oblong, 8–20 mm. long, the upper spikes approximate and short-peduncled, the lower ones strongly separate and long-peduncled; scales broadly or narrowly ovate, the tips obtuse to acute or acuminate or often aristate with a scabrid awn, all these variations occurring within the same spike, the scales reddish-brown-tinged with green center and narrow hyaline margins; lowest bract extending beyond the tip of the culm; perigynia obovoid, at first greenish or straw-colored, becoming whitish and minutely granular.

Along streams, in wet meadows, and in bogs: northern California, south in mountains to Tulare County, rare and local along coast from Humboldt County to Monterey County, Panamint Mountains, and mountains of southern California; north to Alaska and across continent to Atlantic.

2. Carex salinaeformis Mackenzie, Bull. Torrey Club 36:477. 1909. Deceiving sedge. Fig. 87.

Closely related to *Carex Hassei* Bailey, from which it is distinguished by its slightly larger perigynia, which taper slightly at the apex and are yellowish green or straw-colored; in general, the plants are smaller than those of *C. Hassei* and the leaves usually overtop the culms; the pistillate scales consistently have long or short scabrid awns. The species is doubtfully distinct from *C. Hassei,* and perhaps should be considered as a variety, but this cannot be determined from the limited number of collections at present available.

Near the coast in Humboldt, Mendocino, Sonoma, and Santa Cruz counties.

3. Carex obnupta Bailey, Proc. Calif. Acad. Sci. II. 3:104. 1891. Slough sedge. Fig. 88.

Plants forming large clumps with long, stout rhizomes; culms 3–15 dm. tall, sharply triangular; lower leaves bladeless or nearly so; leaf blades 3–8 mm. wide, channeled above and keeled toward base; upper 2 or 3 spikes staminate, often curving; pistillate spikes 2–4, erect, spreading or strongly drooping, the lower spikes short-peduncled (or sometimes long-peduncled), the upper ones usually nearly sessile, oblong to usually linear-cylindric, 3–10 cm. long, usually staminate at apex, the scales ovate-lanceolate, acute to acuminate, dark brown to purplish black with lighter center and narrow hyaline margins, longer than perigynia; lowest bract leaflike, extending beyond the tip of the culm; perigynia oval-ovoid, 2–3.5 mm. long, yellowish green when young, in age brown, smooth, marginally 2-ribbed, abruptly short-apiculate.

Marshes and wet meadows: San Luis Obispo County to Del Norte County; north to British Columbia. Jepson reports it as "local in the Sacramento Valley," but we have not seen any authentic material from this area.

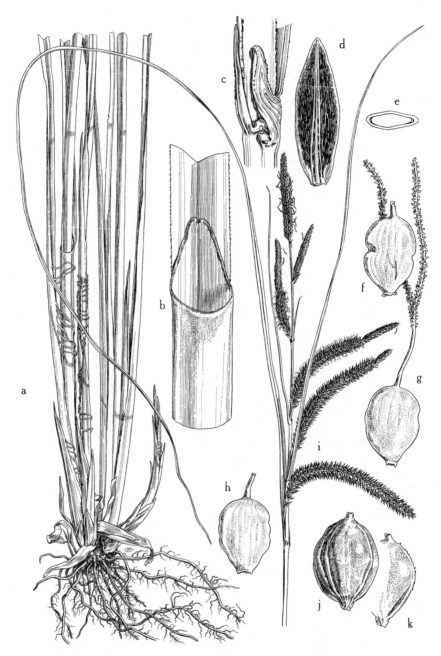

Fig. 88. *Carex obnupta: a,* habit, showing base of plant with filamentose sheaths of sterile shoots, × ⅖; *b,* ligule, × 2; *c,* auricles of bract subtending the inflorescences, × 4; *d,* scale, alike in staminate and pistillate flowers, × 8; *e,* achene (cross section), × 8; *f–h,* achenes, showing variation in shape, × 8; *i,* habit, showing upper part of culm with the lower, curving, pistillate spikes (some with staminate tips), the upper staminate spikes, and the long subtending bracts, × ⅖; *j* and *k,* perigynia, showing smooth and toothed shoulder, × 8.

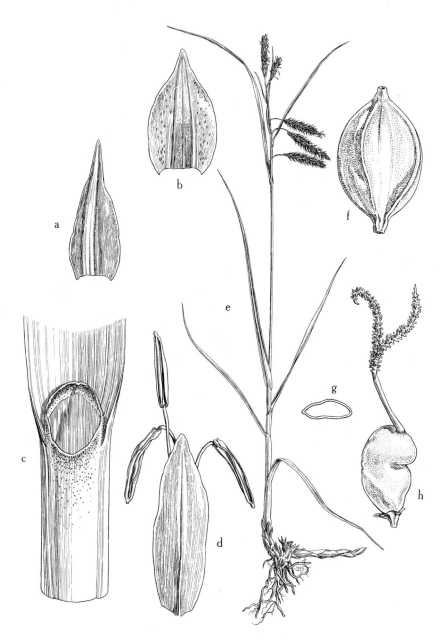

Fig. 89. *Carex Lyngbyei: a,* scale of pistillate flower in upper part of spike, × 8; *b,* scale of pistillate flower in lower part of spike, × 8; *c,* ligule, × 6; *d,* staminate flower and subtending scale, × 8; *e,* habit, showing the flat leaf blades, drooping pistillate spikes, and erect staminate ones at apex, × ⅖; *f,* perigynium, strongly nerved, ovoid, × 12; *g,* achene (cross section), × 12; *h,* pistillate flower, showing achene constricted on one side and style arising at angle, × 12.

Fig. 90. *Carex Barbarae: a,* perigynium, showing the short-toothed beak, $\times$ 8; *b,* scale of pistillate flower, showing rough-awned apex, $\times$ 8; *c,* pistillate flower with perigynium removed, $\times$ 8; *d,* achene (cross section), $\times$ 8; *e,* ligule, $\times$ 3; *f,* upper part of culm, showing subtending bracts, 2 staminate spikes at apex (the lower staminate spike partly pistillate), and the pistillate spikes below, $\times$ ⅖; *g,* base of plant, showing the flat, long-attenuate leaf blades, channeled and keeled toward base, $\times$ ⅖; *h,* staminate flower and subtending scale, $\times$ 8.

4. Carex Lyngbyei Hornem., Fl. Dan. Pl. 1888. 1827. Fig. 89.

Plants forming large clumps with long, stout rhizomes; culms varying from rather slender to very stout, 2.5–9 dm. tall, triangular; lower leaves bladeless or nearly so; leaf blades flat, 2–10 mm. wide; uppermost 1 or 2 spikes staminate, slender-peduncled; lateral spikes 2–4, the upper 1 or 2 often partly staminate above, the lower ones pistillate, drooping on slender peduncles, oblong-cylindric, 2–8 cm. long; the scales ovate to ovate-lanceolate, acute to acuminate, usually longer than the perigynia, reddish brown to brownish black, with lighter center and narrow hyaline margins; lowest bract leaflike, extending beyond the tip of the culm; perigynia oblong-obovoid or ovoid, 2.5–3 mm. long, obscurely to strongly nerved, minutely and abruptly apiculate, minutely granular.

Marshes near the coast: Marin, Mendocino, and Humboldt counties; north to Alaska and circumpolar.

5. Carex Barbarae Dewey in Torr., Mex. Bound. Surv. Bot. 231. 1859. Fig. 90.

Caespitose with long horizontal rhizomes; culms 3–10 dm. tall, stout below, slender but erect above; leaf blades light green, flat above, channeled and strongly keeled toward the base, 3–9 mm. wide, long-attenuate; staminate spikes 1 or 2, if 2, the lower one much shorter, sometimes partly pistillate below; pistillate spikes 3–5, the upper ones often staminate at apex, erect, sessile or nearly so, the lower spikes short-peduncled, oblong- or linear-cylindric, 2.5–8 cm. long; the scales ovate to ovate-lanceolate, usually rough-cuspidate or rough-awned, occasionally merely acute but usually hispid at apex, narrower and from slightly shorter to considerably longer than perigynia; lowest bract leaflike, varying from extending beyond to not reaching the tip of the culm; perigynia oval-orbicular to obovate, 3–4.5 mm. long, marginally 2-ribbed, strongly to obscurely several-nerved on both faces, straw-colored or brownish, red-dotted, puncticulate, rounded and stipitate at base, abruptly short-beaked, the teeth of beak hispidulous within.

Moist places along streams or on slopes, occasionally bordering marshes: mountains of coastal southern California, San Bernardino Mountains, Channel Islands, Coast Ranges from Ventura County north to Humboldt County, Central Valley, Sierra Nevada foothills from Fresno County north to El Dorado County, occasionally ascending to an elevation of 2,500 feet; southern Oregon.

6. Carex Schottii Dewey in Torr., Mex. Bound. Surv. Bot. 231. 1859. Fig. 91.

Caespitose in large clumps with long rhizomes; culms stout below, 1–1.5 m. tall, sharply triangular, very rough above; abaxial surface of lower leaf sheaths sparingly hispidulous, somewhat septate-nodulose and strongly keeled; staminate spikes about 3, 5–14 cm. long; pistillate spikes 3 or 4, the upper part usually staminate, sessile or nearly so, elongate-linear; scales narrowly lanceolate, obtuse or acute, narrower than but usually longer than perigynia, purplish

Fig. 91. *Carex Schottii: a,* perigynium, oval and flattened, × 12; *b,* scale of pistillate flower, showing the light center, × 8; *c,* pistillate flower with perigynium removed, × 12; *d,* achene (cross section), × 12; *e,* basal part of plant, showing filamentose lower sheaths, × ⅖; *f,* upper part of culm, the staminate spikes at apex, the pistillate spikes (with their tips staminate) below, the bracts subtending the spikes leaflike, × ⅖; *g,* staminate flower with subtending scale, × 8; *h,* ligule, × 2.

Fig. 92. *Carex senta: a,* staminate flower with subtending scale, × 10; *b* and *c,* scales of pistillate flowers, showing variation, × 10; *d,* pistillate flower with perigynium removed, × 12; *e,* achene (cross section), × 12; *f,* perigynium, broadly ovate, × 12; *g,* ligule, showing the long auricles, × 6; *h,* lower part of plant, showing filamentose lower sheaths, × ⅖; *i,* upper part of culm, with short subtending bracts, the lower spike pistillate, the upper spikes pistillate below and staminate above, × ⅖.

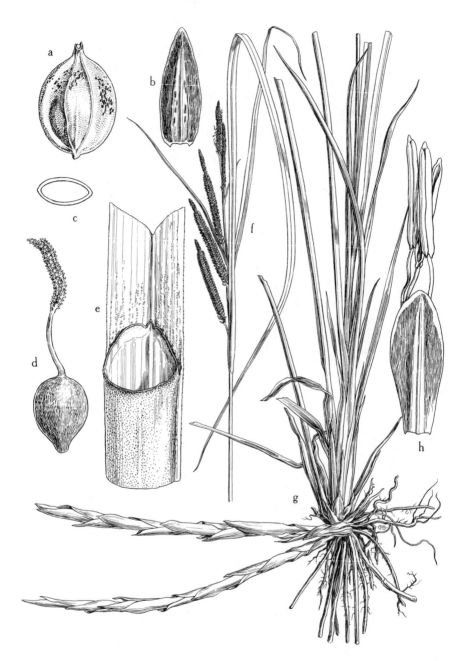

Fig. 93. *Carex sitchensis: a,* perigynium, ovate and puncticulate, × 10; *b,* scale of pistillate flower, × 10; *c,* achene (cross section), × 12; *d,* pistillate flower with perigynium removed, × 12; *e,* ligule, × 4; *f,* upper part of culm, showing the erect, linear-cylindric lower spikes, which are pistillate below and staminate at apex, and the upper staminate spikes, × ⅖; *g,* lower part of plant, showing the stout rhizomes, caespitose culms, and flat leaf blades channeled at the base, × ⅖; *h,* staminate flower and subtending scale, × 10.

black with lighter center; lowest bract leaflike, sheathless, usually extending beyond but sometimes not reaching the tip of the culm; perigynia flattened, obovate or oval, 2.5–3.5 mm. long, greenish-straw-colored, puncticulate, marginally 2-ribbed, somewhat strongly several-nerved on both faces, short-stipitate at base, abruptly very short-beaked.

Stream banks and marshes: San Diego and Riverside counties to Monterey County.

7. Carex senta Boott, Illus. Carex 4:174. 1867. Rough sedge. Fig. 92.

Closely related to *Carex Schottii* Dewey; the culms, however, more loosely caespitose, slender, 3–10 dm. tall; the leaf sheaths rounded, not sharply keeled dorsally; the pistillate spikes linear to oblong, 2.5–5 cm. long; the scales oblong-ovate or oblong-lanceolate (sometimes lanceolate), usually obtuse, but sometimes acute or short-awned, often hyaline-tipped, the narrow 1- to 3-nerved center usually not extending to apex, usually shorter but sometimes as long as or slightly longer than the perigynia; the perigynia broadly ovate or broadly obovate, the beak usually dark-tinged.

Stream banks and marshes: southern California north to Solano County, Sierra Nevada from Tulare County to Butte County.

8. Carex sitchensis Prescott, Mém. Acad. St. Pétersb. VI. Math. Phys. Nat. 2:169. 1832. Fig. 93.

Caespitose with long rhizomes; culms stout below, sharply triangular, 6–12 dm. tall; leaf blades flat, or channeled toward their base, light green or glaucous green, 3–9 mm. wide, usually septate-nodulose in lower parts of blade and on basal sheaths; leaf sheaths strongly purplish-black-tinged on adaxial side at mouth; staminate spikes 2–4; pistillate spikes 3–5, widely separate, linear-cylindric, 3–9 cm. long, on long slender peduncles (especially the lower spikes), erect, spreading or drooping, the upper pistillate spikes staminate above; scales lanceolate or ovate-lanceolate, acute or obtuse, narrower and from longer than to shorter than perigynia, purplish black with lighter center and white hyaline tip; lowest bract leaflike, extending beyond the tip of the culm; perigynia ovate or oval, 2.5–3.5 mm. long, light greenish or straw-colored in age, puncticulate, minutely granular, marginally 2-ribbed, abruptly apiculate above, the beak darkish-tipped, entire.

Marshes, mostly near the coast: Santa Cruz County to Del Norte County, ascending to an elevation of 4,500 feet in Humboldt County. Jepson (Fl. Calif. 1:236, 1922) states that it ranges east to Butte County, but we have not seen authentic specimens from any interior localities; north to Alaska.

9. Carex nebraskensis Dewey, Am. Jour. Sci. II. 18:102. 1854. Fig. 94.

Caespitose with long, stout, horizontal rhizomes; culms 2–10 dm. tall, papillate, sharply triangular, roughened or smooth above; leaf blades pale green, 3–8 mm. wide, flat, the lower sheaths usually prominently septate-

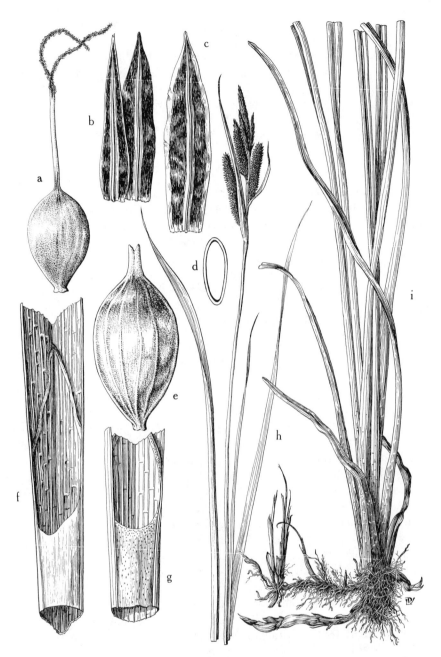

Fig. 94. *Carex nebraskensis: a,* pistillate flower with perigynium removed, × 12; *b,* scales of pistillate flowers, × 12; *c,* scale of staminate flower, × 12; *d,* achene (cross section), × 12; *e,* perigynium, flattened and strongly many-ribbed, × 12; *f* and *g,* ligules, sometimes punctate, × 6; *h,* habit, upper part of plant, showing the leaves, the culm, and the inflorescence with the spikes staminate above and pistillate below, the subtending bracts short, × ⅖; *i,* lower part of plant, showing the stout, horizontal rhizomes, × ⅖.

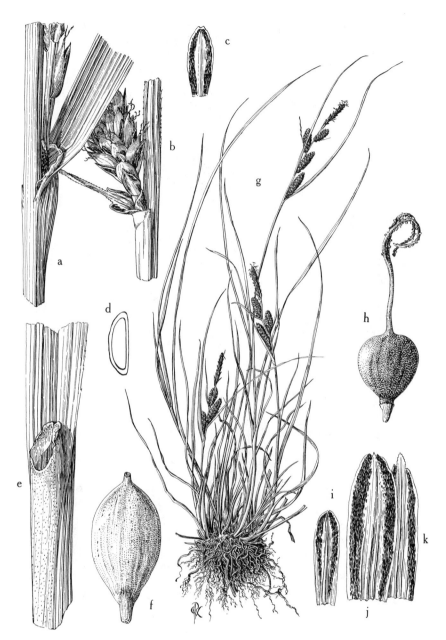

Fig. 95. *Carex Hindsii: a,* lowermost spike of inflorescence, showing leaflike bract and its ligule, × 4; *b,* upper pistillate spike of inflorescence, subtended by awned bract on outer side and spathelike bract on inner side, × 4; *c,* scale of pistillate flower, × 8; *d,* achene (cross section), × 12; *e,* ligule, 2-lobed, × 8; *f,* perigynium, stipitate at base, × 12; *g,* habit, showing the caespitose culms, subtending bracts, and the oblong spikes, the terminal spike staminate, × ⅖; *h,* pistillate flower with perigynium removed, showing stipitate and minutely papillate achene, × 12; *i,* scale of staminate flower from upper part of spike, × 8; *j* and *k,* scales of staminate flowers, from lower part of spike, × 8.

Fig. 96. *Carex Douglasii: a,* habit, staminate plant, × ⅔; *b,* habit, pistillate plant, × ⅔; *c,* staminate flower with subtending scale, × 8; *d,* scale of pistillate flower, × 8; *e,* perigynium, strongly nerved, abaxial view, × 8; *f,* pistillate flower with perigynium removed, × 8; *g,* achene (cross section), × 8; *h,* perigynium, lightly nerved, adaxial view, × 8; *i,* ligule, truncate with ciliate margin, × 8.

nodulose; terminal staminate spike 1.5–4 cm. long, often with 1 or 2 smaller ones near its base, the lateral ones sessile or short-peduncled; pistillate spikes 2–4, erect, the upper one sessile or nearly so, the lower ones short- or long-peduncled, all contiguous or the lower ones somewhat separate, oblong to cylindric, 1.5–5 cm. long; scales lanceolate, obtusish to acute or acuminate, narrower than and from shorter than to longer than perigynia, purplish or brownish black with lighter center and often with narrow hyaline margins; lowest bract leaflike, not sheathing, often dark- or light-auricled, varying from extending slightly beyond to not reaching the tip of the culm; perigynia flattened, oblong-ovate to broadly ovate or obovate, 3–3.5 mm. long, strongly many-ribbed on both faces, greenish to straw-colored or brownish at maturity, abruptly apiculate at apex, the beak often dark-tipped.

Montane meadows and marshes, ascending to an elevation of 10,500 feet: San Diego County, San Jacinto and San Bernardino mountains, Mount Pinos in Ventura County, Argus, Panamint, White, and Inyo mountains, Owens Valley, Sierra Nevada from Tulare and Inyo counties north to Plumas County; mountains of Shasta, Trinity, and Siskiyou counties, Modoc County, rare and local in Lake County and possibly Sonoma County; north to British Columbia, east to Kansas.

10. Carex Hindsii C. B. Clarke, Kew Bull. Misc. Inf., Add. Ser. 8:70. 1908. Fig. 95.

Culms caespitose, 1–5 dm. tall, slender but strict; leaf blades flat above, channeled toward base, 1.5–3 mm. wide, long-attenuate; terminal spike staminate, peduncled or nearly sessile; pistillate spikes 3–6, linear or oblong-linear, 1.5–3.5 cm. long; scales oblong-lanceolate, obtuse or acutish, purplish black with lighter center usually not extending to apex, much shorter and narrower than perigynia; lowest bract leaflike, extending beyond tip of culm; perigynia ovate, flattened, 2.5–3.5 mm. long, about 5-ribbed on both sides, yellowish green, minutely papillate, more or less slenderly stipitate at base, tapering or contracted into a dark-tipped beak.

Local in wet places along the coast: Mendocino, Humboldt, and Del Norte counties; north to Alaska.

11. Carex Douglasii Boott in Hook., Fl. Bor. Am. 2:213, pl. 214. 1839. Fig. 96.

Rhizomes 1–2 mm. thick, tough; culms 6–30 cm. tall, slender but stiff, obtusely triangular, smooth, usually overtopping the leaves, but sometimes shorter; leaf blades 1–2.5 mm. wide, involute above and flat or channeled toward base; heads usually dioecious, the many spikes closely aggregated, but usually distinguishable; pistillate heads suborbicular to oblong, 1.5–5 cm. long, 1–2.5 cm. thick, the scales yellowish brown with broad hyaline margins and green center, acuminate to cuspidate, concealing perigynia; staminate

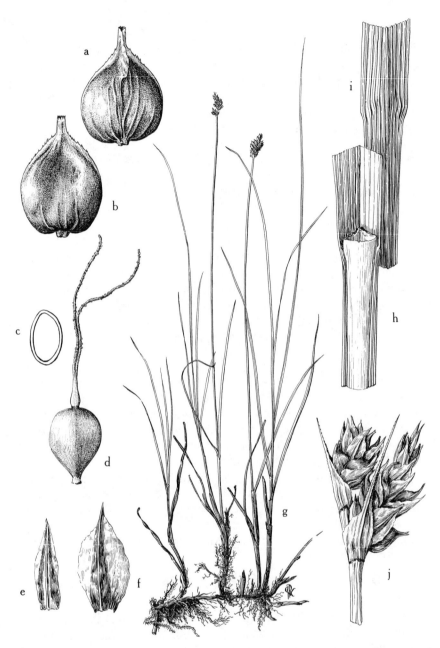

Fig. 97. *Carex simulata: a,* perigynium, abaxial view, × 12; *b,* perigynium, adaxial view, × 12; *c,* achene (cross section), × 12; *d,* pistillate flower with perigynium removed, × 12; *e,* scale of staminate flower, × 12; *f,* scale of pistillate flower, × 12; *g,* habit, showing bractless pistillate heads, × ⅖; *h,* ligule (on adaxial side of blade), × 8; *i,* ligular region of leaf, abaxial view, × 8; *j,* pistillate inflorescence, showing subtending bracts, × 4.

heads similar but somewhat narrower; lowest bract short-cuspidate, not extending beyond tip of culm; perigynia appressed-ascending, ovate-lanceolate, 3–4.5 mm. long, straw-colored to brownish, plano-convex, coriaceous, lightly nerved ventrally, strongly nerved dorsally, rounded and short-stipitate at base, sharp-edged, serrulate above middle, the beak obliquely cut dorsally, in age bidentulate, the apex hyaline.

Wet meadows or dry alkaline flats: Sierra Nevada from Tulare and Inyo counties north to Plumas, Lassen, and Modoc counties, White Mountains (to elevation of 12,500 feet), local in Riverside, Ventura, Santa Barbara, and San Benito counties; Canada to New Mexico.

12. Carex simulata Mackenzie, Bull. Torrey Club 34:604. 1908. Short-beaked sedge. Fig. 97.

Culms 2.5–5.5 dm. tall, sharply triangular and roughened on the angles above, overtopping the leaves; leaf blades 2–4 mm. wide, flat or channeled, light green; spikes densely aggregated into a linear-oblong or oblong-ovoid head 12–25 mm. long, wholly pistillate, wholly staminate or pistillate and partly staminate above, the lower spikes distinguishable; pistillate scales concealing perigynia, cuspidate or short-awned, brown with narrow hyaline margin and prominent, lighter midvein; bracts absent, or if present then shorter than head, cuspidate and enlarged at base; perigynia ascending, unequally biconvex to plano-convex, broadly ovate, smooth, shining, coriaceous, yellowish brown to chestnut-colored, 1.75–2.25 mm. long, rounded and short-stipitate at base, sharp-edged, nerveless ventrally, slenderly few-nerved dorsally, the upper part of the body and beak serrulate (sometimes only sparingly so), the beak obliquely cut dorsally, its apex at length bidentulate, and slightly hyaline.

Wet meadows, swales, or marshes: chiefly on the east side of the Sierra Nevada from Tulare and Inyo counties north, occasional on west slope within this range, mountains of Shasta, Siskiyou, and Modoc counties, rare and local in Marin and Santa Cruz counties; western states.

13. Carex praegracilis W. Boott, Bot. Gaz. 9:87. 1884. Clustered field sedge. Fig. 98.

Culms 2–7.5 dm. tall, sharply triangular, usually overtopping the leaves, sometimes quite strongly; leaf blades 1–3 mm. wide, erect-ascending, flat or somewhat channeled, light green; basal leaf sheaths dark brown or blackish; spikes closely aggregated (but the lowest ones readily distinguishable and somewhat separate) in a linear-oblong or oblong-ovoid head 1–5 cm. long, 6–10 mm. thick; scales with conspicuous hyaline margins and lighter midvein, the lower scales cuspidate, the upper ones acuminate, concealing perigynia; bracts absent or the lower 1 or 2 present, usually shorter than head; perigynia plano-convex, ovate or ovate-lanceolate, straw-colored or at maturity brownish black, coriaceous, dull, nerveless ventrally, lightly several-nerved dorsally,

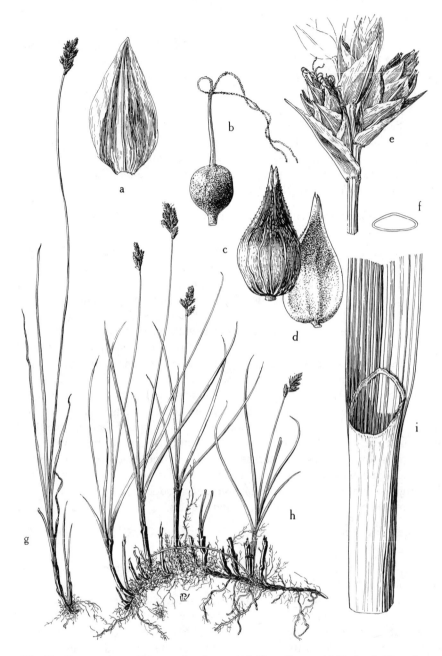

Fig. 98. *Carex praegracilis: a,* scale of upper pistillate flower, $\times$ 10; *b,* pistillate flower with perigynium removed, $\times$ 10; *c,* perigynium, lightly several-nerved, beak obliquely cut, abaxial view, $\times$ 10; *d,* perigynium, adaxial view, $\times$ 10; *e,* lower pistillate spikes with short subtending bracts, $\times$ 4; *f,* achene (cross section), $\times$ 10; *g,* habit, showing the erect-ascending leaf blades, $\times$ ⅖; *h,* habit, showing the dark basal sheaths and the culms extending above the leaves, $\times$ ⅖; *i,* ligule, $\times$ 10.

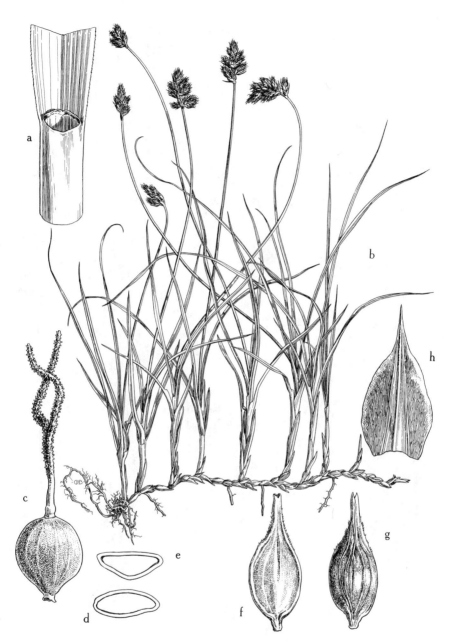

Fig. 99. *Carex pansa: a,* ligule, × 8; *b,* habit, showing rhizome, the recurved-spreading leaf blades, and the closely aggregated pistillate spikes with inconspicuous staminate flowers at apex, × ⅖; *c,* pistillate flower with perigynium removed, showing longitudinally ribbed achene, × 12; *d* and *e,* achenes (cross section), × 12; *f,* perigynium, adaxial view, × 8; *g,* perigynium, abaxial view, × 8; *h,* scale of pistillate flower (those of staminate flowers similar), × 6.

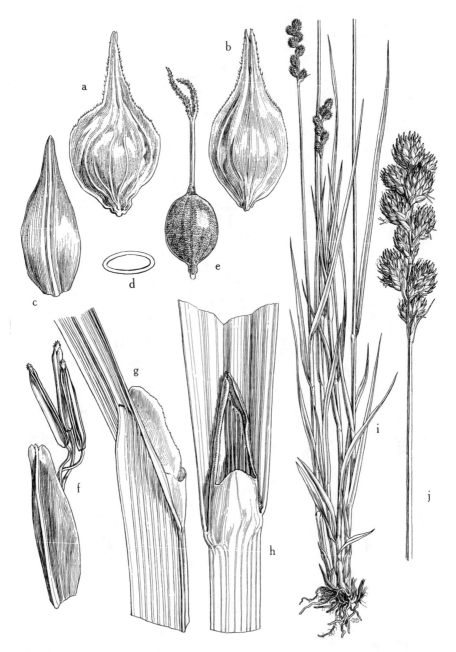

Fig. 100. *Carex feta: a,* perigynium, showing wings, which are serrulate above middle, adaxial view, × 12; *b,* perigynium, abaxial view, × 12; *c,* scale of pistillate flower, × 12; *d,* achene (cross section), × 12; *e,* pistillate flower, showing stipitate achene, × 12; *f,* staminate flower, showing the short, rough awn of the subtending scale, × 12; *g* and *h,* ligule, bilobed, the leaf blade set at an angle, × 6; *i,* habit, showing the short basal leaf blades, the longer culm leaves, and the inflorescences, × ²⁄₅; *j,* inflorescence, showing the short subtending bract and the separate, globose spikes, each one staminate below, pistillate above, × 1⅕.

Fig. 101. *Carex athrostachya: a,* ligule, bilobed, × 8; *b,* lowermost spike of an inflorescence, showing subtending bract with auricled, hyaline base and elongated, serrulate
midvein, and the inconspicuous staminate flowers at base of spike, × 6; *c,* achene (cross
section), × 12; *d,* pistillate flower, showing stipitate achene, × 12; *e* and *f,* scales, showing variation in size and shape, × 12; *g,* perigynium, adaxial view, × 12; *h,* perigynium,
abaxial view, × 12; *i,* habit, showing the slender culms and the closely aggregated spikes,
× ¼; *j,* inflorescence with auricled subtending bracts, the uppermost much reduced, × ⅘.

Fig. 102. (For explanation, see facing page.)

spongy and rounded at base, sharp-edged, serrulate above middle, tapering at apex into a beak ⅓–½ as long as the body, the beak serrulate, obliquely cut dorsally, at length bidentulate, hyaline at tip.

Wet meadows and streams: throughout California at middle to low elevations; western North America.

A widespread species which shows considerable variation throughout its range.

14. Carex pansa Bailey, Bot. Gaz. 13:82. 1888. Sand-dune sedge. Fig. 99.

Closely related to *Carex praegracilis* W. Boott and possibly representing a variant adapted to maritime conditions. The two are distinguished from other related species in having black basal sheaths and blackish stout rhizomes.

Carex pansa differs from *C. praegracilis* not only in respect to the characters noted in the key, but also in having shorter culms (1.5–3 dm. tall), leaf blades which tend to be more recurved-spreading, heads that are shorter (1–2.5 cm. long) and broader (1 cm. thick), and spikes that are closely aggregated and usually not separated (in the lower part of the heads).

Mostly on drifting sands or sandy flats, but sometimes at the edge of marshes along the seacoast: from Monterey County north to Del Norte County; north to Washington.

15. Carex feta Bailey, Bull. Torrey Club 20:417. 1893. Fig. 100.

Culms densely caespitose, 5–12 dm. tall; leaf blades flat, 2–4 mm. wide; head 2–6 cm. long, the spikes 5–12, aggregated or more or less separate, the staminate basal flowers inconspicuous; scales ovate, greenish-straw-colored or light brownish-tinged with hyaline margins and lighter midvein, obtuse or acute, shorter than perigynia; lowest bract setaceous, inconspicuous, not as long as the spikes; perigynia ovate, thickish, 3–4 mm. long, yellowish green with green beak, several-nerved dorsally, obscurely few-nerved ventrally, narrow-winged, serrulate above, tapering or somewhat abruptly narrowed into a flat, reddish-tipped beak about as long as the body.

Wet meadows and marshes of foothills and mountains, at elevations of 100–7,800 feet: San Bernardino Mountains, Sierra Nevada from Madera County to Plumas County and north in mountains of Shasta, Siskiyou, and Trinity counties, south in Coast Ranges to Marin County; San Joaquin Valley in Stanislaus County; north to British Columbia.

Fig. 102. *Carex sub-bracteata: a,* habit, showing the short-creeping rhizome, the flat leaf blades, and the closely aggregated spikes, × ⅖; *b,* pistillate flower with perigynium removed, showing the irregular longitudinal ribs on the achene, the stout, straight style, and the 2 long stigmas, × 10; *c,* achene (cross section), × 10; *d,* lowermost spike of inflorescence, showing the broad-based subtending bracts and the staminate flowers below the pistillate flowers, × 10; *e,* perigynium, abaxial view, showing the many strong veins and the oblique apex, the cut extending nearly to middle, × 10; *f,* perigynium, faintly few-nerved, adaxial view, × 10; *g,* scale of pistillate flower (that of the staminate flower similar), × 8; *h* and *i,* ligule, showing variation, × 6.

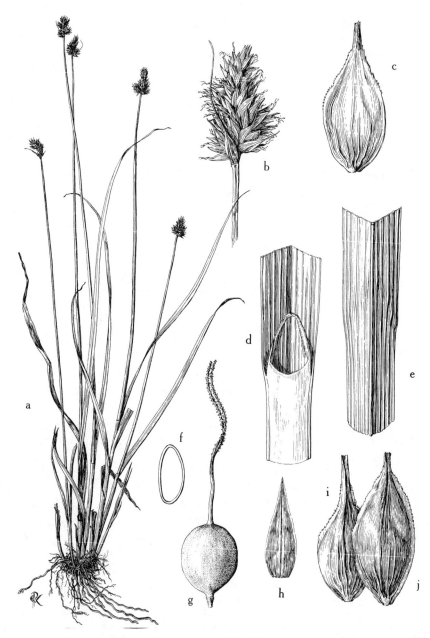

Fig. 103. *Carex Harfordii: a,* habit, × ⅖; *b,* inflorescence, × 4; *c,* perigynium, adaxial view, × 12; *d,* ligule, × 6; *e,* ligular region of leaf, abaxial view, × 6; *f,* achene (cross section), × 12; *g,* pistillate flower with perigynium removed, × 12; *h,* scale, alike in staminate and pistillate flowers, × 8; *i* and *j,* perigynia, showing variation in shape, abaxial view, × 12.

16. Carex athrostachya Olney, Proc. Am. Acad. Arts & Sci. 7:393. 1868. Fig. 101.

Culms caespitose, 1–8 dm. tall; leaf blades flat, 1–3 mm. wide, yellowish green; head ovoid, 1–2.5 cm. long, the spikes 4–20, closely aggregated, the staminate basal flowers inconspicuous; scales ovate or lanceolate-ovate, shorter than perigynia, acute or short-cuspidate, brownish with hyaline margins; perigynia ovate-lanceolate, 3–4 mm. long, light green, becoming straw-colored or brownish, substipitate, ciliate-serrulate above, tapering into a slender, terete, brownish-tipped beak, the margins of the orifice hyaline.

Chiefly wet meadows and copses: San Bernardino Mountains, Sierra Nevada from Tulare County to Lassen County, chiefly at elevations of 4,000–10,000 feet, Modoc County, west in Siskiyou Mountains to Del Norte County and south in Coast Ranges to Marin County at elevations of 300–5,000 feet; western Canada and Pacific states.

17. Carex sub-bracteata Mackenzie, Bull. Torrey Club 43:612. 1917. Fig. 102.

Caespitose, the rhizomes short-creeping, the culms 3–12 dm. tall; leaf blades flat, light green, 2–4 mm. wide; heads 1.5–2.5 cm. long, globose or ovoid, the spikes 5–10, closely aggregated, the staminate basal flowers inconspicuous; scales ovate, obtuse or acute, shorter than perigynia, reddish brown with lighter center and hyaline margins, more or less prominent midvein; perigynia ovate, 3.5–4.5 mm. long, dull green to yellowish brown, substipitate, narrowly margined, serrulate above middle, tapering somewhat abruptly into a beak ⅓ as long as the body, the beak subterete, chestnut-tinged, hyaline at orifice.

Wet meadows, moist woods, streams, and marshes near coast: Santa Barbara County north to Humboldt County, Santa Cruz Island.

This species is very similar to the widely distributed *Carex pachystachya* Cham., and further study may prove that the two species are not distinct.

18. Carex Harfordii Mackenzie, Bull. Torrey Club 43:615. 1917. Fig. 103.

This species is very closely related to *Carex sub-bracteata,* and the distinctions which are used to separate the two species are very slight.

In *C. Harfordii* the spikes are usually more numerous, the perigynia are strongly fine-nerved dorsally, several-nerved ventrally, and the beak of the perigynium is shorter (about ¼ as long as the body).

Brackish marshes near the coast: San Luis Obispo County to Humboldt County.

19. Carex Tracyi Mackenzie, Erythea 8:41. 1922.

Densely caespitose, the rhizomes short, black, fibrillose; culms 2–8 dm. tall, sharply triangular and roughened above, slender to base but strict; leaves with well-developed blades, 4–7 on the lower ⅓ of a fertile culm, the blades flat or channeled with revolute margins, 2–4 mm. wide; leaf sheaths tight, white-

Fig. 104. (For explanation, see facing page.)

hyaline adaxially, thin and prolonged at mouth; spikes 4–7, ovoid or short-oblong, well defined, approximate or slightly separate in a rather stiff, erect, oblong head 1.5–4 cm. long; scales ovate, acute, dull reddish brown; lowest bract short-cuspidate, the upper ones scale-like; perigynia 4–5 mm. long, plano-convex, ovate, with nearly orbicular body, wing-margined to base, serrulate to middle, strongly nerved dorsally and ventrally, abruptly contracted into a beak shorter than the body, serrulate below and terete, nearly smooth and with reddish brown tip.

Wet meadows and low swampy ground: Coast Ranges from Marin County to Humboldt County; north to British Columbia.

20. Carex stipata Muhl. in Willd., Sp. Pl. 4:233. 1805. Fig. 104.

Densely caespitose, the culms 3–12 dm. tall, stout, but weak and flattened in drying, the angles slightly winged and strongly serrulate above; leaf blades 4–8 mm. wide, flat, flaccid, the sheaths septate-nodulose on abaxial surface, the ligule conspicuous, prolonged above base of blade; head 3–10 cm. long, oblong-linear to ovoid, yellowish brown, the lower spikes slightly separate, the staminate basal flowers inconspicuous; scales ovate-triangular, light brownish with light midvein and hyaline margins, acuminate, cuspidate or short-awned, about as long as body of perigynia; lowest bract bristle-like, usually much shorter than inflorescence, or wanting; perigynia yellowish brown at maturity, sharp-edged to base, strongly nerved dorsally, less strongly nerved ventrally, round-cordate, spongy and abruptly stipitate at base, tapering into a serrulate, bidentate beak nearly equaling the body in length or longer, the teeth reddish-brown-tinged.

Marshes and wet meadows: Sonoma County to Del Norte and Siskiyou counties in the Coast Ranges, thence south in the Sierra Nevada to Tuolumne County; north to Alaska, east to Atlantic.

21. Carex densa Bailey, Mem. Torrey Club 1:50. 1889. Dense sedge.

Caespitose, the rhizomes short, stout, black, fibrillose; culms 3–7 dm. tall, sharply triangular, strict and rather stout, smooth or roughened above; leaves with well-developed blades several to a culm, all toward the base, the blades flat, 3–6 mm. wide, the sheaths tight, inconspicuously septate abaxially, and thin, white-hyaline and usually cross-rugulose adaxially, convex and prolonged at mouth; spikes in a close oblong or oblong-ovoid compound head 2–5 cm.

Fig. 104. *Carex stipata:* a, perigynium, abaxial view, showing the strong nerves and the round-cordate, spongy base, × 10; b, achene (cross section), × 10; c, perigynium, adaxial view, the nerves less developed than on abaxial side, × 10; d, achene, showing the spongy base of the perigynium, × 10; e, pistillate flower with perigynium removed, showing the stipitate achene, the short style, the 2 long stigmas, × 10; f–h, scales of pistillate flowers, showing variation in shape of scale and in length of awn, × 10; i, ligule, × 4; j, inflorescence, the spikes not crowded, the bracts bristle-like, × ⅖; k, staminate flower and subtending scale (lowermost flower in spike), × 10; l, habit, showing the conspicuous leaf sheaths and flat, flaccid blades, × ⅖.

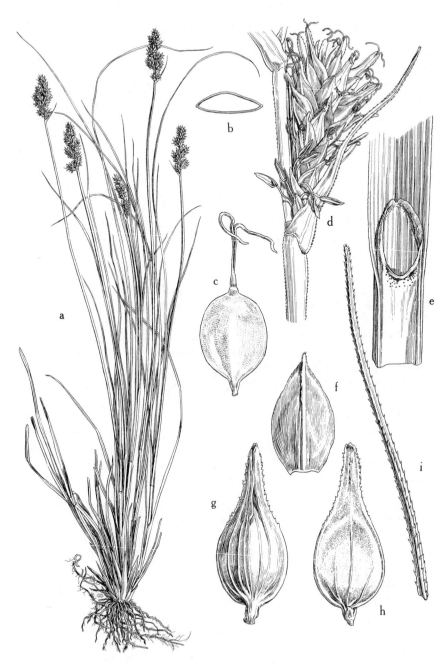

Fig. 105. (For explanation, see facing page.)

long; scales ovate, acute to cuspidate, brownish; lower bracts setaceous; perigynia 3.5–4.5 mm. long, strongly convex dorsally, ovate or ovate-lanceolate, narrowly green-margined, serrulate above middle, the beak about ½ as long as the body.

Boggy ground on slopes or in low, wet meadows: Coast Ranges from Santa Clara County north to Del Norte, Trinity, and Siskiyou counties; Sierra Nevada foothills from Mariposa County north to El Dorado County; southern Oregon.

Carex densa is a member of a complex and variable group of species which includes *C. Dudleyi* Mackenzie and *C. breviligulata* Mackenzie. The last two species occupy habitats similar to those of *C. densa* and overlap its range in the Coast Ranges, but occur less frequently. Typically, *C. Dudleyi* is distinguished from the other species by the elongate awns on all the pistillate scales. The perigynia are also smaller (2.5–3 mm. long, in contrast to 3.5–4.5 mm. long in *C. densa* and *C. breviligulata*). *Carex breviligulata* has been separated from *C. densa* by the much shorter ligule, but this is a trivial difference that appears to have little diagnostic value. *Carex vicaria,* also a member of this complex of species, is separately described below.

22. Carex vicaria Bailey, Mem. Torrey Club 1:49. 1889.

Caespitose, the rhizomes short, stout; culms 2–10 dm. tall, sharply triangular, overtopping the leaves; leaf blades flat, 3–5 mm. wide, long-attenuate; spikes closely aggregated into an oblong or linear head 3–5 cm. long, the staminate flowers apical, inconspicuous, the perigynia 6–15; scales ovate, mucronate, reddish brown with a greenish midrib, as long as or slightly shorter than the perigynia; bracts inconspicuous, slender, abruptly attenuate from a broad hyaline base, only 1 or 2 lower ones developed; perigynia plano-convex, 2–3 mm. long, sharply margined, somewhat nerved dorsally and ventrally, greenish brown or reddish brown, the beak finely toothed, flat, bidentate, and obscurely cleft dorsally.

About springs and in boggy areas: scattered localities in Coast Ranges at middle or low altitudes in Humboldt, Lake, and Santa Cruz counties; north to Washington.

23. Carex arcta Boott, Illus. Carex 4:155, pl. 497. 1867. Fig. 105.

Culms densely caespitose, 1.5–8 dm. tall, sharply triangular, slender but strict, very rough above; leaf blades flat, 1.5–4 mm. wide, light green, very rough toward the long-attenuate apex; spikes 5–15, closely aggregated into an ovoid-oblong or oblong head 1.5–3 cm. long, the upper spikes closely aggre-

Fig. 105. *Carex arcta: a,* habit, showing the slender culms and leaf blades and the ovoid-oblong inflorescences, × ⅖; *b,* achene (cross section), × 16; *c,* pistillate flower with perigynium removed, showing the stipitate achene and the short style, × 16; *d,* lowermost spike of inflorescence, showing toothed bract, the staminate flowers at base, and the pistillate flowers above, × 6; *e,* ligule, × 8; *f,* scale, alike in staminate and pistillate flowers, × 16; *g,* perigynium, abaxial view, showing obliquely cut beak, × 16; *h,* perigynium, adaxial view, × 16; *i,* apical part of leaf, × 8.

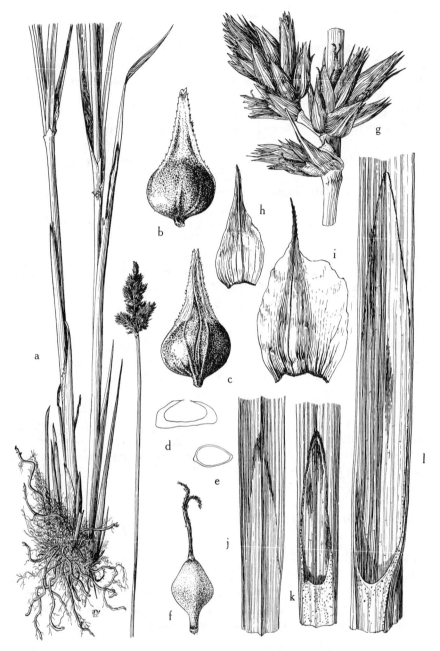

Fig. 106. (For explanation, see facing page.)

gated, the lower ones often slightly separated, the staminate basal flowers inconspicuous; scales ovate, somewhat obtuse to short-cuspidate, shorter than perigynia, hyaline with green midvein, more or less brownish-tinged; lower 1 or 2 bracts developed and from much shorter than to much longer than head; perigynia 2–3 mm. long, sharp-edged, strongly several-nerved dorsally, obscurely nerved at base ventrally, short-stipitate, tapering into a serrulate, flat beak, the beak shallowly bidentate, reddish-brown-tinged.

Swamps and wet woods from sea level to an elevation of 4,600 feet: Del Norte, Humboldt, and Mendocino counties; north to British Columbia, east to Atlantic.

24. Carex Cusickii Mackenzie in Piper & Beattie, Fl. NW. Coast 72. 1915. Fig. 106.

Culms rather densely caespitose, 7–12 dm. tall, sharply triangular; basal leaves bladeless or nearly so; leaf blades flat, the sheaths copper-colored at mouth; inflorescence 2–8 cm. long, the lower branches more or less separate, the upper ones aggregated; spikes closely aggregated, the staminate apical flowers inconspicuous; scales ovate, acute or short-acuminate, chestnut brown with hyaline margins and lighter rough-keeled midvein, usually covering perigynia; lowest bract setaceous, short and inconspicuous; perigynia 3–4 mm. long, brownish black, very thick, strongly convex and nerved dorsally, slightly convex and lightly nerved at base ventrally, spreading in age, truncate and short-stipitate at base, the body ovoid-orbicular, sharp-edged to base, serrate above, abruptly narrowed into a setulose-serrulate, flattened beak, the beak shallowly bidentate and about as long as the body.

Rare and local in wet meadows and marshes: San Luis Obispo County north to Del Norte and Siskiyou counties; south in the Sierra Nevada to Tuolumne County; north to British Columbia, east to Montana.

25. Carex diandra Schrank, Cent. Bot. Ammerk. 57. 1781.

Similar to *Carex Cusickii,* but differing in its narrower leaves, leaf sheaths which are not copper-tinged at mouth, heads which are little interrupted, and smaller perigynia (2–2.75 mm. long) with shining, very plump bodies which are usually not concealed by the scales.

Rare and local in wet meadows: San Bernardino Valley in San Bernardino County, Oriole Lake in Tulare County, also Sonoma, El Dorado, Butte, and Humboldt counties; northern part of North America, Europe.

Fig. 106. *Carex Cusickii: a,* habit, showing the bladeless basal leaves and the inflorescence, × ⅖; *b,* perigynium, adaxial view, × 12; *c,* perigynium, abaxial view, × 12; *d,* perigynium (cross section), showing thick wall, × 12; *e,* achene (cross section), × 12; *f,* pistillate flower with perigynium removed, showing long-stipitate achene, × 12; *g,* part of inflorescence, showing inconspicuous bract, and spikes with pistillate flowers at base, staminate flowers at apex, × 4; *h,* scale of a pistillate flower, × 8; *i,* lowest bract of inflorescence, × 8; *j,* ligular region of leaf, abaxial view, × 2; *k* and *l,* ligules, showing variation in size, × 2.

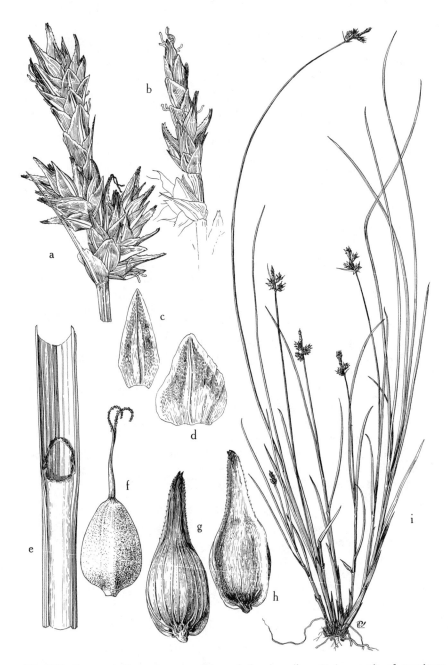

Fig. 107. *Carex phyllomanica: a* and *b,* variation in spikes, × 4; *c,* scale of staminate flower, × 12; *d,* scale of pistillate flower, × 12; *e,* ligule, × 8; *f,* pistillate flower with perigynium removed, × 12; *g,* perigynium, abaxial view, × 12; *h,* perigynium, adaxial view, × 12; *i,* habit, showing densely caespitose, slender culms, × ⅖.

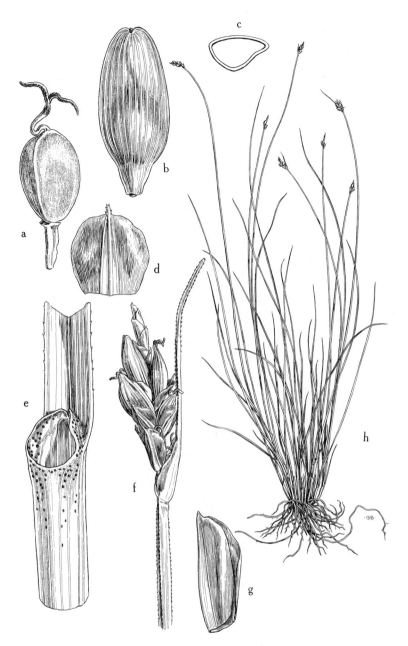

Fig. 108. *Carex leptalea: a,* pistillate flower with a part of the stipitate perigynium attached, the achene obscurely trigonous, the style flexuous, and the stigma bifid, × 16; *b,* perigynium, many-striate and beakless, × 12; *c,* achene (cross section), × 16; *d,* scale of pistillate flower, × 12; *e,* ligule, × 24; *f,* spike subtended by bractlike scale, the inconspicuous staminate flowers at apex, × 6; *g,* scale of staminate flower, × 16; *h,* habit, showing the very slender, densely tufted culms extending above the leaves, × ⅖.

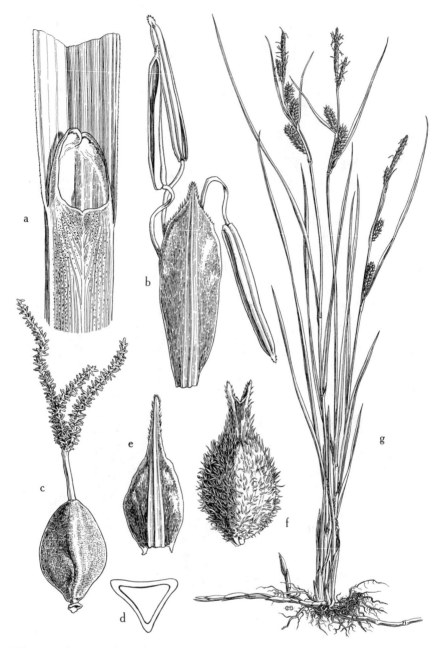

Fig. 109. *Carex lanuginosa: a,* ligule, × 6; *b,* staminate flower, the subtending scale acuminate and ciliate toward apex, × 12; *c,* pistillate flower with perigynium removed, showing the trigonous achene, × 12; *d,* achene (cross section), × 12; *e,* scale of pistillate flower, awned and ciliate upward (shorter than staminate scale), × 12; *f,* perigynium, showing the deeply bidentate beak and the dense, ascending pubescence, × 12; *g,* habit, showing the rhizome, the basal leaf sheaths which become filamentose in age, and the inflorescences with staminate spikes at apex, × ⅖.

26. Carex phyllomanica W. Boott in Wats., Bot. Calif. 2:233. 1880. Fig. 107.

Culms densely caespitose, slender but strict, 2.5–6 dm. tall; leaf blades flat, 1–2.75 mm. wide; spikes 3 or 4, approximate, forming a head 1.5–3.5 cm. long, the terminal one with pistillate flowers above and short-clavate at base, the lateral ones suborbicular, with 8–15 widely spreading perigynia; scales ovate, obtuse to acute, chestnut brown with lighter center and hyaline margins, the midvein obscure at apex; lowest bract setaceous, usually not reaching the tip of the culm but sometimes extending beyond it; perigynia round-truncate at base, conspicuously striate on both faces, green or green tinged with brown, sharp-edged to base, tapering into a serrulate beak about ½ as long as the body, the tip of beak chestnut-tinged, bidentate.

Bogs and marshes near the coast: Santa Cruz County to Del Norte County; north to Alaska.

Carex ormantha (Fern.) Mackenzie is very similar to this species, but differs from it primarily in having widely separate spikes, which form a head 2–6 cm. long, the terminal spike long-clavate. It occurs mostly in boggy places in the mountains at elevations of 4,000–9,500 feet, ranging from southern California north to Oregon. In the Coast Ranges it is usually found at elevations of 2,000–6,000 feet, but in Del Norte County it is found at 500 feet or lower.

27. Carex leptalea Wahl., Sv. Vet.-Akad. Handl. II. 24:139. 1803. Fig. 108.

Densely caespitose, the culms very slender, 1.5–6 dm. tall, obtusely triangular, smooth or slightly roughened, mostly extending beyond the leaves; leaf blades 0.5–1.25 mm. wide, thin, flat or grooved; spikes 4–15 mm. long, bractless or sometimes with 1 or 2 bracts, the staminate part varying from inconspicuous to occupying nearly the whole spike; pistillate scales ovate-obovate, very obtuse to short-cuspidate, shorter than the perigynia (the lowest scales as long as or longer than the perigynia), reddish-brown-tinged with hyaline margins and green center; perigynia erect-ascending, more or less strongly overlapping, oval-elliptic, 2.5–5 mm. long, finely many-striate, yellowish green or light green, substipitate, rounded and beakless at apex; style flexuous, the stigmas 2; achene somewhat trigonous.

Local in bogs and marshes: Humboldt, Marin, and Trinity counties; north to Alaska, then across the continent and south to Florida.

28. Carex lanuginosa Michx., Fl. Bor. Am. 2:175. Woolly sedge. Fig. 109.

Caespitose, with long, stout rhizomes; culms 3–10 dm. tall, sharply triangular and rough above; leaf blades flat, 2–5 mm. wide; basal leaf sheaths becoming filamentose; staminate spikes 2–4 mm. wide, to 4 cm. long; pistillate spikes 1–3, oblong-cylindric, 1–4 (or –5) cm. long, sessile or short-peduncled; scales lanceolate to lanceolate-ovate, acuminate or awned, ciliate upwards, narrower than perigynia, reddish brown with green center and hyaline margins; bracts sheathless or very short-sheathing, the blade of the lowest bract usually

Fig. 110. *Carex livida: a* and *b,* perigynia, showing variations in shape, × 8; *c* and *d,* achenes (cross section), showing variation, × 8; *e* and *f,* scales of pistillate flowers, × 8; *g,* staminate flower with subtending scale, rounded at apex, × 8; *h* and *i,* ligules, × 6; *j,* habit, showing a slender horizontal rhizome, well-developed basal leaf blades, and the inflorescences with the terminal spikes staminate, × ⅔; *k,* pistillate flower with perigynium removed, showing achene borne on long stipe, × 8.

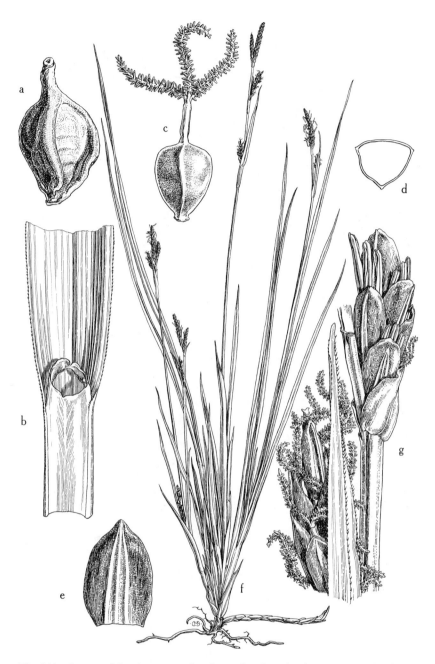

Fig. 111. *Carex californica: a,* perigynium, showing the long, narrow beak, × 10;
b, ligule, × 6; *c,* pistillate flower with perigynium removed, showing the sharply trigonous
achene, × 10; *d,* achene (cross section), × 10; *e,* scale of pistillate flower, × 10; *f,* habit,
showing the horizontal rhizomes, the nearly bladeless basal leaves, the stiff, flat blades,
and the inflorescences with terminal staminate spikes, × ⅖; *g,* tip of inflorescence, showing pistillate and staminate spikes and upper part of the scabrous subtending bract, × 6.

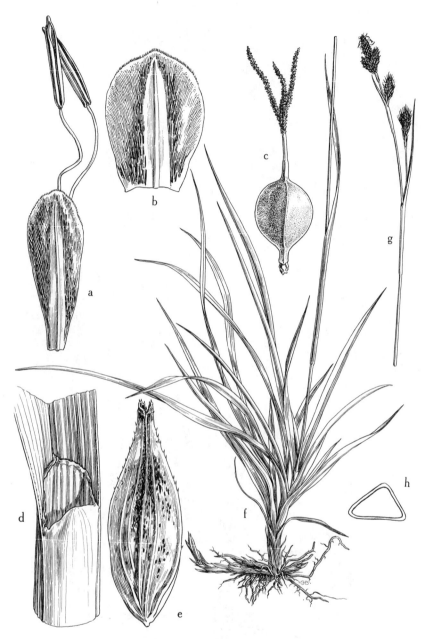

Fig. 112. *Carex luzulina: a,* staminate flower and subtending bract, × 12; *b,* scale of pistillate flower, × 12; *c,* pistillate flower with perigynium removed, showing sharply trigonous, apiculate achene, jointed to the short, slender style, × 12; *d,* ligule, × 4; *e,* perigynium, showing the bidentate beak, × 12; *f,* habit, showing the short rhizome and the stiff, spreading leaf blades, × ⅖; *g,* inflorescence, the terminal spike staminate, × ⅖; *h,* achene (cross section), × 12.

extending beyond the tip of the culm; perigynia broadly ovoid, 2.5–3 mm. long, the nerves nearly hidden by ascending pubescence, abruptly short-beaked, the beak deeply bidentate.

Wet meadows or marshes almost throughout California, but only occasional in the Coast Ranges; British Columbia to Nova Scotia, eastern states, New Mexico.

29. Carex livida (Wahl.) Willd., Sp. Pl. 4:285. 1805. Fig. 110.

Plants with elongate, slender, horizontal rhizomes; culms 0.5–5 dm. tall, smooth; basal leaves with well-developed blades; leaf blades 0.5–3.5 mm. wide, glaucous green; terminal spike 0.7–3 cm. long, short-peduncled; pistillate spikes 1 or 2, contiguous, sessile or short-peduncled (sometimes 3, with the lower one more remote, long-peduncled, oblong, 1–2 cm. long); scales ovate, obtuse to acute, shorter than perigynia, dark brown with broad, green center and hyaline margins; bracts short-sheathing, the lowest one not reaching the tip of the culm to extending beyond it; perigynia 3.5–4.5 mm. long, oblong-ovoid, obtusely triangular, glaucous green, faintly nerved, strongly puncticulate, tapering to a broadly stipitate base; style 3-cleft; achene ochre-yellow to russet brown and with silky shine, minutely beaded, 2–2.5 mm. long, globose to ovoid, trigonous, somewhat irregular, sessile on stipe.

Local in sphagnum bogs: near Mendocino City in Mendocino County; Alaska to Labrador and northeastern United States.

30. Carex californica Bailey, Mem. Torrey Club 1:9. 1889. California sedge. Fig. 111.

Rhizomes stout, elongate; culms 2–5 dm. tall, smooth or nearly so, strongly purple-tinged at base; basal leaves bladeless or nearly so, blades of culm leaves 1.5–5 mm. wide, green, stiffish, flat; sheaths prolonged upward beyond base of blade and continuous with the conspicuous ligule; terminal spike staminate (often with an additional shorter spike at the base), 1.5–3 cm. long, peduncled; pistillate spikes lateral, 1–3, erect, widely separate, sessile or short-peduncled, linear-oblong, 1–4 cm. long; the scales ovate, obtuse to acute, shorter than perigynia, purplish brown with lighter, hispidulous midvein and narrow hyaline margins; bracts leaflet-like, sheathing, the lowest usually not reaching the tip of the culm; perigynia 3.5–4 mm. long, the body broadly ovoid, brownish yellow, several-nerved, obtusely triangular, minutely granulose, sessile and rounded at base.

Rare and local in meadows and bogs: along the coast of Mendocino County; coastal Oregon and southern Washington.

31. Carex amplifolia Boott in Hook., Fl. Bor. Am. 2:228, pl. 226. 1839.

Caespitose with long, stout, blackish, scaly rhizomes; culms stout, erect, 5–10 dm. tall, winged-triangular, rough on angles above, more or less strongly purplish-tinged toward base; leaves with well-developed blades, the lower

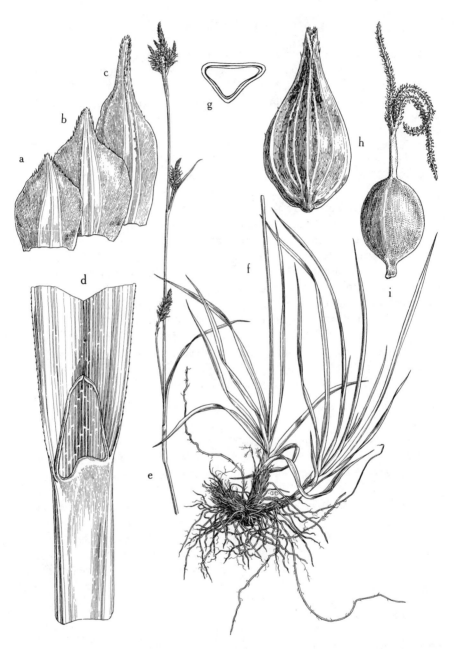

Fig. 113. *Carex sonomensis: a–c,* scales of pistillate flowers, showing variation in size and shape (those of the staminate flowers are similarly variable), × 10; *d,* ligule, × 6; *e,* upper part of culm, showing inflorescence, the tips of spikes staminate, × ⅖; *f,* habit, the long culm cut off, showing rhizome and spreading leaves, × ⅖; *g,* achene (cross section), × 12; *h,* perigynium, × 12; *i,* pistillate flower with perigynium removed, the achene trigonous, stigmas 3 (sometimes 4), × 12.

leaves clustered, the upper ones regularly disposed, light green or glaucous green, 8–18 mm. wide, smooth above except on veins toward the tip; sheaths more or less hispidulous on abaxial surface; staminate spike solitary, terminal; pistillate spikes 3–6, lateral, linear-cylindric, 3.5–14 cm. long; scales lanceolate and acuminate or awned to ovate and acute or mucronate, brownish purple with lighter center; lowest bract leaflike, extending beyond the inflorescence, little sheathing, the upper ones much reduced; perigynia 3 mm. long, obtusely triangular, brownish green, smooth or nearly so, somewhat rugose, 2-ribbed, abruptly contracted into a conic, more or less excurved beak of nearly the length of the body of the perigynium, the orifice oblique, hyaline, becoming bidentulate or bidentate.

Marshy sites and along streams: Coast Ranges from San Mateo County north to Del Norte County, Sierra Nevada from Tulare County north to Butte County, also Siskiyou and Modoc counties; north to British Columbia.

32. Carex luzulina Olney, Proc. Am. Acad. Arts & Sci. 7:395. 1868. Fig. 112.

Densely caespitose with short rhizomes; culms 1.5–7 dm. tall, obtusely triangular; leaves clustered near base, approximately ½ as long as the culms (often shorter), the blades spreading, stiff, light green, 3–7 mm. wide, roughened near apex; spikes 3–6, the terminal spike staminate (or often with a few perigynia at base), the upper ones clustered, the 1–3 lower spikes widely separate on slender, long or short peduncles; pistillate spikes oblong, 7–20 mm. long; scales ovate, obtuse or acutish, the midvein extending either to or nearly to apex; bracts long-sheathing, the blades not reaching the tips of the culms; perigynia lanceolate, 3.5–4.5 mm. long, somewhat triangular, yellowish green, purplish-tinged, few-nerved, tapering at apex into a dark, purplish-tinged, shallowly bidentate beak.

Meadows and bogs: Coast Ranges from Marin County to Humboldt and Siskiyou counties; southwestern Oregon.

This species is closely related to *Carex Lemmonii* Boott and *C. ablata* Bailey. *Carex Lemmonii,* which occurs in the Sierra Nevada, North Coast Ranges south to Lake County, and San Bernardino Mountains, can be distinguished from *C. luzulina* by its widely separated lower spikes; in *C. luzulina* the spikes tend to be congested near the apex. *Carex ablata,* which is not easily distinguished from *C. luzulina,* occurs at higher elevations in the North Coast Ranges, Modoc County, and south to Nevada County. All these species and the closely related *C. sonomensis* Stacey need further detailed study to aid in clarifying their relationships.

33. Carex sonomensis Stacey, Leafl. West. Bot. 2:63. 1937. Fig. 113.

Closely related and similar to *Carex luzulina* Olney, from which it may be distinguished by its almost completely hyaline scales. The spikes are less closely aggregated than in *C. luzulina,* but the uppermost pistillate spikes are more

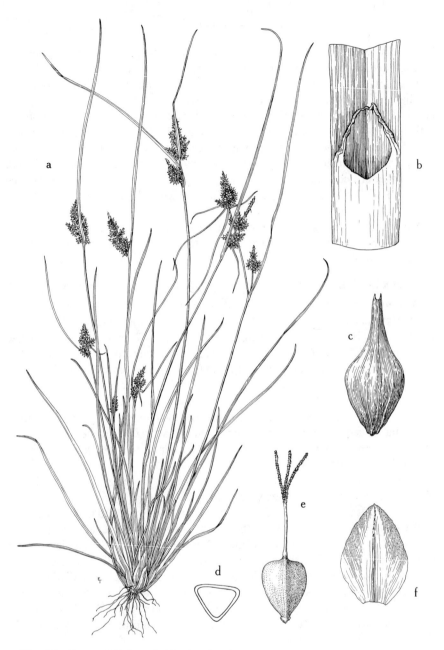

Fig. 114. *Carex viridula: a,* habit, showing the densely caespitose culms and leaves, and the long bracts subtending the inflorescences, × ⅖; *b,* ligule, × 6; *c,* perigynium, × 12; *d,* achene (cross section), × 12; *e,* pistillate flower with perigynium removed, × 12; *f,* scale, × 12.

closely aggregated than they are in the closely related *C. Lemmonii* of the Sierra Nevada. The uppermost spike varies in having the pistillate flowers either at the apex or the base or in being wholly pistillate, whereas in related species it is mostly staminate; the perigynia are usually hyaline- or greenish-tipped rather than dark purplish-tipped as in *C. luzulina,* although this character varies and some perigynia of *C. sonomensis* have brownish tips as well as hyaline ones within the same spike.

Known only from two localities in Sonoma County: Pitkin Marsh, five miles north of Sebastopol (type locality), and marsh on Perry's "Twin Pines Ranch," also near Sebastopol.

34. Carex viridula Michx., Fl. Bor. Am. 2:170. 1803. Green sedge. Fig. 114.

Culms densely caespitose, 0.6–4 dm. tall, smooth, bluntly triangular; leaf blades dull green, grooved, 1.5–3 mm. wide; terminal spike usually staminate, sessile or short-peduncled, 3–15 mm. long; pistillate spikes 2–6, closely aggregated and sessile or the lower ones separate and short-peduncled, oblong or globose-oblong; scales broadly ovate, shorter than perigynia, obtuse, acute or short-cuspidate, hyaline with greenish midvein, reddish-brown-tinged; bracts leaflike, the lower one extending well beyond the tips of the culms, strongly sheathing; perigynia 2–3 mm. long, obovoid, many-nerved, yellowish green, tapering at base, abruptly beaked, the beak minutely bidentate, reddish-tinged.

Marshes and bogs: coast from Mendocino County to Del Norte County; north to Alaska, across continent in Canada and northern United States.

35. Carex gynodynama Olney, Proc. Am. Acad. Arts & Sci. 7:394. 1868.

Culms densely caespitose, 2–7 dm. tall, obtusely triangular; leaves mostly clustered near base, the blades flat, light green, sparsely soft-pubescent; terminal spike staminate or with a few perigynia, sessile or short-peduncled, 1–2 cm. long, often overtopped by the 1 or 2 uppermost pistillate spikes; pistillate spikes 2–4, if more than 2, the 2 upper ones approximate, short-peduncled, the lower 1 or 2 widely separate from those above, erect, on slender, usually long-exserted, sparingly hairy peduncles, oblong-cylindric, 1–3 cm. long; scales broadly ovate or obovate, more or less soft-hairy above, the lower abruptly short-cuspidate, the upper ones often obtuse, reddish brown with green center and conspicuous hyaline margins, wider but shorter than perigynia; bracts strongly sheathing, shorter than the inflorescence; perigynia oblong-ovoid, 4–5 mm. long, finely nerved, membranaceous, yellowish brown and red-dotted, short-pilose toward apex, round-tapering at base, rounded and abruptly beaked at apex, the beak shallowly bidentate.

Moist places: Coast Ranges and near coast from Monterey County to Del Norte County; southern Oregon.

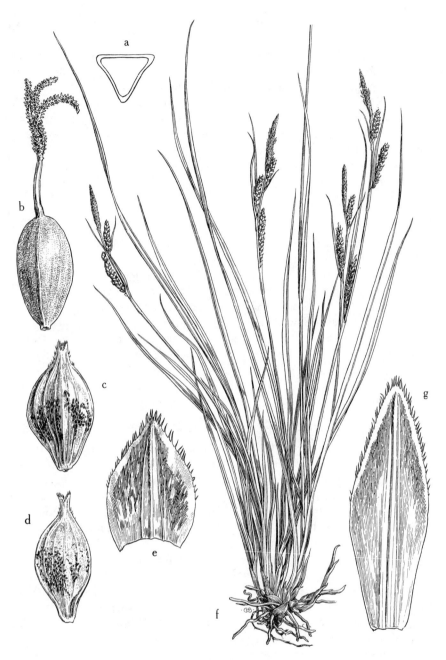

Fig. 115. *Carex mendocinensis: a,* achene (cross section), × 12; *b,* pistillate flower with perigynium removed, showing the shiny, minutely reticulate achene, × 12; *c,* perigynium, abaxial view, × 8; *d,* perigynium, adaxial view, × 8; *e,* scale of pistillate flower, soft-hairy above, × 12; *f,* habit, showing the spreading leaves and culms and the inflorescences of several pistillate spikes and 1 terminal staminate spike, × ⅖; *g,* scale of staminate flower, × 12.

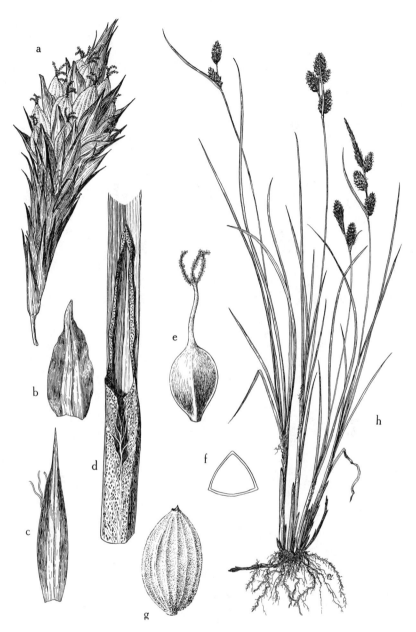

Fig. 116. *Carex Buxbaumii: a,* terminal spike, with staminate flowers below, pistillate flowers above, × 4; *b,* scale of pistillate flower, × 8; *c,* scale of staminate flower, showing filaments, × 8; *d,* ligule and leaf sheath, the latter becoming filamentose, × 6; *e,* pistillate flower with perigynium removed, × 12; *f,* achene (cross section), × 12; *g,* perigynium, finely many-nerved and papillose, × 12; *h,* habit, showing slender rhizome, filamentose sheaths, slender leaf blades, and culms bearing only a few spikes, × ⅖.

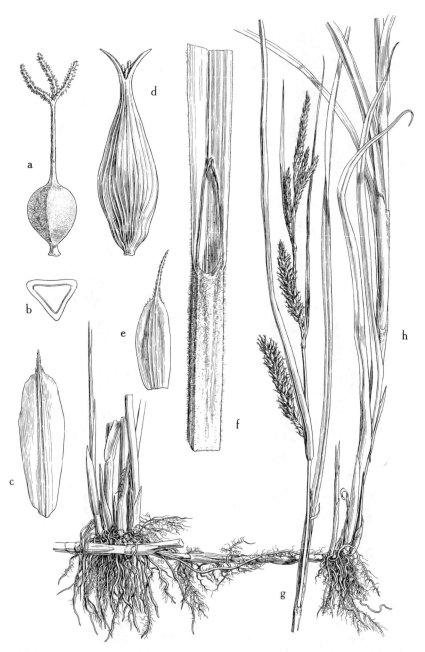

Fig. 117. *Carex atherodes: a,* pistillate flower with perigynium removed, showing short, stipitate achene, × 6; *b,* achene (cross section), × 6; *c,* scale of staminate flower, the awn variable in size, × 6; *d,* perigynium, showing the many strong ribs and the spreading, bidentate beak, × 6; *e,* scale of pistillate flower, the awn variable in size, × 6; *f,* ligule and pubescent leaf sheath, × ⅘; *g,* inflorescence, showing the leaflike sheathing bracts, the pistillate spikes below, and the staminate ones above, × ⅖; *h,* lower part of plant, showing the stout rhizomes, the flat leaf blades, and the pubescent sheaths, × ⅖.

36. Carex mendocinensis Olney ex W. Boott in Wats., Bot. Calif. 2:249. 1880. Fig. 115.

Culms caespitose, 3–8 dm. tall, slender, somewhat nodding, strongly over-topping the leaves, obtusely triangular; basal leaves bladeless or nearly so; blades of culm leaves erect, flat toward apex, channeled toward base, the sheaths sparsely hispidulous on abaxial surface; terminal spike staminate or with a few perigynia, long- or short-peduncled, 1.5–3.5 cm. long; pistillate spikes 2 or 3, the lower one strongly separate (if spikes 3, either 1 or 2 strongly separate), the uppermost erect, short-peduncled or nearly sessile, the lower 1 or 2 weakly erect, slender-peduncled, slender, linear, 1.5–5 cm. long; scales ovate, sharply keeled, obtuse to cuspidate, minutely ciliate at apex, cinnamon brown with lighter midvein and hyaline margins, about the width of but shorter than perigynia; bracts leaflet-like, the blade of the lowest extending beyond the tip of the spike; perigynia oblong-lanceolate, 3–4 mm. long, triangular, light green, membranaceous, puncticulate, obscurely few-nerved, tapering at base and substipitate, abruptly contracted at apex into a hyaline bidentate beak, short-ciliate at mouth.

Wet meadows and marshes: Marin County north to Del Norte County; southwestern Oregon.

This is the species that was formerly known as *Carex debiliformis* Mackenzie (Leafl. West. Bot. 6:157–162, 1951). For many years the name *C. mendocinensis* Olney was applied to plants which are now believed to be hybrids between the present species and *C. gynodynama* Olney. This confusion arose largely from the fact that the type collection of *C. mendocinensis* contains a few plants of the hybrid. This hybrid may be distinguished by the fact that it has sessile or short-peduncled staminate spikes, as well as broader perigynia which are sometimes sparsely pubescent. In addition, fertile plants of the putative hybrid have never been collected, and it has not been found except where the two putative parents grow together. The hybrid has been found at a few scattered localities from Mendocino County south to Marin County.

37. Carex Buxbaumii Wahl., Sv. Vet.-Akad. Handl. II. 24:163. 1803. Fig. 116.

Caespitose, rhizomatous; culms 2–10 dm. tall, slender but stiff, sharply angled, rough above; basal leaf sheaths breaking and becoming conspicuously filamentose; leaf blades 1.5–4 mm. wide, light green, more or less glaucous, flat, channeled toward base, sharply keeled, long-attenuate; spikes 2–4, erect, 1–4 cm. long, sessile or short-peduncled, the lateral ones pistillate, ovoid or oblong-ovoid; scales ovate, longer than the perigynia, long-acuminate or aris-tate, purplish black or purplish brown with light midvein; bracts scale-like, the lowest one not reaching the tip of the culm to extending slightly beyond it; perigynia oblong-obovoid, 3–4 mm. long, glaucous green, papillose, margin-ally 2-ribbed and finely many-nerved, short-stipitate, abruptly very minutely beaked, the beak minutely bidentate, purplish-tipped.

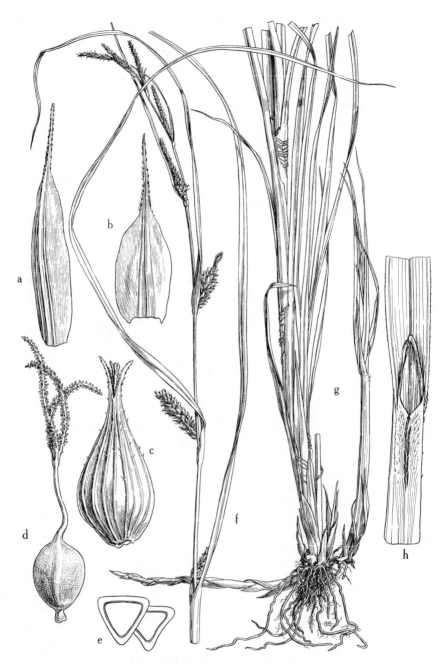

Fig. 118. (For explanation, see facing page.)

Rare and local in bogs or swales: Sierra Nevada in Tuolumne County at elevations of 4,600–8,000 feet, and in Inyo County at Rock Creek Lake Basin at 10,700 feet, also near the coast in Marin, Sonoma, and Humboldt counties; circumpolar.

38. Carex atherodes Spreng., Syst. Veg. 3:828. 1826. Awned sedge. Fig. 117.

Rhizomes stout; culms 3–12 dm. tall; lower leaves bladeless or nearly so, sharply triangular, smooth; blades flat, 3–12 mm. wide, long-attenuate, more or less pubescent beneath (at least toward the base), often prominently septate-nodulose; basal leaf sheaths breaking and becoming filamentose; staminate spikes 2–6, rarely with a few perigynia, 2–5 cm. long; pistillate spikes 2–4, widely separate, cylindric, sessile or the lower ones short-stalked, 3–10 cm. long; scales ovate-lanceolate, abruptly rough-awned, narrower than perigynia, the lower scales longer than and the upper ones rather shorter than perigynia, dull reddish brown with green center and hyaline margins; bracts leaflike, strongly sheathing, the lowest one reaching about to the tip of the culm or extending well beyond it; perigynia ascending to slightly spreading, lanceolate to ovate-lanceolate, yellowish green or light brownish, strongly many-ribbed, somewhat inflated, rounded and substipitate at base, 7–10 mm. long, tapering into a very deeply bidentate beak, the teeth ascending to widely spreading, 1–3 mm. long.

Swamps: Trinity and Modoc counties; circumpolar.

39. Carex Sheldonii Mackenzie, Bull. Torrey Club 42:618. 1915. Fig. 118.

Similar to *Carex atherodes* Spreng., but differs from it in that the lower leaves have well-developed blades; the leaf blades are on the average narrower (3.5–6 mm. wide); the staminate spikes are fewer in number (2 or 3); the pistillate spikes also are fewer in number (2 or 3), oblong-cylindric, and shorter (2–5 cm. long); the perigynia are broader (ovoid to ovate-lanceolate), soft-pubescent, and 5–6 mm. long; the beaks tend to be tipped with purplish red and are as much as 2 mm. long.

Marshy spots and along streams: Warner Mountains in Modoc County, Baxters in Placer County; Pacific states.

40. Carex spissa Bailey, Proc. Am. Acad. Arts & Sci. 22:70. 1886. San Diego sedge. Fig. 119.

Rhizomes stout, elongate; culms very stout, 1–2 m. tall, obtusely triangular, smooth; leaf blades glaucous green, 5–12 mm. wide, flat, with revolute, strongly

---

Fig. 118. *Carex Sheldonii: a,* scale of staminate flower, × 8; *b,* scale of pistillate flower, × 8; *c,* perigynium, pubescent toward the spreading, bidentate beak, × 6; *d,* pistillate flower with perigynium removed, showing the stipitate achene, × 8; *e,* achenes (cross section), showing slightly concave sides, × 8; *f,* upper part of culm, showing the leaflike sheathing bracts and solitary spikes, the staminate spikes at the apex, × ⅖; *g,* lower part of plant, showing rhizome, well-developed basal leaf blades, and sheaths becoming filamentose, × ⅖; *h,* ligule and soft-pubescent leaf sheath, × 1½.

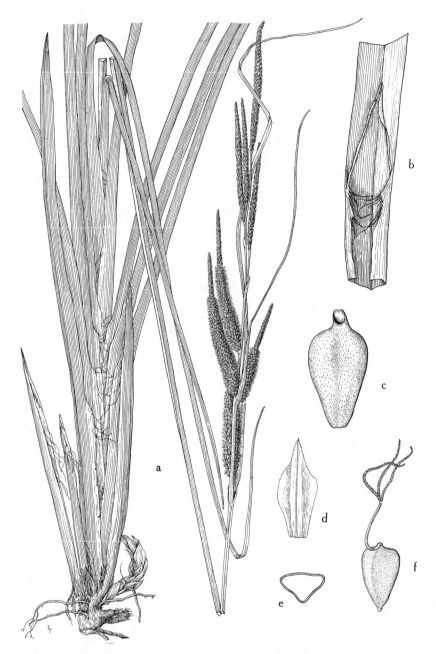

Fig. 119. *Carex spissa: a*, habit, showing the rhizome, the filamentose basal sheaths the stout culms, and the inflorescence with long, leaflike subtending bracts, × ⅖; *b*, ligule and leaf sheath, the sheath breaking into fibers, × 1⅕; *c*, perigynium, × 6½; *d*, scale, × 6½; *e*, achene (cross section), × 6½; *f*, pistillate flower with perigynium removed, showing the substipitate, trigonous achene and the abruptly bent style, × 6½.

serrate margins; basal sheaths brownish on adaxial surface, breaking and becoming conspicuously filamentose; staminate spikes 3–6, sometimes with a few perigynia, 4–10 cm. long; pistillate spikes 3–6, separate, usually staminate at apex, the upper ones sessile, the lower ones more or less long-peduncled, linear-cylindric, 6–15 cm. long; scales ovate-lanceolate, rough-awned, usually longer but narrower than perigynia, brownish with green center and hyaline margins; bracts leaflike, the lowest one very short-sheathing, extending beyond the tip of the culm; perigynia spreading at maturity, obtusely triangular, becoming inflated, 3–4 mm. long, obovoid, straw-colored, striate-dotted with red, glabrous, abruptly short-beaked, the beak 0.5 mm. long, the tip dark-tinged.

Banks of streams and marshy ground: coastal southern California to San Luis Obispo County; into Arizona and Baja California, Mexico.

41. Carex hystricina Muhl. in Willd., Sp. Pl. 4:282. 1805. Porcupine sedge.

Caespitose, the rhizomes mostly short, but some long and slender; culms 3–9 dm. tall, rough above; leaf blades 2–10 mm. wide, more or less septate-nodulose; basal sheaths breaking and becoming filamentose; staminate spike solitary, slender-peduncled, usually with a conspicuous, slender bract at base, 1–5 cm. long; pistillate spikes 1–4, approximate or widely separate, oblong or oblong-cylindric, 1.5–5 cm. long, the upper ones short-peduncled, erect; scales obovate or oblanceolate, serrulate above, hyaline or tinged with brown, with green center prolonged into a long, scabrid awn, the body of the scale much shorter than perigynium; bracts leaflike, short-sheathing or not sheathing, the lowest one extending well beyond the tip of the culm; perigynia lanceolate-ovate, inflated, 5–7 mm. long, light green or greenish-straw-colored at maturity, closely tapering or somewhat contracted into a slender bidentate beak, the slender teeth rigid, erect.

Wet soil: Trinity County; east to Atlantic.

42. Carex comosa Boott, Trans. Linn. Soc. London 20:117. 1846. Bristly sedge. Fig. 120.

Culms caespitose, stout, 5–15 dm. tall, very sharply angled; leaf blades 5–15 mm. wide, strongly septate-nodulose; basal sheaths not becoming filamentose; staminate spike solitary, 3–7 cm. long, slender-peduncled, often bearing a few perigynia; pistillate spikes 2–6, approximate or the lower ones slightly separate, oblong-cylindric, 2–7 cm. long, the upper ones short-peduncled and erect, the lower ones slender-peduncled and often strongly nodding; the scales narrowly lanceolate, serrulate above, hyaline or tinged with reddish brown, with green center prolonged into a long, scabrid awn, the body of the scale much shorter than perigynium; bracts leaflike, little sheathing, the lowest extending well beyond the tip of the culm; perigynia lanceolate-ovate, slightly inflated, 5–7 mm. long, greenish in color or brownish-tinged, closely many-ribbed, spreading or reflexed at maturity, tapering into a very deeply bidentate beak.

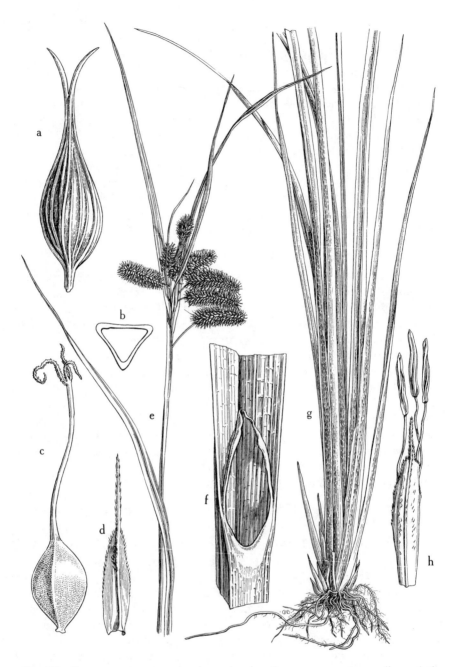

Fig. 120. *Carex comosa: a,* perigynium, showing the numerous strong ribs and the spreading, bidentate beak, × 8; *b,* achene (cross section), × 12; *c,* pistillate flower with perigynium removed, showing the very long style and the 3 short stigmas, × 12; *d,* scale of pistillate flower, showing the long, scabrid awn, × 12; *e,* upper part of culm, showing the leaflike bracts and the nodding pistillate spikes, × ¼; *f,* ligule, × 2; *g,* lower part of plant, showing the stout, erect culms and leaf blades, × ¼; *h,* staminate flower, the subtending scale scabrid, × 6.

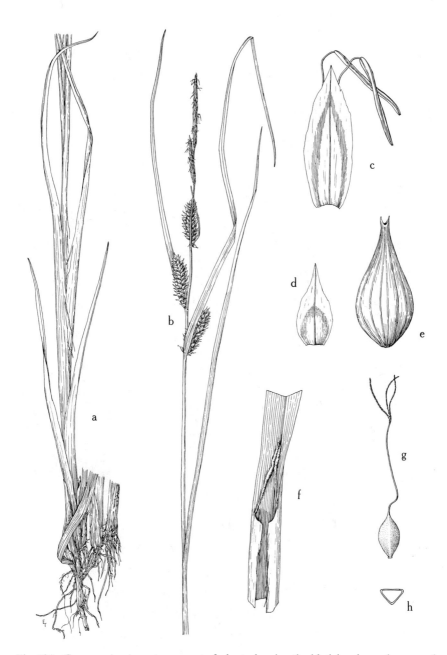

Fig. 121. *Carex vesicaria: a,* lower part of plant, showing the bladeless lower leaves and the rhizomatous base of plant, × ⅖; *b,* upper part of culm, showing the long bracts, the sessile pistillate spikes below, and the terminal staminate spikes above, × ⅖; *c,* staminate flower and subtending scale, × 5; *d,* scale of pistillate flower, × 5; *e,* perigynium, showing bidentate beak with erect teeth, × 5; *f,* ligule, × 1⅗; *g,* pistillate flower with perigynium removed, showing the very long and flexuous style, × 5; *h,* achene (cross section), × 5.

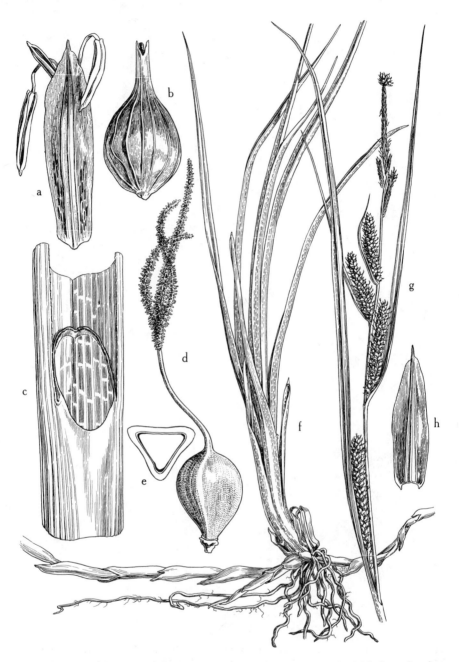

Fig. 122. *Carex rostrata: a,* staminate flower and subtending scale, × 8; *b,* perigynium, showing the slender, erect, bidentate beak, × 8; *c,* ligule, × 6; *d,* pistillate flower with perigynium removed, showing the substipitate achene and curved style, × 12; *e,* achene (cross section), × 12; *f,* habit, lower part of plant, showing the long, horizontal rhizomes, × ⅖; *g,* upper part of culm, the lower spikes pistillate, the staminate spikes terminal, some of the staminate spikes bearing perigynia at apex, × ⅖; *h,* scale of pistillate flower, × 8.

Marshes and ponds: San Bernardino Valley, along coast from Santa Cruz County north to Lake County, Big Lake in Shasta County, San Joaquin delta near Holt in San Joaquin County; east to Atlantic.

43. Carex vesicaria L., Sp. Pl. 979, excluding var. β. 1753. Inflated sedge. Fig. 121.

Rhizomes short-creeping, stout; culms 3–9 dm. tall, sharply triangular and rough above, the lower leaves more or less bladeless; leaf blades flat, 2–6 mm. wide, more or less strongly nodulose on abaxial surface; staminate spikes 2–4, 2–4 cm. long, the upper one peduncled, the lateral ones sessile; pistillate spikes 1–3 (usually 2), 2–5 cm. long, oblong-cylindric, sessile or short-peduncled, widely separate; scales ovate-lanceolate, acute, acuminate or short-awned, reddish-brown-tinged with green center and narrow hyaline margins, ½ to nearly as long as the perigynia; bracts not sheathing, the lowest extending well beyond the tip of the culm; perigynia ovoid, inflated, 4–8 mm. long, yellowish green or brownish, ascending or ascending-spreading at maturity, the smooth, bidentate beak 1.5–2 mm. long, the teeth erect, 0.5 mm. long.

Wet meadows: Coast Ranges from Mendocino County to Siskiyou County, Sierra Nevada from Tulare County to Siskiyou County chiefly at elevations of 4,000–9,000 feet, Modoc County; cosmopolitan.

An occasional form of this species found at high elevations (up to 10,000 feet) throughout the range of the species is characterized by having smaller and narrower pistillate spikes (1.5–3 cm. long) with smaller, less inflated perigynia (3–5 mm. long).

44. Carex rostrata Stokes in With., Arrang. Brit. Pl., 2d ed., 2:1059. 1787. Beaked sedge. Fig. 122.

Closely resembling *Carex vesicaria,* with which it possibly intergrades, but differing, at least in its typical aspect, in the following features: the rhizomes produce long horizontal stolons; the culms are obtusely angled; the lower leaves have well-developed blades; the leaves are more or less strongly septate-nodulose (at least on the sheaths), the blades 2–12 mm. wide; the basal sheaths are little if at all filamentose; the perigynia are 4–6 mm. long, at maturity ascending-spreading or spreading, the lowest sometimes reflexed.

Montane marshes or bogs: Sierra Nevada from Tulare County north to Plumas County, Mono and Inyo counties, San Bernardino Mountains, North Coast Ranges from Glenn County north to Siskiyou County, rare and local in Marin County; north to Alaska and east to Atlantic.

J. T. Howell (Leafl. West. Bot. 6:110–111, 1951) has pointed out some distinct differences in growth habit between *C. rostrata* and *C. vesicaria* which may be distinguished when the two species are seen growing together in the field, but which are not observable in dried material.

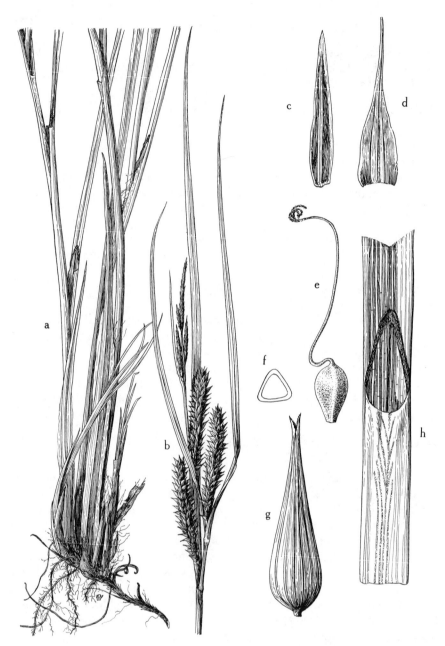

Fig. 123. *Carex exsiccata: a,* lower part of plant, × ⅖; *b,* upper part of culm, showing the long, flat, leaflike bracts, the large perigynia in long, cylindric spikes, and the slender terminal staminate spikes, × ⅖; *c,* scale of staminate flower, × 6; *d,* scale of pistillate flower, × 6; *e,* pistillate flower with perigynium removed, showing the trigonous achene and abruptly bent, long, slender style, × 6; *f,* achene (cross section), × 6; *g,* perigynium, gradually tapering into bidentate beak, × 6; *h,* ligule, × 4.

45. Carex exsiccata Bailey, Mem. Torrey Club 1:6. 1889. Fig. 123.

Similar to *Carex vesicaria* L., from which it differs mainly in having longer perigynia (7–10 mm.), which are more tapered and less abruptly contracted into the beak, and lanceolate to lanceolate-ovate rather than ovate, the beak 2–3 mm. long, with teeth 0.5–1.5 mm. long.

Marshes along the coast: Santa Cruz County to Del Norte County; north to Alaska, east to Montana.

CLADIUM

1. Cladium californicum (Wats.) O'Neill in Tidestr. & Kittell, Fl. Ariz. & N. M. 773. 1941. Saw grass.

Cladium mariscus var. *californicum* Wats., Bot. Calif. 2:224. 1880.
Mariscus californicus (Wats.) Fern., Rhodora 25:52. 1923.

Perennial; culms stout, 1–2 m. tall, subtrigonous to subterete; leaves 1–2 m. long, flat, 7–10 mm. wide, the margins serrate; umbels axillary, compound, the spikelets in glomerules of 3–6; spikelets oblong, acute, 3 mm. long, reddish brown, bearing 1 terminal perfect floret and several staminate florets, the lowermost scales empty, the scales spirally imbricated; bristles none; stamens 2; style bifid to trifid; achene ovoid, smooth, about 2 mm. long, without a tubercle.

Swampy places: coastal and desert southern California, north to San Luis Obispo and Fresno counties; southwestern states and northern Mexico.

CYPERUS

Annual or perennial sedges. Culms in ours simple, trigonous, striate, leafy at base. Involucral leaves 1 to several, extending well beyond the inflorescence. Inflorescence umbellate or capitate, the rays commonly bearing divaricate clusters of spikelets. Spikelets flat or subterete, many-flowered. Spikelets falling away from the head, or, if persistent, then the scales deciduous. Rachis straight, offset, or zigzag, unwinged, or frequently bearing at each node a pair of wings which are the decurrent bases of the next distal scale. Scales 2-ranked, keeled, all fertile, or the lower ones empty. Florets perfect. Perianth absent. Stamens 1–3. Style bifid or trifid; the style base deciduous from the summit of the achene. Achene trigonous or lenticular, naked or clasped by the rachis wings.

Corcoran, Sister Mary Lucy, A revision of the subgenus *Pycreus* in North and South America. Contr. Biol. Lab. Catholic Univ. Am. No. 37. 1941.

Horvat, Sister Mary Liguori, A revision of the subgenus *Mariscus* found in the United States. Contr. Biol. Lab. Catholic Univ. Am. No. 33. 1941.

Kükenthal, Georg, Cyperaceae–Scirpoideae–Cypereae, in Engler, Das Pflanzenreich 4[20]:1–671. 1935.

Fig. 124. *Cyperus niger* var. *capitatus: a,* scale, showing keel and obtuse apex, × 16; *b,* flower with scale removed, showing the bifid style and the 2 stamens, × 16; *c,* mature achene, showing puncticulate surface, × 16; *d,* capitate inflorescence and the involucral leaves unequal in length, × ⅔; *e,* habit, showing the short rhizome, and the slender, erect culms and leaf blades, × ⅕; *f,* spikelet, with lower scales removed to show the zigzag rachis, × 4.

McGivney, Sister M. Vincent de Paul, A revision of the subgenus *Eucyperus* found in the United States. Contr. Biol. Lab. Catholic Univ. Am. No. 26. 1938.

O'Neill, Hugh, The status and distribution of some Cyperaceae in North and South America. Rhodora 44:43–64, 77–89. 1942.

Style bifid; achene lenticular.
  Achene laterally flattened.
    Scales ochre-brown; inflorescence capitate . . . . 1. *C. niger* var. *capitatus*
    Scales reddish brown; inflorescence umbellate . . . . 2. *C. rivularis*
  Achene dorsally flattened . . . . 3. *C. laevigatus*
Style trifid; achene trigonous.
  Spikelets persistent on the spike; scales deciduous.
    Rachis not winged.
      Perennial with short rhizomes.
        Culms sharply trigonous, upwardly scabrellate on the edges . . . . . . 4. *C. virens*
        Culms bluntly trigonous, smooth . . . . 5. *C. Eragrostis*
      Annual with fibrous roots.
        Scales acuminate or awned, the tip recurved.
          Low annuals with celery-scented herbage . . . . 6. *C. aristatus*
          Taller plants without a marked odor . . . . 7. *C. acuminatus*
        Scales obtuse . . . . 8. *C. difformis*
    Rachis winged with a pair of inner hyaline appendages at each node.
      Perennials; achene obtuse (or mucronulate in *C. Parishii*).
        Stoloniferous with tubers; leaves about as long as the culms.
          Scales dull brown or yellow-brown . . . . 9. *C. esculentus*
          Scales shining reddish brown . . . . 10. *C. rotundus*
        Short-rhizomatous; leaves much shorter than the culm . . . . 11. *C. Parishii*
      Robust annual; achene distinctly mucronate . . . . 12. *C. erythrorhizos*
  Spikelets disarticulating above the basal pair of scales, or breaking up into 1-fruited joints at maturity; scales persistent.
    Spikelets disarticulating above the sterile basal pair of scales; rachis wings thin and hyaline.
      Culm swollen at the base into a corm; rhizomes absent . . . . 13. *C. strigosus*
      Culm not swollen at the base; short rhizomes present . . . . 11. *C. Parishii*
    Spikelets breaking up into 1-fruited joints; rachis wings firm and brown; plants annual
                                                  14. *C. ferax*

1. Cyperus niger var. **capitatus** (Britt.) O'Neill, Rhodora 44:86. 1942. Fig. 124.

Cyperus melanostachys H.B.K., Nov. Gen. & Sp. 1:207. 1815.

Perennial with short rhizome, fibrous roots, and caespitose culms; culms slender, smooth, erect, 15–50 cm. tall; leaves 2 on a culm, shorter than culm, narrow, smooth, the sheaths reddish brown; involucral leaves 3, to 15 cm. long; inflorescence a capitate cluster of spikelets; spikelets lanceolate, acutish, 5–12 mm. long, 2–2.5 mm. wide, the rachis zigzag; scales ovate, acutish to obtuse, keeled, 3- to 5-nerved, ochre-brown, deciduous; stamens 2; style bifid; achene lenticular, oblong, short-stipitate, short-mucronate, 1–1.3 mm. long, brown or gray, the surface puncticulate.

Wet places: coastal southern California, Inyo County, Sierra Nevada foothills from Tuolumne County to Butte County; Central Valley, Coast Ranges,

Fig. 125. *Cyperus rivularis: a,* ray of inflorescence, $\times$ 3; *b,* tip of spikelet, the lower
scales and achenes shed, but the stamens persistent on rachis, $\times$ 12; *c,* habit, showing the
low stature and the loosely umbellate inflorescences, $\times$ ⅔; *d,* spikelet, $\times$ 6; *e,* achene,
somewhat compressed, but not trigonous, $\times$ 20; *f,* keeled scale, lateral view, $\times$ 16.

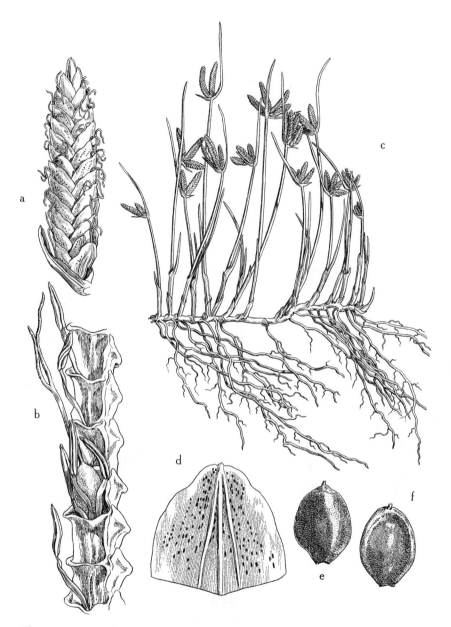

Fíg. 126. *Cyperus laevigatus: a,* compressed spikelet, × 6; *b,* rachis, showing the persistent stamens and an achene with bifid style, × 20; *c,* habit, the culms arising singly from a horizontal rhizome, × ⅘; *d,* obtuse scale, × 3; *e* and *f,* achenes, showing minutely reticulate surface, abaxial and adaxial views, × 12.

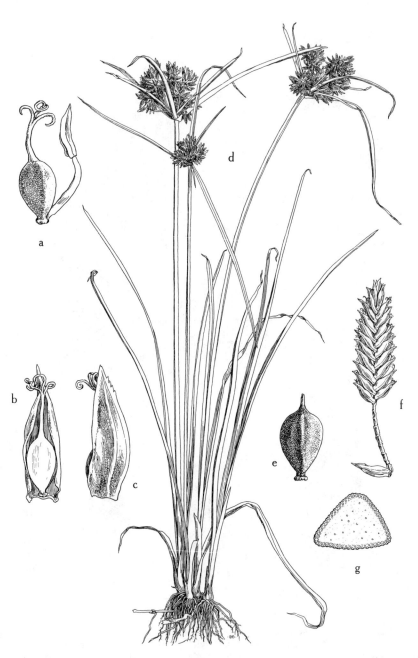

Fig. 127. *Cyperus Eragrostis: a,* young achene with trifid style and single stamen, × 16; *b* and *c,* achene and enclosing scale, front and lateral views, × 16; *d,* habit, showing the compact, globose inflorescences with numerous involucral leaves unequal in length, × ⅕; *e,* sharply trigonous achene, × 16; *f,* flat spikelet, showing the straight rachis, × 3; *g,* culm (cross section), × 4.

there intergrading northward with *Cyperus rivularis;* southwestern states, Colombia.

2. Cyperus rivularis Kunth, Enum. Pl. 2:6. 1837. Shining cyperus. Fig. 125.

Similar to *Cyperus niger,* but of annual habit and lower stature (4–40 cm. tall); inflorescence umbellate and looser; scales deep reddish brown.

Wet soil: North Coast Ranges, Siskiyou Mountains, El Dorado County; east to Atlantic. Intergrading with *C. niger* var. *capitatus.*

3. Cyperus laevigatus L., Mant. 1:179. 1771. Fig. 126.

Perennial with horizontal rhizomes; culms smooth, 15–50 cm. tall; leaves 2–5 on a culm, shorter than culm, sometimes reduced to basal sheaths; involucral leaves 1 or 2, very unequal in length; inflorescence a capitate cluster of 1 to several spikelets, appearing lateral; spikelets linear-oblong, compressed, 6–12 mm. long, 2–3 mm. wide; scales ovate, obtuse, about 1.5 mm. long and 1 mm. wide, whitish on base and keel and chocolate brown on sides; stamens 2 or 3, style bifid; achene lenticular, compressed parallel to rachis, elliptic, obtuse, substipitate, the surface minutely cellular.

About springs and wet flats: throughout desert and coastal southern California, north to Inyo County; widespread in the tropics.

4. Cyperus virens Michx., Fl. Bor. Am. 1:28. 1803.

Similar to *Cyperus Eragrostis* but differing in its sharply trigonous culms, which are upwardly scabrellate. The achenes, as noted by O'Neill (Leafl. West. Bot. 2:108, 1938), are oblong-elliptic, with a short stipe that is not at all broadened at the base (broad and flangelike in *C. Eragrostis*).

Apparently known in California from only two collections, both from Fresno County; southeastern United States, tropical America.

5. Cyperus Eragrostis Lam., Tabl. Encycl. 1:146. 1791. Not Vahl. Fig. 127.
Cyperus vegetus Willd., Sp. Pl. 1:283. 1797.

Perennial sedge with short, thick rhizomes and coarse, fibrous roots; culms trigonous, smooth, erect, 10–90 cm. tall, slightly swollen at base; leaves basal, 6–10, about as long as the culm, scabrellate on the margin and sometimes on midrib; involucral leaves 5–8, unequal in length, to 50 cm. long, scabrellate on margins and midrib; inflorescence a compact, globose compound umbel, the clusters of spikelets almost sessile or on rays to 12 cm. long; spikelets flat, 10–20 mm. long, 3–3.5 mm. wide, the rachis straight; scales ovate, acute, keeled, 3-nerved, straw-colored, falling off with the achenes enclosed; stamen 1; style trifid; achene sharply trigonous, obovoid, stipitate, mucronate, brown, puncticulate.

Our most common species of *Cyperus* on moist land: coastal southern California, Coast Ranges, Sierra Nevada foothills, Modoc County; Mexico, South America.

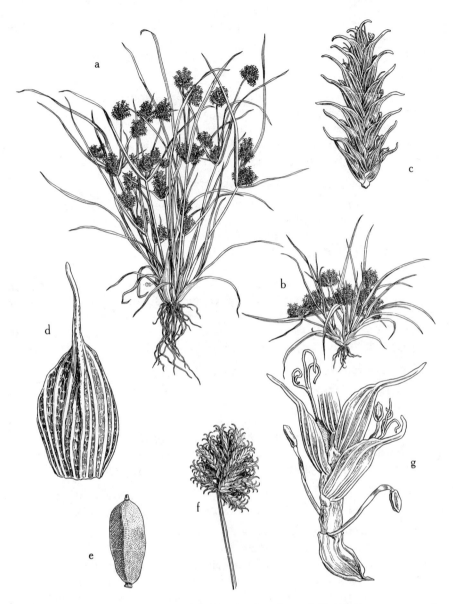

Fig. 128. *Cyperus aristatus: a* and *b,* habit, showing the umbellate inflorescences, each ray bearing a capitate cluster of spikelets, × ⅔; *c,* compressed spikelet, showing the recurved awns of scales, × 8; *d,* scale, showing the strong nerves, × 32; *e,* mature achene with puncticulate surface, × 24; *f,* ray of inflorescence, showing capitate arrangement of spikelets, × 1½; *g,* rachis, showing persistent stamens, arrangement of achenes, and the trifid styles, × 20.

Fig. 129. *Cyperus acuminatus: a,* spikelet, showing the recurved tips of scales, × 8; *b,* scale, 3-nerved, the surface cellular-reticulate, × 20; *c,* trigonous achene, × 20; *d,* habit, showing the globose heads of spikelets on rays of unequal length, × ⅓; *e,* flower without scale, × 20; *f,* ray of inflorescence, showing globose head of spikelets, × 1⅕.

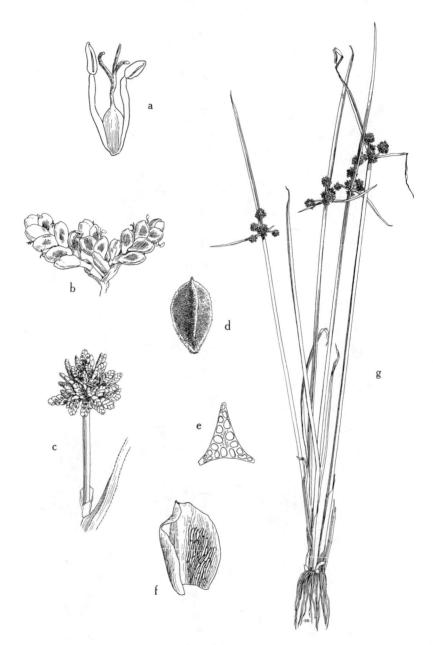

Fig. 130. *Cyperus difformis: a,* flower without scale, × 40; *b,* group of spikelets, × 8; *c,* ray of inflorescence, showing globose head of spikelets and part of scabrellate margin of involucral leaf, × 2; *d,* trigonous achene, showing the minutely cellular surface, × 28; *e,* culm (cross section), × 6; *f,* scale, × 40; *g,* habit, showing the umbellate inflorescences with involucral leaves of unequal length, × ⅓.

6. Cyperus aristatus Rottb., Descr. & Icon. 4:23. 1773. Fig. 128.

Cyperus inflexus Muhl., Descr. Gram. 16. 1817.

Small annual sedge with fibrous roots and celery-scented herbage; culms caespitose, slender, smooth, 1–20 cm. tall; leaves 2 or 3 on a culm, longer or shorter than culm, flat, 0.5–3 mm. wide; involucral leaves extending much beyond the inflorescence; inflorescence umbellate, the umbels capitate or the rays 2 cm. long or longer, bearing capitate clusters of spikelets; spikelets linear-oblong, 4–10 mm. long, compressed; rachis straight, deciduous after the scales; scales lanceolate, awned, the awn recurved, keeled, strongly several-nerved, green to light brown, deciduous; stamen 1; style trifid; achene trigonous, obovoid or oblong, obtuse, mucronulate, 0.7–1 mm. long, brown, the surface puncticulate.

Wet soil: almost throughout the nondesert parts of California; across the continent.

7. Cyperus acuminatus Torr. & Hook., Ann. Lyc. N. Y. 3:435. 1836. Fig. 129.

Annual sedge with fibrous roots and caespitose culms; culms slender, 7–40 cm. tall; leaves 2–4 on a culm, about as long as the culms, flat, 0.5–2.5 mm. wide, scabrellate on margins; involucral leaves 3 or 4, unequal in length, to 18 cm. long; inflorescence umbellate with 2–5 unequal rays, bearing globose heads of spikelets; spikelets ovate-oblong, obtuse, compressed, 4–10 mm. long; rachis straight, unwinged; scales ovate, acuminate, with a recurved tip, 3-nerved, pale green to light brown, the surface cellular-reticulate; stamen 1; achene trigonous, oblong, 0.6–0.9 mm. long, short-stipitate, short-mucronate.

Wet soil: San Joaquin Valley, Sierra Nevada, Coast Ranges; east to Georgia.

8. Cyperus difformis L., Cent. Pl. 2:6. 1755. Fig. 130.

Cyperus lateriflorus Torr., Mex. Bound. Surv. Bot. 226. 1859.

Annual sedge with fibrous roots and caespitose culms; culms smooth, 15–50 cm. tall; leaves 2–4 on a culm, about as long as the culm, 1–4 mm. wide, scaberulous on margins near apex; involucral leaves 2 or 3, unequal in length; inflorescence umbellate, the globose heads of spikelets sessile or on rays to 7 cm. long; spikelets linear, obtuse, subcompressed, 4–8 mm. long; rachis straight, unwinged; scales roundish, obtuse, 0.6–0.8 mm. long, membranous, green with brown sides, readily deciduous; stamens 1 or 2; achene trigonous, obovate, minutely mucronulate, 0.5 mm. long, pale greenish brown, the surface minutely cellular.

Common weed in rice fields: San Joaquin and Sacramento valleys, Sierra Nevada, Marin County. Native of Asia.

9. Cyperus esculentus L., Sp. Pl. 45. 1753. Yellow nut grass. Fig. 131.

Perennial sedge with scaly stolons terminating in tubers; culms stout, smooth, 15–50 cm. tall; leaves numerous, about as long as the culms, flat, 3–10 mm.

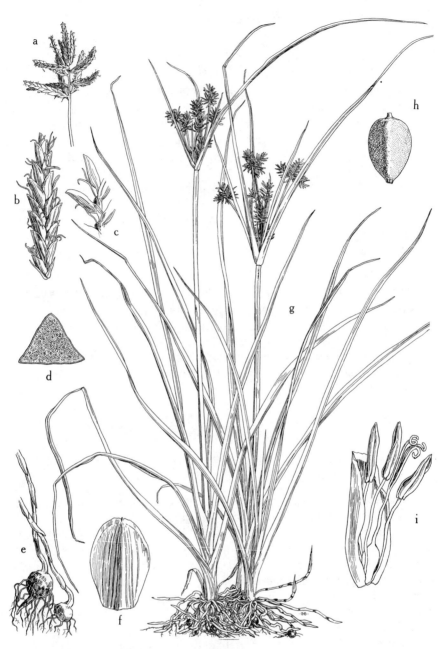

Fig. 131. *Cyperus esculentus: a,* ray, showing remote, divaricate spikelets, × ⅘; *b,* spikelet, × 3; *c,* part of spikelet, with some flowers removed, showing the hyaline, persistent wings of rachis, × 3; *d,* culm (cross section), × 2; *e,* stolons terminating in tubers, × ⅘; *f,* ovate scale, showing mucronulate apex, × 8; *g,* habit, showing stolons, tubers, numerous flat leaves, and umbellate inflorescences with broad, ascending involucral leaves, × ⅘; *h,* achene, showing surface puncticulate, × 12; *i,* flower, × 12.

Fig. 132. *Cyperus rotundus: a,* ray, showing linear spikelets, × ⅘; *b,* part of a spikelet, 1 scale and wing removed to show wing on rachis continuous on either side of scale, × 12; *c,* scale, × 8; *d,* puncticulate, trigonous achene, × 12; *e,* habit, showing the large, umbellate inflorescence and short, scabrellate involucral leaves, × ⅖; *f,* spikelet, showing the very long, filiform, trifid styles, × 3; *g,* stolon with tubers, × ⅘; *h,* flower, × 11.

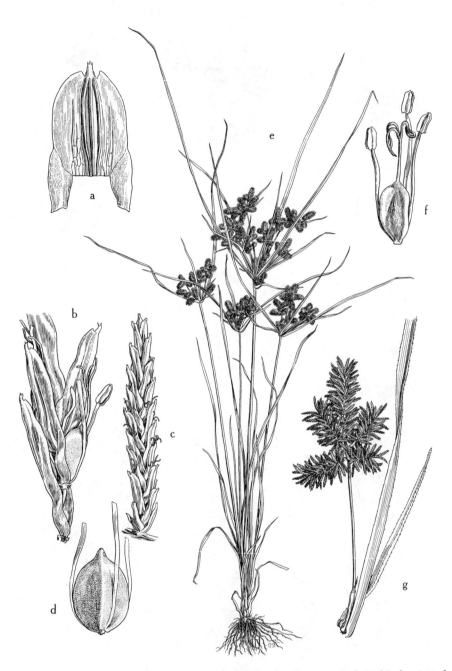

Fig. 133. *Cyperus erythrorhizos: a,* scale with rachis wings attached, × 20; *b,* part of a spikelet, with some scales removed to show inner hyaline membranes forming wings on rachis, × 20; *c,* linear spikelet, × 8; *d,* mature achene, trigonous and with surface finely cellular, × 28; *e,* habit, showing the compound umbels and their numerous involucral leaves, which are unequal in length, × ⅛; *f,* flower, × 20; *g,* ray of inflorescence, showing branches of divaricate spikelets and scabrellate involucral leaves, × ⅘.

wide, smooth; involucral leaves 2–6; inflorescence umbellate, with 5–10 rays, to 12 cm. long, bearing spikes of numerous remotely divaricate spikelets, some of the spikes sessile; spikelets linear, 6–30 mm. long, 2–3 mm. wide, flat; rachis with narrow, hyaline, persistent wings; scales ovate, obscurely mucronulate, 7- to 9-nerved, light brown; stamens 3; style trifid; achene trigonous, oblong, obtuse, 1.3–2 mm. long, the surface puncticulate.

Wet soil: coastal southern California, San Joaquin Valley, Sierra Nevada, less common in northern California; widely distributed.

10. Cyperus rotundus L., Sp. Pl. 45. 1753. Nut grass. Fig. 132.

Similar to *Cyperus esculentus,* but differing in the shining reddish brown color of its scales; tuberous.

Weed introduced from the eastern United States, tropical America, and Europe. Naturalized in scattered locations in coastal southern California, north to Ventura County, and in the San Joaquin Valley (Turlock in Stanislaus County, Stratford in Kings County).

11. Cyperus Parishii Britt. ex Parish, Bull. So. Calif. Acad. Sci. 3:52. 1904.

Cyperus sphacelatus sensu Britt., Bull. Torrey Club 13:212. 1886. Not Rottb., 1773.

Perennial sedge with short rhizomes and fibrous roots; culms subtrigonous, smooth, 10–25 cm. tall; leaves several, much shorter than the culm, 3–5 mm. wide, minutely scabrellate on the margins and midrib; involucral leaves 3 or 4, scabrellate; inflorescence umbellate, the rays 0.5–5 cm. long; spikelets linear, acute, 12–20 mm. long, about 2 mm. wide; rachis with a pair of hyaline wings at each node, these early-deciduous; scales ovate, acute, 2–3 mm. long, strongly several-nerved, the keel green and the sides reddish brown; stamens 3; style trifid; achene trigonous, obovoid-ellipsoid, 1–1.2 mm. long, mucronulate, nearly black.

Wet meadows: southern California; Arizona, New Mexico.

12. Cyperus erythrorhizos Muhl., Descr. Gram. 20. 1817. Fig. 133.

Annual sedge with fibrous roots and caespitose culms; culms bluntly trigonous, smooth, 10–100 cm. tall; leaves several, about as long as the culm, flat, 2–10 mm. wide, margins and midrib scabrellate, the sheaths loosely enveloping culm; involucral leaves 4–10, unequal in length, margins scabrellate; inflorescence a compound umbel or sometimes simple, the rays 1–30 cm. long, bearing numerous divaricate spikelets; spikelets linear, 3–10 mm. long, 1–1.5 mm. wide; rachis with a pair of inner hyaline wings at each node; scales oblong-obovate, mucronulate, keeled, with green midrib and satiny golden brown sides; stamens 3; style trifid; achene sharply trigonous, oblong, not stipitate, distinctly mucronate, grayish white, the surface smooth or very finely cellular under magnification.

Wet soil: coastal southern California, Imperial Valley, Central Valley, Coast Ranges; east to Atlantic.

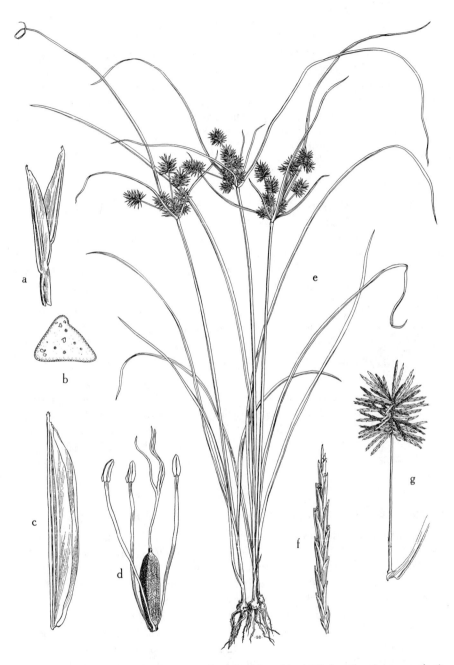

Fig. 134. *Cyperus strigosus: a,* part of a winged rachis, × 6; *b,* culm (cross section), × 5; *c,* scale, strongly nerved, × 12; *d,* linear, puncticulate achene, with trifid style and 3 stamens, × 12; *e,* habit, showing corms swollen at base, the umbellate inflorescences, and the involucral leaves, which are unequal in length, × ⅕; *f,* linear spikelets, × 2½; *g,* ray of inflorescence, showing loose, divaricate cluster of spikelets, × ⅔.

A dwarf form of this species has been collected in Lake County. In this plant the culms are only 4–10 cm. tall.

13. Cyperus strigosus L., Sp. Pl. 47. 1753. Fig. 134.

Perennial sedge with swollen corms at the base of the culms; culms smooth, 10–90 cm. tall; leaves 1 to several, longer than or shorter than the culm, flat, scabrellate on margins and midrib; involucral leaves 3 to several, scabrellate; inflorescence umbellate, the rays unequal in length, to 20 cm. long, terminating in loose, divaricate clusters of spikelets; spikelets linear, flat, 5–25 mm. long, 1–2 mm. wide, disarticulating above the sterile basal pair of scales when ripe; rachis somewhat zigzag, having at each node a pair of wide hyaline wings which clasp the achene; scales oblong-lanceolate, subacute, strongly several-nerved, straw-colored to golden brown; stamens 3; style trifid; achene trigonous; linear-oblong, 1.5–2 mm. long, 0.5–0.6 mm. wide, mucronate, purplish brown, the surface puncticulate.

Moist soil: Central Valley, Shasta and Humboldt counties; east to Atlantic.

14. Cyperus ferax L. C. Rich., Act. Soc. Hist. Nat. Paris 1:106. 1792. Coarse cyperus. Fig. 135.

Plants annual; culms 1 to several, stout, smooth, glabrous, triangular, the angles rounded, 20–50 cm. tall; leaves several to many, as long as or longer than the culms, the blade plicate, channeled above, 3–8 mm. wide, the margin beset with minute, stout prickles; involucral leaves several, 5–30 cm. long, very unequal in length, plicate; inflorescence simulating a compound spike, the rays arising from the top of the culm and umbellately spreading, branching into secondary rays each in the axil of an involucel leaf, the rays becoming enlarged in age by a gibbous, succulent tissue filling the bases of the involucral and involucel leaves; spikelets linear, subterete, 10–25 mm. long, about 2 mm. wide, disarticulating above the sterile basal pair of scales when ripe, the rachis succulent, becoming corky on ripening and breaking up into 1-seeded joints enclosing achene; florets deeply recessed in the joint of the rachis and almost enclosed by its winglike margins; scales ovate, obtuse, sheathing the rachis, strongly 7- to 9-nerved, yellowish brown; achene trigonous, obovoid, obtuse, 1–1.2 mm. long, 0.7–0.8 mm. wide, brown, the surface puncticulate.

A grainfield weed often infesting rice and barley fields: coastal southern California, Imperial Valley, Central Valley; widespread.

DULICHIUM

1. Dulichium arundinaceum (L.) Britt., Bull. Torrey Club 21:29. 1894.

Tall perennial with rhizomes; culms stout, erect, terete, hollow-jointed, 0.3–1 m. tall; leaves numerous, the basal leaves reduced to brown sheaths, the cauline leaves green, flat, linear, 2–8 cm. long, 4–8 mm. wide, spreading or ascending;

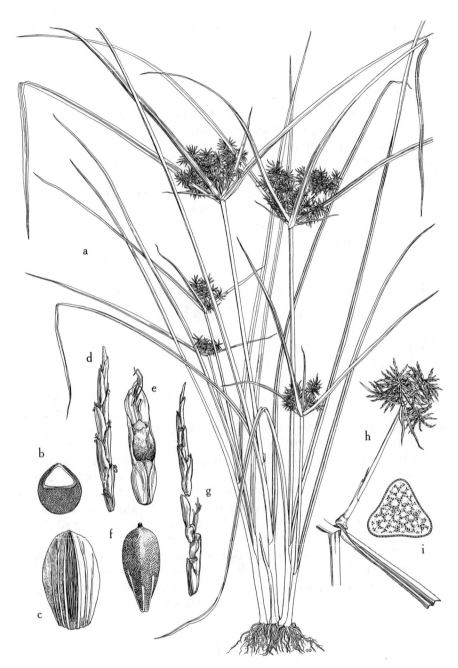

Fig. 135. (For explanation, see facing page.)

peduncles of the axillary racemes 0.4–2.5 cm. long; spikelets 2-ranked, about 6 to a raceme, linear, flat, jointed, spreading, 1–2.5 cm. long, 6- to 12-flowered; scales lanceolate, carinate, strongly nerved, decurrent on the proximal joint, greenish brown, the lowermost scale of each spikelet sterile; florets perfect; bristles about twice as long as the body of the achene; stamens 3; style bifid, the style base persisting on the apex of the achene as a beak; body of the achene lenticular, smooth, about 3 mm. long, the subulate beak as long as the body.

Swamps (rare in California): Swamp Lake and Tiltill Valley in Yosemite National Park, Tuolumne County, Ward Lake in Fresno County, Bald Mountain in Humboldt County; across the continent.

ELEOCHARIS. SPIKE RUSH

Annual or perennial herbs with rhizomes, stolons, or fibrous root systems. Culms simple, terete or subterete, usually striate. Leaves reduced to basal leaf sheaths, sometimes with an apiculate tip, rarely linear or acicular. Spike solitary, terminal, erect, several- to many-flowered, not subtended by an involucral leaf. Scales ovate to lanceolate, spirally imbricate, rarely 2- or 3-ranked. Bristles 8 to 1, or wanting, downwardly barbed. Florets perfect. Stamens 2 or 3. Style bifid or trifid. Achene lenticular or trigonous. Style base persistent, forming a tubercle on the apex of the achene. A group of usually small plants with a long history of taxonomic confusion.

Svenson, H. K., [Monographic studies in the genus *Eleocharis*], appearing in Rhodora in parts from 1929 through 1939. 31:121–135, 152–163, 167–191, 199–219, 224–242. 1929. 34:193–203, 215–227. 1932. 36:377–389. 1934. 39:210–231, 236–273. 1937. 41: 1–19, 43–77, 90–110. 1939.

Culms when fresh terete or subterete but not 4-angled.
  Style trifid; achenes plump or trigonous.
    Achenes with several longitudinal ridges and many fine horizontal lines.
      Dwarf annuals with fibrous roots; montane marshes . . . . 1. *E. bella*
      Perennial with rhizomes or stolons; lowland marshes.
        Culms spongy, striate; stamens 2; scales straw-colored . . . . 2. *E. radicans*
        Culms capillary, furrowed; stamens 3; scales brown . . . . 3. *E. acicularis*
    Achenes not longitudinally ribbed.
      Spikelets 2- to 7-flowered.
        Culms 2–7 cm. tall.
          Achenes smooth; bristles usually as long as or longer than achene
                                                  4. *E. parvula*

Fig. 135. *Cyperus ferax: a,* habit, showing plicate leaves, the involucral leaves, channeled above, and the compound inflorescences, × ⅕; *b,* achene enclosed by the corky rachis (cross section), × 12; *c,* scale, showing the strong nerves, × 12; *d,* spikelet, × 4; *e,* disarticulating joints of rachis, the achenes tightly enclosed, × 8; *f,* mature achene, the surface puncticulate, × 16; *g,* mature spikelet, breaking up into joints, × 4; *h,* primary ray with gibbous, succulent base, bearing secondary rays, each arising from involucel leaf, × ⅔; *i,* culm (cross section), × 3.

Achenes with pitted surface; bristles rudimentary or absent

5. *E. coloradoensis*

Culms 7–14 cm. tall.

Culms slender, usually less than 1 mm. in diameter.

Culms erect .... 6. *E. pauciflora*

Culms arching .... 6a. *E. pauciflora* var. *bernardina*

Culms coarse, 1–2 mm. wide .... 6b. *E. pauciflora* var. *Suksdorfiana*

Spikelets 10- to many-flowered.

Scales loosely 2-ranked; tubercles 3-lobed below, the lobes decurrent on apex of achene .... 7. *E. pachycarpa*

Scales not 2-ranked; tubercles not 3-lobed.

Tubercles broad and flat, apiculate .... 8. *E. Bolanderi*

Tubercles beaked, not flat.

Tubercles subulate, continuous with apex of achene ...... 9. *E. rostellata*

Tubercles conic or pyramidal, constricted at base.

Spikelets linear-lanceolate, acute; achenes smooth (or finely pitted under magnification) .... 10. *E. Parishii*

Spikelets ovoid, blunt; achenes pitted-reticulate under magnification

11. *E. montevidensis*

Style bifid; achenes lenticular.

Achenes shining black; tubercle depressed, but not lamelliform.

Apex of sheath conspicuously white-membranous .... 12. *E. flavescens*

Apex of sheath firm, not membranous.

Achenes 0.5 mm. long .... 13. *E. atropurpurea*

Achenes 1 mm. long .... 14. *E. geniculata*

Achenes brown or olivaceous.

Annuals with fibrous roots; tubercles lamelliform.

Tubercle ⅓ to nearly ½ as high as body of achene; bristles usually exceeding achene .... 15. *E. obtusa*

Tubercle very low, not more than ¼ as high as body of achene; bristles as long as achene or rudimentary .... 16. *E. Engelmannii*

Perennials with rhizomes; tubercles pyramidal, constricted at base

17. *E. macrostachya*

Culms 4-angled, plants 20–50 cm. tall .... 18. *E. quadrangulata*

1. Eleocharis bella (Piper) Svenson, Rhodora 31:201. 1929.

Eleocharis acicularis var. *bella* Piper in Piper & Beattie, Fl. Palouse Reg. 35. 1901.

Eleocharis acicularis of Jepson, Man. Fl. Pl. Calif. 148, as to fig. 135. 1923. Not Roem. & Schult.

Dwarf annual with fibrous roots and caespitose culms, often forming dense round tufts 5–10 cm. in diameter; culms capillary, furrowed, 2–6 cm. tall, light green; basal leaf sheaths loose, obliquely truncated; spikelets 1–3 mm. long, 8- to 10-flowered, or sometimes 2- or 3-flowered; scales with purplish brown sides and green midrib; bristles none; stamens 2; achene with about 30 fine transverse lines; tubercle compressed-conical.

Montane meadows: Sierra Nevada from Tulare County north to Siskiyou Mountains in Siskiyou County, east of the Sierra Nevada crest from Mono County north to Modoc County, North Coast Ranges, where it descends to an elevation of 300 feet, southern California; western states.

Except for its annual habit and smaller average stature, this species closely resembles *Eleocharis acicularis*.

2. Eleocharis radicans (Poir.) Kunth, Enum. Pl. 2:142. 1837.

Eleocharis Lindheimeri Svenson, Rhodora 31:199. 1929.

Perennial with long, creeping rhizomes; culms soft and spongy, striate, 3–8 cm. tall; basal leaf sheaths closely investing the culm; spikelets ovate, acute, 2–4 mm. long, 6- to 12-flowered; scales with scarious straw-colored sides and green midrib; bristles 4, longer than achene; stamens 2; achene narrowly obovate-oblong, with 30–40 fine traverse lines; tubercle narrowly conical (not compressed).

Marshy places below elevation of 4,500 feet: coastal southern California, Fresno County in San Joaquin Valley, San Francisco, Napa, and Sonoma counties; North and South America.

In habit like *Eleocharis coloradoensis,* but rhizome without tubers.

3. Eleocharis acicularis (L.) Roem. & Schult., Syst. Veg. 2:154. 1817. Fig. 136.

Eleocharis acicularis var. *occidentalis* Svenson, Rhodora 31:190. 1929.

Perennial with filiform stolons; culms matted, capillary, furrowed, 2–20 cm. long; basal leaf sheaths truncate, inconspicuous; spikelets ovate to linear, acute, 2–7 mm. long, 5- to 10-flowered; scales ovate-lanceolate, with brown sides, green midrib, and hyaline margins; bristles 3 or 4 as long as achene or wanting; stamens 3; style trifid; achene obovate-oblong with about 40 close transverse lines; tubercle compressed-conical.

Muddy river banks, meadows, vernal pools, and marshes almost throughout California except in the desert regions, ascending in the mountains to an elevation of 8,000 feet; widely distributed, Northern Hemisphere.

Svenson (Rhodora 31:190–191, 1929) has assigned most of the California material to *Eleocharis acicularis* var. *occidentalis,* separating it from the more northern and eastern typical form on the basis of its more rigid culms, brown-margined scales, more depressed tubercles, and constant lack of bristles. There is, however, such a wide range of variation in all of these characters in the California plants that *E. acicularis* var. *occidentalis* is perhaps better regarded as only a minor form of a variable species.

4. Eleocharis parvula (Roem. & Schult.) Link ex Bluff, Nees, & Schauer in Bluff & Fingerhuth, Comp. Fl. Germ., 2d ed., 1:93. 1836. Fig. 136.

Scirpus nanus Spreng., Pugill. 1:4. 1813.

Caespitose annual with fibrous roots, often with minute tuberous stolons; culms capillary, 2–7 cm. tall, greenish or straw-colored; spikelets 2–3.5 mm. long, broadly ovate, 2- to 9-flowered; scales ovate, obtuse or acute, green to yellowish, often reddish brown on the sides; bristles 6, as long as or longer than the achene; stamens 3; style trifid; achene obovate, 1–1.4 mm. long, straw-

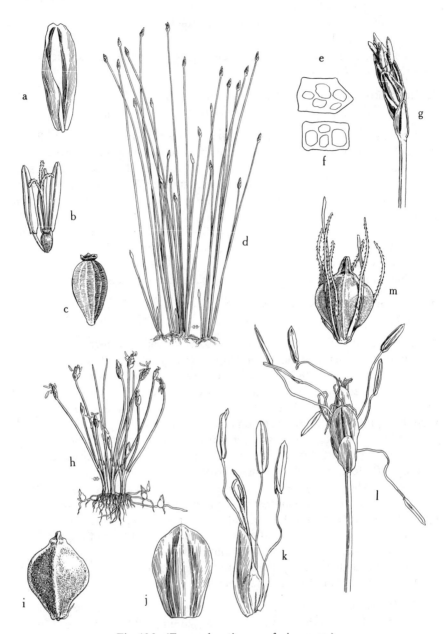

Fig. 136. (For explanation, see facing page.)

colored, smooth and shining, under high magnification sometimes lightly striate; tubercle very small, triangular, confluent with apex of achene, greenish.

Salt marshes: San Luis Obispo and Humboldt counties; cosmopolitan. Similar in habit to *Eleocharis coloradoensis.*

5. Eleocharis coloradoensis (Britt.) Gilly, Am. Midl. Nat. 26:66. 1941. Dwarf spike rush. Fig. 136.

Eleocharis leptos Svenson, Rhodora 31:176. In part. 1929. Not *E. lepta* Clarke, 1900.

Dwarf sedge with fibrous roots, and often with stolons bearing minute tubers; culms capillary, usually somewhat arched, 2–3 cm. tall; basal leaf sheaths loose, or closely investing the culm, obliquely truncate, light brown; spikelets about 3 mm. long, 4- to 6-flowered, ovate, acute; scales ovate, acute, with brown or purple sides, green midrib, and hyaline margins; bristles 3 and short, or none; style trifid; achene subtrigonous, the convex surface bearing an obscure keel, 1–1.5 mm. long, light brown, the surface papillose; tubercle pyramidal.

Alkaline mud flats: Modoc County, Plumas County, Merced County in San Joaquin Valley, Imperial County (infrequently collected); western states.

This species bears a close habital resemblance to *Eleocharis parvula,* with which it has often been confused. Svenson, in his monographic treatment of the genus (Rhodora 35:386–389, 1934), considers it to be a variety of *E. parvula;* but Gilly (Am. Midl. Nat. 26:65–68, 1941) gives substantial evidence that the two plants in question are distinct species. The important specific differences are to be found in the achenes. In *E. coloradoensis* the bristles are 3 in number and very short, or entirely wanting, and the achene surface is papillose, dark brown, and bears a pyramidal tubercle. In *E. parvula* there are 6 bristles, which are usually longer than the achene, the achenes are whitish to pale brown, essentially smooth-surfaced, and the apex tapers to a conical tubercle.

6. Eleocharis pauciflora (Lightf.) Link, Hort. Berol. 1:284. 1827.

Scirpus pauciflorus Lightf., Fl. Scot. 2:1078. 1777.

Perennial with filiform rhizomes bearing small leafy tubers; culms capillary, grooved, erect, 7–14 cm. tall, or sometimes 40 cm. tall; basal leaf sheaths 2–3 cm. long, truncate; spikelets 4–7 mm. long, ovate, 2- to 7-flowered; scales

---

Fig. 136. *Eleocharis. a–g, E. acicularis: a,* scale, showing the hyaline margins, × 12; *b,* young flower with scale removed, bristles wanting, × 12; *c,* mature achene, showing the longitudinal ribs, the close transverse lines, and the compressed, conical tubercle, × 20; *d,* habit, showing the matted, capillary culms, × ⅔; *e* and *f,* culms (cross sections of dried specimens), × 24; *g,* spikelet, × 6. *h–l, E. coloradoensis: h,* habit, showing the somewhat arching capillary culms and the stolons bearing tubers, × 1; *i,* mature subtrigonous achene, showing the minutely papillose surface and the pyramidal tubercle, × 16; *j,* scale, showing broad, hyaline margins, × 12; *k,* flower with short bristles, × 12; *l,* spikelets, × 8. *m, E. parvula:* mature achene, the 6 bristles extending beyond the achene and its conical tubercle, × 16.

Fig. 137. *Eleocharis rostellata: a,* habit, showing the wiry culms, some procumbent and rooting at the tips, × ⅕; *b,* mature, obtusely trigonous achene with surface finely reticulate, the tubercle subulate and continuous with the apex of the achene, × 12; *c,* spikelet, × 4; *d,* flower, × 8; *e,* scale, × 8.

lanceolate, acuminate, purplish brown; bristles 2–6, shorter than to as long as or longer than the achene; style trifid; achene trigonous, the surface finely reticulate, yellowish brown, 2 mm. long; tubercle a subulate beak merging into the dark base of the style.

High mountain meadows: Sierra Nevada at elevations of 7,500–12,000 feet, North Coast Ranges at an elevation of 5,000 feet, San Bernardino and San Jacinto mountains; cosmopolitan.

6a. Eleocharis pauciflora var. **bernardina** (Munz & Johnst.) Svenson, Rhodora 31:174. 1929.

Scirpus bernardinus Munz & Johnst., Bull. Torrey Club 52:221. 1925.

Similar to the species, but culms arching and densely caespitose. Probably only a minor form of the species.

Forming large, dense, grayish green turfs at elevations of 7,500–8,500 feet: San Bernardino Mountains, and Mount Pinos in Ventura County.

6b. Eleocharis pauciflora var. **Suksdorfiana** (Beauverd) Svenson, Rhodora 31:174. 1929.

Eleocharis Suksdorfiana Beauverd, Bull. Soc. Bot. Genève II. 13:267. 1922.

Similar to the species, but culms compressed and conspicuously grooved, 1 mm. wide; spikelets 9- to 12-flowered.

Mountain meadows: Shasta, Modoc, Lassen, and Inyo counties, also rare and local in Marin and Lake counties; Oregon, Washington.

7. Eleocharis pachycarpa Desv. in C. Gay, Fl. Chil. 6:174. 1853.

Perennial with thickened, descending rootstocks; culms caespitose, slender to nearly filiform, quadrangular-sulcate, 10–40 cm. tall, erect to recurved, frequently proliferous; basal leaf sheaths straw-colored, brownish, or slightly purplish-tinged, the apex firm, acute, appressed; spikelets ovate, 5–10 mm. long; scales loosely arranged in 2 ranks, ovate-lanceolate, obtuse to acute, purplish brown; bristles as long as the achene or frequently absent; style trifid; achene obtusely trigonous, ovate to obovate, truncate, yellowish white, the surface smooth or very finely reticulate; tubercle pyramidal, acute to acuminate, its trilobed base decurrent on the angles of the achene.

Known from four stations in Humboldt County, where it grows in moist soil near salt marshes, and from fresh-water pools on serpentine barrens in the Sierra Nevada foothills in El Dorado County, also from one and a half miles northeast of Buena Vista, Amador County. These are first records of the species in California, it previously having been reported as new for North America from Ormsby County, Nevada, on the east side of Lake Tahoe (correspondence, Svenson, 1950). Possibly introduced from Chile.

This species bears a close superficial resemblance to *Eleocharis Bolanderi* Gray, but the basal leaf sheaths and the tubercles readily distinguish the two species.

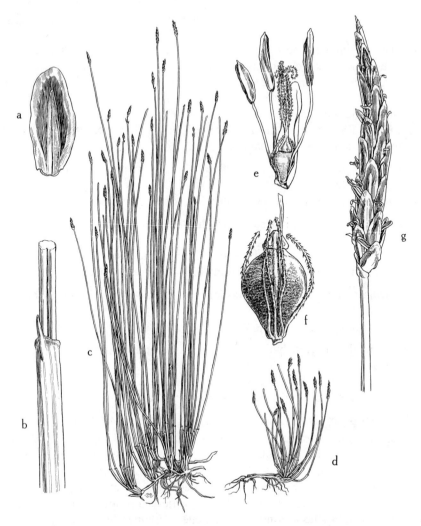

Fig. 138. *Eleocharis Parishii: a,* scale, × 10; *b,* toothed, truncate leaf sheath, × 6; *c* and *d,* habit, showing average and small plants, the slender, erect, fascicled culms, and the creeping rhizomes, × ⅖; *e,* flower, × 10; *f,* trigonous, obovoid achene, the surface faintly reticulate, the tubercle conic, and the subtending bristles longer than the achene, × 20; *g,* spikelet, linear-lanceolate, acute, × 6.

8. Eleocharis Bolanderi Gray, Proc. Am. Acad. Arts & Sci. 7:392. 1868.

Perennial with short, woody rhizomes; culms caespitose, slender, striate, 10–30 cm. tall, glaucous green; basal leaf sheaths obliquely truncate, with a minute tooth, slightly inflated at an indurated purplish summit; spikelets ovate, 3–8 mm. long; scales ovate, acute, dark brown, with narrow hyaline margin (sometimes obscure); bristles 3 or 4, ½–¾ as long as the achene; style trifid; achene trigonous, plump, truncate, yellowish brown, the surface smooth; tubercle flat, shortly apiculate.

Meadows in the Sierra Nevada foothills from Madera County north to El Dorado County; east of the Sierra Nevada crest in Mono County.

9. Eleocharis rostellata Torr., Fl. N. Y. 2:347. 1843. Walking sedge. Fig. 137.

Perennial from a short caudex; culms wiry, coarse, grooved, erect or somewhat arched, 0.25–1.5 m. long, often some culms procumbent and rooting at tips; basal leaf sheaths 2–7 cm. long, obliquely truncate; spikelets oblong, 6–12 mm. long, 10- to 20-flowered; scales ovate-lanceolate; bristles 4–8, about as long as the achene; style trifid; achene obtusely trigonous, plump, 1.2–2 mm. long, the surface finely reticulate; tubercle a subulate beak continuous with the apex of the achene.

Springs or alkaline marshes: Orange County, San Gabriel and San Bernardino mountains, north in desert localities to Mono County, in Coast Ranges in Ventura County, Marin and Sonoma counties; east to Atlantic.

10. Eleocharis Parishii Britt., Jour. N. Y. Micr. Soc. 5:110. 1889. Fig. 138.

Perennial (or sometimes annual?) with slender, creeping, reddish rhizomes; culms slender, striate, erect, 1–3 dm. tall, in fascicles or tufted; leaf sheaths reddish brown at base, usually becoming straw-colored at apex, the apex obliquely truncate, usually with a minute tooth; spikelets linear-lanceolate, acute, 10–15 mm. long, many-flowered; scales ovate-oblong, acute to obtuse, chestnut brown or dark brown, with a short hyaline tip; bristles 6 or 7, as long as, longer than, or shorter than the achene; style trifid; achene trigonous, ellipsoid or obovoid, yellow to light brown, smooth or faintly reticulate under magnification; tubercle short-subulate to conic.

Moist ground in coastal and desert southern California, ascending the mountains to an elevation of 7,000 feet, Coast Ranges from Ventura County north to Siskiyou County, Madera County at an elevation of 7,500 feet, Butte and Lassen counties.

Eleocharis disciformis Parish (Bull. Calif. Acad. Sci. 3:81, pl. 6, 1904) is very similar to *E. Parishii,* but is an annual species with fibrous roots. The former species is apparently known only from the area of its type locality at the eastern base of the San Jacinto Mountains, along the borders of the Colorado Desert in southern California. Its retention as a distinct species will depend upon the study of additional material.

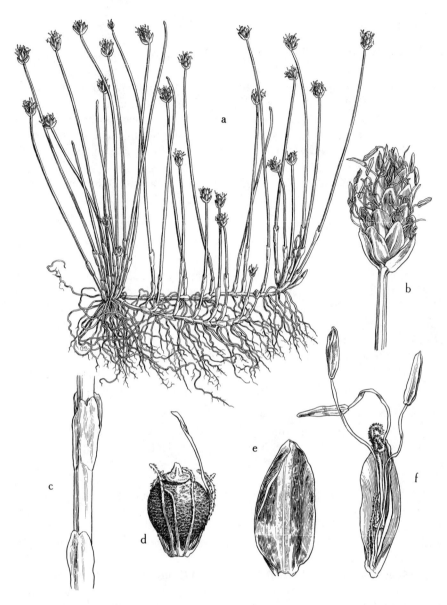

Fig. 139. *Eleocharis flavescens: a,* habit, showing a tufted plant and rhizomes with single culms arising from the nodes, × ⅘; *b,* ovate, obtuse spikelet, × 6; *c,* emarginate basal leaf sheaths, × 6; *d,* mature achene, the tubercle conic, acute, the subtending bristles as long as or slightly longer than achene, × 24; *e,* elliptic scale with pale midvein, × 20; *f,* flower, showing the bifid style and the 3 stamens, × 20.

11. Eleocharis montevidensis Kunth, Enum. Pl. 2:144. 1837.

Eleocharis arenicola Torr. in Engelm. & Gray, Boston Jour. Nat. Hist. 5:237. 1847.
Eleocharis montana of western American authors, not Roem. & Schult.

Perennial with slender, creeping, reddish rhizomes; culms slender, striate, 1–3 dm. tall, in fascicles; leaf sheaths reddish brown, usually becoming straw-colored toward the apex, the apex truncate, often with a minute tooth; spikelets narrowly ovoid to oblong, obtuse, 4–15 mm. long, many-flowered; scales ovate, obtuse, brownish or yellowish with a hyaline margin; bristles 4–6, as long as or shorter than the achene; style trifid; achene trigonous, obovoid, yellowish brown to brown, minutely pitted or reticulate; tubercle conic, short.

Moist ground throughout coastal southern California and extending to the western edge of the deserts; north in the Coast Ranges at scattered stations to Siskiyou County; also reported from serpentine in Lake County; to Texas, Mexico, South America.

Except for its narrowly ovoid to oblong and obtuse spikelets, and often somewhat lighter-colored scales, this species bears a very close resemblance to *Eleocharis Parishii*. The two species apparently intergrade where their ranges are contiguous.

12. Eleocharis flavescens (Poir.) Urban, Symb. Ant. 4:116. 1903. Fig. 139.

Eleocharis flaccida (Reichb.) Urban, Symb. Ant. 2:165. 1900.

Perennial; culms light green, 0.5–4 dm. long, striate; leaf-sheath apexes conspicuously membranous, usually consisting of 2 white, inflated and some-what emarginate lobes; spikelets 2–6 mm. long, ovate, acute or obtuse; scales elliptic to oblong-lanceolate, purplish brown with paler midvein; bristles 6 or 7, as long as or slightly longer than the achene; stamens 3; style bifid; mature achene shining, purplish brown, 0.9–1.1 mm. long, obovate, the surface minutely punctate; tubercle green, conic, acute.

Merced County in San Joaquin Valley; probably introduced from the eastern United States; South America, Europe.

H. K. Svenson, who has studied the collection upon which this record is based, comments that the California material has coarser culms and larger achenes than the tropical material. It may also be noted that the scales are darker in our material.

13. Eleocharis atropurpurea (Retz.) Kunth, Enum. Pl. 2:151. 1837. Fig. 140.

Annual with fibrous roots and caespitose culms; culms filiform, striate, 3–12 cm. tall; basal leaf sheaths loose, obliquely truncate, with an attenuate tooth; spikelets ovoid, many-flowered; scales ovate, obtuse, with purple-brown sides, a green midrib, and a very narrow scarious margin; bristles 2–4 or none, slender, shorter than achene; stamens 2 or 3; style bifid; achene lenticular, obovoid, 0.5 mm. long, shining black, the surface smooth; tubercle minute, depressed, constricted at base, conic, white.

Fig. 140. *Eleocharis atropurpurea: a,* spikelet with mature, shiny, black achenes, × 8; *b,* habit, showing the tufted culms, each with a loose, obliquely truncate, toothed leaf sheath, × ⅘; *c,* flower, style bifid, × 20; *d,* striate culm with ovoid, many-flowered spikelet, × 8; *e,* obtuse scale, × 20; *f,* mature achene, the tubercle minute and conic, the subtending bristles short, × 30.

Marshes, rice fields: San Bernardino County, Glenn and Butte counties, Fresno County; widespread in tropical areas.

Identification of achenes of this species in the stomach contents of ducks first called attention to its presence in California. The species, which is found principally along flyways, is much more abundant than herbarium records would indicate. The achene, which is diagnostic for the species, is avidly sought by ducks.

14. Eleocharis geniculata (L.) Roem. & Schult., Syst. Veg. 2:150. 1817. Not of recent authors. Fig. 141.

Eleocharis capitata R. Br., Prodr. Fl. Nov. Holl. 225. 1810.
Eleocharis caribaea Blake, Rhodora 20:24. 1918.

Annual with fibrous roots and caespitose culms; culms subfiliform, striate, 5–25 cm. tall, or sometimes 40 cm. tall; basal leaf sheaths loose, obliquely truncate, with an attenuate tooth; spikelets ovoid, obtuse, much thicker than the culm, many-flowered; scales ovate, obtuse, pale brown with a scarious margin; bristles 6–8, as long as achene; stamens 2 or 3; style bifid; achene lenticular, obovate, 1 mm. long, shining black, the surface smooth; tubercle depressed, constricted at base, apiculate, spongy, whitish.

Marshes and watercourses in southern California: San Bernardino, Riverside, and Imperial counties; widespread.

15. Eleocharis obtusa (Willd.) Schult., Mant. 2:89. 1824. Fig. 142.

Annual with fibrous roots; culms slender, erect, 10–50 cm. tall; basal leaf sheaths obliquely truncate with a minute tooth; spikelets ovoid to ovoid-oblong, obtuse, many-flowered, 4–12 mm. long; scales obovate to nearly orbicular, obtuse, brown, scarious-margined; bristles 6–8, usually much longer than the achene; style bifid; achene lenticular, obovate, about 1 mm. long, shining brown, the surface smooth; tubercle deltoid, lamelliform, broad, covering the top of the achene and varying from ⅓ to nearly ½ as long as the achene.

Meadows and ponds: North Coast Ranges from Marin County north to Humboldt County; Klamath Mountains; east to Atlantic.

16. Eleocharis Engelmannii Steud., Syn. Pl. Glum., pt. 2—Cyp. 79. 1855.

Eleocharis monticola Fern., Proc. Am. Acad. Arts & Sci. 34:496. 1899.
Eleocharis Engelmannii Britt. in Abrams, Illus. Fl. Pacif. States 1:263. 1923. In part, not as to fig. 628.

Annual with fibrous roots; culms 1–4 dm. long; upper leaf sheaths obliquely truncate or with a minute tooth; spikelets brownish, oblong-cylindric, 5–16 mm. long, obtuse to acute; scales ovate, obtuse to acute, with narrow scarious margins; bristles about 6, approximately as long as the achene, sometimes rudimentary or wanting; style bifid (sometimes trifid); achene lenticular, obovate, about 1 mm. long, shining brown or greenish, the surface smooth; tubercle deltoid, low, broad, covering the top of the achene and less than ¼ as long as the achene.

Meadows and ponds: Sierra Nevada from Tuolumne County north to Plumas County, Modoc County, Central Valley, Marin and Lake counties; east to Atlantic.

Svenson, in correspondence, reports that *Eleocharis Engelmannii* and *E. obtusa* tend to intergrade. Whereas in the eastern United States *E. Engelmannii* may be readily distinguished by virtue of the pale elongate spikelets with low

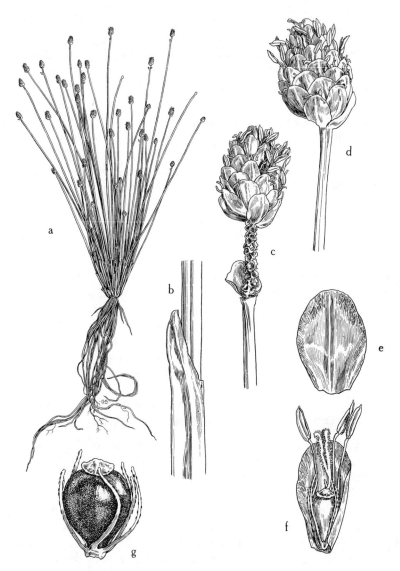

Fig. 141. *Eleocharis geniculata: a,* habit, showing the tufted, filiform culms and the loose leaf sheaths, × ⅖; *b,* obliquely truncate leaf sheath with attenuate tooth, × 12; *c,* older spikelet, the ovoid shape preserved as the lower flowers drop off, × 6; *d,* young, ovoid spikelet, × 6; *e,* ovate, obtuse scale, × 16; *f,* flower, style bifid, × 16; *g,* mature achene, the tubercle depressed and spongy, and the subtending bristles about as long as the achene, × 24.

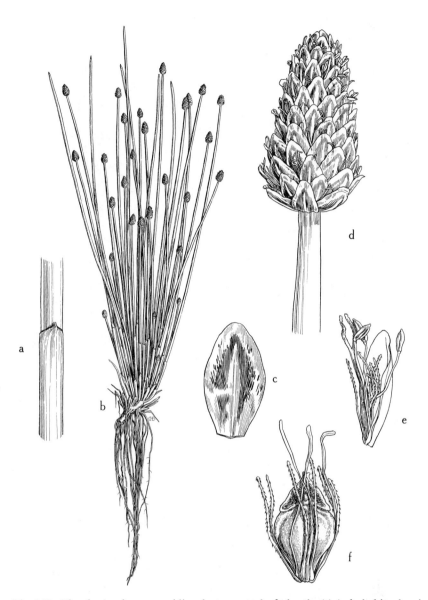

Fig. 142. *Eleocharis obtusa: a,* obliquely truncate leaf sheath, × 4; *b,* habit, showing the erect, tufted culms, basal leaf sheaths, and ovoid, tapering spikelets, × ⅖; *c,* scale, × 12; *d,* spikelet, × 6; *e,* flower, showing the bifid style and the long bristles, × 12; *f,* lenticular, obovate, smooth achene, the tubercle broad and deltoid, the subtending bristles retrorsely barbed, extending beyond the tubercle, × 16.

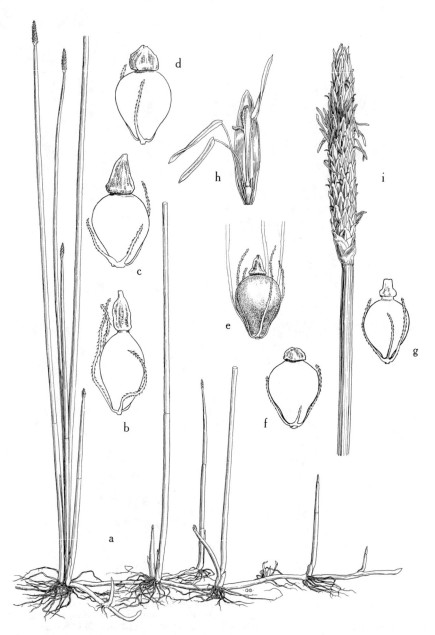

Fig. 143. *Eleocharis macrostachya: a,* habit, showing the tall, erect culms with truncate basal leaf sheaths, and the creeping rhizomes, × ⅖; *b–g,* variations in form and size of achene, tubercle, and subtending bristles, × 12; *h,* flowers, showing the lanceolate scale, × 6; *i,* terminal spike, the lower scales empty, × 2.

tubercles on the achene, in the West these differences are not so clear. Gilly reduces *E. Engelmannii* to a variety under *E. obtusa*.

17. Eleocharis macrostachya Britt. in Small, Fl. SE. U. S. 184, 1327. 1903. Common spike rush, creeping spike rush. Fig. 143.

Eleocharis palustris of American authors, not *Scirpus palustris* L.
Eleocharis uniglumis Schult., Mant. 2:88. 1824.
Scirpus (Heleocharis) mamillata Lindb. f., Acta Soc. Fauna & Fl. Fenn. 23:4, 7. 1902.

Perennial sedge with long, creeping rhizomes; culms loosely or densely caespitose, terete or oval in cross section (becoming flat and ribbon-like upon drying in some plants), stout or slender, striated, erect, 0.5–1 m. tall, pale green to dark green; leaves reduced to basal sheaths, obliquely truncated or sometimes mucronate; terminal spikes 0.5–2.5 cm. long, subtended by 2 or 3 empty scales or uniglumate; fertile scales lanceolate, brown to purple or green, with a green midrib and a scarious margin; bristles 4, or sometimes wanting, about as long as the achene; stamens 2 or 3; style mostly bifid; achene lenticular, yellowish brown, 1.5 mm. long (including tubercle); tubercle broadly to narrowly pyramidal or cone-shaped, constricted at base.

Common and widespread in marshes, vernal pools, ditches, and flooded lands throughout California; east to Atlantic.

This is an extremely variable species consisting of a number of perplexing forms. Several segregates have been proposed, but it is very difficult to delimit these, because of numerous intergradations of characters. Our treatment follows the recent work of Svenson (Rhodora 49:61–67, 1947).

Field observations suggest that many of the variants of *Eleocharis macrostachya* may be genetic responses to local environmental conditions. In brackish marshes near the coast, for example, the plants develop rather broad and soft culms with large spikelets and dark purple to black scales. Plants from interior localities may have culms as broad as this maritime form, or they may be rounder and firmer, but the scales are usually much lighter in color. In plants with broad culms, the culm often approximates the width of the spikelet, at least in its upper part, and often may become conspicuously flattened and ribbon-like in dried specimens. The broad culm, however, may represent only a seasonal response to moisture conditions. Occasionally it would appear that early-season plants develop the broad culms, whereas those developing under conditions of summer drought and a lowered water table develop slenderer and more terete culms. Considerable variation also occurs in size of spikelet, size and shape of the tubercle (which may be as long as it is wide, or longer), and the nature of the sheath apex.

18. Eleocharis quadrangulata (Michx.) Roem. & Schult., Syst. Veg. 2:155. 1817. Fig. 144.

Perennial, often bearing tubers; culms sharply 4-angled, coarse, 3–4 mm. wide on a side, 0.5–1 m. tall; basal leaf sheaths membranous, reddish brown;

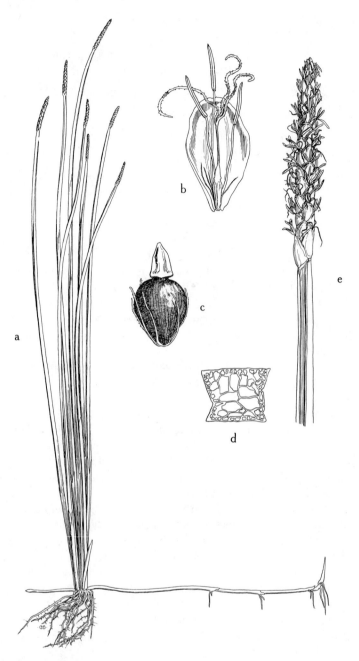

Fig. 144. *Eleocharis quadrangulata: a,* habit, showing the rhizome, the basal leaf sheaths, the tall, 4-angled culms, and the cylindric spikelets, × ⅕; *b,* flower, showing the rounded scale, the 3 stamens, and the trifid style, × 6; *c,* mature achene, the tubercle elongated and triangular, and the slender subtending bristles, × 8; *d,* culm, showing the sharp angles (cross section), × 6; *e,* spikelet, × 1½.

spikelets cylindric, 3–4 mm. in diameter, 2–5 cm. long; scales 4-ranked, elliptic, rounded or somewhat acute, not keeled, 5 mm. long, straw-colored; bristles 4–6, as long as the achene; stamens 3; achene lenticular, obovate, yellowish brown, shining, the surface finely marked with numerous rows of transverse linear cells; tubercle elongated, triangular, constricted at summit of achene.

In shallow water at edge of ponds: Merced County, probably introduced from the eastern United States; Atlantic coast.

ERIOPHORUM

1. Eriophorum gracile Koch in Roth, Catal. Bot. 2:259. 1800. Slender cotton grass.

Perennial with rhizomes; culms slender, smooth, subterete, 30–60 cm. tall; leaf blades triangular, channeled, 2–30 cm. long; involucral leaf solitary, 1–2 cm. long; inflorescence a cluster of 2–5 spikelets; spikelets 6–8 mm. long; scales gray to almost black; bristles numerous, 1–2 cm. long, exserted beyond the scales, remaining attached to the achene at maturity; stamens 3–1; style trifid; achene oblong, 2–3 mm. long.

Mountain meadows in the Sierra Nevada; and, according to Jepson, cool swamps from San Francisco County and Sonoma County northward; Northern Hemisphere.

Fernald designated as *Eriophorum gracile* var. *caurianum* the California material in which the scales are straw-colored or brownish rather than lead-colored or blackish as in the typical form. Since the color of the scales of our plants is so variable, we do not differentiate between the species and the variety.

Fernald, M. L., The North American species of *Eriophorum*. Rhodora 7:81–92, 129–136. 1905.

FIMBRISTYLIS

Perennial or annual sedges. Culms erect, leafy at base. Involucral leaves 1 to several. Inflorescence a loose umbel or capitate cluster of spikelets. Scales spirally imbricated. Florets all perfect. Bristles none. Stamens 1–3. Style bifid or trifid, pubescent or glabrous, the style base swollen and in 1 species persistent as a tubercle. Achene trigonous or lenticular.

Achene trigonous, bearing a conic tubercle, this constricted at the base ... 1. *F. capillaris*
Achene lenticular and mucronate, but not bearing a tubercle.
Inflorescence a simple capitate cluster of 3–8 spikelets; style glabrous; dwarf annuals
2. *F. Vahlii*
Inflorescence a more or less loose umbel of several to many spikelets; style pubescent;
taller perennials .... 3. *F. thermalis*

1. Fimbristylis capillaris Gray, Man. 530. 1848.

Stenophyllus capillaris (L.) Britt., Bull. Torrey Club 21:30. 1894.

Annual sedge with fibrous roots and caespitose culms; culms filiform, grooved, 5–25 cm. tall; leaves basal, the blades filiform, shorter than the culm, the sheaths pubescent with long hairs; involucral leaves 1–3, short and bristly; inflorescence a terminal umbel of 2 to several spikelets, or sometimes a solitary spikelet; spikelets oblong, 5–8 mm. long; scales oblong, obtuse, keeled, puberulent, with deep brown sides and a green midrib; stamens 2; style trifid, glabrous; achene trigonous, truncate, about 1 mm. long, white to light brown, the surface transversely wrinkled; tubercle conic, constricted at base.

Montane meadows: Sierra Nevada, in Mariposa and Tuolumne counties; east to Atlantic.

2. Fimbristylis Vahlii (Lam.) Link, Hort. Berol. 1:287. 1827.

Annual sedge with fibrous roots and caespitose culms; culms filiform, striate, 2–10 cm. tall; leaves basal, the blades nearly filiform, as long as or longer than the culms; involucral leaves 3–5, much longer than the inflorescence; inflorescence a simple capitate cluster of 3–8 spikelets; spikelets many-flowered, 4–8 mm. long; scales lanceolate, acuminate, pale greenish brown; stamen 1; style bifid, papillose above, glabrous below; achene lenticular, mucronate, 0.5 mm. long, white, cancellate with about 14 vertical rows of 15–20 rectangular pits.

Occasional throughout the San Joaquin Valley; east to Atlantic.

3. Fimbristylis thermalis Wats. in King, Geol. Expl. 40th Par. 5:360. 1871.

Perennial sedge with rhizomes; culms slender, grooved, erect, 20–70 cm. tall; leaves basal, the blades flat, 1–2.5 mm. wide, shorter than the culm, somewhat pubescent near the sheath; involucral leaves few, short, narrow; inflorescence a rather loose simple or compound umbel, the rays to 6 cm. long; spikelets oblong, 8–15 mm. long, many-flowered; scales oblong, mucronate, light brown; style bifid, pubescent all along its length, more heavily so above; achene lenticular, obovate, mucronate, about 1.5 mm. long, slate-colored, the surface with numerous rows of very fine pits.

Hot springs: San Bernardino, Kern, Inyo counties; Nevada.

Fimbristylis miliacea (Thunb.) Vahl is listed for California by Abrams (Illus. Fl. Pacif. States 1:268, 1923), on the basis of a collection by A. Wood in 1868 from San Francisco. Apparently it has not since been collected and must be regarded as a transient species.

HEMICARPHA

Dwarf annuals with fibrous roots and caespitose culms. Culms erect, filiform, grooved. Leaves basal, the blades convolute, shorter than the culms. Involucral leaves 1–3, one much longer than the inflorescence. Inflorescence a solitary spikelet or a capitate cluster of 2 or 3 spikelets. Scales spirally imbricated. Florets all perfect. Bristles absent. A single hyaline perianth part is developed between the floret and the rachis of the spikelet, but sometimes is wanting. Stamens 1–3. Style bifid, short, not swollen at base. Achene oblong, finely pitted.

Friedland, Solomon, The American species of *Hemicarpha*. Am. Jour. Bot. 28:855–861. 1941.

Scales mucronate or with short awns 0.2–0.5 mm. long . . . . 1. *H. micrantha*
Scales with awns about 1–1.5 mm. long, the scale tapering to the awn . . 2. *H. occidentalis*

1. Hemicarpha micrantha (Vahl) Pax var. **minor** (Schrad.) Friedland, Am. Jour. Bot. 28:860. 1941.

Annual with fibrous roots; culms erect or somewhat recurved, grooved, 1–3 cm. tall, or sometimes –10 cm. tall; leaves basal, the sheaths weakly united below, open above, with loose hyaline margins; the blades convolute, filiform, and grooved, shorter than the culms; involucral leaves 2 or 3, the blades convolute, one of them much longer than the inflorescence, 0.5–2 cm. long; inflorescence a solitary terminal spikelet or a capitate cluster of 2 or 3 spikelets; spikelets ovoid, obtuse, minute, about 2 mm. long, many-flowered; scales obovate, acuminate, with brown sides and a green midrib, this produced into a short recurved awn 0.2–0.5 mm. long in all but the lowermost scale; awn of the lowermost scale 1–4 mm. long, simulating an involucral leaf; perianth member hyaline, broad and fimbriate-tipped or capillary, 0.5 mm. long, persistent on the rachis; stamen 1; style short; achene oblong, obtuse, mucronulate, 0.5–0.8 mm. long, white to brown or black, the surface finely pitted.

Moist places: middle altitudes in the central Sierra Nevada, Central Valley, North Coast Ranges; warm areas in tropical America.

2. Hemicarpha occidentalis Gray, Proc. Am. Acad. Arts & Sci. 7:391. 1868.

Similar to *Hemicarpha micrantha;* culms 0.3–4 cm. tall; scales ovate-lanceolate, with brown sides and a green midrib, the latter produced into a long recurved awn about 1–1.5 mm. long or subequal to the body of the scale in length.

Wet places at middle altitudes: central Sierra Nevada, Humboldt County, San Jacinto and San Bernardino mountains; Pacific states.

RYNCHOSPORA. BEAKED RUSH

Ours perennial sedges with rhizomes and caespitose culms. Culms erect, trigonous to subterete, leafy. Leaves flat or involute. Inflorescences cymose or capitate, axillary, containing 6–20 spikelets. Scales spirally imbricated, thin, with 1 midvein usually produced as a mucro, the basal 2 to several scales empty, the middle 1–5 scales perfect, and the apical scales subtending staminate florets. Bristles 5–12 in ours, downwardly or upwardly barbed. Stamens 3. Style bifid; the style base persistent on the achene as a tubercle. Achene lenticular, the surface wrinkled or smooth, detaching from the rachis of the spikelet with a persistent gynophore to which the bristles remain attached; tubercle attenuate and subulate to compressed and deltoid.

Gale, Shirley, *Rynchospora*, section *Eurynchospora*, in Canada, the United States and the West Indies. Rhodora 46:89–134, 159–197, 207–249, 255–278. 1944.

Achenes obovate with a subulate tubercle; bristles downwardly barbed (except *R. californica*).
    Scales whitish; bristles 10–12 . . . . 1. *R. alba*
    Scales brown; bristles 5–7.
        Achenes smooth . . . . 2. *R. capitellata*
        Achenes rugulose; bristles upwardly barbed . . . . 3. *R. californica*
Achenes subglobose with a compressed tubercle; bristles upwardly barbed
                                                           4. *R. globularis*

1. Rynchospora alba Vahl, Enum. Pl. 2:236. 1806.

Culms trigonous, slender, erect, 7–70 cm. tall, leafy; leaves linear, shorter than the culm; inflorescences consisting of 1–4 axillary capitate clusters of 6–20 spikelets; spikelets oblong, acute at both ends, 4–6 mm. long, bearing 1 or 2 achenes; scales ovate to ovate-lanceolate, mucronate, whitish; bristles 10–12, stiff, downwardly barbed, about as long as the achene; achene lenticular with an obscure margin, obovate, faintly rugulose, about 1.7 mm. long, light brown; tubercle subulate, 0.6–1.2 mm. long.

Bogs: Yosemite Valley, Pitkin Marsh in Sonoma County, Mendocino County, White Mountains in Inyo County; Northern Hemisphere.

2. Rynchospora capitellata (Michx.) Vahl, Enum. Pl. 2:235. 1806.

Schoenus capitellatus Michx., Fl. Bor. Am. 1:36. 1803. In part.

Culms subtrigonous, slender, erect, 20–100 cm. tall; leaves short, flat, 1.5–3.5 mm. wide; inflorescences consisting of 1–5 axillary capitate clusters of 6–20 spikelets; spikelets oblong, acute at both ends, 3.5–5 mm. long, bearing 2–5 achenes; scales ovate-lanceolate, mucronulate, brown, caducous; bristles 6, downwardly barbed, about as long as the achene; achene lenticular, with a narrow margin, the surface smooth, brown; tubercle subulate, 1–1.6 mm. long.

Bogs: Pitkin Marsh in Sonoma County, Trinity Center in Trinity County, Buckeye Ridge in Nevada County; east to Atlantic.

3. Rynchospora californica Gale, Rhodora 46:272. 1944.

In habit similar to *Rynchospora capitellata;* spikelets ovate; scales ovate, mucronate, caducous; bristles 6 or 7, upwardly barbed, extending beyond the tubercle; achene lenticular with an obscure margin, obovate, light brown, lightly rugulose, 2 mm. long; tubercle deltoid, 1 mm. long. Possibly a hybrid of *R. capitellata* and *R. globularis.*

Known only from bogs and marshes in Sonoma and Marin counties.

4. Rynchospora globularis (Chapm.) Small, Man. SE. Fl. 184. 1933.

Culms 4–10 dm. (rarely 1.5 dm.) tall, robust, trigonous and erect to attenuate, wiry and flexuous; branchlets of the cymes usually terminating in dense or small knobby glomerules, setaceous bracts conspicuous or inconspicuous; spikelets 3–8, 2.5–4 mm. long, bearing 1–3 achenes; bristles 5 or 6, ½ to ⅓ as long as the achene; achene reticulate to striate, transversely ridged, 1.2–1.5 mm. wide, 1.3–1.6 mm. long; tubercle compressed, conical. The larger specimens from Pitkin Marsh have been designated *Rynchospora globularis* var. *recognita* Gale.

Bogs: Pitkin, Perry, and Cunningham marshes in Sonoma County; central and south Atlantic states, Central America.

SCHOENUS

1. Schoenus nigricans L., Sp. Pl. 43. 1753. Black sedge.

Perennial, with stiff quill-like culms and leaves; culms 20–70 cm. tall; leaves basal, blades convolute, some of leaves nearly as long as culms, basal sheaths shining chestnut brown; involucral leaves 2, stiff, sharp-pointed, one elongated to 8 cm. long; spikelets flattened, in a terminal capitate cluster; scales of spikelets in 2 series, lanceolate, carinate, acuminate, chestnut brown, the 3–5 basal scales empty; florets 5–8, perfect; bristles 6, plumose, longer than the achene; stamens 3; style trifid; achene trigonous, white, without tubercle, about 2 mm. long.

Hot springs: San Bernardino Mountains, Death Valley.

SCIRPUS. BULRUSH, CLUB RUSH

Perennial or rarely annual herbs. Culms erect, triangular to terete, leafy or the leaves reduced to basal sheaths. Involucre of several blade-bearing leaves, or reduced to a solitary leaf, or rarely absent. Inflorescence an open umbel or close cluster of numerous spikelets, or a solitary terminal spikelet. Scales spirally imbricated. Florets perfect. Bristles 6–1, or wanting, barbed or ciliate, usually included within the scales, in one species exserted. Stamens 2 or 3. Style bifid or trifid. Achene lenticular or trigonous.

Beetle, A. A., *Scirpus.* N. Am. Fl. 18:481–504. 1947.

Bristles much exserted, upwardly barbed .... 1. *S. criniger*
Bristles included within scales, downwardly barbed (or absent in *S. carinatus* and *S. cernuus*).
  Inflorescence subtended by involucral leaves.
    Involucral leaves 2–5, usually longer than the inflorescence; culms leafy.
      Spikelets small, 0.3–0.6 cm. long.
        Stamens 2; achene lenticular; style bifid .... 2. *S. microcarpus*
        Stamens 3; achene trigonous; style trifid .... 3. *S. Congdonii*

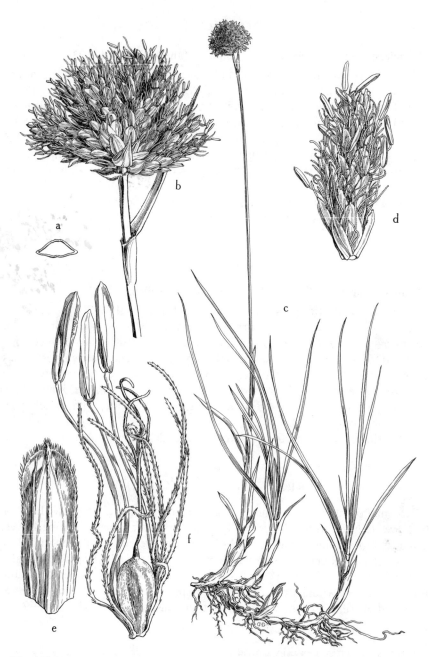

Fig. 145. *Scirpus criniger: a,* achene (cross section), × 12; *b,* capitate inflorescence, showing short, scabrous involucral leaf, × 1½; *c,* habit, the trigonous culm extending much above the leaves, × ⅖; *d,* spikelet, × 4; *e,* scale, showing fimbriate margins and toothed midrib, × 12; *f,* flower, showing the trigonous achene and the much exserted, often irregularly curved, barbed subtending bristles, × 12.

Spikelets larger, 1–4 cm. long.
  Inflorescence loose–umbellate; achene trigonous; style trifid; bristles nearly as long as the achene.
    Leaves 8–16 mm. wide .... 4. *S. fluviatilis*
    Leaves usually less than 5 mm. wide .... 5. *S. tuberosus*
  Inflorescence capitate with 1 to several elongated rays; achene lenticular; style usually bifid; bristles ½ as long as achene .... 6. *S. robustus*
Involucral leaf solitary, often appearing as a continuation of the culm.
  Culm leafy, triangular or subterete; spikelets few, 1–12.
    Perennials with rhizomes.
      Culm subterete, striate; scales not awned.
        Achenes reticulate, shining olive-green, 2 mm. long ..... 7. *S. nevadensis*
        Achenes longitudinally ribbed and horizontally striate, brownish, 1 mm. long .... 8. *S. setaceus*
      Culm sharply triangular; scales short-awned.
        Leaf blades convolute; involucral leaf long, 3–10 cm. long; plant small, to 1.1 m. tall .... 9. *S. americanus*
        Leaf blades flat; involucral leaf short, 1–3 cm. long; plant larger, to 2.2 m. tall.
          Achenes light brown or gray, the surface minutely pitted .. 10. *S. Olneyi*
          Achenes dark brown, the surface horizontally rugose; rare
                                                           11. *S. mucronatus*
    Annuals with fibrous roots.
      Scales only slightly keeled at tip, obtuse .... 12. *S. cernuus*
      Scales sharply keeled, acute or acuminate .... 13. *S. koilolepis*
  Culm leaves reduced to basal sheaths with short blades, to 8 cm. long; culm stout and terete; spikelets numerous, in umbels.
    Bristles broad, often dark red, ciliate or plumose but never barbed
                                                           14. *S. californicus*
    Bristles filiform, downwardly barbed.
      Scales much longer than the achene; style bifid or trifid; common
                                                           15. *S. acutus*
      Scales as long as or only very slightly longer than the achene; style bifid; rare
                                                           16. *S. validus*
Inflorescence not subtended by involucral leaves, these wanting; lowermost scale of spikelet produced into a long, subtending awn .... 17. *S. Clementis*

1. Scirpus criniger Gray, Proc. Am. Acad. Arts & Sci. 7:392. 1867. Fig. 145.

Perennial sedge with rhizomes; culms slender, lightly scabrous above, erect, trigonous, striate, 20–100 cm. tall (10 cm. or less at high altitudes); leaves flat, linear-lanceolate, scabrous, 4–10 cm. long; involucral leaves 2–5, short, 4–8 mm. long; inflorescence a capitate head composed of 5–10 spikelets; spikelets 3–10 mm. long; scales dark to light brown, lanceolate, fimbriate; bristles about 6, upwardly barbed, much exserted, 3–7 mm. long, caducous; filaments elongated, 6–7 mm. long; style 3-cleft; achene trigonous, oblong-obovate, mucronate, 2–3 mm. long, dark brown.

High mountain meadows: Sierra Nevada, Klamath Mountains, North Coast Ranges.

This species shares characters of both *Scirpus* and *Eriophorum* and has

Fig. 146. *Scirpus microcarpus: a,* achene and retrorsely barbed subtending bristles, × 12; *b,* achene (cross section), × 12; *c,* primary ray of umbel, × ⅘; *d,* habit, showing the spreading leaf blades and their basal sheaths, the culm cut off, × ⅕; *e,* upper part of culm, with entire leaf sheaths and compound inflorescence, the involucral leaves extending beyond the inflorescence, × ⅕; *f,* spikelet, × 8; *g,* young leaf sheath, × 2; *h,* ovate, acute scale with prominent midrib, × 16; *i,* flower, the style 2-cleft, × 16.

recently been transferred to the latter genus by Beetle, mainly on the basis of its achenes, which are as long to twice as long as those of other species in a related section in *Scirpus.* The conspicuously exserted bristles give it the superficial appearance of an *Eriophorum,* but in the present species the bristles are barbed, whereas in species of *Eriophorum* they are smooth. The bristle characters have been commonly used to separate this anomalous species of *Eriophorum,* and this distinction is considered valid in the present treatment.

2. Scirpus microcarpus Presl, Rel. Haenk. 1:195. 1830. Fig. 146.

Perennial with stout rhizomes; culms stout, erect, leafy, subterete, 0.7–1.7 m. tall; leaves flat, broad, 1–2 cm. wide, margins scabrous, the blades acuminate, often overtopping the stem; involucral leaves 2–5, the longer ones usually extending beyond the heads; inflorescence a loose, spreading, compound umbel, the primary rays to 10 cm. long; scales green to brown, acute, ovate, with a prominent midrib, not awned; bristles 4, downwardly barbed, somewhat longer than the achene; stamens 2; style 2-cleft; achene whitish, ovate, lenticular, with an obscure dorsal crest, mucronate, small, 1 mm. long.

Along streams and about springs: Coast Ranges, to coastal southern California, Sierra Nevada, there passing into *Scirpus Congdonii* Britt., east of the Sierra Nevada crest from Inyo County north to Modoc County; north to Alaska, east to Atlantic.

3. Scirpus Congdonii Britt., Torreya 18:36. 1918. Fig. 147.

Perennial with stolons and short, stout rhizomes; culms slender, trigonous, leafy, 4–5 dm. tall; basal leaves to 8 mm. broad, the culm leaves fewer and progressively shorter, the blades scabrous on the margins and midrib below, abruptly acuminate; involucral leaves 3–7 cm. long, as long as or shorter than the inflorescence; inflorescence usually abbreviated, the spikelets subcapitate; scales brown, ovate to ovate-lanceolate, abruptly acuminate; bristles white, obscurely scabrous, longer than achene; stamens 3; style 3-cleft; achene white, ovate, trigonous, prominently apiculate, 1–1.25 mm. long.

Montane meadows mostly at middle altitudes, sometimes higher: throughout the Sierra Nevada.

This species bears some resemblance to *Scirpus microcarpus* Presl, and possibly intergrades with it at lower elevations. It may be distinguished, however, by its shorter culms, narrower leaves, which are shorter than or of the same length as the culm, the somewhat larger, acuminate spikelets and scales, the slender bristles curled at the ends, 3 stamens, the trifid style, and a trigonous, more prominently mucronate achene.

4. Scirpus fluviatilis (Torr.) Gray, Man. 527. 1848. River bulrush. Fig. 148.

Perennial sedge with horizontal rhizomes forming tubers; culms stout, sharply triangular, erect, 1–1.5 m. tall; leaves 8–16 mm. wide; involucral leaves 3–5, unequal in length, to 20 cm. long; inflorescence umbellate; spike-

Fig. 147. (For explanation, see facing page.)

lets acute, 1.6–2.5 cm. long; bristles 6, unequal in length, nearly as long as the achene; style trifid; achene usually sharply triangular, angled on back, 4 mm. long.

Widely distributed in interior valleys: Sacramento Valley, Lassen County and Modoc County, occasional near coast in Lake and San Mateo counties; east to Atlantic.

5. Scirpus tuberosus Desf., Fl. Atlant. 1:50. 1798.

Culms sharply trigonous, erect, to 5 dm. tall, from swollen nodes on slender black rhizomes; sheaths covering about ½ the length of the culms, the margin of the orifice conspicuously raised, hyaline; leaf blades mostly less than 5 mm. broad, the margins and midrib scabrous at the tip; involucral leaves unequal in length, the margins scabrous; inflorescence compound-umbellate, of many spikelets, 1–2 cm. long; scales pubescent, light brown to dark brown, the apex notched, the midrib mucronate; bristles 6, unequal in length, shorter than the achene, downwardly barbed; stamens 3; style trifid, or, rarely, both bifid and trifid styles in the same spikelet; achene 2.5 mm. long, dark brown, trigonous, slightly apiculate.

Known in California from at least three stations: south of Willows in Glenn County, north of Gridley in Butte County, east of Norman in Colusa County; native to North Africa and southern Eurasia, from the Mediterranean to India.

6. Scirpus robustus Pursh, Fl. Am. Sept. 1:56. 1814. Fig. 149.

Scirpus paludosus Nels., Bull. Torrey Club 26:5. 1899.

Perennial sedge with horizontal rhizomes forming tubers; culms erect, sharply triangular, 0.5–1.5 m. tall; leaves typically 4–6 mm. wide, sometimes to 15 mm. wide, the involucral leaves 2–5, unequal in length, to 30 cm. long; inflorescence capitate, or with 1 to several elongated rays; spikelets ovate, 1–2.5 cm. long, sometimes cylindric, to 4 cm. long; scales reddish brown to pale straw-colored; bristles 1–6, ½ as long as achene; style usually bifid, sometimes trifid; achene lenticular, 3–4 mm. long.

Common in salt, fresh-water, or alkaline marshes: coastal and desert southern California, Coast Ranges, Central Valley, east side of the Sierra Nevada from Inyo County to Modoc County; widespread across the continent.

There is considerable variation in spikelet and floral characters in this species. A coastal form is characterized by having rather dark reddish brown,

Fig. 147. *Scirpus Congdonii: a,* spikelets, × 8; *b,* flower, showing 3-cleft style and the long, filiform bristles, × 16; *c,* scale, ovate, abruptly acuminate, and with margins minutely fimbriate across the top, × 16; *d,* achene (cross section), × 20; *e,* achene, trigonous and prominently apiculate, the subtending bristles filiform, obscurely scabrous, very long and irregular, × 20; *f,* habit, showing stout rhizomes, the culm leaves progressively shorter and less spreading near the inflorescence, × ⅖; *g,* inflorescence, showing the primary rays and the subcapitate spikelets, × ⅖; *h,* young leaf sheath, × 4; *i,* part of basal leaf, showing ligule and remains of sheath, × 4.

Fig. 148. *Scirpus fluviatilis: a–d*, achenes, showing variation in shape (cross sections), × 6; *e*, flower, style slender and trifid, the bristles unequal in length, × 3; *f* and *g*, awned scales, × 4; *h*, achene, the subtending bristles unequal in length, × 4; *i*, spikelet, × 1⅓; *j*, rhizome, tubers, and sharply triangular culms, × ⅖; *k*, inflorescence with nearly sessile rays and longer primary rays, × ⅖; *l*, habit, showing rhizomes, tubers, sheathing culm leaves, and umbellate inflorescence with the involucral leaves unequal in length, × ⅙.

Fig. 149. *Scirpus robustus: a,* habit, showing the horizontal rhizome-forming tubers, the triangular culms, and the culm leaves, × ⅛; *b,* lenticular achene, the subtending bristles ½ as long as the achene, × 6; *c,* achene (cross section), × 6; *d,* inflorescence, showing the involucral leaves of unequal length, × ⅖; *e,* awned scale, × 4.

Fig. 150. *Scirpus nevadensis: a,* obtuse scale, its margins fimbriate across the top, × 6; *b,* flower, the style 2-cleft, × 6; *c,* mature achene with punctate surface, the subtending bristles short, × 10; *d,* achene (cross section), × 10; *e,* single spikelet with part of a long involucral leaf, × 2½; *f–h,* variations in habit, × ⅖; *i,* capitate cluster of spikelets with one long involucral leaf and several minute ones, × ⅖.

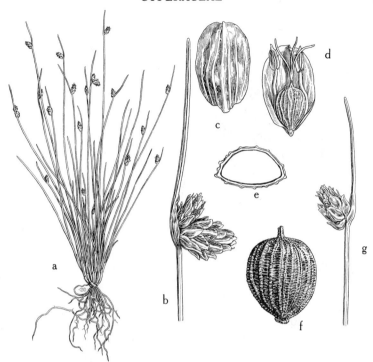

Fig. 151. *Scirpus setaceus: a,* habit, culms and leaf blades filiform, × ⅔; *b,* solitary involucral leaf with paired spikelets, the lower one larger, × 4; *c,* obtuse scale, strongly keeled near apex, × 16; *d,* flower, × 16; *e,* achene (cross section), × 32; *f,* mature achene, showing prominent longitudinal ribs and fine horizontal striations between ribs, × 32; *g,* solitary spikelet, × 4.

ovate spikelets, often correlated with a trifid style and a more or less obtusely trigonous achene. Toward the interior, the plants have more elongate, cylindrical spikelets, possibly because of intermixture with the closely allied *Scirpus fluviatilis* (Torr.) Gray, and have somewhat narrower leaves, paler scales (strawcolored in the more arid parts of its range), bifid styles, and lenticular achenes. This interior form has been recognized by several authors as a distinct species, *S. paludosus* Nels. In our studies, however, we have found no consistent characters which will serve to maintain the two forms as distinct entities. For example, it has been observed that coastal plants bearing the ovate spikelets may also produce the more elongate spikelets, either on the same culm or from another culm on the same rhizome. Although a trifid style often occurs in the coastal plants, individuals are not infrequently found in which bifid and trifid styles are intermixed within the same spikelet or the spikelet may have only bifid styles.

Fig. 152. *Scirpus americanus: a,* single spikelet, × 4; *b,* inflorescence with the solitary involucral leaf, × ⅘; *c,* scale, showing the cleft apex and its short awn, × 6; *d,* obtusely trigonous achene (cross section), × 6; *e,* achene, the subtending bristles unequal in length, × 6; *f,* habit, showing the rhizomes, the triangular culms, and the keeled, convolute, narrow leaf blades, × ¼.

7. Scirpus nevadensis Wats. in King, Geol. Expl. 40th Par. 5:360. 1871. Nevada club rush. Fig. 150.

Perennial sedge, with slender rhizomes, bearing slender, erect, or sometimes arched, subterete culms, 6–45 cm. tall; leaf blades narrow, about 1 mm. wide, convolute, to 15 cm. tall; involucral leaf commonly solitary or 1 long and 1–3 short ones, 1–10 cm. long; inflorescence a capitate cluster of 1–8 spikelets; spikelets narrow, acuminate, 0.5–2 cm. long; scales shining chestnut brown, obtuse, not awned; bristles unequal, from very minute to nearly ½ as long as the achene; style 2-cleft; achene convex on one surface, cuneate, shining olive-green and punctate, 2 mm. long.

Wet alkaline ground: Modoc County to Mono County; Nevada.

8. Scirpus setaceus L., Sp. Pl. 1:49. 1753. Fig. 151.

Perennial sedge with very slender horizontal rootstocks, bearing filiform, erect culms, 6–12 cm. tall; leaf blades filiform, about 0.5 mm. wide, to 8 cm. long; involucral leaf solitary, 4–10 mm. long; spikelets solitary or 2, if 2, the lower one is often larger; spikelets narrow-ovate, 2–4 mm. long; scales dark brown with a broad green midvein, obtuse, strongly keeled near apex, not awned; achene somewhat angled on one side, prominently longitudinally ribbed, finely and horizontally striate between the ribs, brown, 1 mm. long.

Moist places: Humboldt and Del Norte counties; widely distributed in Northern Hemisphere.

In general habit and in spikelet characters, this species bears a close resemblance to *Scirpus cernuus*. It can be readily distinguished from *S. cernuus*, however, by its perennial habit and its distinctive achene markings.

9. Scirpus americanus Pers., Syn. Pl. 1:68. 1805. Three-square. Fig. 152.

Perennial with horizontal rhizomes; culms erect or arched, sharply triangular, stiff and slender, 0.3–1.1 m. tall; leaf blades to 18 cm. long, keeled, convolute, narrow, 2–3 mm. wide; involucral leaf solitary, 3–10 cm. long; inflorescence a capitate cluster of 1–7 spikelets; spikelets oblong, acuminate, 8–12 mm. long; scales pale brown to chocolate brown, cleft at apex, short-awned; bristles 2–6, downwardly barbed, unequal in length, from slightly longer than to only ½ as long as the achene; style 2- or 3-cleft; achene lenticular, or obtusely trigonous, mucronate, 3 mm. long.

Widely distributed in wet ground: along coast from Ventura County to Del Norte County, occasional in San Bernardino County and Imperial County; Inyo, Mono, Lassen, and Modoc counties; San Joaquin Valley; occasional (?) in Sacramento Valley.

A small form bearing minute tubers and few-flowered spikelets is frequently found growing with a larger and stouter form. Whether these two forms are genetic races or edaphic forms cannot at present be determined.

Scirpus americanus var. *polyphyllus* (Boeckl.) Beetle was distinguished from

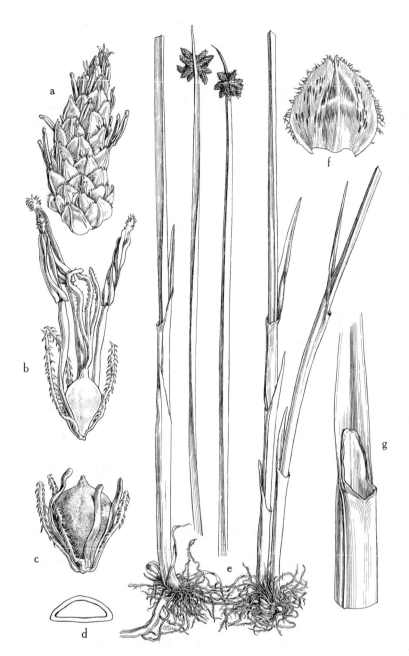

Fig. 153. *Scirpus Olneyi: a,* spikelet, $\times$ 5; *b,* flower without scale, the style 3-cleft, $\times$ 12; *c,* achene, minutely pitted, mucronate, the subtending bristles unequal in length, $\times$ 10; *d,* achene (cross section), $\times$ 10; *e,* habit, showing the rhizomes, the stout, sharply triangular culms, the basal leaf sheaths bearing short blades, and the inflorescences of clustered spikelets with short, solitary involucral leaves, $\times$ ⅖; *f,* short-awned scale with cleft apex, ciliate margins, and broad, rounded base, $\times$ 8; *g,* leaf sheath, the ligule emarginate, $\times$ 1⅕.

the typical form as having a trifid style, 3 or more blade-bearing sheaths, and solid, chocolate brown scales. Although most of the California plants have trifid styles, occasional collections also show bifid styles. The number of blade-bearing leaves and the color of the scales are more variable characters. A 2-leaved condition seems to be of most common occurrence in the California plants. Plants with solid, chocolate brown scales appear to be restricted to collections from along or near the coast, whereas plants from farther inland show pale brown scales somewhat flecked with red. Similar variations in these characters are noted by Beetle for collections of the species from other regions in North America. It therefore appears that there are no consistent differences which will distinguish the California material from typical *S. americanus*.

10. Scirpus Olneyi Gray, Boston Jour. Nat. Hist. 5:238. 1845. Fig. 153.

Perennial sedge with long rhizomes; culms stout, sharply triangular, the sides concave, 0.5–2.2 m. tall; leaf blades short, 2–13 cm. long; involucral leaf short, erect, 1–3 cm. long; inflorescence a capitate cluster of 5–12 spikelets; spikelets ovoid, 5–8 mm. long; scales flecked with brown, short-awned; bristles 4–6, downwardly barbed, unequal in length, from shorter than to as long as the achene; style 3-cleft; achene lenticular, mucronate, light brown or gray, minutely pitted, 2.5 mm. long.

Widely distributed in marshes and occasionally around hot springs: southern Sacramento Valley, east side of the Sierra Nevada from Inyo County north to Modoc County, Coast Ranges north to Napa County; coastal and desert southern California; east to Atlantic.

Alan A. Beetle (Am. Jour. Bot. 30:397, 1943) tentatively suggests that *Scirpus chilensis* Nees & Meyen (Nova Acta Acad. Leop.-Carol. 19, Suppl. 2:93, 1843) may be the correct name for this species. A further discussion of the nomenclature of this species is given by Fernald (Rhodora 45:390–392, 1943).

11. Scirpus mucronatus L., Sp. Pl. 1:50. 1753. Fig. 154.

Culms thick, 4–9 dm. tall, clustered, sharply trigonous, the base with leafless, obliquely truncate sheaths; spikelets oblong, sessile, closely crowded into a lateral, spherical head, overtopped by the deflexed (at length) continuation of the culm; scales longitudinally striate, buff-colored, ovate, obtuse, mucronate; achene transversely wrinkled, mucronulate.

Introduced as a weed in rice fields in Glenn and Butte counties. Native to Eurasia.

This species is similar to *Scirpus Olneyi* Gray, but the spikelets are somewhat longer (8–10 mm. long), the scales are broader and shorter-tipped, and the achenes are dark brown and horizontally rugose.

Fig. 154. (For explanation, see facing page.)

12. Scirpus cernuus Vahl var. **californicus** (Torr.) Beetle, Rhodora 46:145. 1944. Low club rush. Fig. 155.

Annual with fibrous roots; culms caespitose, nearly filiform, 4–20 cm. tall; 1 basal leaf sheath bearing a convolute blade 2–5 cm. long, the others without blades or with short blades to 3 mm. long; involucral leaf solitary, appearing as a continuation of the culm, 2–5 mm. long; spikelet solitary, ovoid, 2–5 mm. long; scales broad, obtuse, keeled at tip, pale brown to deep brown with green midrib; bristles none; stamens 2 or 3; style trifid; achene trigonous, punctate, white turning brownish at maturity, 1 mm. long.

Coastal marshes and tidal flats: San Diego County to Del Norte County, southern California inland to western Riverside and San Bernardino counties.

13. Scirpus koilolepis (Steud.) Gleason, Rhodora 44:479. 1942. Fig. 155.

Isolepis koilolepes Steud., Syn. Pl., pt. 2—Cyp. 318. 1855.
Scirpus carinatus Gray, Proc. Am. Acad. Arts & Sci. 7:392. 1868.

Annual with fibrous roots; culms caespitose, 4–20 cm. tall, filiform; leaf blades to 4 cm. long, obtuse; involucral leaf solitary, appearing as a continuation of the culm, to 2.5 cm. long, obtuse; spikelets 1–3 on a culm, ovate, 2–5 mm. long; scales strongly carinate, greenish brown, acute to acuminate, with broad hyaline margins, the broad, green midrib shortly mucronate; bristles none; style trifid; achene sharply trigonous, puncticulate, light brown to dark brown, 1.5 mm. long.

Coastal marshes: Monterey County north to Mendocino County; across southern United States.

Similar to *Scirpus cernuus* Vahl, but differing in the longer involucral leaves and the greenish, sharply keeled, acute to acuminate scales.

According to Beetle (N. Am. Fl. 18:496, 1947), the correct spelling of this name is "koilolepis."

14. Scirpus californicus (C. A. Mey.) Steud., Nom. Bot., 2d ed., 2:538. 1841. California bulrush. Fig. 156.

Perennial sedge with stout, subterete to triangular culms to 4 m. tall; leaves reduced to basal sheaths; involucral leaf solitary, erect, shorter than inflorescence; inflorescence loosely umbellate; spikelets narrow, acute, 5–10 mm. long; scales ovate, reddish brown; bristles 2–4, dark red, or sometimes pale red, broad and ciliate or plumose, not barbed; style bifid; achene lenticular, 2 mm. long.

Fig. 154. *Scirpus mucronatus: a,* achene, mucronulate, the surface transversely wrinkled (rugose), the subtending bristles unequal in length, × 12; *b,* achene (cross section), × 12; *c,* spikelet, × 6; *d,* head of closely crowded, sessile spikelets overtopped by deflexed continuation of culm, × ⅘; *e,* mature spikelet, the lower scales and achenes shed, × 6; *f,* flower, × 8; *g,* scale, mucronate, longitudinally striate, the margins finely toothed near apex, × 8; *h,* habit, showing the leafless, obliquely truncate basal sheaths, the sharply trigonous culms, and the inflorescences with solitary involucral leaves, which are sometimes horizontally deflexed, × ⅕.

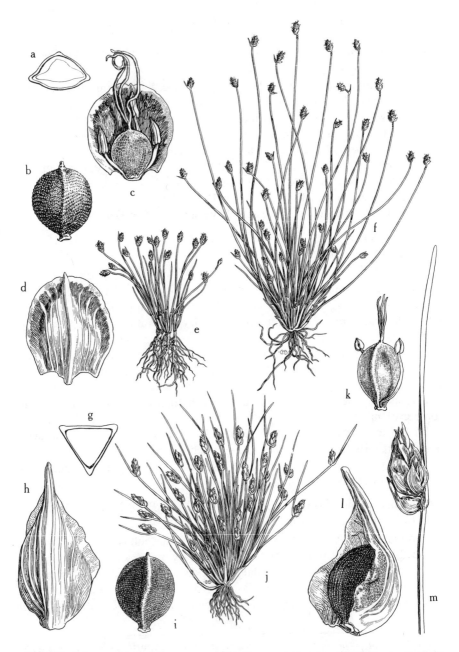

Fig. 155. *Scirpus. a–f, S. cernuus* var. *californicus: a,* trigonous achene (cross section), × 16; *b,* mature achene, its surface minutely papillose, × 16; *c,* flower, × 16; *d,* scale, broad, rounded, and keeled, × 16; *e* and *f,* habit variations, some of the basal leaf sheaths bearing blades, × ⅔. *g–m, S. koilolepis: g,* sharply trigonous achene (cross section), × 12; *h,* scale, showing broad, hyaline margins, × 16; *i,* mature achene, its surface puncticulate, × 12; *j,* habit, × ⅘; *k,* young achene, the style trifid, × 12; *l,* mature achene, tightly enclosed by the strongly carinate scale, × 16; *m,* culm with single spikelet and the solitary, obtuse involucral leaf, × 4.

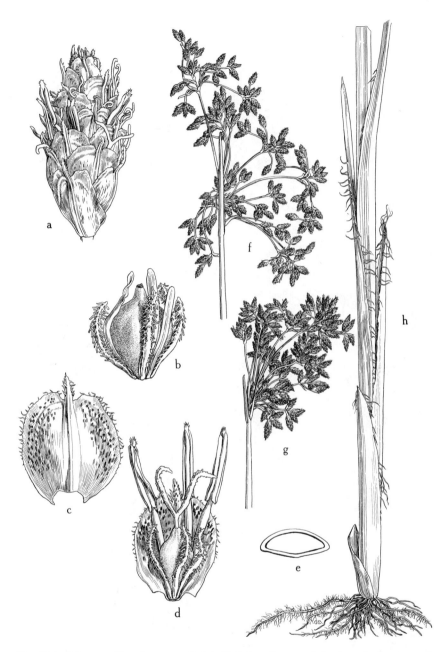

Fig. 156. *Scirpus californicus: a,* spikelet, × 6; *b,* achene and the broad, plumose subtending bristles, × 10; *c,* ovate, awned scale, the margins fimbriate, × 8; *d,* flower, style
bifid, × 8; *e,* achene (cross section), × 10; *f* and *g,* loosely umbellate inflorescences,
showing the solitary, short, erect involucral leaves, × ⅖; *h,* basal leaf sheaths and lower
part of the triangular culm, × ⅖.

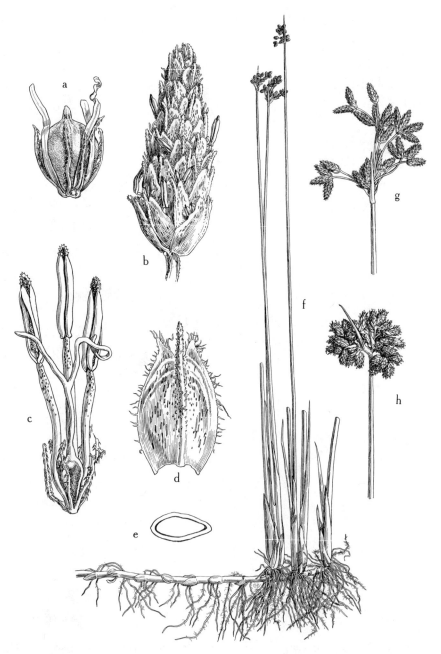

Fig. 157. *Scirpus acutus: a,* mature achene, the subtending bristles with conspicuous retrorse barbs, × 8; *b,* spikelet, × 4; *c,* flower without the scale, × 12; *d,* carinate scale, showing the short awn and the cleft, ciliate apex, × 8; *e,* achene (cross section), × 8; *f,* habit, showing stout rhizome, basal sheaths, and erect culms, × ⅕; *g* and *h,* inflorescences, showing variation, × ⅖.

Common in marshes along coast from San Diego County to Napa County; Central Valley, occasional in the Mojave Desert (Lancaster, Needles).

This species is similar in aspect to *Scirpus acutus* Muhl. It can be distinguished, however, not only by the characters given in the key, but also by its subterete to triangular culms (most noticeable in the upper parts); in *S. acutus* the culms are terete throughout their length. Also, the spikelets of *S. californicus* are smaller, and the scales are more consistently reddish brown.

15. Scirpus acutus Muhl. ex Bigel., Fl. Boston. 15–16. 1814. Tule, great bulrush. Fig. 157.

Perennial with stout rootstocks and thick brown rhizomes; culms stout, to 2 cm. thick, erect, to 5 m. tall; leaves reduced to basal sheaths with blades to 8 cm. long; involucral leaf solitary, terete, shorter than the inflorescence, and appearing as a continuation of the culm; inflorescence dense, capitate, to lax and umbellate; spikelets ovate or cylindric, 8–18 mm. long; scales ovate, carinate, with short awn, flecked with red to varying degrees, so that the spikelet varies in hue from pale brown to reddish brown; bristles slender, downwardly barbed, usually 6 in number, varying in length from ⅓ to fully as long as the achene; achene 2–3 mm. long and well enclosed by scales, lenticular with bifid style, or sometimes trigonous with trifid style.

Very abundant in marshes: Central Valley, widespread throughout Coast Ranges, Sierra Nevada, Modoc County, San Francisco Bay area; east to Atlantic.

Typical *Scirpus acutus* Muhl., with large, lenticular achenes, bifid styles, and densely clustered inflorescences, is mainly interior in distribution. It intergrades in southern California with *S. heterochaetus* Chase, which has predominantly trigonous achenes and trifid styles. A maritime race occurring from Marin County to Humboldt County is distinguished from the typical form by the deeper reddish brown of the spikelets.

The number of divisions of the style is a variable character within this species. Although many plants, particularly those from the interior valleys, have bifid styles and lenticular achenes, frequently plants are found having trifid styles and trigonous achenes, often the two types of styles and achenes occurring within the same spikelet. The southern California *S. heterochaetus* Chase usually has trifid styles, but individual specimens sometimes have bifid styles. In the characterization of a coastal form, *S. rubiginosus,* Beetle (Am. Jour. Bot. 28:697, 1941) has pointed out the variable number of styles, a character used to distinguish this species from *S. acutus.* However, these two plants are so insufficiently defined on this as well as other characters that it seems best to retain this coastal form under *S. acutus.*

16. Scirpus validus Vahl, Enum. Pl. 2:268. 1806.

Perennial with slender, scaly rhizomes; culms stout, to 2 cm. thick, erect, to 5 m. tall; leaves with reduced blades, the upper blades stiff, 4 mm. to 1 dm.

long; involucral leaf 0.7–7 cm. long, rigid; spikelets ovate, 5–10 mm. long; scales pale brown, viscid-dotted, the margins entire or minutely to strongly fimbriate, with a slightly recurved, mucronate tip; bristles downwardly barbed, usually 6, as long as the achene, tortuous; style bifid; achene 2.5 mm. long, deep gray at maturity, lenticular, apiculate.

The status of this species in California is doubtful. We have seen no fully authenticated specimens, but a single collection from Oro Fino, Siskiyou County (*G. D. Butler 137*), compares favorably with extra-California collections of *Scirpus validus* in having smaller, more broadly ovate spikelets (larger and more ovate-cylindric in *S. acutus*), with scales as long as or very slightly longer than the achene. The species is distributed across the continent.

17. Scirpus Clementis Jones, Contr. West. Bot. No. 14:21. 1912.

    Scirpus yosemitanus Smiley, Univ. Calif. Publ. Bot. 9:108. 1921.

Perennial sedge with caespitose culms; culms slender, grooved, 3–10 cm. tall; basal leaf sheath bearing a linear blade 0.5–3 cm. long; true involucral leaf wanting, but lowermost scale of spikelet simulating an involucre by the production of its green midrib into an awn 2–4 cm. long; spikelets ovate, 2- to 4-flowered; scales ovate, obtuse, brown; bristles 6, shorter than achene; stamens 3; style trifid; achene trigonous, the surface minutely apiculate, about 1.5 mm. long.

High montane meadows: Sierra Nevada from Fresno County to Tuolumne County and Mono County to Inyo County.

ARACEAE. ARUM FAMILY

Stout perennial glabrous herbs. Leaves large, often fleshy, basal or alternate and 2-ranked on the stem, simple. Flowers perfect or unisexual, densely crowded on a cylindrical spadix, enclosed or subtended by a usually colored spathe. Perianth of 4–6 scale-like segments or wanting. Stamens in ours 2–6. Ovary 1- to several-celled. Ovules 1 to several in each cell. Fruit in ours a berry.

Leaves cauline as well as basal, swordlike, 2-ranked .... *Acorus*
Leaves all in a basal cluster, broad and often fleshy.
    Leaves cordate-ovate or sagittate-ovate; spathe white or creamy white ... *Zantedeschia*
    Leaves oblanceolate or elliptic; spathe lemon yellow .... *Lysichitum*

ACORUS

1. Acorus calamus L., Sp. Pl. 324. 1753. Sweet flag.

Herbaceous perennial with stout, creeping, aromatic rootstock; leaves 2-ranked, sword-shaped, erect, 6–18 dm. long, 2.5 cm. wide or less, sharp-pointed and sharp-edged, with a rigid midvein running the entire length, the leaves closely sheathing each other and the scape below; scapes 3-angled, keeled on the back; spathe a tapering leaflike continuation of the scape, projecting 2–8 dm. beyond the spadix; spadix cylindrical, spikelike, yellow-green, appearing lateral on the scape, 5–10 cm. long, about 1 cm. in diameter; flowers perfect, densely covering entire spadix; perianth of 6 greenish yellow segments; stamens 6; ovary 2- or 3-celled with several ovules; fruit a gelatinous berry with 1–4 seeds.

Widespread, but known to us from only one collection, this from the border of a marshy pond on the southern edge of the town of Blue Lake, Humboldt County. This collection, *Nobs & Smith 1242,* was in sterile condition, apparently a rather frequent occurrence in the species. Muenscher notes, in Aquatic Plants of the United States (p. 175, 1944), that plants growing in partially drained, marshy fields that become relatively dry in summer may not flower or fruit for years, whereas plants growing in shallow water or in spring-fed marshes usually produce fruit and viable seeds in abundance.

Buell, M. F., *Acorus calamus* in America. Rhodora 37:367–369. 1935.

LYSICHITUM

1. Lysichitum americanum Hultén & St. John, Sv. Bot. Tidskr. 25:455. 1931. Yellow skunk cabbage. Fig. 158.

Lysichitum camtschatcense of American authors, in part. Not Schott.

Robust perennial herb from a fleshy rootstock, glabrous; leaves erect or spreading, 3–15 dm. long, glabrous, entire, oblanceolate or elliptic, subsessile or short-petioled, blade and petiole developing after flowers; spathe 10–18 cm. long, pale lemon yellow, elliptic, boat-shaped, obtuse or shortly acuminate, sheathing at base; spadix at first partly enclosed by the spathe, becoming long-exserted upon a stout peduncle, cylindric or narrowly fusiform-cylindric, 3.5–11 cm. long, 10–25 mm. thick; flowers yellow-green, fetid, numerous, densely aggregated, perfect; perianth segments 4, 3–4 mm. long, fleshy, concave, imbricate; stamens 4, the anthers exserted; ovary 2-celled, ovoid-conical, fleshy, constricted at the base, 1 ovule in each cell; fruit berry-like, the apex prominently green, the lower part white, sunk in spadix.

Swampy woods or bogs: near the coast from Santa Cruz County to Del Norte County; north to Alaska, Pacific coast to Japan.

The Asiatic *Lysichitum camtschatcense,* which has been considered conspecific with *L. americanum,* differs in having a white spathe, odorless flowers, a somewhat smaller spadix, and smaller perianth segments (2–3 mm. long).

Hultén, E., and H. St. John, The American species of *Lysichitum.* Sv. Bot. Tidskr. 25: 453–464. 1931.

———, Notes on *Lysichitum americanum.* Sv. Bot. Tidskr. 28:362. 1934.

Fig. 158. *Lysichitum americanum: a,* young flower, top view (stamens emerging one at a time), $\times$ 8; *b,* habit, showing the glabrous, elliptic leaves and the boat-shaped spathes enclosing spadices, $\times$ ⅓; *c,* fruit, enclosed at base by persistent perianth, $\times$ 3; *d,* the fleshy rootstock, $\times$ ⅖.

ZANTEDESCHIA

1. Zantedeschia aethiopica Spreng., Syst. Veg. 3:765. 1826. Calla of gardeners.

Calla aethiopica L., Sp. Pl. 968. 1753.

Robust, perennial, rhizomatous herb; leaves basal, long-petioled, smooth, shining green, cordate-ovate or sagittate-ovate with slender cusp at apex; spathe white or creamy white, 11.5–23 cm. long, with an open, broad, flaring limb and a long recurved cusp; spadix prominent but much shorter than spathe, pistillate on lower part, staminate on upper part; flowers naked; stamens 2 or 3, free; ovary with 1 or more cells, the ovules usually 4 in each cell; fruit berry-like.

This commonly cultivated species, *Zantedeschia aethiopica,* is sometimes found as an escape in wet or marshy places near the coast. Native of South Africa.

The true calla, *Calla palustris* L., native of the northeastern states, was reported for California from Glen Blair, Mendocino County, by Alice Eastwood (Zoe 5:58, 1900). The collection on which this report was based was destroyed in the San Francisco earthquake and fire of 1906, and the species has not been collected since.

LEMNACEAE. DUCKWEED FAMILY

Diminutive floating aquatics from the surface or just below the surface of ponds, ditches, or quiet water, or occasionally in wet seepage places. Plant consisting of leafless, flat, disclike or spheroid stems, rootless or bearing 1 or more simple roots from the lower surface. Flowers from a saclike spathe in a pouch at the basal margin of the frond or breaking through the surface to one side of the spathe, consisting of a single stamen or a single pistil, often 2 staminate flowers and 1 pistillate flower to a spathe.

Plants without roots; fronds floating at or just below the surface of the water, or a part of the frond breaking through the surface-tension layer of the water; flowers breaking through upper surface of frond.
 Fronds of the plant ligulate, often curved like the segment of a band; solitary or united in stellate colonies .... *Wolffiella*
 Fronds of the plant ellipsoid, minute, appearing as though turgid, often less than 1 mm. long; usually only mother and daughter fronds connected .... *Wolffia*
Plants with evident roots; fronds floating on the surface of the water; flowers developing in the reproductive pouch.
 Roots solitary on each segment of the frond; fronds 1–3 mm. long, undersurface green or streaked with brown .... *Lemna*
 Roots several from each segment of frond; fronds 3–6 mm. long, undersurface usually reddish .... *Spirodela*

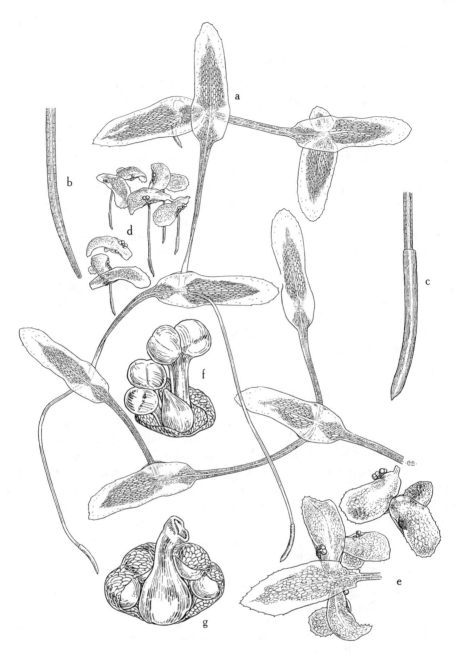

Fig. 159. *Lemna trisulca: a,* habit, vegetative plant, showing mother and daughter fronds remaining attached by long, attenuate stipes, some fronds with solitary root, × 4; *b,* root tip without the rootcap, × 20; *c,* root with rootcap, × 20; *d,* habit, flowering plants, × 2; *e,* flowering plants, showing flowers emerging from lateral pouches, × 4; *f,* spathe at time of development of staminate flowers, × 40; *g,* spathe at time of development of pistillate flowers, × 40.

LEMNA

Diminutive floating aquatics or growing on wet surfaces. Plants consisting of leafless, stemlike tissue with an evident or inconspicuous stipe, bearing on either side a meristematic pouch in which are produced vegetative and flower buds. Vegetative buds usually disarticulate to form independent plants (in *Lemna trisulca* often remaining attached). Flowers monoecious, produced in a membranous spathe, the staminate flowers usually 2 to a spathe, each flower consisting of a single stamen, the pistillate flowers usually 1 to a spathe and having a single pistil. Ovary 1- to 3-ovuled. Seeds usually early ribbed and containing a distinct operculum.

Fronds frequently remaining attached in chainlike colonies, blades elliptic to lanceolate, often serrulate, on long stipes .... 1. *L. trisulca*
Fronds soon separating, 2–5, rarely more, temporarily joined and these sessile or on very short stipes, obovate to oblong, their margins entire.
  Spathe of inflorescence urn-shaped or campanulate, not open on one side; ovules 1 or more, amphitropous; fruits broad at the shoulder, sometimes winged; air chambers within the frond in 2 layers.
    Fruits winged at the shoulders; seeds usually 2; air chambers within the frond conspicuous from below, often inflated to form a gibbous frond ...... 2. *L. gibba*
    Fruits broad but not winged at the shoulders, seed 1; air chambers not conspicuous and never inflated .... 3. *L. minor*
  Spathe open down one side, often kidney-shaped; ovules solitary, orthotropous, but sometimes oblique within the ovary; air chambers within the frond usually in a single layer.
    Fronds obscurely to strongly 3-nerved .... 4. *L. perpusilla*
    Fronds nerveless or very obscurely 1-nerved.
      Fronds thin and flat, without papules, surface texture uniform throughout
                                        5. *L. valdiviana*
      Fronds thick, with a low median ridge bearing 2 or more papules, usually with a thin margin .... 6. *L. minima*

1. Lemna trisulca L., Sp. Pl. 970. 1753. Stan duckweed. Fig. 159.

Plants often forming dense masses, usually floating just below the surface or growing on wet, mossy shores, mother and daughter fronds often remaining attached for several generations by long, attenuate stipes; blade of fronds 6–10 mm. long, lanceolate-elliptic to oblanceolate, attenuate below into a slender stipe, entire or more frequently denticulate toward the apex, each with a solitary root or in some fronds the root lacking; reproductive pouches lateral near the basal end; spathe open on one face; staminate flowers 2; pistillate flower 1; ovule amphitropous; ribs of the seed conspicuously corky-thickened.

Ponds: occasional in the mountains and east of the Sierra Nevada from Owens Valley to the northeast basin of Modoc and Lassen counties, San Bernardino Mountains; north to Alaska, east to the Atlantic, Old World.

In plants as much reduced as the Lemnaceae, differences as great as those between *Lemna trisulca* and the other species of *Lemna* might justify separate generic status. Certainly these differences are fully as great as those between

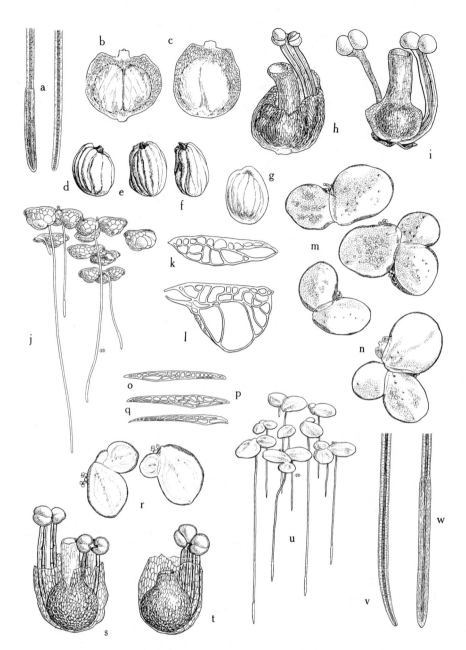

Fig. 160. (For explanation, see facing page.)

Spirodela and *Lemna*. Flowering material from Sage Hen Creek in northern Nevada County indicates that at the time of flowering the fronds break through the surface-tension layer of water just before anthesis. These fertile plants differ markedly in form from the usual vegetative plants and more nearly resemble the other species of the genus in being shorter and more rounded on the dorsal surface. However, many transitional forms between extremes were found flowering. The reproductive pouch of *Lemna trisulca* is consistently more lateral on the frond than in the other species.

2. Lemna gibba L., Sp. Pl. 970. 1753. Inflated duckweed, wind bags. Fig. 160.

Fronds solitary or few in a group, orbicular-obovate, 2–5 mm. long, 2–4 mm. wide, thick, with 2 layers of air spaces within, dark green above and often suffused with red or purple, with a slight ridge and 1- to 3-nerved above and conspicuously round, inflated-gibbous below because of the enlargement of the lower tier of air spaces, or these not much enlarged and the frond merely convex below; spathe saclike, completely surrounding the inflorescence; staminate flowers 2; pistillate flower 1; ovary flattened, winged at the shoulders, the style elongate, the ovules 2, amphitropous; fruit strongly winged above, broader than high; seeds with corky-thickened ribs and a conspicuous operculum.

Frequent in ponds, marshes, and slow streams, occasional on wet banks throughout the valleys and the coastal area to middle altitudes in the mountains. Widely distributed throughout the world, but apparently absent from northeastern United States.

Lemna gibba occurs in two forms almost throughout its range. One is conspicuously gibbous and the other is essentially nongibbous. Whether these are distinct genetic races or mere ecological modifications of the same race is not known. Although both forms flower profusely, fruiting specimens of the gibbous form are more frequently encountered. This suggests the possibility that the nongibbous form is a sterile hybrid. By and large, the nongibbous form is much more abundant than the gibbous form. The relation of these two forms to each other is in need of clarification.

3. Lemna minor L., Sp. Pl. 970. 1753. Water lentil. Fig. 161.

Fronds solitary or clustered, round to elliptic-obovate, 2–4 mm. long, 1.5–3 mm. wide, thick, with 2 layers of air spaces within, these not inflated, obscurely

---

Fig. 160. *Lemna gibba: a,* roots, with and without rootcap, × 12; *b* and *c,* winged fruits, the ovules 1–3, × 12; *d–f,* mature seeds, showing corky ribs and conspicuous operculum, × 16; *g,* young seed, × 16; *h,* staminate and pistillate flowers surrounded by saclike spathe, × 23; *i,* flowers with spathe removed, showing the pistillate flower with winged shoulders, and a pair of staminate flowers, × 25; *j,* flowering and fruiting fronds, with inflated, gibbous lower side, and solitary roots, × 2; *k* and *l,* fronds (longitudinal section), showing variation in the air spaces, × 8; *m,* habit, top view of flowering fronds, × 4; *n,* habit, top view of fronds, the parent frond with fruit, × 4. *o–w,* nongibbous plants: *o–q,* fronds (longitudinal section), × 8; *r,* habit, flowering fronds, × 4; *s* and *t,* pistillate and staminate flowers surrounded by spathe, × 28; *u,* habit, showing group of plants, × 2; *v* and *w,* roots, with and without rootcap, × 20.

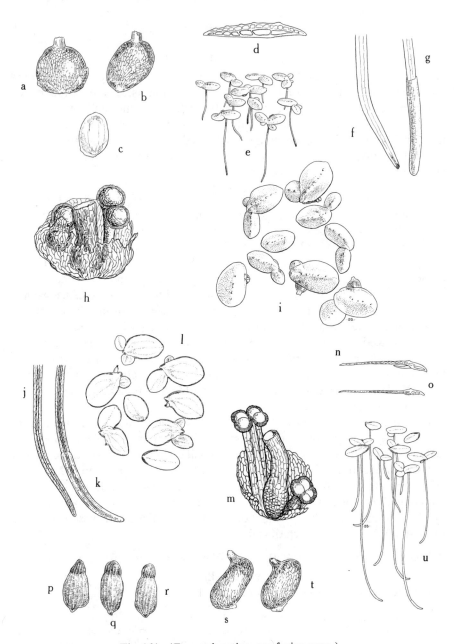

Fig. 161. (For explanation, see facing page.)

1-nerved above, dark green surface often suffused with red or purple, low median ridge often terminated with a conspicuous papule, sometimes a row of papules along the median ridge; spathe saclike, short; ovule solitary, amphitropous; fruit usually asymmetrical, with broad shoulders, these sometimes slightly winged; seed with corky-thickened ribs and a conspicuous operculum.

Common throughout the mountains and valleys of California; general over the continent, cosmopolitan.

Lemna minor tends to resemble small plants of the nongibbous form of *L. gibba,* from which it may be readily distinguished by the lack of a conspicuous wing on the fruits and the less conspicuous air chambers in the frond, as well as by the solitary seed. The spathe is comparatively shorter and more flaring.

4. Lemna perpusilla Torr., Fl. N. Y. 2:245. 1843. Fig. 161.

Lemna paucicostata Hegelm. in Gray, Man., 5th ed., 681. Jan., 1868.

Fronds solitary or in small clusters, obovate to orbicular-obovate, somewhat oblique, 2–3.5 mm. long, 1–2.5 mm. wide, obscurely 3-nerved, fairly thick, with large air spaces within, these in 1 layer, the apical papule usually prominent, sometimes a row of papules along mid-nerve; spathe open on one side; style long; ovule orthotropous but the fruit somewhat asymmetrical; seed ribbed and cross-striate between ribs, the ribs not corky-thickened.

Occasional and widely distributed in California: Central Valley, coastal area; east to the Atlantic.

Lemna perpusilla may be distinguished by its 3-nerved fronds, the single layer of air spaces within, and the open spathe. Apparently in California it has long been confused with *L. minor.*

5. Lemna valdiviana Phil., Linnaea 33:239. 1864.

Lemna cyclostasa of western American authors, and of Thompson, Ann. Mo. Bot. Gard. 9:15. 1897. Not Chev., Fl. Paris 2:256. 1827.
Lemna minor var. *cyclostasa* Ell., Bot. S. C. & Ga. 2:518. 1824.

Fronds narrowly elliptic to oblong or oval, 1–5 mm. long, 0.5–2.5 mm. wide, symmetrical to slightly falcate, obscurely 1-nerved, or nerveless, the upper surface of conspicuously uniform texture throughout, often translucent; spathe open, reniform; fruit long-exserted, narrowed to the style; seed orthotropous,

Fig. 161. *Lemna. a–i, L. minor: a* and *b,* shouldered fruits, *a* showing remains of spathe, × 16; *c,* seed, × 16; *d,* frond (longitudinal section), × 8; *e,* group habit, showing the papules along median ridge and the solitary roots, × 2; *f* and *g,* roots, with and without rootcap, × 28; *h,* pistillate and staminate flowers in short, saclike spathe, × 36; *i,* habit, top view showing young, flowering, and fruiting fronds, × 4. *j–u, L. perpusilla: j* and *k,* roots, with and without rootcap, × 12; *l,* habit, showing top view of fruiting fronds, × 4; *m,* pistillate and staminate flowers, the surrounding spathe open on one side, × 40; *n* and *o,* fronds (longitudinal section), the air spaces in a single layer, × 8; *p–r,* mature seeds, ribbed, cross-striate between ribs, × 20; *s* and *t,* fruits, asymmetrical, × 20; *u,* group habit, the roots long, × 2.

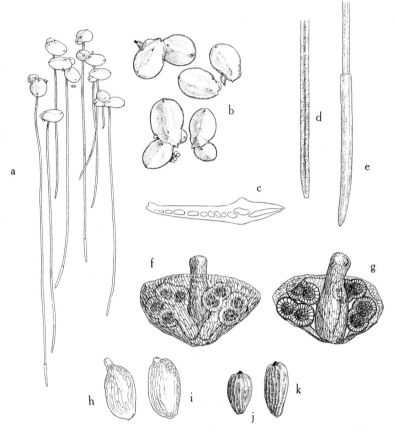

Fig. 162. *Lemna minima: a,* habit, showing flowering fronds with long, solitary roots, × 2; *b,* habit, top view of fronds with flowers and fruits, × 4; *c,* frond (longitudinal section), the air spaces in a single layer, × 12; *d* and *e,* root tips, with and without rootcap, × 20; *f* and *g,* pistillate and staminate flowers, the broad enclosing spathe reniform and open on one side, × 48; *h* and *i,* fruits, slightly flattened, × 16; *j* and *k,* mature seeds, longitudinally ribbed and cross-striate, × 16.

the ribs not corky-thickened, conspicuously striate between ribs, the operculum prominent.

Common in ponds and ditches and about springs in the valleys and mountains; Western Hemisphere.

Lemna valdiviana may be distinguished from other species by the usually thin, oblong, 1-nerved or nerveless fronds with a uniform surface texture.

6. Lemna minima Phil., Linnaea 33:239. 1864. Fig. 162.

Fronds solitary or in very small clusters, oblong to elliptical or somewhat obovate, 1.5–4 mm. long, 1–2.5 mm. wide, usually small, only occasional members large, thick, the upper surface convex, commonly nerveless or with an obscure nerve and a row of papules along the middle, usually with a thin margin around the frond, this becoming hyaline near the base, the air chambers in 1 layer; spathe open, reniform, and large; ovule orthotropous; fruit usually symmetrical to asymmetrical; seeds oblong, ribbed and cross-striate, the ribs not corky-thickened, the operculum pointed.

Common throughout California; Western Hemisphere.

This is the smallest species of *Lemna* in our flora and may be distinguished by the nerveless or obscure single-nerved frond with papules down the middle.

SPIRODELA

Fronds solitary or in clusters of 2–5 with 3 to several roots fascicled below. Reproductive pouches 2, 1 on either side of the basal end. Inflorescence arising from pouch which consists of a saclike spathe enclosing 1 pistillate and 2 or 3 staminate flowers. Ovary somewhat winged on the shoulders. Fruit a utricle.

Roots 2 or 3; fronds oblong to oblong-obovate, 2–4 mm. long, not conspicuously nerved
  .... 1. *S. oligorrhiza*
Roots 4 to several; fronds orbicular-obovate, 3–8 mm. long, almost as wide, conspicuously
  several-nerved, the nerves radiating from stipe base .... 2. *S. polyrrhiza*

1. Spirodela oligorrhiza (Kurz) Hegelm., Lemnaceen 147. 1868. Fig. 163.

Lemna oligorrhiza Kurz, Jour. Linn. Soc. London 9:267. 1867.

Fronds with 2 or 3 roots, oblong-obovate, 2–4 mm. long, 1.5–2 mm. broad, closely resembling in size and shape those of some species of *Lemna;* flowers in membranous spathe open on one side; pistillate flowers 1 and staminate flowers 2 to each spathe, each flower consisting of a single pistil or a single stamen; fruit 1- or 2-seeded, often asymmetrical; seed ribbed, minutely reticulate.

In California known only from the vicinity of Berkeley; Missouri, Southern Hemisphere.

Spirodela oligorrhiza can be readily mistaken for any of several species of *Lemna,* but can be easily identified by its 2 or 3 roots. The plants flower abundantly at Berkeley.

2. Spirodela polyrrhiza (L.) Schleid., Linnaea 13:392. 1839. Fig. 164.

Fronds conspicuously orbicular-ovate, almost as broad as long, dark glossy green above, often reddish or purple below, with 5–11 conspicuous nerves radiating from the base of the reproductive pouch, giving the appearance of being peltate, the stipe marginal or submarginal and the reproductive pouches on either side; roots several, these with a single vascular strand and a large pointed root cap; plants reproducing chiefly by asexual budding, the buds arising from the reproductive pouches and growing to nearly mature size before

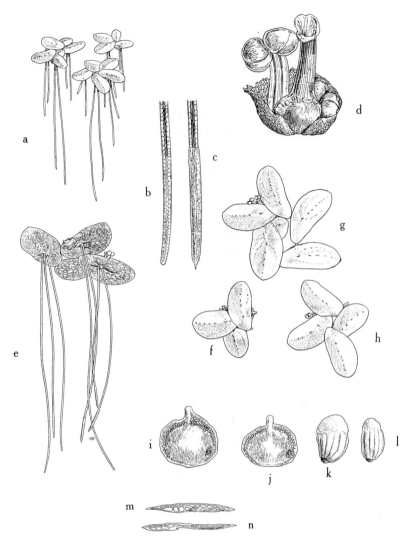

Fig. 163. *Spirodela oligorrhiza: a,* habit, × 2; *b* and *c,* root tips, with and without root-cap, the rootcap 1.2–1.9 mm. long, × 20; *d,* pistillate flower and pair of staminate flowers enclosed by short spathe open on one side, × 40; *e,* habit, lower surface, showing the large, netlike air spaces, and the flowers and fruit arising from pouch, × 4; *f,* fronds, top view, with flowers and fruit, × 4; *g* and *h,* fronds, remaining attached to one another by slender stipes, × 4; *i* and *j,* fruits, showing the winged shoulders, × 16; *k* and *l,* seeds, ribbed and minutely reticulate, × 16; *m* and *n,* frond (longitudinal section), showing air chambers, × 8.

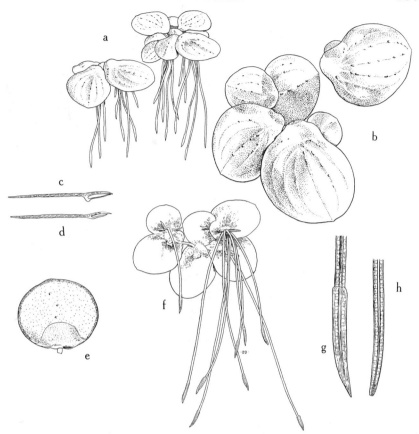

Fig. 164. *Spirodela polyrrhiza: a,* habit, showing group of fronds, each with numerous roots, × 2; *b,* fronds producing buds, top view, showing the conspicuous nerves radiating from the reproductive pouches, × 4; *c* and *d,* fronds (longitudinal section), showing air chambers, × 4; *e,* winter bud, × 12; *f,* fronds, lower surface, showing the slender stipes by which they are attached to one another, and the clustered roots, × 3; *g* and *h,* root tips, with and without rootcap, the rootcap 0.9–1.5 mm. long, × 20.

separating from the parent; flowers borne in a saclike spathe enclosing the pistillate flower, which consists of a solitary pistil, and 2 or 3 staminate flowers, each consisting of a solitary stamen; fruit wing-margined above, 1- or 2-ovuled.

This is the largest of the surface-floating duckweeds. It is often present as scattered, large fronds in masses of *Lemna* and *Wolffiella* and ocasionally in almost pure stands. Flowers have not yet been reported on California material. The plants carry through the winter by producing winter buds that are dense and sink to the bottom of the pond.

Frequent at low and middle altitudes throughout California, most abundant in the Central Valley, where it sometimes covers the entire surface of a pond; North and South America, Europe.

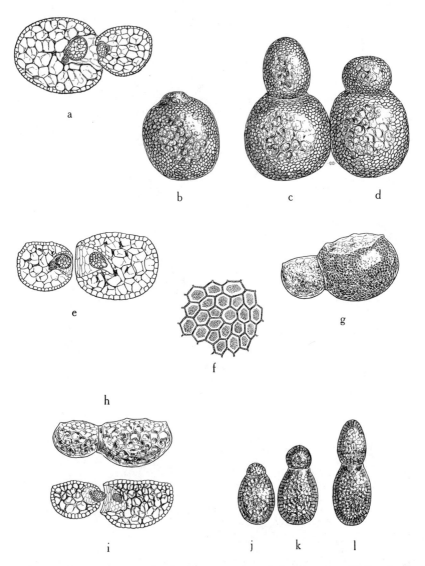

Fig. 165. *Wolffia. a–d, W. columbiana: a,* parent and daughter fronds (longitudinal section) at the point of division, showing the small epidermal cells, × 20; *b–d,* habit of plants at various stages, dorsal view, × 20. *e–g, W. arrhiza: e,* parent and daughter fronds (longitudinal section), showing the large epidermal cells and flattened dorsal surface, × 20; *f,* chromoplasts in epidermal cells; *g,* habit, side view, showing flattened dorsal surface with conspicuously raised distal cells (note that chromoplasts are lacking on the dorsal surface), × 20. *h–l, W. cylindracea: h,* habit, side view, showing long, ellipsoid shape of fronds and their flattened dorsal surface with several low papules, × 20; *i,* parent and daughter fronds (longitudinal section), showing small air chambers in the upper half and larger chambers in the lower half, × 20; *j–l,* habit of plants at various stages, × 20.

WOLFFIA

Diminutive floating rootless plants, scarcely visible to the naked eye as individuals and often forming uninterrupted green masses on the surface of the water as a thin green scum. Plant 0.3–0.9 mm. long, spheroid to ellipsoid, sometimes flattened above, with a funnel-shaped reproductive pouch at one end bearing asexually successive daughter fronds, ours not punctate but sometimes bearing masses of red pigment bodies in each epidermal cell. Inflorescence breaking through the upper surface of the frond, composed of 1 staminate flower consisting of a single stamen and 1 pistillate flower consisting of a single pistil. Ovule orthotropous but often lying oblique in the utricle.

In five collections of members of this genus recently made in California so much variation is evident that it is at present impossible to be certain which species are represented in our flora. A brief history of the problem as it presented itself is in order, to explain our tentative identifications. In June, 1947, a collection was made of plants growing on the surface of the water of Dune Lake in San Luis Obispo County. These were very abundant but sterile. In the hope of finding flowering material a return trip was made the following year, but only with extreme diligence in searching were any plants found. These were likewise sterile. They were identified as *Wolffia columbiana*. However, at Oso Flaco Lake, to the south, an abundance of material was found which appeared to be identical with the European *W. arrhiza*. In 1949 none was found in Oso Flaco Lake, but in sloughs along the San Joaquin River wash northeast of Fresno another *Wolffia* collection was made which consisted of a mixture of what appeared to be *W. columbiana* and specimens about half its size that were flat-topped and contained a few low papules. These resembled *W. cylindracea*. A further collection of this small material was taken from a ditch in northeastern San Joaquin County near the Amador County border, southwest of Ione.

The plants in these collections have distinctive characters. The Dune Lake material was made up of spheroid and ellipsoid fronds that were not flat-topped. The Oso Flaco Lake material was slightly flat-topped, and the upper epidermal cells were about ½ as large as similar cells of the Dune Lake material. The plants themselves were about the same size. Whereas the Oso Flaco Lake specimens averaged 33–39 cells lengthwise across the top, the Dune Lake specimens averaged 16–18 cells. In cross section, plants of both of these collections exhibited large air spaces bounded by thin-walled cells. The small plants from the San Joaquin Valley were about ½ as wide as the coastal plants, but averaged nearly as long. They were very much flatter than any of the coastal material and differed from both of the other collections in that the air spaces within the fronds were bounded by thick-walled cells in the upper half of the frond, and by thin-walled cells in the lower half. No flowers have been observed on any of the California specimens.

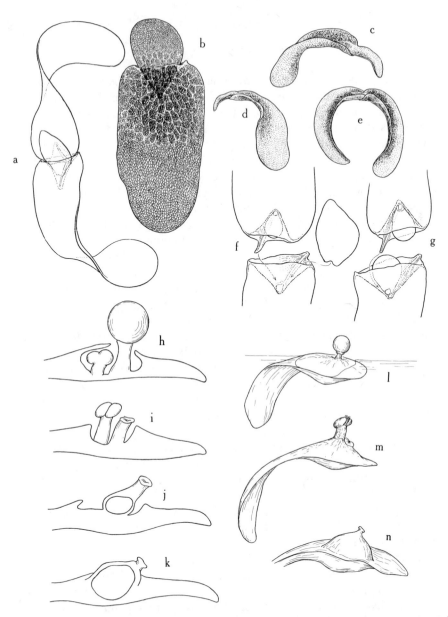

Fig. 166. *Wolffiella lingulata: a,* parent and daughter fronds forming recurved band, *× 8; b,* parent frond and young daughter frond, top view, *× 4; c–e,* side view of fronds, *× 8; f* and *g,* parent and daughter fronds dividing, *× 8; h–k,* diagrammatic representation of development of flower and fruit; *l,* frond, side view, showing stigma with a spherical globule of liquid; *m,* frond, side view, the single anther protruding beyond the stigma; *n,* fruit embedded in frond.

If we follow a philosophy that in dealing with extremely reduced plants minor morphological characters assume proportionately greater significance, it would seem that we are justified in regarding these three collections of *Wolffia* as distinct.

Air spaces within the frond bounded by thin-walled cells, fronds 0.7–1.5 mm. long, ellipsoid to spheroid, rounded at distal end and slightly flattened on the upper surface.
    Epidermal cells of the upper surface small, numbering 25–30 cells in a lengthwise row on the frond, fronds ellipsoid, scarcely or a little flattened above
                                            1. *W. columbiana*
    Epidermal cells of the upper surface of frond large, numbering 16–18 cells in a lengthwise row, fronds ellipsoid, somewhat flattened above and bearing a few low papillae
                                            2. *W. arrhiza*
Air spaces within the upper half of frond bounded by thick-walled cells, fronds 0.3–1.3 mm. long, tapering to distal end, conspicuously flattened above and roughened
                                            3. *W. cylindracea*

1. Wolffia columbiana Karsten, Bot. Unters. 1:103. 1865. Fig. 165.

Fronds spheroid to ellipsoid, 1¼–1½ times as long as wide, 0.7–1.5 mm. long, scarcely if at all flattened, the epidermal cells small (about ⅓–⅔ as large as those of *Wolffia arrhiza*), 25–30 cells in a lengthwise row on upper surface of frond; stomata few (1–6); no red-pigmented cells present; air chambers bordered by thin-walled cells, relatively uniform throughout frond.

Forming a thin green scum on the surface of ponds in our region: known only from Dune Lake, San Luis Obispo County, and from sloughs along the San Joaquin River wash northeast of Fresno in Fresno County; eastern states.

This material is characterized by the almost complete absence of flattening of the upper surface and by its small epidermal cells. No flowers have as yet been observed.

2. Wolffia arrhiza (L.) Wimm., Fl. Schles. 140. 1857. Fig. 165.

Fronds spheroid to ellipsoid, 1¼–1½ times as long as broad, 0.7–1.5 mm. long; flattened on the upper surface, the distal cell of the flattened area often conspicuously raised, the epidermal cells large (nearly twice as large as those of *Wolffia columbiana*), 16–18 cells in a lengthwise row on the frond; stomata many, large; pigment cells lacking in the epidermis (but reddish chromoplasts developing in all epidermal cells late in the season); air chambers all thin-walled, ranging from large in the lower half of frond to somewhat smaller above.

In California known only from Oso Flaco Lake in San Luis Obispo County; world-wide in temperate zones.

The large diameter and flat top, in combination with the large cells and numerous stomata, are distinctive.

3. Wolffia cylindracea Hegelm., Lemnaceen 123. 1868. Fig. 165.

Frond ellipsoid, strongly flattened and roughened as though with several low papules, about twice as long as wide, 0.3–1.3 mm. long with many small stomata, no pigment cells in the epidermis; air chambers in the upper half of

Fig. 167. *Eichornia crassipes: a,* flower, $\times$ ⅔; *b,* flower (longitudinal section), showing the irregularly adnate stamens, $\times$ ⅘; *c,* leaf, showing its orbicular, leathery blade and
inflated petiole, $\times$ ⅖; *d,* ovary (cross section), $\times$ 8; *e,* habit, showing the loose terminal
spike of flowers, the floating leaves, and the roots, $\times$ ⅖.

the frond small and bounded by thick walls, those of the lower half of the frond larger and bounded by thin walls.

Ponds and ditches of the San Joaquin Valley: ditch in northeastern San Joaquin County near the Amador County line just southeast of Ione, slough of the San Joaquin River northeast of Fresno in Fresno County; Africa.

The narrow frond with the much-flattened top and the dimorphic character of the air spaces are distinctive.

WOLFFIELLA

1. Wolffiella lingulata (Hegelm.) Hegelm., Bot. Jahrb. 21:303. 1895. Fig. 166.

Rootless floating plants, growing below or breaking through the surface tension of the water; fronds broadly oblong to linear, symmetrically or asymmetrically falciform, length 1½–7 times the width, 5–10 mm. long, 1–5 mm. wide; surface concave or channeled, the parent and daughter fronds often recurved and together appearing like a segment of a band; vegetative pouch solitary, triangular, a stipe scar at or near one corner of the pouch, either to the right or the left; inflorescence breaking through the surface to one side of the pouch; flowers without a spathe, monoecious, the pistillate flower appearing before the staminate one; stamen 1; pistil 1; fruit an indehiscent, bladderlike, 1-seeded utricle.

In still water throughout the valleys: Butte County to Kern County, south along the coast from San Francisco; Mexico.

A single species in California.

Wolffiella oblonga has been reported from California, but all the material so labeled in herbaria seems to be *W. lingulata*. For a discussion of this problem see Mason, The Flowering of Wolffiella lingulata (Hegelm.) Hegelm. (Madroño 4:241–251, 1938).

PONTEDERIACEAE. PICKEREL-WEED FAMILY

Perennial aquatic herbs, wholly submersed, emersed, or floating. Flowers perfect, regular or irregular, solitary or in spikes, subtended by a leaflike spathe. Perianth 6-parted, united only at base or forming a tube. Stamens 3 or 6, inserted on the tube or at the base of the perianth, usually dissimilar or unequal in size. Ovary superior, 1- or 3-celled; style 1; stigma 3-lobed or 6-toothed. Fruit a many-seeded capsule, or a 1-celled, 1-seeded utricle.

Leaves with orbicular blades and bulbular petioles; plants free-floating or rooting in mud
Eichornia

Leaves linear or ovate and cordate at base.
    Leaves linear, thin and grasslike; perianth segments united into a tube; plants chiefly
        submersed .... *Heteranthera*
    Leaves ovate and cordate at base; perianth segments free except at base; plants erect
Monochoria

Fig. 168. *Heteranthera dubia: a,* capsule, sessile in leaf axil and enclosed by spathe, × 1½; *b,* mature seed, finely cross-striate and with membranous longitudinal ribs, × 12; *c,* habit, showing the ribbon-like leaves, sessile flowers and fruit, and roots at the nodes, × ⅖; *d,* spathe with flower in leaf axil, × 2; *e* and *f,* leaf sheaths, showing the stipule-like appendages, × 4.

EICHORNIA

1. Eichornia crassipes (Mart.) Solms-Laubach in DC., Monogr. Phan. 4:527. 1883. Water hyacinth. Fig. 167.

Piaropus crassipes Britt., Ann. N. Y. Acad. Sci. 7:241. 1892.

Ours an herb averaging 1–6 dm. tall, with rootstocks floating or rooting in the mud; submersed leaves when present long and narrow, those above the water clustered at the nodes and with ovate to orbicular, leathery blades 3–8 cm. wide, and wholly or partly inflated petioles; flowers in a loose, terminal spike, bluish purple and showy, slightly irregular and somewhat 2-lipped; perianth segments 6, united into a short tube below; stamens 6, irregular, 3 included, 3 exserted, the filaments irregularly adnate to the perianth; ovary ovoid, 3-celled.

Locally abundant at a few localities in the San Joaquin Valley, extending north in the Sacramento Valley to Clarksburg, Yolo County, also on Warner Creek, San Bernardino County; native of tropical America.

This species shows considerable variation in size, the plants ranging from a few centimeters to nearly a meter in height.

HETERANTHERA

1. Heteranthera dubia (Jacq.) MacM., Metasp. Minn. Valley 138. 1892. Water star grass. Fig. 168.

Ours a submersed, grasslike herb, with slender, branching stems often rooting at the nodes; leaves linear or ribbon-like, thin, sessile, finely parallel-veined and without a distinct midvein, the sheaths thin, tipped on either side with small, acute, stipule-like appendages; spathe terminal, 1-flowered, exposed above the water; perianth pale yellow with an elongated filiform tube (usually less than 25 mm. long in our specimens) and a rotate, 6-parted limb, the segments linear-lanceolate; stamens 3, equal in size, the filaments dilated below; stigma several-lobed; capsule 1-celled, with 3 parietal placentae, many-seeded; seeds oblong-ovoid, finely crosslined, with prominent, raised, membranous, longitudinal ribs.

Known by us only from Butte Sink in Butte County and Pit River in Modoc County, in still or flowing waters. A collection from Mendocino County is cited by Sereno Watson (Bot. Calif. 2:187, 1880). A specimen from "Near San Francisco," collected by Vasey, is in the Gray Herbarium. Isolated localities in western states, widespread in midwestern and eastern states, Mexico.

In all the California material that we have seen, the perianth tube is only about 10–25 mm. long, but in plants from the eastern states it may be as long as 10–15 cm. Collections studied from localities in Oregon were found to have perianth tubes as long as those in the eastern plants.

Fig. 169. *Juncus acutus* var. *sphaerocarpus: a,* single panicle cluster, showing mature capsules, × 6½; *b,* capsule, × 6½; *c,* part of stem, enclosed by basal sheath, showing the terete, stemlike leaf, × ⅕; *d,* capsule valve, showing the tailed seeds, × 6½; *e* and *f,* seeds, × 12; *g,* stamens, × 6½; *h,* lower part of plant, showing basal sheaths and tufted stems, × ⅖; *i,* upper part of plant, showing leaves, stem, and the compound panicle, × ⅖.

MONOCHORIA

1. Monochoria vaginalis (L.) Presl, Rel. Haenk. 1:128. 1830.

Erect, emersed aquatic with hollow stems to 3 dm. tall; leaves ovate, cordate-based, to 10 cm. long; inflorescence a spike; flowers few, blue; perianth segments 6, joined only at the base; stamens 6, one of them long, the others short; ovary 3-celled, many-ovuled; fruit a loculicidal capsule.

Locally established in experimental rice plots at Biggs Rice Station in Butte County, California; native of the Eastern Hemisphere.

In India the young plants are eaten as greens.

JUNCACEAE. RUSH FAMILY

Perennial or sometimes annual, grasslike, usually tufted herbs, commonly growing in moist or wet places. Stems simple, rarely branching, leafy or naked and scapelike. Leaf blades terete, or laterally or vertically flattened, the blades arising from sheaths or the sheaths sometimes bladeless. Inflorescence in terminal or sometimes apparently lateral heads, spikes, corymbs, or panicles. Flowers greenish or brownish, with 6 distinct, similar, glumelike segments. Stamens 6 or sometimes 3. Ovary superior, 3- or sometimes 1-celled. Stigmas 3. Fruit a loculicidally 3-valved capsule. Seeds several to many, usually distinctly reticulate or ribbed, often tailed.

The only genus in the family which concerns us in the present work is *Juncus*. The species of the related genus *Luzula* are usually inhabitants of moist woods and meadows, but *L. campestris* var. *congesta* (Thuill.) Buchenau has been found occasionally in the vicinity of marshes near the ocean. It may be distinguished from *Juncus* species by its closed leaf sheaths and hollow stems. In *Luzula campestris* var. *congesta* the inflorescence consists of several spikes congested in a pyramidal or conical head, and the perianth segments are usually rather dark brown.

JUNCUS

Inflorescence apparently lateral; the lowest involucral leaf resembling a continuation of the stem, terete.
Basal leaf sheaths with terete, stout, sharp-pointed blades.
Perianth segments 3–4 mm. long ....1. *J. acutus* var. *sphaerocarpus*
Perianth segments 5–6 mm. long .... 2. *J. Cooperi*
Basal leaf sheaths bladeless (or with slender, compressed blades in *J. mexicanus*).
Perianth segments usually more than 3 mm. long.
Leaf blades wanting.
Perianth segments usually dark brown, 5–7 mm. long .... 3. *J. Leseurii*
Perianth segments green, straw-colored, or brownish-tinged, 3.5–5 mm. long.
Stems less than 3 mm. thick; woody tissue absent or poorly developed
4. *J. balticus*
Stems 3–5 mm. thick; woody tissue well developed .... 5. *J. textilis*

Leaf blades usually well developed on the uppermost basal sheaths
6. *J. mexicanus*
Perianth segments 2–3 mm. long.
Stamens 3, opposite the outer perianth segments.
Perianth segments with green or pale brown, scarious margins
7a. *J. effusus* var. *pacificus*
Perianth segments with distinct, dark brown lateral bands
7b. *J. effusus* var. *brunneus*
Stamens 6 .... 8. *J. patens*
Inflorescence terminal; the lowest involucral bract not resembling a continuation of the stem (or, if so, conspicuously channeled along the inner side).
Small annuals.
Capsule oblong; perianth segments 4–5 mm. long .... 9. *J. bufonius*
Capsule oblong-ovoid or subglobose; perianth segments 3–4 mm. long
10. *J. sphaerocarpus*
Perennials.
Leaf blades with partial or complete internal septa.
Leaf blades terete or somewhat compressed, the septa complete.
Earliest leaves filiform, submersed .... 11. *J. supiniformis*
Leaves all of one kind, not filiform or submersed.
Epidermis of the leaves and stems transversely rugosely roughened
12. *J. rugulosus*
Epidermis of the leaves and stems smooth.
Capsule slender-subulate, golden brown .... 13. *J. Torreyi*
Capsule broader, varying from ovate-lanceolate to ovate-oblong.
Stamens 3.
Capsule clavate-oblong .... 14. *J. Bolanderi*
Capsule narrowly ovate-lanceolate .... 15. *J. acuminatus*
Stamens 6.
Anthers shorter than the filaments; perianth segments 2–3 mm. long
16. *J. articulatus*
Anthers longer than the filaments; perianth segments 3–4 mm. long
17. *J. nevadensis*
Leaf blades gladiate, the septa incomplete.
Heads few (1–3), in large, glomerate clusters; perianth segments dark reddish brown.
Stamens 3; anthers as long as or shorter than the filaments .. 18. *J. ensifolius*
Stamens 6; anthers usually much longer than the filaments
19. *J. phaeocephalus*
Heads numerous, smaller, in large compound panicles; perianth segments straw-colored or paler reddish brown.
Capsule narrowly oblong, attenuate, exserted well beyond the perianth
20. *J. oxymeris*
Capsule oblong, acute or obtuse.
Style often long-exserted; anthers as long as or longer than the filaments
19a. *J. phaeocephalus* var. *paniculatus*
Style usually included; anthers usually shorter than the filaments
21. *J. xiphioides*
Leaf blades without septa, flat and grasslike, inserted with flat surface facing the stem.
Leaf blades 1.5–3 mm. wide .... 22. *J. falcatus*
Leaf blades 3–7 mm. wide .... 23. *J. orthophyllus*

1. Juncus acutus var. **sphaerocarpus** Engelm. in Rothrock, Rep. U. S. Geol. Surv. W. 100th Merid. 6:376. 1878. Spiny rush. Fig. 169.

Stems in large tufts, 6–12 dm. tall, stout, pungent, terete or slightly compressed; leaves from basal sheaths terete, nearly as long as and resembling the stems; involucral bract 5–15 cm. long, stout, pungent; panicle compound, with branches very unequal in length, the longer ones 10–20 cm. long; flower clusters 2- to 4-flowered; perianth 3–4 mm. long, yellowish brown, the outer segments broadly lanceolate, acute, scarious-margined, the inner ones shorter, retuse at the very broad, scarious-margined apex; stamens slightly shorter than the perianth, the anthers much longer than the filaments; capsule broadly obovate or subglobose, apiculate, brown, extending well beyond the perianth; seeds acute at each end or slightly tailed, finely reticulate.

Coastal salt marshes: San Diego County north to Santa Barbara County, Santa Catalina Island, and extending inland on alkaline sinks into the Colorado Desert; South America, Europe.

Jepson gives San Francisco as the northernmost range of this species, and Abrams cites San Luis Obispo County as the northernmost limit. However, I have seen no verifying collections from either of these two localities.

2. Juncus Cooperi Engelm., Trans. Acad. Sci. St. Louis 2:590. 1868.

Stems in large tufts from stout, much-branched rootstocks, 4–8 dm. tall, stout, pungent, terete, finely striate; leaves from basal sheaths with terete, stout, pungent blades, short or nearly as long as the stems; involucral bract 5–10 cm. long, stout, pungent; panicle compound, with branches very unequal in length, the longer ones to 10 cm. long; flowers 2 to several in a cluster; perianth pale green or straw-colored, 5–6 mm. long, the segments oblong-lanceolate, broadly hyaline-margined, the outer ones prominently cuspidate and longer than the inner ones; stamens 6, about as long as the inner perianth segments, the anthers much longer than the filaments; capsule ovate-oblong, acute, extending slightly beyond the perianth; seeds with a white appendage at each end, slightly margined on the side, finely reticulate.

Alkaline flats in the Colorado and Mojave deserts; into southern Nevada.

3. Juncus Leseurii Boland., Proc. Calif. Acad. Sci. 2:179. 1862. Salt rush. Fig. 170.

Stems usually stout, smooth, either erect, tortuous, or distinctly arching, 3–9 dm. tall, arising from stout, creeping rootstocks; basal leaf sheaths bladeless, shining or dull brown; involucral leaf often as long as or much shorter than the stem, pungent; panicle lateral, often forming a compact, headlike inflorescence in plants with arching stems, but in plants having the stems usually erect, the inflorescence more open, with 3 or 4 rays, the rays 2–3 cm. long; perianth 5–7 mm. long, the outer segments lanceolate-acuminate, the inner ones a little shorter and acute, with green midribs and membranous, mostly

Fig. 170. (For explanation, see facing page.)

dark brown margins; stamens about ⅔ as long as the segments, the anthers much longer than the filaments; capsule oblong-ovoid, acute, shorter than the perianth, brown; seeds ovoid, smooth or faintly reticulate.

Salt marshes or sand dunes along the coast: Ventura County north to Del Norte County; north to British Columbia.

There are two forms of this species which, in general appearance, are quite distinct from each other. One form, which shows a preference for sand-dune habitats, has tortuous or very prominently arching stems with the flowers grouped in a compact, headlike inflorescence. This is probably the form designated as *Juncus Leseurii* var. *Tracyi* Jepson. The other form, which is often found in salt marshes or at their borders, has erect stems, and the inflorescence is an open panicle. The plants commonly grow together.

This species is sometimes confused with *J. balticus* Willd., from which it may be distinguished by its longer and usually darker-colored perianth segments.

The species name is variously spelled "Lescurii" or "Lesueuerii" in the literature, but its spelling in the original description is that given above.

4. Juncus balticus Willd., Mag. Gesell. Nat. Freunde Berlin 3:298. 1809. Fig. 171.

Stems in small clusters or arising singly from creeping rootstocks, 2–9 dm. tall, strict, terete or compressed, moderately stout; basal leaf sheaths bladeless; panicle lateral, lax, few- or many-flowered, its branches disposed to be secund; perianth segments 3.5–5 mm. long, lanceolate, the outer segments acuminate, the inner ones acute and slightly shorter, greenish or straw-colored or brownish with a green midrib, the hyaline margins usually rather broad and well developed on the inner segments; stamens about ⅔ as long as the perianth, the anthers much longer than the filaments; capsule as long as or slightly shorter than the perianth, oblong-ovoid, mucronate, pale or dark brown; seeds oblong-cylindric, faintly reticulate, often with a whitish, membranous surface.

Common and widespread on wet ground from sea level to an elevation of 10,000 feet: throughout California; cosmopolitan.

5. Juncus textilis Buchenau, Abh. Naturw. Ver. Bremen 17:336. 1902. Basket rush.

Juncus Leseurii var. *elatus* Engelm. in Wats., Bot. Calif. 2:205. 1880.

Stems arising from creeping rootstocks, 10–20 dm. tall, 3–5 mm. thick, stout, rigid, very finely striate, pale green; basal leaf sheaths bladeless; panicle lax, 5–10 cm. high, many-flowered; perianth segments about 4 mm. long, lan-

Fig. 170. *Juncus Leseurii: a*, habit, showing the erect stems and open inflorescences, × ⅕; *b*, flowers, × 5; *c*, seeds with membranous coat removed, × 20; *d*, seed, membranous coat partially removed, × 20; *e*, seed, membranous coat intact, × 20; *f*, capsule (cross section), × 4; *g*, mature capsule, × 4; *h*, stamen, × 4; *i*, inner perianth segment, × 4; *j*, outer perianth segment, × 4; *k*, habit, showing type with tortuous stems, × ⅖; *l*, stem, showing sheath at base of inflorescence, × 4; *m*, stem, showing apex of basal sheath, × 4; *n*, habit, showing arching type, with the headlike inflorescences, × ⅖.

Fig. 171. (For explanation, see facing page.)

ceolate, the outer segments short-acuminate or acute, the inner ones a little shorter and acute to obtuse, pale green with broad hyaline margins (especially the inner segments) somewhat tinged with pale brown; stamens ½–⅔ as long as the segments, the anthers much longer than the filaments; capsule as long as or slightly shorter than the perianth, oblong-ovoid, mucronate, brown; seeds oblong to narrowly obovoid, finely reticulate.

Marshes and along streams below an elevation of 4,000 feet: coastal and montane southern California, north to San Luis Obispo and Kern counties.

Juncus textilis is closely related to *J. balticus* Willd., and might well be regarded as one of the more conspicuous variants of that species. In its typical aspect, however, the plant is distinguished by its much stouter habit, its paler green stems, and its somewhat larger, more numerous-flowered panicle. Internally also, the stems usually have well-developed fascicles of strengthening cells, whereas in *J. balticus* such tissue is entirely lacking or poorly developed. As a consequence, the stems of *J. textilis* are more woody in texture, retain their terete shape when dried, and do not tend to flatten as do those of *J. balticus*.

6. Juncus mexicanus Willd. in Schult., Syst. Veg. 7:178. 1829.

Stems from creeping rootstocks, 2–6 dm. tall, slender, sometimes twisted, compressed; basal leaf sheaths brown or straw-colored, few or many of them bearing slender blades 5–20 cm. long; inflorescence loosely 5- to many-flowered, 2–6 cm. high; perianth segments 4–5 mm. long, the outer segments lanceolate, acuminate, the inner ones a little shorter and acute, pale brown with broad hyaline margins; stamens ½–⅔ as long as the segments, the anthers much longer than the filaments; capsule as long as the perianth, narrowly ovoid, mucronate, brown; seeds finely and irregularly reticulate.

Moist or alkaline places: coastal and montane southern California, occasional in the deserts, north in the coastal counties to Marin County and in the Sierra Nevada to Plumas County; Mexico.

This species is closely related to *Juncus balticus* Willd. Parish (Muhlenbergia 6:118, 1910), on the basis of field studies, points out that the presence of compressed and usually contorted stems, as well as the production of scapelike leaf blades, are not constant characters in this species, and that various types of forms intermediate with *J. balticus* are found. In its typical aspect, however, the plant is fairly easily distinguished from *J. balticus* by its much slenderer habit, its compressed stems, and the presence of slender leaf blades on at least some of the basal sheaths.

Fig. 171. *Juncus balticus: a*, capsule valve, showing seeds, × 5; *b*, outer perianth segment, × 5; *c*, inner perianth segment, × 5; *d*, stamen, × 5; *e*, mature capsule, × 5; *f*, outer perianth segment, × 5; *g*, inner perianth segment, × 5; *h*, part of inflorescence, × 5; *i*, flower, × 5; *j–l*, seeds, some with and some without membranous coat, × 16; *m*, simple inflorescence, × ⅖; *n*, habit, upper part of plant, showing inflorescences, × ⅖; *o*, habit, basal part of plant, showing rootstock and sheaths, × ⅖; *p* and *q*, variation in inflorescences, × ⅖; *r*, habit, showing creeping rootstock, × ⅖; *s* and *t*, enclosing sheath of inflorescence, × 3; *u*, habit variation, × ⅖.

Fig. 172. (For explanation, see facing page.)

7a. Juncus effusus var. **pacificus** Fern. & Wieg., Rhodora 12:89. 1910. Common rush, bog rush, soft rush. Fig. 172.

Juncus effusus of western American authors, not L.

Stems moderately stout but soft, 1.5–3 mm. (rarely 4 mm.) wide, from stout, branching rootstocks, 5–15 dm. tall; basal sheaths bladeless or the inner ones tipped with a short awn, the edges usually overlapping nearly to the subtruncate or emarginate tip, the veins converging at the tip, dull chocolate brown or chestnut-colored at base or throughout, chartaceous, the inner sheaths dark (or occasionally pale) toward the summit, 6–15 cm. long; inflorescence a many-flowered, loosely clustered panicle (sometimes congested), 2–10 cm. long; perianth segments 2–3 mm. long, green or with pale brown scarious margins, lanceolate, acute to acuminate; stamens 3, ½ or less as long as the perianth, the anthers as long as or shorter than the filaments; capsule oblong-obovoid, of about the same length as the perianth (or sometimes longer), obtuse or retuse; seeds reticulate.

A common species in boggy places on hillsides or valley flats: Coast Ranges from Monterey County north to Del Norte County (intergrading with the coastal *Juncus effusus* var. *brunneus* throughout this range), Sierra Nevada foothills at medium elevations from Fresno County north to Siskiyou County, coastal southern California and at low elevations in the mountains, San Joaquin Valley and lower Sacramento Valley; Pacific states.

This is the most widely distributed variety of the species in California. It is the plant most commonly referred to *Juncus effusus* L., but according to Fernald and Wiegand (Rhodora 12:84, 1910) the typical form of the species does not occur in America.

7b. Juncus effusus var. **brunneus** Engelm., Trans. Acad. Sci. St. Louis 2:491. 1868. Fig. 172.

This variety may be distinguished from *Juncus effusus* var. *pacificus* by its more compact inflorescence, darker flowers, and looser, paler sheaths. Where the ranges of these two varieties are contiguous, however, there seems to be much intermixing of characters.

Marshes and moist places: near the coast from San Luis Obispo County north to British Columbia.

Two additional varieties, *J. effusus* var. *gracilis* Hook. and *J. effusus* var.

Fig. 172. *Juncus effusus. a–j, J. effusus* var. *pacificus: a,* habit, showing the awned basal sheaths, × ⅖; *b,* habit, showing upper part of stems with loosely clustered panicles, × ⅖; *c,* basal sheaths, showing awned apex, × 1½; *d,* inner perianth segment, × 12; *e–g,* mature seeds, × 20; *h,* flowers, × 4; *i,* mature capsule, × 12; *j,* stamen and perianth segments, × 12. *k–n, J. effusus* var. *gracilis: k* and *l,* basal sheaths, × 1½; *m,* inner perianth segment, × 12; *n,* inflorescence, × ⅖. *o–t, J. effusus* var. *brunneus: o,* inner perianth segment, × 12; *p,* basal sheath, × 1½; *q* and *r,* variations in inflorescence, × ⅖; *s,* habit, lower part, showing rootstock and basal sheaths, × ⅖; *t,* habit, upper part, showing inflorescences, × ⅖.

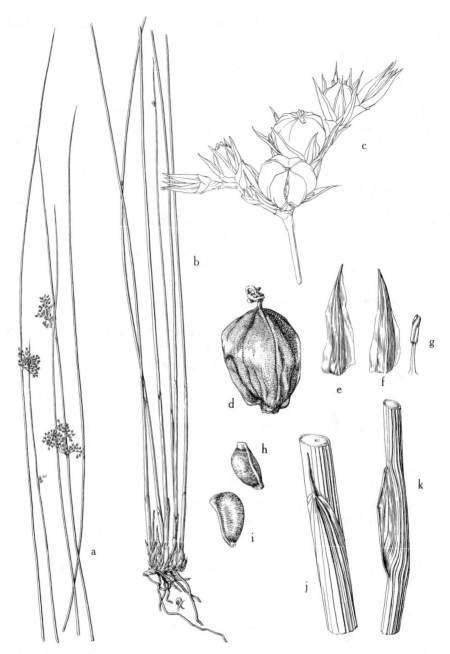

Fig. 173. *Juncus patens: a,* habit, upper part of stems, showing panicles, × ⅖; *b,* habit, lower part of plant, showing rootstock and basal leaf sheaths, × ⅖; *c,* part of inflorescence, × 5½; *d,* mature capsule, × 12; *e,* outer perianth segment, × 12; *f,* inner perianth segment, × 12; *g,* stamen, × 12; *h* and *i,* mature seeds, × 24; *j,* basal leaf sheath, showing awned tip, × 5½; *k,* sheath of inflorescence, × 5½.

exiguus Fern. & Wieg., are chiefly montane in their distribution and are infrequently collected. Upon further study they may prove to be only minor forms of *J. effusus* var. *pacificus.*

8. Juncus patens E. Mey., Syn. Luzul. 28. 1823. Fig. 173.

Stems loosely or densely tufted, light green, from stout, branching rootstocks, 5–10 dm. tall, 1.5–2.5 mm. thick; basal leaf sheaths bladeless, with or without short awns, the uppermost leaf sheaths dark reddish brown throughout or with upper parts greenish or straw-colored, often separating from the stem and in dried specimens suggesting blades; inflorescence an open or somewhat compact panicle, many-flowered, 2–7 cm. long; perianth segments with a green midrib, the margins brownish-tinged or hyaline, lanceolate or broadly subulate, acuminate, 2.5–3 mm. long, spreading in fruit; stamens 6, the anthers shorter than the filaments; capsule subglobose, a little shorter than the perianth, obtuse and apiculate at apex; seeds somewhat obliquely oblong, faintly reticulate.

Common at borders of marshes, along streams, or on springy ground: Santa Cruz and Santa Rosa islands, on the mainland near the coast, and in the Coast Ranges from Ventura County north to Del Norte County; southern Oregon.

9. Juncus bufonius L., Sp. Pl. 328. 1753. Common toad rush. Fig. 174.

Annual, branching at the base, the stems erect, 1.5–2 dm. or 2.5–3.5 dm. tall; leaves narrow, usually revolute and bristle-like, 1–3 on the stem; inflorescence a dichotomous cyme, the flowers inserted singly on the branches and remote to closely secund or even subcapitate (varying even within a single colony); perianth 4–5 mm. long, the segments lanceolate, long-acuminate, greenish with white scarious margins, the outer segments longer; stamens usually 6 (rarely 3), ½ or less as long as the perianth, the anthers shorter than the filaments; capsule oblong, obtuse or somewhat truncate, mucronate, shorter than the perianth; seeds oblong, faintly reticulate.

Widely distributed along streams or in dried pools: throughout California except in the desert (occasional there) and high mountain regions; cosmopolitan.

This species is very variable in size and habit of plant and in the arrangement of the flowers. In its typical form, the flowers are usually scattered and secund on elongate branches, and the perianth segments are acute to acuminate. There are, however, two variant forms that occur occasionally and appear to merge rather imperceptibly with the typical form. *Juncus bufonius* var. *halophilus* Buchenau & Fern. (Rhodora 6:39, 1904) is a plant of small stature, with the ultimate flowers closely approximate, the inner perianth segments distinctly shorter than the outer ones and obtuse or rounded at the apex, and the seeds truncate at the apex. It has been collected occasionally on the coast at the edges of salt marshes. *Juncus bufonius* var. *congestus* Wahlb. (Fl. Goth. 38, 1820) has been collected somewhat more frequently throughout the range of the species. The distinctive feature of this variety is that the flowers tend to be

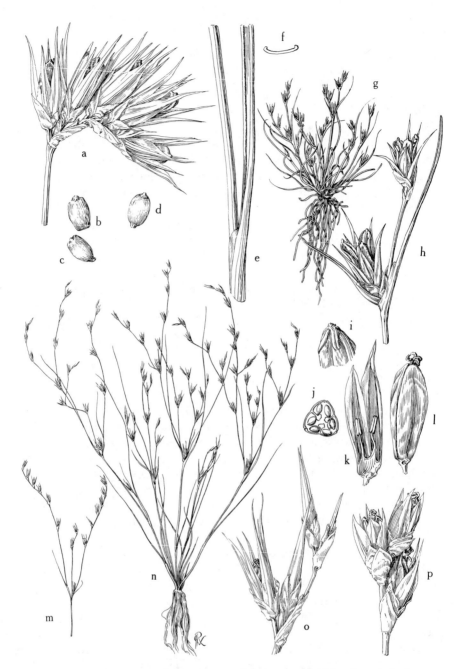

Fig. 174. *Juncus bufonius: a,* inflorescence forming scorpioid head, × 4; *b–d,* mature seeds, × 20; *e,* basal leaf sheath and blade, × 4; *f,* leaf (cross section), × 8; *g,* habit, condensed type, × ⅘; *h,* inflorescence, × 4; *i,* apex of capsule, the anthers appressed, × 8; *j,* capsule (cross section), × 6½; *k,* stamens and perianth segments, × 6½; *l,* capsule, × 6½; *m,* variation in inflorescence, × ⅖; *n,* habit, × ⅖; *o* and *p,* part of inflorescence, showing variation in perianth segments, × 4.

scorpioid and closely approximate in headlike clusters at the ends of the branches. This form is often found growing intermixed with the typical plant, and it is probable that the two forms intergrade.

10. Juncus sphaerocarpus Nees in Funk, Flora 1:521. 1818. Round-fruited toad rush.

Annual, branching at the base, 4–20 cm. tall, the branches filiform; leaves narrow, the blades flat or involute, 1–3 on the stem; inflorescence usually occupying more than half the length of the stem; flowers inserted singly on the branches and more or less remote; perianth 3–4 mm. long, the segments lanceolate, acuminate, greenish with white scarious margins, subequal in length (or the outer segments slightly longer), spreading at maturity; stamens 6, about ½ as long as the segments, the anthers shorter than the filaments; capsule oblong-ovoid or subglobose, about ⅔ as long as the perianth; seeds oblong, faintly reticulate.

Occasional in dried-up pools and along streams in coastal and montane southern California, the Sierra Nevada, San Joaquin Valley, and Coast Ranges; cosmopolitan.

This species is very similar in many of its characters to *Juncus bufonius,* but can usually be distinguished by its slender, smaller habit, smaller flowers, the length and arrangement of the inflorescence, and its broader capsule.

11. Juncus supiniformis Engelm., Trans. Acad. Sci. St. Louis 2:461. 1868.

Stems tufted, from slender rootstocks, 1–4 dm. high; early leaves elongated and capillary, floating; stem leaves terete, much shorter than the floating ones, more or less noticeably septate, as long as or sometimes longer than the stems; inflorescence a panicle of a few 3- to 9-flowered heads; perianth greenish or reddish-brown-tinged, 4–5 mm. long, narrowly lanceolate, acute to acuminate, nerved; stamens 3, about ½ as long as the perianth, the anthers shorter than the filaments; capsule narrowly oblong, as long as or longer than the perianth, light brown, acute at summit and stoutly beaked.

Rare and local in ponds on the Mendocino County coast at Caspar, Mendocino City, and Noyo. On the authority of Buchenau, Jepson (Fl. Calif. 1:254, 1921) includes Humboldt County within the range of the species, but we have not seen any collections from that region.

12. Juncus rugulosus Engelm., Bot. Gaz. 6:224. 1881.

Stems stout, 4–10 dm. tall, from stout, creeping rootstocks, conspicuously transverse-rugose, as are also the sheaths and blades; basal sheaths bladeless, often purplish-tinged; leaf sheaths with a conspicuous auricle, the blades terete, distinctly septate, attenuate into a flagellate-filiform tip; panicle 6–25 cm. long, diffuse, decompound, with many heads, these few-flowered; perianth 2.5–3 mm. long, light greenish brown or brown, the segments narrowly lanceolate,

Fig. 175. *Juncus Torreyi: a,* mature seed, × 24; *b,* leaf, with a part of it removed to show septum, × 3; *c,* flower, × 8; *d,* ligulate, auricled leaf sheath, × 3; *e,* habit, × ⅖; *f,* habit, lower and upper parts of plant, the inflorescences of globose heads, × ⅖; *g,* inflorescence, more branched type, × ⅖; *h,* outer perianth segment, × 8; *i,* inner perianth segment, × 8; *j,* stamen, × 8; *k,* capsule, × 8; *l,* basal part of plant, showing slender rootstock and tuber-like thickenings, × 1⅕.

acuminate; anthers shorter than to as long as the filaments; capsule narrowly oblong, tapering into a beak, longer than the perianth; seeds reticulate.

Wet places or marshes: coastal southern California and mountains at elevations below 6,500 feet, north in the Coast Ranges to Monterey County.

13. Juncus Torreyi Cov., Bull. Torrey Club 22:303. 1895. Fig. 175.

Stems stout, 2–6 dm. tall, arising singly from tuber-like thickenings on slender rootstocks; leaves terete, the blades more or less abruptly divergent from the stem, 2–5 mm. thick, auricled; inflorescence terminal with 1 to many many-flowered heads forming a condensed panicle, the entire cluster subtended by a long-pointed sheath; perianth segments light brown, lanceolate-subulate, 4–5 mm. long, the outer segments longer than the inner ones; stamens 6, about ½ as long as the perianth; capsule subulate, golden brown, as long as the perianth; seeds reticulate.

Wet places at elevations below 5,000 feet: almost throughout coastal and desert southern California, Modoc County; east to Atlantic.

On the authority of Coville, Jepson (Fl. Calif. 1:255, 1921) includes the upper San Joaquin Valley within the range of this species. However, I have not seen any verifying collections from this area.

14. Juncus Bolanderi Engelm., Trans. Acad. Sci. St. Louis 2:436. 1866. Fig. 176.

Stems from creeping rootstocks, 4–8 dm. tall; basal sheaths bladeless; sheaths of stem leaves with conspicuous auricles; the blades slightly compressed to terete, 10–20 cm. long, distinctly septate; inflorescence of 2–4 large (7–10 mm. broad), globose or hemispherical, many-flowered heads, the heads usually in a cluster, but sometimes solitary; perianth usually dark brown but tending to be lighter brown on plants from interior localities, 3.5–4 mm. long, the segments narrowly lanceolate, setaceously acuminate; stamens 3, the anthers shorter than the filaments; capsule clavate-oblong, shorter than or as long as the segments, obtuse and apiculate at apex; seeds reticulate.

Stream banks and marshes: near the coast and in Coast Ranges from Marin County north to Siskiyou County; Oregon.

Jepson (Fl. Calif. 1:255, 1921) describes *Juncus Bolanderi* var. *riparius* as having smaller, lighter-colored heads arranged in a loose panicle, and less setaceous perianth segments. Plants from interior localities tend to have these features, but the differences are not always well marked or consistently uniform.

15. Juncus acuminatus Michx., Fl. Bor. Am. 1:192. 1803. Fig. 177.

Stems caespitose, 2.5–10 dm. tall, from short, inconspicuous rootstocks; leaves distinctly auriculate and strongly septate; inflorescence an open panicle 5–15 cm. long, the branches stiffly ascending; heads 5–50, turbinate to nearly spherical, 5- to 20-flowered; perianth light brown to straw-colored, 2.5–3.5 mm. long, the segments lanceolate-subulate or acuminate, nearly equal in size;

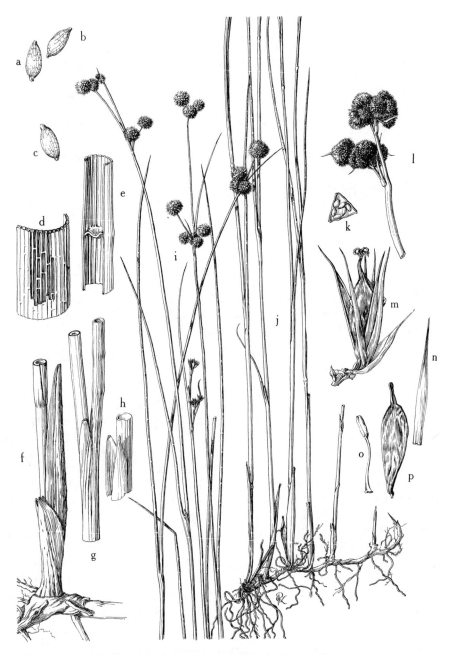

Fig. 176. *Juncus Bolanderi: a–c,* mature seeds, × 20; *d,* part of lower bladeless sheath, the epidermis removed in one area, × 3; *e,* leaf blade with part removed to show septum, × 3; *f,* basal bladeless sheaths, × 1½; *g* and *h,* auricled sheaths of septate stem leaves, × 1½; *i,* upper part of plant, showing inflorescences of globose heads, × ⅖; *j,* lower part of plant, showing the creeping rootstock, × ⅖; *k,* capsule (cross section), × 8; *l,* variation in inflorescence, × ⅖; *m,* flower, × 8; *n,* perianth segment, × 8; *o,* stamen, × 8; *p,* capsule, × 8.

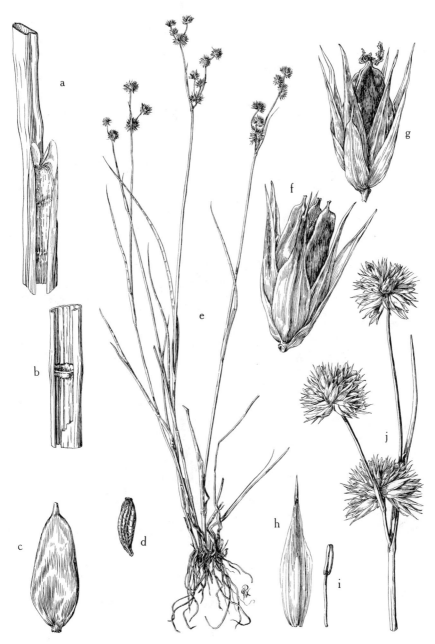

Fig. 177. *Juncus acuminatus: a,* auricles of leaf sheath, $\times$ 2½; *b,* leaf, with a part removed to show septum, $\times$ 2½; *c,* capsule, showing beak, $\times$ 10; *d,* mature seed, $\times$ 40; *e,* habit, showing the septate leaves, caespitose stems, and open panicles, $\times$ ⅖; *f,* mature capsule, after dehiscence, $\times$ 10; *g,* perianth and mature capsule, $\times$ 10; *h,* perianth segment, $\times$ 10; *i,* stamen, $\times$ 10; *j,* inflorescence, showing the spherical heads of flowers, $\times$ 1½.

Fig. 178. *Juncus articulatus: a,* mature seed, × 40; *b,* perianth segment, × 10; *c,* stamen, × 10; *d* and *e,* habit, showing the septate leaves and the loose inflorescence, × ⅖; *f,* mature capsule, after dehiscence, × 10; *g,* perianth and mature capsule, × 10; *h,* inflorescence, showing the heads on stiffly spreading branches, × 3; *i,* leaf sheath and part of septate leaf, × 3.

stamens 3, ½ or more as long as the perianth, the filaments longer than the anthers; capsule narrowly ovate-lanceolate, broadly acute, beaked, slightly shorter than or as long as the perianth; seeds oblong, often tapering at each end, reticulate.

Borders of marshes and streams and in meadows in scattered localities: Central Valley and Coast Ranges; north to British Columbia, east to Atlantic.

This species has not previously been listed for California, although it has been known in Oregon as far south as Josephine County.

16. Juncus articulatus L., Sp. Pl. 327. 1753. Jointed rush. Fig. 178.

Stems erect or spreading from short rootstocks, 2–6 dm. tall; stem leaves with rather loose sheaths and strongly septate, terete blades 5–10 cm. long; inflorescence 3–10 cm. high, loose, the branches stiffly spreading; heads hemispheric to top-shaped, 3- to 12-flowered; perianth 2–3 mm. long, the segments nearly equal in size, lanceolate, acuminate, reddish brown with a greenish midrib; stamens 6, ½–¾ as long as the perianth, the anthers shorter than the filaments; capsule dark brown, shining, longer than the perianth, 3-angled, ovate, sharply acute, tapering to a conspicuous tip; seeds reticulate.

Streamsides near the coast: Mendocino County to Del Norte County; cosmopolitan.

17. Juncus nevadensis Wats., Proc. Am. Acad. Arts & Sci. 14:303. 1879. Fig. 179.

Stems rather slender, arising singly or loosely caespitose from slender, creeping rootstocks, 1.5–6 dm. tall; leaves with long, narrow, somewhat truncate auricles on sheaths, the blades 1–1.5 mm. wide, somewhat compressed, completely but often obscurely septate; heads often 1 or 2 and much congested or 4–10 and few-flowered; perianth 3–4 mm. long, dark brown, the segments about equal in size, lanceolate, acuminate; stamens 6, the anthers linear, much longer than the filaments; stigmas often long-exserted; capsule ovate-oblong, abruptly acute and mucronate at apex, slightly shorter than the perianth, dark brown; seeds acute at each end, reticulate.

Montane meadows: San Bernardino Mountains, Sierra Nevada from Tulare County north to Siskiyou and Trinity counties, east of the Sierra Nevada crest from Inyo County north to Modoc County; north to Washington. Probably intergrading at higher altitudes with *Juncus mertensianus* Bong.

In its typical aspect, this species is distinguished by its slender habit, its very narrow, erect leaves, its dark-colored perianth, and its prominent ligules. Several collections, however, chiefly from east of the Sierra Nevada crest, are much stouter plants, often with conspicuously septate-nodulose sheaths and noticeably septate blades. The capsule in these plants tends to be more abruptly acute at the tip, and the perianth segments are lighter-colored.

Fig. 179. *Juncus nevadensis: a* and *b,* inner perianth segments, × 6½; *c,* outer perianth segment, × 6½; *d,* auricled leaf sheath, × 4; *e* and *f,* mature seeds, × 32; *g,* mature capsule, typical shape, × 6½; *h,* mature capsule (typical of material from east of the Sierra Nevada), × 6½; *i,* habit, leaves noticeably septate (typical of material from east of the Sierra Nevada), × ⅖; *j,* flower, × 10; *k,* auricled leaf sheath and conspicuously septate leaf, × 4; *l,* habit, typical form, showing creeping rootstock, × ⅖.

Fig. 180. *Juncus ensifolius: a,* habit, showing the slender rootstock, the equitant septate leaves, and the densely flowered heads, × ⅖; *b* and *c,* mature seeds, × 24; *d,* perianth and mature capsule, × 10; *e,* capsule (cross section), × 8; *f,* leaf, with a part of surface removed to show septum, × 3; *g,* inner perianth segment, × 14; *h,* outer perianth segment and stamen, × 14; *i,* leaf, equitant and septate, × ⅕.

Fig. 181. (For explanation, see facing page.)

18. Juncus ensifolius Wiks., Sv. Vet.-Akad. Handl. III. 1823:274. 1823. Three-stamened rush. Fig. 180.

Stems from slender rootstocks, 2–5 dm. tall, compressed; leaves distinctly equitant, 2–5 mm. wide, incompletely septate, the upper ones often nearly equaling the inflorescence in height; heads 1 to 3-glomerate, densely flowered, dark reddish brown; perianth 2.5–3.5 mm. long, the segments nearly equal in size, lanceolate, acuminate; stamens 3, ½–⅔ as long as the perianth, the filaments longer than the anthers; capsule dark reddish brown, longer than the perianth, oblong, obtuse or shortly acute at the summit with a short mucro; seeds sharply reticulate.

Wet ground near the coast: Humboldt County and north in western states to Alaska.

The species as here defined is somewhat less inclusive than in other treatments of the California material, but I believe that such a treatment accords best with the original concept of the species. It is frequently confused with *Juncus xiphioides,* but the few, densely flowered, glomerate, and dark-colored heads distinguish it from that species.

19. Juncus phaeocephalus Engelm., Trans. Acad. Sci. St. Louis 2:484. 1868. Fig. 181.

Juncus phaeocephalus var. *glomeratus* Engelm., *loc. cit.*

Stems flattened, 2-edged, 2–9 dm. tall, from stout, creeping rootstocks; leaves flattened laterally, without auricles, the blades 2–4 mm. (rarely 5 mm.) wide, more or less distinctly ribbed by transverse septa; inflorescence of 1 to a few large globose or hemispherical heads; perianth dark brown or purplish brown, 4–5 mm. long, the segments lanceolate, about equal in size, the outer segments acuminate or with subulate tips, the inner ones acute; stamens 6, ⅔ as long as the perianth, the anthers usually much longer than the filaments; style long, the stigmas often long-exserted; capsule oblong, abruptly acuminate, with a prominent beak; seeds reticulate.

Coastal meadows and borders of marshes: San Luis Obispo County north to Del Norte County.

Occasional collections of this species suggest a possible hybridization with *Juncus Bolanderi* Engelm. Although *J. phaeocephalus* and *J. Bolanderi* resemble each other closely in habit and occupy similar ranges, *J. Bolanderi* may be distinguished by having narrower, terete, and completely septate leaves,

Fig. 181. *Juncus phaeocephalus: a,* mature capsule, beak prominent, × 8; *b,* capsule (cross section), × 4; *c,* inner perianth segment, × 8; *d,* outer perianth segment, × 8; *e,* stamen, × 20; *f* and *g,* inflorescences, × ⅖; *h,* leaf (cross section), × 4; *i,* leaf sheath, × 1½; *j,* sheath of inflorescence, × 1½; *k,* leaf, with a part removed to show septum, × 3; *l,* mature seed, × 24; *m,* flowers, showing long-exserted stigmas, × 6½; *n,* habit, showing the creeping rootstock, the flattened, septate leaves, and the globose heads of inflorescence, × ⅖; *o,* inner perianth segment, × 8; *p,* inflorescence, × ⅖.

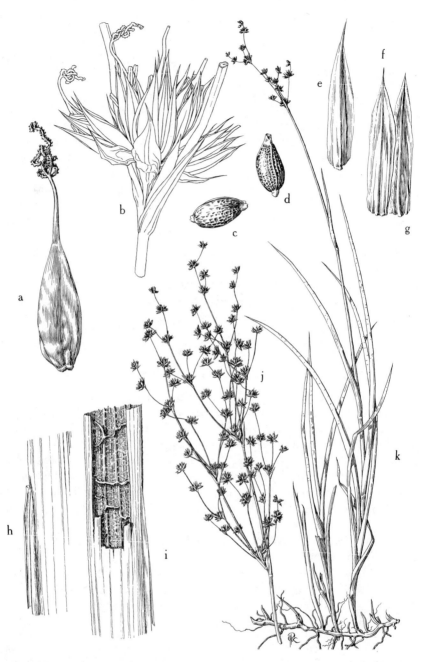

Fig. 182. *Juncus oxymeris: a,* mature capsule, × 12; *b,* single flower head, × 8; *c* and *d,* mature seeds, × 32; *e,* outer perianth segment, × 12; *f* and *g,* inner perianth segments, × 12; *h,* part of leaf, showing sheath and small auricle, × 4; *i,* leaf, with a part removed to show septa, × 8; *j,* inflorescence, × ⅖; *k,* habit, showing the creeping rootstock and the flattened, equitant, septate leaves, × ⅖.

conspicuous auricles on the leaf sheaths, narrowly lanceolate and setaceously acuminate perianth segments, three stamens in which the anthers are shorter than the filaments, and a clavate, oblong capsule as long as or slightly longer than the perianth. Certain collections from Humboldt and Del Norte counties, where these two species are found growing together, display the leaf and auricle characters of *J. Bolanderi,* though in somewhat modified form, but have floral characters more like those of *J. phaeocephalus.* Additional field studies are needed to help in determining the extent of intergradation, if any, between these two species.

19a. Juncus phaeocephalus var. **paniculatus** Engelm., Trans. Acad. Sci. St. Louis 2:484. 1868.

This variety differs from the glomerate coastal form in having numerous few-flowered heads, arranged in a loose-flowered panicle. Also, the perianth tends to be paler reddish brown or sometimes only dark-tinged at the tips of the segments. There is also a tendency toward shorter perianth segments (averaging 3–4 mm. in length), but in plants from coastal localities the segments are usually longer. The variety is found intergrading with the species near the coast, but it apparently extends inland as far as the lower Sacramento Valley, and it also occurs throughout coastal southern California at low altitudes and is occasional in the mountains.

20. Juncus oxymeris Engelm., Trans. Acad. Sci. St. Louis 2:483. 1868. Pointed rush. Fig. 182.

Stems compressed, 3–6 dm. tall, arising from stout, creeping rootstocks; leaves flattened laterally, equitant, imperfectly septate, 5–20 cm. long, equaling the stem in height or shorter, 3–7 mm. wide, the sheaths loose, the auricles small or wanting; panicle usually ample, 6–20 cm. long, compound with numerous few-flowered heads, the branches ascending; perianth light brown or green, 2.5–3 mm. long, the segments equal in size, lanceolate, acute to narrowly acuminate; stamens 6, ½–⅔ as long as the perianth, the anthers ½ as long as to longer than the filaments; capsule brown, usually longer than the perianth, narrowly oblong, 3-sided, tapering into a slender beak; seeds finely reticulate.

Chiefly montane meadows: southern California, Sierra Nevada from Tulare County north, and to Shasta County in Siskiyou Mountains, occasional in the Central Valley; north to Washington.

Sierra Nevada collections which have been referred to *Juncus phaeocephalus* var. *paniculatus* are believed to be either *J. oxymeris* or possibly, in some instances, *J. nevadensis.*

Juncus oxymeris Engelm. closely resembles *J. phaeocephalus* var. *paniculatus,* and the two are sometimes difficult to distinguish. However, in *J. oxymeris* the branches of the inflorescence tend to be slenderer (nearly filiform),

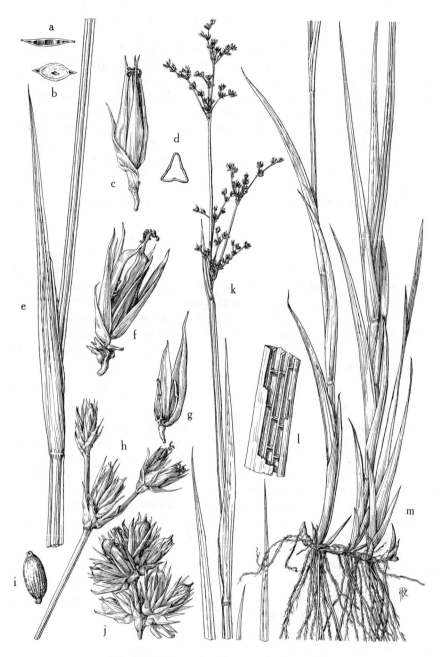

Fig. 183. (For explanation, see facing page.)

the heads are smaller, the outer as well as the inner perianth segments have conspicuous subulate or attenuate tips, and the capsule is more narrowly oblong and is tapered more gradually into a prominent beak which is exserted well beyond the perianth at maturity.

21. Juncus xiphioides E. Mey., Syn. Junc. 50. 1822. Fig. 183.

Stems compressed, acutely 2-edged, 4–8 dm. tall, from stout, creeping root-stocks; leaves flattened laterally, the sheaths without auricles, the blades 3–10 mm. wide, more or less distinctly ribbed by transverse septa; inflorescence variable, commonly of numerous heads in a loose or compact compound panicle, but sometimes with a few relatively large heads; perianth brownish or reddish-tinged, 2.5–3 mm. long, the segments lanceolate, acuminate (prominently subulate in immature plants); stamens 3 or 6, ½ (or sometimes more) as long as the perianth, the anthers shorter than or of about the same length as the filaments; style usually included; capsule oblong, shortly acute or slightly tapering below the mucronation, as long as to slightly longer than the perianth; seeds reticulate.

Streams, meadows, and marshes: almost throughout California; Arizona and northern Baja California, Mexico.

This is an extremely variable species, consisting of many intergrading forms. The broad, gladiate leaves, which have been considered a distinctive feature of the species, are characteristic only of plants from coastal localities, whereas plants from the interior, such as in the Sierra Nevada, have much narrower leaves and are not nearly as robust or tall as the coastal plants. The Sierra Nevada plants also tend to have smaller, fewer-flowered heads. The several varieties recognized by Engelmann for California do not have sufficiently well-marked characters to merit recognition.

22. Juncus falcatus E. Mey., Syn. Luzul. 34. 1823.

Stems 10–30 cm. tall, arising from slender, stoloniferous rootstocks; basal leaves flat and grasslike, falcate, ⅔ as long as to longer than the stems, 1.5–3 mm. wide; stem leaves solitary, near the middle of the stem, 3–4 (or –9) cm. long; heads solitary or sometimes 2 or 3, 5- to 25-flowered; lowest bracts 8–30 mm. long; perianth 5–6 mm. long, dark brown, the segments usually with a greenish midrib, finely roughened, the outer segments short-acuminate or acute, the inner ones varying from acute to obtuse; stamens ½ or more as long as the perianth, the anthers much longer than the filaments; capsule obovoid, fully

---

Fig. 183. *Juncus xiphioides: a,* leaf (cross section), × 1½; *b,* stem (cross section), × 1⅕; *c,* flower, × 10; *d,* capsule (cross section), × 10; *e,* leaf, the sheath without auricles, × ⅘; *f,* perianth and mature capsule, × 10; *g,* inner and outer perianth segments and stamens, the anthers shorter than filaments, × 10; *h,* branch of inflorescence, × 4; *i,* mature seed, × 24; *j,* variation in inflorescence, the heads larger, × 4; *k,* habit, upper part of plant, showing inflorescence, × ⅖; *l,* leaf, with a part of epidermis removed to show septa, × 1½; *m,* habit, lower part of plant, showing the stout, creeping rootstock and the flattened leaf blades, × ⅖.

Fig. 184. *Juncus orthophyllus:* a, inflorescence, × ⅖; b, outer perianth segment, × 5½;
c, inner perianth segment, × 5½; d, mature capsule, × 5½; e, stamen, × 5½; f and
g, mature seeds, × 20; h, basal part of plant, showing bladeless sheath, × 2½; i, minutely
auricled sheath, × 2½; j, sheath without auricles, × 2½; k, habit, showing the creeping
rootstock and the grasslike leaves, × ⅖; l, heads of inflorescence, × 3.

or nearly as long as the perianth, dark brown, rounded or depressed at the apex, mucronate; seeds longitudinally reticulate, with a whitish coat.

Wet places or marshes along sand dunes near the coast: Monterey County north to Humboldt County; widespread.

A closely related species, *Juncus Covillei* Piper (*J. falcatus* var. *paniculatus* Engelm.), may possibly intergrade with *J. falcatus,* but the former occurs along streams and in meadows farther inland in the Coast Ranges and in the central Sierra Nevada. It differs from *J. falcatus* chiefly in its fewer-flowered (3- to 5-flowered), paniculately arranged heads; its lighter brown, more obtuse perianth segments; and its narrower, more oblong capsule, which often is longer than the perianth at maturity.

23. Juncus orthophyllus Cov., Contr. U. S. Nat. Herb. 4:207. 1893. Fig. 184.

Juncus latifolius Buchenau, Monogr. Juncac. 425. 1890. Not Wulf, 1789.

Stems from creeping rootstocks, 1.5–4 dm. tall, pale green; leaves mostly basal, grasslike, ¼ to nearly as long as the stem, 1–7 mm. broad, many-nerved, spreading or erect, auricles absent or weakly developed; stem leaves none (or sometimes 1 or 2 above the middle, 2–5 cm. long); inflorescence a panicle of 2–7 heads, these usually 6- to 10-flowered; perianth 4–6 mm. long, the segments lanceolate, with a green midrib bordered by brown, hyaline-margined, finely roughened on the back, the inner segments usually longer than the outer ones, acute, bristle-tipped; stamens 6, about ⅔ as long as the perianth, the anthers longer than the filaments; capsule oblong-ovoid, obtuse, mucronate, about as long as the perianth; seeds longitudinally reticulate.

Chiefly montane meadows: San Bernardino Mountains, Sierra Nevada, higher Coast Ranges north to Siskiyou County; east of the Sierra Nevada from Inyo County to Modoc County; north to Washington.

A related species, *Juncus longistylis* Torr., which is found chiefly east of the Sierra Nevada crest, is sometimes confused with *J. orthophyllus.* It differs from *J. orthophyllus* in having smooth and shining perianth segments, with the inner ones as long as or shorter than the outer ones. Also, the leaves of *J. longistylis* have well-developed auricles, and the stem leaves vary in number from 1 to 3. It may possibly intergrade with *J. orthophyllus.*

LILIACEAE. LILY FAMILY

Herbs from bulbs or rootstocks. Flowers perfect, regular. Perianth of 6 segments, the outer and inner ones essentially alike or the outer segments sepaloid and the inner ones petaloid. Stamens 6, or sometimes reduced to fewer. Ovary superior, 3-celled. Fruit a capsule or berry.

Fig. 185. *Calochortus uniflorus: a,* petal, denticulate across the apex, showing gland and hairs on basal part, × 2; *b,* stamen, × 4; *c,* habit, showing bulblets, the leaves extending beyond the flowers, and the nodding capsule, × ⅔; *d,* gland of petal, × 12.

Styles 3, distinct; plants from rhizomes (except *Zygadenus,* which is from bulbs).
  Leaves equitant.
    Stamens glabrous; flowers in heads, white .... *Tofieldia*
    Stamens densely woolly; flowers racemose, yellow .... *Narthecium*
  Leaves not equitant.
    Stems glabrous; leaves basal from a tunicated bulb .... *Zygadenus*
    Stems pubescent, from rhizomes, leafy throughout; leaves plaited .... *Veratrum*
Style 1, sometimes 3-lobed; plants bulbous.
  Stems from a tunicated bulb or a corm.
    Perianth not conspicuously differentiated into an inner and an outer series.
      Flowers borne in umbels or heads.
        Perianth segments distinct to base; herbage with odor of onions ...... *Allium*
        Perianth segments united below; stamens inserted on perianth tube ... *Brodiaea*
      Flowers borne in racemes or panicles.
        Raceme dense; perianth 6–10 mm. long, white .... *Schoenolirion*
        Raceme lax; perianth 20–25 mm. long, blue, rarely white .... *Camassia*
    Perianth conspicuously differentiated into narrow outer sepal-like segments and
      broader inner petal-like segments; flowers without bracts .... *Calochortus*
  Stems from scaly bulbs; anthers versatile; style entire; perianth segments essentially
    alike .... *Lilium*

ALLIUM

1. Allium validum Wats. in King, Geol. Expl. 40th Par. 5:350. 1871. Swamp onion.

Herbaceous alliaceous plant arising from membranous-coated bulbs which are connected by stout rhizomes; bulb coats white or red-tinged, the reticulations vertical, fine; scape 5–9 dm. tall; leaves flat, 6–15 mm. broad, 3–8 dm. long; flowers in erect, terminal, many-flowered umbels on pedicels 10–15 mm. long; perianth segments all similar, petaloid, long-acuminate, 5–10 mm. long, rose to nearly white; stamens 6–15 mm. long, usually longer than the perianth; style usually exserted, 7–8 mm. long; capsule subglobose, not crested, loculicidal.

Wet alpine meadows: Sierra Nevada, Coast Ranges south to Lake County; Pacific states.

BRODIAEA

Erect scapose herb arising from a corm. Leaves few, linear. Flowers in a bracteate umbel. Pedicels jointed beneath the perianth. Perianth segments united. Stamens 6, or 3 alternating with 3 staminodia; filaments slender or winged and extended beyond the anther as teeth or appendages. Capsule 3-celled, loculicidal, many-seeded.

Filaments filiform; pedicels 5–20 cm. long .... 1. *B. peduncularis*
Filaments deltoid, somewhat united at base; pedicels 1–5 cm. long .... 2. *B. hyacinthina*

Fig. 186. *Camassia quamash* subsp. *linearis: a,* fruiting raceme, × ⅘; *b,* fruit (cross section), × 2½; *c,* habit, showing bulb and leaves, × ⅖; *d,* seed, × 8; *e,* upper part of scape and flowering raceme, × ⅖.

1. Brodiaea peduncularis (Lindl.) Wats., Proc. Am. Acad. Arts & Sci. 14:237. 1879.

Triteleia peduncularis Lindl., Bot. Reg. 20: pl. 1685. 1834.

Scape 2–9 dm. tall; leaves 2 or 3, often longer than the stem, 5–10 mm. wide; umbels 4- to 17-flowered; pedicels 5–20 cm. long; perianth 10–25 mm. long, rose-purple, rarely white with deep purple band on the back, the tube shorter than the lobes; stamens 6 in 2 rows; anthers 3 mm. long; ovary yellow; capsule 5–6 mm. long on stipe of about the same length.

Along streams and in wet meadows: North Coast Ranges.

2. Brodiaea hyacinthina (Lindl.) Baker, Gard. Chron. III. 20:459. 1896.

Hesperoscordum hyacinthinum Lindl., Bot. Reg. 15: pl. 1293. 1829.

Scapes 3–8 dm. tall; leaves usually 2, often shorter than the stem, 3–12 mm. wide; umbels 10- to 40-flowered; pedicels 1–5 cm. long; perianth open-campanulate, 10–15 mm. long, cleft below the middle, white to lavender, each segment with a green midvein; filaments 2 mm. long, deltoid, the anthers yellow or purple; capsule stipitate.

Bogs and wet meadows in the mountains: northern California; north to British Columbia and Idaho.

CALOCHORTUS

1. Calochortus uniflorus Hook. & Arn., Bot. Beechey Voy. 398, pl. 94. 1841. Large-flowered star tulip. Fig. 185.

Stems 5–25 mm. tall, arising from membranous corms, the bulblets 1–4, beneath ground; basal leaves longer than the inflorescence, 8–12 mm. wide; bracts long, leaflike; umbels 1–3, usually 1- to 4-flowered; flowers open-campanulate; sepals ovate-lanceolate, greenish; petals cuneate, denticulate, naked above, sparsely hairy below, the gland truncate above, fringed, with dense hairs on upper border, the lower edge irregularly convex; capsule nodding, elliptic, 10–15 mm. long, septicidal.

Wet meadows: North Coast Ranges, South Coast Ranges to Monterey County.

CAMASSIA

Slender scapose herbs from tunicated edible bulbs. Leaves linear, basal. Inflorescence a simple terminal raceme. Flowers blue to white. Pedicels jointed at base of flower. Perianth of 6 distinct, spreading, linear-oblong segments, withering-persistent. Stamens 6, inserted on base of perianth segments; anthers versatile. Style filiform with 3-lobed stigma, the base persistent. Capsule 3-lobed, loculicidal.

Fig. 187. *Narthecium californicum: a,* mature beaked capsule, × 4; *b,* habit, showing rootstock, basal and stem leaves, and part of fruiting inflorescence, × ⅖; *c,* seed, showing long tail at each end, × 6; *d,* flower, showing the woolly filaments, × 3; *e,* mature seed, × 12; *f,* flowering raceme, × ⅘; *g,* creeping rootstock and basal leaves, × 1⅕.

Perianth segments regular; fruiting pedicels spreading, curved near the tip; capsule erect
..... 1. *C. Leichtlinii*
Perianth segments slightly irregular (1 segment offset); fruiting pedicels erect or ascending
and curved from the base .... 2. *C. quamash*

1. Camassia Leichtlinii subsp. **Suksdorfii** (Greenm.) Gould, Am. Midl. Nat. 28:723. 1942.

Scape 3–8 dm. tall; raceme many-flowered; bracts shorter than pedicels; flowers creamy white to dark blue; perianth segments 20–35 mm. long, usually 3- to 5- or 7-nerved, connivent and twisted into a single group in fruit; capsule oblong, about 15 mm. long, many-seeded.

Wet meadows and bogs: Sierra Nevada, North Coast Ranges; north to British Columbia. The typical phase of the species is locally confined to southern Oregon.

2. Camassia quamash (Pursh) Greene, Man. Bot. Bay Reg. 313. 1894. Common camass. Fig. 186.

Bulbs black, ovoid, 15–35 mm. in diameter; scapes stout, 3–10 dm. tall; leaves somewhat shorter than the scape, 3–20 mm. wide; raceme 5- to 35-flowered, 5–30 cm. long, the pedicels short; bracts about as long as the flowers; flowers white to blue; perianth segments 15–25 mm. long, linear-oblong, slightly unequal in length, usually separating into 2 groups or remaining separate, twisting in drying.

Wet meadows and bogs: North Coast Ranges; north to British Columbia, east to Rocky Mountains.

Two California subspecies of *Camassia quamash, C. quamash* subsp. *linearis* and *C. quamash* subsp. *breviflora,* have been proposed by Gould; the former is confined to the North Coast Ranges of California, but the range of the latter extends from the northern Sierra Nevada north to Washington.

LILIUM

Tall, leafy perennial herbs with a scaly bulb or rootstalk. Leaves narrow, sessile. Flowers 1 to many, solitary or in a terminal raceme, showy. Perianth segments 6, spreading or recurved, deciduous, a nectar-bearing groove near the base. Stamens 6, shorter than the perianth segments and lightly attached near the base. Style 1; stigma 3-lobed. Capsule loculicidal. Seeds many.

Flowers erect or ascending; upper ⅓ of perianth segments spreading, bright orange
..... 1. *L. parvum*
Flowers nodding; perianth segments strongly recurved to below middle
..... 2. *L. pardalinum*

Fig. 188. *Schoenolirion album:* *a* and *b,* mature seeds, black, shining, and wrinkled, × 6; *c,* habit, showing the fibrous bulb, the flat leaves, and the many-flowered raceme, × ⅖; *d,* flower and bract, × 6; *e,* mature capsule, short-stalked, × 5.

1. Lilium parvum Kell., Proc. Calif. Acad. Sci. 2:179, fig. 52. 1863. Small tiger lily.

Stems 4–20 dm. tall; rhizome thick with short, thick scales, each with 3 or 4 joints; leaves in 1–3 whorls or scattered, lanceolate to linear, 5–15 cm. long, 5–30 mm. broad; inflorescence few- to many-flowered, erect or ascending, funnelform, upper ⅓ spreading and somewhat revolute, 2–4 cm. long, yellowish orange, often spotted with purple; stamens 20–25 mm. long; anthers 3 mm. long.

Bogs and swamps: high Sierra Nevada; mountains of southern Oregon.

2. Lilium pardalinum Kell., Proc. Calif. Acad. Sci. 2:12. 1859. Leopard lily.

Stems 9–20 dm. tall, from branched rhizome covered with 2-jointed fleshy scales, the rhizomes forming large matlike masses; leaves in 3 or 4 whorls of 9–15, linear-lanceolate, 6–25 mm. wide; flowers whorled or racemose on long spreading pedicels, nodding; perianth segments 5–8 cm. long, 12–18 mm. wide, revolute to below the middle, orange with purple spots on lower half; stamens shorter than perianth segment, the anthers red, 10–15 mm. long; capsule oblong, 3 cm. long, acutely angled.

Springs, bogs, stream banks at low altitudes: northern California, Sierra Nevada and North Coast Ranges, central California; southern Oregon.

Several other species of *Lilium* are found along moist banks and in wet meadows. They include: *L. parryi* Wats., *L. maritimum* Kell., *L. occidentalis* Purdy.

NARTHECIUM

1. Narthecium californicum Baker, Jour. Linn. Soc. London 15:351. 1876. Bog asphodel. Fig. 187.

Abama californicum Heller, Cat. N. Am. Pl. 3. 1898.

Perennial from a creeping rootstalk; stem simple, erect; basal leaves linear, equitant, acute, 10–25 cm. long, 3–4 mm. wide, the stem leaves distant, short; raceme loose, the pedicels 6–12 mm. long; perianth segments 6, distinct, yellow, narrow, acute, 6–8 mm. long; stamens 6, the anthers red, the filaments woolly; ovary sessile, stigma obscurely 3-lobed; capsule beaked; seeds many, with long tail at each end.

About springs and bogs in mountain meadows: northern California to Fresno County and Mendocino County; southern Oregon.

SCHOENOLIRION

Stem and linear leaves from tunicated bulb. Flowers in dense raceme, often 1 or 2 secondary racemes beneath. Bracts small, scarious. Perianth segments

Fig. 189. *Tofieldia occidentalis: a* and *b,* mature seeds, showing the loose outer coat
and slender appendages, × 12; *c,* habit, showing the equitant, linear leaves, the naked
stem, and the headlike inflorescence, × ⅖; *d,* variation in inflorescence, × ⅖; *e,* mature
capsule, × 4; *f,* bud and flower, × 5.

distinct, 5–10 mm. long, white, scarious in age. Stamens 6, united to perianth segments at base. Ovary ovate with a short stipe; stigma 3-cleft. Capsule loculicidal. Seeds black.

Perianth segments 3–6 mm. long, outer bulb-coats somewhat fibrous .... 1. *S. album*
Perianth segments 8–12 mm. long, outer bulb-coats membranous ...... 2. *S. bracteosum*

1. Schoenolirion album Durand, Jour. Acad. Nat. Sci. Phila. II. 3:103. 1855. Fig. 188.

Outer bulb-coats often fibrous, the bulb 12–18 mm. broad; stems tall, glabrous, 4–15 dm. tall; leaves flat, 4–12 mm. wide, 2–6 dm. long; raceme many-flowered, the bracts acuminate, 2–3 mm. long; perianth segments 3–6 mm. long, white, often tinged with pink, lilac, or green; stamens longer than perianth segments; ovary remotely 6-lobed, the style 3-cleft; capsule short-stalked, 6 mm. long, broadly ovoid; seeds oblong, 4 mm. long.

Stream banks and wet meadows: North Coast Ranges and northern Sierra Nevada, Siskiyou Mountains; southern Oregon.

2. Schoenolirion bracteosum (Wats.) Jepson, Fl. Calif. 1:268. 1922.

Outer bulb-coats membranous, the bulb 20–30 mm. broad; raceme with flowers on pedicels 2–3 mm. long; perianth segments 8–12 mm. long, white; stamens about ½ as long as the perianth segments.

Wet meadows and bogs: Del Norte County; southern Oregon.

TOFIELDIA

1. Tofieldia occidentalis Wats., Proc. Am. Acad. Arts & Sci. 14:283. 1879. Fig. 189.

Perennial herb arising from a slender rootstock; stems simple, naked above, 2–6 dm. tall; leaves equitant, linear, 5–25 cm. long, 2–6 mm. wide; raceme 1–3 cm. long, the flowers in groups of 3's, yellowish white or greenish white; perianth segments 6, subequal in length, distinct, persistent, 3–6 mm. long; stamens 6, the filaments filiform; capsule obovate, 6–8 mm. long, tipped with spreading persistent styles, septicidal; seeds with a spongy, loose outer coat with a slender appendage about as long as the seed.

Mountain springs: Sierra Nevada, northern California, North Coast Ranges to Mendocino County; north to Alaska.

VERATRUM

Tall, leafy, perennial herb from short, thick, often poisonous rootstocks. Stem and inflorescence pubescent. Leaves broad, plaited, prominently nerved, clasping. Flowers in terminal panicles, polygamous, white or greenish. Peri-

anth segments similar, distinct, often adnate to base of ovary. Stamens 6, opposite perianth segments, free, short, curved, the anthers cordate, with confluent
pollen sacs. Styles 3, persistent. Capsule 3-celled, 3-lobed, many-seeded.

Perianth segments fimbriate; capsule 8 mm. long . . . . 1. *V. fimbriatum*
Perianth segments entire or serrulate; capsule 20–30 mm. long . . . . . . 2. *V. californicum*

1. Veratrum fimbriatum Gray, Proc. Am. Acad. Arts & Sci. 7:391. 1868.

Tall, stout, leafy perennial; leaves narrowed at base, 1–5 dm. long, 2–10 cm.
wide, strongly nerved; pedicels 6–10 mm. long; perianth segments 5–10 mm.
long, rhombic-ovate, the upper margins irregularly fimbriate, the broad base
bearing 2 glandular spots; capsule emarginate at apex, about 8 mm. long;
seeds 6 mm. long, 5–7 in each of the 3 cavities, green, scarcely margined.

Common in wet meadows near the coast: Sonoma and Mendocino counties.

2. Veratrum californicum Durand, Jour. Acad. Nat. Sci. Phila. II. 3:103. 1855.

Plant stout, leafy, 1–2 m. tall; leaves sheathing at base, 2–5 dm. long, 1–2
dm. wide, ovate or the upper ones narrower, lightly pubescent; panicle 2–5 dm.
long, tomentose; pedicels 2–6 mm. long; flowers dull white; perianth segments
8–15 mm. long, 5–8 mm. wide, greenish-margined, with a greenish spot at
base; capsule 20–30 mm. long; seeds wing-margined.

Wet meadows: North Coast Ranges, Sierra Nevada, high mountains of
southern California; western states, Baja California, Mexico.

ZYGADENUS

1. Zygadenus venenosus Wats., Proc. Am. Acad. Arts & Sci. 14:279. 1879.
Deadly zygadene, death camass.

Perennial glabrous herbs; tunicated bulbs dark, 1–2 cm. in diameter; stem
2–6 dm. tall; leaves linear, 4–6 mm. broad, carinate, folded, mostly basal;
raceme usually simple, 10–15 cm. high; pedicels 6–12 mm. long; flowers
numerous, greenish white; perianth free from the ovary, the segments triangular-ovate to elliptical, 4–6 mm. long, clawed, subcordate at base, withering-
persistent, each segment bearing a gland near the base; stamens 6, as long as
the perianth and slightly adnate to the claw; capsule 3-celled, 8–12 mm. long;
seeds usually 2 in each cell.

Wet meadows: Sierra Nevada, Coast Ranges; western states and north to
British Columbia.

Called "death camass," also "hog potatoes" in Sonoma County.

IRIDACEAE. IRIS FAMILY

Perennial herbs with equitant, 2-ranked leaves and bracted flowers. Flowers in ours regular, perfect. Perianth of 6 members essentially alike or in 2 unlike series of 3 each. Stamens 3. Ovary inferior, 3-celled. Fruit a capsule.

Perianth segments in 2 unlike series; styles petal-like .... *Iris*
Perianth segments essentially alike; styles filiform .... *Sisyrinchium*

IRIS

1. Iris missouriensis Nutt., Jour. Acad. Nat. Sci. Phila. 7:58. 1834. Western iris.

Perennial herb with stout, creeping rootstock; stem slender, simple, terete, 3–5 dm. tall; equitant leaves usually basal, shorter than or as long as the stem, 4–10 mm. wide; bracts scarious, 4–7 cm. long, acute; pedicels 2–8 cm. long; perianth segments 6-clawed, united below into a tube 5–8 mm. long, the 3 outer segments broad, spreading or reflexed, 5–7 cm. long, glabrous, without crest, the 3 inner segments somewhat shorter, erect, narrow, white to blue, often with darker veins; capsule 3–7 cm. long, 6-angled; seeds 4 mm. long, obovate.

Wet meadows: Sierra Nevada, north to Modoc County, San Bernardino Mountains, Inner Coast Ranges; Great Basin, north to British Columbia.

Iris longipetala Herbert, which is similar to *I. missouriensis,* sometimes occurs on seepage bogs in the San Francisco Bay region.

Another species that should be mentioned is *I. pseudacorus* L., an Old World species, widely cultivated in America, which has escaped in Merced County and is apparently moving down the watercourses. It is a large plant, 8–12 dm. tall, with conspicuous yellow flowers.

SISYRINCHIUM

Perennial tufted herbs with flattened winged stems and narrow equitant leaves. Flowers 2 to several from a pair of deeply keeled spathelike bracts. Perianth of 6 similar members. Stamens wholly or partially united by their filaments, the anthers free. Ovary inferior, 3-celled; style 3-lobed. Fruit a capsule. Seeds several.

Herbage becoming black on drying, leaving a purple stain; stems broadly winged; perianth
    segments 12–18 mm. long .... 1. *S. californicum*
Herbage remaining green on drying, without stain; stems narrowly winged; perianth seg-
    ments 8–12 mm. long .... 2. *S. Elmeri*

Fig. 190. *Sisyrinchium californicum: a,* mature capsule, × 1½; *b,* habit, showing the broadly winged stems, spathelike bracts, equitant leaves, flowers, and capsules, × ⅖; *c,* flower, the filaments partially united, × 1½; *d,* mature seed, adaxial view, × 12; *e,* capsule (cross section), × 3; *f,* mature seed, abaxial view, × 12.

1. Sisyrinchium californicum Dryand., Hort. Kew., 2d ed., 4:135. 1812. Golden-eyed grass. Fig. 190.

Hydastylus californicus Salisb., Trans. Hort. Soc. London 1:310. 1812.

Stems 15–60 cm. tall, glaucescent, 2–6 mm. wide, broadly winged; leaves equitant, 8–30 cm. long, 2–6 mm. wide; bracts unequal in length, the outer ones longer; flowers 3–7; perianth segments yellow with dark veins, 12–18 mm. long, obtuse; capsule 6–12 mm. long.

Bogs and swamps near the coast: Monterey County north; southern Oregon.

2. Sisyrinchium Elmeri Greene, Pittonia 2:106. 1890. Sierra golden-eyed grass.

Hydastylus Elmeri Bickn., Bull. Torrey Club 27:380. 1900.

Stems 10–30 cm. tall, 1–2 mm. broad, narrowly winged; leaves 6–15 cm. long; bracts subequal in length; perianth yellow with dark veins, the segments 8–12 mm. long; capsule 7 mm. long.

Borders of streams and in bogs at middle altitudes: Sierra Nevada, San Bernardino Mountains.

ORCHIDACEAE. ORCHID FAMILY

Perennial herbs with sheathing entire leaves. Flowers perfect, irregular. Perianth of 6 segments, the 3 outer segments (sepals) regular, 2 of the inner segments (petals) alike, the third often very dissimilar, being developed into a spur, a sac, or an elongate lip. Stamens variously united to the style. Ovary inferior, often long and twisted; style often extended into a beak between the anthers. Fruit a 3-valved capsule. Seeds numerous, minute.

Lower perianth lip saclike, ovoid; anthers 2 . . . . *Cypripedium*
Lower perianth lip open-ligulate or concave; anther 1.
  Leaves plicate; flowers in a lax raceme . . . . *Epipactis*
  Leaves not plicate; flowers in erect spikes.
    Perianth with a spur . . . . *Habenaria*
    Perianth without a spur . . . . *Spiranthes*

CYPRIPEDIUM

1. Cypripedium californicum Gray, Proc. Am. Acad. Arts & Sci. 7:389. 1867. California lady's slipper. Fig. 191.

Fibrous-rooted perennial from short rhizome; stems erect, 3–7 dm. tall, glandular-pubescent; leaves 7–10 cm. long, ovate-lanceolate, acute, the upper ones smaller and often acuminate; flowers 3–7, solitary and recurved in the axils of bractlike leaves; sepals broadly elliptic-oval, the lower ones united, 10–15 mm. long; petals oblong, as long as or shorter than the sepals, the lower

Fig. 191. *Cypripedium californicum: a,* habit, × ⅖; *b,* inflorescence, × ⅘; *c,* upper lip of flower, face view, × 6; *d,* flower (longitudinal section), × 1½.

one forming a pouchlike, ovate-globose sac or "slipper," open above, white or rose, 18–20 mm. long, pubescent within; fertile stamens 2, a third, sterile, forming a fleshy hood or flap arched over the stigma; stigma rough, 4 mm. long; capsule reflexed, 20 mm. long.

Bogs or wet soil in mountains: Marin County and Placer County; north to southern Oregon. Not common.

Cypripedium montanum Dougl. and *C. fasciculatum* Kell. also occur in our area on moist soil or in riparian habitats.

EPIPACTIS

1. Epipactis gigantea Dougl. in Hook., Fl. Bor. Am. 2:202, pl. 202. 1839. Stream orchis. Fig. 192.

Stem 3–10 dm. tall, leafy, from a mass of branched, creeping rootstocks; leaves ovate-lanceolate, plicate, clasping, 5–15 cm. long; flowers in a raceme with leaflike bracts; perianth segments nearly equal in length, spreading, greenish or purplish, free, sessile; lip strongly constricted at middle, the basal part deeply concave, the distal part dilated and petal-like; anther solitary on a slender jointed base, sessile behind a broad truncate stigma which functions as an operculum to the anther, the pollinia 2-parted, becoming attached to the gland capping the beak of stigma; capsule reflexed at maturity.

Growing in tufts at the margins of streams, mountains: throughout California; north to Washington, Montana, and British Columbia, south to Baja California, Mexico, east to Texas.

HABENARIA

Stems leafy, often tall, from fleshy fibrous or tuberous roots. Leaves linear or oblong-lanceolate. Flowers spicate or racemose, not showy. Perianth bilabiate; sepals and petals similar, the lower sepals often spreading, the lip spreading, entire or 3-lobed, a slender spur at its base. Anther sacs somewhat divergent. Stylar column short. Capsule erect.

Flowers white; lip rhombic-lanceolate; stem thick .... 1. *H. leucostachys*
Flowers greenish; lip linear; stem slender .... 2. *H. sparsiflora*

1. Habenaria leucostachys (Lindl.) Wats., Bot. Calif. 2:134. 1880. White bog orchid.

Limnorchis leucostachys Rydb., Mem. N. Y. Bot. Gard. 1:105. 1906.

Stem stout, erect, 2–7 dm. tall, from a fascicle of elongate tubers; basal leaves 1–2 dm. long, the upper leaves shorter, oblanceolate to lanceolate, acute; spike dense, 1–3 dm. long; bracts shorter than flowers; flowers glistening white;

Fig. 192. *Epipactis gigantea: a,* habit, showing creeping rootstocks, stems, and clasping leaves, × ⅖; *b,* habit, upper part of stem, showing raceme of flowers with their leaflike bracts, × ⅔; *c,* anther and stigma, × 4; *d,* flower, face view, × 1⅕; *e,* mature capsule, × 1⅕.

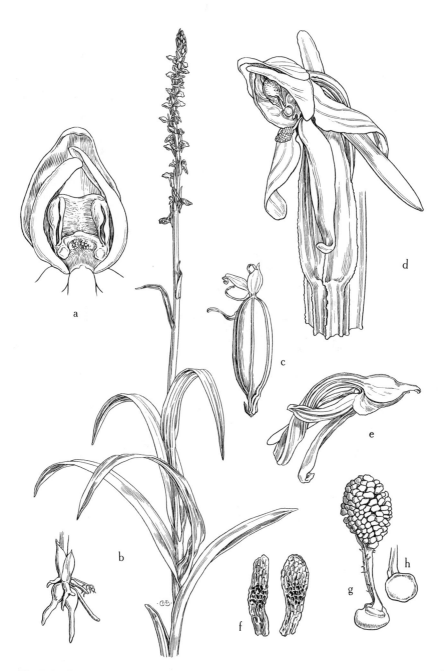

Fig. 193. *Habenaria sparsiflora: a,* upper sepal and petals, stigma and empty anther sacs, × 6; *b,* habit, × ⅓; *c,* mature capsule, × 2; *d,* bract and flower, 1 anther still enclosed in anther sac, × 4; *e,* flower, lateral view, × 3; *f,* seeds, × 30; *g,* caudicle and massula (pollinium), × 16; *h,* base of caudicle, × 16.

Fig. 194. *Spiranthes Romanzoffiana: a,* base of plant, showing linear leaves and fleshy roots, × ⅔; *b,* spike, × 1⅕; *c,* single whorl of flowers, × 1⅕; *d,* flower, lateral view, × 4; *e,* flower (longitudinal section), × 4; *f,* back of column, showing anther, × 6; *g,* front of column, showing stigma, × 6; *h,* lower lip, × 6.

sepals oblong, unequal in size, 4–6 mm. long, the upper one broadly ovate, the lateral ones lanceolate; lateral petals erect-oblique, lanceolate; lip ovate, attenuate, subequal to or shorter than the clavate spur; capsule 8–12 mm. long.

Common about bogs and springs: North Coast Ranges in Marin and Sonoma counties, mountains of southern California and the Sierra Nevada; north to Alaska.

This species exhibits considerable variation in the density of the spike.

2. Habenaria sparsiflora Wats., Proc. Am. Acad. Arts & Sci. 12:276. 1877. Green bog orchid. Fig. 193.

Limnorchis sparsiflora Rydb., Bull. Torrey Club 28:631. 1902.

Stem slender, erect, 3–6 dm. tall; leaves lanceolate, 8–10 cm. long, 10–15 mm. wide; spike slender, 1–2 dm. long, relatively few-flowered; bracts subequal to or longer than the flowers; flowers greenish, 10–12 mm. long; upper sepal ovate, the lower ones lanceolate, spreading; upper petals oblique at base, erect; lip narrowly linear, the spur 6–8 mm. long, slightly longer than lip; capsule 12 mm. long.

About springs and bogs, middle and higher elevations in the mountains: North Coast Ranges, Sierra Nevada; to Baja California, Mexico.

SPIRANTHES. SPIRAL ORCHID

Stems from a cluster of fleshy roots, erect, often stout but sometimes slender, leafy. Flowers in a twisted 2- or 3-ranked spike. Perianth segments all erect and connivent, united at base into a short tube, white to greenish white; lip sessile or with short claw, the base clasping the stylar column and bearing a protuberance on each side, the margins undulate-crisped. Stylar column short, obliquely inserted with a sessile, short-stalked anther on back and the stigma in front. Capsule erect.

Lip ovate to orbicular at base, narrowed above middle but strongly dilated at apex, protuberances at base of lip blade ridgelike, minute or obsolete ... 1. S. Romanzoffiana
Lip oblong at base, lanceolate above, protuberances at base of lip blade nipple-like and prominent .... 2. S. porrifolia

1. Spiranthes Romanzoffiana Cham. & Schl., Linnaea 3:32. 1828. Hooded spiral orchid. Fig. 194.

Ibidium Romanzoffianum House, Muhlenbergia 1:129. 1906.

Stem glabrous, 15–45 cm. tall; leaves linear, 5–30 cm. long; spike 5–15 cm. long, dense; flowers greenish white, 6–8 mm. long; bracts conspicuous, 10–12 mm. long; lip recurved.

Wet meadows, mostly of higher mountains: Sierra Nevada, Coast Ranges, rare in southern California; across northern part of North America.

Fig. 195. *Anemopsis californica: a,* flower-like inflorescence subtended by petaloid bracts, × ⅘; *b,* seed, × 20; *c,* spike, showing receptacle (longitudinal section), × 1⅓; *d,* single bracteate flower, × 8; *e,* habit, × ⅖.

2. Spiranthes porrifolia Lindl., Gen. & Sp. Orchid. 467. 1840. Western spiral orchid.

Ibidium porrifolium Rydb., Bull. Torrey Club 32:610. 1905.

Stem slender, 2–4 dm. tall; leaves linear or oblanceolate, 1–2 dm. long; spike densely flowered, 5–10 cm. long; flowers cream-colored or greenish white; lip lanceolate, 2 nipple-shaped callosities at base, the apex dilated.

Bogs and marshes: Coast Ranges, Sierra Nevada, southern California; north to Washington and Idaho, and east to Colorado.

SAURURACEAE. LIZARD'S-TAIL FAMILY

Erect perennial herbs with alternate, simple, stipulate leaves. Flowers without perianth, perfect, in dense, slender, peduncled spikes or racemes, subtended by bracts. Stamens 6–8 or sometimes fewer, usually hypogynous. Ovary 3- or 4-celled. Fruit a succulent follicle or capsule.

ANEMOPSIS

1. Anemopsis californica (Nutt.) Hook. & Arn., Bot. Beechey Voy. 390. 1841. Yerba mansa. Fig. 195.

Colonial plant arising from stolons and rhizomes; stem and herbage spicy-aromatic; leaves basal and cauline, the basal leaves 5–15 cm. long, petioled, the blade elliptic-oblong to broadly spatulate, truncate to cordate at base, entire, glabrous to puberulent or pubescent, the cauline leaves sessile and clasping at the nodes with 1–3 small, petioled leaves fascicled in the axils; inflorescence scapelike with 1 or 2 leafy-bracted nodes and usually a solitary terminal flower-like spike, subtended by a whorl of white or reddish petaloid bracts; spike closely packed with bracteate flowers, the floral bracts white, obovate, with a narrow claw; true perianth none; stamens 6–8; ovary sunk in rachis of spike, ovules 6–8; fruit a dehiscent capsule.

Alkaline floodlands of Central Valley; to Arizona and Baja California, Mexico.

SALICACEAE

Trees and shrubs with simple leaves. Flowers dioecious, borne in catkins in the axils of bracts, the staminate flowers having 1 to many stamens, the pistillate ones having a single pistil, the perianth absent, usually a gland or a disc at the base of the flower. Fruit a capsule, dehiscent into 2 valves. Seeds densely long-comose.

Bracts of flowers fimbriate; disc present .... *Populus*
Bracts of flowers entire; disc absent or reduced to an elongate gland .... *Salix*

POPULUS. POPLARS AND ASPENS

Trees with resinous buds. Leaves usually with long petioles and broad, del-toid or orbicular blades, occasionally these lanceolate or ovate, the margins toothed or entire. Flowers in catkins, each flower subtended by a fimbriate bract and inserted in a cup-shaped or spreading disc. Staminate catkins lax and usually pendulous; stamens in ours 6–60. Pistillate catkins pendulous, erect or spreading, at length racemose, the flowers sessile; style short or none; stig-mas 2–4, entire or lobed. Capsule 2- to 4-valved. Seeds comose.

Leaves longer than wide, finely crenulate or serrulate, gradually to abruptly acuminate; petioles terete.
  Leaf blades rhombic-lanceolate to ovate, usually cuneate at the base, abruptly acumi-nate at the apex, green and glabrous on both surfaces; capsule glabrous
  1. *P. acuminata*
  Leaf blades broadly ovate to ovate-lanceolate, cordate or truncate at base, tapering above, green on upper surface, glaucous on undersurface; capsule pubescent
  2. *P. trichocarpa*
Leaves nearly as wide as or wider than long; coarsely serrate, petioles flattened laterally.
  Leaf blades broadly ovate to nearly orbicular, crenate-serrate with glandular teeth on the margin except at the base, 3–5 cm. long; stamens 6–12 ..... 3. *P. tremuloides*
  Leaf blades broadly deltoid, coarsely serrate-dentate, without glandular teeth, 4–7 cm. long; stamens 60 or more.
    Leaves and petioles glabrous; buds elongate .... 4. *P. Fremontii*
    Leaves and petioles pubescent; buds short, blunt .... 5. *P. Macdougalii*

1. Populus acuminata Rydb., Bull. Torrey Club 20:50. 1893. Smooth-barked cottonwood.

Tree to about 20 m. tall, with brownish or gray trunk; petioles 3–7 cm. long, terete (not laterally flattened); leaf blades rhombic-lanceolate to ovate, commonly less than twice as long as wide, somewhat abruptly acuminate at apex, finely crenate, usually rounded or broadly short-cuneate at the base, both surfaces green; capsule glabrous, the pedicels in fruit (in ours) about 3 mm. long.

Riverbanks: known only from Lone Pine Creek in Inyo County; Saskatche-wan, Texas, Arizona, Montana, Nebraska, and South Dakota.

2. Populus trichocarpa Torr. & Gray in Hook., Icon. Pl. 9: pl. 878. 1852.

Tree to 60 m. (occasionally to 70 m.) tall; winter buds resinous, long-pointed; leaves deltoid to lanceolate or ovate-attenuate, very variable in shape, truncate, rounded or cordate at base, the margins finely serrate, 5–20 cm. long, dark lustrous green above, glaucous to silvery or rufous below; stamens 40–60; ovary ovoid, densely pubescent; capsule subglobose, densely hairy, sessile or subsessile.

Swamps and stream banks, widely ranging in area and altitude throughout California; north to Canada, east to Rocky Mountains.

The narrow-leaved forms have been variously treated as *Populus trichocarpa* var. *ingrata* Jepson, or *P. angustifolia* James. The latter is a Great Basin species ranging from Arizona and New Mexico northward to Saskatchewan. It has been reported from several widely scattered localities in California on the basis of sterile specimens bearing only mature foliage. It has been impossible to verify these identifications, because critical identification requires a count of stamens and observation of the fruit. However, fertile specimens from these same localities have been identified as *P. angustifolia*. It is a matter of interest that *P. angustifolia* is intermediate in some of its characters between *P. tremuloides* and *P. trichocarpa*.

3. Populus tremuloides Michx., Fl. Bor. Am. 2:243. 1803. Quaking aspen.

Slender trees, often in groves or thickets; bark greenish white, becoming black in age; leaves orbicular to deltoid, rounded or tipped with an abrupt point, cordate, truncate or cuneate at base, denticulate to entire, the blades 3–8 cm. long, on slender petioles; stamens 6–12; bract lacerate, the lobes bristle-tipped; ovary ovoid; stigmas 2, each 2-lobed; capsule glabrous.

Bogs and swamps, and in seasonally wet meadows: throughout California at elevations of 5,000–10,000 feet; similar localities throughout North America.

Our material has been variously treated as *Populus tremuloides* var. *aurea* or as *P. aurea,* chiefly because of the fall coloring.

4. Populus Fremontii Wats., Proc. Am. Acad. Arts & Sci. 10:350. 1875.

Spreading tree to 35 m. tall; winter buds acute, resinous, with green scales; leaves 4–7 cm. long, on slender, flattened petioles, the blades deltoid, acute or acuminate, truncate or cordate at base, coarsely serrate-dentate, especially about midway on margins, green on both surfaces, glabrous; stamens 60 or more; ovary glabrous; capsule 3-valved, 8–12 mm. long, glabrous.

Swamps and stream banks at low altitudes: Central Valley and surrounding foothills, Coast Ranges, southern California; south and west to Nevada, Arizona, and Baja California, Mexico.

5. Populus Macdougalii Rose, Smithson. Misc. Coll. 61 (Art. 12):1. 1913.

Populus Fremontii var. *Macdougallii* Jepson, Man. Fl. Pl. Calif. 268. 1923.

Similar to *Populus Fremontii* in many details, but the leaves, petioles, and twigs pubescent; leaves blue-green, and the winter buds ovoid and obtuse.

Lowlands along the Colorado River and its delta.

SALIX. WILLOW

Trees or shrubs with simple, alternate, deciduous leaves bearing stipules which may or may not persist, the buds covered by 1 scale, the catkins usually erect and appearing either before or with the leaves. Flowers of the catkins each in the axil of a scale, the staminate flowers consisting of 1–5 free or united stamens and a basal gland, the pistillate flowers consisting of a single pistil and with a gland at base of ovary. Fruit a capsule, dehiscent, bearing many seeds covered by long, silky hairs.

Of the large numbers of willows in California, only those that play a role in bogs and swamps are regarded as significant to our problem. This group includes all the species that are typically trees and most of those that sometimes assume treelike proportions. It also includes those that typically form thickets in wet, boggy, or swampy places.

Stamens 4–9; typically trees (except *S. caudata*) having a single trunk 6–15 m. tall when
    mature; bark thick and deeply fissured (see also *S. Hindsiana, S. lasiolepis, S. Coul-*
    teri, and *S. Scouleriana* below, which are also sometimes treelike).
  Petioles with wartlike glands just below the blade.
    Leaves green below; shrub .... 1. *S. caudata*
    Leaves glaucous below; tree .... 2. *S. lasiandra*
  Petioles without wartlike glands below the blade.
    Leaves glaucous beneath, broadly lanceolate, acute .... 3. *S. laevigata*
    Leaves green on both sides, narrowly lanceolate and long-pointed .. 4. *S. Gooddingii*
Stamens 2 or 1; mature plants typically shrubs, having several main stems from the base
    and usually less than 6 m. tall; bark smooth or roughened but not deeply fissured;
    often forming thickets.
  Stamen filaments hairy.
    Filaments distinct to base.
      Leaves short-petioled, linear to linear-lanceolate.
        Twigs puberulent to pubescent; leaves pubescent above and below
                      5. *S. Hindsiana*
        Twigs pruinose, glabrous; leaves becoming glabrate, at least above .. 6. *S. exigua*
      Leaves sessile or subsessile, lanceolate to lanceolate-elliptic.
        Herbage densely silky-lanate above and below .... 7. *S. Parksiana*
        Herbage closely felty-pubescent, not silky-lanate, sometimes becoming glabrate
                      8. *S. melanopsis*
    Filaments united at base .... 9. *S. Geyeriana*
  Stamen filaments glabrous (or sometimes pilose at base).
    Stamens 2.
      Filaments united at base.
        Leaves broadly obovate, glabrous and glaucous below.
          Leaves elliptic, acute or obtuse at apex .... 10. *S. Piperi*
          Leaves obovate, rounded at apex .... 11. *S. Tracyi*
        Leaves oblanceolate to oblong-acute, pubescent to glaucescent below
                      12. *S. lasiolepis*
      Filaments free at base.
        Capsule glabrous or thinly pubescent.
          Leaf blade glabrous below.
             Leaf glaucous below.
               Leaves 2–5 cm. long; twigs yellow .... 13. *S. lutea*
              Leaves 5–10 cm. long; twigs brown or green .... 14. *S. Mackenziana*
            Leaf green below .... 15. *S. pseudocordata*

Leaf blade hairy below.
  Catkins sessile, without leaves on pedicel .... 16. *S. Hookeriana*
  Catkins peduncled, bearing leaves .... 17. *S. commutata*
Capsule hairy.
  Leaves glabrate at maturity (see also *S. Scouleriana*).
    Plants 1–5 m. tall; leaves 4–9 cm. long .... 18. *S. Lemmonii*
    Plants less than 1 m. tall; leaves 2–3 cm. long .... 19. *S. monica*
  Leaves densely pubescent, at least below.
    Branchlets pruinose .... 20. *S. subcoerulea*
    Branchlets never pruinose.
      Style evident.
        Leaves tomentose beneath.
          Floral gland very short .... 21. *S. Jepsonii*
          Floral gland elongate.
            Leaves linear-lanceolate .... 22. *S. Breweri*
            Leaves oblong-obovate .... 23. *S. delnortensis*
        Leaves tomentose on both sides.
          Catkins on leafy peduncles .... 24. *S. Eastwoodiae*
          Catkins subsessile .... 25. *S. orestera*
      Style none .... 26. *S. Scouleriana*
Stamen 1 .... 27. *S. Coulteri*

1. Salix caudata (Nutt.) Heller, Muhlenbergia 2:186. 1906.

A shrubby species very close in its form and structure to *Salix lasiandra* and perhaps best treated as of varietal status in that species.

Along streams and in marshy thickets: scattered in the Siskiyou Mountains and in the Sierra Nevada, especially on Great Basin slopes; north to Alberta and east to South Dakota.

2. Salix lasiandra Benth., Pl. Hartw. 335. 1857. Yellow willow tree.

Tree to 15 m. tall with rough, fissured bark; twigs from conspicuously yellow to red; winter buds keeled, blunt; stipules conspicuous; petioles with wart-like glands beneath the blades, the blades lanceolate to lanceolate-acuminate; staminate catkins 2–6 cm. long, on a very short, leafy shoot, the scales yellow, the stamens 4–9, free to base; pistillate catkins on a short, leafy shoot, the scales hairy at base, glabrous above, brown, deciduous in fruit, the ovary and capsule glabrous, the style short.

Swamps and stream courses: at low and middle altitudes throughout California, reaching higher elevations in southern California; north to Alberta, south to New Mexico.

Distinguishable by its many stamens, treelike habit, and wartlike glands on petiole.

3. Salix laevigata Bebb, Am. Nat. 8:202. 1874. Red willow.

Tree to 15 m. tall, the bark rough, the twigs reddish brown, glabrous; winter buds pointed; stipules minute, caducous; petioles 4–10 mm. long, the blades oblong-elliptic, acute at both ends or obtuse at base, green above, glaucous below; staminate catkins lax, erect, 3–10 cm. long, the peduncle leafy, the stamens 4–7, the filaments free; pistillate catkins slender, 2–5 cm. long, the peduncle leafy; capsule ovoid.

Swamps and stream courses: Coast Ranges, southern California; east to Utah and Arizona.

Distinguishable by its treelike habit, broadly elliptic mature leaves, absence of glands on petiole, and 4–6 stamens.

4. Salix Gooddingii Ball, Bot. Gaz. 40:376. 1905.

Salix nigra of western American authors, not Marshall.
Salix nigra var. *vallicola* Dudley in Abrams, Fl. Los Angeles 100. 1904.

Tree to 12 m. tall with rough bark, the branchlets pubescent; stipules often glandular above; petioles 6–10 mm. long; blades narrowly lanceolate-attenuate, finely glandular-serrulate, gray-green on both sides; staminate catkins 4–8 cm. long, the stamens 3–5, the filaments free; pistillate catkins 4–8 cm. long, on leafy branchlets; capsule ovoid-attenuate, pilose when young, becoming glabrate.

Swamps and stream courses: at low altitudes throughout California; east and south to Nevada, Arizona, and Baja California, Mexico.

The common black willow, distinguished by its gray, narrow, lanceolate-attenuate leaves, and 3–5 free stamens.

5. Salix Hindsiana Benth., Pl. Hartw. 335. 1857. Sand-bar willow.

Salix sessilifolia var. *Hindsiana* Anderss., Oefv. K. Vet.-Akad. Förhandl. 15:117. 1858.

Shrub, sometimes treelike, often forming dense thickets, 2–9 m. tall, the bark gray and furrowed but not thick, the twigs silvery-tomentose when young; leaves linear to linear-lanceolate, entire to denticulate, silky-tomentose above and below, sometimes becoming glabrate in age; staminate catkins on leafy peduncles, 2–5 cm. long, the bracts green-tipped, the stamens 2, the filaments free; pistillate catkins on leafy peduncles, 2–5 cm. long, the ovary sessile, silky-pubescent, the stigma deeply cleft but style evident.

Common in the summer-warm areas along rivers and sloughs to an elevation of 5,000 feet, often at sites of former Indian habitation; southwestern Oregon to Baja California, Mexico.

This was one of the common basket willows of the Indians.

6. Salix exigua Nutt., Sylva 1:75. 1843.

Salix argophylla of Jepson, Man. Fl. Pl. Calif., in part. Not Nutt.
Salix exigua var. *Parishiana* Jepson, Man. Fl. Pl. Calif. 264. 1923.

Shrub 2–4 m. tall; twigs pruinose to silky-tomentose; leaves without stipules, the blades tapering to a short petiole, 5–12 cm. long, linear to linear-lanceolate, remotely denticulate, canescent to silky-pubescent on both surfaces, becoming glabrate in age; staminate catkins on long peduncles, 2–4 cm. long, the stamens 2, the filaments free, pubescent; pistillate catkins 3–6 cm. long, on leafy peduncles, the scales lanceolate, acute, white-pilose, the stigma sessile; capsules subsessile to short-pediceled, ovoid-attenuate, glabrous to sericeous and becoming glabrous.

Stream banks and swamps: Great Basin slopes, San Joaquin Valley to southern California; western North America.

7. Salix Parksiana Ball, Univ. Calif. Publ. Bot. 17:400. 1934.

Shrub to 3 m. tall; twigs brown, densely pilose-tomentose, becoming glabrate; stipules, when present, lanceolate-auriculate, the petiole short or wanting, the blades narrowly to broadly elliptic to oblanceolate or oblong-elliptic, acute at both ends, 4–12 cm. long, irregularly glandular-denticulate, densely to thinly pilose-tomentose below; catkins appearing with the leaves on peduncles 2–5 cm. (sometimes –10 cm.) long, bearing 3–5 leaves; staminate catkins 4–8 cm. long, lax, the scales oblong to obovate, pale, pilose on lower half, the stamens 2, the filaments free, hairy below; pistillate catkins 4–8 cm. long, the scales oblong to ovate, 3-veined, pilose-tomentose, becoming glabrate, the stigma lobes short; capsule narrowly ovoid, sessile or subsessile, glabrous to thinly pilose.

Gravel bars and river bars: Humboldt and Del Norte counties; north to Oregon.

8. Salix melanopsis Nutt., Sylva 1:78. 1843. Dusky willow.

Salix melanopsis var. *Bolanderiana* Schn., Bot. Gaz. 67:338. 1919.

Shrub 3–5 m. tall, the twigs dark, lustrous brown; stipules lanceolate to semicordate, dentate, the leaf blades oblanceolate to elliptic, acute, 4–8 cm. long, closely denticulate, dark green and glabrous above, glaucescent below; catkins 3–4 cm. long; staminate catkins on long peduncles, the stamens 2, the filaments free, pubescent; pistillate catkins with erose scales, glabrous to thinly pilose, often striate with 3–5 nerves; capsule ovoid, glabrous, sessile, 4–5 mm. long.

Floodplains and stream courses almost throughout California; east and north to Rocky Mountains, southern Canada.

9. Salix Geyeriana Anderss., Oefv. K. Vet.-Akad. Förhandl. 15:122. 1858.

Shrub to 1 m. tall, the twigs black and pruinose, glabrous to pubescent; leaves exstipulate, linear-oblanceolate to elliptic, acute at each end, 2–6 cm. long, dark green above, glaucous below, thinly to densely pubescent on both sides, the margins entire to revolute; staminate catkins oblong, 1 cm. long, leafy-peduncled, the stamens 2, the filaments united at base; pistillate catkins subglobose to oblong, 1–2 cm. long, the scales with red tips, the style short or none; capsule 5–7 mm. long, pubescent.

In wet meadows and along stream borders, middle altitudes and above: Sierra Nevada, Siskiyou Mountains; east to Rocky Mountains, north to Oregon.

Salix Geyeriana var. *argentea* Schn. of the central Sierra Nevada differs in its silvery, pilose leaves.

10. Salix Piperi Bebb, Gard. & For. 8:482. 1895.

Shrub 5–6 m. tall; twigs glabrous; leaves short-pediceled, broadly elliptic, oblanceolate or obovate, acute or acuminate, serrulate, glabrous, dark green above, glaucous below; catkins appearing before or with the leaves; staminate catkins 3–5 cm. long on short peduncles, the scales long-pilose, the stamens 2, the filaments united at base, glabrous; pistillate catkins 4–10 cm. long, on short, leafy peduncles, the style conspicuous; capsule glabrous or thinly pubescent, 6–7 mm. long, pedicellate.

In swamps and along streams: Humboldt and Del Norte counties; north to British Columbia.

11. Salix Tracyi Ball, Univ. Calif. Publ. Bot. 17:403. 1934.

Shrub to 6 m. tall; twigs yellow to brown, puberulent to glabrate; leaves with auriculate, subdentate to entire stipules, the blades oblanceolate to obovate, rounded and often apiculate at tip, entire to crenate-serrulate, at length glabrate, green above and glaucous below; some catkins appearing before leaves, the peduncles short, leafy; staminate catkins 2–4 cm. long, the stamens 2, the filaments united at base, glabrous; pistillate catkins 2–6 cm. long, the ovary glabrous, the stigma equal to the style in length; capsule pediceled.

Gravel bars and riverbanks: Eel River in Humboldt County and north to southern Oregon.

Closely related to *Salix lasiolepis,* but differing in its broadly obovate leaves, which are glabrous and glaucous below.

12. Salix lasiolepis Benth., Pl. Hartw. 335. 1857. Arroyo willow.

Shrub, often forming thickets, rarely treelike, to 12 m. tall; twigs pubescent; leaves oblanceolate to oblong-acute, green and glabrous above, pubescent or glaucescent beneath, obscurely serrulate, the petioles 2–15 mm. long; catkins appearing before leaves, the scales dark, silky-tomentose; staminate catkins 2–4 cm. long, the stamens 2, the filaments united at base, glabrous; pistillate catkins 2–3 cm. long, the style evident; capsule glabrous or pubescent.

The most common thicket willow of low and middle altitudes along both active and vernal streams, in swamps, and about springs; western United States, Mexico.

13. Salix lutea Nutt., Sylva 1:63. 1843. Yellow willow.

Shrub 2–5 m. tall; twigs puberulent, becoming glabrous, yellow to reddish brown; stipules ovate to lunate, serrulate to entire, the petioles short, lanceolate or oblanceolate to elliptic, acute, 4–10 cm. long, serrulate, glaucous beneath; catkins subsessile, leafy-bracted; staminate catkins 2–3 cm. long, the stamens 2, the filaments free; pistillate catkins 2–4 cm. long, the scales pilose, crisp, the styles evident; capsule ovate-attenuate, glabrous.

Wet meadows and along streams at elevations of 4,000–7,000 feet: Sierra Nevada, San Bernardino Mountains; east to Wyoming, north to British Columbia.

14. Salix Mackenziana (Hook.) Barratt in Anderss., Sv. Vet.-Akad. Handl. IV. 6 (Monogr. Salix):160. 1867.

Salix cordata var. *Mackenziana* Hook., Fl. Bor. Am. 2:149. 1839.
Salix cordata of Jepson, Man. Fl. Pl. Calif. 266. 1923.

Shrub, sometimes treelike, 2–6 m. tall; branchlets dark brown to yellow; leaves stipulate, petioled, the blades lanceolate to lanceolate-acuminate, rounded to subcordate at base, glandular-serrulate, glabrous, glaucous beneath; catkins appearing at same time as leaves, on leafy-bracted peduncles; staminate catkins 2–3 cm. long, dense, the stamens 2, the filaments free, glabrous; pistillate catkins 2–6 cm. long, lax, the style evident; capsule ovoid, glabrous.

Wet meadows and stream courses at low altitudes: northern California; to northern Canada.

15. Salix pseudocordata Anderss., Sv. Vet.-Akad. Handl. IV. 6 (Monogr. Salix):161. 1867.

Shrub to 3 m. tall; twigs lustrous brown; petioles short, the blades oblanceolate to oblong, 3–6 cm. long, rounded to subcordate at base, glandular-serrate, dark green above, green beneath; staminate catkins 2–3 cm. long, the stamens 2, the filaments free, glabrous; pistillate catkins leafy-bracted, the scales thinly white-pilose, the style evident; capsule green, ovoid-attenuate.

Forming thickets in marshy ground at high altitudes: Sierra Nevada, southern California.

Differs from *Salix lutea* in that its leaves are not glaucous beneath.

16. Salix Hookeriana Barratt in Hook., Fl. Bor. Am. 2:145. 1839.

Shrub, sometimes treelike, 4–10 m. tall; twigs densely pubescent; leaves short-petioled, oblong-lanceolate to elliptic-oval, crenate-serrate, dark green and glabrous above, densely tomentose below; catkins appearing before the leaves; staminate catkins stout, 2–5 cm. long, the scales black, with long white hairs, the stamens 2, the filaments free, glabrous; pistillate catkins 4–7 cm. long, the style longer than stigma; capsule ovoid-attenuate, glabrous.

Rare: coastal area of Humboldt and Del Norte counties; north to Washington.

17. Salix commutata Bebb, Bot. Gaz. 13:110. 1888.

Shrub to 3 m. tall; twigs dark, tomentose to glabrate; leaves broadly elliptic to obovate, often rounded at apex, sometimes obtuse, entire, densely silky-pubescent on both sides or becoming glabrate in age; catkins appearing with the leaves, 2–5 cm. long, the peduncles leafy, the scales oblanceolate, obtuse, long-hairy, the stamens 2, the filaments free, glabrous; capsule glabrous or somewhat pubescent, 5–7 mm. long.

Swamps or bogs and wet stream banks: higher mountains of Siskiyou and Modoc counties; north to Alaska.

18. Salix Lemmonii Bebb in Wats., Bot. Calif. 2:88. 1880.

Shrub to 5 m. tall; twigs glabrous to silky or glaucous; leaves stipulate, lanceolate to elliptic-lanceolate, green and glabrous above, glaucous beneath, the petioles 2–5 mm. long, the stipules caducous; catkins appearing with leaves, on short, leafy peduncles, scales dark, silky-pubescent; staminate catkins 1–3 cm. long, the stamens 2, the filaments free, pilose only at base; pistillate catkins 2–4 cm. long, the style short, the stigma trifid; capsule 6–8 mm. long, silky-tomentose.

Forming thickets in swamps and bogs or along streams: in the mountains at elevations of 6,000–10,000 feet, a common middle-altitude species; Nevada, Oregon.

19. Salix monica Bebb in Wats., Bot. Calif. 2:90. 1880.

Salix phycifolia var. *monica* Jepson, Man. Fl. Pl. Calif. 265. 1923.

Shrub to 1 m. tall; twigs brown, glabrous; leaves short-petioled to subsessile, 2–4 cm. long, elliptic, ovate to obovate, entire, glabrous or nearly so; catkins appearing with leaves, sessile; staminate catkins 1–2 cm. long, the stamens 2, the filaments free, glabrous; pistillate catkins 1–3 cm. long, the scales clothed with long shaggy hairs; capsule ovoid-attenuate, pubescent.

Stream margins and bogs: high Sierra Nevada; Rocky Mountains.

20. Salix subcoerulea Piper, Bull. Torrey Club 27:400. 1900.

Shrub to 3 m. tall; twigs dark brown, glabrous, conspicuously pruinose; leaves without stipules, short-petioled, the blades oblong-lanceolate to oblanceolate, 3–7 cm. long, entire or crenulate, green and puberulent above, densely silky or felty-sericeous-pubescent beneath; staminate catkins leafy-bracted, subsessile, 1–4 cm. long, the stamens 2, the filaments free, glabrous; pistillate catkins short-peduncled, 1–4 cm. long, the style very conspicuous; capsule sericeous, ovoid-attenuate.

Forming thickets in bogs and streams: local in high southern Sierra Nevada (Onion Valley); eastern Oregon and Washington, east to Utah and New Mexico.

21. Salix Jepsonii Schn., Jour. Arnold Arb. 1:89. 1919.

Shrub to 3 m. tall; twigs brown, pubescent but not pruinose; otherwise very similar to *Salix subcoerulea* and possibly only subspecifically distinct from it.

Forming thickets in bogs and streams: central Sierra Nevada at elevations of 7,000–10,000 feet.

22. Salix Breweri Bebb in Wats., Bot. Calif. 2:89. 1880.

Shrub to 1 m. tall; branchlets tan to light brown, pubescent to glabrate; stipules inconspicuous; petioles short; leaf blades oblong to linear-lanceolate, acute, 3–6 cm. long, entire, somewhat tomentose on midrib above, gray-tomentose below; staminate catkins 1–2 cm. long, the stamens 2, the filaments free, glabrous; pistillate catkins 2–4 cm. long, the style longer than stigma; capsule subsessile, ovoid-attenuate, tomentose.

Along vernal watercourses in serpentine outcrops: Inner Coast Ranges.

23. Salix delnortensis Schn., Jour. Arnold Arb. 1:96. 1919.

Salix Breweri var. *delnortensis* Jepson, Man. Fl. Pl. Calif. 267. 1923.

Low shrub to 1 m. tall; twigs gray-tomentose; leaves obovate-oblong to elliptic, 2–3 cm. long, entire, tomentose to glabrate above, densely tomentose below; staminate catkins 1–3 cm. long, the stamens 2, the filaments free, glabrous; pistillate catkins 3–5 cm. long, slender, the scales silky-tomentose; capsule ovoid, silky-pilose.

Bogs and stream banks: local in Del Norte County.

24. Salix Eastwoodiae Cockerell in Heller, Cat. N. Am. Pl., 2d ed., 89. 1910.

Salix californica Bebb in Wats., Bot. Calif. 2:89. 1880. Not Lesquereaux.
Salix commutata of Jepson, Man. Fl. Pl. Calif. 267. 1923. In part.

Shrub to 2 m. tall, often forming thickets; branchlets brown; leaves elliptic-lanceolate to elliptic-oblong or obovate when young, 4–6 cm. long, acute, rounded, truncate or obtuse at base, sparsely tomentose on both sides, the margins usually callous-toothed or glandular-denticulate, these sometimes obsolete; catkins appearing with the leaves, the scales broadly oblanceolate, tomentose, the peduncles leafy; staminate catkins 1–3 cm. long, the stamens 2, the filaments free at base, pilose below; pistillate catkins 2–5 cm. long; capsule tomentose, 5–7 mm. long.

Marshy thickets, bogs and stream banks: at higher elevations of the Sierra Nevada; east to Nevada, north to Washington.

25. Salix orestera Schn., Jour. Arnold Arb. 1:164. 1920.

Salix glauca var. *villosa* of California authors, not Bebb.
Salix glauca var. *orestera* Jepson, Man. Fl. Pl. Calif. 267. 1923.

Thicket-forming shrub 1–3 m. tall; twigs yellow-brown, tomentose to glabrate; leaves oblanceolate to narrowly elliptic, acute, entire to glandular-serrate, silky-villous on both sides, becoming glabrescent and glaucescent beneath; staminate catkins 2–5 cm. long, the scales silky-pilose, the stamens 2, free or partly united, sometimes pilose at base; style longer than stigma; capsule silky-villous, pedicellate.

Common, high montane, stream-bank and bog species: Sierra Nevada, San Bernardino Mountains; northeastern Nevada.

26. Salix Scouleriana Barratt in Hook., Fl. Bor. Am. 2:145. 1839.

Shrub to 5 m. tall, occasionally treelike to 8 m.; leaves obovate or oblanceolate, rounded or obtuse at apex, thick, lustrous green above, silvery to glaucous and glabrate or short-hairy below; catkins sessile, appearing before leaves, the scales black at tip, long-hairy; staminate catkins short, the stamens 2, long-exserted, the filaments free, glabrous; pistillate catkins sessile or very short-peduncled, the ovary white-silky, the style obsolete, the stigma sometimes notched; capsule tomentose.

Swamps and bogs: almost throughout California; north to Alaska, east to New Mexico.

27. Salix Coulteri Anderss., Oefv. K. Vet.-Akad. Förhandl. 15:119. 1858.

Salix sitchensis var. *Coulteri* Jepson, Man. Fl. Pl. Calif. 265. 1923.

Shrub, or sometimes treelike, to 8 m. tall; leaves thick, obovate to oblanceolate, entire or serrulate, dark green and glabrous or glabrate above, variously pubescent below, from densely white-tomentose to silky, the stipules reniform, the petioles 4–8 mm. long; catkins appearing with leaves, sessile or on short peduncles; staminate catkins 3–7 cm. long, the scales long-hairy, the stamen 1; pistillate catkins 5–10 cm. long, the scales dark at tip, densely long-hairy, the style evident; capsule ovoid.

Swamps, bogs, and streams: near the coast from Santa Barbara County north; north to Washington.

Differs from the typical form of *Salix sitchensis* Sanson in leaf shape and pubescence.

MYRICACEAE. BAYBERRY FAMILY

Trees or shrubs. Leaves simple, alternate, deciduous or evergreen, entire, serrate, or crenate-dentate, lanceolate to oblanceolate or spatulate; stipules wanting. Flowers monoecious or dioecious, in short spikes or heads borne in the axils of the leaves of the same year. Perianth none. Staminate flowers in cylindrical spikes with usually 4–16 stamens borne in the axil of a deltoid bract; filaments distinct or united; anthers ovate, 2-celled, dehiscent by a longitudinal slit. Pistillate flowers in ovoid to subglobose spikes, each with a solitary, 1-celled ovary subtended by 2 bractlets; ovule 1, orthotropous; styles 2, linear. Fruit in ours a small, oblong, 1-celled, 1-seeded drupe or nut, the flesh often waxy. Seed erect; endosperm none.

MYRICA. WAX MYRTLE

Leaves evergreen, flowers monoecious; plants of coastal areas . . . . 1. *M. californica*
Leaves deciduous, flowers dioecious; plants local in the central Sierra Nevada
2. *M. Hartwegii*

1. Myrica californica Cham. & Schl., Linnaea 6:535. 1831.

Small tree or shrub 2–10 m. tall; bark thin, dark gray, usually smooth; leaves alternate, dark green on both under and upper surfaces, lanceolate to oblong-lanceolate, acute, narrowed to a short petiole, the margins remotely serrate to subentire, glabrous; spikes monoecious, the lower flowers staminate, the upper ones pistillate; fruit covered with a waxy, dark blue coat.

Bogs, swamps, and stream courses: coastal California; north to Washington.

2. Myrica Hartwegii Wats., Proc. Am. Acad. Arts & Sci. 10:350. 1875.

Shrub 1–2 m. tall; twigs pubescent, at length glabrate; leaves deciduous, oblanceolate to oblong-spatulate, 4–8 cm. long, the margins serrate to crenate-serrate, narrowed to a short petiole, light green, pubescent on margins and veins; flowers appearing before leaves, dioecious; staminate catkins 1–2 cm. long, the stamens 2–4, shorter than subtending scale; pistillate spike subglobose; nutlets smooth and appearing as though winged by the adnate fleshy scales.

Margins of streams at middle altitudes: Sierra Nevada from Yuba County to Fresno County.

BETULACEAE. BIRCH FAMILY

Trees and shrubs with leaves alternate, deciduous, pinnately veined and variously serrate-margined. Flowers monoecious, appearing before the leaves, the staminate flowers in elongate lax catkins, the pistillate flowers solitary, or in short, globose or elongate clusters. Staminate flowers with a 2- to 4-membered perianth; stamens 1–10. Pistillate flowers with inferior ovary; stigma 2-lobed. Fruit winged.

Pistillate catkins in racemes, subglobose to somewhat elongate; the scales thick, woody, persistent .... *Alnus*
Pistillate catkins solitary; the scales thin, 3-lobed, deciduous .... *Betula*

ALNUS. ALDER

Trees or thicket-forming shrubs. Leaves alternate, simple, petioled, pinnately veined, doubly serrate. Flowers monoecious, the staminate in lax catkins, the pistillate in short, oblong, woody spikes. Staminate flowers 3–6 per scale; perianth 4-lobed; 1–4 stamens. Pistillate flowers 2 in each axil, subtended by 2–4 bracteoles. Nutlet flat, winged.

Catkins appearing on twigs of previous season, blooming before the leaves appear.
    Leaf margins narrowly revolute; coastal species .... 1. *A. oregona*
    Leaf margins plane, not revolute; interior and montane species.
        Leaves cuneate at base; stamens 2 or 3; mostly riparian trees ..... 2. *A. rhombifolia*

Leaves truncate to cordate at base; stamens 4; large thicket-forming shrubs
3. *A. tenuifolia*

Catkins appearing on twigs of current season with the leaves; stamens 5 or 6
4. *A. sinuata*

1. Alnus oregona Nutt., Sylva 1:28. 1842. Red alder.

Alnus rubra Bong., Mém. Acad. St. Pétersb. VI. Math Phys. Nat. 2:162. 1832. Not Marsh.

Tree to 25 or 30 m. tall; leaves ovate, 7–15 cm. long, doubly serrate, dark green and glabrate above, often rusty-pubescent below, the margin narrowly revolute; staminate catkins 10–15 cm. long, the scales glabrous, the stamens 4; pistillate cones ovoid, 20–25 mm. long; nutlets narrowly membrane-margined.

Swamps and stream margins in the coastal strip: central California north to Alaska.

2. Alnus rhombifolia Nutt., Sylva 1:33. 1842. White alder.

Riparian tree 15–25 m. tall; leaves ovate, 5–8 cm. long, variously doubly serrate to nearly entire, dark green above, lighter and puberulent below, the margins plane; petioles flattened, pubescent; staminate catkins pendulous, 3–5 at the ends of the fertile panicle; pistillate cones 4–10 on the lower branches of the fertile panicle; cones 8–12 mm. long; nutlets narrow-margined.

Common along live streams at low and middle altitudes: west side of the Sierra Nevada to the coast; north to British Columbia, south to Baja California in Mexico.

3. Alnus tenuifolia Nutt., Sylva 1:32. 1842. Mountain alder.

Shrub, forming thickets, 1–3 m. tall; leaves 5–10 cm. long, ovate to oblong, doubly serrate, truncate or cordate at base, dark green above, lighter below, glabrate; cones 8–12 mm. long, the scales 3-lobed at summit; nutlets scarcely winged.

Bogs, stream banks, and lake borders: Sierra Nevada, upper altitudes of the northern Trinity and Siskiyou mountains; north to Alaska.

4. Alnus sinuata (Regel) Rydb., Bull. Torrey Club 24:190. 1897. Sitka alder.

Alnus viridis var. *sinuata* Regel in A. DC., Prodr. 16²:183. 1868.
Alnus sitchensis Sarg., Silva 14:61. 1902.

Thicket-forming shrub, ours rarely more than 3–6 m. tall; leaves ovate, unequally doubly glandular-serrate, glabrous or hairy on the veins; catkins appearing with the leaves on current-season twigs; stamens long-exserted, the scales ovate-apiculate; pistillate cones 12–15 mm. long, on leafy-bracted peduncles; nutlets broadly winged.

Bogs and stream banks, high montane: Trinity and Siskiyou counties; north to Alaska.

BETULA. BIRCH

Ours shrubs, rarely treelike with thin bark. Leaves short-petioled, the blades orbicular, toothed. Catkins solitary, the staminate ones elongate. Flowers 3 to each bract, the lateral pair subtended by a bracteole adnate to bract. Stamens 2; filament branched, each branch with a pollen sac. Pistillate spikes shorter than staminate ones. Bracts 3-lobed, deciduous. Nutlet flattened, winged.

Shrubs usually less than 1 m. tall; leaves about 2 cm. long, lobes of bracts parallel
1. *B. glandulosa*
Shrubs usually more than 1 m. tall; leaves 3–6 cm. long, lobes of bracts divergent
2. *B. fontinalis*

1. Betula glandulosa Michx., Fl. Bor. Am. 2:180. 1803.

Low spreading shrub; twigs glandular, warty; leaves obovate to orbicular, rounded at apex, cuneate at base, serrate above base, 2–3 cm. long; staminate catkins 2–3 cm. long, lax; pistillate catkins 1–2 cm. long, erect; nutlets scarcely winged.

Bogs: high mountains of Lassen and Modoc counties; circumpolar.

2. Betula fontinalis Sarg., Bot. Gaz. 31:239. 1901. Water birch.

Betula occidentalis f. *inopina* Jepson, Fl. Calif. 1:349. 1909.

Shrub 1–10 m. tall; twigs glandular-dotted; leaves ovate to oblong-ovate, cuneate, truncate, or cordate at base, 3–5 cm. long, sharply singly or doubly serrate, resinous-glandular above when young, glandular-dotted below; catkins spreading or pendulous; bracts ciliate, the lateral lobes shorter than the middle ones; seeds winged.

Mountain streams: chiefly east of the Sierra Nevada crest, North Coast Ranges in Humboldt and Siskiyou counties; north to Alaska.

URTICACEAE

Rhizomatous or stoloniferous perennial herbs, less commonly simple annuals; herbage with stinging hairs. Leaves simple, opposite in ours, petioled, 3- to 7-nerved, stipulate, coarsely simple-serrate, dentate, or more commonly double-serrate. Flowers small, greenish, monoecious or occasionally dioecious, in axillary clusters or short, simple or branching spikes. Staminate flower with 4 sepals, 4 opposite stamens, and a rudimentary pistil. Pistillate flower in ours with unequal sepals, the outer sepals short and narrow, the inner ones long and broad, at length investing the flattened achene; petals none; ovary superior, 1-celled; style 1, the stigma sessile; ovule orthotropous with a straight embryo, the endosperm oily. Fruit an achene enclosed by calyx. Species exceedingly variable.

URTICA. NETTLE

Stems gray, densely pubescent, often velvety or sometimes strigose.
   Flower clusters subequal in length to the leaves; stem and leaves coarsely velvety-
      pubescent; seeds smooth .... 1. *U. holosericea*
   Flower clusters much shorter than leaves; stems densely strigose but becoming glabrate;
      seeds tuberculate .... 2. *U. Breweri*
Stems nearly glabrous; leaves dark green above, gray-pubescent beneath, becoming gla-
   brate in age .... 3. *U. californica*

1. Urtica holosericea Nutt., Jour. Acad. Nat. Sci. Phila. II. 1:183. 1848.

   Urtica gracilis var. *holosericea* Jepson, Fl. Calif. 1:367. 1909.

Stems simple, from horizontal, perennial rhizomes, 1–3 m. tall; herbage bristly and finely pubescent; leaves ovate to lanceolate, 5–10 cm. long, the upper ones much shorter, on short petioles, densely soft velvety-pubescent beneath, cinereous above, coarsely serrate; stipules oblong-lanceolate, 6–10 mm. long; clusters of staminate flowers loose, of nearly the same length as the leaves; clusters of pistillate flowers denser and shorter; inner sepals densely hispid, as long as the ovate achene.

Banks of streams and ditches and in swamps and marshes in the valleys and mountains from sea level to an elevation of about 10,000 feet: throughout California except in the deserts; Washington to Baja California, Mexico.

2. Urtica Breweri Wats., Proc. Am. Acad. Arts & Sci. 10:348. 1875.

Stems simple, stout, from horizontal perennial rhizomes, 1–2 m. tall; herbage glabrous to grayish with short strigose hairs and scattered bristles; leaves 4–10 cm. long, on very bristly petioles, the blade ovate to ovate-lanceolate, rounded or subcordate below the tip, acute, the margin coarsely serrate, glabrate to sparsely strigose above, finely velvety-pubescent beneath; flower clusters much shorter than leaves; achenes shorter than inner sepals.

Streams and marshes, ditch banks: southern California; east to Texas, Mexico.

3. Urtica californica Greene, Pittonia 1:281. 1889.

Stems often branched from the base, erect, from shallow, stolon-like perennial rhizomes; herbage hispid and finely strigose-pubescent; leaves broadly ovate, cordate, coarsely serrate, 7–15 cm. long, short-pubescent beneath and often gray, dark green and glabrate above; stipules oblong, 8–12 mm. long; flower clusters paniculately branched or sometimes simple, longer than petioles; achenes puncticulate, as long as inner sepals, ovate.

Low moist places near the coast of central California.

Urtica californica tends to resemble the northern *U. Lyallii* Wats., which as yet has not been reported in California. The pistillate flower clusters of *U. californica* are paniculate, and the leaves are not prolonged into an entire tip, whereas in *U. Lyallii* the clusters are unbranched and more compact and the leaf tip is prolonged into an entire segment.

POLYGONACEAE. BUCKWHEAT FAMILY

Annual or perennial herbs with simple leaves. Flowers small, regular, perfect or monoecious, often mixed in same inflorescence, occasionally dioecious. Perianth of 1 whorl, 4- to 6-cleft. Stamens 3–9, inserted on perianth. Ovary superior; style 1, entire or 2- to 4-parted. Fruit a lenticular or trigonous achene.

Perianth segments 5 or 4, equal in length; leaves mostly cauline; stems often bent or swollen at nodes .... *Polygonum*
Perianth segments 6, very unequal in size, the inner ones much enlarged; leaves mostly basal, sometimes some cauline and reduced; stems usually erect, not bent at nodes
Rumex

POLYGONUM. SMARTWEED

Aquatic, terrestrial, or amphibious herbs. Leaves alternate, entire with scarious, sheathing (stipular sheath), often conspicuously venose stipules. Flowers on jointed pedicels clustered in the axils of leaves or bracts or more often in terminal spikelike racemes which may be solitary and terminal, in pairs, or in groups of 1–9 at the ends of branches, or occasionally reduced in the axils of leaves. Perianth united below, pink, green, or white, the lobes erect in fruit, often closely investing the achene, usually with an evident glandular disc lining the lower part (this frequently not evident in dried material). Stamens 3–9, often unequally inserted, some in the sinuses of the lobes, others below on the tube or occasionally some on the margin of the gland. Ovary superior; style 2- or 3-cleft or -parted; stigmas capitate. Achene trigonous or lenticular, sometimes both kinds on the same plant, light tan through reddish brown to black.

I am especially indebted to Mr. Malcolm Nobs, whose personal interest and very thorough field observations have made possible the present understanding of *Polygonum* in California. The conclusions, however, are my own.

Leaves articulate with the sheath, short-petiolate; stems wiry-striate.
    Branches elongate, 10–70 cm. long; inflorescence not imbricate-leafy; if internodes short, the leaves barely longer than the flowers.
        Flowers borne in axils of ordinary foliage leaves; achenes striate .... 1. *P. aviculare*
        Flowers borne in axils of much reduced, scale-like leaves; achenes smooth
                                                    2. *P. argyrocoleon*
    Branches rarely more than 6–7 cm. long; inflorescence conspicuously imbricate-leafy; leaves twice as long as the flowers or longer .... 3. *P. esotericum*
Leaves not articulate with the sheath; stems not wiry-striate, often succulent and geniculate.
    Stems simple; leaves dimorphic, the basal leaves long-petioled and several, the cauline ones sessile, cordate, few; flowers white to ivory white .... 4. *P. bistortoides*
    Stems branched; leaves of essentially the same shape throughout, all cauline and short-petioled, rarely truncate or cordate below; flowers pink, green, or greenish white.

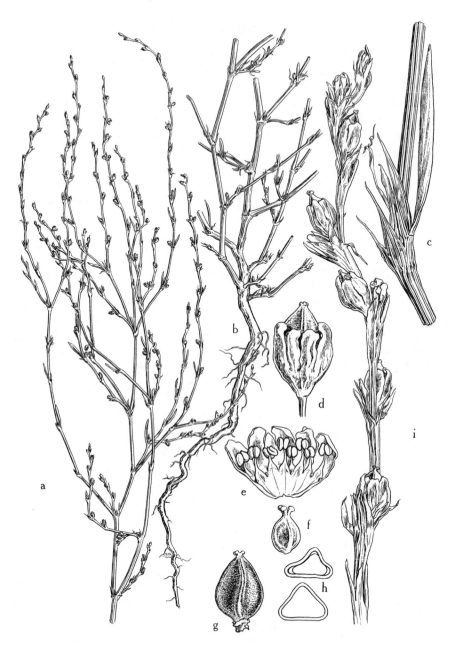

Fig. 196. *Polygonum argyrocoleon: a,* habit, upper part of plant, × ⅖; *b,* habit, lower part of plant, × ⅖; *c,* leaf and sheath, × 3; *d,* mature achene, partially enveloped by perianth, × 8; *e,* perianth, spread open, showing reduced glands between stamens, × 8; *f,* pistil, × 8; *g,* achene, × 6; *h,* achenes (cross section), × 6; *i,* tip of inflorescence, × 4.

Plants dimorphic, floating when aquatic, often coarse, stout and erect when terrestrial; flowers borne in spikelike racemes which are solitary or in pairs, terminal on stout peduncles, thick, densely flowered; perianth bright rose-red.
 Spikes 5–10 cm. long; leaves lanceolate to ovate-lanceolate, attenuate at apex; sheaths in terrestrial plants never herbaceous-margined .... 5. *P. coccineum*
 Spikes 1–4 cm. long; leaves oblong-elliptic, obtuse or acute at apex; sheaths herbaceous, green in terrestrial plants .... 6. *P. natans*
Plants not dimorphic whether aquatic or terrestrial, erect or decumbent; spikelike racemes 2 to many, arranged in a panicle, slender on thin peduncles; perianth pink, green, or white.
 Perianth conspicuously glandular-punctate.
 Achenes shiny black or dark reddish brown, smooth; plants perennial, decumbent, abundantly rooting at nodes; spikes mostly elongate
7. *P. punctatum*
 Achenes dull brown, minutely rugose-striate; plants annual, usually simple and erect at base, occasionally rooting at lower nodes; spikes elongate and some in lower leaf axils short .... 8. *P. Hydropiper*
 Perianth not glandular-punctate.
 Perianth 5-cleft.
 Peduncles not glandular-stipitate; stipular sheaths coarsely ciliate, the cilia much coarser than pubescence of sheath.
 Spikes slender, linear, 3–8 cm. long; flowers pink, white, or greenish; perennials .... 9. *P. hydropiperoides*
 Spikes short, oblong, 1–3 cm. long; flowers at length dull rose to white; annuals .... 10. *P. Persicaria*
 Peduncles glandular-stipitate; stipular sheaths ciliate or glabrous, if ciliate then the cilia not clearly differentiated from hairs on back of sheath.
 Spikes, at least the longest, interrupted, the lowest internodes elongate and bearing fascicles of flowers on the well-separated nodes; bracts of inflorescence ciliate .... 11. *P. mexicanum*
 Spikes compact, not interrupted; bracts of inflorescence glabrous or with few cilia .... 12. *P. pennsylvanicum*
 Perianth 4-cleft; spikes elongate and drooping; veins of perianth segments conspicuously dichotomous and recurved at apex .... 13. *P. lapathifolium*

1. Polygonum aviculare L., Sp. Pl. 362. 1753. Wire grass.

Prostrate or decumbent, diffusely branched annual; stems tough, wiry, the internodes striate; leaves oblong to linear-oblanceolate, 5–35 mm. long, dark bluish green with a short petiole jointed to the sheath; stipular sheaths hyaline, splitting and becoming lacerate in age; flowers 2–5 in fascicles in the leaf axils, the pedicels very short, included in sheath; perianth 5-cleft, 2–3 mm. long, the lobes green, margined with white or pink; stamens 8, the filament bases dilated and glandular; style 3-cleft; achene trigonous, closely enveloped by the perianth, microscopically glandular-roughened, dull dark brown.

A common weed of many habitats, often on the borders of marshes and ponds, recently emerged floodlands, and the borders of tidal salt marshes; from Europe.

Fig. 197. *Polygonum bistortoides: a,* stipular leaf sheath, × 2; *b,* habit, showing rhizome, basal leaves, and stem with cauline leaves and solitary spike, × ⅓; *c* and *d,* mature achenes (cross sections), × 6; *e,* mature achene, × 6; *f,* part of flowering spike, showing the long-pediceled flowers exserted from hyaline, sheathlike bracts, × 4; *g,* rootstock, × ⅖; *h,* pistils, the style 3-cleft, × 4; *i,* perianth, spread open, showing the exserted stamens arising from glandular disc just below sinuses, or from individual glandular lobes, × 4.

2. Polygonum argyrocoleon Steud. ex Kuntze, Linnaea 20:17. 1847. Persian wireweed. Fig. 196.

Glabrous annual usually branched from base, the branches slender, minutely striate, erect or ascending, occasionally decumbent in dry, open situations, dark bluish green; leaves linear-lanceolate, entire, 1–6 cm. long, pale green and soon-deciduous, the petioles very short, jointed to sheath; stipular sheaths delicate, hyaline, silvery above when young, becoming lacerate in age, yellow-brown below; inflorescence spikelike, the flowers 3–10, clustered in the axils of reduced, bractlike leaves, the pedicels exserted from sheaths; perianth lobes 5 or 6, the margins pink or white; stamens 6–8; disc little developed; achene trigonous, smooth, brown, shiny, microscopically punctate, closely enveloped by perianth but exserted when mature.

Widely scattered in marshy ground or on the margins of temporary ponds: interior of southern California in Imperial, Riverside, and San Diego counties, to the San Joaquin and Sacramento valleys; east to Arizona. Native of Persia.

Plants which may prove to be *Polygonum scoparium* Req. ex Loisel of Corsica have been collected on Ryer Island in Suisun Bay, Solano County.

3. Polygonum esotericum Wheeler, Rhodora 40:310. 1938.

Annual, 5–10 cm. tall, branched and spreading above, glabrous; leaves linear, 5–15 mm. long, distinctly jointed to sheath, those of the inflorescence somewhat reduced and imbricate, with a broad, scarious white margin, enfolding the bud, the margin becoming reflexed in anthesis; sheaths deeply lacerate, hyaline; flowers 2–5 in each leaf axil on short pedicels; perianth segments about 2 mm. long, white to pinkish, with a conspicuous green nerve; stamens 5–7, the filaments of the inner 3 conspicuously dilated at base, the others not dilated; style branches nearly sessile; achene obovoid-attenuate, dark brown to almost black, conspicuously longitudinally striate with a fine reticulum, not shiny.

Summer beds of vernal pools: Lassen and Modoc counties; north to Klamath County, Oregon.

An ecologically associated form in northern Modoc County, with smaller, light brown, shiny achenes, needs further study. From what is now known of it, it appears to mature earlier. Wheeler (Rhodora 40:311, 1938) calls attention to stamen differences. Material collected is from the southern limit of the known range of *Polygonum esotericum,* and differs from the typical form in its shorter spike.

4. Polygonum bistortoides Pursh, Fl. Am. Sept. 1:271. 1814. Snakeweed. Fig. 197.

Plant from a horizontal rhizome, perennial; basal leaves on petioles as long as the blades, the blades oblong-lanceolate, acute or obtuse, tapering to the petioles, glabrous and sometimes glaucous; cauline leaves sessile, 3–15 cm.

Fig. 198. *Polygonum coccineum: a,* stem of terrestrial plant, the mature branch glabrous, the young branch puberulent, × ⅖; *b,* part of flowering spike, showing the short-pediceled, fasciculate flowers in axils of hairy, sheathing bracts, × 4; *c,* habit, terrestrial plant, pubescent, × ⅖; *d,* mature achene, × 6; *e,* stipular leaf sheaths, terrestrial plant, × ⅘; *f,* lower part of stem, aquatic form, showing roots at the swollen nodes, × ⅖; *g,* habit, aquatic plant, glabrous, × ⅖; *h,* flower, spread open to show the conspicuous glandular disc and regular stamen insertion, × 4.

long, lanceolate, cordate to subcordate at base; stipular sheaths cylindric, oblique at apex; spike solitary, oblong, densely flowered, 1–6 cm. long; perianth 5-cleft, in ours white; stamens exserted; achene trigonous, brown, smooth, shining.

Bogs and wet meadows: Sierra Nevada, North Coast Ranges, mountains of southern California; Rocky Mountains and north.

5. Polygonum coccineum Muhl. ex Willd., Enum. Hort. Berol. 428. 1809. Fig. 198.

Polygonum Muhlenbergii Wats., Proc. Am. Acad. Arts & Sci. 14:295. 1879.

Dimorphic, amphibious, aquatic, or terrestrial perennial, rooting at nodes; terrestrial plants erect or decumbent; aquatic plants with floating tips and spreading or floating leaves, or at length with erect, branched, aerial stems; stems at length swollen above the nodes, glabrous to puberulent or finely tomentose; leaves lanceolate-attenuate to oblong or ovate, 8–15 cm. long, slightly unequal at the cuneate, truncate, or cordate base, finely silky-pubescent, the margins somewhat undulate; petioles 2–4 cm. long, the lower half decurrent and flanked by a stipular sheath extending as a cylinder around the stem for 10–15 mm. above its junction with the petiole; stipular sheath about 12-nerved and truncate across the top, with a few hairs of unequal lengths along the nerves or scattered between them; inflorescence of 1 or 2 elongated spikes 2–8 cm. long on a stout, red, densely glandular-pubescent peduncle 1–3 cm. long; flowers fascicled in the axils of hairy, stipulate bracts on short, glabrous pedicels, 1 flower of each fascicle blooming at a time, the flowers thus blooming in succession over entire spike; perianth 5-membered, petaloid, bright rose-pink, about 5 mm. long, the lobes free above, united below, ovate; stamens inserted just below the sinuses of the perianth lobes, exserted; anthers versatile; glandular disc attached to the base of the perianth tube, orange-red, 5-lobed, the tips of lobes free; ovary flattened; styles elongate, bifid to middle; stigmas capitate; achene lenticular, beaked with the persistent style base, deep reddish brown, shiny, minutely punctate.

In ponds and swamps of valleys and to middle altitudes in the mountains: southern California; north to British Columbia, east to the Atlantic.

6. Polygonum natans (Michx.) Eat., Man. Bot., 3d ed., 400. 1822. Floating knotweed. Fig. 199.

Polygonum amphibium of Jepson, Man. Fl. Pl. Calif. 288. 1923. Not L.

Dimorphic amphibious perennial, predominantly aquatic, sometimes terrestrial on adjacent banks or after recession of water; stems from creeping rhizomes, submersed or floating on surface, rooting from nodes, or erect and terrestrial; leaves 5–10 cm. long, oblong to elliptic, obtuse or acute at apex, rounded to subcordate at base, entire, glabrous above and below, rarely somewhat scabrous above; stipular sheaths cylindric, membranous, 10–20 mm. high,

Fig. 199. *Polygonum natans: a,* achene (cross section), × 6; *b,* mature achene, × 6; *c,* flower, showing exserted stamens, × 6; *d,* perianth, spread open, showing the prominent glandular disc and the regularly inserted, stout filaments, × 6; *e,* terrestrial plant, showing habit, scabrous pubescence, × ⅖; *f,* stipular sheath of terrestrial plant, scabrous, the broad margin ciliate, × 3; *g,* stipular sheath of aquatic plant, glabrous and membranous, × 2; *h,* habit of aquatic plant, showing the rooting nodes, the glabrous floating leaves, and the short, dense flowering spikes, × ⅖; *i,* lower part of spike, showing the bilobed sheathing bracts and the flowers on short, glabrous pedicels, × 4.

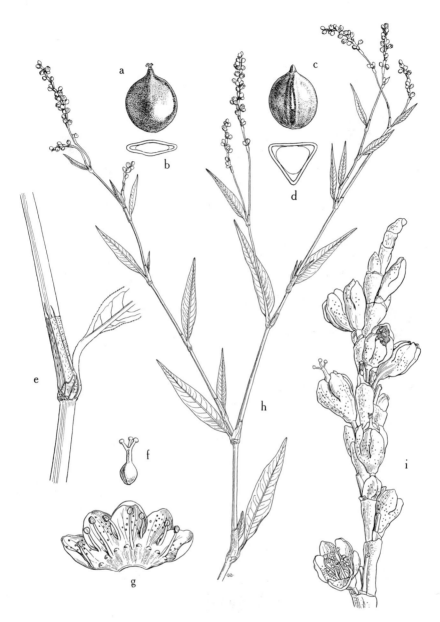

Fig. 200. *Polygonum punctatum: a* and *b,* achene, lenticular type (cross section), × 5; *c* and *d,* achene, trigonous type (cross section), × 5; *e,* stipular leaf sheath, glandular-dotted and bristly-ciliate, × 2; *f,* pistil, × 6; *g,* perianth, spread open, showing the small glands of the disc and stamen insertion on two levels, × 6; *h,* habit, showing the nearly flat, acuminate leaves and the somewhat interrupted spikes, × ⅖; *i,* upper part of spike, showing the gland-dotted perianth and the funnelform sheathing bracts, × 4.

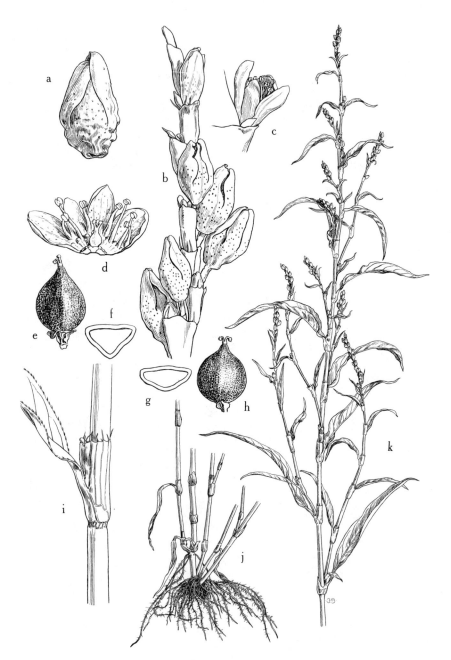

Fig. 201. (For explanation, see facing page.)

those of floating or submersed stems glabrous and entire, those of emersed stems scabrous and ciliate-margined; inflorescence of 1 or 2 densely flowered, terminal spikes 10–35 mm. long; pedicels short, thick, glabrous, included in sheath; sheathing bracts bilobed, tapering to a margin on opposite side of peduncle, glabrous or ciliate-margined; perianth petaloid, bright rose-pink, 4–5 mm. long, 5-cleft to below middle, the base of tube with prominent red-orange, glandular disc; stamens 5, inserted below sinuses in the lobes of the disc; style deeply 2-cleft to below middle, exserted; achene lenticular, minutely roughened, dark reddish brown.

Lakes and ponds, especially in the mountains, but widely scattered in the lowlands of California; cosmopolitan.

7. Polygonum punctatum Ell., Bot. S. C. & Ga. 1:445. 1817. Perennial smart-weed. Fig. 200.

Polygonum acre H.B.K., Nov. Gen. & Sp. 2:179. 1817. Not Lam. 1778.

Aquatic to amphibious perennial, rarely behaving as an annual; stems 3–10 dm. tall, erect or decumbent at base, rooting at nodes, simple or much-branched, green above, reddish below, slightly swollen above nodes, often glandular-punctate; leaves lanceolate to ovate-lanceolate, acuminate, cuneate at base, 5–15 cm. long, glabrous except on veins; petioles short; stipular sheaths cylindric, 9–17 mm. high, membranous, expanding with nodes, at length splitting, scabrous, glandular-dotted, truncate above and bristly-ciliate; flowers in elongate, somewhat interrupted spikes, the pedicels glabrous, exserted; sheathing bract narrowly funnelform, truncate, entire or rarely sparsely ciliate, glandular-dotted; perianth green, 5-parted to below middle, conspicuously glandular-punctate, green to greenish white, jointed to pedicels; stamens 6–8, inserted in sinuses or 1–3 inserted on lower half of tube, the glands green, evident in young fresh material, becoming obsolete in age; style 2- or 3-cleft; stigmas capitate; achene lenticular or trigonous, microscopically roughened, black and shining.

Marshes, ponds, and ditches in shallow water: throughout California at low and middle elevations; southern Canada, south to Argentina.

Polygonum punctatum has been divided by Fassett (Brittonia 6:369–393, 1949) into twelve varieties, several of which range into California, where they intergrade to such an extent that no clear morphological and geographic lines

Fig. 201. *Polygonum Hydropiper: a,* perianth enclosing mature fruit, × 6; *b,* tip of spike, showing the glandular-punctate flowers and the funnelform, ciliate sheathing bracts, × 4; *c,* young flower, × 6; *d,* opened perianth, showing the glands of the disc and the longer filaments inserted between them, × 6; *e,* mature achene, trigonous type, × 6; *f,* mature achene, trigonous type (cross section), × 6; *g,* mature achene, semilenticular type (cross section), × 6; *h,* mature achene, semilenticular type, × 6; *i,* stipular leaf sheath, showing coarsely ciliate apex, × 2½; *j,* habit, basal part of plant, showing the roots and the stems swollen above the nodes, × ⅖; *k,* habit, upper part of plant, showing the successively smaller leaves and the variable spikes, × ⅖.

Fig. 202. *Polygonum hydropiperoides: a,* perianth, spread open, showing irregular stamen insertion, × 6; *b,* pistil, × 6; *c,* mature achene, × 8; *d,* mature achene (cross section), × 8; *e,* upper part of spike, showing the ciliate sheathing bract subtending the flowers, × 4; *f,* stipular leaf sheaths, coarsely strigose, ciliate at apex, × 1⅙; *g,* habit, showing roots at lower nodes and slender, interrupted spikes, × ⅖.

can be drawn separating them. Just what may be the cause of intergradation is not entirely clear. It may result from the fact that, in the history of private game management, seed of many aquatic plants has been purchased from eastern sources and introduced into ponds and marshes at gun clubs. This material, once established, may then have crossed with native stock and with other introduced stock to produce a very confusing taxonomic situation.

8. Polygonum Hydropiper L., Sp. Pl. 361. 1753. Annual smartweed. Fig. 201.

Erect amphibious annual, 2–10 dm. tall, branched from the base or from above base, sometimes rooting at lower nodes; stems glabrous or occasionally sparsely glandular-punctate, slightly swollen above nodes, reddish green to purple; leaves lanceolate to oblong-lanceolate, 2–9 cm. long, becoming reduced upward, acuminate, tapering to a cuneate base, from nearly glabrous to glandular-punctate on both surfaces, sparsely strigose along veins, the margins ciliate to strigose; petioles 1–3 mm. long; stipular sheaths cylindrical, membranous, splitting in age, 6–12 mm. long, and coarsely ciliate, truncate across top, scabrous to nearly glabrous; flowers clustered in most leaf axils, becoming aggregated in spikes at the ends of branches, the spikes thus very variable in size, shape, and position, often appearing as though interrupted, the tips sometimes drooping; sheathing bracts funnelform, 1- to 3-flowered, bearing punctate glands, the margins ciliate; perianth 3-, 4-, or 5-cleft to below the middle, growing with fruit, conspicuously punctate with glandular dots, the lobes green with white or rose margins; stamens 4–7, inserted in sinuses or between lobes of the disc; style 2- or 3-cleft to base; stigmas capitate; achene lenticular or trigonous, microscopically pitted, dark dull brown.

In shallow water or on wet banks: northern California, Coast Ranges from Mendocino County northward, south in the Sierra Nevada to Tuolumne and Inyo counties; north to British Columbia, east to the Atlantic.

The species is outstanding for its wide range of variation in the inflorescences on an individual plant.

9. Polygonum hydropiperoides Michx., Fl. Bor. Am. 1:239. 1803. Water smartweed. Fig. 202.

Low, decumbent aquatic or amphibious perennial, much-branched, rooting at nodes; stem glabrous to sparsely scabrous on upper branches, swollen above the nodes, usually reddish; leaves linear-lanceolate to broadly lanceolate, 5–15 cm. long, tapering to short petioles, strigose on the lower veins and strigose-ciliate on the margins, often resinous-dotted below; stipular sheaths 1–2 cm. long, coarsely strigose, truncate above, the upper margin coarsely ciliate; inflorescence a terminal panicle of spikes, these slender, interrupted, erect or sometimes nodding at tip, the peduncles scabrous to nearly glabrous; sheathing bracts funnelform, extending to a rounded tip, glabrous to strigose, greenish at base to rose-pink at the ciliate margin; pedicels glabrous, exserted; perianth

Fig. 203. *Polygonum Persicaria: a,* mature achene, lenticular, × 6; *b,* mature achene, trigonous type, × 6; *c,* perianth, spread open, showing stamens inserted irregularly, × 8; *d,* tip of spike, showing flowers and ciliate sheathing bracts, × 4; *e,* stipular leaf sheaths, strigose, apices bristly-ciliate, × 1½; *f,* habit, basal part of plant, showing some roots at nodes, × ⅖; *g,* habit, showing the flat, lanceolate leaves and the inflorescence of stout, densely flowered spikes, × ⅖.

petaloid, rose-pink to white or greenish, 2–3 mm. long, 5-cleft to below middle, the lobes campanulately spreading at anthesis; stamens 5–8, inserted in sinuses and between some lobes of the glandular disc; disc 5- to 9-lobed, yellow-green; style 2- or 3-cleft; stigmas capitate; achene trigonous to lenticular, black and shiny to light brown.

Shallow ponds and marshes at lower elevations: along coast and in interior valleys to southern California, as varieties; Oregon.

Polygonum hydropiperoides var. *asperifolium* Stanford differs from the species in being strigose, with coarse, appressed hairs on both upper and lower surfaces of leaves: southern California; Mexico to Oregon.

Polygonum hydropiperoides var. *persicarioides* (H.B.K.) Stanford differs from the species in typically having five stamens and a lenticular, reddish brown achene. These characters are all inconstant; the plants otherwise resemble typical *P. hydropiperoides* so closely as to make identification difficult. It occurs in southern California, especially in the lowlands along the Colorado River; thence south to Mexico and east to Texas and Louisiana. Greene described the Colorado River plants in the vicinity of Needles as *P. fusiforme*.

10. Polygonum Persicaria L., Sp. Pl. 361. 1753. Lady's thumb. Fig. 203.

Erect or decumbent amphibious annual, rooting at nodes, often in large clumps; stems glabrous, 2–9 dm. tall, diffusely branched or occasionally simple and erect, glabrous and at length swollen at nodes; leaves lanceolate to linear-lanceolate, 3–15 cm. long, acuminate, tapering to the short petiole, sparsely strigose to nearly glabrous, sometimes glandular-dotted; stipular sheaths extending 10–20 mm. beyond junction of petiole, strigose, the apex truncate and bristly-ciliate; inflorescence of a few short, stout, densely flowered spikes, these erect, 8–25 mm. long, on tips of terminal and lateral branches; sheathing bracts membranous, ciliate; perianth petaloid, deep dull rose to white, 5-parted to near middle, the base lined with a 5-lobed, yellowish green, glandular disc; stamens 6, some inserted in the sinuses and 1–3 between the glands of the disc; style short, 2- or 3-branched; stigmas capitate; achene lenticular to trigonous, microscopically pitted, black and shining.

Common in marshes and wet fields: throughout California. Naturalized from Europe.

11. Polygonum mexicanum Small, Bull. Torrey Club 19:356. 1892. Giant smartweed. Fig. 204.

Erect, robust annual, simple or little-branched below, usually much-branched above, 1–3 m. tall; stems stout, red, glabrous below, densely clothed with stalked glands above, strongly swollen above the nodes; leaves linear-lanceolate to lanceolate, 15–20 cm. long, tapering to the short petiole, strigose throughout above, or only on the veins, densely glandular-dotted below, the margins strigose-ciliate; stipular sheaths cylindric, membranous, 10–20 mm.

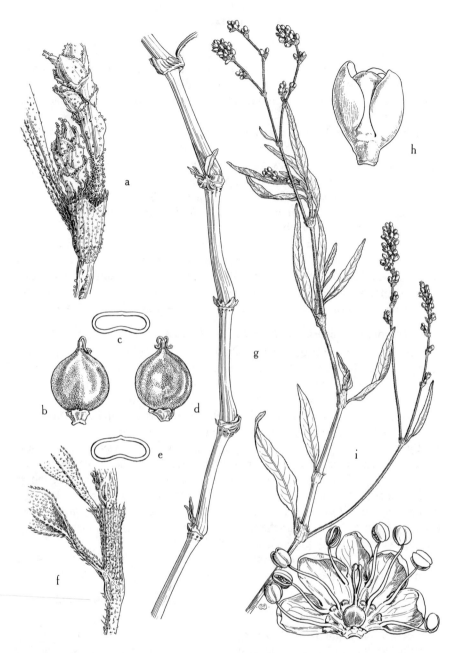

Fig. 204. *Polygonum mexicanum: a,* lower part of young spike, showing the glandular-stipitate internodes and the sheathing bracts, × 4; *b–e,* achenes, concave and convex sides, and cross sections, × 6; *f,* stipular sheath and leaf, both strigose and glandular-dotted, × 2½; *g,* lower part of stem, glabrous and swollen above the nodes, × ⅖; *h,* young flower, × 8; *i,* habit, showing the upper, glandular-stipitate part of stem, × ⅖; *j,* flower, spread open to show glandular disc and irregular stamen insertion, × 6.

long, strigose, truncate with fimbriate-ciliate margin; inflorescence an erect panicle, the branches spikelike with flowers fascicled at the nodes, the internodes glandular-stipitate, the lower internodes elongate, those above short, the spike thus often conspicuously interrupted; sheathing bracts fimbriate-margined, often glandular on back; perianth petaloid, rose-pink, 3–4 mm. long, cleft to below middle, lined with an inconspicuous 5- to 7-lobed glandular disc; stamens 8 or 9, 5 inserted in the sinuses, 3 or 4 irregularly inserted near the glandular disc; style 2-cleft to near base; stigmas capitate; achene lenticular, tapering to a short beak, microscopically pitted, black, shiny.

Marshes and along stream courses, little known in California: 3 miles south of Tupman in Kern County, Bigwater Gun Club in Merced County; Mexico.

12. Polygonum pennsylvanicum L., Sp. Pl. 362. 1753. Pinkweed.

Stout, erect annual, 3–10 dm. tall, branching above; stems mostly glabrous below to stipitate-glandular in the inflorescence, reddish; leaves 4–22 cm. long, lanceolate to broadly lanceolate, acuminate, abruptly tapering to short petioles, sparsely punctate above and below, the veins somewhat strigose, the margins strigose-ciliate; stipular sheaths cylindric-funnelform, glabrous, truncate and without bristles, membranous and soon fracturing, 10–15 mm. long; spikes mostly compact, rarely interrupted, 2–4 cm. long, erect, the peduncles densely stipitate-glandular; sheathing bracts funnelform, the tip acute, glabrous to sparsely glandular, the margins with a few minute cilia toward apex, otherwise entire; pedicels exserted, glabrous; perianth rose to white, 3–4 mm. long, 5-parted to below middle; stamens 8 or less, some in the sinuses of perianth, others from between lobes of the glandular disc; disc prominent, 8-lobed; style 2- or 3-cleft to below middle; stigmas capitate; achene lenticular to trigonous, microscopically roughened, black and shiny.

Low marshy ground: Kenwood in Sonoma County, Lake Hodges in San Diego county; introduced from eastern United States.

Closely related to *Polygonum mexicanum* but differing in being less robust and in having more compact, uninterrupted spikes. A collection from Lake Hodges, San Diego County (*Smith & Nobs 635*), is placed here provisionally. Its very robust character suggests *P. mexicanum,* but its uninterrupted spikes, less coarse, stipitate-glandular peduncles, and less conspicuously ciliate bracts suggest *P. pennsylvanicum.*

13. Polygonum lapathifolium L. Sp. Pl. 360. 1753. Willow smartweed. Fig. 205.

Stout erect annual to 15 dm. tall; stems glabrous, conspicuously swollen above nodes; leaves linear-lanceolate to oblong-lanceolate, acuminate, tapering to short petioles, the upper surface nearly glabrous, with strigose scabrous veins and margins, the undersurface densely glandular-punctate; stipular sheaths membranous, cylindric, 10–20 mm. high, glandular on margins, at

Fig. 205. *Polygonum lapathifolium: a,* habit, showing the stout stem, swollen above the nodes, and the drooping, spikelike racemes, × ⅓; *b,* perianth, spread open, showing glandular disc and stamen insertion, × 8; *c,* perianth enclosing mature achene (note the forked, recurved veins, which have become prominent), × 8; *d,* tip of spike, the sheathing bracts obliquely tapered, their margins glandular, × 6; *e,* mature achene, × 8; *f,* young stipular sheaths, showing glandular or ciliate margins and fractured sheath enclosing the swollen part above the node, × 1⅕.

length splitting and becoming truncately fractured; inflorescence an elongate, drooping, spikelike raceme, densely flowered, the peduncles glabrous to glandular; sheathing bracts funnelform, obliquely tapered, 2 mm. long, glandular on margins; perianth pink to white, 4- or 5-cleft to below middle, the tube lined with a thin, 5-lobed glandular disc, the veins of perianth segments dichotomously forked and recurved at tips, becoming prominent in age; stamens 6, inserted on tube and in sinuses; style 2-parted nearly to base; stigmas capitate; achene lenticular, minutely roughened, dark shining brown.

Marshy areas and shallow water: widely distributed except at high elevations; introduced from Europe.

Polygonum lapathifolium var. *salicifolium* Sibth. (Fl. Oxon. 129, 1794) differs from the species in its leaves, which are white-tomentose beneath. It is widely scattered in northern California: San Francisco, Clear Lake in Lake County, Alturas in Modoc County, Lake Tahoe.

RUMEX

Annual or perennial herbs, glabrous or a few pubescent. Leaves simple, alternate, entire or crisped, the upper ones reduced; stipules united into a scarious cylindrical sheath (ochrea), which is often evanescent. Flowers perfect or dioecious (sometimes polygamous), pediceled, borne in interrupted or dense glomerules (clusters) along the branches of a paniculate inflorescence. Perianth 6-lobed, the 3 outer segments linear, unchanged in fruit, the 3 inner segments (valves or wings) growing with the fruit and often with a callous grain on the abaxial surface. Stamens 6, on short filaments. Styles 3, stigmas tufted. Fruit a 3-sided achene enclosed by the enlarged inner perianth segments.

Rechinger, K. H., Jr., The North American species of *Rumex*. Field Mus. Nat. Hist., Bot. Ser. 17:1–151. 1937.

Flowers dioecious or polygamous; callous grains lacking.
  Leaves with hastate basal lobes; plants with rhizomes . . . . 1. *R. Acetosella*
  Leaves narrowed at base, never with hastate lobes; plants from fusiform roots, occasionally somewhat rhizomatous . . . . 2. *R. paucifolius*
Flowers always (or for the most part) perfect; leaves never hastate at base; callous grains present or absent.
  Stems erect, procumbent, or ascending, branched below inflorescence.
    Mature inner perianth segments more than 20 mm. wide, without callous grains, longer than pedicels . . . . 3. *R. venosus*
    Mature inner perianth segments less than 15 mm. wide, shorter than pedicels.
      Callous grains absent from inner perianth segments.
        Segments shallowly dentate; stems slender, with leafy branches numerous on lower half of plant; leaves linear-lanceolate; panicles large and open
                                                     4. *R. californicus*
        Segments crenulate near base to entire; stems usually stout, leafy branches usually arising only near base of plant; leaves lanceolate; panicles small, compact . . . . 5. *R. utahensis*

Fig. 206. *Rumex paucifolius: a,* habit, staminate plant, × ⅖; *b,* staminate inflorescence, × 5; *c,* pistillate flowers, × 4; *d,* habit, pistillate plant, × ⅖.

Callous grains on 1 or more of the inner perianth segments.
  Margins of mature inner perianth segments narrower than callous grains.
    Segments 4–5 mm. long; leaf blade 2–3 times as long as wide .. 6. *R. crassus*
    Segments 2–4 mm. long; leaf blade 3–6 times as long as wide.
      Callous grains on 1 segment only; achenes shorter than inner perianth segments, 1.8–2 mm. long .... 7. *R. salicifolius*
      Callous grains usually on 3 segments; achenes subequal to the inner perianth segments in length, 2.5 mm. long .... 8. *R. transitorius*
  Margins of mature inner perianth segments broader than callous grains.
    Segments 2–2.5 mm. long; leaves oblong-elliptic, papillose-puberulent (terrestrial form) .... 9. *R. lacustris*
    Segments 3–4 mm. long; leaves linear-lanceolate, glabrous
                                                    10. *R. triangulivalvis*
Stems mostly erect, unbranched below inflorescence.
  Mature inner perianth segments without distinct callous grains.
    Segments about 10 mm. long .... 11. *R. fenestratus*
    Segments 4–5 mm. long .... 12. *R. occidentalis*
  Mature inner perianth segments with 1 or more distinct callous grains.
    Segments essentially entire.
      Lower leaves flattish, truncate, or subcordate; glomerules remote; callous grain subequal to perianth margin in width .... 13. *R. conglomeratus*
      Lower leaves crisped or undulate, gradually narrowed to base, glomerules approximate; callous grain much narrower than perianth margin
                                                    14. *R. crispus*
    Segments dentate.
      Plants perennial; leaf blade up to 2½ times as long as wide.
        Pedicels thick, not longer than mature perianth, articulate at or below middle; leaves relatively small, characteristically fiddle-shaped ... 15. *R. pulcher*
        Pedicels slender, 1–2½ times the length of the mature perianth, articulate near base; leaves relatively large, not fiddle-shaped .. 16. *R. obtusifolius*
      Plants mostly annual; leaves 3–6 times as long as wide, or longer.
        Pedicels short, stout; perianth segments dentate; leaves widest above middle, almost 3 times as long as broad .... 17. *R. violascens*
        Pedicels long, slender; leaves linear-lanceolate, 5–7 or many times as long as broad.
          Mature inner perianth segments triangular; callous grains fusiform, narrow; widespread in diverse habitats .... 18. *R. fueginus*
          Mature inner perianth segments elliptic or ovate; callous grains thick, oblong; local in salt marshes .... 19. *R. persicarioides*

1. Rumex Acetosella L., Sp. Pl. 338. 1753.

Perennial with slender, running rhizomes; stems numerous, slender, wiry, erect or decumbent at base, usually unbranched, 2–4 dm. tall, scabrous; leaves linear or lanceolate, 2.5–8 cm. long, hastate, the basal lobes usually large, the petioles of lower leaves often longer than the blades; panicles many-branched, the glomerules without subtending leaves; flowers dioecious, the pistillate flowers turning red in age; pedicels as long to twice as long as mature perianth, not articulated; mature inner perianth segments entire, not enlarging in fruit, about 1 mm. long, scarcely as long as the achene, lacking callous grains and without

Fig. 207. *Rumex venosus: a,* flowering branch, × ⅖; *b,* flower in fruit, showing the enlarged, membranous inner perianth segments, × 1⅕.

distinct nervation; achene 1+ mm. long, almost as broad, the surface smooth and shiny, mahogany red.

Waste places, often in swampy areas: throughout California; north to British Columbia, east to Atlantic. Native of Europe.

2. Rumex paucifolius Nutt. apud Wats., Jour. Acad. Nat. Sci. Phila. 7:49. 1834. Fig. 206.

Perennial; stems 3–7 dm. tall from a stout, simple or branched, fusiform root, somewhat rhizomatous; leaves mostly basal, linear to oblong-lanceolate, 4–10 cm. long, 1–3 cm. wide, entire, narrowed to long petioles; panicles leafy-bracted only at base, usually dense and contracted; flowers dioecious or polygamous, becoming reddish; pedicels short, 2 mm. long, imperceptibly articulated below the middle; mature inner perianth segments 2.9–3.8 mm. long and as wide, cordate, finely veined, without callous grains; achene smooth, 1–1.8 mm. long, glossy brown, widest below middle.

High mountain meadows and streams: Sierra Nevada; north to British Columbia, east to Rocky Mountains.

3. Rumex venosus Pursh, Fl. Am. Sept. 2:733. 1814. Fig. 207.

Perennial herb from a woody rootstock, glabrous and rather pallid, the stems stout, procumbent or ascending, rarely suberect, 1.5–4.5 dm. tall, simple or usually few-branched from lower leaf axils, somewhat flexuous; leaves ovate to obovate-elliptic, 3–10 cm. long, 4–5 cm. wide, acute or acuminate at apex, cuneate at base, more or less coriaceous; branches of panicle 1 to several, erect, more or less interrupted; pedicels shorter than mature perianth, articulate near middle; flowers perfect, reddish; mature inner perianth segments 14–18 mm. long, 24–30 mm. wide, the base deeply cordate, the apex rounded or shortly acute, margins entire, surface finely reticulate, without callous grains; achene 5–7 mm. long, smooth and shining, the sides concave, the angles margined, brown.

Sandy river valleys: northeastern California; north to eastern Oregon and Washington, northeast to Saskatchewan and Montana.

This species has the largest fruiting perianth segments in the genus.

4. Rumex californicus Rech. f., Rep. Sp. Nov. 40:297. 1936.

Rumex salicifolius of California authors, in part.

Perennial; stems many, finely sulcate-striate, ascending or suberect, 3–6 dm. tall, with many leafy branches arising below the middle of the plant in the axils of the leaves; leaves linear-lanceolate, the lower ones to 10 cm. long and to 0.5 cm. wide, the petiole about as long as the blade is wide; panicle large and open, the simple branches arcuately diverging from the stem (or somtimes appressed), the lower glomerules remote, the upper ones nearly approximate, or all of them approximate, contiguous in fruiting state; flowers perfect; pedicels articulate in

Fig. 208. *Rumex salicifolius: a,* habit, × ⅖; *b,* single whorl of inflorescence, × 3; *c* and *d,* flower in fruit, showing well-developed callous grain on 1 inner perianth segment, × 7.

their lower third or fourth; mature inner perianth segments about 3 mm. long and 2.5 mm. wide, broadly triangular, acute, membranous, dark, irregularly and shallowly denticulate toward base, prominently reticulate-nerved with conspicuous midvein, without callous grains; achene dark brown to black, about 2 mm. long and 1.3 mm. wide.

Mountain meadows and moist ground along creeks: southern California north through the Sierra Nevada to Plumas County, occasional in Coast Ranges (Glenn County, San Benito County).

5. Rumex utahensis Rech. f., Rep. Sp. Nov. 40:298. 1936.

Perennial; stems many, usually stout, thinly striate, glabrous, smooth, erect, rarely subflexuose or arcuate-ascending, 1.5–4 (–6) dm. tall, the leafy branches arising mostly from lower nodes; leaves elongate, scabrous, plane or somewhat undulate, the petioles as long as the blade is broad, or sometimes longer; lower leaves lanceolate, 4–5 times as long as wide, tapering at both ends; branches of panicle simple, erect, small, compact, the lower ones often subtended by a leaf, the glomerules approximate, in fruit contiguous; flowers perfect; pedicels slender, imperceptibly articulate near their base and broadened at tip, shorter than mature perianth; mature inner perianth segments 2.5–3 mm. long, ovate to deltoid, margins crenulate to subentire, face reticulately nerved with midvein prominent, without callous grains; achene about 2 mm. long and 1.3 mm. wide, dark brown to black, acuminate at both ends, slightly more so at apex.

Along rivers and wet margins of lakes: Mono County; adjacent Nevada, north to Alberta and east to Colorado and Utah.

6. Rumex crassus Rech. f., Rep. Sp. Nov. 40:295. 1936.

Rumex salicifolius of western American authors, in part.

Perennial; stems many, procumbent or flexuose-ascending, with many leafy branches from lower nodes, 2–5 dm. tall, strongly striate, glabrous, smooth, the petioles about as long as the blade is broad, or longer; lower leaves ovate-lanceolate or oblong-lanceolate, 6–8 cm. long, 2–2.5 cm. broad, usually 2½–3½ times as long as wide, the base narrow or broadly cuneate, the apex acute; panicle branches short, usually simple, small and compactly formed, only the lower one subtended by a leaf, the glomerules usually approximate and contiguous in fruit; flowers perfect; pedicels articulate on their lower third or fourth, 1–2 times as long as mature perianth; mature inner perianth segments 4–5 mm. long, 3–4 mm. wide, ovate or deltoid, ligulate, coriaceous when dry, brownish purple, margin minutely and irregularly crenulate to denticulate, reticulately veined, the callus of anterior segment ovate, swollen, about 4 mm. long, 2.5 mm. wide, large and prominent, almost covering grain, other 2 segments without callous grain or rarely with minute grain; achene about 2.5 mm. long and 2 mm. wide, broadest just below middle, the apex short-acuminate.

Edge of salt marshes: coastal northern California; north to Oregon.

Fig. 209. *Rumex transitorius: a,* inflorescence, × ⅖; *b,* part of lower stem, showing well-developed leaves, × ⅖; *c,* flower in fruit, showing callous grains on the 3 inner perianth segments and pedicel articulated near its base, × 6½; *d,* part of inflorescence, × 3.

7. Rumex salicifolius Weinm., Flora 4:28. 1821. Willow dock. Fig. 208.

Perennial from a stout taproot; stems 3–7 dm. long, spreading or erect, with leafy branches, striated, glabrous; leaves glabrous, glaucous, bright green, flat, entire, linear-oblong to lanceolate, 6–12 cm. long, 2–2.5 cm. wide, 3–6 times as long as wide, gradually and equally narrowed at base and apex, short-petioled; panicle branches congested above, the lower whorls leafy-bracted; pedicels 1½–2 times as long as the mature perianth, articulate near their base; flowers perfect; mature inner perianth segments deltoid, 2–3 mm. long, 1–1.5 mm. wide, usually only the anterior segment with a callous grain, the callus large, about 2 mm. long and 1 mm. wide, almost covering the segment, very prominent; achene 1.8–2 mm. long, 1–1.3 mm. wide, slightly wider below base, the apex more or less acuminate, dark brown.

Wet places in valleys, foothills, and high mountains: throughout California; north to British Columbia.

This species, as here delimited, is characterized by the comparatively small fruiting perianth segments, with only 1 segment bearing a callous grain, this almost covering the segment.

8. Rumex transitorius Rech. f., Rep. Sp. Nov. 40:296. 1936. Fig. 209.

Rumex salicifolius of western American authors, in part.

Perennial; stems solitary or more often several, branching from lower nodes, arcuate-ascending or suberect, flexuous, strongly striate, glabrous, smooth, 2.5–6 dm. tall; leaves thickish, glabrous, smooth, the lower leaves lanceolate, 6–12 cm. long, 2–2.5 cm. wide, plane, 3½–6 times as long as wide, tapered at both ends but more shortly at the base, the petiole about as long as the blade is broad; inflorescence forming an open panicle, the branches spreading, the glomerules rather densely approximated, only the lower branches subtended by a leaf; pedicels 1–1½ (–2) times as long as the mature perianth, articulate on lower quarter; flowers perfect; mature inner perianth segments 2.5–3 mm. long, 2–2.3 mm. wide, ovate or lanceolate-ovate, rigid-membranous, sordid yellow-brownish to purple, margin entire or subentire, obscurely reticulate-veined, usually each segment with a callous grain, these smooth, globular, large, almost covering the segments; achene blackish brown, about 2.5 mm. long, 1.3 mm. wide, the apex long-acuminate.

Marshes: central California; north to British Columbia.

This species is characterized by the fact that all three fruiting perianth segments bear large callous grains which almost cover the segments.

9. Rumex lacustris Greene, Erythea 3:63. 1895. Fig. 210.

Rumex salicifolius var. *denticulatus* of western American authors, in part.

Perennial; stems 5–9 dm. tall, branching from lower nodes, finely striate, the aquatic stems subglabrous with broad leaves, the terrestrial stems thick at

Fig. 210. *Rumex lacustris: a,* habit, × ⅖; *b* and *c,* flower in fruit, showing the well-developed callous grains on the 3 inner perianth segments, × 8.

the base, papillose, with usually narrow leaves; terrestrial leaves somewhat fleshy, papillose-puberulent, becoming glabrescent, oblong-elliptic, 5–7 cm. long, 1.5–2 cm. wide, the apex round-acuminate, the margin minutely crenulate-crisped, petioles often much longer than the larger blades are wide; panicle branches short to elongate, simple, divergent, at least the lowermost leafy-bracted, the glomerules remote to contiguous; pedicels slender, short, at most equaling the mature perianth in length, obscurely articulate below middle; flowers perfect; mature inner perianth segments 2–2.5 mm. long, 1.5 mm. wide, broadly ovate-linguiform, narrow and rounded at base, acute at apex, the margin entire, membranous, pale brown with raised reticulate nerves, all with small callous grains, these slightly rugulose or subsmooth, 1.5–2 mm. long, 0.5–0.6 mm. wide; achene brownish black, 2–2.2 mm. long, 1–1.1 mm. wide, acuminate at both ends, slightly more so at apex.

Margins of lakes: Sierra County to Modoc County; north into southeastern Oregon.

There are two forms of this species, an aquatic type and a terrestrial one. We have encountered only the terrestrial form. The small fruiting perianth segments are very characteristic of this species.

10. Rumex triangulivalvis (Danser) Rech. f., Rep. Sp. Nov. 40:297. 1936.

Rumex salicifolius of many authors, in part.
Rumex salicifolius var. *denticulatus* Torr., Mex. Bound. Surv. Bot. 178. 1859.

Perennial; stems solitary or many from lower nodes, elongate, flexuous, at length decumbent, striate, glabrous, 4–10 dm. tall; lower leaves thin in texture, glabrous, pale green, plane or rarely subundulate, linear-lanceolate, 12–15 cm. long, about 5 times as long as broad, short-acuminate at both ends, slightly more so at apex, the petiole frequently shorter than the blade is wide; inflorescence branched, the glomerules usually approximate, only the lowermost branches subtended by a leaf; pedicels usually shorter than mature perianth, occasionally 1½ times as long, articulate near base; flowers perfect; mature inner perianth segments 3–4 mm. long, 2.5–3 mm. wide, rigid-membranous, pale brown, triangular, subcordate or truncate at base, the apex more or less acute, the margin entire or minutely crenulate toward base, the surface reticulate, nervose, the callous grains usually subequal in size (rarely 1 or 2 segments with calluses diminutive or wanting), prominently fusiform, rounded basally, somewhat acute at both ends, 1.8–2.5 mm. long, 0.6–0.9 mm. wide, usually scrobiculate-rugulose, small in proportion to segment; achene 2 mm. long, about 1.3 mm. wide, very dark brown to black, widest below middle, acuminate at both ends, slightly more so at apex.

Meadows and marsh areas at middle altitudes: Sierra Nevada (Placer County) and Great Basin (Mono County); north to Canada, east to Atlantic.

11. Rumex fenestratus Greene, Pittonia 4:306. 1901.

Rumex occidentalis of American authors, in part.

Perennial; stems unbranched, erect, to 2 m. tall, pale reddish yellow, striate; basal leaves membranous in texture, narrowly oblong-triangular or somewhat wider, deeply cordate, more or less acute, the margins crisp-undulate or almost plane, papillose-scabrous on the nerves, the blade longer than or subequal to petiole, the cauline leaves oblong-triangular, often more than 3 times as long as wide, short-petioled; panicles broad, dense, usually compound, not leafy-bracted, the glomerules many-flowered, contiguous in fruit; pedicels filiform, 1½ times as long as mature perianth, obscurely articulate below middle; flowers perfect; mature inner perianth segments 10 mm. long, 7–9 mm. wide, the base more or less narrowly cordate, the apex bluntly acuminate, the margin subentire or more often erose-denticulate (especially toward base), pale fleshy brown, finely reticulate-nerved, without callous grains; achene black, 3.5–4 mm. long, about 2 mm. wide, almost equally acuminate at both ends.

Coastal marshes: San Francisco north in Coast Ranges, Lassen County; north to Alaska, northeastern Canada.

This differs from *Rumex occidentalis* in that its fruiting perianth segments and its mature achenes are almost twice as large.

12. Rumex occidentalis Wats., Proc. Am. Acad. Arts & Sci. 12:253. 1876. Western dock.

Perennial from a stout taproot; stems usually simple, stout, erect, striated, glabrous, 4–20 dm. tall, reddish or suffused with purple; petioles of lower leaves from ⅓ to nearly as long as the blade is wide, the blades from oblong-triangular to ovate-triangular, 15–40 cm. long, 2–2½ times as long as wide, somewhat crisped on margins, cordate or subcordate at base, obtuse to acute at apex; panicles dense, strict, 3–6 dm. long, leafless or with only a few small leaves below; pedicels 1–2 times as long as mature perianth, obscurely articulate toward base; flowers perfect; mature inner perianth segments round-ovate, 4–5 mm. long, 5–6 mm. wide, prominently reticulate-veined, subcordate, the margins erose or serrulate, without callous grains (or rarely with one); achene brown, smooth, shining, about 2.5–3 mm. long, 1.5 mm. wide, acuminate at both ends, slightly more so at apex.

Bogs and marshes near fresh or brackish water: coastal central California to Modoc and Lassen counties; north to Alaska, east to Texas.

A highly polymorphic species, the vegetative characters showing great variation; under this species Rechinger groups all closely related grainless forms which are characterized by relatively small fruiting perianth segments and small achenes, thus excluding *Rumex fenestratus*.

13. Rumex conglomeratus Murr., Prodr. Fl. Göttingen 52. 1770.

Perennial; stems slender, erect, glabrous, 1–2 m. tall from a cluster of fusiform, woody roots or a stout taproot, lacking leafy branches; lower leaves glabrous, light green, ovate to oblong or oblong-lanceolate, 10–20 cm. long, flat or very slightly undulate, truncate or subcordate below; panicle of many long, loose, slender, erect or divergent branches, the glomerules remote, not crowded, nearly all with a lanceolate, subtending leaf; pedicels as long as mature perianth, articulate near base; mature inner perianth segments oblong, 2.5–3 mm. long, reticulate-veined, entire, each with a callous grain, these large, ovate-oblong, nearly as wide as the perianth segment; achene reddish brown, shining, smooth, 2 mm. long.

Weed of moist valley floodlands to middle altitudes in the mountains, common in Pacific states; naturalized from Europe.

14. Rumex crispus L., Sp. Pl. 335. 1753. Curly-leaved dock. Fig. 211.

Coarse perennial; stems stout, erect, straight, without axillary branches, 5–15 dm. tall, from a deep taproot, glabrous, dark bluish green; lower leaves elliptical to oblong-lanceolate, 10–30 cm. long, prominently undulate and crisped on margin, cuneate at base, long-petioled, the upper smaller leaves all cordate or obtuse at base; panicles usually strict, of wandlike branches with few leaves, the glomerules usually dense and approximate; pedicels 1½–2 times as long as the mature perianth, articulate below middle; flowers perfect; mature inner perianth segments about 5 mm. long and 3–4 mm. wide, round-ovate, subcordate, the margin entire to scarcely erose, all segments with smooth, oblong callous grains much narrower than perianth margin, occasionally 1 or 2 segments naked; achene smooth, glossy, reddish brown, acute, 2 mm. long.

Waste ground and floodlands throughout the valleys and foothills; naturalized from Europe.

15. Rumex pulcher L., Sp. Pl. 336. 1753. Fiddle dock.

Perennial; stems erect, 5–8 dm. tall, rather slender, with rigid, divaricately spreading branches; lower leaves long-petioled, oblong or some of them fiddle-shaped, 3–15 cm. long, obtuse at apex, cordate at base, scabrous beneath, the upper leaves narrowed at both ends; panicles loose, with elongate, intricate, divergent branches, the clusters rather remote, with or without subtending leaves; pedicels thick, about as long as the mature perianth, articulate at or below middle; flowers perfect; mature inner perianth segments ovate or oblong-ovate, 4–6 mm. long, 2.5–4.5 mm. wide, usually with 5–10 awnlike or short teeth on each margin, truncate at base, usually all 3 segments bearing small callous grains, these often of unequal size; achene about 3–4 mm. long, smooth, shining, mahogany red, the apex elongated, the base obtuse.

Widespread in moist, waste places: throughout California; north to Oregon, east to Atlantic and Gulf states. Native of Europe..

Very polymorphic; represented in California by *Rumex pulcher* subsp. *eu-pulcher* Rech. f., which has leaves contracted above the base, and fruiting perianth segments longer than broad, with the teeth relatively long, and by

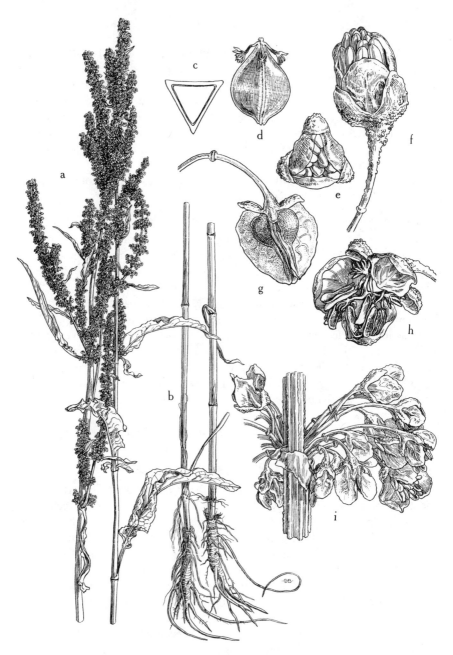

Fig. 211. (For explanation, see facing page.)

R. pulcher subsp. *divaricatus* (L.) Murb., which has leaves usually not contracted above the base, and perianth segments about as long as broad, with the teeth relatively short.

16. Rumex obtusifolius L., Sp. Pl. 335. 1753. Bitter dock.

Perennial; stems from a stout taproot, simple or sparingly branched at base, tall, slender, 6–12 dm. tall; lower leaves ovate-oblong to broadly oblong-lanceolate, 10–35 mm. long, somewhat undulate, margins slightly crisped, usually deeply cordate at base, somewhat papillate on lower surface, glabrous, dark green, on long petioles; panicle strict, leafless or with a few leaves at base, open, the glomerules not crowded, the lower ones remote; pedicels slender, 1–2½ times as long as the mature perianth, articulate near base; flowers perfect; mature inner perianth segments deltoid-ovate, 5–6 mm. long, with 3–5 pronounced, spinose, or subulate teeth on each margin, 1 segment with a small callous grain, sometimes the other 2 with very small grains; achene 2 mm. long, reddish brown, shining.

Wet grounds and stream banks, widely distributed. Naturalized from Europe.

A very polymorphic species. Most of the American plants can be referred to the western European subspecies, *Rumex obtusifolius* subsp. *agrestis* (Fries) Danser.

17. Rumex violascens Rech. f., Rep. Sp. Nov. 39:171. 1936.

Annual or biennial (or occasionally perennial); stems stout, erect, 3–8 dm. tall, from an elongated taproot, dark green, usually tinged with red, usually simple below; lower leaves narrowly oblanceolate to narrowly obovate, 4–10 cm. long, ½–¼ as wide, flat or obscurely crisped, glabrous, cuneate to truncate at base; panicle narrow, leafy-bracted basally, ascending, 5–12 cm. long, the glomerules many, remote, only the apical contiguous in fruit; pedicels articulate below the middle, somewhat swollen below the joint, about as long as mature perianth; flowers perfect; mature inner perianth segments 2.5–3 mm. long, 2–3 mm. wide, narrowly ovate-triangular, the apex acute, the margins irregularly and acutely toothed toward base, each bearing an ovate, prominent callous grain 1.5–2 mm. long, about 0.75 mm. wide, somewhat unequal in size, more or less impressed-punctate and sometimes transversely rugulose above; achene about 1.7 mm. long, about 1.2 mm. wide, brown, smooth, shining.

Fig. 211. *Rumex crispus: a,* habit, upper part of plant, showing the undulate leaves and the wandlike panicle branches, × ⅕; *b,* habit, lower part of stem and the deep taproots, × ⅕; *c,* achene (cross section), × 8; *d,* mature achene, showing the reflexed styles with tufted stigmas, × 8; *e,* young flower, top view, × 8; *f,* young flower, showing inner and outer perianth segments, × 8; *g,* mature fruit, the achene enclosed by inner perianth segments which bear smooth callous grains, × 4; *h,* flower, showing tufted stigmas and anthers after dehiscence, × 8; *i,* whorl of flowers, showing the sheathing stipules (in older plants, only long fibers remain), × 3.

Fig. 212. *Rumex fueginus: a,* habit, × ⅖; *b* and *c,* flowers in fruit, showing the inner perianth segments with slender teeth and long, narrow callous grains, × 8.

Low land and ditch banks: Imperial and San Joaquin valleys; east to Texas and New Mexico, south to Mexico.

In American botanical literature this species has often been confused with *Rumex Berlandieri* Miesn.

18. Rumex fueginus Phil., Anal. Univ. Chile 91:493. 1895. Fig. 212.

Rumex persicarioides of most western American authors, not L.

Annual or occasionally biennial; stems erect or ascending, 1.5–6 dm. tall, usually strict, slender or stout, more or less striate, papillose-scabrous, glabrescent or glabrous, becoming brownish or sometimes purplish; lower leaves membranous or subcoriaceous, the margin more or less undulate-crisped, glabrous and smooth or scabrous-pubescent, linear-lanceolate, the blade 5–7 times as long as wide, more or less cordate or truncate at base and widened above base, the apex acute; petiole shorter than blade; upper leaves progressively smaller, narrower; panicle broad, the glomerules many-flowered, contiguous and compact above, remote below, leafy-bracted; pedicels slender, articulate near base, 1–2 times as long as mature perianth; flowers perfect; mature inner perianth segments triangular, 1.7–2 mm. long, 0.7–0.9 mm. wide exclusive of the teeth, subcoriaceous, the apex ligulate, acute, the margins each with 2 (or 3) divergent, setaceous-subulate teeth, each segment with a callous grain, these fusiform, cellular-punctate, prominent, about 1–1.4 mm. long, 0.5–0.7 mm. wide, the apex obtuse, narrowing into midrib; achene brown, 1.3–1.4 mm. long, 0.5–0.7 mm. wide, the ends usually subequally acuminate.

Lake margins and marshy ground: scattered localities throughout California; north to Canada, throughout United States except southeastern states, South America.

According to Rechinger, the differences in the mature inner perianth segments (see discussion under *Rumex persicarioides*) are adequate to distinguish *Rumex fueginus* from *R. persicarioides* in North America when considered along with distribution. *Rumex fueginus* is the more widespread species in North America, whereas *R. persicarioides* is limited to salt marshes and saline situations and is found mainly in the New England states and southeastern Canada.

19. Rumex persicarioides L., Sp. Pl. 335. 1753. Golden dock.

Annual; stems erect, straight or angularly flexuose, 1–5 dm. long, slender, minutely scabrous-pubescent; leaves oblong or lanceolate, 5–10 cm. long, 5–7 times as long as wide, truncate or subcordate at base, short-petioled, somewhat crisped-undulate, the apex acute; panicles ample, broad, the glomerules loose or crowded in an interrupted spike, each subtended by a leaf; pedicels slenderly filiform, about 1½ times as long as mature perianth, articulate at base; mature inner perianth segments 2–3 mm. long, about 1 mm. wide exclusive of teeth, narrowly lingulate-triangular, attenuate above, with 2 or 3 subulate-setaceous

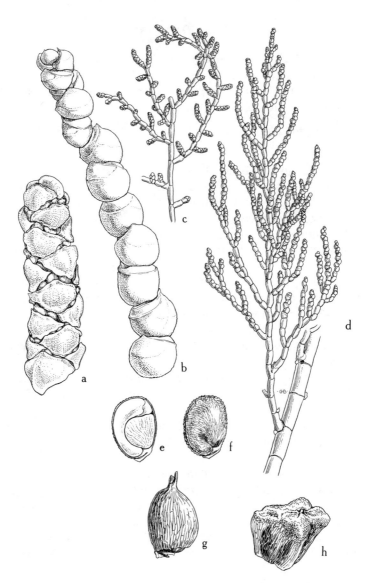

Fig. 213. *Allenrolfea occidentalis: a,* flowering spike, showing stamens protruding between scales, × 8; *b,* jointed stem, showing the alternate, scalelike leaves, × 6; *c,* inflorescence, × ⅘; *d,* vegetative branch, × ⅘; *e,* seed (longitudinal section), showing the marginal, curved embryo, × 20; *f,* seed, × 20; *g,* pericarp enclosing the seed, × 20; *h,* spongy calyx enclosing the fruit, × 16.

teeth on each side, subcoriaceous, each segment with a thick, prominent, cellular-punctate, oblong callous grain about 1 mm. long and almost as wide as segment and rounded at both ends; achene brown, about 1.1 mm. long, 0.5–0.6 mm. wide, acuminate at both ends, slightly more so at apex.

Salt marshes and saline areas: local in San Mateo County; also local in Oregon; New England states and southeastern Canada.

This species is altogether like *Rumex fueginus* in vegetative aspects; its fruiting material differs from that of *R. fueginus* in the proportionately somewhat narrower mature inner perianth segments, these elliptic rather than triangular in outline, and in the thick, swollen callous grains, which are not narrowed into the midrib at the apex and which cover almost the entire perianth segment.

CHENOPODIACEAE. GOOSEFOOT FAMILY

Annual or perennial herbs or shrubs; herbage often mealy-scurfy or succulent. Leaves simple, alternate or occasionally opposite, sometimes reduced and scale-like, without stipules. Flowers small, usually inconspicuous, perfect or unisexual, often mixed in the same inflorescence. Calyx persistent, 2- to 5-lobed, the lobes rarely reduced to 1 or none on pistillate flowers. Petals none. Stamens equal to the sepals in number and opposite them, sometimes fewer. Ovary superior, 1-celled, 1-seeded; styles 1–3. Fruit a utricle. Embryo curved or bent.

Leaves all functional; stems not jointed; flowers not embedded in fleshy joints.
  Leaves alternate.
    Stamen 1; sepal 1 .... *Monolepis*
    Stamens and sepals more than 1.
      Pistillate flowers enclosed in 2 sepaloid bracts .... *Atriplex*
      Pistillate flowers not enclosed in 2 sepaloid bracts.
        Calyx lobes without hooks or spines.
          Leaves with distinct blade and petiole.
            Flowers in the axils of bracts .... *Beta*
            Flowers not in the axils of bracts .... *Chenopodium*
          Leaves sessile, linear, terete or flattened.
            Embryo annular; stems erect from a compact, persistent woody base
                                              Kochia
            Embryo coiled in a flat spiral; stems annual or perennial, when woody, stems openly branched.
              Flowers monoecious, the staminate ones in terminal spike ... *Sarcobatus*
              Flowers perfect, or occasionally some pistillate ones intermixed, in axils of leaves.
                Leaves and bracts not spine-tipped .... *Suaeda*
                Leaves and bracts spine-tipped .... *Salsola*
        Calyx lobes in fruit bearing hooks or spines .... *Echinopsilon*
  Leaves opposite .... *Nitrophila*

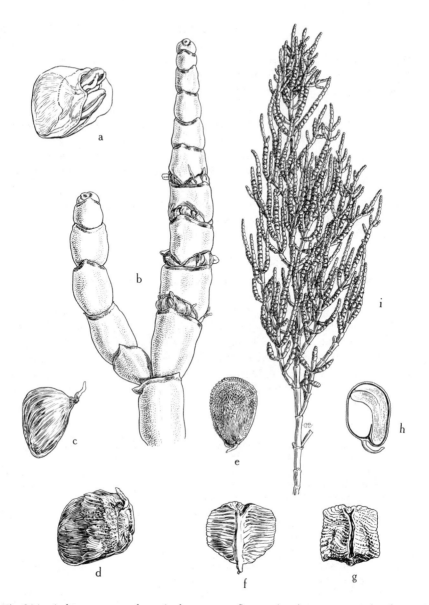

Fig. 214. *Arthrocnemum subterminale: a,* young flower, showing stamens and style protruding from terminal slit of calyx, × 12; *b,* branch, showing the perfoliate ring of reduced leaves and the subterminal flowering spike, × 4; *c,* mature fruit, showing lateral beak, × 12; *d,* corky calyx around fruit, lateral view, × 12; *e,* mature seed, showing funiculus, × 12; *f,* calyx enclosing fruit, adaxial view, × 12; *g,* calyx enclosing fruit, abaxial view, × 12; *h,* seed (longitudinal section), showing L-shaped embyro, × 12; *i,* habit, branches numerous and wide-spreading, × ⅖.

Leaves reduced to the scale-like tips of fleshy, moniliform joints of stems.
 Leaves opposite; annual, erect herbs or prostrate or decumbent perennials.
  Fruiting spikes all terminal; seeds hairy .... *Salicornia*
  Fruiting spikes usually subterminal, the tips proliferating; seeds glabrous
                                                            Arthrocnemum
 Leaves alternate; erect shrubs .... *Allenrolfea*

ALLENROLFEA

1. Allenrolfea occidentalis (Wats.) Kuntze, Rev. Gen. Pl. 2:545. 1891. Iodine
    bush. Fig. 213.

Erect shrub, 3–16 dm. tall; branches jointed, green, fleshy; leaves scale-like,
reduced to an obtuse fleshy point, alternate, glabrous, clasping the stem; fruiting spikes terminal on sessile or subsessile branchlets, the scales alternate or
spirally arranged, the joints broader than long; flowers 3 to a bract; sepals 4,
fleshy, carinate, coalesced over the fruit or free, the tips meeting at the orifice,
becoming spongy in fruit; stamens 2, exserted; ovary 1-ovuled; style erect,
2- or 3-lobed; seed enclosed in a membranous pericarp, minutely rugose;
embryo marginal and curved, enclosing copious endosperm.

Alkaline floodlands: San Joaquin Valley to Inyo County, eastern Mojave
Desert, Imperial Valley; east to Texas.

ARTHROCNEMUM

1. Arthrocnemum subterminale (Parish) Standley, Jour. Wash. Acad. Sci.
    4:399. 1914. Fig. 214.

Salicornia subterminalis Parish, Erythea 6:87. 1898.

Rhizomatous perennial, often widely spreading; stems erect or decumbent,
2–3 dm. tall, fleshy, usually dying to the ground during winter; leaves reduced
to a perfoliate ring, sometimes evident only as a pair of opposite obtuse points
on margin of ring; flowering spikes terminal, but often becoming subterminal
because of continued vegetative growth and branching above the fertile part,
the fertile part 1–5 cm. long, the joints of the spike usually longer than broad,
the flowering basal part about as long as the sterile upper part; flowers usually
3, occasionally 5, sessile in the axil of the obtuse bract; calyx saccate, open by
a terminal slit, becoming corky in fruit, or sometimes, in the coastal area,
merely shriveling around the fruit; stamens 2, usually simultaneous in anthesis;
pistil 1, the style persistent as a lateral beak on the fruit; fruit glabrous; seed
glabrous, the embryo L-shaped, in copious endosperm.

Common on alkaline floodlands: central San Joaquin Valley south through
the western Mojave Desert to Riverside and San Diego counties, thence north
along the coast to Los Angeles County; San Clemente Island.

We have followed Standley in regarding *Arthrocnemum* as distinct from *Salicornia* because of the indeterminate growth of the flowering shoot, the presence of abundant endosperm in the seed, the L-shaped embryo, and the wholly glabrous seeds. It is much more abundant in the interior than *Salicornia*, and frequently covers large areas in the alkaline plains of the San Joaquin Valley.

ATRIPLEX. SALTBUSH

Plants herbaceous or shrubby, often white-scurfy, to green and glabrous throughout. Leaves alternate or opposite, simple, sessile or petioled. Flowers dioecious or monoecious in axillary glomerules or these aggregated into terminal spikelike panicles. Staminate flower with a 3- to 5-parted calyx; stamens 3–5. Pistillate flower enclosed between 2 bracts which are free or variously united. Perianth none. Stigmas 2. Fruit a utricle with a membranous pericarp.

Plants annual, wholly herbaceous.
  Leaves green, sparsely scurfy or mealy only when young.
    Leaves cordate to hastate below.
      Plants spreading, decumbent or prostrate to erect; saline and salt marshes
                                          1. *A. patula*
      Plants erect and compact; Central Valley .... 2. *A. Phyllostegia*
    Leaves all narrowed at base .... 3. *A. Serenana*
  Leaves gray-scurfy.
    Leaves cordate, truncate or subhastate at base.
      Leaves 5–15 mm. long, cordate or subcordate at base.
        Leaves tomentose; fruiting bracts united to near the middle ... 4. *A. cordulata*
        Leaves mealy; fruiting bracts united to near the top .... 5. *A. vallicola*
      Leaves 15–40 mm. long, truncate or subhastate at base .... 6. *A. truncata*
    Leaves rounded or cuneate at base (occasionally some subhastate in *A. argentea*).
      Leaves ovate to triangular-ovate, 2–5 cm. long .... 7. *A. argentea*
    Leaves ovate-lanceolate to oblanceolate-elliptic.
      Plants erect, not decumbent or spreading.
        Leaves 20–60 mm. long; plants rounded, bushy .... 8. *A. rosea*
        Leaves 5–20 mm. long.
          Plants slender, erect, often simple or virgate .... 9. *A. tularensis*
          Plants much-branched, low and rounded .... 10. *A. pusilla*
      Plants prostrate, decumbent or spreading, if erect then only the ultimate branches
        erect.
        Bracts broadest at or near the middle; seeds brown to amber.
          Leaves deltoid or ovate; bracts united to the middle ...... 11. *A. coronata*
          Leaves broadest above the middle, oblanceolate or elliptic; bracts united
            nearly to the summit .... 12. *A. pacifica*
        Bracts broadest near the base; seeds dark brown to black .... 13. *A. Parishii*

Plants perennial, herbaceous above, sometimes woody at base but not shrubby.
  Leaves entire; plants erect or spreading, 1–4 dm. high.
    Leaves of fruiting bracts prominently 3-nerved; coastal salt marshes
                                                            14. *A. Coulteri*
    Leaves of fruiting bracts not prominently nerved; interior alkaline floodlands
                                                            15. *A. fruticulosa*
  Leaves repand-dentate, some entire; plants prostrate; fruiting bracts fleshy, the free
    margin denticulate to entire .... 16. *A. semibaccata*

1. Atriplex patula L., Sp. Pl. 1053. 1753.

Erect, spreading, or prostrate annual herb, 2.5–10 dm. high, the branches ascending or spreading or the lower ones decumbent; herbage green or sometimes white-farinaceous, becoming green and glabrate in age; leaves petioled, the blades linear to lanceolate, ovate or deltoid, narrowed to the base to truncate or rounded, or cordate to hastate, entire, crenulate to irregularly sinuate-dentate, the lower leaves opposite and the upper ones alternate, or all leaves opposite; bracts of the pistillate flowers sessile or stalked, united only near the base, deltoid to linear, hastate to cuneate at base, 3–7 mm. long, nearly as wide as long, the margins free, entire or sparsely dentate, the faces smooth or with enations or tubercles, sometimes veined; seeds 1–3 mm. long, dark brown or black.

Common in salt or brackish marshes along the coast and around bogs.

Several variants have been segregated, among which *Atriplex patula* var. *hastata* Gray is often accorded specific status because of the hastate leaf bases. See figure 215.

Following is a key to the varieties of *A. patula,* each of which is accorded specific status in some of the works on California botany.

Margins of the fruiting bracts narrow-toothed, the teeth small or large but not abundant,
    faces often with tubercles.
  Leaf blades triangular-hastate; bracts broadly rounded or truncate at base
                                                            A. patula var. *hastata*
  Leaf blades lanceolate or oblong-linear, not hastate; bracts broadly cuneate at base
                                                            A. patula var. *patula*
Margins of fruiting bracts wide, entire or with an occasional tooth, the face smooth or
    rarely appendaged.
  Leaves ovate-rhombic, irregularly sinuate-dentate; bracts small, about 3 mm. long, the
    face ribbed or furrowed .... *A. patula* var. *spicata*
  Leaves linear to linear-lanceolate, entire or sometimes toothed; bracts 4–20 mm. long,
    margins entire .... *A. patula* var. *obtusa*

2. Atriplex Phyllostegia (Torr.) Wats., Proc. Am. Acad. Arts & Sci. 9:108. 1874.

Erect annual herb 2–6 dm. tall; branches close and compact, pubescent or glabrate; lowermost pair of leaves opposite, the remainder alternate, petioled or sessile, from lanceolate to broadly triangular, cordate or hastate below, the tip attenuate, 1–3 cm. long, succulent, scurfy or glabrate, becoming papery on

Fig. 215. *Atriplex patula* var. *hastata: a,* pistillate inflorescence, $\times$ 4; *b,* habit, showing the spreading branches, the hastate leaves, and the spikelike inflorescences, $\times$ ⅖; *c* and *d,* young staminate flowers, $\times$ 12; *e,* seed, showing curved embryo, $\times$ 12; *f,* single pistillate flower, showing tuberculate bracts, $\times$ 8.

drying; bracts of the pistillate flowers sessile or stalked, joined only near base, lanceolate or lanceolate-oblong, cordate or hastate or cuneate at base, the tips free, elongate, 5–20 mm. long, entire or laciniate, widely separating, usually 3-ribbed, sometimes with facial appendages; seeds brown.

Alkaline floodlands: San Joaquin Valley; east to Great Basin and north to Oregon.

3. Atriplex Serenana Nels., Proc. Biol. Soc. Wash. 17:99. 1904.

Atriplex bracteosa Wats., Proc. Am. Acad. Arts & Sci. 9:115. 1874. Not Trautv. 1870.

Erect or decumbent annual herb, 3–10 dm. tall, often in dense, tangled masses; herbage sparsely scurfy to glabrate; leaves alternate, sessile or subsessile, 2–8 cm. long, lanceolate to narrowly ovate or elliptic, narrowed to base, acute or acuminate at tip, sparingly dentate; bracts of pistillate flowers sessile or subsessile, united to about the middle, the base cuneate, 2–3 mm. long, the margins acutely dentate, the faces tuberculate or smooth, nerved; seeds brown.

Alkaline floodlands: Sacramento Valley south through southern California; south to northern Baja California in Mexico, and east to Nevada.

Atriplex Serenana var. *Davidsonii* (Standley) Munz, which has short, globose, terminal, staminate clusters, occurs on alkaline or saline soils along the southern California coast.

4. Atriplex cordulata Jepson, Pittonia 2:304. 1892.

Erect, much-branched, rigid annual, 1–4 dm. tall, the branches stout; herbage mealy; leaves alternate except the lower ones, sessile, cordate, clasping, entire, thick, tomentose, 5–15 mm. long, white-mealy; bracts of pistillate flowers 4–5 mm. long, sessile or subsessile to short-stalked, united to near the middle, broadened below the middle, the margins free, green, and acutely dentate, the faces smooth, scurfy, occasionally with a few protuberances; seeds deep reddish brown to nearly black, shiny.

Alkaline floodlands: Sacramento and San Joaquin valleys.

5. Atriplex vallicola Hoover, Leafl. West. Bot. 2:130. 1938.

Erect, much-branched annual 1–4 dm. tall; herbage mealy; leaves cordate or subcordate at base, 2–7 mm. long; bracts of pistillate flowers united nearly to the top, irregular in shape, the upper margin free, undulate, irregularly toothed.

Alkaline floodplains: San Joaquin Valley.

This species is closely related to *Atriplex cordulata*.

6. Atriplex truncata (Torr.) Gray, Proc. Am. Acad. Arts & Sci. 8:398. 1872.

Erect, sparsely branched annual, 2–12 dm. tall; stems dull white or reddish; leaves alternate, the lower ones opposite, mostly sessile, triangular to ovate, truncate to subhastate at base, entire, undulate, gray-scurfy, 1–4 cm. long;

bracts of pistillate flowers short-stalked, broadly cuneate, truncate across the top, 2–3 mm. long, the faces smooth or with a few minute tubercles, conspicuously veined and reticulate; seeds light brown, shiny.

Alkaline basins: east side of the Sierra Nevada; east to Colorado and New Mexico.

7. Atriplex argentea Nutt., Gen. 1:198. 1818. Silver saltbush.

Erect, globose annual, 2–6 dm. tall; branches grayish when young; leaves alternate, the lower ones opposite, sessile to subsessile, or the lower ones petioled, lanceolate-ovate or deltoid, 2–5 cm. long, cuneate to subhastate at base; bracts of pistillate flowers sessile or subsessile, united to the middle, more or less orbicular, 4–8 mm. long, the margins foliaceous, subentire to laciniate, the faces smooth or variously appendaged; seeds brown.

Alkaline floodlands: Central Valley and Mojave Desert, northeastern California; east to North Dakota.

A form with strictly sessile upper leaves has been segregated as *Atriplex argentea* subsp. *expansa* (Wats.) Hall & Clements.

8. Atriplex rosea L., Sp. Pl., 2d ed., 1493. 1763.

Erect, rounded, bushy annual, 2–10 dm. tall; leaves alternate, the lowermost opposite, sessile or petioled, ovate to lanceolate, cuneate or rounded at base, acute and mucronulate at apex, 2–6 cm. long, coarsely toothed, gray-scurfy; fruiting bracts sessile, united at base, rhombic to ovate, the base broad, 4–12 mm. long, firm and indurated in age, the margins dentate, the faces tuberculate; seeds dark brown, dull.

Weed introduced in alkaline areas: southern California to northeastern Washington.

9. Atriplex tularensis Cov., Contr. U. S. Nat. Herb. 4:182. 1893.

Atriplex cordulata var. *tularensis* Jepson, Fl. Calif. 436. 1914.

Slender, erect annual with rigid, brittle branches, 1.5–4 dm. tall; herbage coarsely white- or gray-scurfy; leaves alternate except the lowermost, sessile or subsessile, ovate to lanceolate, rounded at base, acute or acuminate above, entire, thin, 1- or 2-nerved, 1–2 cm. long; bracts of pistillate flowers sessile, united to near summit, ovate or rhomboid, then somewhat narrowed below, 2–3.5 mm. long, the free margin thin, few- to many-toothed, or somewhat erose, the tip usually a prominent tooth, the faces white, mealy; seeds dark brown, shiny.

Alkaline plains: southern end of San Joaquin Valley.

10. Atriplex pusilla (Torr.) Wats. Proc. Am. Acad. Arts & Sci. 9:110. 1874.

Much-branched erect annual, 5–25 cm. tall; herbage sparsely mealy, often reddish; leaves numerous, imbricated above, sessile, 3–8 mm. long, ovate to oblong-lanceolate, acute or acuminate, entire; fruiting bracts sessile, ovate, 1–2 mm. long, acute or acuminate, entire, the faces smooth.

Alkaline floodlands: northeastern California; north to Oregon, east to Nevada.

11. Atriplex coronata Wats., Proc. Am. Acad. Arts & Sci. 9:114. 1874.

Spreading annual, the branches erect, 1–3 dm. tall; herbage gray-scurfy, glabrate in age; leaves alternate except the lower ones, petioled except the upper ones, ovate-elliptic, acute at apex, entire, 5–25 mm. long; fruiting bracts sessile, united to above the middle, the margins green, foliaceous, irregularly dentate or laciniate, the faces smooth to tuberculate; seeds purplish brown to dark amber, shiny.

Alkaline floodlands: Central Valley, southern California.

12. Atriplex pacifica Nels., Proc. Biol. Soc. Wash. 17:99. 1904.

Atriplex microcarpa (Benth.) D. Dietr., Syn. Pl. 5:536. 1852. Not Waldst. & Kit. 1812.

Prostrate or procumbent annual, forming tangled mats 1–4 dm. broad, the branches erect; leaves alternate and sessile, or the lowermost opposite and petioled, oblanceolate or elliptic, narrowed to base, sometimes mucronulate at the acute or obtuse tip, entire, thin, gray-scurfy, 5–20 mm. long; bracts of pistillate flowers sessile or short-stalked, scarcely compressed, united to near the summit, the base narrowed and the summit broad, obtuse and with 1 to several minute teeth across top, the faces smooth or with a few tubercles, 1-nerved; seeds light brown.

Borders of salt marshes: coastal southern California; south into northern Baja California, Mexico, and adjacent islands.

13. Atriplex Parishii Wats., Proc. Am. Acad. Arts & Sci. 17:377. 1882.

Low, horizontally spreading annual, 5–20 cm. tall; stems thick, brittle; herbage white-scurfy, somewhat reddish in age; leaves opposite and alternate, the upper ones imbricate, sessile, ovate to ovate-lanceolate, acute at tip, entire, 4–10 mm. long; fruiting bracts sessile, united to above the middle, ovate or rhombic or subhastate, acute, 2–3.5 mm. long, entire or with a few prominent teeth, the faces smooth or muricate; seeds brown to black.

Alkaline floodlands: Central Valley to southern California.

14. Atriplex Coulteri (Moq.) D. Dietr., Syn. Pl. 5:537. 1852.

Spreading perennial; herbage 1–10 dm. tall; stems slender, glabrate, sometimes reddish; leaves alternate, on short petioles or the upper leaves sessile, narrowly elliptic to lanceolate or oblong, entire, gray to greenish, 1–2 cm. long, 1-nerved; fruiting bracts sessile, united to the middle, obovate, 2.5–3 mm. long, the margins free, sharply dentate, the faces prominently 3-nerved with a few sharp tubercles or these absent; seeds brown.

Margins of salt marshes: southern California; northern Baja California, Mexico, and adjacent islands.

15. Atriplex fruticulosa Jepson, Pittonia 2:306. 1892.

Erect or spreading, woody-based perennial, 1–3 dm. tall; herbage coarse, scurfy to glabrate at maturity, reddish; leaves alternate, short-petioled or the upper ones sessile, elliptic to linear-lanceolate, 5–15 mm. long, entire, densely gray-scurfy on both faces, 1-nerved; fruiting bracts sessile or subsessile, scarcely compressed, united to above the middle, ovoid to subglobose, 3–4 mm. long, the margins free, acutely dentate, the faces cristate or muricate or sometimes smooth, indurate; seeds dark brown.

Alkaline floodlands: Central Valley.

16. Atriplex semibaccata R. Br., Prodr. Fl. Nov. Holl. 406. 1810.

Prostrate perennial, woody below, forming mats 14 dm. wide; herbage mealy, at length glabrate; leaves alternate, short-petioled, elliptic to elliptic-oblanceolate, 1–4 cm. long, irregularly repand-dentate, gray-scurfy to greenish, 1-nerved; fruiting bracts sessile or stalked, convex and fleshy, becoming flat when dry, rhombic, 3–6 mm. long, united below the middle, the margins free, denticulate or entire, the faces 3- to 5-nerved, becoming reddish; seeds brown or black.

Common on alkaline floodlands and on sea bluffs: San Joaquin Valley, southern California; naturalized from Australia.

BETA

1. Beta vulgaris L., Sp. Pl. 222. 1753. Common beet.

Robust, glabrous biennial herb; leaves alternate, large, long-petioled, reduced and subsessile in the inflorescence; flowers perfect, greenish white, in sessile, axillary, 2- or 3-flowered clusters, disposed in panicled spikes, cohering in fruit by the enlarged bases of the calyx; calyx 5-parted, becoming indurated and closing over the fruit; stamens 5, perigynous; filaments sometimes connate at base; ovary embedded in the succulent base of the calyx; styles 2 or 3, short, stigmatose on the inside; embryo annular.

Naturalized in marshes, from gardens: Petaluma in Sonoma County, Suisun in Solano County, Alvarado in Alameda County, Monterey in Monterey County, and San Bernardino in San Bernardino County.

CHENOPODIUM. GOOSEFOOT, PIGWEED

Annual or perennial herbs, often white-mealy or glandular. Leaves alternate, petioled. Flowers perfect, greenish to reddish, sessile in clusters which are simple or aggregated into spikes. Calyx 5-parted, rarely 1- to 3- or 4-parted, persistent, at length enclosing the fruit. Corolla none. Stamens 5 or fewer. Ovary flattened; ovule either horizontally disposed or erect in the ovary; styles 2–4. Seed with coiled embryo and a membranous pericarp (ripened ovary wall).

Herbage glabrous or mealy, not strong-scented.
    Fruiting calyx dry, never reddish; achene usually horizontal in pericarp; spikes slender, often paniculate . . . . 1. *C. album*
    Fruiting calyx fleshy, often reddish; achene usually erect in pericarp; flowers in compact glomerules or dense spikes . . . . 2. *C. rubrum*
Herbage glandular-pubescent and strong-scented . . . . 3. *C. ambrosioides*

1. Chenopodium album L., Sp. Pl. 219. 1753. Lamb's quarters.

Erect annual 6–15 dm. tall; herbage usually much-branched, light green or white; leaves mealy, broadly ovate, sinuate-dentate below, somewhat attenuate and entire above; flowers clustered in dense spikes which are arranged in a panicle; sepals strongly carinate; fruit a 1-seeded utricle.

Common on floodlands and in waste places near marshes: throughout western North America except Arctic regions; native of Eurasia.

This species is commonly grazed by ducks and geese and represents an important item of green food.

Another species, *Chenopodium murale* L., which has dark green herbage, sometimes occurs in similar habitats and is also grazed by waterfowl.

2. Chenopodium rubrum L., Sp. Pl. 218. 1753. Red goosefoot.

Erect annual with angled stem, 4–10 dm. tall; herbage dark green, the petioles and stems and sometimes the inflorescence often reddish; leaves oblong-lanceolate to ovate, coarsely sinuate, 2–6 cm. long; flowers clustered in short axillary spikes; sepals 2–4, fleshy; stamens 1 or 2; achene shiny, the margin acute.

Often locally abundant about alkaline or saline marshlands: Central Valley, coastal and southern California; naturalized from Europe.

Chenopodium rubrum var. *humile* Wats., which has prostrate or ascending stems, is found in the marshes of northern California.

3. Chenopodium ambrosioides L., Sp. Pl. 219. 1753. Mexican tea.

Erect stout annual, or short-lived perennial, 5–15 dm. tall, much-branched; herbage glabrous or when young tomentose, somewhat glandular, aromatic; leaves short-petioled, oblong to lanceolate, repand-toothed or sometimes nearly entire, 5–15 cm. long; flowers in dense, axillary, leafy, spikelike clusters; sepals obtuse, slightly carinate, appressed over utricle; styles 3 or 4; pericarp deciduous; seeds shiny, reddish.

Common in floodlands of alkaline or salt marshes and along interior streams: throughout California; Pacific states and across the continent. Native of tropical America.

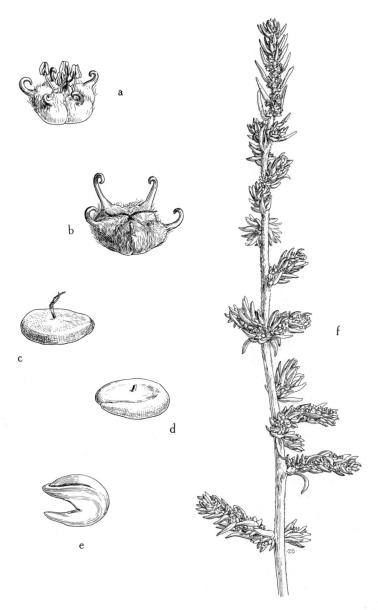

Fig. 216. *Echinopsilon hyssopifolium: a,* flower, the broadly ovate sepals armed with hooked spines, × 8; *b,* mature fruit containing a single, horizontally placed seed, × 8; *c* and *d,* mature seeds, × 8; *e,* curved embryo, × 8; *f,* stem and inflorescences, × 1.

Seeds are produced in abundance.

Chenopodium ambrosioides var. *anthelminticum* (L.) Gray, which tends to be spreading in habit and has more finely cut leaves and slender spikes which appear leafless, has much the same range as the species but is most frequent in the Sacramento Valley.

ECHINOPSILON

1. Echinopsilon hyssopifolium (Pall.) Moq. in A. DC., Prodr. 13^2:135. 1849. Fig. 216.

Bassia hyssopifolia Kuntze, Rev. Gen. Pl. 547. 1891.

Erect, decumbent, or prostrate annual, 3–5 dm. tall; herbage pilose throughout; leaves alternate, sessile, linear-lanceolate, entire, 2–4 cm. long; flowers in axillary glomerules, villous to tomentose; calyx globose, the sepals broadly ovate, obtuse, 1 mm. long, each armed with a hooked spine; fruit a utricle, enclosed in calyx; seed horizontal, free from pericarp; embryo annular.

Common in alkaline floodplains and in alkaline vernal pools: Sacramento and San Joaquin valleys, Owens Valley, and Orange County. Recently introduced and spreading rapidly; native of Europe.

KOCHIA

Perennial herbs, woody at base, somewhat silky-pubescent. Leaves linear, terete or flat, entire, alternate or opposite. Flowers perfect or with some pistillate ones intermixed, solitary or few in the axils, without bracts. Calyx herbaceous, subglobose, 5-lobed, persistent over fruit, at length developing 5 horizontal wings. Stamens 5, usually exserted. Ovary depressed; styles 2 or 3, filiform. Seed with membranous pericarp; embryo nearly annular, green; endosperm none.

Flowering stems essentially simple, pubescence white; leaves linear-terete
1. K. americana
Flowering stems much-branched, pubescence somewhat rufous-silky; leaves linear to elliptic, flat .... 2. *K. californica*

1. Kochia americana Wats., Proc. Am. Acad. Arts & Sci. 9:93. 1874.

Perennial from a compact, branched, woody crown; seasonal branches 1–4 dm. tall, simple, erect, slender, white-silky-pubescent; leaves linear, almost filiform, terete, 1–3 cm. long; flowers embedded in white wool; calyx 5-lobed, each lobe developing in age a conspicuous, cuneate, horizontal wing; stamens 5; ovary 1-celled; utricle glabrous.

Alkaline floodlands: Modoc plateau, south along east side of the Sierra Nevada to Mojave Desert; north to Oregon, east to Wyoming and Colorado.

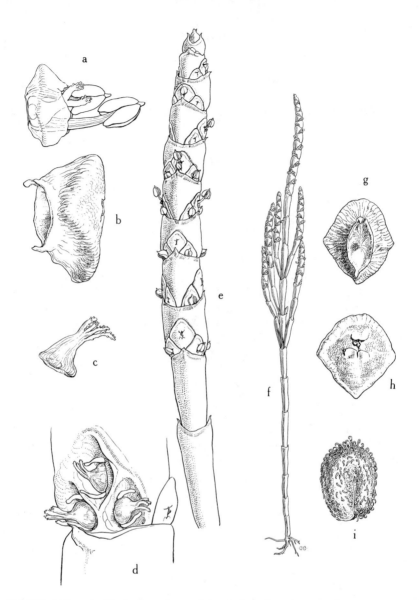

Fig. 217. *Salicornia Bigelovii: a,* flower, lateral view, the stigmas and stamens exserted from the slitlike orifice of calyx, × 12; *b,* calyx, lateral view, with the mature seed removed from its cavity, × 12; *c,* pistil, showing variation in shape of ovary and number of style branches, × 12; *d,* triad of flowers in their cavities, the calyces removed, × 8; *e,* spike, showing the angular triads of flowers and mucronate bracts, × 3; *f,* habit, the few branches appressed and erect, × ⅖; *g,* calyx, adaxial view, showing cavity of ovary, × 8; *h,* calyx, abaxial view, the lobes of orifice having closed around the remains of the stigmas, × 8; *i,* mature seed with retrorse, appressed, hooked hairs, × 16.

2. Kochia californica Wats., Proc. Am. Acad. Arts & Sci. 17:378. 1882.

Kochia americana var. *californica* Jones, Contr. West. Bot. No. 11:19. 1903.

Perennial from a compact, woody, branched base; seasonal branches paniculately much-branched, 2–5 dm. tall; herbage densely covered with rufous-silky hairs; leaves linear to elliptic, flat, 1–2 cm. long, 2–5 mm. wide; flowers and fruit densely rufous-silky.

Alkaline floodlands and desert water holes: southern San Joaquin Valley to Mojave Desert.

MONOLEPIS

1. Monolepis Nuttalliana (Schult.) Greene, Fl. Fran. 168. 1891. Patata.

Low annual, mealy when young, becoming glabrate; branches erect or ascending; leaves alternate, fleshy, attenuate above, narrowed to short petiole, hastate; flowers polygamous, sessile, clustered in the upper axils; sepal 1, entire, bractlike, persistent; stamen 1; styles 2, filiform; achene with thin, minutely pitted, persistent pericarp; embryo annular, surrounded by copious endosperm.

Alkaline floodplains: southern California north to Sacramento Valley, Sierra Nevada; north to Alberta, east to the Great Plains.

Monolepis spathulata Gray and *M. pusilla* Torr. occur in alkaline soils of the higher mountains.

NITROPHILA

1. Nitrophila occidentalis (Moq.) Wats. in King, Geol. Expl. 40th Par. 5:297. 1871.

Low, perennial, rhizomatous, glabrous herb; leaves sessile, fleshy, linear, pungent, 1–3 cm. long, reduced upward, opposite; flowers axillary, perfect; sepals 5 to 7, imbricate, carinate; petals none; stamens 5, united at base into a thin yellowish disc; style longer than the subglobose ovary; stigmas 2; achene beaked by the persistent style, included within the connivent sepals, the pericarp membranous.

Moist alkaline soils: Central Valley, and east side of the Sierra Nevada; south to Baja California, Mexico.

SALICORNIA. PICKLEWEED

Erect, spreading, or prostrate herbs, sometimes rooting at nodes, 5–40 cm. tall; stems fleshy, jointed, glabrous. Leaves reduced to 2 opposite perfoliate cusps, often somewhat mucronate. Flowers in decussate terminal spikes 1–10 cm. long, the joints as long as or longer than the flowers, the flowers sessile in opposite triads, the middle flower of each triad usually the longest and

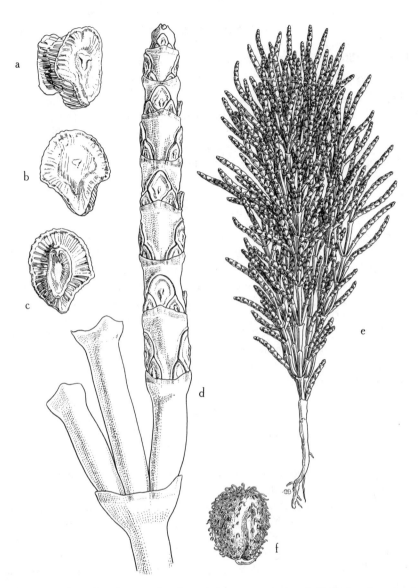

Fig. 218. *Salicornia rubra: a,* calyx, lateral view, ovary enclosed, × 12; *b,* calyx, abaxial view, the orifice completely closed, × 12; *c,* calyx, adaxial view, showing cavity of seed, × 12; *d,* spike, showing rounded apex of much higher middle flower of triad, × 4; *e,* habit, plant very much branched and bushy, × ⅔; *f,* mature seed, showing the retrorse, hooked hairs, × 16.

higher on stem. Calyx completely saccate, 4-membered, free at tips, around a slitlike orifice, becoming corky in fruit. Stamens 2, arising above and below pistil, the connective terminating in a cusp. Styles 2- to 4-branched, terminal and erect; ovary membranous. Fruit 1-celled and 1-seeded, dehiscent by an irregular circumscissile degeneration of the ovary and the base of the calyx tube, these shed separately from the seeds or sometimes the exposed seed adhering to the calyx base. Seed without endosperm, clothed with hooked or straight, retrorse, appressed hairs, sometimes becoming glabrate, the embryo bent so that the cotyledons are reflexed against the radicle.

Plants annual, erect or spreading.
  Stems few-branched; spikes 3–5 mm. in diameter; bracts of spike mucronate
                                       1. *S. Bigelovii*
  Stems many-branched; spikes slender, 2–3 mm. in diameter.
    Plants distinctly bushy, the many branches virgately erect . . . . 2. *S. rubra*
    Plants low-spreading, the branches decumbent and divaricate . . . . 3. *S. depressa*
Plants perennial, erect or trailing, often rooting at the nodes . . . . 4. *S. pacifica*

1. Salicornia Bigelovii Torr., Mex. Bound. Surv. Bot. 184. 1859. Fig. 217.

Salicornia mucronata Bigel., Fl. Boston., 2d ed., 2. 1824. Not Lag.
Salicornia europea L. of western American authors.

Erect or occasionally ascending annual; stems usually simple and erect at base, with few to several appressed branches, glabrous; spikes stout, 2–8 cm. long, 4–5 mm. thick, the joints 1–2 times as long as the longest flower of each triad, the bracts mucronate, usually widely spreading in mature plants; middle flower of triad much higher than lateral flowers; stamens 2, exserted from the slitlike orifice of the calyx; styles 2- to 4-branched; stigma exserted; ovary membranous; fruit dehiscent; seeds falling separately, often adhering to the cavity of the flower on the spike.

Salt marshes: southern California.

2. Salicornia rubra Nels., Bull. Torrey Club 26:122. 1899. Fig. 218.

Very bushy plants 1–4 dm. tall; branches many from the base to the tip, virgately erect; spikes 2–4 cm. long, thicker than the stem, 2–3 mm. thick, the joints extending slightly beyond the middle flower of each triad, the middle flower much higher than the lateral flowers; fruit sparsely pubescent with retrorse appressed hairs to glabrate.

In high spots in the salt marshes bordering San Francisco Bay, as well as in alkaline and saline floodlands in Modoc County; east to Wyoming and Minnesota.

3. Salicornia depressa Standley, N. Am. Fl. 21:85. 1916. Fig. 219.

Tufted, spreading or decumbent annual, divaricately branched from the base or the lower nodes, glabrous; branches several, 3–20 cm. long; spikes 1–6 cm. long, 2–3 mm. thick; joints cuneiform, 1–1½ times as long as the longest

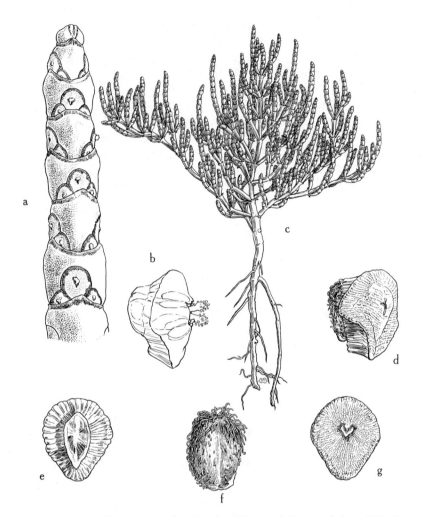

Fig. 219. *Salicornia depressa: a,* spike, showing the rounded apex of the middle flower of a triad and the nearly obtuse bracts, × 6; *b,* flower, lateral view, the styles barely exserted, × 12; *c,* habit, showing the tufted, spreading branches, × ⅖; *d,* calyx, enclosing seed, × 12; *e,* calyx, adaxial view, showing cavity of seed, × 12; *f,* seed, showing the long, retrorsely appressed, hooked hairs, × 16; *g,* calyx, abaxial view, the orifice closed, × 12.

flower; bracts obtuse, sometimes not evident; middle flower of triad much higher than the lateral ones; anthers barely exserted; seeds pubescent.

Salt marshes and alkaline floodlands, widely dispersed but infrequently collected in California: San Francisco Bay, Tolay Creek in Sonoma County, east of Palo Alto in Santa Clara County, Monolith in Kern County, San Diego County; southward into Baja California, Mexico.

4. Salicornia pacifica Standley, N. Am. Fl. 21:83. 1916. Common pickleweed. Fig. 220.

Suffruticose perennial, erect, decumbent, or prostrate, usually from a horizontal rhizome, occasionally rooting along the decumbent or prostrate branches, or the individual plant solitary and erect; joints constricted at nodes, 2–5 mm. thick; leaves reduced to a perfoliate collar with opposite cusps, these often obscure, glabrous, glaucous or green; spikes 1–10 cm. long, terminating the ultimate branches, the joints 1.5–2.5 mm. long, usually wider than long, the middle flower of each triad only slightly higher than the lateral ones; sepals 4 or 3, fused or sometimes those of lateral flowers nearly free; stamens 2, not simultaneous in anthesis; seeds covered with white, stiff, appressed hairs, falling free from calyx on dehiscence or adhering to it and falling with it.

Salt marshes along the coast: from Baja California, Mexico, north to British Columbia, and sparingly in wet saline or alkaline floodlands in the interior.

Specimens from the interior are characterized by a thicker spike and a stronger tendency to root along the decumbent or prostrate stems. These have been described as *Salicornia pacifica* var. *utahensis* (Tidestr.) Munz (Man. So. Calif. Bot. 142, 1935). However, along the lower San Joaquin River, they merge completely with coastal types.

SALSOLA

1. Salsola Kali L. var. **tenuifolia** Tausch, Flora 11:326. 1828. Russian thistle, tumbleweed.

Intricately branched, low, rounded herb with stiff stems; leaves rigid-linear or subulate-spinescent; flowers perfect, solitary, sessile and axillary, each subtended by a rigid spinescent bract and 2 bractlets; sepals becoming horizontally winged on back; stamens 5; ovary hollowed above like a shallow cup; styles 2; seed horizontal; embryo coiled into a conic spiral; endosperm none.

Alkaline floodlands and some agricultural areas: abundant weed in southern California, also San Joaquin Valley and Solano, Lassen, and Modoc counties; native of Eurasia.

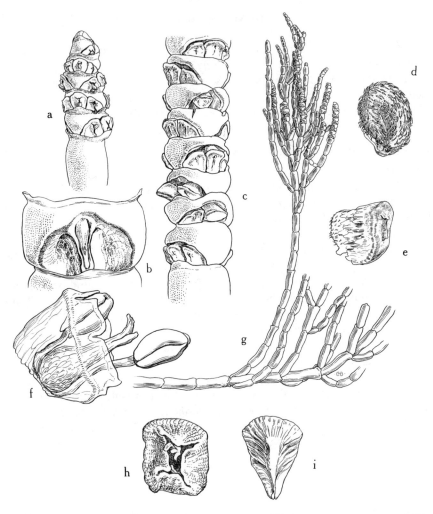

Fig. 220. *Salicornia pacifica: a,* young flowering spike at anthesis, showing cusplike bracts and the broad flower triads and joints, × 4; *b,* joint, showing cavity of flower triad, × 8; *c,* mature spike, the calyces falling with seeds, × 4; *d,* mature seed with nearly straight, appressed hairs, × 16; *e,* calyx enclosing seed, lateral view, × 8; *f,* young flower, the conspicuous calyx lobes wide open, × 20; *g,* habit, × ⅖; *h,* calyx, abaxial view, × 12; *i,* calyx, the seed enclosed, adaxial view, × 12.

SARCOBATUS

1. Sarcobatus vermiculatus (Hook.) Torr. in Emory, Notes Mil. Reconn. 150. 1848. Black greasewood.

Erect, divaricately branched shrub with thorny branches; leaves alternate, linear, sessile, entire; flowers monoecius or dioecious, without bracts; staminate flowers in terminal catkin-like spikes subtended by spirally arranged scales, the perianth none, the stamens 2–5, under a stipitate, peltate scale, the filaments short; pistillate flowers axillary and commonly solitary, sessile, the ovary adnate to a saclike calyx, the style short, the stigmas 2, spreading horizontally, the calyx laterally margined by a narrow border which becomes in fruit a broad, circular, horizontal, wavy, membranous wing.

Alkaline clay soil of desert valleys: Colorado and Mojave deserts, and east of Sierra Nevada crest from Inyo County to Lassen and Modoc counties; east to New Mexico and Colorado, and north to Washington.

SUAEDA

Annual or perennial herbs or subshrubs. Leaves alternate, linear, flat or terete, succulent. Flowers usually clustered in the axils of the uppermost leaves, sometimes simulating a spike. Sepals 4 or 5, valvate, fleshy, sometimes with a narrow hyaline margin, united below. Petals none. Stamens 4 or 5, the filaments short. Styles usually 2. Fruit a 1-seeded utricle enclosed by the connivent sepals; embryo coiled in a flat spiral.

Calyx in age becoming corniculately fleshy-appendaged, appendages forming a horizontal wing around the fruit.
    Leaves erect; stems virgately branched .... 1. *S. minutiflora*
    Leaves spreading; stems divaricately branched, erect or decumbent.
        Leaves, at least in inflorescence, dilated at base; stems in larger plants stout and
            coarse .... 2. *S. depressa*
        Leaves not dilated at base, stems and leaves very slender .... 3. *S. occidentalis*
Calyx lobes not corniculately appendaged, sepals flat, terete, or at most carinate.
    Leaves of the inflorescence often reduced, 1–3 times as long as the flowers, the internodes clearly exposed; calyx and herbage glabrous to minutely puberulent.
        Leaves flattened; herbage dark green .... 4. *S. Torreyana*
        Leaves terete; herbage blue-green .... 5. *S. fruticosa*
    Leaves of the inflorescence 3–5 times as long as the flowers, usually crowded and concealing the short internodes; calyx and herbage pubescent, often densely villous or tomentose, occasionally glabrate .... 6. *S. californica*

1. Suaeda minutiflora Wats., Proc. Am. Acad. Arts & Sci. 18:194. 1883.

Plant glaucous and glabrous, erect, stout, virgately branching, annual or perennial; leaves linear to linear-lanceolate, 2–5 cm. long, erect to somewhat appressed, terete to semiterete, often those of the inflorescence somewhat

Fig. 221. *Suaeda fruticosa: a,* habit, showing erect, terete leaves, × ⅖; *b,* flower, × 8; *c,* part of inflorescence, showing flower and subtending leaves, × 6.

broader and shorter, usually extending beyond the flower clusters; inflorescence of short, slender, appressed, axillary branches, rarely unbranched; flowers densely crowded, 1 or 2 of the sepals corniculately appendaged; seeds horizontal, dark reddish brown.

Salt marshes and alkaline plains: coastal area from Santa Barbara County south to Riverside and San Diego counties, inland in San Bernardino and Riverside counties.

2. Suaeda depressa (Pursh) Wats. in King, Geol. Expl. 40th Par. 5:294. 1871.

Plant green to somewhat glaucous, glabrous, stout, erect or decumbent annual or short-lived perennial, 2–6 dm. tall, often much-branched; leaves linear to lanceolate-linear, semiterete, 1–3 cm. long, reduced in the inflorescence, these broader at base and often narrowly ovate, 2–3 times as long as the flowers; flowers crowded in dense spikes, 1–5 in each cluster, 1 to several of the sepals corniculately appendaged below middle; seeds vertical or horizontal, black.

Alkaline plains: northeastern California and east of the Sierra Nevada to coastal southern California; western states.

Suaeda depressa var. *erecta* Wats. is an erect annual form, simple to sparingly branched. It occurs in the general region of the species.

3. Suaeda occidentalis Wats., Proc. Am. Acad. Arts & Sci. 9:90. 1874.

Glaucous and glabrous erect or spreading annual, 1–3 dm. tall, much-branched with very slender branches; leaves narrowly linear, appearing as though filiform when dry, 10–25 mm. long, 1 mm. wide, not much reduced in inflorescence; flowers 1–3 to each axil, forming an interrupted spike; sepals becoming transverse-winged in age below the middle, the wing entire; seeds black.

Alkaline floodplains: Modoc plateau south to Lassen County (rare); north to Oregon and Washington, east to Colorado.

4. Suaeda Torreyana Wats., Proc. Am. Acad. Arts & Sci. 9:88. 1874.

Suaeda Torreyana Wats. var. *ramosissima* Munz, Man. So. Calif. Bot. 44. 1935.

Plants green, essentially glabrous or sometimes puberulent or sparsely pubescent above, much-branched, erect or ascending, 3–10 dm. tall; twigs and branches usually slender, the internodes usually conspicuous; leaves 2–3 cm. long, linear, very evidently flat, those of the inflorescence becoming reduced to 1–3 times as long as the flower cluster; flowers 1–5 in a cluster in the leaf axils, the clusters separated by the slender, often wiry, internode; calyx deeply cleft, the lobes rounded on back; seeds black, minutely tuberculate.

Alkaline floodlands: east of the Sierra Nevada, also Colorado River area and Imperial Valley; north to eastern Oregon, east to Utah and New Mexico.

The plants most obviously pubescent have been segregated as *Suaeda ramosissima* Standley, but intergradations seem to make clear definition difficult between *S. ramosissima* and *S. Torreyana*.

5. Suaeda fruticosa (L.) Forsk., Fl. Aegypt. 70. 1775. Fig. 221.

Suaeda Moquinii Greene, Pittonia 1:264. 1889.

Stems erect or ascending, annual or more commonly perennial, often shrublike with woody base; herbage glabrous and glaucous; leaves terete to subterete, linear, 1–3 cm. long, those of the inflorescence usually somewhat reduced and rarely more than 3 times as long as the flowers; flowers in clusters of 1–7, the clusters separated by a slender, often wiry, internode; calyx deeply cleft, the lobes rounded to somewhat keeled on back; seeds black.

Alkaline and saline habitats: coastal southern California, Central Valley from central California southward, also east of the Sierra Nevada; north to Canada, south to northern Mexico.

6. Suaeda californica Wats., Proc. Am. Acad. Arts & Sci. 9:89. 1874.

Suaeda californica var. pubescens Jepson, Fl. Calif. 447. 1914.

Perennial, often villous-pubescent to glabrate in age, rarely if ever wholly glabrous; stems 2–10 dm. tall, decumbent or ascending, usually densely long-leafy throughout; leaves 1–4 cm. long, linear-terete to subterete, those of the inflorescence crowded and very little reduced; flowers 1–4 in an axil, the clusters close by virtue of the crowded leaf bases; calyx densely pubescent to glabrous and glaucous, the lobes rounded on back; seeds black.

Coastal salt marshes: central and southern California; south to Baja California, Mexico.

The more highly pubescent forms have been segregated as *Suaeda taxifolia* Standley, but pubescence is too inconstant to be of great significance. A short-leaved type has been described as *Dondia brevifolia* Standley [*Suaeda taxifolia* subsp. *brevifolia* (Standley) Abrams].

AMARANTHACEAE

Ours mostly annual herbs. Leaves alternate or opposite, without stipules. Flowers inconspicuous, regular, usually crowded into dense spikes or racemes and subtended by a membranous bract and 2 similar hyaline bractlets. Perianth of 3–5 dry or membranous sepals, free or united at the base. Petals wanting. Stamens usually 5, opposite the calyx lobes. Pistil 1, the ovary superior, 1-celled, the styles 1–3. Fruit usually a utricle, nutlet, or circumscissile capsule, dehiscent or indehiscent.

AMARANTHUS

1. Amaranthus californicus (Moq.) Wats., Bot. Calif. 2:42. 1880.

Amaranthus carneus Greene, Pittonia 2:105. 1890.

Prostrate annual, much-branched herb, the branches numerous, stout, 10–30 cm. long; leaves obovately lanceolate, obtuse, mucronate, pale green, glabrous, frequently white-margined, 4–15 cm. long, petiolate; flowers in axillary clusters, monoecious, green or reddish green; bracts lanceolate, subulate at tip, about 1 mm. long; sepals of staminate flowers 3, membranous; stamens 3; sepals of pistillate flowers usually 1 or wanting, sometimes 2 or 3; utricle smooth; seeds reddish, rotund.

Moist soils and floodlands, dried-up ponds: southern California north to Washington.

Amaranthus Palmeri Wats. occurs as a "careless weed" of the bottomlands of southern California and desert regions eastward.

Amaranthus graecizans L. is a common tumbleweed in cultivated areas and in floodlands throughout the state.

BATIDACEAE

A monotypic family consisting of the following genus and species.

BATIS

1. Batis maritima L., Syst. Nat., 10th ed., 2:1380. 1759.

Woody, perennial, maritime plant, 1–10 dm. tall, the branches erect or ascending; leaves without stipules, opposite, entire, fleshy, linear-oblanceolate, 1–3 cm. long, subterete; flowers unisexual in sessile or subsessile, axillary, spikelike catkins subtended by bracts; staminate spikes ovoid-cylindric, 5–10 mm. long; staminate flowers with calyx shallowly 2-lipped, the stamens exserted, 4 or 5, alternating with and longer than the triangular staminodia; pistillate spikes short-pedunculate, the pistillate flowers lacking calyx and corolla but subtended by small, nonimbricated, deciduous bracts; pistil 1; ovary superior, 4-loculed, each locule bearing 1 ovule; style 1; stigma capitate; fruit fleshy, compound, consisting of several united ovaries.

Coastal salt marshes: Los Angeles and south along the coast; tropics and subtropics of the Western Hemisphere, Galapagos and Hawaiian Islands.

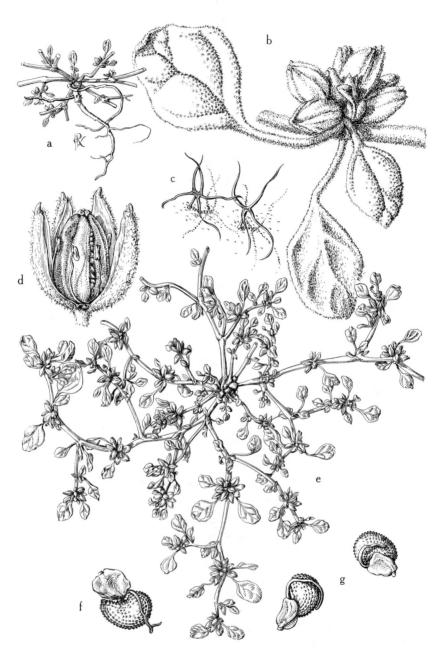

Fig. 222. *Glinus lotoides: a,* basal part of plant, showing roots, × ⅔; *b,* leaves and glomerule of flowers, the pubescence of stellate hairs, × 3; *c,* stalked, stellate hairs, × 40; *d,* flower, 1 sepal removed to show ovary and stamens, × 6½; *e,* plant, as seen from above to show prostrate, spreading habit, × ⅔; *f* and *g,* mature seeds, showing strophiole and funiculus, × 20.

AIZOACEAE. MESEMBRYANTHEMUM FAMILY

Ours annual or perennial herbs with prostrate or trailing stems. Leaves commonly opposite but sometimes whorled or alternate, often succulent. Flowers solitary or clustered in the leaf axils, regular, perfect, hypogynous, perigynous, or epigynous. Sepals 4 or 5. Petals none or many. Stamens 3 to many. Styles 2 to several. Fruit in ours chiefly capsular and dehiscent, sometimes fleshy.

Leaves whorled, more than 2 at a node; capsules loculicidally dehiscent.
    Plants glabrous; stems threadlike; flowers pedicellate .... *Mollugo*
    Plants densely soft-pubescent; stems stout; flowers in glomerules .... *Glinus*
Leaves opposite, 2 at a node; capsules circumscissile.
    Upper part of the capsule corky-thickened, sometimes bilobed; plants fleshy; paired
        leaves unequal in size .... *Trianthema*
    Upper part of the capsule hyaline-membranous.
        Paired leaves equal in size or nearly so, fleshy; hyaline petiole margins entire; styles
           3–5 .... *Sesuvium*
        Paired leaves unequal in size, not fleshy, the larger leaf with a petiole nearly twice
          as long as that of the smaller one; hyaline petiole margins laciniate; styles 2
                            Cypselea

CYPSELEA

1. Cypselea humifusa Turpin, Ann. Mus. Hist. Nat. Paris 7:219, pl. 121. 1806.

Prostrate, matted annual 2–10 cm. across; leaves opposite, glabrous, the opposing members of a pair unequal in size, the larger one with blade and petiole to 12 mm. long, the smaller one about ½ that length, at successive nodes on a given stem the smaller leaf occurring on alternate sides and bearing a short lateral stem in its axil; petioles with conspicuous, hyaline, laciniate margins at the base; flowers 2 mm. long, solitary in the angle between the short lateral stem and the main stem, on a short pedicel, subtended by a scarious, laciniate bract; calyx of 4 or 5 lobes, unequal, scarious-margined, free from the ovary; petals none; stamens 1–3; ovary globose, 1-celled; style 2-cleft; seeds numerous, reddish brown.

Locally established on low ground: San Joaquin Valley and Santa Cruz County; native of the West Indies.

GLINUS

1. Glinus lotoides L., Sp. Pl. 463. 1753. Fig. 222.

Prostrate, or sometimes ascending, much-branched annual; herbage soft-tomentose with branched hairs; stems 10–20 cm. long; leaves whorled, slender, petioled, orbicular to obovate, the apex rounded or acute, the blade 6–15 mm. long; flowers in dense glomerules in upper leaf axils, pediceled or subsessile; calyx free from ovary, the sepals 5, distinct, broadly oblong, 5–7 mm. long; petals none; stamens 5–10, rarely more; stigmas 3, short, sessile; ovary 3-celled;

capsule loculicidally dehiscent; seeds blackish with a strophiole, the funiculus conspicuous, coiled around seed.

Locally established in moist flats and along marsh and lake margins: Central Valley, Lake County, and Santa Clara County; native of Europe.

MOLLUGO

1. Mollugo verticillata L., Sp. Pl. 89. 1753.

Prostrate, glabrous annual with numerous, slender, unequally forked stems 10–20 cm. long, not succulent; leaves whorled, 5 or 6 in each whorl, entire, unequal in size, spatulate; pedicels slender; sepals 5, white inside, 1.5–2 mm. long, oblong; petals none; stamens 3–5, hypogynous; ovary 3-celled, superior; stigmas 3, short, nearly sessile; capsule loculicidally dehiscent, many-seeded; seeds not strophiolate.

Waste places, cultivated ground, lake margins, and lowlands: throughout California; native of tropical America.

SESUVIUM

1. Sesuvium sessile Pers., Syn. Pl. 2:39. 1807. Lowland purslane.

Glabrous, fleshy, prostrate, freely branching herb; leaves fleshy, opposite, the members of a pair nearly equal in size, broadly spatulate, 1–4 cm. long, the petiole with broad hyaline entire margin at base; flowers solitary, sessile or short-stipitate, subtended by caducous bract; calyx tube adnate to ovary, the calyx lobes 5, scarious-margined, 6 mm. long, pink within, the central part extended into a linear fleshy horn near the apex; petals none; perigynous stamens numerous, the filaments red, united at base; ovary 3- to 5-celled, half-superior; capsule circumscissile; seeds numerous, smooth.

Low alkaline flats and floodplains: Sacramento Valley to southern California; western and central United States, south to Mexico, Brazil.

TRIANTHEMA

1. Trianthema Portulacastrum L., Sp. Pl. 223. 1753.

Decumbent, succulent annual herb, branching much as in *Cypselea;* leaves opposite, simple, petioled, the blades orbicular-ovate, the opposing members of a pair unequal in size; flowers solitary, sessile, inconspicuous, borne under the axillary sheath at every node; sepals lanceolate, mucronate; petals none; stamens 5 or 6; ovary 2-celled or with 1 cell aborted; capsule several-seeded, circumscissile, the thickened, often bilobed, crestlike apical part containing a single embedded seed falling with the capsule valve.

Weed along irrigation ditches: Imperial Valley; to Texas and Florida, Baja California in Mexico, and the West Indies.

PORTULACACEAE. PURSLANE FAMILY

Succulent glabrous herbs with simple, entire, opposite, alternate, or basal leaves. Flowers perfect, regular, usually hypogynous. Sepals 2, free or cohering at base. Petals 3–16. Stamens few to many and opposite petals. Anthers versatile. Ovary superior or partly inferior. Fruit a dehiscent capsule.

Capsule 2- or 3-valved .... *Montia*
Capsule circumscissile .... *Portulaca*

MONTIA

Plants annual, or perennial by rhizomes or stolons, succulent, glabrous, often glaucous. Leaves basal, opposite, or alternate. Inflorescence paniculate or racemose. Pedicels spreading or recurved in fruit. Flowers white or pinkish lavender. Sepals 2, subequal in size, persistent. Petals 3–5, often unequal in size. Stamens 3 or 5. Ovary 1-celled; styles 3; ovules 3. Capsule globose or ovoid, 3-valved. Seeds 1–6.

Petals about 2 mm. long, unequal in size, more or less united at base; annual
1. *M. fontana*
Petals 5–10 mm. long, equal in size, free or united at base; perennial.
Stem with more than 1 pair of opposite leaves; petals slightly emarginate to entire
2. *M. Chamissoi*
Stem with 1 pair of opposite leaves; petals conspicuously emarginate ... 3. *M. sibirica*

1. Montia fontana L., Sp. Pl. 87. 1753. Fig. 223.

Low, delicate, procumbent or ascending annual, rooting at the nodes, aquatic or terrestrial; stems 3–10 cm. long; leaves opposite, subsessile, linear-spatulate to obovate, 5–15 mm. long; inflorescence 3- to 9-flowered, leafy; sepals 2, 1–3 mm. long; corolla white, 3- to 5-lobed, slightly longer than the calyx; stamens 3–5; ovary 1-celled; ovules 3; capsule 3-lobed, 1–2 mm. high; seeds black, dull, covered with muricate tubercles.

Muddy stream margins and floating in pools, where it may become terrestrial as the pools dry: throughout California; cosmopolitan.

2. Montia Chamissoi (Ledeb.) Durand & Jack., Index Kew. Suppl. 1:282. 1903. Fig. 224.

Perennial by bulblets produced at the ends of runners; stems prostrate or ascending, 5–30 cm. long, leafy particularly at tip; leaves opposite, petiolate, oblanceolate, 1–5 cm. long; inflorescence axillary or terminal, 3- to 8-flowered, 1 or 2 bracts at base of raceme or bractless; pedicels recurved in fruit; sepals

Fig. 223. (For explanation, see facing page.)

orbicular, about 2 mm. in diameter; petals white or pink, entire, or barely retuse; stamens 3–5; ovules 3; capsule 1–1.5 mm. long; seeds black, shining, with low tubercles.

Wet meadows, stream banks, bogs: throughout California; north to Alaska, east to Minnesota, and southeast to northern Mexico.

3. Montia sibirica (L.) Howell, Erythea 1:39. 1893. Fig. 225.

Stems several, erect, 1.5–5 dm. tall, the short crown or slender rootstock persisting; basal leaves several, the blades 2–6 cm. long, ovate-truncate or attenuate at base, the apex acute, the petioles 8–20 cm. long; stem leaves similar but sessile, opposite, 2–5 cm. long; raceme open, 6–30 cm. long, occasionally branched, 10- to 30-flowered, bracteate; pedicels 2–6 mm. long; petals 6–10 mm. long, pale pink-lavender or white, often with pink-purple venation, notched at apex; capsule about as long as the calyx; seeds dark, shining, 2–2.5 mm. long, low-tuberculate.

Moist shady places and bogs: along the coast from Santa Cruz County northward in California; north to Alaska and Siberia, east to Idaho and Montana.

PORTULACA

1. Portulaca oleracea L., Sp. Pl. 445. 1753.

Fleshy, glabrous, prostrate, annual herb; stems 10–20 cm. long; leaves alternate, cuneate to obovate, 5–20 mm. long, clustered at ends of branches, often with a lacerate membrane in the axil; flowers sessile, yellow, 2–6 mm. broad; sepals 2, united below, partly adnate to the ovary; petals 4–6, inserted on the calyx, 3 mm. long, opening only in bright sunlight; stamens 7 to many; style 3- to 6-parted; capsule globular, 1-celled, many-seeded, circumscissile.

Moist, waste areas, floodlands: throughout California. Naturalized from Europe.

CARYOPHYLLACEAE. CHICKWEED FAMILY, PINK FAMILY

Annual or perennial herb with nodose stems. Leaves opposite, simple and entire. Flowers hypogynous to perigynous-epigynous. Sepals 4 or 5, free or united. Petals as many as the sepals or fewer, sometimes absent, entire, notched or 2-lobed. Stamens 1–2 times as many as the petals, sometimes fewer. Styles

---

Fig. 223. *Montia fontana: a,* young pistil, showing the papillose stigmas, × 20; *b,* habit, showing roots at the nodes, the subsessile leaves, and the nodding inflorescences, × ⅘; *c,* mature seed, × 20; *d,* mature capsule before dehiscence, × 8; *e,* dehisced capsule with seeds, × 8; *f,* empty capsule, the valves having closed up by a sudden twist, thus discharging the seeds, × 8; *g,* flower, top view, × 12; *h,* flower, spread open, showing the stamens inserted on the small inner petals, × 12; *i,* flower, × 12; *j,* corolla, the calyx removed, × 12; *k,* inflorescence with buds, × 4.

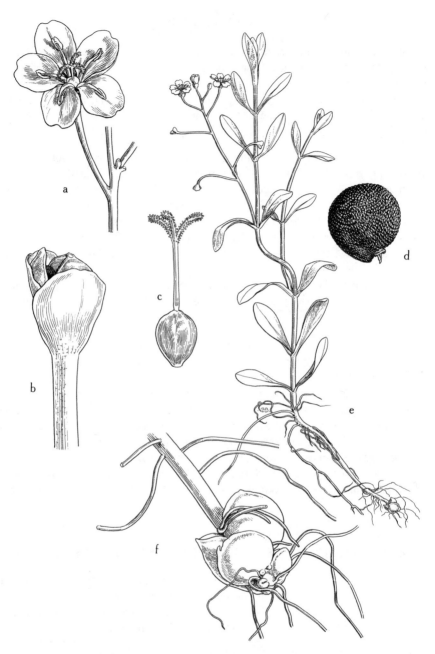

Fig. 224. *Montia Chamissoi: a,* flower, × 2½; *b,* calyx and capsule at time of dehiscence, × 12; *c,* pistil, showing the long, papillose stigmas, × 8; *d,* mature seed, the tubercles beadlike, × 16; *e,* habit, × ⅔; *f,* bulblet at end of runner, × 3.

Fig. 225. *Montia sibirica:* *a* and *b,* flower at successive stages of anthesis, × 2½; *c,* mature seed, × 8; *d,* calyx and young capsule, × 4; *e,* valves of capsule after dehiscence and discharge of seeds, × 6; *f,* habit, × ⅖.

Fig. 226. *Arenaria paludicola: a,* calyx and mature capsule, × 6; *b,* habit, basal part of plant, showing roots at the nodes, × ⅘; *c,* habit, upper part of stem, showing flowers solitary in leaf axils, × ⅘; *d,* mature seed, globose, shiny, × 32; *e,* flower, × 3; *f,* connate leaf bases, × 8.

2–5, free or united. Ovary 1- to 5-celled. Fruit a dehiscent capsule with entire or bifid valves. Seeds with curved embryos.

Leaves without stipules.
  Petals 2-cleft or -parted .... *Stellaria*
  Petals entire, emarginate, or absent.
    Styles as many as sepals .... *Sagina*
    Styles fewer than sepals .... *Arenaria*
Leaves with scarious stipules .... *Spergularia*

A species which grows in California but is not specially treated here is the weed *Saponaria officinalis* L. This plant is occasionally encountered in floodlands and along irrigation ditches in the northern part of the state. It may be recognized by its coarse habit and its several to many stems, each topped by a broad cyme of pink flowers.

ARENARIA

Annual or perennial herbs, low, often tufted or matted. Leaves linear, subulate, or ovate. Flowers small, solitary or cymose. Sepals 5. Petals 5, entire. Stamens 10. Styles 3. Capsule globose, dehiscent by 3 valves.

Annual; leaves ovate; flowers cymose; capsule valves entire .... 1. *A. serpyllifolia*
Perennial; leaves linear; flowers solitary; capsule valves 2-toothed or 2-cleft
                                                         2. *A. paludicola*

1. Arenaria serpyllifolia L., Sp. Pl. 423. 1753.

Low, branching annual, 5–15 cm. tall; stems several from the base, retrorsely puberulent; leaves ovate, acute, 2–4 mm. long; flowers in open, leafy-bracted cymes; pedicels 4–10 mm. long; sepals 5, ovate-lanceolate, 3 mm. long; petals entire or nearly so, ½ as long as the sepals; stamens 10; styles 3; capsule globose, the valves 2-toothed or 2-cleft.

Along river bars: North Coast Ranges; native of Europe.

2. Arenaria paludicola Robins., Proc. Am. Acad. Arts & Sci. 29:298. 1894. Fig. 226.

Glabrous, flaccid perennial; stems commonly simple, weak, sulcate, rooting at the lower nodes, 3–10 dm. long, or longer when growing in swamps among tall marsh plants; leaves linear or lanceolate, 1–3 cm. long, 2–6 mm. wide, somewhat connate at the base; peduncles solitary in leaf axils; pedicels 1.5–4 cm. long; sepals herbaceous, 3–4 mm. long; petals white, 6 mm. long; capsule shorter than the calyx.

Swamps, widely distributed but local along the coast: southern California, Oso Flaco Lake in San Luis Obispo County; north to Puget Sound, Washington.

SAGINA

Small, often inconspicuous herbs, annual or perennial. Leaves subulate-filiform to linear, opposite, scarious-connate at base. Flowers terminal, usually solitary and long-pediceled. Sepals free, 4 or 5 or none. Petals usually shorter than sepals, entire, white. Stamens 3–10. Styles 4 or 5. Capsule ovoid or spheroid. Seeds many, smooth.

Sepals 4.
 Petals none; lower leaves ciliate-margined .... 1. *S. apetala*
 Petals present; lower leaves not ciliate .... 2. *S. procumbens*
Sepals 5.
 Stems filiform.
  Annual, without sterile rosette of leaves at base .... 3. *S. occidentalis*
  Perennial, with sterile rosette at base .... 4. *S. saginoides*
 Stems stout, with basal rosette of leaves .... 5. *S. crassicaulis*

1. Sagina apetala Ard., Animad. Bot. Sp. Alt. 2:22, pl. 8. 1764.

Slender, erect annual, 3–5 cm. tall; herbage minutely glandular-pubescent to nearly glabrous; leaves linear, ciliate toward base, the tip with a minute bristle; sepals 4; petals none; capsule slightly longer than calyx.

Wastelands and roadsides in wet places; native of Europe.

Our material has been segregated as *Sagina apetala* var. *barbata* Fenzl ex Ledeb. (Fl. Ross. 1:338, 1842).

2. Sagina procumbens L., Sp. Pl. 128. 1753.

Annual or perennial herb, branching from base; stems 3–8 cm. tall; leaves often crowded, longer than internodes, linear, glabrous, 2–8 mm. long; flowers solitary on terminal capillary peduncles; sepals 4, an occasional plant with a few of its flowers having 5 sepals; petals shorter than sepals, rarely none; capsule barely longer than the sepals, narrowly ovoid.

Marshy spots along coast: Mendocino County north to British Columbia; native of Eurasia.

3. Sagina occidentalis Wats., Proc. Am. Acad. Arts & Sci. 10:344. 1875.

Low, glabrous annual, 5–10 cm. tall; stems decumbent or ascending; lower leaves filiform, 6–12 mm. long, the upper ones subulate, 4–6 mm. long; flowers terminal, the pedicels 6–12 mm. long; sepals 5, 1.5 mm. long; petals 5, as long as sepals or slightly shorter, white; stamens 3–10; capsule 4- or 5-valved, many-seeded, 2 mm. long.

Moist ground, borders of salt marshes: throughout California at medium altitudes; north to Canada, east to Idaho.

4. Sagina saginoides (L.) Britt., Mem. Torrey Club 5:151. 1894.

Sagina Linnaei Presl, Rel. Haenk. 2:14. 1835.

Tufted perennial, 2–10 cm. tall, glabrous; leaves linear, subulate, 4–10 mm. long; flowers terminal or occasionally 1 or 2 in axils of upper leaves; sepals 5; petals shorter than sepals; stamens 10; capsule ovoid, longer than sepals.

Mountain meadows and bogs: throughout California; Eurasia.

Our material has been segregated as *Sagina saginoides* var. *hesperia* Fern. (Rhodora 27:131, 1925).

5. Sagina crassicaulis Wats., Proc. Am. Acad. Arts & Sci. 18:191. 1883.

Perennial herb from stout root, glabrous and succulent; stems distinctly not filiform, 4–15 cm. tall; leaves in a basal rosette and cauline, linear, 15–30 mm. long, with broad, connate, scarious bases; sepals 5; petals subequal to sepals in length, white; stamens 10; capsule ovoid.

Marshes and seepage places along coast: central California to Alaska.

SPERGULARIA

Annual or perennial herbs of alkaline or saline soils. Leaves linear, subulate, or semiterete, often fleshy. Stipules scarious. Flowers pink or whitish in terminal racemes or cymes. Sepals 5. Petals 5 or fewer, distinct. Stamens 2–10. Styles 3 (rarely 5), distinct to base. Capsule 3-valved.

Leaves densely fascicled; stipules lanceolate-acuminate; commonly perennial.
    Plants prostrate; roots fibrous; seeds reticulate .... 1. *S. rubra*
    Plants erect or ascending; roots fleshy; seeds smooth .... 2. *S. macrotheca*
Leaves not densely fascicled; stipules deltoid; annual .... 3. *S. marina*

1. Spergularia rubra (L.) J. & C. Presl, Fl. Cech. 94. 1819.

Short-lived, prostrate perennial or annual from fibrous roots; stems usually unbranched, 6–25 cm. long, wiry, many from a matlike base; herbage glabrous or slightly glandular; leaves linear, 3–10 mm. long; stipules ovate, scarious, 4 mm. long, conspicuous; pedicels slender, 4–10 mm. long; sepals oblong, 3–4 mm. long; petals red, as long as the sepals; stamens 6–10; capsule acute, about as long as the calyx.

Waste places, floodplains: central and northern California; introduced and slowly spreading. Native of Eurasia.

2. Spergularia macrotheca (Hornem.) Heynh., Nom. 2:689. 1840. Fig. 227.

Stout perennial, 1–3 dm. tall, erect or ascending from a thick root; herbage glandular, pubescent; leaves narrowly linear, 2–5 cm. long; stipules 5–6 mm. long; flowers in terminal cymes, the branches racemose; pedicels 4–14 mm. long; sepals scarious-margined, 6–8 mm. long; petals about as long as the sepals, pink or rose; capsule about as long as the calyx.

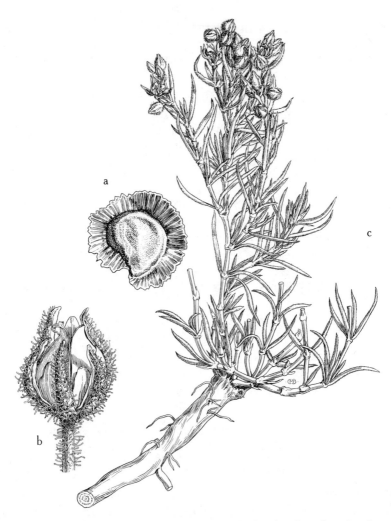

Fig. 227. *Spergularia macrotheca: a,* mature seed, showing broadly winged margin, × 20; *b,* calyx with dehisced capsule, showing the persistent petals, × 4; *c,* habit, showing the thick root, the glandular, pubescent herbage, and the flowers in terminal cymes, × ⅔.

Margins of salt marshes near the coast: Washington to Baja California, Mexico.

A variable species with several named varieties.

3. Spergularia marina (L.) Griseb., Spicil. Fl. Rumel. 1:213. 1843.

Spergularia salina J. & C. Presl, Fl. Cech. 95. 1819.

Herbage fleshy, nearly glabrous; stem branching, erect, occasionally prostrate, 7–15 cm. long; leaves usually shorter than the internodes, narrowly linear, 2–4 cm. long; flowers in terminal leafy cymes or racemes; sepals ovate to ovate-lanceolate, 2–5 mm. long; petals only slightly shorter than sepals, pink; stamens 2–5; capsule slightly longer than the calyx.

Alkaline places, salt marshes, seashore: Sacramento and San Joaquin valleys, also coastal areas to southern California; north to Washington, east to Atlantic.

STELLARIA

Low, diffusely branched, herbaceous perennials. Leaves opposite, linear to lanceolate to ovate. Flowers in the axils of upper leaves and bracts or sometimes in terminal cymes or cymose panicles, hypogynous. Sepals 5. Petals 5. white, cleft or divided to near base. Stamens 3–10. Ovary 1-celled; styles 3–5. Capsule ovoid to spheroid, dehiscent, splitting from above into 6–10 segments. Seeds smooth or variously roughened.

In addition to the species herein treated, *Stellaria Jamesiana, S. umbellata,* and *S. crispa* are to be found in California. These latter species may be expected in bogs in the mountains.

Flowers in the axils of foliage leaves.
    Petals much shorter than the sepals or sometimes absent; herbage essentially glabrous but sometimes pilose . . . . 1. *S. calycantha*
    Petals equal or subequal to the sepals in length; herbage villous-pubescent
                                                          2. *S. littoralis*
Flowers in the axils of much reduced, scarious bracts . . . . 3. *S. longipes*

1. Stellaria calycantha (Ledeb.) Bong., Mém. Acad. St. Pétersb. VI. Math. Phys. Nat. 2:127. 1832.

Stellaria borealis Bigel., Fl. Boston., 2d ed., 182. 1824.

Plant from slender, perennial rhizomes; stems many, slender, erect or ascending, 1–3 dm. tall, simple or branched, glabrous or somewhat pilose; leaves sessile, linear-lanceolate to lanceolate, ciliolate at base, 7–30 mm. long; flowers solitary in the upper axils, on slender pedicels 2–3 cm. long, or occasionally in a terminal leafy-bracted cyme; sepals lanceolate, acute, scarious-margined; petals in ours shorter than sepals, occasionally absent, 2-cleft to near base; capsules ovoid; seeds brown, reticulate.

Fig. 228. *Nuphar polysepalum: a,* habit, showing rootstock and fruit, × ⅖; *b,* seed, × 4; *c,* flower, showing the conspicuous, thick sepals, the numerous arching stamens, and the radiate stigma, × ⅖; *d,* flower (longitudinal section), the short petals concealed by the stamens, × ⅖.

Marshy ground: chiefly in the mountains but reaching the coast in Mendocino, Humboldt, and Del Norte counties; circumpolar.

An exceedingly variable species which has been subdivided into many subspecies.

2. Stellaria littoralis Torr., Pacif. R. R. Rep. 4:69. 1857.

Low, dichotomously branched, villous, pubescent perennial; stems 2–5 dm. long; leaves sessile, crowded, ovate-lanceolate, acute, 15–40 mm. long; flowers in a terminal, compound cyme; bracts foliaceous; pedicels 6–15 mm. long; sepals 5, distinct, lanceolate with broad scarious margins, 4–6 mm. long; petals as long as or slightly longer than the sepals, bifid nearly to the base, white; stamens 10, hypogynous; ovary 1-celled with numerous ovules, styles 3; capsule longer than the calyx, ovoid or oblong, dehiscent by 6 valves; seeds brown, reticulate.

Bogs and marshes near the seacoast: Humboldt County to San Francisco County.

3. Stellaria longipes Goldie, Edinb. Philos. Jour. 6:327. 1822.

Tufted perennial; stems erect or ascending, 1–3 dm. tall, simple or branched; herbage essentially glabrous; leaves sessile, linear-lanceolate; flowers in terminal bracteate cymes or solitary on long terminal pedicels; bracts scarious and reduced; sepals lanceolate, acute, scarious-margined; petals 2-cleft, longer than sepals; capsule ovoid, dark-colored, longer than calyx; seeds smooth.

Bogs and seepage areas: in the mountains; New Mexico to Alaska and east to Atlantic.

NYMPHAEACEAE. WATER-LILY FAMILY

Perennial aquatic herbs from stout rhizomes. Leaves long-petioled, the blades floating, peltate or cordate. Flowers perfect, solitary in leaf axils, usually showy, hypogynous or perigynous. Perianth segments 4–6 or numerous, the sepals and petals often intergrading and, in some, petals and stamens intergrading. Anthers introrse, numerous. Carpels 8 or more, united or embedded in an expanded torus. Stigmas often radiate or peltate. Ovules 1 to many.

NUPHAR

1. Nuphar polysepalum Engelm., Trans. Acad. Sci. St. Louis 2:282. 1865. Yellow pond lily. Fig. 228.

Nymphaea polysepala Greene, Bull. Torrey Club 15:84. 1888.

Rhizomatous, aquatic, perennial herb; leaves long-petioled, the blade floating or raised above the surface, broadly oval, 15–50 cm. long, with a deep, wedge-shaped sinus; flowers long-peduncled, usually raised above the leaves,

Fig. 229. *Brasenia Schreberi: a,* fruit (longitudinal section), × 2; *b,* flowering branch, all the submersed parts clothed with a thick, translucent, gelatinous coating, × 4/5; *c,* mature seed, × 4; *d,* roots, × 4/5; *e,* flower, × 3; *f,* mature fruits, × 1 1/5; *g,* mature fruit, × 2; *h,* habit, showing peltate leaves, × 1/5.

sometimes floating on surface; sepals conspicuous, thick, 6–12, yellow, sometimes tinged with orange-red, persisting on the fruit until they decay, 3–5 cm. long; petals numerous, inconspicuous, usually concealed by the stamens; stamens numerous; anthers reddish; pistil 1, urceolate, flat-topped with a radiate stigma; fruit ovoid-baccate, 3–5 cm. long; seeds numerous, with endosperm.

Common in fresh water: throughout northern California; western states to Alaska.

CABOMBACEAE

Perennial aquatic herbs. Stems rhizomatous and ascending, thickly coated with mucilage. Leaves floating, peltate or sometimes the submersed leaves dissected. Flowers in leaf axils, solitary, perfect, hypogynous, regular, the perianth segments 6. Petaloid stamens 3 to numerous, the anthers opening lengthwise. Carpels free, 2 to numerous, 1-celled, 1- to 3-seeded.

Members of the Cabombaceae are often included in Nymphaeaceae, but the free, simple carpels and extrorse stamens clearly separate the Cabombaceae from that family.

BRASENIA

1. Brasenia Schreberi J. F. Gmelin, Syst. Veg. 1:853. 1796. Fig. 229.

Aquatic, rooted in mud, the stems 3–20 dm. long; herbage reddish brown, the submersed parts clothed with a thick translucent gelatinous coating; leaves alternate, long-petioled, floating, the blades oval, peltate, 4–12 cm. long, 3–8 cm. broad, the sides unrolling in vernation; flowers solitary in leaf axils, emersed during anthesis, submersed at all other times; sepals 3; petals 3, scarcely differentiated, reddish purple, linear-oblong, 10–15 mm. long; stamens 12–18, the filaments slender, the anthers extrorse; pistils 4–18, separate; fruit indehiscent, with a stout beak.

Ponds and slow streams: widely scattered from San Joaquin and Butte counties in the Central Valley to Lake and Mendocino counties in the North Coast Ranges, and to Fresno, Tuolumne, and El Dorado counties in the Sierra Nevada; north to British Columbia, east to Atlantic.

Tokura reports that in Japan the flowers of this species each open twice, on two successive days (Jour. Jap. Bot. 13:829–839, 1939, English summary). On the first day the flower is pushed above the water about 6 A.M. and opens within an hour, and during this period pollination is effected. The flower closes at about 9 A.M. and is then withdrawn under the water. On the next day the same flower is again pushed above the water, this time higher above the surface. The stamens then dehisce, and after a few hours the flower is again withdrawn, never to emerge again.

We depart from the current local practice of treating this genus as a member

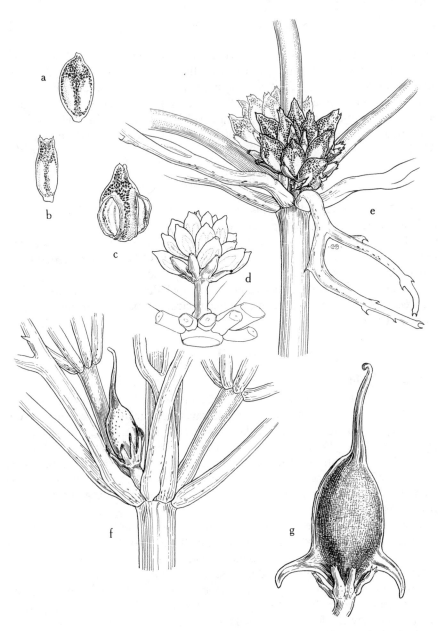

Fig. 230. *Ceratophyllum demersum: a–c,* stamens at various stages of development, × 8; *d,* staminate flower, the involucre calyx-like, × 4; *e,* node with leaves, branches, and a pair of staminate flowers arising on opposite sides of stem, × 4; *f,* node with leaves, branches, and solitary pistillate flower, × 8; *g,* mature fruit, the basal spines recurved, × 6.

of the water-lily family (Nymphaeaceae), in which it has been placed in the subfamily Cabomboideae. Its free carpels suggest a close affinity to Ranalean groups, whereas the fruit of the water-lily family suggests close affinity with the Rhoedalean or poppy order. We therefore regard Cabombaceae as a separate family.

CERATOPHYLLACEAE

Perennial, rootless, aquatic plants. Leaves sessile, whorled, dichotomously divided into filiform or linear segments. Flowers usually unisexual, small, regular, subtended by an 8- to 12-cleft, calyx-like involucre, the perianth lacking; staminate flowers with 10–20 stamens, the pistillate flowers with a single 1-celled ovary. Fruit a nutlet with a persistent style.

CERATOPHYLLUM

1. Ceratophyllum demersum L., Sp. Pl. 992. 1753. Hornwort. Fig. 230.

Submersed or at length free-floating aquatic; stems to 3 m. long, densely branched; leaves verticillate, 9 or 10 in a whorl from a swollen node, 15–25 mm. long, 2- to 4-forked, the lobes conspicuously serrate, dentate on the outer and lower margin, each tooth ending in a bristle-like cusp; flowers monoecious or sometimes dioecious, the staminate flowers in pairs on opposite sides of the stem, short-stalked and bearing a calyx-like involucre of 8–12 short, stout, oblong segments, each 2- or 3-toothed at the apex, the pistillate flowers usually solitary and consisting of a unilocular ovary and a long style; fruit ellipsoid, flattened, 4–6 mm. long, 1-celled and beaked by the persistent style and developing 2 recurved lateral basal spines.

Common in quiet, fresh-water pools and slow streams throughout California; world-wide.

Exceedingly variable in the form and toothing of its foliage.

RANUNCULACEAE

Herbs with alternate or basal leaves. Flowers usually perfect, the parts distinct. Sepals 3 to about 15, usually 5. Petals 3–16 or rarely none, usually 5. Stamens usually many, occasionally reduced to few. Pistils 3 to many, 1-celled, maturing to an achene, follicle, or berry.

Fruit a several-seeded follicle.
  Flowers regular.
    Petals without spurs .... *Caltha*
    Petals with spurs .... *Aquilegia*

Fig. 231. *Caltha biflora: a,* habit, × ⅖; *b,* young fruiting head, × 4; *c,* mature follicle, showing dehiscence, × 3; *d,* mature seed, × 12.

Flowers irregular.
 Upper sepal spurred . . . . *Delphinium*
 Upper sepal produced as a hood . . . . *Aconitum*
Fruit a 1-seeded achene or follicle.
 Sepals with spurs, white and petaloid . . . . *Myosurus*
 Sepals not spurred, usually green . . . . *Ranunculus*

The genera *Thalictrum* and *Kumlienia,* not treated here, are also found in California, often in wet habitats.

ACONITUM

1. Aconitum columbianum Nutt. in Torr. & Gray, Fl. N. Am. 1:34. 1838.

Perennial plant resembling *Delphinium* in habit and foliage; sepals 5, the upper one hooded or helmet-shaped; petals 2–5, the upper pair hooded, long-clawed, and often concealed in the sepal hood, the 3 lower petals minute; stamens many; pistils 3; fruit a follicle.

Mountain stream banks and bogs: high Sierra Nevada; north to British Columbia, east to the Rocky Mountains.

AQUILEGIA

1. Aquilegia eximia Van Houtte ex Planch., Fl. Serres 12:15. 1857.
 Aquilegia Tracyi Jepson, Fl. West. Middle Calif., 2d ed., 165. 1911.

Caespitose plant, 5–10 dm. tall, glandular-pubescent throughout; leaves chiefly basal, triternately compound; flowers red with yellow markings; sepals 5, regular, petaloid; petals 5, spurred, the orifice cut backward and not erect, the lamina thus obsolete; stamens many, some without anthers; carpels 5; fruit follicular, many-seeded.

Stream banks, chiefly in serpentine outcrops: Coast Ranges of central California.

CALTHA

1. Caltha biflora DC. subsp. **Howellii** (Huth) Abrams, Illus. Fl. Pacif. States 2:175. 1944. Fig. 231.

Glabrous, succulent perennial herb, from a short, vertical rootstock bearing fibrous roots; leaves simple, reniform-orbicular, the basal leaves long-petiolate, obscurely repand-crenate, 5–8 cm. broad, the upper leaves smaller; flower solitary on a 1-leaved scape; sepals 6–9, large, white, petaloid, 10–15 mm. long; petals absent; stamens numerous; carpels several to many; fruit a follicle opening by a ventral suture, many-seeded; seeds shiny brown with irregular ridges.

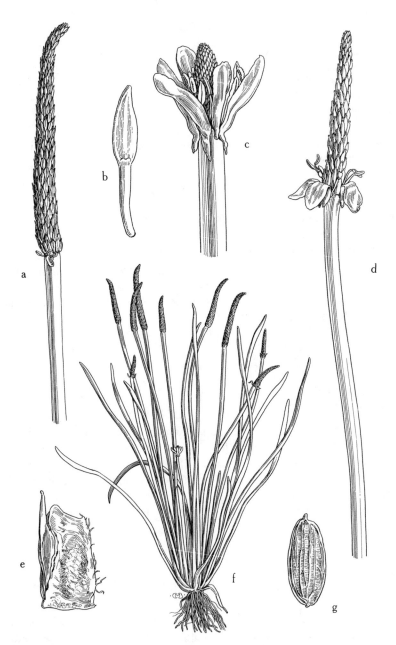

Fig. 232. *Myosurus minimus: a,* mature fruiting spike, $\times$ 2; *b,* petal and terete claw, showing nectariferous pit, $\times$ 20; *c,* flower, the spurs of sepals short, $\times$ 6; *d,* young fruiting spike, $\times$ 4; *e,* mature achene, its beak parallel with back, $\times$ 16; *f,* habit, showing the slender, clavate scapes and the linear, spatulate leaves, $\times$ ⅔; *g,* mature seed, $\times$ 20.

Bogs and marshes: Sierra Nevada, Humboldt County, Siskiyou Mountains; north to British Columbia.

DELPHINIUM

Ours herbaceous perennials with erect stems and cauline and basal, palmately lobed leaves. Flowers showy, irregular. Sepals 5, petaloid, the upper one spurred. Petals 2–4, the upper and lower pairs unlike. Stamens many. Pistils 1–3. Fruit a follicle. Seeds often winged or sculptured.

Leaf blades cuneate at base; plants 4–6 dm. tall; North Coast Ranges .. 1. *D. uliginosum*
Leaf blades not cuneate at base; plants 8–25 dm. tall; high montane ... 2. *D. scopulorum*

1. Delphinium uliginosum Curran, Bull. Calif. Acad. Sci. 1:151. 1885.

Erect perennial herb from woody, fibrous roots, 4–6 dm. tall; herbage nearly glabrous; basal leaves 2–6 cm. broad, cuneate, 3-cleft, segments entire or toothed; flowers in a strict raceme; pedicels shorter than spur; sepals blue, 8–12 mm. long, the one with a spur 12–15 mm. long; upper petals white-tipped, the lower ones violet, ciliate on margins; fruit of 3 follicles, 8–10 mm. long; seeds beset with minute, blunt processes.

Margins of streams, rivulets, and boggy spots: on serpentine outcrops in the North Coast Ranges.

2. Delphinium scopulorum Gray var. **glaucum** (Wats.) Gray, Bot. Gaz. 12:52. 1887.

Stems several to many, from a stout, woody root system, 8–25 dm. tall, glabrous to glaucous, leafy; leaves 8–15 cm. broad, 5- to 7-parted, into cuneate, incised segments, the central segment of each group prominent; flowers in prominent racemes; sepals blue to purple, 8–12 mm. long, the spine stout, equal to the lobes in length; upper petals notched, purple-tipped, the lower petals cleft; follicles 8–10 mm. long; seeds winged at summit.

Mountain streams and wet meadows: throughout California; western states and north to Alaska.

Delphinium scopulorum Gray is a very variable species, in which several subspecies and varieties have been segregated.

MYOSURUS

Small acaulescent annuals (most frequently 3–15 cm. tall), with entire, linear, basal leaves. Flowers solitary, sessile or more often borne on short to elongate scapes. Sepals petaloid, usually 5 (3–8), spurred at base, with 1-, 3-, or 5-nerved blades; petals inconspicuous, 5, sometimes fewer, with a nectariferous pit at the junction of the narrow, terete claw and the short, flat, linear blade. Stamens 5–25. Pistils numerous (10–400), spirally arranged on an elongate

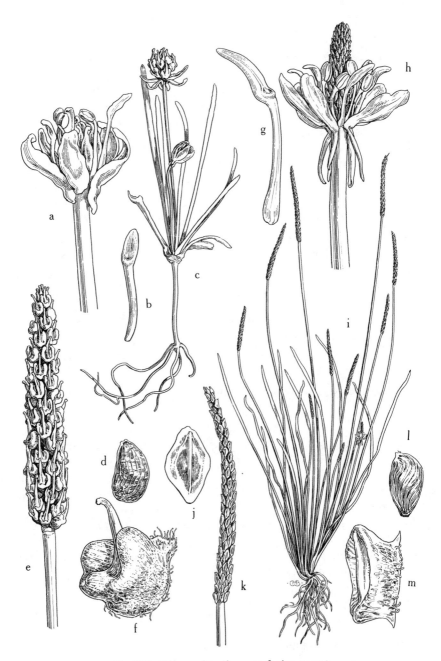

Fig. 233. (For explanation, see facing page.)

receptacle. Achenes with a persistent style, appearing as an appressed or divaricate beak at maturity.

The manuscript for this treatment was revised by Donald E. Stone.

Mature fruiting spikes with the beaks of the fruit absent or closely appressed to the spike, that is, directed in a plane parallel to the axis of the spike, the spike thus appearing smooth.
 Fruiting spikes extending beyond the leaves.
  Mature achenes truncate at lower margin, usually 1.1 mm. or longer (–1.5 mm.), beak of achene appressed, extending slightly beyond body . . . . . 1. *M. minimus*
  Mature achenes rounded at lower margin, usually 1 mm. long or shorter, beak of achene short or lacking . . . . 1a. *M. minimus* var. *filiformis*
 Fruiting spikes shorter than or as long as the leaves, sessile or with stalk of intermediate length . . . . 1b. *M. minimus* var. *apus*
Mature fruiting spikes with divergent beaks, that is, with the beak of each fruit spreading from the plane of the axis of the spike.
 Fruiting spikes extending beyond the leaves.
  Achenes thick, square (that is, as long as broad), with horseshoe-shaped basal and lateral margin surrounding a large beak . . . . 2. *M. cupulatus*
  Achenes longer than broad, lacking a horseshoe-shaped margin.
   Sepal 1-nerved, spikes short (8–10 mm. long) . . . . 3. *M. aristatus*
   Sepal faintly 3-nerved, spikes long (20–50 mm. long)
                3a. *M. aristatus* subsp. *montanus*
 Fruiting spikes shorter than or as long as the leaves.
  Spikes sessile or nearly so . . . . 4. *M. sessilis*
  Spikes not sessile, scape very stout . . . . 4a. *M. sessilis* subsp. *alopecuroides*

1. Myosurus minimus L., Sp. Pl. 284. 1753. Fig. 232.

Plants 3–15 cm. tall; leaves linear-spatulate, 3–10 cm. long; scapes slender, 5–15 cm. long, often slenderly clavate; sepals 5, oblong, the spur linear to short-triangular; petals 5, or occasionally fewer or wanting, the claw terete, the blade elliptic, subequal to the claw in length; stamens 5 to several; spike slender, 1–4 cm. long, parallel-sided or slightly tapering toward apex; achenes many, rounded on back, the persistent style developed into an erect, short, appressed beak parallel with back of achene, the sides of achene with a few hairs, membranous; seed ellipsoid to ovate, microscopically sculptured.

Vernal pools and vernally wet meadows, widely scattered: chiefly in Central Valley and valleys of Coast Ranges to middle altitudes; east to Atlantic, north to Alaska, Europe.

Fig. 233. *Myosurus. a–f, M. cupulatus:* a, flower, the spurs and petals variable in size, × 16; *b,* petal, showing the short blade and the shallow nectariferous pit, × 16; *c,* habit, showing leaves and flowering scapes, × 4; *d,* mature seed, × 20; *e,* mature fruiting spike, × 4; *f,* mature achene, showing the divergent beak arising between thickened margins, × 20. *g–m, M. minimus* var. *filiformis: g,* petal, showing long claw and rounded, nectariferous pit, × 16; *h,* flower, showing long, slender spurs and numerous petals, × 6; *i,* habit, the leaves, scapes, and spikes very slender, × ⅔; *j,* mature achene, abaxial view, × 16; *k,* mature fruiting spike, × 2; *l,* mature seed, longitudinally striate, flattened, and twisted, × 20; *m,* mature achene, lateral view, showing short beak, × 20.

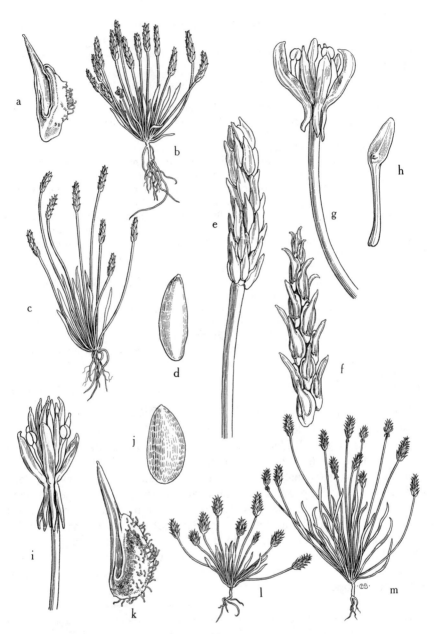

Fig. 234. *Myosurus. a–h, M. aristatus* subsp. *montanus: a,* mature achene with divergent beak and strongly nerved lateral angles, × 8; *b* and *c,* habit variations, scapes extending above the leaves, × ⅖; *d,* mature seed, × 12; *e,* young fruiting spike, × 4; *f,* part of a mature fruiting spike, × 4; *g,* flower, stamens few, × 12; *h,* petal, the nectariferous pit rounded, × 16. *i–m, M. aristatus: i,* flower, the calyx spurs long and slender, × 8; *j,* mature seed, × 16; *k,* mature achene, the beak long and divergent, × 12; *l* and *m,* habit variations, the fruiting spikes bristly, × ⅖.

1a. Myosurus minimus var. **filiformis** Greene, Bull. Calif. Acad. Sci. 1:277. 1885. Fig. 233.

Myosurus lepturus var. *filiformis* Greene in Abrams, Illus. Fl. Pacif. States 2:195. 1944.

Differing from the species in its very slender scape and spike and in the very short to inconspicuous beak of the achene: southern California, San Joaquin Valley, Sacramento Valley, Mendocino County; Baja California, Mexico.

Specimens with very long, slender, conical spikes have been treated as *Myosurus major* Greene [Pittonia 3:257, 1898; *M. minimus* subsp. *major* (Greene) Campbell, El Aliso 2:396, 1952]. The type came from Siskiyou County and is characterized by a pronounced keel on the back of the carpel, with an attendant suppression of the lateral angles.

1b. Myosurus minimus var. **apus** Greene, Bull. Calif. Acad. Sci. 1:277. 1885.

Tufted annual, 2–6 cm. tall; leaves linear, terete to subspatulate, 2–6 cm. long, extending beyond the spike; spike sessile to short-stalked, cylindrical to subconical, 1–2 cm. long; stalk when present very stout; sepals 5–10 (5), the spur blunt to blunt-tapered; beak of achene closely appressed to spike, approximately ½ as long as the body.

This is part of a stabilized hybrid complex that is intermediate between the putative parents *Myosurus minimus* and *M. sessilis,* although characterized by the more or less erect beak of *M. minimus.*

Vernal pools and alkaline marshes: San Diego County and coastal southern California north through the San Joaquin Valley, also Lake County.

2. Myosurus cupulatus Wats., Proc. Am. Acad. Arts & Sci. 17:362. 1882. Fig. 233.

Tufted annual, often very diminutive, 3–8 cm. tall; leaves linear to linear-spatulate, 1–5 cm. long; sepals 5 or 6, oblong, the spur variable, often short and "blunt-tapered," sometimes as much as ⅓ as long as the blade; petals linear-filiform with a narrow, elongate claw 4–5 times as long as the narrow, short-oblong blade; scapes 3–8 cm. long; fruiting spike slender, 2–5 cm. long; achenes with a thickened margin producing a cuplike depression on back from which protrudes the divergent beak; seed short-oblong, flattened.

Vernally wet spots in the desert mountains and mountains bordering the deserts: southern California; east to Texas.

This species may readily be distinguished by the loose-fitting achenes which at maturity are separated by obvious spaces.

3. Myosurus aristatus Benth. ex Hook., London Jour. Bot. 6:458. 1847. Fig. 234.

Tufted annual; leaves linear to linear-spatulate, 1–5 cm. long, with a broad, almost membranous base; scapes very slender, 2–5 cm. tall; sepals oblong, erect or spreading, with a slender spur from ½ as long to fully as long as the

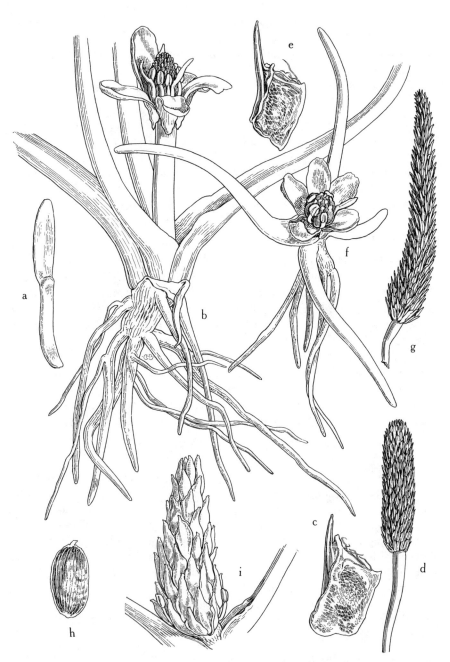

Fig. 235. *Myosurus. a–d, M. sessilis* subsp. *alopecuroides: a,* petal, × 16; *b,* habit, showing the flowering scape and the basal parts of leaves, × 6; *c,* mature achene, its beak divergent, × 16; *d,* mature fruiting spike, × 2. *e–i, M. sessilis: e,* mature achene, its beak long and divergent, × 12; *f,* habit, showing young flower, sessile among the linear leaves, × 8; *g,* mature fruiting spike, × 2; *h,* mature seed, × 20; *i,* young fruiting spike, × 6.

blade; petals present, or at maturity often none; fruiting spike 5–10 mm. long; achenes somewhat quadrate in outline, the back sharply keeled, extending to the elongate, divergent, sometimes falcate beak.

Myosurus aristatus may be readily distinguished by the elongate, divergent beaks of the achene, which are fully as long as the body of the achene, and by the very short spike, which, because of the achene, appears somewhat bristly.

Modoc plateau in northeastern California; north to British Columbia, east to the Rocky Mountains. Reported also from the mountains of southern California.

3a. Myosurus aristatus subsp. **montanus** (Campbell) Stone, comb. nov. Fig. 234.

Myosurus minimus subsp. *montanus* Campbell, El Aliso 2:349. 1952.

Leaves filiform-spatulate, the earliest ones almost terete, 1–6 cm. long; scapes extending much beyond the leaves, 5–15 cm. long; spikes relatively short, 2–22 mm. long; sepals 5, the spur linear-filiform; petals none to 5, the blade much shorter than the claw; stamens 5–10; carpels with a conspicuous beak divergent from the axis of back of achene; mature achene strongly keeled on back above and into beak, rounded below, lateral angles conspicuously nerved, the interfaces and sides thin-membranous, in no sense becoming rigid on ripening; seed linear-ovoid to fusiform.

Wet margins of ditches and ponds, vernal pools in the mountains: Sierra Nevada, North Coast Ranges, San Bernardino Mountains; east to Arizona and South Dakota, north to Saskatchewan.

The relationship of *Myosurus aristatus* subsp. *montanus* to *M. aristatus* is evidenced by the conspicuous nerving of the lateral angles of the carpel, the divergent beak of the mature fruit, and the high frequency of apetalous flowers. The subspecies differs from *M. aristatus* primarily in the longer flower spikes, and also in the faintly 3-nerved sepal blades, which are conspicuous at maturity. The variability of the achene shape, particularly beak angle, is quite noticeable when the achene is compared to the stable *M. aristatus* fruit. From *M. minimus* this subspecies differs in having fewer achenes per given length of spike, in its more conspicuous divergent beak, and in its very short petal blade. The fact that *M. aristatus* subsp. *montanus* is common in the meadows and ponds about Sierra Valley led to its frequently being identified as *M. lepturus* Greene, which was described from this area. *Myosurus lepturus,* however, is a synonym of *M. minimus.*

4. Myosurus sessilis Wats., Proc. Am. Acad. Arts & Sci. 17:362. 1882. Fig. 235.

Myosurus minimus var. *sessiliflorus* (Huth) Campbell, El Aliso 2:397. 1952.

Tufted annual, 4–10 cm. tall; leaves linear-spatulate, subequal to or extending beyond the spike; flowers sessile to subsessile, or very short-petioled; sepals

Fig. 236. *Ranunculus repens: a,* mature achenes, showing variation in the recurved beak, × 12; *b,* flower, × 1½; *c,* single petal, showing the broad scale covering the nectariferous pit, × 3; *d,* habit, showing the trailing stems, the coarsely serrate, divided leaves, and the large flowers, × ⅖.

5, with a short, blunt spur, or occasionally the spur approaching linear; petals 1–5, rarely absent, the blade subequal to the claw in length; stamens 5 to several; achene flat, keeled or rounded on back with a conspicuous spongy margin and a prominently divergent beak; seeds broadly ellipsoidal, microscopically striate; spike 1–3 cm. long, 2–3 mm. thick, usually conspicuously tapered.

Alkaline vernal pools: Sacramento, San Joaquin, Livermore, and San Benito valleys; north to Oregon.

The California specimens are often much stouter than the northern material which stands as the type. Nevertheless, there seem to be no character differences significant enough to warrant separation. The Jepson material upon which *Myosurus aristatus* var. *sessiliflorus* Huth was based is a very close match in stature for the Howell material used by Watson as the type of *M. sessilis*.

4a. Myosurus sessilis subsp. **alopecuroides** (Greene) Stone, comb. nov. Fig. 235.

Myosurus alopecuroides Greene, Bull. Calif. Acad. Sci. 1:278. 1885.

Scapes stout, short, thickened upward; spikes thick, cylindrical to strongly conical; beaks of achenes usually divergent, rarely somewhat appressed.

The name *Myosurus alopecuroides* was applied to plants which are intermediate in stature and other characters between *M. sessilis* and *M. minimus* and which usually occur with them. Greene's type included a single specimen of *M. sessilis*. These plants are of hybrid origin and thus are included in the hybrid complex with *M. minimus* var. *apus*.

RANUNCULUS

Annual or perennial herbs, caespitose or stoloniferous, terrestrial or aquatic. Basal leaves usually long-petioled, the petioles of the cauline leaves usually shorter or sometimes obsolete; leaves alternate, rarely opposite, the petioles dilated in stipular region. Flowers from a branched inflorescence or terminal on scapelike pedicels. Sepals 3–5, usually soon-deciduous, rarely marcescent. Petals glossy yellow or a few white, usually having a nectariferous pit which is covered by a minute scale. Stamens 10 to many. Pistils 5 to many; ovule solitary. Fruit an achene.

For a more detailed account and more complete synonymy see Lyman Benson, "Ranunculus" (in Abrams, Illus. Fl. Pacif. States 2:197–215, 1944).

Fig. 237. *Ranunculus Bloomeri: a,* habit, showing the succulent, ascending stems and the roundish leaves, × ⅖; *b,* mature achene, the beak long and slender, × 8; *c,* flower, × 1½; *d,* flower, lateral view, showing the short, reflexed sepals, × 1½; *e,* petal, the nectariferous pit covered by a narrow, emarginate scale, × 3; *f,* fruiting head, × 2.

Flowering stems erect or prostrate or trailing and rooting at the nodes, if floating at or below the surface of water, then flowers yellow; achenes not transversely ridged (except *R. sceleratus*).
 Pericarp of achene thick and firm, not striate or nerved.
  Achenes smooth or hairy, not spiny.
   Leaves (either the cauline or the basal) lobed, parted, divided, or dissected.
    Achene beaked with a persistent style, without corky thickenings on keel or margins.
     Receptacle glabrous; beak of achene strongly hooked .... 1. *R. repens*
     Receptacle hairy; beak of achene straight .... 2. *R. Bloomeri*
    Achene practically beakless, or if beaked, with corky-thickened margins.
     Achene without beak, stigma sessile; face of achene with transverse ridges or punctate .... 3. *R. sceleratus*
     Achene with well-developed beak and with corky-thickened margins
                  4. *R. flabellaris*
   Leaves entire, crenate, or serrate, sometimes undulate.
    Petals 5–10.
     Stems erect or reclining, never rooting at the nodes; cauline leaves lanceolate, oblanceolate, or linear .... 5. *R. alismaefolius*
     Stems trailing, often stolon-like, rooting at the nodes; cauline leaves ovate or ovate-lanceolate.
      Stems fistulose-inflated, 1.5–4 mm. thick; blade broadly ovate to suborbicular or elliptic .... 6. *R. hydrocharoides*
      Stems filiform, less than 1 mm. thick.
       Blade lanceolate to linear-spatulate or oblanceolate .... 7. *R. flammula*
       Blade broadly ovate to deltoid .... 8. *R. Gormanii*
    Petals 1–3, minute, 1–1.5 mm. long; annuals.
     Sepals 5; achenes papillate on faces; upper cauline leaves sessile
                   9. *R. pusillus*
     Sepals 3; achenes reticulate on faces; upper cauline leaves petioled
                  10. *R. alveolatus*
  Achenes spiny on the faces .... 11. *R. muricatus*
 Pericarp of achene thin and fragile, striate and 3-nerved; receptacle becoming much elongated in fruit .... 12. *R. Cymbalaria*
Floating aquatics; achenes roughly transverse-ridged; petals white, never glossy, sometimes the claw yellow.
 Style 2–3 times as long as ovary; receptacle glabrous .... 13. *R. Lobbii*
 Style ½ as long as ovary; receptacle hairy .... 14. *R. aquatilis*

1. Ranunculus repens L., Sp. Pl. 554. 1753. Fig. 236.

Stems trailing, rooting at the nodes; basal leaves long-petioled, often densely coarse-pilose, the blades 3–10 cm. broad, bi- or triternately parted or divided, the ultimate lobes coarsely serrate; flowers large, 10–25 mm. broad; calyx not reflexed; petals nearly twice as long as the sepals, deep orange-yellow; achenes flat-faced, the keel flanked by a thickened margin or ridge, the beak stout, recurved or hooked.

Along stream courses, in marshes and wet springy spots: central California; sometimes escaping into lawns, then wholly prostrate and the petioles short; native of Europe. Sometimes there is a light-colored spot in the sinuses of the

Fig. 238. *Ranunculus sceleratus: a,* flower, the hairy sepals reflexed, × 4; *b,* petal, showing the open nectariferous pit at the constricted base of the petal, × 20; *c,* mature achene, × 20; *d,* young upper leaf and auricle of sheath, × 4; *e,* variation in achene, × 20; *f,* habit, showing the cluster of basal leaves and the flowering and fruiting heads, × ⅖; *g,* habit variation, hardly any basal leaves present, × ⅖.

clefts of the leaves. A double-flowered form of *Ranunculus repens* is in common cultivation.

2. Ranunculus Bloomeri Wats., Bot. Calif. 2:426. 1880. Fig. 237.

Somewhat succulent, erect or ascending perennial, 3–5 dm. tall; basal leaves on long petioles, the petioles glabrous or hairy, the blades 2–5 cm. long, 3-foliate or sometimes simple, roundish in outline, the segments coarsely dentate or sometimes obscurely 3-lobed, glabrous or somewhat hairy; flowers few, large, 2–3 cm. broad; sepals reflexed, deciduous; petals 5–8, much longer than the sepals, 10–17 mm. long, emarginate at apex, usually with a greenish spot near the base, the scale narrow, the nectary large, emarginate; achenes many, plump, the beak slender, equal or subequal to the body.

Marshy or wet meadows: Coast Ranges and coastal valleys of central California.

3. Ranunculus sceleratus L., Sp. Pl. 551. 1753. Cursed buttercup. Fig. 238.

Erect, somewhat fistulose, succulent annual, usually with erect branches 1–3 dm. tall; leaves 2–5 cm. broad, the lower leaves long-petioled, the blades parted into 3–5 cuneate segments, these often again cleft into coarsely toothed segments, glabrous, the upper leaves nearly sessile, the blades with linear, entire segments; flowers 6–10 mm. broad; sepals reflexed; petals slightly shorter than to longer than sepals, pale yellow; receptacle elongating in fruit, 3–10 mm. long; achenes very numerous, scarcely beaked.

Common in muddy shallow pools, at lake margins, and on stream banks, often thriving in brackish and alkaline sites: San Joaquin Valley to Modoc County; north to British Columbia, south to Arizona, east to the Atlantic. Native of Europe.

Ranunculus sceleratus L. is said to possess very acrid juice which raises blisters on the skin. The leaves resemble those of celery. Because the California material has more luxuriant foliage and some "pinprick" depressions on the face of the achene, it has been referred to *Ranunculus sceleratus* var. *multifidus* Nutt. (in Torr. & Gray, Fl. N. Am. 1:19, 1838).

4. Ranunculus flabellaris Raf., Am. Mo. Mag. 2:344. 1818.

Ranunculus delphinifolius Torr. in Eat., Man. Bot., 2d ed., 395. 1818.

Aquatic perennial; stems flaccid and fistulose, glabrous, submersed, the lower nodes rooting; leaves reniform, all cauline, finely dissected, the ultimate segments flat; sepals deciduous in anthesis; petals 5–8, yellow, about twice as long as the sepals; receptacle elongating in fruit, 7–10 mm. long; achenes glabrous, smooth, very numerous, corky-margined below, the beak as long as body, hooked, winged-margined.

Shallow water or wet mud: northeastern and northwestern California; north to British Columbia, east to Atlantic.

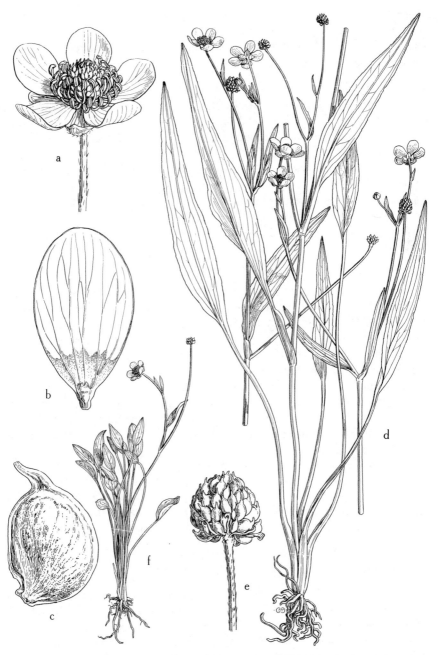

Fig. 239. *Ranunculus. a–e, R. alismaefolius: a,* flower, × 2; *b,* single petal, showing the bilobed scale covering nectariferous pit, × 5; *c,* mature achene, × 12; *d,* habit, showing the long, serrulate leaf blades, the flowers, and the fruiting heads, × 2/5; *e,* fruiting head, × 2. *f, R. alismaefolius* var. *alismellus:* habit, showing small stature, short, ovate leaf blades, and single-flowered stems, × 2/5.

Readily distinguishable from *Ranunculus aquatilis* by its yellow flowers and the flat leaf segments.

5. Ranunculus alismaefolius Geyer ex Benth., Pl. Hartw. 295. 1848. Fig. 239.

Fistulose, erect perennial, 2–5 dm. tall, branched; lower leaves long-petioled, the upper ones short-petioled or sessile, the blades oblanceolate to lanceolate-elliptic, 2–20 cm. long, 1–3 cm. broad, entire or obscurely serrulate; flowers 5–20 mm. wide; calyx pubescent, not reflexed; petals longer than sepals; achenes 30–50, smooth, plump, with a short, slender beak.

Swamps: Del Norte, Humboldt, and Mendocino counties; north to British Columbia.

The most common form in California is *Ranunculus alismaefolius* var. *alismellus* Gray. It differs from the species in its smaller stature (it is only 10–25 cm. tall), its 1-flowered stems, and its leaf blades, which are rarely more than 2 cm. long. It is the common form throughout the Sierra Nevada and the higher mountains of southern California. See figure 239.

6. Ranunculus hydrocharoides Gray, Mem. Am. Acad. Arts & Sci. II. 5:306. 1855.

Aquatic perennial; stems fistulose-inflated, 1.5–4 mm. thick, from an erect stolon-producing caudex, 8–30 cm. tall; stolons rooting at the nodes; basal leaves long-petioled, the blades ovate to cordate, 1–3 cm. long, somewhat fleshy, the cauline ovate to spatulate; flowers 5–7 mm. wide; sepals spreading, deciduous; petals 5 mm. long, straw yellow, narrowly ovate; achenes 20–25, 1 mm. long, the beak fleshy, with a hooklike tip.

In marshy places about springs: Owens Valley; south to Arizona and Mexico.

7. Ranunculus flammula var. **ovalis** (Bigel.) Benson, Bull. Torrey Club 69:305. 1942. Creeping buttercup. Fig. 240.

Stoloniferous perennial rooting at the nodes, the slender stolons often somewhat arched, usually less than 1 mm. thick, often intricately interlaced; leaves usually erect, the petioles as long as or longer than the blade, rarely wanting, the blades lanceolate or oblanceolate to linear-spatulate, 1–4 cm. long, entire; flowers 4–10 mm. broad, on leafy or naked erect stems or terminating naked or bracteate stolons; petals much longer than the sepals; achenes few, broad, terminated by a short apiculate beak.

Very common on margins of slow streams, lakes, or ponds: throughout mountains and higher valleys of California; north to Alaska, east to Atlantic.

There are a few records which show that the stolons have sometimes carried the plants into deep water, where they tended to float on the surface.

A form with broader leaf blades, from northeastern California and Oregon, has been segregated as *Ranunculus flammula* var. *samolifolius* (Greene) Benson (Bull. Torrey Club 69:306, 1942).

Fig. 240. *Ranunculus flammula* var. *ovalis: a,* petals, showing the shallow nectar-iferous pit, × 8; *b,* flower, the petals much longer than the sepals, × 3; *c,* habit, showing the slender, arching stolons and the erect leaves, × ⅔; *d,* horizontal stolon, × ⅖; *e,* ma-ture achenes, showing variation in size and shape of beak, × 20; *f,* mature fruiting head, × 6.

8. Ranunculus Gormanii Greene, Pittonia 3:91. 1896.

Prostrate, glabrous perennial with filiform, stolon-like stems, rooting at some of the nodes; cauline and basal leaves essentially alike, on slender petioles, the blades broadly ovate to deltoid, 20–30 mm. long, entire, 3-nerved; flowers on scapelike pedicels; sepals spreading, ovate, 2–3 mm. long; petals 5–6 mm. long, bright straw yellow; achenes 6–15, 1.5 mm. long, with an inconspicuous margin, the beak slender, curved or hooked at tip.

Boggy mountain streams: Del Norte, Humboldt, and Siskiyou counties.

9. Ranunculus pusillus Poir. in Lam., Encycl. 6:99. 1804. Dwarf buttercup.

Slender-stemmed annual, with reclining stems rooting at the nodes, 1–5 dm. long, glabrous except on the dilated, elongate petioles; blades of basal and lower cauline leaves round-ovate, toothed or entire, 5–20 mm. long, the upper cauline leaves elliptic to lanceolate, entire or denticulate, sessile; flowers very small; sepals not reflexed, 1–1.5 mm. long; petals 1–3, obovate, barely 1–1.5 mm. long; achenes numerous, smooth or papillate, with a minute beak.

Wet ground or margins of pools: outer North Coast Ranges from Santa Cruz Mountains to Humboldt County; southeastern United States.

10. Ranunculus alveolatus Carter apud Benson & Carter, Am. Jour. Bot. 26: 555. 1939.

Semiaquatic annual; stems many from the base, decumbent and rooting at the nodes, ultimately erect, glabrous to sparsely pubescent; lower leaves on petioles 3–10 cm. long, the cauline petioles 6–35 mm. long, the blades simple, ovate to ovate-lanceolate, entire to somewhat dentate; sepals 3, with margins membranous at base, ovate, 2–2.5 mm. long; petals 2 or 3, yellow, ovate, 2–2.5 mm. long, the nectary scale truncate, glabrous; receptacle glabrous, 2–3.5 mm. long; achenes 15–25, flattened, oval, the surface alveolate, the beak obscure.

Marshy areas, margins of ponds and small streams: foothills of the northern Sierra Nevada.

11. Ranunculus muricatus L., Sp. Pl. 555. 1753. Spiny buttercup. Fig. 241.

Erect annual, 6–20 cm. tall, branched from the base; herbage succulent, light green, glabrous; basal leaves on long broad petioles, the blades 2–5 cm. broad, reniform to cordate-orbicular, often 3-lobed or 3-cleft, the cauline simple or divided, short-petioled or sessile; flowers 6–14 mm. broad; sepals thin, 4–5 mm. long, reflexed; petals light yellow, longer than sepals; achenes flat, thickened around the margin, the depressed faces beset with several long prickles.

Frequent in marshy ground: Central Valley, coastal area of central California; north to Oregon and Washington. Native of Europe.

Fig. 241. *Ranunculus muricatus: a,* young flower, showing reflexed sepals, × 2½;
b, petal, showing scale covering nectariferous pit, × 6; *c,* flower, × 2½; *d,* mature muricate achene, side view, × 4; *e,* habit, plant branched from the base, the petioles succulent,
× ⅖; *f,* mature achene, marginal view, × 4.

Fig. 242. *Ranunculus Cymbalaria* var. *saximontanus: a,* petal, gland covered by a scale, × 12; *b,* flower, the petals shorter than the sepals, × 4; *c,* mature achenes, showing variation in shape, × 20; *d* and *e,* variations in habit, × ⅔.

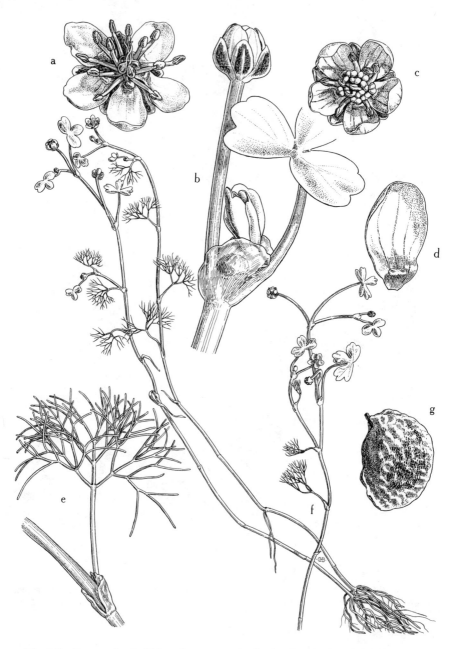

Fig. 243. *Ranunculus Lobbii: a,* flower at anthesis, showing the long styles, × 4; *b,* buds in axil of floating leaf, the enlarged petiole base auricled, × 4; *c,* young flower, × 4; *d,* petal, showing the broad, short scale covering the nectariferous pit, × 8; *e,* submersed, dissected leaf and sheath, × 3; *f,* habit, showing submersed and floating leaves, × ⅔; *g,* mature achene, rugose and black-dotted, the basal part of style forming a beak, × 12.

12. Ranunculus Cymbalaria var. **saximontanus** Fern., Rhodora 16:162. 1914. Desert buttercup. Fig. 242.

Small stoloniferous perennial, rooting at nodes; flowering stems 3–15 cm. tall, 1- to 3-flowered; leaves on long petioles, the blades round-ovate to reniform, often cordate at base, 1–2 cm. long; flowers 8–12 mm. wide; sepals not reflexed, as long as or slightly longer than petals; receptacle elongate, cylindrical in fruit, 4–12 mm. long; achene longitudinally striate on sides, the beak stout.

Common in alkaline marshes and along watercourses: San Joaquin Valley and coastal southern California, thence east through the Mojave Desert and north along the east side of the Sierra Nevada to Lassen and Modoc counties; north to Alaska, east to Nebraska, and south to Mexico.

13. Ranunculus Lobbii (Hiern) Gray, Proc. Am. Acad. Arts & Sci. 21:364. 1886. Fig. 243.

Floating annual aquatic, occasionally with filiform, dissected, submersed leaves or these often absent, the floating leaves 3- or sometimes 5-lobed or -parted, the lobes oblong or ovate, often notched; flowers emersed; sepals 5, 2 mm. long; petals white, withering-persistent, longer than sepals; stamens 5–16; style filiform, about 3 times as long as the ovary; achenes few, rugose, often covered with minute black dots.

Common in vernal pools: mountains and valleys of central coastal California; north to British Columbia.

This species may be readily distinguished from *Ranunculus aquatilis* by the very long style, and by the commonly 3-lobed floating leaves, the usual absence of dissected submersed leaves, and the annual habit.

14. Ranunculus aquatilis var. **capillaceus** (Thuill.) DC., Prodr. 1:26. 1824. Aquatic buttercup. Fig. 244.

Floating perennial aquatic; submersed leaves filiform-dissected, the floating leaves when present 3-lobed, the lobes toothed at the summit; flowers 6–10 mm. broad; sepals deciduous; petals white, sometimes greenish yellow at base; styles subulate; achenes usually numerous, transversely rugose, sometimes hispidulous, the beak almost obsolete.

Common in permanent pools and slow streams or ditches: almost throughout California; widely distributed in North America and Europe.

A considerable amount of variation is exhibited in the size of the flowers, the rigidity of the dissected leaves, and the pubescence of the base of the petioles. The plants with floating leaves have been distinguished as *Ranunculus aquatilis* var. *hispidulus* Drew (Bull. Torrey Club 16:150, 1889). See figure 244.

Fig. 244. *Ranunculus aquatilis. a–d, R. aquatilis* var. *hispidulus: a,* habit, showing submersed and floating leaves, the flowers and fruiting heads on long, slender peduncles, × ⅘; *b,* flower, × 4; *c,* mature achene, × 12; *d,* transitional forms of floating leaves, showing toothed summit, × 1½. *e–h, R. aquatilis* var. *capillaceus: e,* habit, submersed leaves only, the peduncles short and stout, × ⅘; *f,* single petal, showing the low scale surrounding the shallow nectariferous pit, × 8; *g,* achene from head comprised of about 35 achenes, × 12; *h,* achene from a head comprised of few achenes, × 12.

CALYCANTHACEAE. SWEET-SHRUB FAMILY

Aromatic shrubs with opposite, entire leaves and no stipules. Flowers large, solitary, terminating the branches. Bracts, sepals, and petals undifferentiated and arranged in a close spiral on the outer rim of a thickened, cuplike receptacle. Stamens many, inserted on rim of the receptacle, the inner ones sterile. Pistils many, distinct, nearly enclosed in the receptacle, becoming achenes.

CALYCANTHUS

1. Calycanthus occidentalis Hook. & Arn., Bot. Beechey Voy. 340, pl. 84. 1840. Sweet bush, spicebush. Fig. 245.

Aromatic deciduous shrub 1–3 m. tall, with many-branched stems from base; leaves ovate to oblong-lanceolate, acute, rounded at base, 6–30 cm. long, on short petioles; flowers with sepals and petals red, becoming tan at tips, linear-spatulate, unequal in length, 10–30 mm. long; stamens distinct, the filaments short; pistils many, distinct, inserted within the cuplike receptacle, which enlarges in fruit, enclosing the many achenes.

Common on margins of streams at low altitudes: northern Sierra Nevada, North Coast Ranges.

LAURACEAE. LAUREL FAMILY

Trees and shrubs with alternate, entire leaves without stipules, aromatic. Flowers usually bisexual, regular, the perianth usually of 6 undifferentiated sepal-like segments. Stamens in 4 whorls of 3 stamens each, the innermost stamens usually sterile, the filaments free, the anthers basifixed. Pistil 1, the ovary superior, 1-celled, the ovule solitary. Fruit a drupe.

UMBELLULARIA

1. Umbellularia californica (Hook. & Arn.) Nutt., Sylva 1:8. 1842. California laurel, Oregon myrtle.

Aromatic evergreen tree 10–20 m. tall; leaves simple, lanceolate to lanceolate-elliptic, acute or rarely obtuse, glabrous; flowers in axillary and terminal umbels, perfect, regular; sepals 6; petals none; stamens in 4 whorls, the 2 outer whorls of 3 stamens each appearing as a simple whorl, the inner whorl of 3 sterile filaments and the remaining whorl of 3 stamens, each with a pair of large orange glands on the filament; anthers 4-celled, dehiscent by 4 valves, those of the outer stamens dehiscing inward and those of the inner stamens outward;

Fig. 245. *Calycanthus occidentalis: a,* achene, velvety-hirsute, $\times$ 3; *b,* flowering branch, $\times$ ⅖; *c,* capsule, $\times$ ⅕; *d,* flower, top view, $\times$ ⅔; *e,* flower (longitudinal section), showing pistils deep in the receptacle, $\times$ 3.

ovary 1-celled, 1-ovuled; fruit a drupe, subtended by the thickened base of the calyx, the drupe dark purple when ripe.

Often a riparian tree but by no means confined to riparian habitats: San Diego County and north in the Coast Ranges and the Sierra Nevada; to southern Oregon.

CRUCIFERAE

Annual, biennial, or perennial herbs with acrid juice. Leaves chiefly alternate. Flowers in racemes, occasionally in corymbs, hypogynous. Sepals 4. Petals 4 or absent. Stamens 6, usually in 2 whorls, the outer whorl of 2 stamens, the inner whorl of 4 stamens, occasionally only 4 or 2 stamens, often with 4 green glands at base of stamens. Ovary 2-celled, with a partition between the cells, the placentae in 2 rows in each cell on opposite margins of the valve. Pod dehiscent or indehiscent.

Pods short, 1–2 times as long as broad.
    Pods globose, not tipped by winglike lobes .... *Subularia*
    Pods flattened at right angles to the partition, tipped by a 2-lobed wing .... *Lepidium*
Pods elongate, 5 to many times as long as broad.
    Valves of pod nerved by a conspicuous vein.
        Seeds from opposite placentae arranged in the cell of the pod in a single row
                                                                   Barbarea
        Seeds from opposite placentae arranged in the cell of the pod in 2 rows ... *Rorippa*
    Valves of pod not nerved.
        Basal leaves not from a subterranean tuber .... *Cardamine*
        Basal leaves from a scaly tuber .... *Dentaria*

BARBAREA

1. Barbarea americana Rydb., Mem. N. Y. Bot. Gard. 1:174. 1900. American winter cress.

Barbarea vulgaris of Jepson, Man. Fl. Pl. Calif. 424. 1925.

Erect, glabrous biennial, often with many stems from base, 3–6 dm. tall; basal leaves lyrate-pinnatifid with an orbicular terminal lobe and 2–4 pairs of linear lateral lobes, the cauline leaves reduced above; petals yellow; pod 3–5 cm. long, obscurely 4-angled with a stout, short beak.

Moist slopes, springs, and bogs at low and middle altitudes almost throughout California; north to British Columbia, east to New England, south to Mexico.

CARDAMINE

Annual or perennial herbs. Leaves variously lobed or sometimes entire. Flowers in ours white, in racemes or corymbs. Stamens 6 or 4. Pod flattened parallel to the partition, the valves without nerves. Seeds appearing as though in a single row, wingless.

Most of our species occur at high altitudes.

Leaves simple, entire, or shallowly lobed or with a few lateral teeth.
    Leaves ovate or elliptical; stems 3–10 cm. tall .... 1. *C. bellidifolia*
    Leaves cordate or reniform; stems 20–60 cm. tall .... 2. *C. Lyallii*
Leaves, at least those of the stem, pinnatifid.
    Basal leaves simple, the cauline leaves 3- to 5-foliate .... 3. *C. Breweri*
    Basal leaves pinnatifid, with 5–15 leaflets.
        Perennials with rhizomes .... 4. *C. Gambelii*
        Annuals.
            Leaflets linear to oblong; pods with 20–30 seeds .... 5. *C. pennsylvanica*
            Leaflets mostly orbicular; pods with 8–20 seeds .... 6. *C. oligosperma*

1. Cardamine bellidifolia L., Sp. Pl. 654. 1753.

Tufted perennial from a branched caudex; stems 3–12 cm. tall, glabrous; leaves long-petioled, ovate to elliptic, with a few lateral teeth; flowers 1–5; petals white to pink, spatulate, 3–4 mm. long; pod erect, 2–4 cm. long.

Moist places: high elevations in northern California; north to Alaska, east to New England, Europe, Asia.

2. Cardamine Lyallii Wats., Proc. Am. Acad. Arts & Sci. 22:466. 1887.

Rhizomatous perennial; stems erect, 2–6 dm. tall, simple or branched, glabrous to sparsely pilose; leaves on long petioles, the blades cordate-reniform, shallowly sinuate; petals white; pod erect or spreading, 2–4 cm. long, with a short, stout beak.

Montane: northern Sierra Nevada to Siskiyou County; north to Oregon, Nevada.

3. Cardamine Breweri Wats., Proc. Am. Acad. Arts & Sci. 10:339. 1875.

Rhizomatous perennial; stems erect or decumbent, 2–3 dm. tall, glabrous to pubescent; leaves pinnate, with 3 or occasionally 7 leaflets, sometimes the basal leaves simple and reniform-cordate; petals white; pod 2–3 cm. long, ascending or erect.

Along streams, middle altitudes: Sierra Nevada, coastal areas of Humboldt County; north to British Columbia, east to Rocky Mountains.

4. Cardamine Gambelii Wats., Proc. Am. Acad. Arts & Sci. 11:147. 1876.
    Swamp cress.

Rhizomatous or stoloniferous, often somewhat coarse perennial, rooting at the nodes; stems 3–10 dm. tall, glabrous to soft-villous; leaves with 2–6 pairs of leaflets, these toothed; racemes dense, elongating in age; petals white; pod on divaricately spreading pedicel, erect or ascending.

Swamps: southern California except in the deserts, Santa Barbara County; south to Mexico.

5. Cardamine pennsylvanica Muhl. ex Willd., Sp. Pl. 3:486. 1800.

Fibrous-rooted annual or biennial; stems 2–10 dm. tall, glabrous to sparsely pubescent; leaflets 7–13, oblong to linear or suborbicular, toothed or entire; petals white; pod slender, 2–3 cm. long.

Swamps: Nevada County; north to British Columbia, east to Florida and Newfoundland.

6. Cardamine oligosperma Nutt. in Torr. & Gray, Fl. N. Am. 1:85. 1838.

Delicate, fibrous-rooted annual; stems simple, erect, 1–4 dm. tall, sparsely hirsute to glabrous; leaves in a basal rosette and cauline, the leaflets 5–11, sub-equal in size, each with a minute petiolule, oval to orbicular, shallowly lobed or toothed; racemes few-flowered; petals white, longer than sepals; pod erect, slender, 2–3 cm. long.

Moist places at low and middle altitudes: almost throughout California except in the deserts; north to Alaska, east to Atlantic.

DENTARIA. TOOTHWORT

Perennial herbs from a horizontal, tuber-like rootstock. Basal leaves springing directly from rootstock and often appearing as though not a part of the plant that bears the flowering stem, the blades simple, the stem leaves simple or compound. Flowers in simple or corymbose racemes. Petals with slender claws, the blades ovate and spreading, longer than sepals. Pod flattened parallel to the partition, linear, the valves nerveless to faintly nerved, elastically dehiscent.

Tubers orange-yellow; petals deep purple . . . . 1. *D. gemmata*
Tubers white; petals white to pink.
  Basal leaf coarsely 5- to 9-toothed above middle; stem leaves usually simple
                                          2. *D. pachystigma*
  Basal leaf entire or sinuately lobed; stem leaves commonly 3- to 5-foliate, rarely simple
    in one variety . . . . 3. *D. integrifolia*

1. Dentaria gemmata (Greene) Howell, Fl. NW. Am. 1:49. 1897.

Tubers ovoid, orange-yellow, 7–10 mm. thick; stems stout, simple, erect, 10–25 cm. tall; basal leaves 3- to 5-foliolate, the lobes mucronate, the stem leaves 3- to 7-foliolate, narrowly lanceolate; racemes short; petals deep purple, 10–15 mm. long; pod 3–5 cm. long, on pedicel 15–20 mm. long.

Stream banks and shallow water: Del Norte County; southwestern Oregon.

2. Dentaria pachystigma Wats. in Gray, Syn. Fl. N. Am. 1¹:155. 1895.

Stems stout, simple, erect, 10–20 cm. tall, glabrous; basal leaves simple, the blades orbicular, cordate, crenately toothed, the stem leaves 2 or 3, the simple blades cordate-reniform, coarsely dentate to sinuate or crenate; racemes nearly sessile, short, often nearly corymbose, then the lower pedicels much elongate; petals white to pale pink; pod 2–5 cm. long, the beak short and stout.

Wet places: Sierra Nevada, North Coast Ranges.

Two varieties have been distinguished. *Dentaria pachystigma* var. *corymbosa* Abrams (*D. corymbosa* Jepson), from the North Coast Ranges and the southern Sierra Nevada, is based on the corymbose inflorescence and narrower style. *Dentaria pachystigma* var. *dissectifolia* Detling, based upon the pinnate cauline leaves, occurs on serpentine outcrops in the northern Sierra Nevada.

3. Dentaria integrifolia Nutt. in Torr. & Gray, Fl. N. Am. 1:88. 1838.

> *Dentaria californica* Nutt., *loc. cit.*
> *Dentaria integrifolia* var. *californica* (Nutt.) Jepson, Fl. West. Middle Calif. 222. 1901.
> *Dentaria californica* var. *integrifolia* Detling, Am. Jour. Bot. 23:576. 1936.

Stems simple, erect, sometimes tinged with purple, glabrous; basal leaves simple or trifoliate, the blades orbicular, ovate or reniform, the stem leaves 3- to 5-foliolate, the leaflets broadly lanceolate to linear; racemes mostly simple and terminal; sepals green to reddish; petals white to pinkish; pod 3–4 cm. long, the valves not nerved.

Vernally wet meadows or moist woods, chiefly at low and middle altitudes: central California.

Several geographic groups have been recognized. *Dentaria integrifolia* var. *californica* Jepson occurs chiefly in moist woods and is characterized by less succulent foliage than the species. *Dentaria integrifolia* var. *Tracyi* Jepson (*D. californica* var. *sinuata* Detling) has simple basal leaves with sinuate margins and occurs from Mendocino County northward.

LEPIDIUM

Erect or spreading herbs, occasionally somewhat woody; annuals, biennials, or perennials. Leaves entire to pinnatifid. Flowers in racemes. Petals white or greenish white, rarely wanting. Stamens 6 or less. Style none or, in a few species of drier habitats, slender and elongate. Pod a silicle, oblong or obovate, flattened at right angles to the partition and winged at apex. Seeds 1 in each cell.

A genus of about twenty species in California, of which the following occur in beds of vernal pools in alkaline floodplains.

Inner free margins of the wings of the pod essentially parallel; pods conspicuously reticulate-veined.
    Wings subequal to body of pod in length .... 1. *L. latipes*
    Wings much shorter than body of pod .... 2. *L. dictyotum*
Inner free margins of wings divergently spreading; pods inconspicuously reticulate-veined.
    Wings much shorter than body of pod .... 3. *L. oxycarpum*
    Wings ½ to fully as long as body of pod .... 4. *L. acutidens*

1. Lepidium latipes Hook., Icon. Pl. 1: pl. 41. 1837.

Stout-stemmed annual, branched from base; stems procumbent or erect, 2–8 cm. long; leaves linear and entire or with a few linear teeth, 5–10 cm. long, pubescent; racemes dense, 1–4 cm. long, the pedicels conspicuously flattened; petals greenish, longer than the sepals; pod oval, the wings subequal to the body in length, the sinus narrow and parallel-sided, the body strongly reticulate-veined.

Common in former beds of alkaline pools: San Diego County to Sacramento Valley.

2. Lepidium dictyotum Gray, Proc. Am. Acad. Arts & Sci. 7:329. 1867.

Decumbent to ascending annual, branching from the base, pubescent; leaves linear, entire or with a few teeth or lobes; racemes dense, the pedicels flattened; petals none, or if present then shorter than the calyx, white; pod ovate, tipped by a pair of rounded lobes, the sinus between lobes narrow.

Wet alkaline soils at low altitudes: southern California; north to Washington and Idaho.

3. Lepidium oxycarpum Torr. & Gray, Fl. N. Am. 1:116. 1838.

Minutely hirsute annual; stems several from base, 10–15 cm. long, erect or ascending, culminating in elongate racemes; leaves linear, entire or with a few triangular teeth or linear acute lobes; pedicels flattened, spreading; sepals falling early, very unequal in size; petals none; stamens 2; pod finely reticulate, broadly ovate, tipped with 2 triangular, attenuate teeth.

Margins of salt marshes: San Francisco Bay region.

4. Lepidium acutidens (Gray) Howell, Fl. NW. Am. 1:64. 1897.

Lepidium dictyotum var. *acutidens* Gray, Proc. Am. Acad. Arts & Sci. 12:54. 1876.
Lepidium oxycarpum var. *acutidens* Jepson, Man. Fl. Pl. Calif. 441. 1925.

Decumbent or ascending annual, branched from base, pubescent; leaves entire or remotely denticulate, 2–5 cm. long; racemes lax, the pedicels flattened and appressed; pod strongly reticulate, short-ovate, tipped by a pair of triangular, lanceolate teeth, the sinus between teeth divergent.

Saline and alkaline spots and the beds of vernal pools: northeastern California south through the valleys to southern California; north to eastern Oregon.

RORIPPA

Herbs with simple or pinnate leaves. Flowers in racemes or sometimes in small panicles, yellow or white. Sepals spreading in anthesis. Petals narrowed to the base. Stamens 6. Stigma capitate, from obscurely lobed to evidently 2-lobed. Pods short-cylindrical to ovoid or spheroid, sometimes curved, the seeds in 2 rows in each cell.

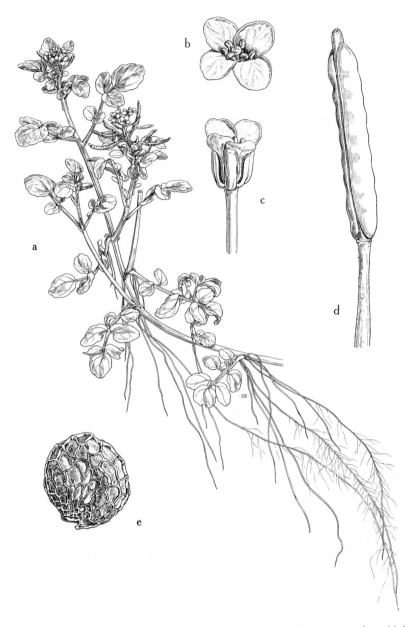

Fig. 246. *Rorippa Nasturtium-aquaticum: a,* habit, × ⅔; *b,* flower, top view, × 4; *c,* flower, side view, × 4; *d,* pod, × 4; *e,* seed, × 20.

This group of highly variable plants of wet habitats is not too well understood. The plants have been variously treated under the generic names *Radicula, Nasturtium,* and *Rorippa.* The name *Rorippa* is technically correct under the International Rules of Botanical Nomenclature if all the species in the following treatment are to be retained in the genus. If the common water cress is to be segregated it would be placed under the genus *Nasturtium.*

Flowers white; leaves pinnately parted into rounded lobes; plants often rooting at nodes
                                     1. *R. Nasturtium-aquaticum*
Flowers yellow; leaves pinnatifid to pinnately lobed, occasionally some of the basal leaves simple.
  Plants perennial.
    Pods 4–6 times as long as broad, straight to slightly curved . . . . 2. *R. sinuata*
    Pods 1–2 times as long as broad . . . . 3. *R. subumbellata*
  Plants annual or biennial.
    Pods linear, 4–8 times as long as broad; style less than 1 mm. long.
      Plants spreading; pods curved . . . . 4. *R. curvisiliqua*
      Plants erect; pods straight or slightly curved . . . . 5. *R. palustris*
    Pods globose to ovoid, 1–3 times as long as broad.
      Stems erect, sparsely to densely hispid; leaves pinnatifid, hirsute on veins
                                          6. *R. hispida*
      Stems diffusely branched from base; leaf lobes rounded, glabrous . . . 7. *R. obtusa*

1. Rorippa Nasturtium-aquaticum (L.) Schinz & Thell., Fl. Schweiz, 3d ed., 240. 1909. Water cress. Fig. 246.

    Radicula Nasturtium-aquaticum (L.) Britten & Rendle, List Brit. Seed Pl. 3. 1907.

Perennial herb with ascending or prostrate stems, rooting at nodes; leaves pinnately parted, the leaflets 3–9, ovate to orbicular (or in some, lanceolate to linear in the upper part of plant), glabrous; flowers in short racemes, becoming elongate in fruit; pedicels 1–2 cm. long; petals white, 3–4 mm. long; stigma capitate, subsessile to short-styled; pod divaricately spreading, 1–3 cm. long.

This is the common water cress used in salads.

Margins of small streams, bogs, and springs: common throughout California; native of Eurasia.

2. Rorippa sinuata (Nutt.) Hitchc., Spring Fl. Manhattan 18. 1894.

    Radicula sinuata (Nutt.) Greene, Leafl. Bot. Obs. & Crit. 1:113. 1905.

Perennial from slender, creeping rootstock; stems decumbent or prostrate, the branches ascending; leaves 5–8 cm. long, deeply and regularly sinuately pinnately parted, the lobes linear, oblong or deltoid, entire or toothed; flowers 6–8 mm. long; petals yellow, longer than the sepals; pod thick-cylindric or lanceolate, straight or slightly curved, 8–15 mm. long, tipped by the persistent style.

Stream banks and borders of lakes and marshes: widely scattered throughout California; northeast to Saskatchewan, east to Texas.

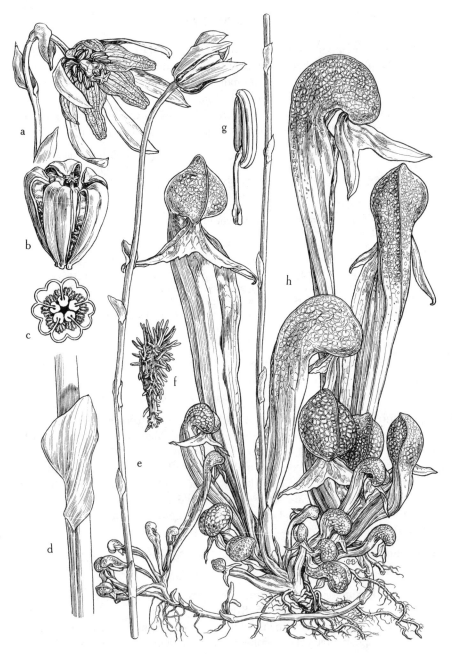

Fig. 247. *Darlingtonia californica: a*, flower, × ⅔; *b*, capsule, × ⅘; *c*, capsule (cross section), × ⅘; *d*, stem leaf, × 2; *e*, flowering stalk, × ⅖; *f*, seed, × 8; *g*, stamen, × 3; *h*, habit, showing specialized basal leaves with tubular petioles, × ⅖.

3. Rorippa subumbellata Rollins, Contr. Dudley Herb. 5:177. 1941.

Perennial herb from slender rhizomes; stems decumbent, branched, pubescent with simple hairs; leaves short-petioled to sessile, 1–3 cm. long, oblong to oblanceolate, pinnately lobed into rather regular segments, pilose to glabrous; inflorescence at length racemose; sepals nonsaccate, persistent, glabrous to sparsely pilose; petals yellowish to white; stigma not expanded; style elongate, becoming stout in fruit; pod orbicular-oblong, glabrous, 3–5 mm. long.

Known only from the shores of Lake Tahoe in El Dorado County, California, where it has been abundantly collected.

This species is very close to *Rorippa Columbiae* Suksd., from which it differs most noticeably in its glabrous capsules and its unexpanded stigma.

4. Rorippa curvisiliqua (Hook.) Bessey, Mem. Torrey Club 5:169. 1894.

Radicula curvisiliqua Greene, Leafl. Bot. Obs. & Crit. 1:113. 1905.

Plant annual, or sometimes behaving as biennial near the coast, diffusely branched; branches ascending, 1–3 dm. tall, glabrous; leaves pinnatifid, the segments oblong, obtuse; flowers in short racemes, the pedicels 2–4 mm. long, ascending or spreading; petals yellow; pod 8–12 mm. long, curved, beaked by a short, persistent style.

Common in wet places at low and middle altitudes: almost throughout California; north to British Columbia, south to Baja California in Mexico, east to Rocky Mountains.

5. Rorippa palustris (L.) Bess., Enum. Pl. Volh. 27. 1821.

Radicula palustris Moench, Meth. Pl. Marburg. 263. 1794.

Biennial herb; stems erect, branching above, essentially glabrous; leaves lyrate-pinnatifid, the terminal lobe usually large, coarsely toothed; petals 2 mm. long; pod linear to linear-oblong, straight or slightly curved, the style about 1 mm. long.

Widely scattered in our area in marshy places; cosmopolitan in the Northern Hemisphere.

Rorippa palustris subsp. *occidentale* (Wats.) Abrams (*R. islandica* var. *occidentalis* Butters & Abbe) is the common form of *R. palustris* in California.

6. Rorippa hispida (Desv.) Britt., Mem. Torrey Club 5:169. 1894.

Plant annual or biennial; stems stout, erect, branched above base, 3–12 dm. tall, beset with sparse or dense, rather stiff hairs; leaves lyrate-pinnatifid, the lower surface, petioles, and veins often hirsute; pod ovoid, 4–6 mm. long, glabrous, tipped with a persistent style.

Marshes and bogs: northern California; north to Alaska, east to Atlantic.

The material from northeastern California is nearly glabrous.

Fig. 248. *Drosera rotundifolia: a,* habit, × 1⅕; *b,* leaf blade, showing stalked glands,
× 4; *c,* flower at anthesis, showing nonspreading petals and sepals, × 6; *d,* flower, spread
open to show stamens and pistil, × 6; *e,* flower with maturing capsule, × 7; *f,* scape and
raceme of flowers with maturing capsules, × 2.

7. Rorippa obtusa (Nutt.) Britt., Mem. Torrey Club 5:169. 1894.

Diffusely branched, glabrous annual; stems decumbent at base, ascending, 1–3 dm. tall; leaves pinnatifid, with obovate, rounded lobes, or blades simple and entire; pedicels ascending or spreading, 2–4 mm. long; petals yellow, spatulate, 1 mm. long; pod 4–8 mm. long, ovoid.

Wet soil along streams and ditches: low and middle altitudes in California; Mexico to British Columbia, east to Mississippi Valley.

The entire-leaved forms have been segregated as *Rorippa obtusa* var. *integra* Marie-Victorin. We have collected it near Bridgeport in Mono County.

SUBULARIA

1. Subularia aquatica L., Sp. Pl. 642. 1753. Water awlwort.

Tufted, aquatic, annual herb; leaves terete, 1–3 cm. long, in a basal rosette, glabrous; flowers in a short raceme on a naked scape; petals white; stamens 6; style none; pod short-stipitate, 2–3 mm. long, ovoid or spheroid, the valves 1-ribbed.

In clear cold water in the mountains, usually in ponds or lakes: Sierra Nevada; north to Canada, east to Maine.

SARRACENIACEAE

Herbaceous, insectivorous perennials inhabiting swamps or bogs. Leaves basal, tubular, retrorsely pubescent within, usually in rosettes. Flowers solitary, nodding, regular. Sepals 4 or 5, imbricate. Petals 5 or absent, distinct, hypogynous. Stamens numerous. Pistil 1, 3- to 5-celled; placentation axile; ovules numerous. Fruit a loculicidal capsule. Seeds clavate, covered with slender protuberances.

DARLINGTONIA

1. Darlingtonia californica Torr., Smithson. Contr. 6 (Art. 4):5. 1854. Pitcher plant. Fig. 247.

Herb 1–6 dm. tall; leaves with tubular petioles enlarged upward and terminating in a rounded hood with a circular orifice on one side; blades 2-forked, attached to the hood at top of orifice; flowers pendulous on a scape, solitary; sepals yellowish green, 5, rotate, persistent, 3–5 cm. long; petals 5, dark reddish purple, heavily veined, narrow-ovate, 2–3 cm. long; stamens 13–15, 1 cm. long; ovary superior, 3- to 5-celled, cuneate-obovoid; stigmas 5, rotate on a short style; fruit a loculicidal capsule, obovate to oblong, 3–4 cm. long.

Boggy meadows and stream banks, sea level to an elevation of 6,000 feet:

Fig. 249. *Drosera longifolia:* *a,* habit, × 1⅕; *b,* leaf blade, showing stalked glands, × 4; *c,* part of raceme, showing flowers at anthesis, × 3; *d,* flower, spread open to show stamens and pistil, × 4; *e,* mature capsule, × 6; *f,* seed, × 28.

northern Sierra Nevada, northwestern California in Siskiyou, Trinity, and Del Norte counties; Oregon.

DROSERACEAE

Ours small, perennial herbs, glandular. Leaves long-petioled in a basal rosette. Flowers regular, hypogynous. Ovary 1-celled; styles 1–5; ovules numerous. Capsule 3-valved.

DROSERA

Perennial herbs of sphagnum bogs; the orbicular or oblong leaf blades covered with stalked red glands; flowers in a one-sided raceme on a naked scape, hypogynous; calyx persistent, 5-parted; petals 5; stamens 5; styles 2–5, usually 3, often deeply 2-lobed; ovules numerous; placentation parietal; capsule 3-valved.

Leaf blades orbicular, broader than long, spreading; sepals subequal to the petals in length
1. *D. rotundifolia*
Leaf blades 4–8 times as long as broad, erect or nearly so; sepals ½ as long as the petals
2. *D. longifolia*

1. Drosera rotundifolia L., Sp. Pl. 282. 1753. Round-leaved sundew. Fig. 248.

Plants with flowering stems to 25 cm. (mostly 10–15 cm.) long; leaves in a basal rosette, spreading, long-petiolate, the blades nearly orbicular, about 1 cm. in diameter, the upper surface and the margins bearing long, gland-tipped hairs; flowering stalk slender, erect, with a terminal, 2- to 10-flowered (sometimes to 25-flowered), one-sided raceme; pedicels 2–4 mm. long; sepals 4 mm. long, about as long as the white or reddish petals; seeds numerous, spindle-shaped.

Bogs: North Coast Ranges from Sonoma County to Del Norte County, Sierra Nevada at elevations of 3,000–8,000 feet. from Tulare County north to Siskiyou County; north to Alaska, east to Labrador, Europe, Asia.

2. Drosera longifolia L., Sp. Pl. 282. 1753. Oblong-leaved sundew. Fig. 249.

Plants with flowering stems to 15 cm. tall; leaves in a basal rosette, erect or nearly so, the blades obovate-spatulate, gradually tapering to the long, slender petiole, the margins and upper surface bearing long gland-tipped hairs; flowering stalk slender, erect, bearing a few- to several-flowered, one-sided raceme; pedicels 2–4 mm. long; sepals 4 mm. long, about ½ as long as the white petals; seeds oblong with a loose, rough seed coat.

Bogs: Nevada and Plumas counties in the Sierra Nevada; north to Alaska and east to Atlantic, Europe, Asia.

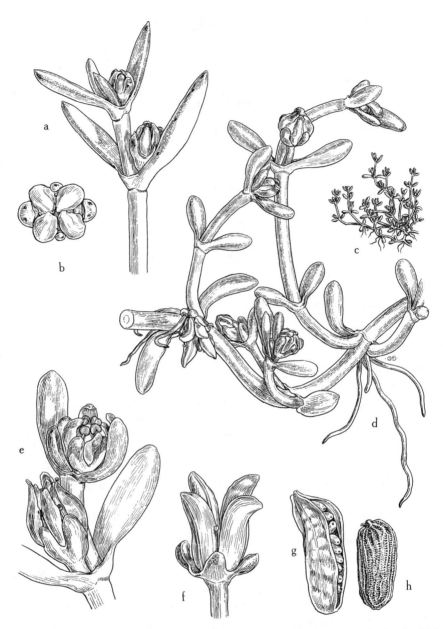

Fig. 250. *Tillaea aquatica: a,* tip of branch, showing flowers solitary in leaf axils, × 6; *b,* flower, top view, × 12; *c,* habit, × ⅔; *d,* part of plant, showing the flowers, and the stems rooting at the nodes, × 6; *e,* flowers, showing different stages in anthesis, × 12; *f,* fruit, × 12; *g,* single carpel, after dehiscence, × 16; *h,* seed, × 60.

CRASSULACEAE

Succulent, glabrous, annual or perennial herbs. Leaves (in ours) opposite, simple, entire, fleshy, lacking stipules. Inflorescence bracteate, cymose, or sometimes solitary. Flowers small, hypogynous, perfect, regular. Sepals and petals of same number, 4- to 30-merous (often 5-merous), distinct. Stamens usually in 2 whorls, as many or twice as many as petals, distinct. Pistils 3 or more, distinct or basally connate; ovary superior, 1-loculed, 1-carpeled; style 1; stigma 1. Fruit of 1- to many-seeded, free or partially united follicles.

TILLAEA

1. Tillaea aquatica L., Sp. Pl. 128. 1753. Fig. 250.

Tillaea aquaticum (L.) Britt., Bull. N. Y. Bot. Gard. 3:1. 1903.

Diminutive, glabrous, succulent annual; stems usually decumbent and rooting at lower nodes, branched, 2–6 cm. long or sometimes becoming free-floating with elongate internodes; leaves entire, oblong, 4–6 mm. long, opposite and connate at base; flowers very small, 4-merous, perfect, sessile or short-pedicellate, solitary in axils; sepals 1 mm. long; petals greenish white, oblong, distinct, twice as long as sepals; carpels distinct, erect, 8- to 10-seeded, somewhat spreading in fruit; styles short.

Wet ground or vernal pools: widely distributed in California; North America, also Europe.

Doubtfully distinct is *Tillaea aquatica* var. *Drummondii* (Torr. & Gray) Jepson, in which the lower pedicels elongate in fruit and become longer than the leaves.

SAXIFRAGACEAE

Perennial herbs with simple, alternate leaves, mostly basal. Flowers perfect, perigynous, borne in racemes, panicles, or cymes, or rarely solitary. Calyx 5-lobed or 5-cleft. Petals commonly 5, sometimes 4–8. Stamens 5 or 10, sometimes 3, or 8–20, or numerous. Carpels 2, rarely 3 or 5, wholly or partially united or rarely distinct. Styles as many as placentae, the latter axile or parietal. Fruit a capsule or follicle. Seeds with fleshy endosperm.

Seeds borne on placentae attached to the ovary wall.
  Flowers clustered in the axils of the upper leaves; calyx lobes normally 4; petals absent; stamens 8 . . . . *Chrysosplenium*
  Flowers in more or less elongated racemes; calyx lobes 5; petals normally present.
    Petals filiform, entire; calyx tube irregular, stamens 3 . . . . *Tolmiea*
    Petals pinnately dissected; calyx tube regular, stamens 5 . . . . *Pectiantia*

Seeds borne on a central placenta.
    Stamens 5.
        Ovary superior; petals persistent, flower more or less purple .... *Bolandra*
        Ovary inferior; petals deciduous .... *Boykinia*
    Stamens 10.
        Flowers on essentially naked scapes, basal leaves appearing after the flowering stage,
            leaves orbicular-peltate .... *Peltiphyllum*
        Flowers appearing with or after the leaves .... *Saxifraga*

BOLANDRA

1. Bolandra californica Gray, Proc. Am. Acad. Arts & Sci. 7:341. 1868.

Perennial herb with short, bulblet-bearing rootstock; stem 1–3 dm. tall, slender, leafy, glabrous below, glandular-puberulent above; basal and lower cauline leaves petioled, the petioles 3–10 cm. long; leaf blades round-cordate, glabrous, 5- to 7-lobed, the segments rounded and coarsely toothed, upper cauline leaves sessile, ovate, toothed; sepals 5, long-attenuate; petals 5, purplish, recurving; stamens 5, opposite the sepals; styles 2.

In moist places or on wet rocks: central Sierra Nevada from Mariposa County to El Dorado County.

BOYKINIA

Perennial herbs with scaly rootstocks and with leafy lateral flowering shoots. Basal tuft of leaves reniform, lobed, and crenate, the stipules either leafy or reduced to bristles. Inflorescence in a paniculate or corymbose cyme bearing white flowers. Sepals 5, lanceolate or ovate-lanceolate. Petals 5, obovate, usually narrowed to a claw, usually soon-deciduous. Stamens 5, opposite sepals. Ovary 2-celled. Styles 2. Capsule 2-beaked.

Leaves cleft or incised with acute teeth; petals longer than the sepals.
    Stipules often large, foliaceous .... 1. *B. major*
    Stipules small, reduced to brownish bristles .... 2. *B. elata*
Leaves merely rounded-lobed, crenate with broad, mucronate teeth; petals about equal in
    length to the sepals .... 3. *B. rotundifolia*

1. Boykinia major Gray, Bot. Calif. 1:196. 1876.

Stout plant with stems 3–10 dm. tall, somewhat glandular-scabrous with brown hairs; petioles 1–2 dm. long; leaf blades roundish-cordate, 5–20 cm. broad, 5- to 7-cleft, the divisions again cleft and coarsely toothed, the lowest stipules membranous, the upper stipules foliaceous, partly clasping the stem; flowers densely clustered in a cymoid panicle, elongated in fruit, densely glandular-puberulent; calyx tube at first hemispherical, soon urn-shaped or sub-globose, the lobes triangular-acute; petals white, broadly ovate or obovate, clawed, 5–7 mm. long.

Along stream banks and in wet meadows: Sierra Nevada from Madera County north; northeast to Montana.

2. Boykinia elata Greene, Fl. Fran. 190. 1891.

Stems slender, erect, 2–6 dm. tall, more or less brown-hairy below, densely glandular above; petioles of basal leaves 3–6 cm. long, hirsute with brown hairs; stipules at base of petioles reduced to rusty bristles; leaf blades cordate, 2–5 cm. wide, shallowly lobed or incised, serrate; flowers slightly irregular, borne in a panicle of one-sided racemes, densely glandular-puberulent; bractlets usually foliaceous, spatulate, incised; calyx tube campanulate, often purplish and glandular below; calyx in age urn-shaped, the lobes lanceolate-triangular; petals obovate to oblanceolate, obtuse.

Along wooded streams: Coast Ranges and Sierra Nevada; north to British Columbia.

3. Boykinia rotundifolia Parry ex Gray, Proc. Am. Acad. Arts & Sci. 13:371. 1878.

Stems stout and tall, 6–10 dm. tall, densely glandular-hirsute; basal and lower cauline leaves long-petioled, the petioles 1–1.5 dm. long, also densely glandular-villous; blades round-cordate, 1–1.5 dm. in diameter, with many shallow, rounded, crenate lobes; the upper stem leaves short-petioled and orbicular; calyx 10-nerved, hirsutulous, more or less glandular; petals obovate, spatulate.

Along shaded banks of canyons: mountains of southern California.

CHRYSOSPLENIUM

1. Chrysosplenium glechomaefolium Nutt. in Torr. & Gray, Fl. N. Am. 1:589. 1840. Pacific water carpet.

Glabrous perennial herb, somewhat succulent, dichotomously branched, with a slender rootstock; stems mostly ascending, often rooting at the lower nodes, 8–25 cm. long; leaves opposite, 8–15 mm. broad; petioles 1–2 cm. long; flowers solitary, small, greenish yellow; calyx rotate, 4-lobed; petals absent; stamens 8; capsule at length exserted.

Wet places, coastal region: North Coast Ranges from Mendocino County north; north to British Columbia.

PECTIANTIA

Low perennials with scaly rootstocks. Stems slender, scapiform, terminating in a spikelike raceme of small flowers. Calyx tube partially adnate to ovary, 5-lobed, lobes triangular or ovate, reflexed. Petals 5, pinnately divided. Sta-

Fig. 251. *Peltiphyllum peltatum: a,* basal part of plant, showing stout, fleshy rootstock, × ⅖; *b,* peltate leaves arising from rootstock, × ¼; *c,* inflorescence, × ⅖; *d,* flower, × 2; *e,* pistil, showing the 2 carpels, × 6; *f* and *g,* seeds, × 12; *h,* fruit, × 4.

mens 5, filaments very short. Styles 2, short. Seeds numerous, obovoid, smooth and shiny.

Stamens opposite to the petals .... 1. *P. pentandra*
Stamens opposite to the sepals .... 2. *P. ovalis*

1. Pectiantia pentandra (Hook.) Rydb., N. Am. Fl. 22:93. 1905.

Mitella pentandra Hook., Bot. Mag. 56: pl. 2933. 1829.

Plant 1–3 dm. tall; scapes minutely glandular-puberulent; petioles of basal leaves 5–10 cm. long, hairy; leaf blades broadly cordate, coarsely toothed, with 9–11 rounded lobes, sparingly hairy on both sides; calyx green, purplish inside; petals divided into 5–7 slender divisions; stamens 5, opposite petals.

Montane streams and bogs: northern Coast Ranges and southern Sierra Nevada; Rocky Mountains, north to Alaska.

2. Pectiantia ovalis (Greene) Rydb., N. Am. Fl. 22:94. 1905.

Mitella ovalis Greene, Pittonia 1:32. 1887.

Plant 1–3 dm. tall; scape and petioles hirsute with reflexed long hairs; leaf blades oval with a cordate base, with 5–9 rounded, shallow lobes, broadly toothed; coarsely hirsute on both sides; flowers yellowish green; stamens 5, opposite to sepals.

In wet places in coastal regions: North Coast Ranges from Marin County north; north to British Columbia.

PELTIPHYLLUM

1. Peltiphyllum peltatum (Torr.) Engler in Engler & Prantl, Nat. Pflanzenf. 3²ᵃ:61. 1891. Fig. 251.

Perennial, coarse herb, with stout, fleshy rootstocks, 3–10 dm. tall; scapes naked; leaves all basal, tufted; petioles long and hairy; leaf blades peltate, orbicular, 1–4 dm. broad, many-lobed, toothed, glabrous or nearly so; flowers white (appearing before the leaves), in terminal, simple or paniculate, compound cymes; sepals 5, 3–4 mm. long, ovate to oblong-ovate, reflexed in age; petals 5, broad, sessile, 5–7 mm. long; stamens 10; follicles 8–11 mm. long, more or less spreading.

In swift streams and along stream banks: Sierra Nevada north from Tulare County, North Coast Ranges north from Humboldt County; southwestern Oregon.

SAXIFRAGA

Mostly perennial herbs, the simple leaves mainly in a basal cluster. Flowers in ours white, usually paniculately cymose. Calyx adnate to at least the base of the ovary. Sepals 5. Petals perigynous. Stamens 10, inserted with petals. Capsule 2-beaked. Seeds small, numerous.

Leaves long-petioled; inflorescence open.
  Leaf blades orbicular, cordate, the margins coarsely crenate.
    Petiole with stipule-like wing at base; leaf margins with larger crenations in turn dentate; aspect of inflorescence green .... 1. *S. Mertensiana*
    Petiole without stipule-like wing at base; leaf margins simply and deeply crenate; aspect of inflorescence usually purplish .... 2. *S. arguta*
  Leaf blades oblong to ovate, the margins dentate .... 3. *S. Marshallii*
Leaves sessile to mostly broadly winged to the base or short-petioled; flower clusters compact .... 4. *S. oregana*

1. Saxifraga Mertensiana Bong., Mém. Acad. St. Pétersb. VI. Math. Phys. Nat. 2:141. 1832.

Acaulescent perennial with a short, thick, scaly rootstock; leaf blades orbicular, cordate, 2–7 cm. broad, glabrous or hirsutulous, the margins crenate, the crenation usually with 3 minor crenations, on petioles 4–12 cm. long; flower stalk 10–30 (–40) cm. long, terminating in an open, paniculately cymose inflorescence; sepals 1.5–2 mm. long, abruptly recurved after anthesis; petals narrowly ovate, white, twice as long as the sepals; filaments broadened above, petaloid; mature fruits to 6 mm. long and nearly as broad.

Moist woods and along stream banks: North Coast Ranges from Sonoma County north, occasional in the Sierra Nevada north of Mariposa County; north to Alaska.

2. Saxifraga arguta D. Don, Trans. Linn. Soc. London 13:356. 1822. Brook saxifrage.

Acaulescent perennial with a long rootstock; leaf blades orbicular, cordate, the margins coarsely and deeply crenate, glabrous, the petioles 4–12 (–20) cm. long; flower stalks 5–40 cm. long, terminating in an open, paniculately cymose inflorescence; sepals 1.5–2 mm. long, often purple, sharply recurved; petals suborbicular, abruptly and shortly clawed, about twice as long as the sepals; filaments broadened above, petaloid, conspicuous because of the early falling of the petals; mature fruits to 10 mm. long, twice as long as broad.

Along streams at high elevations: Sierra Nevada and extreme northern area of North Coast Ranges; north to British Columbia, east to Rocky Mountains, south to Arizona and New Mexico.

3. Saxifraga Marshallii Greene, Pittonia 1:159. 1888. Spotted saxifrage.

Acaulescent perennial from a short rootstalk; leaf blades oblong to ovate, 1–5 cm. long, with dentate or crenate-dentate margins, glabrous above, often rusty-long-pubescent below, the petioles 1–10 cm. long; flower stalks 7–40 cm. long, terminating in an open, paniculately cymose inflorescence; sepals reflexed, 2–4 mm. long; petals oval to oblong, slightly longer than the sepals and with

2 yellow spots below the middle; filaments broadened above, petaloid; mature fruits 2.5–4 mm. long, the 2 carpels spreading above the middle.

Moist, rocky places along rivers and creeks: North Coast Ranges north from Mendocino County; north to Washington and Idaho.

4. Saxifraga oregana Howell, Erythea 3:34. 1895. Bog saxifrage.

Saxifraga integrifolia Hook. var. *sierrae* Cov., Proc. Biol. Soc. Wash. 7:78. 1892.

Large, coarse, acaulescent perennial from a long and very stout rootstock; leaves oblong-spatulate, remotely and shallowly toothed, glabrous, 8–30 cm. long, the blades gradually tapering to form a broadly winged petiole; flower stalks 30–90 cm. tall, terminating in a paniculately cymose inflorescence, the cymules densely flowered; sepals 2–5 mm. long, reflexed, equaling the obconic calyx tube in length; petals 2–4 mm. long, elliptic to obovate, sessile; filaments subulate; fruits 3–7 mm. long and as broad, the beaks spreading at right angles.

Mountain meadows: Sierra Nevada from Tulare County north, and in the northern part of North Coast Ranges; north to Washington, east to Idaho and Nevada.

TOLMIEA

1. Tolmiea Menziesii (Pursh) Torr. & Gray, Fl. N. Am. 1:582. 1840.

Perennial herb with long-petioled basal leaves and lateral leafy, racemose flowering branches; petioles 0.5–2 dm. long; leaf blades round-cordate or cordate-ovate, lobed, irregularly serrate, 2–12 cm. broad; racemes elongate, bracteate, 20- to 60-flowered; calyx tubular with 5 lobes unequal in size (2 large, 3 small), deeply cleft on one side; petals 5 or 4, filiform, elongated, inserted in the sinuses of the calyx opposite the upper and lateral lobes; ovary oblong, the 2 valves spreading in age, the fruit protruding through the slit on the lower side of the persistent calyx.

Along mountain streams: North Coast Ranges north from Mendocino County; north to southern Alaska.

PARNASSIACEAE. GRASS-OF-PARNASSUS FAMILY

Glabrous perennial herb with short rootstock and scapelike stem. Leaves basal, entire and petioled, the single cauline leaf or bract sessile. Scape 1-flowered. Sepals 5. Petals 5, white or pale yellow, veined and deciduous. Stamens 5, hypogynous or perigynous, alternate with petals and with clusters of more or less united gland-tipped staminodia, the filaments subulate. Ovary 1-celled, superior or half-inferior, with 3 or 4 parietal placentae. Stigmas commonly 4, sessile. Capsule 3- or 4-valved. Seeds numerous, winged.

PARNASSIA

1. Parnassia californica (Gray) Greene, Pittonia 2:102. 1890.

Parnassia palustris var. *californica* Gray, Bot. Calif. 1:202. 1876.

Scape 3–6 dm. tall; basal leaves round-ovate to elliptic, 2–4 cm. wide; petioles 2–15 cm. long; sepals 4–6 mm. long, oval to oblong; petals 10–15 mm. long, entire, 3- to 7-veined; sterile filaments capillary, 20–24 in a set united to the middle.

Wet meadows: San Bernardino Mountains, Sierra Nevada, and from San Benito County north in Coast Ranges in widely scattered localities; to Josephine County, Oregon.

PLATANACEAE

Large trees with bark falling in thin plates. Leaves large, alternate, palmately lobed. Stipules thin, sheathing, entire or toothed. Flowers small, greenish, the staminate and pistillate flowers in separate, dense, pedunculate heads. Sepals and petals minute. Stamens with long anthers and very short filaments. Ovary 1-ovuled. Style 1. Fruit a small nut.

PLATANUS

1. Platanus racemosa Nutt., Sylva 1:47. 1842. California plane tree, California sycamore.

Tree 10–25 m. tall; leaves palmately 3- to 5-lobed, usually cordate or truncate below the middle, 10–15 cm. broad, scarcely as long as broad, the lobes usually entire, tomentose when young, usually glabrate in age; petioles shorter than leaf blade; staminate heads 8–10 mm. in diameter; pistillate heads 2–2.5 cm. in diameter in fruit.

Along river and stream banks in the foothills and valleys; upper Sacramento Valley and interior valleys of Coast Ranges, south to coastal southern California; northern Baja California, Mexico.

ROSACEAE

Herbs, shrubs, or trees, with alternate, simple, often pinnately dissected or compound leaves; stipules usually present. Flowers usually perfect, solitary or in clusters, perigynous or epigynous. Sepals normally 5, rarely 4, or 6–9, often subtended by bractlets. Petals as many as sepals or absent. Stamens 1 to many, commonly in 3 series. Pistils distinct, 1 to many. Ovary 1-celled with 1 to several ovules. Fruit achenes, follicles, or drupelets.

Fruit of 1–5 dehiscent follicles.
  Flowers perfect; carpels 1–5, when more than 1, united below; shrub .... *Physocarpus*
  Flowers dioecious; carpels usually 5, distinct; tall herbs .... *Aruncus*
Fruit of indehiscent achenes or drupelets.
  Pistils more than 1.
    Pistils becoming drupelets; leaves compound or simple .... *Rubus*
    Pistils becoming dry achenes; leaves compound or pinnately dissected.
      Style deciduous from achene; herb .... *Potentilla*
      Style persistent on achene .... *Geum*
  Pistil only 1; petals absent .... *Sanguisorba*

ARUNCUS

1. Aruncus vulgaris Raf., Sylva Tell. 152. 1838. Goat's beard.

Aruncus sylvester Kostel., Ind. Pl. Hort. Prag. 151. 1844.

Tall perennial herb with erect stems, 1–2 m. tall, glabrous; leaves 2- or 3-pinnate, the leaflets ovate-lanceolate, acuminate, irregularly serrate, 3–12 cm. long, more or less hairy on both surfaces; flowers dioecious in a terminal compound panicle composed of many long slender spikes; sepals 5; petals 5, 1 mm. long, white; stamens 15–30, long-exserted; pistils 3–5, distinct; style short; follicles reflexed in fruit, about 3 mm. long.

Along shady streams and in bogs: North Coast Ranges; north to Alaska and east to Atlantic, Europe, Asia.

GEUM

1. Geum macrophyllum Willd., Enum. Hort. Berol. 557. 1809. Big-leaf avens.

Perennial herb with stout stems, erect, bristly-pubescent, 3–10 dm. tall; stipules broad, foliaceous; basal leaves petioled, the leaflets incised and serrate, the terminal leaflets very large, round-cordate, 6–10 cm. broad, the lateral leaflets oval or obovate, with smaller ones interspersed, the stem leaves reduced; flowers in open cyme; bractlets linear, minute; sepals 3–5 mm. long; petals yellow, 4–8 mm. long; fruiting heads globose or slightly elongate, 5–12 cm. long; achenes small with a hooked beak.

Wet meadows in the mountains: southern California north; north to Alaska, east to Montana, Newfoundland, and New Hampshire, eastern Siberia.

PHYSOCARPUS

1. Physocarpus capitatus (Pursh) Kuntze, Rev. Gen. Pl. 2:219. 1891. Ninebark.

Erect or straggly shrub with exfoliating bark, 1–5 m. tall; leaves round-ovate, 3- to 5-lobed, the lobes irregularly serrate, glabrous or stellate-pubescent

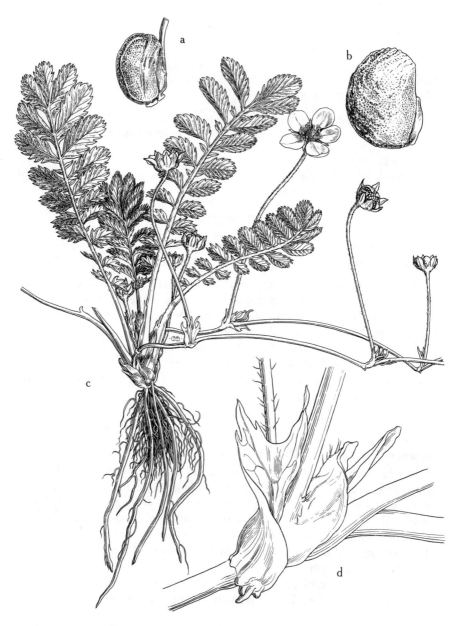

Fig. 252. *Potentilla Anserina: a,* young achene, showing basal part of style, × 12;
b, mature achene, × 12; *c,* habit, × ⅔; *d,* stipules, × 4.

below, 3–6 cm. long; petioles 1–2 cm. long; flowers in hemispherical, densely flowered terminal corymbs; calyx tube 5-lobed, stellate-pubescent; petals 5, white, 3 mm. long; stamens 20–40; follicles 3–5, more or less hairy.

Along streams or on wet banks: Sierra Nevada, Santa Barbara County north through Coast Ranges; north to southern British Columbia and Idaho.

POTENTILLA

Perennial herbs with pinnately compound leaves. Flowers yellow or purple in terminal cymes, or sometimes caespitose. Calyx tube (hypanthium) saucer-shaped, campanulate, or cup-shaped, bearing 5 lobes and 5 alternate bractlets. Petals 5. Stamens usually 20, the filaments filiform or dilated. Pistils many, in fruit becoming dry achenes, the style basal, lateral, or terminal or nearly so.

Leaves pinnately compound.
  Petals purple, inconspicuous; stems stout, ascending; leaves 5- to 7-foliate, green above
    and below .... 1. *P. palustris*
  Petals yellow, conspicuous; stems slender, creeping and rooting at nodes; leaves many-
    foliolate, white-silky below.
    Stems and pedicels glabrous or nearly so; achenes without a groove ... 2. *P. pacifica*
    Stems and pedicels pubescent with ultimately spreading hairs; achenes with a deep
      dorsal groove .... 3. *P. Anserina*
Leaves palmately compound, 3-foliolate; flowers inconspicuous; petals pale yellow
                                                              4. *P. millegrana*

1. Potentilla palustris (L.) Scop., Fl. Carn., 2d ed., 1:359. 1772. Marsh poten-
    tilla.

Stems stout, ascending from long, creeping rootstocks, 2–5 dm. tall, often rooting at the decumbent base, glabrous below, puberulent above; leaves 5- to 7-foliolate, the lower leaves long-petioled, the leaflets 2–6 cm. long, oblong or oval, serrate, obtuse or acute, the stipules membranaceous; sepals ovate, acu-minate, 10–15 mm. long, wine-purple inside; petals dark purple, smaller than sepals.

Swamps and bogs: northern Sierra Nevada, North Coast Ranges; north to Alaska.

2. Potentilla pacifica Howell, Fl. NW. Am. 1:179. 1898. Pacific silverweed.
  Potentilla Anserina var. *grandis* Torr. & Gray, Fl. N. Am. 1:444. 1840.

Very similar to *Potentilla Anserina* in habit, except for its more or less gla-brous herbage and achenes which are neither corky nor grooved.

Coastal salt-water and fresh-water marshes and moist sand flats: Ventura County north to Humboldt County; north to Alaska.

3. Potentilla Anserina L., Sp. Pl. 495. 1753. Silverweed. Fig. 252.

Leaves and peduncles in a basal tuft from a cluster of roots (main stem almost none) producing numerous runners 3–6 dm. long; leaves 1–2 dm. long, pinnate with 9–31 larger leaflets, and smaller ones interposed, spreading or flat on the ground, silky and green above, white-silky and tomentose beneath; flowers 1–2 cm. in diameter, on pedicels 3–10 cm. long; bractlets simple and lanceolate, or often broader and ovate-lanceolate, toothed or divided, generally a little longer than the broadly ovate sepals; petals oval, 7–10 mm. long; achenes numerous, corky, grooved at upper end.

Marshy or springy places: cismontane southern California; east to Atlantic, Europe, Asia.

4. Potentilla millegrana Engelm. ex Lehm., Ind. Sem. Hort. Hamb. 1849: Add. 12. Collect. No. 19. 1849.

Decumbent, many-branched annual or biennial, soft-pubescent or glabrate; leaves 3-foliolate, the leaflets oblong-cuneate; cyme spreading, with many inconspicuous flowers, its leaves bractlike; flowers 6–8 mm. broad, the pale yellow petals half as long as the sepals; stamens usually 10; achenes smooth, light-colored.

River bottomlands: in scattered localities throughout California; north to Washington, east to Dakota.

RUBUS

1. Rubus spectabilis Pursh, Fl. Am. Sept. 1:348. 1814. Salmonberry.

Shrub, stems erect, 2–5 dm. tall, with reddish brown or yellowish, shreddy bark, unarmed, or with short, straight prickles; leaves usually 3-foliolate, the petioles 4–6 cm. long, the leaflets thin, ovate, doubly serrate, often more or less lobed, sparingly pubescent on both sides, the terminal leaflet 4–10 cm. long, acuminate at apex, truncate or cuneate at base; flowers 2–4 or usually solitary; sepals ovate, about 1 cm. long; petals reddish purple, elliptical, 15–20 mm. long; berry ovoid, red or yellow, 15–20 mm. long.

Stream banks: Santa Cruz County and north in Coast Ranges; east to Idaho and north to the Aleutian Islands.

Material from Santa Cruz to Sonoma County is sometimes segregated taxonomically as *Rubus spectabilis* var. *Menziesii* Wats. because of the densely pilose undersurface of leaves.

SANGUISORBA

1. Sanguisorba microcephala Presl, Epimil. Bot. 202. 1849.

Glabrous perennial with slender stem, 1–2 m. tall, simple or branched above; leaves unequally pinnate, the lower stipules narrowly lanceolate, adnate to the petioles, the upper stipules expanded above with a rounded, foliaceous, toothed blade, the leaflets oblong-ovate, 1–4 cm. long, 1–1.25 cm. wide, rounded at the tip, cordate at the base, dark green above, paler beneath, toothed; spike oblong-cylindric, dense, 15–25 mm. long; sepals dark purple, oval, 2–2.5 mm. long; achene enclosed in 4-winged hypanthium.

Swamps: northern California; north to Alaska.

LEGUMINOSAE

Herbs, shrubs, or trees (ours herbs). Leaves alternate, stipulate, usually compound. Flowers irregular, perfect, hypogynous. Calyx synsepalous, 5-toothed or 2-lipped, usually persistent. Corolla with 5 clawed petals, these (in ours) papilionaceous in arrangement (upper banner enclosing the 2 lateral wings, which in turn enclose 2 lower petals that are fused to form a keel, the whole "butterfly-like"). Stamens 10, the filaments usually united into a sheath around the ovary (monadelphous) or the upper stamen free (diadelphous). Pistil 1, simple, with many ovules and parietal placentation. Fruit a legume (2-valved pod) with a single row of seeds along ventral suture, usually opening by both dorsal and ventral sutures, but sometimes indehiscent.

Leaves pinnate, rachis terminating in a tendril .... *Lathyrus*
Leaves pinnate or palmate, rachis not terminating in a tendril.
  Calyx deeply 2-lipped; stamens monadelphous .... *Lupinus*
  Calyx 5-toothed; stamens diadelphous.
    Perennial herbs with heavy odor and dark glands .... *Psoralea*
    Annual or perennial herbs without heavy odor and dark glands.
      Leaflets 3.
        Leaves pinnately 3-foliolate, terminal leaflets distinctly petiolulate; flowers in racemes or spikes .... *Melilotus*
        Leaves palmately 3-foliolate; flowers in a head or very short spike .. *Trifolium*
      Leaflets 5 to many.
        Leaflets 20–70, evenly pinnate; pods septate between seeds .... *Sesbania*
        Leaflets 5–11, odd-pinnate; pods nonseptate .... *Lotus*

LATHYRUS

Herbaceous vines, ours perennial. Leaves pinnate, tendril-bearing, the stipules conspicuous. Flowers showy, axillary, usually in one-sided racemes. Calyx with teeth of unequal size, the 2 upper teeth much reduced. Corolla with large, broad banner. Stamens diadelphous. Style flat, hairy along upper side next the free stamen. Pods flat.

Stems slender; leaflets 2 or 3 pairs; inflorescence 2- to 6-flowered .... 1. *L. palustris*
Stems stout; leaflets 4–12 pairs; inflorescence 6- to 15-flowered .... 2. *L. Jepsonii*

Fig. 253. *Lathyrus Jepsonii: a,* fruit, $\times$ ⅘; *b,* habit, $\times$ ⅖; *c,* seed, $\times$ 4; *d,* flower, $\times$ 2; *e,* stem (cross section), $\times$ 2; *f,* diadelphous stamens, and the style hairy on one side near tip, $\times$ 2.

1. Lathyrus palustris L., Sp. Pl. 733. 1753.

Climbing perennial herb to 1 m. tall; stems winged on angles; herbage glabrous to sparsely pubescent; leaflets 4–6, linear-oblong, 2–7 cm. long; tendrils branched or simple; stipules sagittate; peduncles longer than leaves, 3- to 6-flowered; corolla blue or purple, 10–14 mm. long; pods glabrous, 3–5 cm. long.

Coastal marshes: Humboldt and Del Norte counties; north to Alaska, east to New England, Europe.

2. Lathyrus Jepsonii Greene, Pittonia 2:158. 1890. Fig. 253.

Climbing perennial herb, 1–3 m. tall; stems winged on the angles; herbage glabrous; leaflets 8–12, linear-lanceolate to oblong, mucronulate. 2–7 cm. long; stipules sagittate, both lobes lanceolate or oblong; peduncles shorter than leaves; racemes 6- to 15-flowered; corolla deep rose, 16–20 mm. long; pods 7–12 cm. long.

Margins of sloughs, often among tules: delta region of the Central Valley, Suisun marshes, San Pablo Bay.

LOTUS

Annual or perennial herbs, sometimes suffrutescent. Leaves alternate, pinnately compound, leaflets 3 to many, the stipules foliaceous or reduced to glands. Flowers solitary or umbellate, usually leafy-bracteate. Calyx with 5 teeth subequal in size. Corolla yellow to white, sometimes marked with red or pink. Stamens diadelphous. Pod flat or cylindrical, straight to curved, 1- to many-seeded, dehiscent or indehiscent.

Stems from a thick root crown; petal claws long, well exserted from calyx
                                                                1. L. formosissimus
Stems from slender, branching rootstocks; petal claws short, scarcely or not exserted from
      calyx .... 2. L. oblongifolius

1. Lotus formosissimus Greene, Pittonia 2:147. 1890. Fig. 254.

Hosackia gracilis Benth., Trans. Linn. Soc. London 17:365. 1837.

Herbaceous perennial; stems several from a thick root crown, decumbent, 1–4 dm. long; herbage glabrous; leaves with 5–9 leaflets, these elliptic to obovate, retuse, 7–20 mm. long, equally distributed on opposite sides of rachis, the stipules membranous; peduncles 4–8 cm. long, about equaling or exceeding the leaves in height; umbels 4- to 6-flowered, subtended by a 3- to 5-foliate, petioled bract; calyx tube 3 mm. long, the teeth nearly equal in size, 2 mm. long, subulate; petals 12–14 mm. long, their claws long and exserted from calyx, the banner yellow, reflexed, the keel and wings rose-red, the wings spreading; pods straight, dehiscent, not beaked, remaining erect, 3 cm. long, 2 mm. wide.

Wet places in coastal areas: Monterey County north; north to Washington.

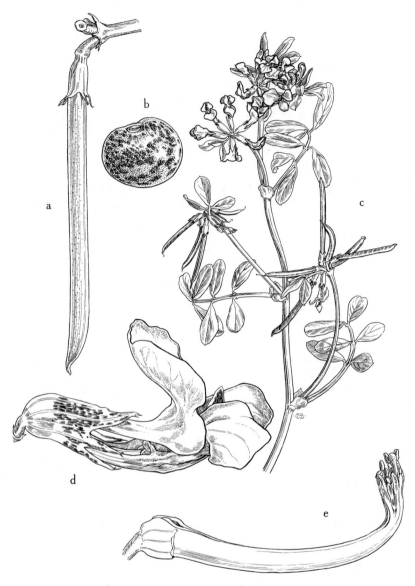

Fig. 254. *Lotus formosissimus: a,* fruit, × 2; *b,* seed, × 12; *c,* habit, × ⅔; *d,* flower, × 4; *e,* diadelphous stamens and tip of style, × 6.

2. Lotus oblongifolius Greene, Pittonia 2:146. 1890. Stream lotus.

Hosackia oblongifolia Benth., Pl. Hartw. 305. 1848.

Herbaceous perennial; stems several, erect or ascending from slender, branched rootstocks, 1.5–4 dm. long; herbage appressed-pubescent to nearly glabrous; leaves with 7–11 leaflets, these linear-lanceolate to elliptic, acute, 7–20 mm. long, equally distributed on opposite sides of rachis, the stipules membranous; peduncles exceeding the leaves in height; umbels 1- to 5-flowered, closely subtended by a 1- to 3-foliate bract; calyx tube 2–3 mm. long, the teeth narrowly subulate, about as long as the tube; petals 10–12 mm. long, the claws short and little exserted from calyx, the banner yellow or orange, ovate, erect, the keel and wings whitish to yellow, sometimes tinged with red; pods straight, dehiscent, not beaked, remaining erect, 2.5–4 cm. long.

Wet places at elevations of 1,000–6,000 feet: Coast Ranges, Sierra Nevada, southern California.

Also found in wet places are the widespread *Lotus oblongifolius* var. *Torreyi* Ottley of the Coast Ranges and Sierra Nevada (leaflets obovate, wings and keel white) and the localized *L. oblongifolius* var. *cupreus* Ottley of the southern Sierra Nevada (plants glabrous, prostrate).

LUPINUS. LUPINE

1. Lupinus polyphyllus Lindl., Bot. Reg. 13: pl. 1096. 1827.

Perennial herb; stems erect, stout, frequently hollow, little-branched, 5–15 dm. tall; leaves alternate, palmately compound with 7–17 leaflets on petioles 15–30 cm. long, the leaflets oblanceolate or lanceolate, glabrous or somewhat hirsute below, 7–15 cm. long; inflorescences densely racemose, 15–60 cm. long, the peduncles 3–10 cm. long; flowers 12–14 mm. long on pedicels 10–16 mm. long, subverticillate, the bracts ciliate, caducous; calyx bilabiate, the lips entire or somewhat toothed; banner reddish purple, the wings blue, the keel falcate, nonciliate, acuminate; stamens monadelphous below, free above, the anthers alternately large and small; pods dark brown, shaggy, 25–40 mm. long, 7–9 mm. wide, 5- to 9-seeded.

Moist flats to an elevation of 4,800 feet: North Coast Ranges, Sierra Nevada; north to British Columbia.

Closely related to *Lupinus polyphyllus* and sometimes to be found in similar wet situations is *L. superbus* Heller of the San Bernardino Mountains and Sierra Nevada, which plant, according to Jepson (Fl. Calif. 2:255, 1936), is possibly just a less robust state of *L. polyphyllus*. Also to be looked for occasionally in similar habitats is *L. rivularis* Dougl. of the Coast Ranges and Sierra Nevada.

MELILOTUS. SWEET CLOVER

Ours annual herbs. Stems erect, branching. Herbage fragrant on drying. Leaves pinnately 3-foliolate, petioled, alternate. Inflorescence of small yellow or white flowers in spikelike racemes on axillary peduncles. Calyx 5-toothed, the short teeth nearly equal in size. Stamens diadelphous. Pod ovoid, straight, indehiscent or tardily dehiscent, 1- or 2-seeded, longer than the calyx.

Corolla white .... 1. *M. alba*
Corolla yellow .... 2. *M. indica*

1. Melilotus alba Desv. ex Lam., Encycl. 4:63. 1797. White sweet clover.

Stems 3–15 dm. tall; herbage glabrous or somewhat pubescent; leaflets broadly or narrowly oblong, serrate, apically truncate or emarginate, 10–20 mm. long; racemes numerous, narrow, 5–10 cm. long; petals white, 4–6 mm. long, the banner longer than wings; fruiting pedicels reflexed; pod glabrous, 3 mm. long, somewhat wrinkled.

Almost throughout California, especially in river bottoms and moist areas; native of Europe.

2. Melilotus indica (L.) All., Fl. Ped. 1:308. 1785. Yellow sweet clover.

Stems 2–7 dm. tall; herbage glabrous or sparsely pubescent; leaflets narrowly or broadly cuneate-obovate, serrate above the middle, obtuse to truncate, 15–25 mm. long; racemes numerous, 2–10 cm. long; petals yellow, 2–3 mm. long; pods coarsely reticulate, glabrous, 2 mm. long.

Common throughout California, more often in dry situations but sometimes in boggy places; native of Europe.

PSORALEA

Ours perennial herbs. Herbage with strong odor and dark glands. Leaves 1- to 5-foliolate, alternate, the stipules large. Flowers in pedunculate spikes or racemes. Calyx 5-lobed, the lobes nearly equal in size. Petals purple to white; banner ovate to orbicular, clawed; keel broad, obtuse. Stamens in ours diadelphous. Pod ovoid, indehiscent, usually included within the calyx.

Peduncles and petioles erect from prostrate, creeping stems .... 1. *P. orbicularis*
Peduncles and petioles borne on erect stems .... 2. *P. macrostachya*

1. Psoralea orbicularis Lindl., Bot. Reg. 23: pl. 1971. 1837.

Stems prostrate, creeping, rooting at nodes; petioles and peduncles erect; leaves 3-foliolate, orbicular-ovate, 3–8 cm. long, glabrous to finely pubescent, the petioles 1–5 dm. long; peduncles 2–7 dm. long; raceme dense, villous with blackish hairs, 5–25 cm. long, its bracts lanceolate, caducous; calyx densely villous; corolla reddish purple, 15 mm. long.

Wet meadows or creek bottoms: Coast Ranges, Sierra Nevada to elevation of 4,000 feet; south to Baja California, Mexico.

2. Psoralea macrostachya DC., Prodr. 2:220. 1825. Leather-root.

Stems erect from woody rootstocks, 5–30 dm. tall; herbage glabrous to tomentose; petioles 3–12 cm. long; leaves 3-foliolate, 2–8 cm. long, rhombic-ovate to lanceolate-ovate; peduncles usually exceeding the leaves in height; racemes spikelike, 5–12 cm. long, the rachis densely white- or black-pubescent, its bracts broadly ovate, attenuate, caducous; calyx densely villous, the lowest tooth as long as or longer than the corolla, the other 4 short; petals purple, the banner 10 mm. long.

Moist areas along streams, in wet meadows, or salt marshes: South Coast Ranges to San Diego County.

SESBANIA

1. Sesbania macrocarpa Muhl., Cat. Pl. Am. 65. 1813. Colorado River hemp.

Glabrous, annual herb; stems striate, 3–10 dm. tall; leaves 5–30 cm. long, exceeding the flowers in height, evenly pinnate, the rachis ending in a setaceous point, the leaflets 20–70, linear-oblong, 10–25 mm. long, entire, obtuse; flowers 1 to several on axillary peduncles; calyx campanulate, 2 deciduous bractlets at base, the lobes short, acute; corolla yellowish, about 15 mm. long, the banner reflexed, streaked with purple; stamens diadelphous; ovary stipitate, with many ovules; pod short-stipitate, linear-elongate, with partitions between the 15–30 seeds, 2-valved, 10–15 cm. long, 3 mm. wide.

River bottoms and overflow lands: lower Colorado River and Colorado Desert; east to Atlantic.

TRIFOLIUM. CLOVER

Annual or perennial herbs. Leaves palmately 3-foliolate (to 5-foliolate), the leaflets usually denticulate; stipules foliaceous, adnate at base, and clasping the petiole. Inflorescence pedunculate, of closed umbels, capitate heads, or short spikes. Calyx 5-toothed, with the teeth usually equal or subequal in size. Corolla white, yellow, purple, or pink, withering-persistent. Stamens diadelphous. Pods usually included within the persistent calyx, 1- to 8-seeded, straight, dehiscent or indehiscent.

Trifolium frequently favors moist habitats, and selection of species for inclusion in this treatment is difficult. However, from among the forty-odd California species, the following are considered as being particularly characteristic of perennially wet places, although in their total distribution they are not necessarily limited to such wet places.

Heads subtended by an involucre; flowers not reflexed in age.
  Annual; stems slender; flowers with corolla 6–8 mm. long .... 1. *T. variegatum*
  Perennial by creeping rootstocks; stems stout; flowers with corolla 10–12 mm. long
                                                                  2. *T. Wormskjoldii*
Heads naked, without involucre; perennial by woody root crown; flowers strongly reflexed
  in age .... 3. *T. Bolanderi*

1. Trifolium variegatum Nutt. in Torr. & Gray, Fl. N. Am. 1:317. 1838.
   White-tip clover.

Annual; stems slender, several from base, many-branched, decumbent or ascending, 2–6 dm. long; herbage glabrous throughout; leaflets obovate, very variable in size, sharply serrate, on slender peduncles, the stipules ovate, laciniate; heads commonly small, irregularly subglobose, 6–12 mm. broad, on axillary peduncles; involucre flat, smaller than head, the lobes toothed; calyx tube 5- to 20-nerved, the teeth subulate-setaceous, 1 tooth sometimes bifid; corolla purple, sometimes white-tipped, 6–8 mm. long, considerably longer than the calyx; pod 2-seeded.

Low, moist or wet places to an elevation of 4,000 feet: throughout California; north to British Columbia.

A very variable and ubiquitous species, represented by many named varieties.

2. Trifolium Wormskjoldii Lehm., Ind. Sem. Hort. Hamb. 1825:17. 1825.
   Cow clover.

   Trifolium involucratum Ortega, Hort. Reg. Bot. Matr. 33. 1797. Not *T. involucratum*
     Lam. 1778.
   Trifolium spinulosum Dougl. in Hook., Fl. Bor. Am. 1:133. 1830.
   Trifolium fimbriatum Lindl., Bot. Reg. 13: pl. 1070. 1827.

Perennial from creeping rootstocks; stems thick, weak, branching from base, decumbent, 1–4 dm. long; herbage glabrous throughout; petioles 2–6 cm. long, the leaflets oblong to obovate, finely serrate, most of them obtuse, 10–25 mm. long; stipules laciniate; peduncles 3–6 cm. long, stout; involucre 15 mm. broad, flat, deeply lobed, the lobes laciniate; heads many-flowered, 20–30 mm. broad, showy; calyx tube 10-nerved, 8–9 mm. long, the teeth subulate, not dilated, slightly longer than the tube; corolla conspicuously inflated, purple to pinkish, about 12 mm. long, the banner deeply emarginate and often lighter-colored; seeds 2–6.

Frequent along streams, near springs, or in wet, saline depressions, to an elevation of 8,000 feet: throughout California; north to British Columbia, south to Mexico.

3. Trifolium Bolanderi Gray, Proc. Am. Acad. Arts & Sci. 7:335. 1868.

Perennial from stout, woody root crown; herbage glabrous throughout, the stems numerous, decumbent or ascending, 1–2 dm. long; leaves sparse, mostly basal, the leaflets obovate to cuneate-oblong, slightly serrate, rounded at apex,

0.8–1.8 mm. long; peduncles mostly terminal, 5–20 cm. long, slender; heads ovoid, 10–15 mm. long, without involucre; flowers on short pedicels, at length completely reflexed; calyx 3–4 mm. long, the teeth lanceolate-acuminate, somewhat shorter than or as long as the tube; corolla pink or lavender, 7–8 mm. long; pod 2-seeded.

Rare endemic, wet mountain meadows at an elevation of about 7,000 feet: Sierra Nevada from Mariposa County to Fresno County. An uncommon species.

CALLITRICHACEAE

Aquatic or terrestrial annual herbs with delicate stems, monoecious or (in *Callitriche marginata*) dioecious. Leaves opposite, entire, without stipules. Flowers unisexual, the perianth lacking, each flower subtended by a pair of falciform or obliquely oval bracteoles (or these wanting), the staminate flowers 1–3 in the axil of a foliage leaf, the pistillate flowers 1 or rarely more, similarly placed. Staminate flower consisting of a single stamen on a slender filament. Pistillate flower of a single pistil of 2 carpels, the styles 2, slender, often much longer than the ovary. Carpels splitting on maturity and usually forming a fruit of 4 nutlets, the flattened nutlets winged, margined or smooth, each bearing 1 seed.

CALLITRICHE

The long list of synonyms associated with many of the species of *Callitriche* attests that it is a group made up of highly polymorphic species, a situation that lends itself to divergent interpretations among taxonomists. The polymorphism may be attributed to several causes. First, several species may be either amphibious or terrestrial, and in each habitat the species in question presents a very different appearance. Second, apomixis prevails in most species, especially in the early flowers, in which no stamens are produced but seed sets regularly. This provides for the preservation of chance genetic strains. Finally, Giuseppe Martinoli (unpublished MS) has demonstrated that certain species involve polyploid complexes.

Fassett, N. C., *Callitriche* in the New World. Rhodora 53:137–155, 161–194, 209–222. 1951.

Fruits sessile in leaf axils (or if pediceled, the pedicels much shorter than leaves); plants normally wholly aquatic.
  Plants with floating ovate leaves at maturity, the submersed leaves linear; flowers with conspicuous bracteoles .... 1. *C. palustris*
  Plants wholly submersed at maturity, the leaves all linear; flowers without bracteoles (fruits 2 in the variety) .... 2. *C. hermaphroditica*

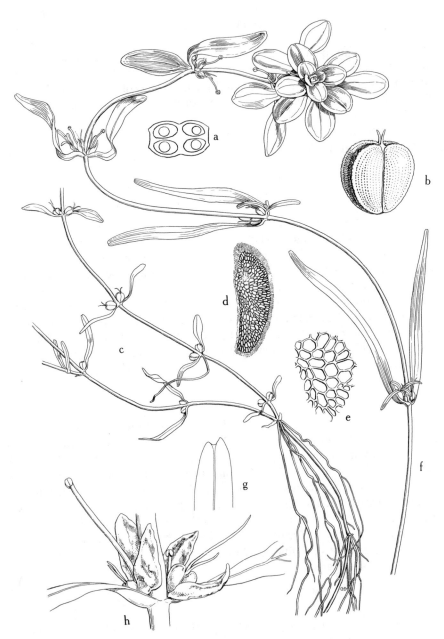

Fig. 255. *Callitriche palustris: a,* fruit (cross section), $\times$ 12; *b,* mature fruit, $\times$ 12; *c,* habit, lower submersed part of plant, showing the short leaves and the sessile fruits, $\times$ 1⅕; *d,* mature carpel, showing the longitudinal wing, $\times$ 20; *e,* reticulation of carpel surface, $\times$ 40; *f,* habit, showing upper part of stem terminating in a rosette of spatulate floating leaves, the submersed leaves linear, $\times$ 1½; *g,* apex of young linear leaf, slightly notched, $\times$ 10; *h,* flowers, showing the winglike bracteoles, $\times$ 8.

Fruits not sessile in leaf axils, at maturity the pedicels from ½ as long as the leaves to
several times as long.
Plants chiefly aquatic; leaves round-ovate, petioled, or the lower submersed ones occasionally linear; monoecious; pedicels long and slender ... 3. *C. longipedunculata*
Plants chiefly terrestrial; leaves all linear-spatulate, thick; dioecious; pedicels usually
about as long as the leaves, stout, sometimes reflexed and hypogeous
4. *C. marginata*

1. Callitriche palustris L., Sp. Pl. 969. 1753. Fig. 255.

Callitriche palustris var. *Bolanderi* Jepson, Fl. Calif. 2:435. 1936.
Callitriche palustris var. *stenocarpa* Jepson, Man. Fl. Pl. Calif. 603. 1925.

Submersed aquatic, usually with a rosette of floating leaves terminating the
longer stems, less frequently terrestrial on the margins of ponds or streams;
stems 1–5 dm. long or occasionally longer; submersed lower leaves linear,
clasping at base, 1-nerved, shallowly notched at tips, the upper submersed
leaves spatulate and petioled, 3-nerved, 1–3 cm. or rarely 5 cm. long; floating
leaves rosulate, the petioles broad, the blades ovate to orbicular, 3- to 5-nerved,
rounded or sometimes notched at apex; bracteoles more conspicuous than in
other species, linear to oblong, or obliquely oval; staminate flowers with filaments 1–3 mm. long; fruits sessile, suborbicular, sometimes slightly longer
than wide, often widening in proportion to length just before dehiscence,
rounded at base, from slightly to deeply notched at apex the carpels with a
longitudinal wing or margin, or this absent.
Quiet water: throughout the state; Northern Hemisphere.

2. Callitriche hermaphroditica L., Cent. Pl. 1:31. 1755. Fig. 256.

Callitriche autumnalis L., Fl. Suec., 2d ed., 2. 1755.

Stems 1–2 dm. long, wholly submersed; leaves all alike, dark livid green,
6–25 mm. long, linear to linear-lanceolate, clasping the stem, 1-nerved, the
apex notched or bifid; flowers 1–3 to a node, bracteoles none; pistillate flowers
with styles deflexed and much longer than ovary; fruits sessile or sometimes
short-pediceled, roundish, the carpels closely appressed, thin, winged.
Ponds and vernal pools: northern California (rare); east to Atlantic.

2a. Callitriche hermaphroditica var. bicarpellaris (Fenley), comb. nov.

Callitriche autumnalis L. var. *bicarpellaris* Fenley in Jepson, Fl. Calif. 2:436. 1936.

This is a form in which two carpels regularly abort. It is known from the
Central Valley and from the valley of the Russian River.

3. Callitriche longipedunculata Morong, Bull. Torrey Club 18:236. 1891.
Fig. 257.

Callitriche marginata var. *longipedunculata* (Morong) Jepson, Man. Fl. Pl. Calif. 603.
1925.

Rooted aquatic with a rosette of floating leaves or sometimes the filiform
stems reclining on wet soil after water recedes; floating leaves spatulate to

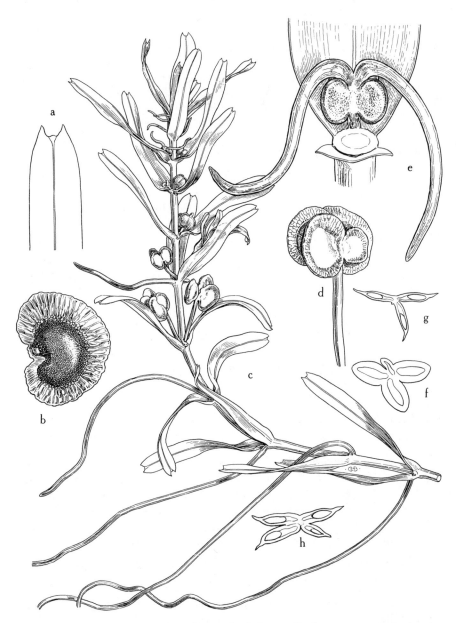

Fig. 256. *Callitriche hermaphroditica: a,* leaf tip, × 12; *b,* mature carpel, showing broad wing, × 20; *c,* habit, wholly submersed plant, showing the linear-lanceolate, notched leaves, the flowers, and the short-peduncled fruits, × 4; *d,* mature fruit, × 8; *e,* young flower, showing the long, deflexed styles, × 40; *f,* young fruit (cross section), × 12; *g* and *h,* mature fruit (cross section), × 8.

Fig. 257. *Callitriche longipedunculata: a,* mature carpel, the wing regular, × 20; *b,* node with young flowers in the leaf axils, the bracteoles large-celled, × 12; *c,* fruit (cross section), × 12; *d,* habit, showing submersed leaves and floating rosette, staminate and pistillate flowers in an axil, and fruits with elongated peduncles, × 4; *e,* mature fruits, × 8.

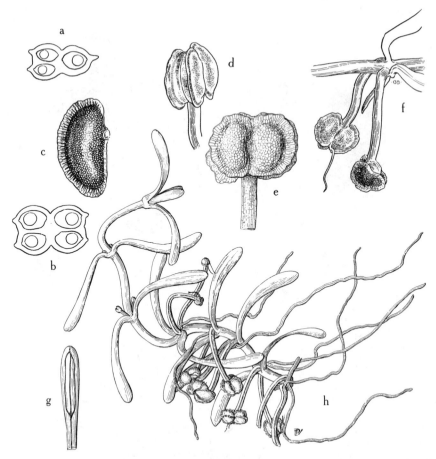

Fig. 258. *Callitriche marginata: a* and *b,* fruit (cross sections), × 12; *c,* mature carpel, wing regular, × 20; *d* and *e,* mature fruits, × 12; *f,* fruits, the peduncles reflexed at maturity, × 8; *g,* leaf, showing 3-nerved venation, × ⅖; *h,* habit, × 4.

round-ovate, the submersed leaves similar to the floating leaves or a few of the lower ones linear; petioles 2–3 mm. long; those in the floating rosette shortest; bracteoles about 1 mm. long, obovate; flowers sessile, the staminate flowers 1 or 2 or absent from some axils, the filaments 1–2 mm. long; fruits narrowly winged, on peduncles 4–25 mm. long.

A very common species in early spring: Central Valley, Coast Ranges, Sierra Nevada foothills, south to coastal southern California. Usually gone by summer.

4. Callitriche marginata Torr., Pacif. R. R. Rep. 4:135. 1857. Fig. 258.

Plant prostrate, chiefly terrestrial; stems 5–15 mm. long; leaves all linear-spatulate, thick, 3-nerved; bracts minute or none; fruiting pedicels stout, spreading, sometimes reflexed and hypogeous, about as long as the leaves; flowers dioecious; fruit dark gray, notched at base and apex, the carpels conspicuously margined, the style persistent, 1–2 mm. long, filiform, reflexed over fruit.

Common in moist, cool places in the valleys and mountains: throughout California; north to British Columbia, east to Atlantic.

LIMNANTHACEAE. MEADOW-FOAM FAMILY

Annual herbs with dissected, alternate, petioled leaves. Flowers perfect, regular, in ours 5-merous, on axillary peduncles. Calyx of distinct sepals, persistent, free from the ovary. Petals withering-persistent. Stamens 10. Carpels 5, united only at the base, where the common style arises; carpels becoming smooth or tuberculate, 1-seeded nutlets.

LIMNANTHES

Flowers whitish or yellow, sometimes tinged with rose. Sepals valvate in the bud. Stamens 10, distinct, those opposite the sepals with glandlike swelling at base. Stigmas 5, capitate.

Mason, Charles T., Jr., A systematic study of the genus *Limnanthes* R. Br. Univ. Calif. Publ. Bot. 25:455–512. 1952. From this study the present treatment has been abstracted.

Petals reflexing during maturation of the nutlets.
    Corolla cup-shaped to campanulate, petals 10–18 mm. long; stamens 5–7 mm. long
        1. *L. Douglasii*
    Corolla funnelform or, if campanulate, the petals less than 10 mm. long; stamens less than 5 mm. long.
        Corolla funnelform; petals 8–15 mm. long, yellow with white tips and usually prominent striations; leaflets entire to incisely toothed, often 3-lobed; Sierra Nevada foothills .... 2. *L. striata*

Corolla campanulate to funnelform; petals 7–9 mm. long, cream-colored with white
   tips, striations not prominent; leaflets entire, rarely 3-lobed; Mendocino County
                                                                         3. *L. Bakeri*
Petals inflexing during maturation of the nutlets.
   Petals conspicuously exceeding the sepals in length at anthesis.
      Petals 7–10 mm. long; stamens 2–4 mm. long; anthers 0.5–1 mm. long.
         Petals aging pink; anthers 1 mm. long . . . . 4. *L. gracilis*
         Petals remaining white in age; anthers 0.5 mm. long . . . . 5. *L. montana*
      Petals 10–15 mm. long; stamens 5–6 mm. long; anthers 2 mm. long . . . . . 6. *L. alba*
   Petals slightly if at all exceeding the sepals in length at anthesis . . . . 7. *L. floccosa*

1. Limnanthes Douglasii R. Br., London Edinb. Philos. Mag. 3:70–71. 1833.

Limnanthes rosea var. *candida* Jepson, Fl. Calif. 2:411. 1936.

Stems 10–50 cm. tall; leaves long-petiolate, 5–25 cm. long, once or twice
pinnately dissected with 5–13 entire, lobed, or parted, often incisely toothed
leaflets, their divisions linear to obovate; flowers 5-merous, cup-shaped to
campanulate; sepals lanceolate, 5–15 mm. long, glabrous or rarely with a few
hairs on the margin, not accrescent; petals 10–18 mm. long, cuneate, obovate,
deeply emarginate to obcordate, white, yellow, or yellow with white tips,
occasionally aging pink, the veins inconspicuous to prominent, rose or cream-
colored, scattered long hairs arising from the veins, 2 rows of short hairs on
the claw; stamens 5–7 mm. long; pistil 5–7 mm. long, including stigmatic
branches up to 1 mm. long; nutlets obovoid, 2.5–5 mm. long, smooth to
strongly tuberculate or ridged, or wrinkled and with a crown of tubercles.
   Vernally wet meadows: coastal areas from San Luis Obispo County north
to Humboldt County, inland in the Central Valley from Madera County north
to Shasta County; Umpqua Valley in Oregon.
   Four entities previously considered as species are treated by C. T. Mason
(*op. cit.*) as varieties of *Limnanthes Douglasii*:

Nutlets smooth, wrinkled, or with scattered tubercles, if covered with tubercles then the
   petals colored; leaflets ovate.
   Petals yellow or yellow with white tips.
      Petals yellow with white tips . . . . 1a. *L. Douglasii* var. *Douglasii*
      Petals yellow . . . . 1b. *L. Douglasii* var. *sulphurea*
   Petals of living plants white; nutlets smooth or crowned with tubercles
                                                                1c. *L. Douglasii* var. *nivea*
Nutlets with high, prominent ridges; petals white with rose veins; leaflets narrow, almost
   linear . . . . 1d. *L. Douglasii* var. *rosea*

1a. Limnanthes Douglasii var. Douglasii.

Limnanthes Howelliana Abrams, Madroño 6:27. 1941.

Leaflets 5–11, entire to incisely toothed, often 3- to 5-lobed; petals cuneate
to obovate, deeply emarginate, yellow with white tips; nutlets 2.5–4 mm. long,
smooth, wrinkled, or tuberculate, often with a crown of short, conic tubercles.

Open, wet fields: coastal valleys of central and northern California; Umpqua Valley in Oregon.

1b. Limnanthes Douglasii var. **sulphurea** C. T. Mason, Univ. Calif. Publ. Bot. 25:477. 1952.

Leaflets 7–13, broadly obovate, incisely toothed or lobed; petals 12–18 mm. long, cuneate to obovate, emarginate, yellow; nutlets 3.5–5 mm. long, smooth or wrinkled, often with a crown of tubercles.

Wet places: Point Reyes peninsula in Marin County.

In geographical range, this variety is limited to a small area near the ocean end of Point Reyes peninsula. There are several populations separated from one another by the rolling terrain. Although colonies of *Limnanthes Douglasii* var. *Douglasii* are situated within a half mile of populations of *L. Douglasii* var. *sulphurea,* the patterning and color show no evidence of introgression.

1c. Limnanthes Douglasii var. **nivea** C. T. Mason, Univ. Calif. Publ. Bot. 25:477. 1952. Fig. 259.

Leaflets 5–11, incisely toothed, often 3- to 5-lobed; petals cuneate, emarginate, white, often with prominent dark purple veins; nutlets 3–4 mm. long, smooth or wrinkled, occasionally with a crown of tubercles.

Wet meadows: Coast Ranges from San Luis Obispo County north to Humboldt County.

Living plants of this variety are readily recognized by their white flowers and smooth nutlets. Confusion in identification is likely to arise, however, when the material is dried. Individuals of some populations develop a yellow-cream color at the base of the petals when pressed, thereby resembling a faded form of *Limnanthes Douglasii* var. *Douglasii.*

1d. Limnanthes Douglasii var. **rosea** (Benth.) C. T. Mason, Univ. Calif. Publ. Bot. 25:480. 1952.

Limnanthes rosea Benth., Pl. Hartw. 302. 1848.

Leaflets 7–11, lobed, or parted with linear segments; petals cuneate to obovate, emarginate to obcordate, white with pink veins, or white with cream base, aging rose-pink, or yellowish with white tips; nutlets 4–5 mm. long, with high, prominent tubercles or ridges, often appearing white.

Wet places: Central Valley, from the Coast Ranges to the Sierra Nevada foothills.

This variety exhibits the greatest diversification in flower color of any member of the species. The white flowers with pink veins, stamens, and pistils are predominantly in the northern part of the range, whereas those with a cream base and with yellow stamens and pistil are in the southern part. Populations with flowers of both types are common. Some variation in the tuberculation of the nutlets is noticeable, the roughest nutlets occurring to the northeast.

Fig. 259. *Limnanthes Douglasii* var. *nivea: a,* flower, before dehiscence of anthers, × 1½; *b,* flower, after dehiscence of anthers, × 1⅕; *c,* flower, showing young nutlets and deflexed, marcescent petals, × 2; *d,* petal, × 3; *e,* mature nutlet, × 8; *f,* habit, × ⅔.

2. Limnanthes striata Jepson, Fl. Calif. 2:411. 1936.

Stems ascending to erect, 10–30 cm. tall; herbage glabrous; leaves 10–15 cm. long, pinnate with 7–11 segments, the leaflets simple or 2- or 3-lobed, entire to incisely toothed; flowers 5-merous, funnelform on peduncles to 12 cm. long; sepals glabrous, 4–6 mm. long, linear-lanceolate, not accrescent; petals 8–15 mm. long, cuneate to spatulate, with truncate to slightly emarginate or undulate white tips, the bases yellow-cream, often marked with 7–9 prominent brown-purple veins, not aging lilac, scattered long hairs arising from the veins and 2 rows of hairs at the base; stamens 3–4 mm. long; pistil 3–4 mm. long, including stigmatic branches 1 mm. long; nutlets 2.5–3 mm. long, obovoid, smooth, wrinkled, or strongly tuberculate, often with a crown of tubercles at the tip.

Borders of small streams and vernal pools: Sierra Nevada foothills from Mariposa County north to El Dorado County.

The leaflet form and the color pattern of the petals of this species resemble those of *Limnanthes Douglasii* var. *Douglasii*. The two entities are readily distinguished from each other by the spatulate shape and lighter yellow base of the petals and the funnelform corolla of *L. striata*.

3. Limnanthes Bakeri J. T. Howell, Leafl. West. Bot. 3:206. 1943.

Stems often branching above the base, ascending or erect; herbage glabrous; leaves sparse, to 10 cm. long, pinnate with 3–9 segments, the leaflets elliptic to ovate, entire, or rarely the lower ones 2- or 3-lobed, the lower leaves dying early; flowers 5-merous, campanulate to funnelform, on peduncles 3–10 cm. long; sepals lanceolate, 5–7 mm. long, glabrous, not accrescent; petals cuneate, truncate with erose tips, 7–9 mm. long, yellow-cream with white tips, with scattered hairs on the petals and 2 rows of hairs at the base; stamens 3–4 mm. long, the anthers 1 mm. long; pistil 3–4 mm. long, including stigmatic branches up to 1 mm. long; nutlets obovoid, 3–3.5 mm. long, with a dense covering of short, broad tubercles.

In wet meadowlands: near Willits in Mendocino County.

This species owes its distinctiveness to its somewhat succulent, obtusely lobed, sparse leaves and its small flowers.

4. Limnanthes gracilis Howell var. **Parishii** (Jepson) C. T. Mason, Univ. Calif. Publ. Bot. 25:490. 1952.

Limnanthes versicolor var. *Parishii* Jepson, Fl. Calif. 2:411. 1936.

Stems 7–30 cm. tall; herbage glabrous; leaves 3–8 cm. long, pinnate with 5–9 segments, the leaflets lanceolate to ovate, entire to 3-lobed; flowers cup-shaped, with distal part of petals recurving; sepals ovate, accrescent; petals white, lacking scattered hairs but with 2 rows of hairs at the base; stamens 2.5–3 mm. long; style 2–3 mm. long; nutlets obovoid, 3 mm. long, crowned with short, broad tubercles.

Fig. 260. *Limnanthes alba: a,* habit, × ⅔; *b,* flower, before dehiscence of anthers, × 1½; *c,* flower, showing young nutlets and inflexed petals, × 2; *d,* mature nutlet, × 6; *e,* petals, × 3.

Wet places: Cuyamaca and Laguna mountains in San Diego County.

Limnanthes gracilis var. *Parishii* is the most southerly member of the family. It is separated from its nearest relatives within the genus by a distance of approximately 300 miles and several mountain ranges. Jepson described the plant as a variety of *L. versicolor* (Greene) Rydb., from which it is easily distinguished by its smaller flowers, stamens, and pistil, and its rough nutlets.

5. Limnanthes montana Jepson, Fl. Calif. 2:412. 1936. Mountain meadow foam.

Stems 10–40 cm. long, ascending to erect; herbage glabrous to sparsely pilose; leaves 3–15 cm. long, pinnate to bipinnate with 7–11 segments, the leaflets linear to ovate, simple to 3-lobed, 3-parted, or 3-divided; flowers 5-merous, campanulate to funnelform, on peduncles to 11 cm. long; sepals lanceolate to ovate-lanceolate, acuminate, 3–6 mm. long, glabrous to sparsely villous, not accrescent; petals cuneate to obovate, truncate, emarginate to repand, 7–10 mm. long, white, or white with a yellowish cream base, some with prominent purple veins, not aging pink, with scattered, long hairs from the veins and 2 rows of hairs at the base; stamens 3 mm. long, the anthers 0.5 mm. long; pistil 3 mm. long, including stigmatic branches 1–1.5 mm. long; nutlets obovoid, 2–3 mm. long, covered with low, broad tubercles.

Springs and bogs, and often bordering small streams: Sierra Nevada from Mariposa County south to Tulare County.

6. Limnanthes alba Benth., Pl. Hartw. 301. 1848. Fig. 260.

Stems ascending to erect, 10–30 cm. tall; herbage glabrous to villous; leaves to 10 cm. long; pinnate with 5–9 segments, the leaflets ovate, simple to 3-lobed or 3-parted; flowers 5-merous, cup-shaped to campanulate, on peduncles to 10 cm. long; sepals lanceolate to ovate, acuminate, 7–8 mm. long, densely villous, not accrescent; petals obovate, truncate to obcordate, 10–15 mm. long, white, some aging lilac-purple or pink at the tips, with a few scattered long hairs from the veins and 2 rows of hairs at the base; stamens 5–6 mm. long, the anthers 2 mm. long; pistil 5–6 mm. long, including stigmatic branches 1–2 mm. long; nutlets obovoid, 3–4 mm. long, very rough with broad ridges.

Herbage (when young) sparsely to densely pilose; sepals densely villous; nutlets with broad ridges .... 6a. *L. alba* var. *alba*
Herbage glabrous; sepals glabrous to sparingly villous; nutlets smooth or with scattered, sharp tubercles .... 6b. *L. alba* var. *versicolor*

6a. Limnanthes alba Benth. var. **alba.**

Limnanthes alba var. *detonsa* Jepson, Fl. Calif. 2:411. 1936.

Wet places and roadside ditches: east side of the Central Valley from Butte County south to Merced County.

6b. Limnanthes alba var. **versicolor** (Greene) C. T. Mason, Univ. Calif. Publ. Bot. 25:497. 1952.

Limnanthes versicolor (Greene) Rydb., N. Am. Fl. 25:99. 1910.

Moist places and along streams: Shasta County south along the Sierra Nevada foothills to Tuolumne County.

7. Limnanthes floccosa Howell, Fl. NW. Am. 1:108. 1897.

Stems decumbent to erect, 3–25 cm. tall; herbage densely villous; leaves 3–8 cm. long, pinnate with 5–11 segments, the leaflets linear to ovate-elliptic, entire, incisely toothed, or lobed; flowers urceolate, campanulate, or deeply cup-shaped, on peduncles 1–8 cm. long; sepals lanceolate to ovate, acuminate, 5–8 mm. long at anthesis, glabrous to woolly, accrescent; petals 6–8 mm. long, obovate, white, not aging pink, completely glabrous or possessing 2 rows of hairs at the base; stamens 2–4 mm. long, the anthers 0.5–1 mm. long; pistil 2–4 mm. long; nutlets obovoid, 3–4.5 mm. long, prominently tuberculate, the tubercles varying from short-conic to long, sharp processes.

Wet places: Butte County, northern California; north to Jackson County in Oregon.

ACERACEAE

Trees or shrubs. Leaves opposite, palmately lobed to palmately compound or pinnately compound, deciduous. Flowers in ours in corymbs or racemes, perfect, or unisexual and then monoecious or dioecious, or perfect and unisexual mixed in the same inflorescence. Sepals 5, united at base. Petals present or absent. Stamens 4–12, inserted on the margin of a perigynous disc, or in some staminate flowers, the flower hypogynous and the stamens centrally inserted. Ovary 2-celled (or 3-celled), laterally compressed, developing a long wing from the summit of each lobe, thus in fruit producing 2 (or 3) samaroid fruitlets which separate at base. Ovules 2 in each cell.

Because of its compound leaves, dioecious flowering, and certain features of stem anatomy, the fourth species below is sometimes treated as a distinct genus, *Negundo*.

ACER

Leaves simple, palmately lobed or parted; petals present (in *A. glabrum* the flowers rarely dioecious).
Flowers in a coarse, elongate raceme; fruits more or less hispid; leaves large, deeply 5-lobed .... 1. *A. macrophyllum*
Flowers in slender corymbs; fruits glabrous.
Leaves shallowly 7- to 9-lobed or lobes sometimes fewer and then again lobed or toothed .... 2. *A. circinatum*
Leaves chiefly 3-lobed or 3-parted .... 3. *A. glabrum*
Leaves compound; flowers dioecious; petals absent .... 4. *A. Negundo*

1. Acer macrophyllum Pursh, Fl. Am. Sept. 1:267. 1814. Big-leaf maple.

Large tree, often approaching 30 m. in height; leaves palmately 5-parted or 3-lobed, 10–25 cm. broad; perfect and staminate flowers in the same raceme; stamens 7–9, the filaments villous below; carpels 2 or occasionally 3; body of fruit bristly.

At margins of streams or in moist woods to an elevation of 5,000 feet: Coast Ranges and Sierra Nevada foothills from southern California northward; north to British Columbia and Alaska.

2. Acer circinatum Pursh, Fl. Am. Sept. 1:267. 1814. Vine maple.

Small tree, or more commonly a large shrub to 6 m. in height; leaves 5–10 cm. broad, 5- to 9-lobed, the lobes equal in size or sometimes basically 3- or 5-lobed and the lobes again lobed or coarsely toothed; flowers 4–12 in a corymb, mostly staminate, 1 or 2 perfect; sepals longer than petals; stamens 6–12, included in perfect flowers, exserted in staminate flowers, the filaments villous below; fruitlets rotately spreading in a plane at right angles to the peduncle.

Stream banks, moist meadows, and canyons to an elevation of 5,000 feet: Humboldt and Del Norte counties, east to Hatchet Mountain in Shasta County and Deer Creek in Butte County.

3. Acer glabrum Torr., Ann. Lyc. N. Y. 2:172. 1828. Sierra maple.

Low, often sprawling shrub 1–3 m. high; leaves 2–8 cm. broad, palmately 3-lobed, or with an additional smaller pair of lobes at base, unequally serrate; flowers 4–9, in corymbs simulating an umbel, monoecious or sometimes dioecious, when monoecious the sexes variously distributed in the corymbs; sepals equaling the petals in length; stamens 7–10, the filaments glabrous; fruits usually several to an inflorescence, the samaras ascending.

Stream banks, springs and moist canyons and wet, rocky slopes: Sierra Nevada and North Coast Ranges at elevations above 6,000 feet; north to Alaska, east to the Rocky Mountains.

4. Acer Negundo var. **californicum** Sarg., Gard. & For. 4:148. 1891. Box elder.

Tree, often with several stems, to 20 m. in height; young stems tomentose to velutinous; leaves pinnately 3- to 5-foliate, the leaflets variously toothed, lobed, or divided; flowers dioecious, the petals absent; staminate flowers on threadlike peduncles, the stamens 4 or 5; pistillate flowers in racemes; fruits finely tomentose.

Along streams and in moist meadows: Central Valley, Coast Ranges to San Bernardino Mountains.

The eastern *Acer Negundo,* which has escaped cultivation in several places in California, may be readily recognized by its glabrous twigs.

Fig. 261. *Hibiscus californicus: a,* mature open capsule, × ⅘; *b,* seed, the tubercles prominent, × 8; *c,* habit, upper part of plant, × ⅖; *d,* stellate hairs on leaf, × 4.

VITACEAE. GRAPE FAMILY

Woody vines climbing by tendrils which are at least once-branched. Leaves in ours simple, alternate. Flowers small, regular, greenish, or whitish, in a compound thyrse. Sepals united at base to form an inconspicuous 5-lobed cup. Petals 5 (4–6), united at tips, separated at base and falling early as a unit. Stamens as many as petals and opposite them. Pistil 1, the ovary superior. Fruit a 2-celled berry.

VITIS

1. Vitis californica Benth., Bot. Voy. Sulphur 10. 1844. California wild grape.

Climbing stems 2–12 m. long; leaf blade round-reniform or cordate, sometimes simple or 3- to 5-lobed, thinly arachnoid below when young, flocculent in age; petals 5; stamens 5; fruit purple, 6–10 mm. in diameter.

Along streams and river courses and in swamps: Coast Ranges, Sierra Nevada foothills, Butte Sink, islands of the Sacramento River delta region.

MALVACEAE. MALLOW FAMILY

Herbs or soft-woody shrubs, usually with stellate pubescence and with mucilaginous sap. Leaves alternate, simple, palmately veined and often palmately lobed, stipulate. Flowers commonly perfect, sometimes dioecious (*Sidalcea*), regular. Calyx 5-lobed, valvate in the bud, often with an involucel of bractlets at base. Petals 5, twisted in the bud, inserted on the base of the stamen tube. Stamens numerous, hypogynous, the filaments united in a column or tube around the pistils. Pistil 1, composed of several to many carpels, the ovary superior, commonly with as many cells as styles or stigmas. Fruit a loculicidal capsule, or composed of a circle of united carpels separating at maturity.

Anthers scattered along the entire stamen column; carpels 5; fruit a loculicidal capsule; shrublike herb 1–2 m. tall .... *Hibiscus*
Anthers in 1 or 2 clusters at the end of the stamen column, lower part of column naked; carpels several, separating on dehiscence.
    Inflorescence racemose or spicate; styles stigmatic lengthwise .... *Sidalcea*
    Inflorescence axillary; style branches with capitate stigmas; flowers cream-white; herbage scurfy-canescent .... *Sida*

HIBISCUS

1. Hibiscus californicus Kell., Proc. Calif. Acad. Sci. 4:292. 1873. Fig. 261.

Stout, canelike herb of shrublike appearance, often many-stemmed from a coarse perennial root or sometimes the stems few, 6–20 dm. tall; herbage soft

Fig. 262. *Sidalcea rhizomata: a,* calyx with mature carpels, × 1½; *b,* habit, basal part of plant, showing the horizontal rhizome and the basal leaves, × ⅖; *c,* stamen column, the inner series of stamens hidden by the outer series, × 4; *d,* mature carpel, × 6; *e,* habit, upper part of plant, the cauline leaves divided, × ⅖.

velvety gray-pubescent; petioles from ½ as long as to fully as long as the blade; stipules minute, caducous; leaf blades ovate-deltoid, cordate, acuminate, crenate-dentate, 5–15 cm. long, densely stellate-pubescent beneath, the hairs of the upper surface of the leaf mostly simple, with a few stellate clusters intermingled; peduncles appearing as though inserted on the petiole 3–5 mm. from its base, apparently adnate to the stipular sheath, with a tumid joint 1–2 cm. below the involucre; involucre cleft into 10 linear-lanceolate lobes as long as or longer than calyx; calyx campanulate, cleft to the midde, conspicuously nerved, the lobes acute, densely gray-stellate-pubescent; corolla white, with a dark purplish red spot in the center, the petals 5–10 cm. long, obovate, broadly rounded at apex, conspicuously veined; stamens numerous in a cylindrical column; style 5-cleft; stigmas capitate; fruit a loculicidal capsule, dehiscent into 5 valves which fill the calyx; seeds numerous, reniform, prominently tuberculate on surface.

Marshes and swamplands: very abundant along slough of lower Butte Creek north and west of Marysville Buttes, thence south along the Sacramento River, and frequent among tules on the delta islands of the San Joaquin and Sacramento rivers, ascending the San Joaquin River course only a short distance.

SIDA

1. Sida hederacea (Dougl.) Torr. ex Gray, Mem. Am. Acad. Arts & Sci. II. 4:23. 1849. Alkali mallow.

Decumbent white-scurfy perennial; branches 1–5 dm. long; leaves petioled, the blades round, reniform, or ovate, dentate, 2–5 cm. broad; flowers in the leaf axils or in small clusters on articulated pedicels; involucel of 1–3 slender, deciduous bractlets; calyx campanulate to turbinate, the lobes ovate to lanceolate; petals 1–2 cm. long, cream-white; carpels 6–10, 1-seeded, triangular, splitting into 2 valves.

Common on alkaline floodlands: throughout the valleys south to coastal southern California; Washington to northern Mexico.

SIDALCEA

Herbs; leaves long-petioled, with orbicular, crenately incised, parted, or divided blades. Flowers in bracteate spikelike racemes, all perfect or with perfect and pistillate flowers or dioecious. Calyx without bracteoles. Petals purple to rose or white. Stamens usually in an upper and a lower series on the stamen column. Fruit of 5–9 carpels.

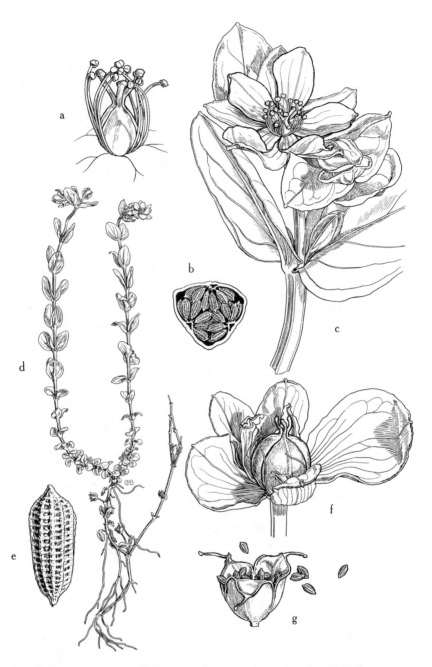

Fig. 263. *Hypericum anagalloides: a,* stamens, ovary, and styles, × 8; *b,* mature capsule (cross section), × 8; *c,* inflorescence, × 3; *d,* habit, the leaves sessile, ovate, × ⅖; *e,* mature seed, × 40; *f,* mature capsule in calyx, the sepals unequal in length, × 6; *g,* capsule, showing septicidal dehiscence, × 6.

Carpels longitudinally grooved and striate on back; stems from a decumbent or a rhizomatous base.

Plant markedly succulent; inflorescence a dense spike . . . . 1. *S. rhizomata*
Plant not succulent; inflorescence a slender, elongate spike . . . . 2. *S. calycosa*
Carpels not longitudinally grooved or striate; stems from a heavy, branched root crown
3. *S. neo-mexicana*

1. Sidalcea rhizomata Jepson, Man. Fl. Pl. Calif. 629. 1925. Fig. 262.

Stems erect or ascending from a perennial horizontal rhizome, 3–5 dm. tall; herbage glabrous to sparingly hirsute-pilose, succulent throughout; leaves long-petioled, the blades of basal leaves 2–10 cm. broad, crenately incised, the stem leaves divided, the segments oblanceolate; stipules 8–16 mm. long, ovate, acuminate or obtuse, serrate; bracts of inflorescence scarious, hairy, 2-lobed, 8–16 mm. long; spikes short, densely flowered; calyx hairy, the hairs simple, the lobes acuminate, scarious, purple-tipped, 6–12 mm. long; corolla light purple, 2–5 cm. long; carpels striate-grooved on back, lightly reticulate on sides.

Marshes: Point Reyes, Marin County.

2. Sidalcea calycosa Jones, Am. Nat. 17:875. 1883.

Stems erect from a decumbent base, sparingly branched, 3–10 dm. high; herbage glabrous to sparsely pilose; leaves long-petioled, the blades of basal leaves crenately incised, the cauline leaves of 6–7 cuneate divisions, each entire or 3-toothed at apex; stipules scarious, linear to ovate, serrate; bracts of inflorescence 2-parted, the lobes ovate-lanceolate; racemes few- to many-flowered; calyx hirsute, the lobes acuminate, purple-tipped, becoming scarious; corolla light purple, 1–2 cm. long; carpels striate-ridged on back, reticulate on sides.

Bogs or marshes: northern Sierra Nevada and north to Shasta County, North Coast Ranges in Napa and Sonoma counties, Sacramento Valley.

3. Sidalcea neo-mexicana var. **parviflora** (Greene) Roush, Ann. Mo. Bot. Gard. 18:186. 1931.

Sidalcea parviflora Greene, Erythea 1:148. 1893.

Stems stout, 4–12 dm. tall, from a perennial, branched root crown; herbage glabrous to stellate-pubescent on leaves; leaf blades crenate-dentate, 5- to 9-cleft, 2–5 cm. wide; raceme slender, the bracts bifid, scarious, 3–4 mm. long; calyx hirsute to stellate-pubescent, the lobes triangular-ovate, acuminate; petals rose-colored, erose, 6–12 mm. long; carpels reticulate, glabrous, the beak short, recurved.

Brackish or alkaline marshes: Inyo County south through Mojave Desert to coastal Los Angeles and Orange counties, South Coast Ranges north to Monterey County.

Fig. 264. *Bergia texana: a,* mature capsule at dehiscence, showing the denticulate sepals, × 6; *b,* habit, prostrate form, × ⅔; *c,* leaf axil with flowers, × 4; *d,* mature seed, shiny brown, obscurely quadrate-reticulate, × 60; *e,* habit, erect form, × ⅔.

HYPERICACEAE

Ours glabrous annual or perennial herbs. Foliage with pellucid or black dots, the leaves opposite, entire, without stipules. Flowers regular, perfect, in terminal or axillary cymes. Sepals 5, herbaceous, persistent. Petals 5, yellow, in ours shorter than sepals, deciduous or withering-persistent. Stamens numerous, usually united by their filaments into 3–5 fascicles. Ovary superior, 1- to 3-celled, placentation parietal, the styles 3, distinct. Fruit a septicidal capsule with numerous seeds.

HYPERICUM

Forming dense mats with ascending or erect stems; leaves 4–12 mm. long; sepals ovate, longer than capsule .... 1. *H. anagalloides*
Stems erect from base; leaves 10–20 mm. long; sepals linear to lanceolate, shorter than capsule .... 2. *H. mutilum*

1. Hypericum anagalloides Cham. & Schl., Linnaea 3:127. 1828. Tinker's penny. Fig. 263.

Weak annual with few to many erect or procumbent stems or forming dense mats with ascending stems; leaves lanceolate to ovate, obtuse, 4–12 mm. long; flowers small, terminal and solitary or in leafy cymose clusters or panicles; sepals ovate, unequal in length, longer than capsule; petals orange-yellow (coppery); stamens 10–20; capsule 1-celled.

Common about fresh-water streams and springs or in bogs: throughout California; north to British Columbia.

2. Hypericum mutilum L., Sp. Pl. 787. 1753.

Erect annual; stems simple or branched, 2–5 dm. tall; leaves ovate, 10–20 mm. long; flowers small, numerous, in leafy cymes; sepals linear to lanceolate, usually shorter than capsule; stamens 6–12; capsule 1-celled.

Shores of lower Sacramento and San Joaquin rivers; introduced from Europe.

ELATINACEAE

Annual or perennial herbs with opposite or verticillate leaves. Flowers small, axillary, sessile or pedicellate, 2- to 5-merous. Sepals as many as the petals or fewer, or under some aquatic conditions the petals not evident. Petals often membranous, campanulately spreading in terrestrial plants, often closely investing the ovary in aquatic individuals. Stamens as many as the petals or twice as many, occasionally reduced to 1, when equal in number either alternate or opposite the petals. Ovary 2- to 5-celled, usually 3-celled; placentation axile. Seeds several to many in a locule, usually with a reticulate surface pattern.

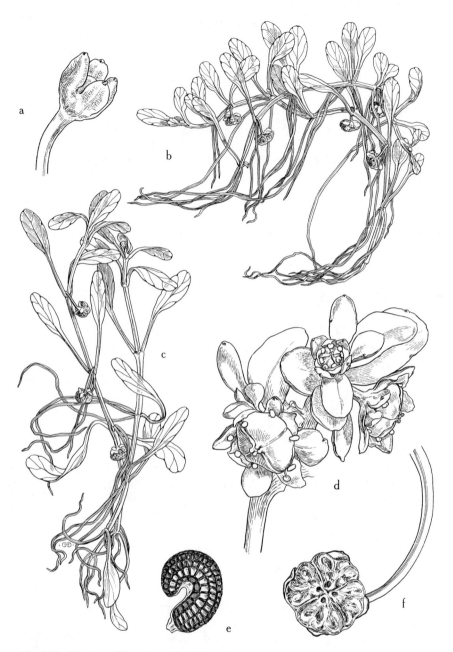

Fig. 265. *Elatine californica: a,* bud, showing calyx, × 12; *b* and *c,* habit, × 2; *d,* terrestrial form, showing apex of branch in flower and fruit, × 8; *e,* mature seed, × 40; *f,* mature capsule, showing 4 carpels, top view, × 12.

Flowers 5-merous; sepals conspicuously scarious-margined, midrib clearly evident; herbage glandular-pubescent .... *Bergia*
Flowers 2-, 3-, or 4-merous; sepals not scarious-margined, without evident midrib; herbage glabrous .... *Elatine*

BERGIA

1. Bergia texana Seub. in Walp., Rep. 1:285. 1842. Fig. 264.

Diffusely branched annual; stems 5–15 cm. long; herbage glandular-pubescent; leaves tapering to a petiole, the blades obovate to oblanceolate or elliptic,
serrulate toward the apex, 1–3 cm. long; stipules deeply serrate; flowers on
short pedicels, solitary or clustered in the leaf axils, 5-merous; sepals 3–4 mm.
long, cuspidate, with broad, scarious margins, denticulate along midvein; petals
white, oblong or obovoid, subequal to the sepals in length; stamens 5–10;
capsule globose, 5-carpeled; seeds many, brown, shiny, obscurely quadrate-
reticulate.

Common on margins of pools or on floodplains where water has stood:
Clear Lake in Lake County, Central Valley, coastal southern California; east
to Texas.

ELATINE

Aquatic, amphibious, or terrestrial annuals from the bottoms of pools or
slow streams or on wet sandy or muddy shores; stems erect or prostrate,
flaccid-succulent, 3–100 mm. long; herbage glabrous. Leaves in ours opposite,
sessile or petioled, with hyaline entire or toothed stipules, the blade linear-
spatulate to oblong or orbicular-obovate, the margin obscurely and remotely
crenate. Flowers 1 or 2 to a node, sessile or pediceled, 2-, 3-, or 4-merous.
Sepals 3 or reduced to 2, equal or unequal in size, in some species withering-
persistent. Petals membranous, usually orbicular in outline, in terrestrial plants
often campanulately spreading, in aquatic plants often closely investing the
ovary or in some not evident. Stamens as many as the petals or twice as many,
or in some aquatic forms reduced to 1 or sometimes none. Ovary 3- or 4-celled,
septicidally dehiscent.

The problems presented by this genus are discussed in a paper by Mason
(Madroño 13:239–240, 1956).

Flowers 4-merous; stamens 8; pedicels elongating in fruit .... 1. *E. californica*
Flowers 3-merous; stamens 3 or varying from 1 to 6.
  Flowers subsessile to distinctly pediceled; fruit turning to one side at maturity; sepals
  3, equal in size .... 2. *E. ambigua*

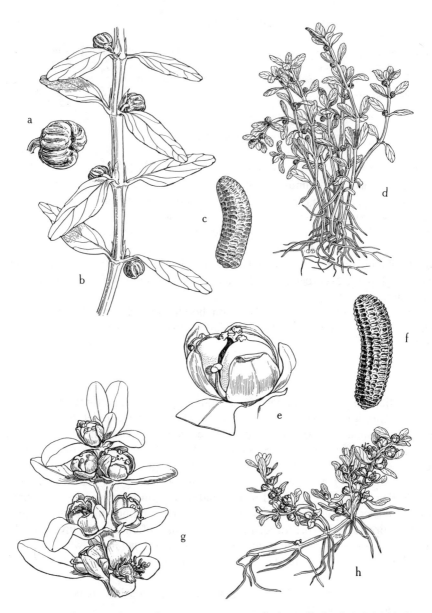

Fig. 266. *Elatine. a–d, E. ambigua: a,* mature capsule, its pedicel reflexed, × 12; *b,* part of fruiting stem, × 4; *c,* mature seed, its sculpture shallow, × 40; *d,* habit, aquatic form, × 1. *e–h, E. rubella: e,* mature capsule at dehiscence, showing stamens alternating with the carpels, × 8; *f,* mature seed, its longitudinal ribs conspicuous, × 40; *g,* habit, upper part of stem, showing leaves and capsules, × 4; *h,* habit, terrestrial form, × 1.

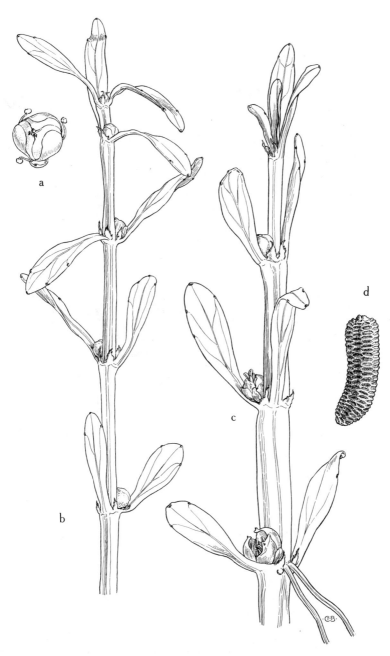

Fig. 267. *Elatine gracilis: a,* flower, × 12; *b,* habit, terrestrial form, × 4; *c,* habit, aquatic form, × 6; *d,* mature seed, showing conspicuous transverse ridges, × 40.

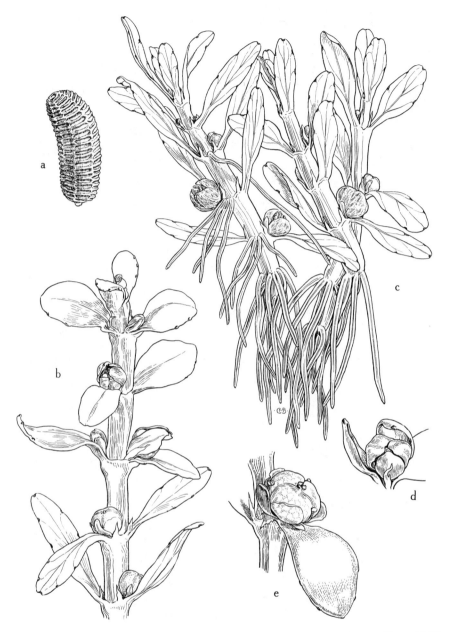

Fig. 268. *Elatine chilensis: a,* mature seed with conspicuous transverse ridges, × 60; *b,* branch, terrestrial form, × 6; *c,* habit, aquatic form, × 4; *d,* young flower, showing the pair of large sepals, × 12; *e,* mature capsule, showing the third, reduced sepal, the petals, and the stamens alternating with the carpels, × 6.

Flowers sessile, erect; sepals 2 or 3, when 3 then 1 of them reduced.
Stamens 3, alternate with the carpels.
Seeds with 16–35 pits (areolae) in a row.
Pits of seeds much broader than long, transverse ridges more prominent than the longitudinal ones, the seeds thus appearing rugose.
Seeds 6–10 in each locule; plants slender, erect; leaves all linear-spatulate .... 3. *E. gracilis*
Seeds 15–40 in each locule; plants coarser; leaves, at least the upper ones, linear-spatulate to orbicular-obovate .... 4. *E. chilensis*
Pits of seeds nearly as long as broad, longitudinal and transverse ridges almost equally distinct .... 5. *E. rubella*
Seeds with 10–15 pits in each row; pits nearly as long as broad.
Leaves 1–2 mm. wide .... 6. *E. brachysperma*
Leaves 3–4 mm. wide, orbicular-obovate .... 7. *E. obovata*
Stamens 3–6, when 3, opposite the carpels................... 8. *E. heterandra*

1. Elatine californica Gray, Proc. Am. Acad. Arts & Sci. 13:361. 1878. Fig. 265.

Matted, prostrate plant of muddy shores, or erect aquatic; leaves short-petioled to subsessile, obovate to oblanceolate, 4–12 mm. long; flowers on short pedicels, the pedicel elongating in fruit to become 1–2 times as long as the fruit; sepals 4, equal or subequal to one another in size, oblong, united at base and growing with the fruit; petals 4, obovate; stamens 8; capsule with 4 carpels; seeds J- or U-shaped, rounded at one end and truncate at the other with a subapiculate base.

Ponds, vernal pools, rice fields, and margins of streams and ditches: widespread in California but not often collected, especially abundant in Sierra Valley, the locality from which the type was taken; north to Washington.

2. Elatine ambigua Wight in Hook., Bot. Misc. 2:103. 1831. Fig. 266.

Erect aquatic annual 1–8 cm. tall, simple or branched from near base; leaves lanceolate to elliptic, or oblong, tapering to a short petiole, the margins obscurely crenate to entire, bright green above to glaucescent below; stipules nearly entire; flowers subsessile to short-pediceled, usually 1 to a node, turning to one side in fruit; sepals 3, all the same size; neither petals nor stamens seen in ours; capsule 3-celled, membranous; seeds 4 or 5 (–10) to a locule, white to pale yellow, sculpturing indistinct, the pits (areolae) shallow, 20–30 in a row. (No terrestrial forms known in California.)

Common in rice fields of the Central Valley or in canals draining these fields; native of Asia.

3. Elatine gracilis Mason, Madroño 13:240. 1956. Fig. 267.

Plant slender, erect, 2–4 cm. tall; leaves opposite, subequal to the internodes in length or slightly longer, linear-spatulate, narrowed to a petiole-like base, with attenuate, lacerate stipules, the margin of blade with a few crenulations, each with a minute callus; flowers usually 1 to a node, sessile; sepals 2, or a

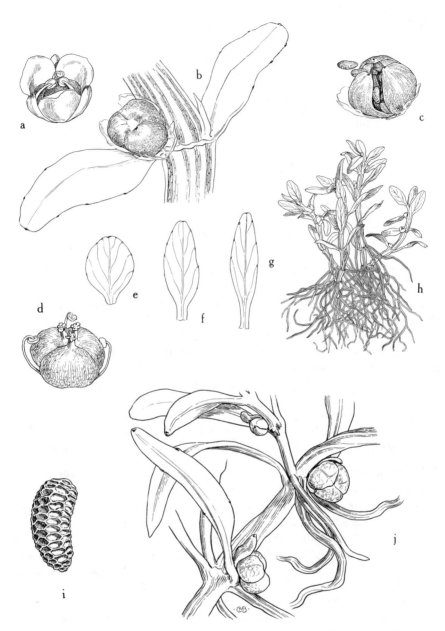

Fig. 269. *Elatine brachysperma: a,* flower, × 10; *b,* capsule in leaf axil, × 10; *c,* mature capsule at dehiscence, × 12; *d,* young capsule, the stamens alternating with the carpels, × 16; *e–g,* leaf variation, × 4; *h,* habit, aquatic form, × 1; *i,* mature seed, × 60; *j,* terrestrial form, part of stem, showing roots, capsules, and slender type of leaf, × 6.

third sepal present and reduced; petals 3, very thin, membranous, almost orbicular, closely investing the growing fruit; stamens 3–1 or sometimes none, alternate with the carpels; carpels 3; seeds straight or slightly curved, 7 or 8 in a locule, with 9 or 10 rows of pits (areolae) and 20–30 pits to each row, the transverse ridges more conspicuous than the longitudinal ones, the seed therefore appearing as though transversely rugose. Aquatic phase of the plant not very different from the terrestrial form.

Known only from vernal pools in the valley of the Little Truckee River in Nevada County, Sierra Nevada.

4. Elatine chilensis C. Gay, Fl. Chil. 1:286. 1845. Fig. 268.

Plant 5–100 mm. long, aquatic or on wet mud, then creeping and rooting at nodes; leaves obovate to broadly spatulate, rounded at summit, 3–4 mm. long, 1–3 mm. broad, narrowed at base to a petiole, with entire, triangular, attenuate, hyaline stipules; flowers solitary in leaf axils, 1 or 2 to a node, sessile; sepals 2, or a third sepal present and reduced, the largest sepal oblong; petals white to pink, orbicular; stamens 3, alternate with the carpels; seeds 24–33 in a locule, borne at the base of the placental axis, erect, cylindric, slightly curved, with 25–35 short, broad pits, the transverse ridges more conspicuous than the longitudinal ones, the seed therefore appearing as though transversely rugose.

Muddy shores of ponds: Sierra Valley, Madeline Plains, associated with *Elatine rubella* and *E. californica;* South America.

5. Elatine rubella Rydb., Mem. N. Y. Bot. Gard. 1:260. 1900. Fig. 266.

Elatine triandra of authors referring to western American plants, not *E. triandra* Schkuhr.

Elatine triandra var. *genuina* Fassett, Rhodora 41:369. 1939. As to western American plants.

Plant prostrate on wet mud or erect and elongate, to 15 cm. high; stems flaccid; herbage often with a reddish pigment; leaves opposite, lanceolate to linear-spatulate, truncate or emarginate at apex, 2–15 mm. long; flowers sessile, 1 or 2 at a node, erect, 3-merous; sepals 2 or 3, frequently very unequal in size, petals broad, thin-membranous; stamens 3, alternate with carpels; seeds 12–30 to each locule, borne along the placental axis, cylindrical to slightly curved, longitudinally ribbed, each rib with 16–25 deep, hexagonal pits.

Ponds, vernal pools, ditches, and rice fields: almost throughout California, but not often collected; western America.

This species has been included under the European species *Elatine triandra*, from which it differs in having sepals that are unequal in size. California material which has been referred by authors to *E. americana* (a species of the eastern United States) belongs under *E. rubella*.

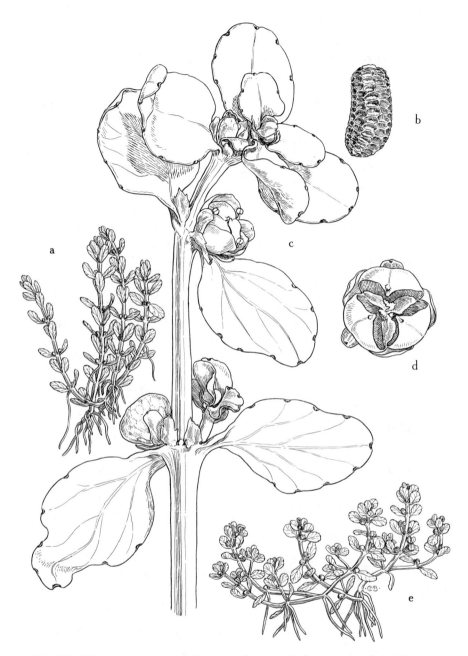

Fig. 270. *Elatine obovata: a,* habit, aquatic form, × ⅘; *b,* mature seed, × 40; *c,* upper part of stem, showing leaves, flowers, and capsule, × 8; *d,* dehisced, empty capsule, showing septa and locules, top view, × 12; *e,* habit, terrestrial form, × ⅘.

6. Elatine brachysperma Gray, Proc. Am. Acad. Arts & Sci. 13:361. 1878. Fig. 269.

Elatine triandra var. brachysperma Fassett, Rhodora 41:374. 1939.

Low, spreading plant or the tip ascending; stems 1–5 cm. long; leaves narrowly oblong to linear-spatulate, narrowed or rounded at tip and often very variable on the same plant, 1–2 mm. wide; stipules linear-lanceolate, irregularly toothed; flowers 3-merous, sessile, 1 or 2 to a node; sepals 2, or a third sepal present and much reduced; petals 3; stamens 3, alternate with carpels; capsule depressed, 3-celled; seeds 11–15 in each locule, short-oblong-ellipsoid, longitudinally ribbed, 9–15 pits in each row.

Shallow water or muddy shores of vernal pools, ponds, or ditches: widely distributed in California but infrequently collected; western and middle-western United States.

7. Elatine obovata (Fassett) Mason, Madroño 13:240. 1956. Fig. 270.

Elatine triandra var. obovata Fassett, Rhodora 41:375. 1939.

Erect, spreading, or prostrate annual; leaves orbicular to obovate, sessile to short-petioled, the margins remotely crenate with a minute callus in the crenation, 3–4 mm. wide; stipules ovate-lanceolate, irregularly toothed; flowers 1 or 2 to a node, usually in alternate leaf axils, sessile or subsessile and erect, 3-merous; sepals 2, or a third sepal present and reduced, the 2 large sepals oblong and subequal to the petals in length; petals membranous, broad; stamens 3, alternate with the carpels; carpels 3; capsule depressed; seeds 11–15 in each locule, short-oblong-ellipsoid, 9–15 pits in a row, the pits hexagonal, deep.

Coastal area and valleys: central California from Lake and Marin counties to Merced County. Fassett reports it from Mexico.

Elatine obovata is clearly related to E. brachysperma, but to date it is known only from the Coast Ranges and valleys at low altitudes in California, whereas E. brachysperma is usually montane. The two species are distinguishable chiefly by the size and shape of their leaves and by the over-all coarseness of E. obovata.

8. Elatine heterandra Mason, Madroño 13:240. 1956. Fig. 271.

Terrestrial plant prostrate, rooting at nodes (aquatic form unknown); leaves 2–4 mm. long, obovate to broadly elliptic-oblong, narrowed to a short petiole or sessile with a few minute callous notches along the margin; stipules hyaline, lanceolate, about 1 mm. long; flowers solitary at nodes and alternate; sepals 2, or a third sepal present and much reduced, the 2 large sepals oblong and subequal to or exceeding the petals in length; corolla campanulately spreading, the petals orbicular; stamens very variable, usually 6 in 2 whorls but varying from 1 to 6, when only 3 stamens, these opposite the carpels; carpels

Fig. 271. *Elatine heterandra: a,* mature seed, × 40; *b,* young flower, × 20; *c,* upper part of a branch, showing the leaves and variability in the flowers, × 8; *d,* mature capsule, the stamens opposite the carpels, × 12; *e,* habit, terrestrial form, × 2.

3; fruit a depressed capsule; seeds 8–12 to a locule, straight or curved, truncate, and apiculate at the hilum, rounded at the opposite end, the pits in 9 or 10 rows, 12–15 in each row.

Sierra Valley and margin of pond 1½ miles east of Calpine, both in Sierra County, Sierra Nevada, also Snow's Lake in Lake County, North Coast Ranges.

FRANKENIACEAE. FRANKENIA FAMILY

Low, perennial herbs or dwarf bushes. Leaves small, opposite, entire, sometimes crowded and fascicled. Flowers perfect, hypogynous, sessile, solitary or pseudocymose. Calyx tubular, 4- or 5-toothed. Petals 4 or 5, with an appendage at the very base of the limb, the appendage decurrent on the claw. Stamens 4–7, exserted from the calyx tube. Ovary 1-celled, with basal or subbasal placentae. Style 1, slender, 2- or 3-cleft, included. Capsule linear, angled, included in the persistent calyx, 2- to 4-valved, the seeds attached by filiform funiculi to the capsule wall, the embryo straight.

FRANKENIA

Style 3-cleft; ovules many; plant herbaceous or nearly so; leaves 5–10 mm. long, linear-oblanceolate to obovate, from more or less expanded to more or less revolute on a given plant .... 1. *F. grandifolia*
Style 2-cleft; ovules 2 or 3; dwarf bush; leaves 2–4 mm. long, all strongly revolute
2. *F. Palmeri*

1. Frankenia grandifolia Cham. & Schl., Linnaea 1:35. 1826. Fig. 272.

Perennial, herbaceous or woody at base, 1–4 dm. tall, glabrous to short-hirsute; leaves often fascicled, sessile to subsessile, obovate to narrowly oblanceolate, 6–10 mm. long, margins from plane to strongly revolute on a single plant, opposite pairs united by a membranous sheathing base; flowers solitary in the leaf axils or appearing as though clustered on short branches; calyx cylindrical with acute teeth; petals 4 or 5, pink, slightly unequal in length, spatulate, with a slightly erose ligular scale at the summit of and decurrent on the claw; stamens 4–7, exserted from calyx tube; style 3-lobed; fruit a capsule; seeds many.

Margins of salt marshes: from central California south to Baja California, Mexico.

2. Frankenia Palmeri Wats., Proc. Am. Acad. Arts & Sci. 11:124. 1876.

Low bush, 1–2 dm. high; herbage canescent; leaves opposite, thickly clothing the stem, sessile to subsessile, 2–4 mm. long, linear-oblong but so strongly revolute as to appear terete, stipulate, the stipules of opposite leaves joined; flowers densely fascicled on a dwarf branch; calyx 3 mm. long; petals white,

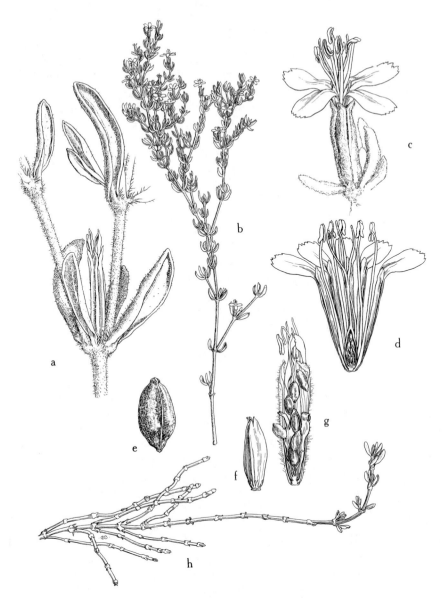

Fig. 272. *Frankenia grandifolia: a,* part of stem, showing flower and leaves fascicled in the axils, the leaf margins revolute, × 3; *b,* habit, upper part of stem, × ⅖; *c,* flower, × 4; *d,* corolla, longitudinally split, the ligulate appendages decurrent on the claws, × 4; *e,* mature seed, × 16; *f,* mature capsule with calyx removed, × 4; *g,* calyx and capsule (longitudinal section), showing the subbasal attachment of the funiculi, × 4; *h,* habit, suffruticose basal part of plant, × ⅖.

limb oblong, with a small ligular scale at summit of claw; stamens 4; style 2-lobed; seeds 1–3.

Borders of salt marshes: San Diego County to Baja California, Mexico.

TAMARICACEAE

Shrubs or small trees with sessile, scale-like alternate leaves. Flowers in slender, catkin-like spikes or racemes, perfect, hypogynous, regular. Sepals imbricate or valvate, 4–6, free. Petals free to connate, equal in number to sepals. Stamens 5–10, free or connate at base to a hypogynous disc; anthers 2-celled. Ovary superior, 1-celled but of 3 or 4 united carpels; styles 2–4, free or united at base. Ovules numerous on a parietal or basal placenta. Fruit a capsule. Seeds with a tuft of hairs at apex or bearded all over. Embryo straight.

TAMARIX. TAMARISK

Leaves and bracts sheathing the stem and giving a jointed appearance; herbage often
    coated with a secreted salt .... 1. *T. aphylla*
Leaves and bracts not sheathing the stem; herbage not coated with salt.
    Flowers 4-merous; racemes lateral on previous season's growth ...... 2. *T. tetrandra*
    Flowers 5-merous; racemes terminal in large panicles .... 3. *T. pentandra*

1. Tamarix aphylla (L.) Karsten, Deutsche Fl. 641. 1882. Salt tree.

Tamarix articulata Vahl, Symb. Bot. 2:48, pl. 32. 1791.

Bushy tree with many slender branches often white with saline secretion from punctate glands; branchlets slender and numerous, of short articulate joints, glaucous, green; leaves sheathing stem, the free part reduced to a minute cusp; bracts flaring abruptly from stem; flowers sessile, white, 5-merous, in spikelike racemes; sepals 1.5 mm. long, orbicular; petals 2 mm. long, oblong; stamens 5, inserted between the lobes of the hypogynous disc; pistil 2.5 mm. long; styles 3; capsule 3.5 mm. long.

An occasional escape from cultivation along stream washes: Central Valley; native of Africa.

2. Tamarix tetrandra Pall. ex Bieb., Fl. Taur.-Cauc. 1:247. 1808. Four-petaled tamarisk.

Tamarix parviflora DC., Prodr. 3:97. 1828.

Shrub or small tree, 2–3 m. tall, with slender, arching branches; leaves and bracts green, scale-like, deltoid, acute to acuminate, gibbous at base, deciduous; racemes slender, rigid, 2–6 cm. long, on old wood of previous year; flowers 4-merous, pink, on pedicels, appearing before leaves; petals spreading, 2–3 mm. long; stamens 4, the filaments dilated at base and attached to angles of the hypogynous disc; pistil 1.5–3 mm. long; styles 3 or 4; capsule 3–4 mm. long.

A common escape: Lake County and the Sierra Nevada foothills south to southern California; native of southeastern Europe.

3. Tamarix pentandra Pall., Fl. Ross. 1²:72. 1788. Desert tamarisk.

Shrub or small tree; herbage glabrous to minutely puberulent; branches often purplish; leaves and bracts lanceolate to ovate-lanceolate or deltoid, acute to acuminate, 3.5 mm. long or less; racemes on stems of current season, often in panicle-like clusters terminating branches; flowers 5-merous, white or pink, persistent, on short pedicels; petals 1.5–2 mm. long; stamens 5, inserted between lobes of hypogynous disc; pistil 1.5–2 mm. long; styles 3; capsule 3–3.5 mm. long.

A common escape along watercourses in the most arid parts of the state; common along the Colorado River.

VIOLACEAE

Ours perennial herbs. Leaves alternate, simple, entire or lobed; stipules persistent. Flowers (in *Viola*) single on axillary peduncles. Sepals 5, persistent, unequal in size, basally auriculate. Petals 5, unequal in size and shape, the corolla thus zygomorphic with 2 upper petals, 2 lateral ones, and 1 lower, spurred petal. Stamens 5, hypogynous, the filaments short and broad and connivent over ovary, each bearing an anther on its inner face. Ovary superior, 1-celled, placentae parietal, style and stigma 1. Fruit a 3-valved capsule bearing several seeds on the middle of each valve and dehiscing loculicidally.

VIOLA

Leaf blades cordate, reniform, or at least truncate at base.
    Petals pale violet; leaf blades 2.5–6 cm. wide .... 1. *V. palustris*
    Petals white; leaf blades 1–2 cm. wide .... 2. *V. Macloskeyi*
Leaf blades rhombic-ovate, tapering to the petiole .... 3. *V. occidentalis*

1. Viola palustris L., Sp. Pl. 934. 1753.

Acaulescent plant with caespitose scapes and leaves from slender rhizome which produces slender stolons; herbage glabrous; leaves on petioles 4–10 cm. long, the blades round-cordate, entire to shallowly crenate, 2–5 cm. wide; petals pale violet to white, 6–12 mm. long, lateral petals sparsely bearded.

Swamps and moist shady woods: coastal Mendocino County; north to Alaska, east to Atlantic, Europe, Asia.

2. Viola Macloskeyi Lloyd, Erythea 3:74. 1895.

Viola blanda var. *Macloskeyi* Jepson, Man. Fl. Pl. Calif. 648. 1925.

Acaulescent, small, delicate plant with caespitose scapes and leaves from a slender rhizome which produces slender stolons; herbage glabrous to tomentose; leaves on slender petioles 1–2 cm. long, the blades orbicular to ovate-cordate or round-reniform, crenulate, 1–2 cm. wide; petals white, 6–8 mm. long, usually beardless or the lateral pair bearded.

Common in bogs and wet meadows: mountains throughout California except the South Coast Ranges and mountains in the deserts; north to British Columbia.

3. Viola occidentalis Howell, Fl. NW. Am. 69. 1897.

Caespitose scapes and leaves from a stout, vertical rootstock, becoming stoloniferous late in the season; herbage glabrous; leaves long-petioled, the blades rhombic-ovate, attenuate to the petiole; petals white, the lower one veined with purple and bearded.

Rare, bogs of Del Norte County; to adjacent southern Oregon.

DATISCACEAE

Perennial herbs with alternate, divided leaves. Flowers dioecious, rarely perfect. Calyx synsepalous. Corolla none. Stamens indefinite. Ovary inferior, 1-celled, with 3 parietal placentae; styles 3, bifid. Fruit a capsule, opening at the top between the styles.

DATISCA

1. Datisca glomerata (Presl) Baill., Hist. Pl. 3:407. 1872.

Stout, glabrous, perennial herb; stems several to many from the base, 4–14 dm. tall; leaves divided and incised, the divisions sharply serrate; flowers in clusters in the axils of leafy branches, apetalous, dioecious or the pistillate flowers with a few stamens; staminate flowers with a very short calyx of 4–9 unequal, united lobes; stamens 8–12, usually 10, the filaments short; pistillate flowers with ovoid, 3-angled and 3-toothed calyx tube; stamens when present 2–4, alternate with calyx teeth; ovary inferior, 1-celled, with 3 parietal placentae; styles 3, each bifid; fruit a capsule with many small seeds.

Dry stream beds or washes, margins of rivers and streams, springy spots in the valleys and foothills: throughout California at elevations of 50–6,500 feet; south to Mexico.

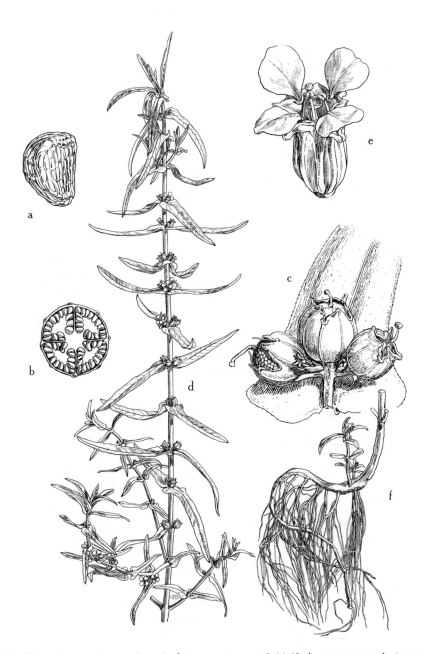

Fig. 273. *Ammannia. a–c, A. auriculata: a,* mature seed, × 40; *b,* mature capsule (cross section), the placentation axile, × 4; *c,* peduncled inflorescence, the subtending leaf auricled at base, × 3. *d–f, A. coccinea: d,* habit, the flowers sessile in leaf axils, × ⅖; *e,* flower, × 4; *f,* habit, basal part of plant, showing roots, × ⅖.

LYTHRACEAE

Glabrous herbs. Leaves simple, entire, opposite or rarely alternate, usually exstipulate. Flowers solitary or clustered, perfect, regular. Calyx tubular, free from the ovary but often the tube enclosing the ovary and capsule, 4- to 6-toothed, usually with accessory teeth in the sinuses. Petals 4–6, inserted with the stamens on the calyx tube. Ovary 2- to 6-celled or rarely 1-celled, placentation axile; style 1; stigma 2-lobed. Capsule 1- to several-celled.

It is to be expected that many more members of this family will eventually make their appearance as rice-field weeds. Species of *Rotala* are especially abundant in rice fields of Asia and the Nile Valley.

Leaves alternate, calyx tube cylindrical, petals 5 or 6 .... *Lythrum*
Leaves opposite, calyx tube campanulate to urceolate, petals 4.
    Flowers usually several to a leaf axil, sessile or in cymes; leaves sessile by an auriculate base; capsules bursting irregularly on dehiscence, the walls without microscopic horizontal striations .... *Ammannia*
    Flowers solitary in the leaf axils; leaves tapering to the sessile or short-petioled base; capsules splitting by valves at the tips, the walls with microscopic horizontal striations .... *Rotala*

AMMANNIA

Herbaceous annuals 1–8 dm. tall, erect, declined, or prostrate. Leaves decussately opposite, base usually dilated and cordate-auriculate. Flowers 1 to several in a leaf axil, sessile, subsessile, pediceled, or on peduncled cymules, typically 4-merous, rarely 5- or 6-merous. Bracts usually fertile. Calyx campanulate or urceolate, becoming hemispherical in fruit, herbaceous, without appendages or these very small. Petals in ours 4, conspicuous. Stamens episepalous, inserted on the lower part of the calyx tube. Capsule globose or ellipsoid, the walls not horizontally striate.

Flowers all sessile or subsessile in the leaf axils; capsule 3–5 mm. in diameter
                                                      1. *A. coccinea*
Flowers pediceled, solitary or in clusters, the clusters peduncled; capsule 2–3 mm. in diameter .... 2. *A. auriculata*

1. Ammannia coccinea Rottb., Pl. Hort. Havn. Descr. 7. 1773. Fig. 273.

Erect or ascending annual, occasionally prostrate; stems to 50 cm. long, square, often conspicuously 4-angled; herbage glabrous; leaves 15–80 mm. long, sessile, auriculately lobed at base, lanceolate to oblong-lanceolate; flowers several, sessile to subsessile in the leaf axils; calyx in anthesis 1–2 mm. long, becoming 3–5 mm. long in fruit, appendages when present minute, sepals low-triangular; petals orbicular, 1–2 mm. long, bright rose-colored, rarely albino; stamens 4–8, exserted; capsule globose, 3–5 mm. in diameter.

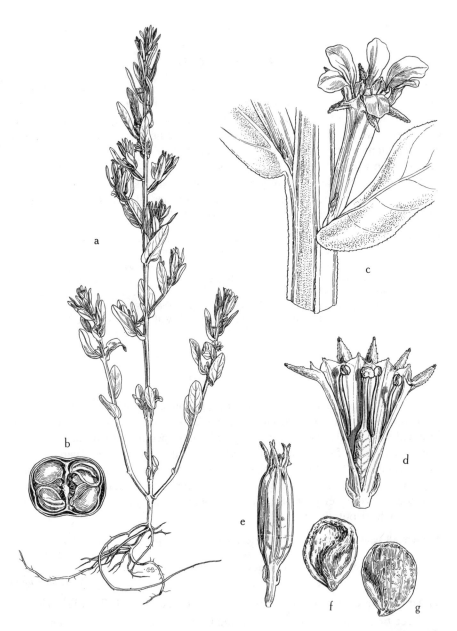

Fig. 274. *Lythrum hyssopifolia: a,* habit, erect plant, the sessile leaves with flowers in their axils, × ⅔; *b,* capsule (cross section), showing axile placentation, × 16; *c,* flower in leaf axil, showing the minute bract near base of calyx tube and the conspicuous, rough calyx appendages, × 8; *d,* calyx, longitudinally split, showing its toothlike, reduced lobes alternating with rough appendages on the outside, and the stamens inserted in the tube, × 8; *e,* calyx enclosing mature capsule, × 10; *f* and *g,* mature seeds, hollow on one side, shiny, brown, striate, × 30.

Common in wet soil near fresh water of floodlands, stream banks, ponds, and marshes, in valleys and foothills to an elevation of 4,000 feet: throughout California; Western Hemisphere.

Readily distinguished from *Ammannia auriculata* by the compact, sessile whorls of flowers and fruits.

2. Ammannia auriculata Willd., Hort. Berol. pl. 7. 1806. Fig. 273.

Erect or ascending annual, the stems to 50 cm. long, square, sometimes conspicuously angled; herbage glabrous; leaves 15–80 mm. long, sessile, linear-lanceolate, auriculately lobed at base, often much reduced in the inflorescence (ours not reduced); flowers 1–15 in a peduncled cluster, each flower with pedicel 3–15 mm. long; calyx in anthesis 1–2 mm. long, becoming 2–3 mm. long in fruit, the appendages when present usually minute, but sometimes nearly as large as the sepal; petals bright rose to white; stamens 4–8, exserted; capsule 2–3 mm. in diameter, rarely more.

Wet lands: Central Valley; south to Mexico, east to south Atlantic states, South America.

The California material of *Ammannia auriculata* usually has larger leaves on the upper part of the plant and larger fruit than seem to be typical for this species.

LYTHRUM

Slender or sometimes coarse herbs, erect or spreading to prostrate. Leaves in ours alternate, sessile, simple, entire. Flowers solitary in the leaf axils. Calyx cylindric to cylindric-funnelform, ribbed, the sepals reduced to minute teeth, the appendages alternate with sepals, usually larger. Petals 5 or 6. Stamens equaling the petals in length, inserted on calyx tube. Capsule oblong to cylindric.

Capsule very slender, broadest at base and tapering upward; petals 1–2 mm. long
1. *L. tribracteatum*
Capsule stout, oblong, cylindrical or elliptical, or sometimes attenuate below; petals 2–8 mm. long.
    Flowers pediceled, petals deep rose-purple to violet, 4–8 mm. long; stems erect, 4–20 dm. tall .... 2. *L. californicum*
    Flowers sessile or subsessile, petals pale pink to lavender, 2–3 mm. long; stems 0.5–10 dm. tall .... 3. *L. hyssopifolia*

1. Lythrum tribracteatum Salzm. ex Spreng., Syst. Veg. 4²:190. 1827.

Stems diffuse or prostrate, 5–50 cm. long, usually branched and often with a purplish pigment, the angles of the stem inconspicuously serrulate; lower leaves opposite, becoming subopposite and at length alternate above, 4–27 mm. long, 1–5 mm. wide, narrowly oblong or oblanceolate, obtuse, attenuate to the base; flowers pediceled, axillary, the lower flowers solitary, the upper

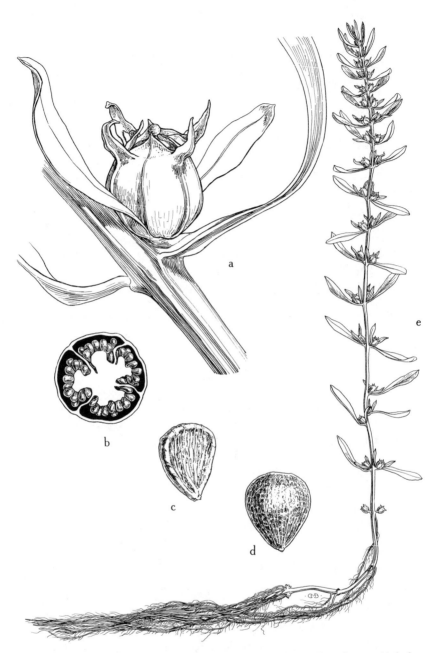

Fig. 275. *Rotala dentifera: a,* fruit in leaf axil, showing the long bracts, × 6; *b,* mature capsule (cross section), showing axile placentation, × 6; *c* and *d,* mature seeds, adaxial and abaxial views, × 40; *e,* habit, showing the flowers solitary in leaf axils, × ⅖.

ones densely spaced on accessory short branches inserted behind a solitary flower; bracts lanceolate, subequal in length to the calyx or longer; calyx cylindric, funnelform, 4–7 mm. long, adpressed to the stem, the lobes minute, the appendages minute and obtuse; petals 1–2 mm. long, deep rose-purple; stamens 6, unequally inserted near base of calyx tube and not extending beyond middle of tube; pistil with capitate stigma, equaling the stamens in length; capsule cylindric, tapering toward apex.

Margins of receding ponds and reservoirs on wet lands: Elmira in Solano County, Dorris Reservoir in Modoc County; native of Asia.

2. Lythrum californicum Torr. & Gray, Fl. N. Am. 1:482. 1840. Common loosestrife.

Plant perennial; stems erect, 3–20 dm. tall, the branches spreading or virgate, sometimes simple below, often rhizomatous; herbage pallid to glaucous, glabrous; leaves linear to linear-lanceolate, sessile, 5–70 mm. long, those of the inflorescence usually much reduced; flowers solitary in the leaf axils on short, bracteate pedicels; bracts minute; calyx cylindric to vase-shaped, the lobes low-triangular, the alternating appendages usually longer than the calyx lobes; petals deep violet, 4–8 mm. long, obovate; stamens 6–8, unequally inserted near base of tube and unequal in length; pistil with large, capitate stigma exserted from calyx tube; capsule ellipsoidal.

About springs on floodlands, irrigation ditches, and marshes: valleys and foothills; southeast to Texas.

A form growing among the tules of the Suisun marshes is particularly striking in that it seems to be adapted to considerable summer salinity in the tidal water that overflows it daily. It is characterized by taller stems (to 2 m. tall), horizontal rhizomes, much larger leaves, and longer peduncles, and the leaves of the inflorescence are often three times as long as the fruiting calyx. There seems to be no clear morphological or physiological basis at present for separating it from *Lythrum californicum*, yet in the extremes the differences are striking.

3. Lythrum hyssopifolia L., Sp. Pl. 447. 1753. Fig. 274.

Lythrum adsurgens Greene, Pittonia 2:12. 1889.

Annual or short-lived perennial, simple below or many-stemmed from the base, often erect or ascending but sometimes prostrate, 1–10 dm. long; herbage glaucous, glabrous; leaves sessile, linear to oblong or lanceolate, 5–30 mm. long; flowers subsessile in the leaf axils; bracts minute; calyx short-cylindric, 4–5 mm. long, strongly ribbed, the alternating appendages rough and more conspicuous than the lobes; petals 4, 2–3 mm. long, pale pink to lavender; stamens 4, opposite sepals, inserted near middle of tube; pistil with small, capitate stigma included in calyx tube; capsule ellipsoidal to cylindric; seeds shiny, brown, striate, hollow on one side.

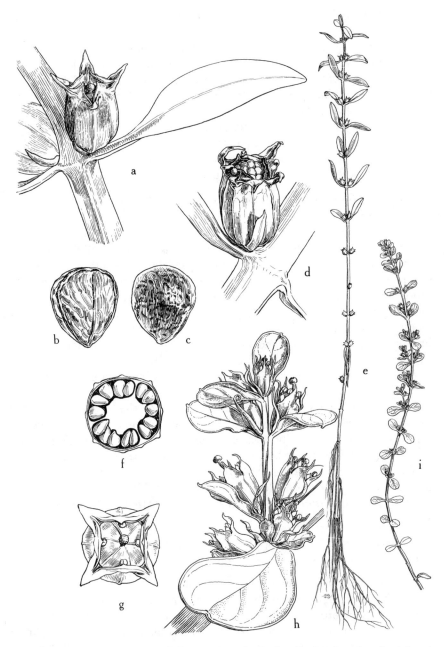

Fig. 276. *Rotala. a–g, R. ramosior: a,* fruit in leaf axil, showing the short bracts, × 6; *b* and *c,* mature seeds, adaxial and abaxial views, × 40; *d,* mature capsule, × 6; *e,* habit, × ⅔; *f,* capsule (cross section), × 8; *g,* maturing capsule, top view, × 8. *h* and *i, R. indica: h,* spikelike branch, the flowers solitary in the leaf axils, the bracts slender, × 6; *i,* habit, × ⅔.

Wet soil, in marshes and at the margins of streams and pools, often a garden weed: throughout California; world-wide.

The perennial plants were referred to *Lythrum adsurgens* by Greene, and Jepson followed this interpretation. We have noted in *Ammannia,* in *Rotala,* and in this species of *Lythrum* a tendency to root at the nodes when the stems are decumbent or prostrate. Where water persists through the dry season, *Lythrum hyssopifolia* tends to become perennial and more robust. As no other characters appear to separate *L. adsurgens* from *L. hyssopifolia,* the former is not retained as a separate entity in this treatment.

ROTALA

Annual herbs, ours with decussately opposite simple leaves. Flowers solitary in the leaf axils or sometimes on axillary branchlets, each subtended by a pair of bracts. Calyx with or without appendages situated just below the sinuses of the sepals. Petals usually inconspicuous and smaller than sepals. Stamens, in ours, 4–6. Stigma capitate. Capsule septicidally dehiscent, 2- to 4-valved, the valves with microscopic horizontal striations. Seeds ovate, flat or concave on one face, strongly convex on the other.

Leaves oblanceolate to linear-oblanceolate, 1–5 cm. long; flowers solitary in the leaf axils, bracts linear-lanceolate; calyx herbaceous, bearing appendages below the sinuses of the lobes, sepals deltoid to short-attenuate; capsule 3- or 4-valved.
Bracts subtending the flower and growing with the fruit, 1–2 times as long as the fruit
1. *R. dentifera*
Bracts subtending the flower not growing with the fruit, rarely more than ½ as long as the fruit . . . . 2. *R. ramosior*
Leaves broadly spatulate to orbicular, obtuse or rounded at apex, 5–15 mm. long; flowers in axillary spikelike branches; calyx thin, becoming scarious, without appendages below sinuses, sepals broadly lanceolate; capsule 2-valved . . . . 3. *R. indica*

1. Rotala dentifera (Gray) Koehne, Bot. Jahrb. 1:161. 1880. Fig. 275.

Stems erect or ascending, 3–50 cm. long; herbage glabrous; leaves lanceolate-elliptic to oblanceolate, narrowed to a very short petiole or subsessile, 2–4 cm. long; flowers sessile and solitary in the leaf axils, the bracts linear to linear-lanceolate, accrescent, longer than the flowers, 1–2 times as long as the fruit; appendages of calyx 2–5 times as long as the sepals, sometimes becoming reflexed, the sepals low-triangular, sometimes cuspidate; petals small, pink to white, elliptic, 1 mm. long; style short or obsolete; ovary 4-celled (sometimes 3-celled).

In shallow water or at the margins of ponds or streams: San Joaquin Valley; south to Mexico.

Rotala dentifera is distinguished from *R. ramosior* by the long, accrescent bracts subtending the flowers and fruits, and by the long appendages on the calyx. Koehne used this latter character to distinguish the two species. It is adequate in extreme examples only. The character of the bracts is more reliable.

2. Rotala ramosior (L.) Koehne in Mart., Fl. Bras. 13²:194. 1875. Fig. 276.

Plant annual; stems simple, erect, or branched from a declined base, or rarely prostrate and diffuse, 4-angled, 3–30 cm. long; leaves 1–5 cm. long, 2–12 mm. broad, oblanceolate to linear-oblanceolate, attenuate to a sessile or subsessile base; flowers solitary in the leaf axils; subtending bracts at the base of the calyx ½ to fully as long as the calyx, rarely more than ½ as long as the fruit, not accrescent; calyx in anthesis 2.5–3 mm. long, in age 5 mm. long, the appendages ½–2 times as long as the lobes, the sepals broadly deltoid; petals as long as the sepals or shorter, white to rose; style shorter than ovary; capsule ovoid to globose, 3- or 4-valved.

Common in wet places almost throughout the valleys and foothills to an elevation of 4,000 feet, usually associated with fresh water and especially abundant as a rice-field weed; Western Hemisphere.

3. Rotala indica (Willd.) Koehne, Bot. Jahrb. 1:172. 1880. Fig. 276.

Stems 4–30 cm. long, usually branched; leaves 4–17 mm. long, 1.5–8 mm. wide, cuneate-obovate to spatulate, obtuse, rarely acute, mucronate or in age the mucro absent and the apex subemarginate, margins cartilaginous; flowers solitary in the leaf axils and on short, foliaceous, spikelike branches; bracts subequal to calyx lobes in length, linear, acute; calyx 2–3 mm. long, broadly campanulate, angled on the nerves but thin and scarious in age, without appendages below sinuses of lobes, the sepals broadly lanceolate; petals much shorter than sepals, persistent, obovate, acute; stamens 4–6, inserted midway on calyx tube, the anthers barely exserted; style as long as the ovary; capsule oblong-ellipsoid, 2-valved; seeds ovate, flat or concave on one face, strongly convex on the other.

Known from a solitary specimen found on the margin of a marshy spot along the road from Biggs to Butte City in Butte County, probably introduced through rice culture; native of southern Asia.

ONAGRACEAE. EVENING-PRIMROSE FAMILY

Herbs or rarely shrubs with simple alternate or opposite leaves. Flowers perfect, axillary or in terminal racemes. Calyx tube adnate to ovary, and with the petals inserted at its summit. Calyx lobes 4. Petals 4 (sometimes 5 or 2). Stamens same number as the petals or twice as many. Ovary 4-celled (sometimes 5- or 2-celled), inferior; style 1; stigma capitate, discoid, or 4-lobed. Fruit a capsule.

Stems prostrate or floating.
   Petals conspicuous, yellow; leaves alternate .... *Jussiaea*
   Petals inconspicuous or absent; leaves opposite .... *Ludwigia*
Stems erect or decumbent.
   Seeds bearing a tuft of long white hairs; inflorescence open-paniculate, sometimes leafy
                                                                          Epilobium
   Seeds naked.
      Flowers small, pink or white; anthers attached at base; inflorescence spicate, con-
         spicuously imbricate-bracteate .... *Boisduvalia*
      Flowers large, yellow; anthers attached at the middle .... *Oenothera*

BOISDUVALIA

Erect annual herbs with alternate, sessile leaves. Flowers small, in leafy-bracted spikes or axillary along the branches. Calyx tube short but evident, the sepals 4, erect. Petals 4, obovate, sessile, emarginate or 2-lobed, magenta-pink to white. Stamens 8, in 2 whorls, the inner whorl shorter, the anthers basifixed. Capsule 4-celled, 4-valved, sessile. Seeds naked.

Petals 8–12 mm. long .... 1. *B. macrantha*
Petals 2–4 mm. long.
   Capsule terete, membranous.
      Bracts of inflorescence ovate.
         Stems 3–15 dm. tall, herbage pubescent; capsule septicidal ...... 2. *B. densiflora*
         Stems 15–25 cm. tall, glabrous; capsule loculicidal .... 3. *B. glabella*
      Bracts of inflorescence linear; capsule loculicidal .... 4. *B. stricta*
   Capsule 4-sided, coriaceous .... 5. *B. cleistogama*

1. Boisduvalia macrantha Heller, Muhlenbergia 2:101. 1905.

Stems simple or branched from base, 7–35 cm. tall, erect; herbage glabrous below, villous in inflorescence; leaves lanceolate to oblanceolate, remotely serrulate, 18–35 mm. long; petals rose, cuneate, notched at apex, 8–12 mm. long; inner whorl of stamens 7 mm. long, the outer whorl 5 mm. long; capsule villous, 15–20 mm. long, tip produced into a short beak; seeds 2 mm. long, triangular-ovate.

Vernally wet lands of the rolling hills and plains: northern end of Sacramento Valley.

2. Boisduvalia densiflora (Lindl.) Wats. in Brew. & Wats., Bot. Calif. 1:233. 1876.

Annual; stems erect, simple below, branched above or sometimes several branches from the base, 3–15 dm. tall; leaves lanceolate, 2–8 cm. long, becoming bractiform in the inflorescence, then ovate, acute and often imbricate, 6–12 mm. long; inflorescence spicate, often elongate; sepals lanceolate on a constricted calyx tube; petals about 4 mm. long, cuneiform, notched at apex, pink to white; capsule 4 mm. long; seeds ovate to triangular-ovate.

Very common in low ground where water has stood: throughout California in the valleys and foothills to an elevation of 6,000 feet; north to British Columbia, east to the Rocky Mountains.

A form with densely imbricate bracts has been designated as *Boisduvalia densiflora* var. *imbricata* Greene (Fl. Fran. 225, 1891). Another form, which has villous herbage and petals with unequally sized lobes, has been designated as *B. densiflora* var. *bipartita* (Greene) Jepson (*B. bipartita* Greene, Erythea 3:119, 1895).

3. Boisduvalia glabella Walp., Rep. 2:89. 1843.

Stems several from the base, 15–25 cm. tall; herbage glabrous and green; leaves lanceolate to ovate, 10–15 mm. long; bracts ovate; petals 4 mm. long, deep pink; stamens of outer whorl shorter than those of inner whorl, the anthers often nearly sessile; capsule nearly straight, pointed, dehiscence loculicidal; seeds fusiform, numerous.

Floodlands and beds of former vernal pools: coastal southern California, north through Coast Ranges to Humboldt County, Central Valley, north to Modoc County; north to British Columbia, east to Nevada.

4. Boisduvalia stricta (Gray) Greene, Fl. Fran. 225. 1891.

Stems simple or diffusely branched, 10–35 cm. tall; herbage canescent and somewhat pilose-pubescent; leaves often sparse, linear, 1–4 cm. long; petals pale violet to white, 2 mm. long; capsule 10–15 mm. long, at length recurved, tardily dehiscent.

Stream beds or beds of vernal pools: San Jacinto Mountains, Coast Ranges, Sierra Nevada; north to Washington.

5. Boisduvalia cleistogama Curran, Bull. Calif. Acad. Sci. 1:12. 1884.

Stems rigidly branched to simple, white, 1–2 dm. tall; herbage glaucescent, glandular-pilose; leaves linear to lanceolate, 2–4 cm. long, remotely denticulate; earliest flowers cleistogamous and not opening; petals 4 mm. long, light pink, bifid at apex; capsule quadrate, sharply angled, puberulent, coriaceous, dehiscence not known.

Vernal pools at low altitudes: delta region of the Central Valley, north to Butte County, east to the Sierra Nevada foothills.

EPILOBIUM. WILLOW HERB

Annual or perennial herbs, rhizomatous or sometimes producing fleshy bulblets or offsets. Stems erect or occasionally decumbent or prostrate. Leaves opposite or alternate. Flowers in racemes or panicles, purple, rose, or white. Calyx lobes sessile on the ovary, rarely from a short tube prolonged beyond the ovary. Sepals 4. Petals 4, sometimes emarginate or bifid. Stamens 8, one whorl shorter than the other. Stigma 4-lobed. Capsule elongate, 4-celled, 4-valved, splitting lengthwise. Seeds minute, each bearing a tuft of long hairs (coma).

Petals entire, showy, 10–20 mm. long, spreading, slightly irregular; calyx tube almost wanting.
    Leaves thin, prominently veined; racemes many-flowered, not leafy
                                                  1. *E. angustifolium*
    Leaves thick, veins inconspicuous; racemes few-flowered, leafy ...... 2. *E. latifolium*
Petals deeply notched to obcordate, strictly regular, 2–12 mm. long; calyx tube short but always evident.
    Plants green and not at all glaucous.
        Rootstocks not producing fleshy, scaly buds; stems simple below, often much-branched above.
            Stems 0.5–3 dm. tall, simple; stoloniferous .... 3. *E. oregonense*
            Stems 3–12 dm. tall, simple below but branched above; not stoloniferous.
                Stems reddish; leaves mainly opposite; petals 6–10 mm. long ... 4. *E. Watsonii*
                Stems white or green; at least the upper leaves alternate; petals 3–5 mm. long
                                          5. *E. californicum*
        Rootstocks producing globose winter buds with thick, fleshy scales; stems simple
                                    6. *E. brevistylum*
Plants pallid, glaucous and glabrous throughout .... 7. *E. glaberrimum*

1. Epilobium angustifolium L., Sp. Pl. 347. 1753. Fireweed.

Stems from a stout perennial crown, erect, 8–20 dm. tall, glabrous or puberulent or sometimes quite pubescent; leaves lanceolate, subentire, 8–15 cm. long, sessile to short-petioled, the veins conspicuous; flowers in terminal, bracteate racemes, showy, reflexed in bud, then ascending; corolla slightly irregular, the petals 1–2 cm. long, entire, spreading; stamens in a single series, the filaments dilated at base, slightly declined; style longer than the stamens, declined, hairy at base; capsule 4–8 cm. long.

Bogs and margins of moist woods, often coming in great abundance after forest and brush fires: throughout the mountain areas, reaching sea level along the north coast in Humboldt and Del Norte counties; Northern Hemisphere.

2. Epilobium latifolium L., Sp. Pl. 347. 1753.

This species differs from *Epilobium angustifolium* in having thicker, glaucous leaves with inconspicuous veins and in having many stems from the base.

Bogs and moist woods: high altitudes in the Sierra Nevada; northern part of Northern Hemisphere.

3. Epilobium oregonense Hausskn., Monogr. Epilob. 276. 1884.

Stoloniferous perennial; stems slender, erect, usually simple, 6–30 cm. tall, usually few; herbage green, not glaucous; leaves erect, opposite or subopposite, linear to ovate-oblong, 6–15 mm. long, sessile, denticulate; calyx tube short; corolla cream-white to pink, the petals deeply emarginate, 2–12 mm. long; style straight; stigma entire or subentire.

Bogs in the higher mountains: throughout California; Pacific states.

Fig. 277. *Epilobium californicum: a,* perianth, longitudinally split, showing the 2 whorls of stamens, × 4; *b,* mature capsule at dehiscence, × ⅔; *c,* flower, showing the long, slender ovary, × 4; *d,* habit, the upper leaves alternate, × ⅖; *e,* mature seed, showing the terminal tuft of hairs (coma), × 6; *f,* seed, abaxial view, × 16; *g,* seed, adaxial view, × 16.

4. Epilobium Watsonii Barbey in Brew. & Wats., Bot. Calif. 1:219. 1876.

Stout, erect perennial; stems simple below, much-branched above, 6–15 dm. tall, often reddish; herbage short-pilose-pubescent, often glandular; leaves usually opposite, subsessile to short-petioled, the blades oblong-lanceolate to narrowly ovate, serrulate, 3–8 cm. long; racemes leafy, very floriferous, compact or open; petals deeply emarginate, 6–10 mm. long, deep magenta-pink.

Common in moist or wet habitats: low altitudes in Coast Ranges from San Luis Obispo County to Sonoma County. The most common species in this area.

A form with glabrous herbage is recognized as *Epilobium Watsonii* var. *franciscanum* (Barbey) Jepson (*E. franciscanum* Barbey in Brew. & Wats., Bot. Calif. 1:220, 1876). It is found in habitats similar to those occupied by the species, and ranges from Monterey County to Del Norte and Siskiyou counties, and north to Washington.

5. Epilobium californicum Hausskn., Monogr. Epilob. 260. 1884. Fig. 277.

Stout, erect perennial; stems simple below, branched above, 6–12 dm. tall; herbage glabrous to variously white-pubescent; leaves opposite or the upper ones alternate, subsessile to short-petioled, the blades lanceolate to oblong-lanceolate, denticulate, 2–8 cm. long, much reduced above; petals white, 3–5 mm. long; capsule slender, short-pediceled.

About springs and along the margins of streams and marshes: coastal southern California north through Central Valley and the mountains; north to Washington. This species seems to be tolerant of both brackish and fresh water.

Epilobium californicum is an exceedingly variable species which represents a large genetic complex with respect to both morphological and physiological characters. The several varietal entities which have been segregated on the basis of pubescence characters do not appear to have significant ecological and geographic distinctness and are not recognized here.

6. Epilobium brevistylum Barbey in Brew. & Wats., Bot. Calif. 1:220. 1876.

Stems perennial from fleshy, scaly winter buds, erect, usually simple, 3–8 dm. tall; herbage glabrous or finely pubescent; leaves sessile, ascending to nearly erect, narrow-ovate to linear-lanceolate, 2–7 cm. long; petals emarginate, 3–5 mm. long, magenta to white.

Wet ground along marshes, ponds, streams, and ditches: throughout the mountains and foothills of California; western states.

The fleshy winter rosettes are sought by waterfowl, particularly geese. One frequently sees areas in which birds have dug up the plants and have eaten the winter buds.

Epilobium brevistylum is an exceedingly variable species, and several subspecific entities, based upon pubescence, flower size, and leaf characters, have been segregated. The frequent great variability within a single colony indicates

Fig. 278. *Jussiaea californica: a,* habit, pubescent form, $\times$ ⅔; *b,* capsule (cross section), $\times$ 4; *c,* seed, enclosed in a thick tough coat, $\times$ 8; *d,* mature seed with the outer coat removed, $\times$ 8; *e,* calyx, showing bractlets, $\times$ 1⅕; *f,* mature capsule, $\times$ 1⅕; *g,* flower, $\times$ 1⅕; *h,* habit, basal part of plant, showing roots, $\times$ ⅖.

that we are dealing with plants evidencing a complex genetic situation that has not as yet become sufficiently stable either geographically or ecologically to warrant recognition.

7. Epilobium glaberrimum Barbey in Brew. & Wats., Bot. Calif. 1:220. 1876.

Slender, erect perennial, 1–12 dm. tall, usually simple, from scaly rhizomes; herbage glabrous and very glaucous; leaves oblong-lanceolate to linear, subentire, 2–7 cm. long, sessile or short-petioled; sepals from a short tube; petals lavender to white, notched, 2–12 mm. long; style straight, stigma entire.

Springy or boggy ground: Modoc County to Del Norte County, Sierra Nevada to Tulare County, mountains of southern California; western states.

JUSSIAEA

1. Jussiaea californica (Wats.) Jepson, Fl. West. Middle Calif. 326. 1901. Fig. 278.

Jussiaea repens var. *peploides* (H.B.K.) Griseb., Cat. Pl. Cubens. 107. 1866.

Perennial, rooted aquatic herb, with floating stems 3–30 dm. long, or the stems extending prostrate onto wet shores; herbage glabrous or sometimes pubescent; leaves alternate, petioled, the blades oblong-elliptic to obovate, 1–3 cm. long; flowers pediceled, solitary in the leaf axils; sepals 5, lanceolate, 1 cm. long, at length deciduous from the mature fruit; petals 5, yellow, broadly obovate, 10–15 mm. long; stamens 10; ovary inferior; fruit hard, corky, 5-celled, cylindric, 12–25 mm. long; seeds with a tough coat.

Often covering ponds, sloughs, small streams, and ditches, to the extent of being an obstruction to navigation in small boat waterways: throughout California at low altitudes.

LUDWIGIA

Aquatic or palustrine perennial herbs, rooting in mud and floating on the water, or more often prostrate on muddy shores; herbage glabrous. Leaves opposite (in ours). Flowers solitary in the leaf axils, sessile or subsessile. Sepals 4. Petals none or minute. Stamens 4, opposite sepals, often with bractlets on the sides. Ovary inferior, 4-celled. Fruit short, often broader than long, enlarging toward the summit or broadest just above the middle, dehiscent by pores or slits.

Fruits with 4 longitudinal green bands; the bracts, when present, basal and minute; fruits as broad as or broader than long .... 1. *L. palustris*
Fruits without green bands; bractlets well above base; fruits longer than broad
2. *L. natans*

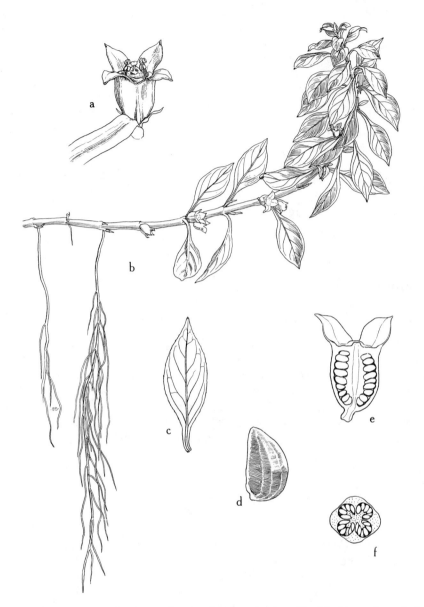

Fig. 279. *Ludwigia palustris: a,* flower, showing the broadly deltoid sepals and the longitudinal bands on ovary, × 4; *b,* habit, × ⅘; *c,* leaf, × 1½; *d,* mature seed, shiny, striate, × 28; *e,* capsule (longitudinal section), × 4; *f,* capsule (cross section), × 4.

1. Ludwigia palustris (L.) Ell., Bot. S. C. & Ga. 1:211. 1821. Fig. 279.

Ours chiefly prostrate, forming small mats 1–6 dm. broad; leaves obovate, acute, narrowed to a petiole, 12–25 mm. long; sepals broadly deltoid; petals none or minute and reddish; ovary with 4 green bands on the sides; bracts minute; capsule quadrate, 2–4.5 mm. long, almost as broad at the top; seeds minute.

Margins of ponds and floodlands in scattered localities: North Coast Ranges and the northern Sierra Nevada foothills; North America, Europe.

2. Ludwigia natans Ell., Bot. S. C. & Ga. 1:581. 1821.

Stems simple or branched, prostrate or floating; leaves elliptic to obovate, obtuse or subacute, 1–4 cm. long, petioled; flowers sessile or subsessile, turbinate; bractlets usually on the sides of the hypanthium; sepals ovate, acuminate; petals minute or wanting; fruit quadrate-campanulate or quadrate-turbinate, 4–6 mm. long, smooth.

Wet mud on the margins of ponds, or sometimes the stems floating on the water: San Bernardino Valley; southeastern United States.

OENOTHERA

Herbs with leaves alternate or often in a basal rosette. Flowers yellow or white fading to brownish or pink. Calyx tube prolonged above ovary. Sepals 4, reflexed. Petals 4, usually showy. Stamens 8. Fruit a 4-valved capsule. Seeds naked.

Stigma 4-lobed; plants biennial, erect, 3–15 dm. tall .... 1. *O. Hookeri*
Stigma capitate; plants perennial, forming a basal rosette .... 2. *O. heteranthera*

1. Oenothera Hookeri Torr. & Gray, Fl. N. Am. 1:493. 1840. Evening primrose.

Tall, erect herb, canescently puberulent to hirsute; leaves ovate to lanceolate or oblanceolate, 10–25 cm. long, the lowermost petioled; flowers sessile in racemose spikes; calyx tube 2–4 cm. long, lobes free in bud; corolla yellow, the petals 2–5 cm. long; anthers versatile, 1–2 cm. long; stigma 4-lobed; capsule quadrangular, woody.

About springs and rivulets: low and middle altitudes throughout California; western states.

2. Oenothera heteranthera Nutt. in Torr. & Gray, Fl. N. Am. 1:507. 1840.

Oenothera subacaulis (Pursh) Jepson, Man. Fl. Pl. Calif. 683. 1925.

Caespitose perennial; herbage glabrous or nearly so; leaves in a basal rosette, 3–12 cm. long, lanceolate to ovate, entire or denticulate (lyrate-pinnatifid in variety) or with 2 or 3 pairs of small lobes at base; flowers sessile on the root

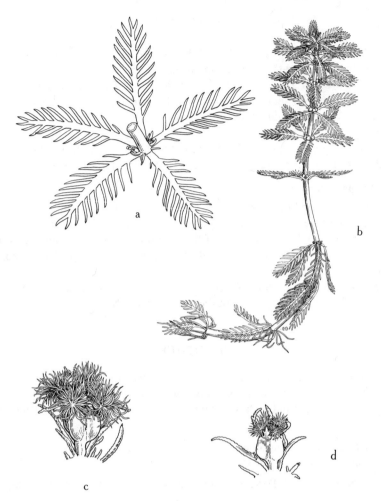

Fig. 280. *Myriophyllum brasiliense: a,* section of stem, showing whorl of leaves with flowers in the axils, $\times$ 2; *b,* habit, the emersed part of plant erect, with flowers in the leaf axils, and the submersed part of plant with roots arising at the nodes, $\times$ ⅘; *c,* pistillate flower, showing bracteoles, $\times$ 8; *d,* young pistillate flower of upper part of plant, showing calyx lobes and bracteoles, $\times$ 8.

crown; the calyx tube elongate, appearing as a peduncle, 2–7 cm. long; petals yellow, about 1 cm. long; capsule oblong, smooth to rugulose; seeds pitted.

Wet mountain meadows: east of the Sierra Nevada crest; western states.

A form designated as *Oenothera heteranthera* var. *taraxacifolia* Jepson (Man. Fl. Pl. Calif. 683, 1925) differs from the species chiefly in the lyrate-pinnatifid leaves. It occurs in moist habitats in Plumas County and north to Siskiyou and Modoc counties.

HALORAGACEAE

Aquatic, annual or perennial herbs. Leaves opposite or whorled, the sub-mersed ones often with capillary divisions. Flowers minute, perfect or unisex-ual (monoecious or dioecious), solitary or clustered, axillary or spicate. Calyx epigynous, 2- to 4-lobed. Petals minute and 2–4 or absent. Stamens 1–8. Ovary inferior, 1- to 4-celled; styles 1–4. Fruit indehiscent, of 2–4 1-seeded carpels, nutlike or drupelike.

Stems erect; leaves simple and entire .... *Hippuris*
Stems lax; submersed leaves finely dissected .... *Myriophyllum*

HIPPURIS

1. Hippuris vulgaris L., Sp. Pl. 4. 1753. Mare's tail.

Stems submersed or partly emersed, erect, simple, 3–6 dm. tall, hollow; herbage glabrous, usually pallid; leaves simple, 6–12 in a whorl, linear-atten-uate, 5–35 mm. long, thick; flowers sessile in middle and upper axils, usually perfect; petals none; stamen 1, inserted on anterior edge of calyx; ovary in-ferior, crowned with the rimlike entire calyx; fruit 2–3 mm. long, 1-celled and 1-seeded.

Ponds and streams: widely scattered from sea level to an elevation of 7,000 feet; Northern Hemisphere.

MYRIOPHYLLUM

Plants wholly submersed, or the upper part of the plant emersed or just the spikes emersed; the emersed leaves opposite, simple or pectinate, the submersed leaves usually divided into capillary segments. Flowers sessile, in the axils of the leaves or in bracteate, interrupted spikes at the ends of branches; monoe-cious or dioecious. Staminate flowers in the upper part of the spike or in leaf axils of dioecious plants. Sepals 4. Petals 4 or none. Stamens 4–8. Pistillate flowers in the lower part of the spike or in leaf axils in dioecious plants. Calyx

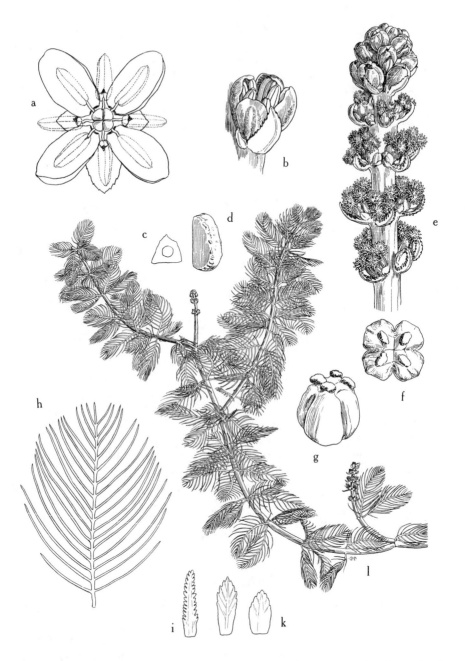

Fig. 281. *Myriophyllum exalbescens: a,* diagram of staminate flower, showing perianth segments in 4's, stamens 8, and 3 subtending bracts, 2 of them shorter than the third, × 10; *b,* young staminate flower, × 8; *c,* nutlet (cross section), × 8; *d,* mature nutlet, rugose on back, × 8; *e,* spike, the staminate flowers at the apex, the pistillate flowers below, × 4; *f,* mature fruit, top view, × 8; *g,* mature fruit, side view, × 8; *h,* submersed leaf, × 2; *i–k,* bracts of inflorescence, × 6; *l,* habit of plant entirely submersed except for the spike, showing whorls of leaves, ⅔.

4-toothed or entire, the tube adnate to the fruit. Stigmas 4, plumose. Fruit splitting into 4 nutlets.

Flowers and fruits in the axils of ordinary emersed foliage leaves; flowers dioecious
1. *M. brasiliense*
Flowers and fruits in terminal bracteate spikes, sometimes the spike proliferating vegetatively at apex; flowers unisexual or sometimes some perfect.
Bracts shorter than flowers or fruits . . . . 2. *M. exalbescens*
Bracts longer than flowers or fruits.
Bracts serrate; spikes not proliferating terminally; stamens 4 . . . 3. *M. hippurioides*
Bracts pectinately lobed; spikes proliferating into a vegetative shoot at apex; stamens 8 . . . . 4. *M. verticillatum*

1. Myriophyllum brasiliense Camb. in St. Hil., Fl. Bras. Merid. 2:252. 1829. Fig. 280.

Aquatic with the upper part of the plant emersed and erect; herbage pallid, glabrous; leaves oblong, stiff, dissected into 10 or more pectinately arranged, filiform divisions; flowers dioecious, the staminate flowers unknown to us, the pistillate ones in the axils of foliage leaves, usually conspicuous as a tuft of white, plumose stigma lobes.

Widely scattered in ponds, ditches, and slow streams: introduced in Marin, Humboldt, San Joaquin, Merced, El Dorado, and San Diego counties; set out in these areas by dealers in aquatics for the purpose of market propagation. In cultivation only the pistillate plants are used. In the trade it is known as parrot's feather.

2. Myriophyllum exalbescens Fern., Rhodora 21:120. 1919. American milfoil. Fig. 281.

Myriophyllum spicatum var. *exalbescens* Jepson, Man. Fl. Pl. Calif. 691. 1925.

Stems simple or forking; herbage often with a brownish purple pigment; leaves in whorls of 3 or 4 with 7–18 pairs of flaccid divisions, 1–4 cm. long; spikes emersed, the lower flowers whorled, pistillate, the upper ones staminate, bright pink-red, subtended by short, entire or serrate bracts which are usually shorter than the flowers or fruits; petals oblong-ovate, 2–3 mm. long; anthers 8; fruits subglobose, 2–3 mm. long; mericarps rounded on back, smooth or rugulose.

Ponds, irrigation ditches, and quiet streams: throughout California; east to Atlantic.

3. Myriophyllum hippurioides Nutt. ex Torr. & Gray, Fl. N. Am. 1:530. 1840. Western milfoil. Fig. 282.

Stems stout, 6–12 dm. long; leaves in whorls of 4–6; emersed leaves linear, the lower ones pinnate, the upper ones from deeply serrate to entire; submersed leaves pinnately dissected into capillary divisions; flowers perfect or the lower ones pistillate and the upper ones staminate; bracteoles acuminate, from

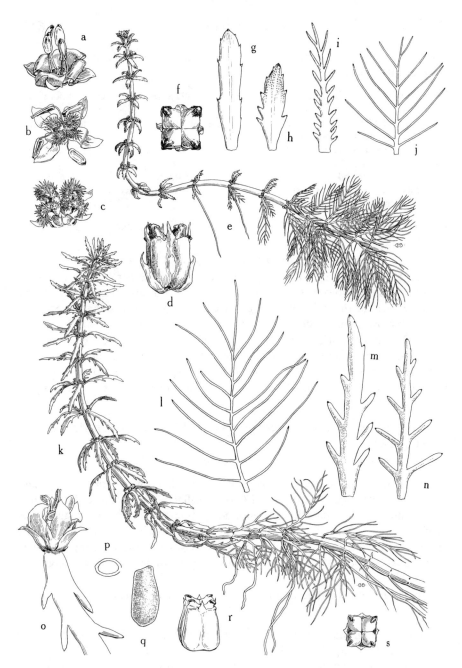

Fig. 282. (For explanation, see facing page.)

coarsely serrate to often only toothed at apex; petals white, 1–2 mm. long; anthers 4; fruit 1.5–2 mm. long; mericarps laterally compressed, with 2 angles along the outer ridge.

Ponds, irrigation ditches, sloughs, and quiet streams: valleys and foothills of northern California, Central Valley.

4. Myriophyllum verticillatum L., Sp. Pl. 992. 1753. Fig. 283.

Aquatic, wholly submersed, or with the spike and upper leaves sometimes emersed; stems simple or branched, to 2.5 m. long, producing winter buds either in the inflorescence or in the leaf axils; leaves in whorls of 4 or 5, the submersed leaves flaccid, 1–4 cm. long, with capillary divisions, the emersed ones smaller and with coarser pectinate divisions; flowers in whorls of 4–6, the lower flowers pistillate and the upper ones staminate, sometimes the middle ones perfect, subtended by pectinately lobed leaves which are longer than flowers; bracteoles minute, palmately 7-lobed; petals spatulate or rudimentary in pistillate flowers; anthers 8; fruit subglobose, deeply furrowed; nutlets light brown, bony.

Ponds: Humboldt, Del Norte, Lake, and Mono counties; Northern Hemisphere.

Myriophyllum verticillatum is apparently rare in California. Our material differs from that reported elsewhere in that the flowering spikes are terminated by a short vegetative shoot. We have too little material to determine whether this is a significant character or not. Jepson's reference (Man. Fl. Pl. Calif. 691, 1925) to 30–40 leaves to a whorl is obviously a misprint.

ARALIACEAE

Herbs, shrubs, trees, or sometimes lianas, with solid, pithy stems and alternate, simple or compound leaves. Flowers regular, mostly unisexual. Calyx adnate to the ovary and sometimes represented by 5 teeth at the apex. Petals 5, inserted at top of hypanthium. Stamens 5, alternate with petals. Ovary 1- to several-celled, the styles as many as the ovary cells. Fruit a berry or a drupe.

Fig. 282. *Myriophyllum hippurioides:* a, staminate flower from upper part of inflorescence, × 10; b, perfect flower, × 10; c, pistillate flower from lower part of inflorescence, × 10; d, fruit, showing sepals and lateral bracteoles, × 10; e, habit, showing emersed and submersed part of plant, × ⅖; f, fruit, top view, × 10; g–i, variation in emersed leaves, × 4; j, submersed leaf, × 2; k, habit, emersed and submersed parts of plant, × ⅖; l, submersed leaf, × 2; m and n, variation in emersed leaves, × 3; o, staminate flower in leaf axil, showing sepals, lateral bracteoles, and pair of dark brown, lateral scales, × 10; p, nutlet (cross section), × 11; q, nutlet, minutely reticulate, × 11; r, fruit, side view, × 8; s, fruit, top view, × 8.

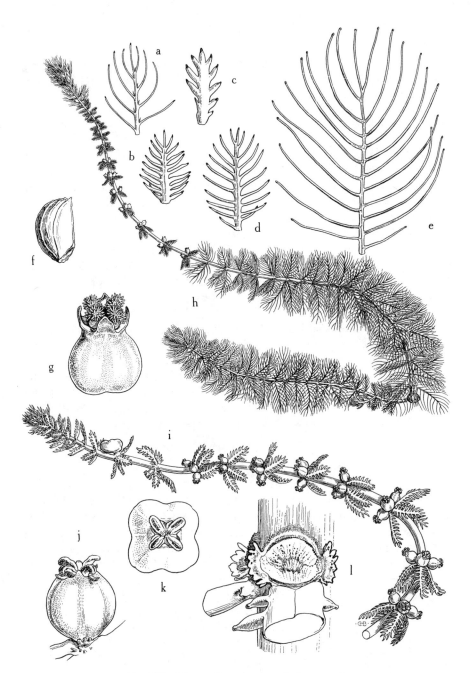

Fig. 283. (For explanation, see facing page.)

ARALIA

1. Aralia californica Wats., Proc. Am. Acad. Arts & Sci. 11:144. 1876. Elk clover.

Coarse perennial herb, 2–3 m. tall; leaves glabrous, very large, 3–15 dm. long, ternately compound, then 3- to 5-pinnate, leaflets ovate, serrate; flowers small, in a large panicle of umbels; petals white, about 2 mm. long, not inflexed; styles 5, united to the middle; ovary inferior, 5-celled; fruit a small, black berry about 4–5 mm. broad.

Stream beds and moist spots in shaded canyons, at elevations of 100–2,500 feet: Coast Ranges, Sierra Nevada, southern California; southern Oregon.

UMBELLIFERAE

Annual or perennial herbs. Stems frequently hollow. Leaves alternate, usually compound; petioles usually basally dilated; stipules minute or absent. Flowers small, arranged in a compound or rarely simple umbel or sometimes in a head, the umbels usually subtended by bracts (involucre), the secondary umbels (umbellets) subtended by bractlets (involucel). Hypanthium completely adnate to ovary. Sepals 5 or obsolete. Petals 5, the tips often inflexed. Stamens 5, alternate with petals, inserted on margin of epigynous disc; anthers versatile. Ovary inferior, 2-celled, 1 ovule in each cell. Styles 2, united below to form a thickened base (stylopodium). Fruit dry, of 2 1-seeded carpels (schizocarp), each carpel (mericarp) with usually 5 longitudinal ribs on the exposed outer (dorsal) surface, some or all of the ribs often winged; mericarps attached to an upward filiform prolongation of the receptacle (carpophore) and separating from the carpophore at maturity along their contiguous inner (ventral) surfaces (commissure); mature mericarp wall (pericarp) frequently with dark oil tubes in spaces or sinuses (intervals) between ribs on dorsal surface and also on commissural side.

This natural family has usually been separated from related families by the distinctive fruit. Moreover, the technicalities separating its various genera and species also lie mainly in the fruit, and to this end, mature fruits are necessary for positive species identification.

Mathias, M. E., and L. Constance, in Abrams, Illustrated Flora of the Pacific States. (Umbelliferae.) 3:215–283. 1951.

Fig. 283. *Myriophyllum verticillatum:* a, leaf, form occurring at tip of stem, × 3; b–d, emersed leaves, × 3; e, submersed leaf, each of its divisions with a globose tip, × 2; f, mature nutlet, × 6; g, young pistillate flower, × 12; h, habit, plant developing from a winter bud, showing the various leaf types, × ⅖; i, habit, upper part of inflorescence, a winter bud in leafy apex, × 1⅕; j, fruit, showing rudimentary petals and subtending bracteoles, × 6; k, fruit, top view, × 6; l, leaf axil, the flower removed to show the palmately lobed bracteoles, × 20.

We are greatly indebted to this excellent recent treatment, as well as to the direct help of its authors, for our account of this family.

Inflorescence forming a distinct umbel.
  Umbels simple or proliferous; leaves simple.
    Leaf blade orbicular .... *Hydrocotyle*
    Leaves bladeless, reduced to hollow, cylindrical, septate phyllodes ...... *Lilaeopsis*
  Umbels compound; leaves compound.
    Ovary and fruit covered with hooked bristles .... *Sanicula*
    Ovary and fruit glabrous or pubescent, but lacking bristles.
      Ribs of fruit not prominently winged; fruit terete or somewhat compressed laterally.
        Stems spotted with purple marks; leaves decompound into very small segments
                              Conium
        Stems not spotted; leaves 1- to 3-pinnate, with larger segments.
          Ribs indistinct in the more or less uniformly corky covering of the pericarp
                              Berula
          At least some of ribs of mature fruit prominent, often corky.
            Involucel lacking; sepals lacking .... *Apium*
            Involucel present; sepals present.
               Sepals minute; involucre conspicuous .... *Sium*
               Sepals prominent; involucre inconspicuous or absent.
                  Styles less than ½ as long as mature fruit; fruit ovoid to globose
                                  Cicuta
                  Styles ½ or more than ½ as long as mature fruit; fruit oblong-
                    cylindric .... *Oenanthe*
      Ribs of fruit winged; fruit more or less compressed dorsally.
        Umbellets capitate .... *Sphenosciadium*
        Umbellets lax, with pedicellate flowers and fruit.
          All ribs winged; roots not fleshy and tuberous.
            Slender plants; leaf divisions small, deeply incised .... *Conioselinum*
            Coarse plants; leaf divisions large, serrate or toothed .... *Angelica*
          Only lateral ribs winged; roots fleshy, tuberous, fascicled .... *Oxypolis*
Inflorescence capitate, forming a true head (see also *Sanicula* and *Sphenosciadium*, in which the umbellets are capitate) .... *Eryngium*

Borderline marsh plants of this family that are not treated herein are: *Eryngium racemosum*, several species of *Perideridia, Lomatium dissectum* var. *multifidum, Apium leptophyllum*, and *Sanicula crassicaulis*.

ANGELICA

1. Angelica genuflexa Nutt. ex Torr. & Gray, Fl. N. Am. 1:620. 1840. Kneeling angelica.

Stout, glabrous to scaberulous perennial; stems leafy, 8–15 dm. tall; leaves thin, ternately pinnate to bipinnate, main divisions widely spreading and reflexed on geniculate rachis, ultimate divisions ovate to ovate-lanceolate, 4–10 cm. long, serrate or coarsely toothed; umbels large, terminal, compound, rays 20–45, unequal in length, 2–7 cm. long, the involucre none, the involucel of linear to filiform, hispidulous bractlets, the pedicels 5–15 mm. long; petals glabrous, white; sepals lacking; fruit orbicular, 3–4 mm. long, glabrous; carpels with dorsal ribs prominent, filiform to narrowly winged, the lateral ribs with broad, thin wings; oil tubes solitary in intervals, 2 on commissure.

Coastal marshes: North Coast Ranges in Humboldt County; north to Alaska, also on the eastern coast of Asia.

APIUM

1. Apium graveolens L., Sp. Pl. 264. 1753. Celery.

Glabrous biennial; roots fibrous; stems erect, 5–15 dm. tall, branching above; basal leaves pinnate, 1–6 dm. long, long-petioled, with 3–5 (–7) divisions 2–4.5 cm. long, these broadly ovate, coarsely toothed and deeply incised, upper leaves with leaflets 3, sessile or short-petioled; umbels compound, opposite the leaves or terminal, 7- to 16-rayed, the rays to 2.5 cm. long; involucre and involucel wanting; sepals wanting; petals white, minute; stylopodium low-conical; fruit glabrous, ovoid, laterally compressed, 1.5 mm. long, the ribs prominent, slightly winged; oil tubes solitary in intervals, 2 on commissural side.

In marshes or along streams; garden plant escaped from cultivation.

BERULA

1. Berula erecta (Huds.) Cov., Contr. U. S. Nat. Herb. 4:115. 1893. Cut-leaved water parsnip.

Aquatic or subaquatic, glabrous perennial; stems erect, stout, freely branched, 2–8 dm. tall; leaves pinnate, with 5–8 pairs of leaflets, these ovate to oblong, subentire to laciniately lobed or variously incised, 1.5–4 cm. long; involucre and involucel conspicuous; umbels compound, fruiting rays 6–15, 1–2 (–3) cm. long, the pedicels 2–5 mm. long; flowers white; sepals minute; fruit glabrous, somewhat flattened laterally, basally emarginate, 1.5–2 mm. long; carpels nearly orbicular, the ribs slender and inconspicuous, the pericarp thick and corky; oil tubes numerous, contiguous, surrounding the seed cavity.

Marshes or streams at elevations of 500–4,000 feet: coastal southern California, Sierra Nevada, Siskiyou Mountains; North America, Europe, Asia.

CICUTA. WATER HEMLOCK

Large glabrous perennials. Tuberous rootstocks horizontal or erect, bearing fibrous or fleshy roots. Stems erect, branching. Leaves large, broad, usually bi- or tripinnate, leaf divisions serrate or incised. Umbels compound, numerous, terminal; involucre present and inconspicuous or absent; rays many, unequal in length; involucel of small bractlets. Sepals prominent, acute. Petals broad, white. Stylopodium depressed or low-conical. Fruit laterally somewhat flattened, ovate to globose, glabrous. Carpel ribs corky, prominent, obtuse, the lateral ones largest; oil tubes solitary in the intervals, 2 on commissure.

Plants of coastal salt marshes; oil tubes large .... 1. *C. Bolanderi*
Fresh-water marshes or streams; oil tubes small .... 2. *C. Douglasii*

1. Cicuta Bolanderi Wats., Proc. Am. Acad. Arts & Sci. 11:139. 1876.

Stems 1.5–3 m. tall; leaves oblong to ovate, bipinnate, 2–4 dm. long, the leaflets linear to lanceolate, serrate, 3–8 cm. long; bracts and bractlets lanceolate; rays 2.5–5 cm. long, unequal in length; pedicels 3–4 mm. long; fruit 3–4 mm. long, the ribs prominent, commissural surface concave; oil tubes large; seeds very oily.

Coastal marshes: southern California to Marin County.

2. Cicuta Douglasii (DC.) Coult. & Rose, Contr. U. S. Nat. Herb. 7:95. 1900.

Cicuta californica Gray, Proc. Am. Acad. Arts & Sci. 7:344. 1868.

Stems 1–2 m. tall; herbage glabrous, glaucous, sometimes purplish; leaves oblong to ovate, bi- or tripinnate, 1–4 dm. long, the leaf segments lanceolate, serrate or dentate-incised, 4–8 cm. long; bracts and bractlets lanceolate; rays 3–6 cm. long; pedicels about 4 mm. long; fruit 2–4 mm. long, suborbicular; carpel ribs broad, low, and corky, the intervals narrow; oil tubes small, dark; seed not very oily.

Fresh-water marshes or flowing streams, valleys or mountains, at elevations of 5–7,000 feet: coastal southern California, eastern Sierra Nevada, North Coast Ranges; north to Alaska.

CONIOSELINUM

1. Conioselinum chinense B.S.P., Prelim. Cat. N. Y. 22. 1888.

Conioselinum pacificum (Wats.) Coult. & Rose, Proc. Am. Acad. Arts & Sci. 11:140. 1876.

Large, glabrous perennial; stems stout, leafy, branching, 3–15 dm. tall; leaves ternate, the divisions pinnately compound, the leaflets ovate, acute, deeply pinnatifid or incised, 1–4.5 cm. long; umbels on stout peduncles, compound, involucre of 2–4 bracts or lacking, the involucel with many bractlets, the rays 14–25, 1.4–4.5 cm. long, the pedicels slender, 5–8 mm. long; sepals lacking; petals white; stylopodium conical; fruit glabrous, oblong, flattened dorsally, 5–7 mm. long; carpels with dorsal ribs acute or narrowly winged, the lateral ribs with corky-thickened broad wings; oil tubes 1 or 2 in the intervals, 2–4 on commissural side.

Marshes, at elevations of 5–1,500 feet: North Coast Ranges, rare; north to Alaska, also on the Atlantic and eastern Asiatic coasts.

CONIUM

1. Conium maculatum L., Sp. Pl. 243. 1753. Poison hemlock.

Large, glabrous, biennial herb; stems 5–30 dm. tall, branching, spotted with purple; herbage with mouselike odor; leaves pinnately decompound and leaflets pinnatifid into small segments; lower leaves with dilated, sheathing petioles, 3–6 dm. or more long including petiole, the upper leaves sessile; umbels large, compound, 3–8 cm. across, the rays 10–15, 2–4 cm. long, the pedicels slender, 4–6 mm. long; involucre and involucel small, of ovate, acuminate bracts and bractlets; flowers white; sepals obsolete; fruit 2–3 mm. long, glabrous, broadly ovate, slightly flattened laterally; carpels with prominent undulate ribs; oil tubes obscure.

European weed of waste places, abundantly established in shady or low, wet ground throughout cismontane California.

ERYNGIUM. BUTTON SNAKEROOT

Creeping to erect, glabrous biennials or perennials. Roots fibrous or with taproot developed. Stems dichotomously branching. Leaves opposite, or the upper leaves sometimes alternate, usually oblanceolate and spinose-serrate, or pinnately or palmately lobed, the basal leaves, if growing in water, sometimes with blades obsolete and petioles septate, fistulous and elongate. Inflorescence usually cymosely branched, of capitate umbels (heads) which are solitary and terminal on the dichotomous branches or on short peduncles in their forks; involucre with bracts spinose, conspicuous, subtending the head; involucel with bractlets usually spinose-tipped, subtending the flowers. Sepals persistent on fruit, conspicuous. Petals white, blue, or purple. Stylopodium lacking. Fruit covered with whitish scales or tubercles. Carpels lacking ribs, oil tubes none or inconspicuous.

Heads obviously blue .... 1. *E. articulatum*
Heads greenish.
  Basal leaves bladeless or the petiole much longer than the small blade.
    Bractlets scarious-margined at base; fruit with scales subequal in size
                                          2. *E. alismaefolium*
    Bractlets prominently scarious-winged at base; fruit with scales unequal in size
                                        3. *E. aristulatum*
  Basal leaves (except in the aquatic stage) with the petioles shorter or little longer than
    the prominent blade .... 4. *E. Vaseyi*

1. Eryngium articulatum Hook., London Jour. Bot. 6:232. 1847.

Eryngium articulatum var. *Bakeri* Jepson, Madroño 1:104. 1923.

Stem erect, stout, 3–10 dm. tall, dichotomously branching above; basal leaves elongate, the petioles fistulous, jointed (septate), 1–4 dm. or more long, much longer than the abortive, lanceolate to ovate, entire to laciniate blades, or blades quite lacking; cauline leaves similar, narrower, sessile; inflorescence cymose, heads 1–2.5 cm. long, many, pedunculate, ovoid, bright blue; bracts stiff, spinose-serrate or toothed, reflexed, scarious-dilated at base, about as long as the heads; bractlets spiny-toothed, about as long as the fruits; sepals entire, ovate-lanceolate; styles shorter than or longer than the sepals; fruit ovoid, 2–3 mm. long, densely covered with flat, white or brown scales.

Swamps and marshes, to an elevation of 4,000 feet: Central Valley, North Coast Ranges; north to Oregon.

2. Eryngium alismaefolium Greene, Erythea 3:64. 1895.

Stems low, numerous, basally clustered and dichotomously branching just above base, 0.5–3 dm. tall, mostly shorter than basal leaves; basal leaves of terete, septate petioles 10–15 (–40) cm. long with blades short or lacking, later leaves developing a lanceolate to ovate, spinosely serrate or pinnatifid blade; cauline leaves similar to later leaves but reduced, often basally connate and scarious-margined; inflorescence cymose, heads many, short-pedunculate, subglobose, 5–10 mm. broad; bracts of involucre subulate-lanceolate, 5–15 mm. long, usually spinose, longer than the heads; involucel bractlets entire to spinose, subulate-lanceolate, slightly longer than the flowers, marginally scarious below; sepals 1–3 mm. long, scarious-margined; styles about as long as sepals or longer; fruit ovoid, about 2 mm. long, densely covered with white, subequal scales.

Wet mountain meadows at elevations of 4,000–5,500 feet: northeastern California; to southern Oregon and northern Nevada.

3. Eryngium aristulatum Jepson, Erythea 1:62. 1893.

Eryngium Parishii Coult. & Rose, Contr. U. S. Nat. Herb. 7:57. 1900.

Stems prostrate or ascending, stout to slender, 1–8 dm. long; basal leaves with septate petioles 3–25 cm. long, the blades spinulose-serrate to incised or lobed, short, sometimes obsolete; cauline leaves opposite, sessile, spinulose-serrate; inflorescence cymose, heads many, pedunculate, globose, 5–12 cm. across; involucral bracts as long as or longer than the heads, spinose and scarious-margined at base; bractlets prominently scarious-winged near base, the inversely sagittate wings enclosing fruit, and spinose in the sinuses; sepals hyaline-margined, lanceolate, 1.5–3.5 mm. long; styles longer than the sepals; fruit ovoid, 1.5–2.5 mm. long, densely covered by narrow scales, those of apex much larger than those of base.

Vernal pools and salt marshes: North Coast Ranges, foothills of the northern Sierra Nevada.

4. Eryngium Vaseyi Coult. & Rose, Bot. Gaz. 13:142. 1888.

Stems erect, stout, 1.5–4 (–6) dm. tall from a fascicle of woody-fibrous roots; basal leaves oblong-lanceolate to ovate, 9–25 cm. long, 2–8 cm. broad, deeply pinnatifid, the segments unequal, narrow or broad, usually remote, spinulose-lobed or again pinnatifid; petioles very short, dilated, 1–4 cm. long; cauline leaves like the basal leaves, the upper ones sessile, opposite; inflorescence corymbose, the heads rather small, numerous, short-pedunculate, the flowers numerous; heads subglobose, 5–10 mm. in diameter; bracts about 8, linear-subulate with lateral spines, shorter than to much longer than the heads; bractlets similar to bracts; sepals lanceolate to ovate, 1–3 mm. long, acute or obtuse, mucronate, scarious-margined, entire or occasionally spinose; petals oblong, 1–1.5 mm. long; styles shorter than or longer than the sepals; fruit ovoid, 2–3 mm. long, densely covered with appressed, white, subequal, lanceolate scales 0.5–1 mm. long, or the calycine scales slightly the longest.

A common inhabitant of vernal pools and roadside ditches: upper Salinas Valley, Central Valley.

The described forms include the following two varieties:

Spines of the bractlets mostly simple .... 4a. *E. Vaseyi* var. *castrense*
Spines of the bractlets themselves spinose .... 4b. *E. Vaseyi* var. *globosum*

4a. Eryngium Vaseyi var. **castrense** (Jepson) Hoover ex Math. & Const., Am. Midl. Nat. 25:387. 1941. Fig. 284.

Eryngium castrense Jepson, Madroño 1:108. 1923.

Plants often stouter; heads subglobose to globose-ovoid, 6–15 mm. long; bracts spreading, 0.8–2.5 cm. long, densely beset with spines on the abaxial surface as well as along the margins, scarious-winged at the base, usually much longer than the heads; bractlets densely spiny and usually with some abaxial spines, 0.5–2 cm. long, much longer than the fruit; sepals usually somewhat spinose.

Vernal pools and roadside ditches: Central Valley.

4b. Eryngium Vaseyi var. **globosum** (Jepson) Hoover ex Math. & Const., Am. Midl. Nat. 25:387. 1941.

Eryngium globosum Jepson, Madroño 1:108. 1923.

Plants often stouter; heads globose to ovoid, 8–18 mm. long; bracts spreading to slightly reflexed, 1–3 cm. long, pinnately spinose and often with some spines on the abaxial surface, much longer than the heads; bractlets pinnately spinose and often with a few abaxial spines, broadly scarious-winged at the base, the wings spinose, longer than the fruit; sepals, or some of them, pinnatifid with 3–8 spiny teeth.

Vernal pools and roadside ditches: southeastern San Joaquin Valley.

Fig. 284. *Eryngium Vaseyi* var. *castrense: a,* petal, showing fimbriate apex, × 20; *b,* inflorescence, × ⅘; *c,* fruit, top view, × 12; *d,* flower, side view, × 6; *e,* floral bractlet, × 6; *f,* habit, aquatic stage of plant, × ⅖; *g,* habit, transitional stage, × ⅖; *h,* habit, terrestrial stage, × ⅖.

HYDROCOTYLE. MARSH PENNYWORT

Low perennial herbs. Stems slender, creeping, rooting at the nodes. Leaves simple, orbicular, peltate or reniform. Umbels simple, sometimes proliferous. Involucre small or wanting. Sepals minute or wanting. Petals white. Fruit strongly flattened laterally, each mericarp with 5 primary ribs. Oil tubes lacking or obscure.

Leaves orbicular, peltate.
    Inflorescence a simple umbel .... 1. *H. umbellata*
    Inflorescence proliferating to give an interrupted spike .... 2. *H. verticillata*
Leaves reniform, not peltate .... 3. *H. ranunculoides*

1. Hydrocotyle umbellata L., Sp. Pl. 234. 1753.

Stems creeping, freely branched; rootstocks tuber-bearing; leaves on erect petioles, the blades suborbicular, centrally peltate, crenate, 1–4 cm. broad; umbels simple, many-flowered; fruit 1–3 mm. in diameter, deeply notched at base and apex, the ribs thick and corky.

Edges of ponds, ditches, marshy ground, at low and medium altitudes: throughout California; east to Atlantic, south to Mexico, southern Africa.

2. Hydrocotyle verticillata Thunb., Diss. Hydrocot. 2. 1798.

Stems creeping in mud; rootstocks tuberous; leaves orbicular, peltate, crenate, 3–5 cm. broad; petioles slender, ascending; inflorescence an interrupted, simple or branched spike, the peduncles shorter than or about as long as the petioles; fruit sessile or subsessile, shallowly notched at apex, rounded or abruptly cuneate at base.

Streams and low, muddy ground, at low altitudes: central and southern California; east to Atlantic states, Mexico, West Indies.

Hydrocotyle verticillata Thunb. var. *triradiata* Fern. (Rhodora 41:437, 1939) differs from the species in having peduncles as long as or longer than the petioles, and fruits with pedicels 1–10 mm. long (fig. 285). It is found in marshes and stream borders from the Central Valley to the San Francisco Bay area.

3. Hydrocotyle ranunculoides L. f., Suppl. 177. 1781. Fig. 286.

Stems floating or creeping in mud; leaves orbicular-reniform, cordate, not peltate, 5–8 cm. across, crenately incised and again more deeply lobed, on elongate petioles 5–10 cm. long; umbels simple, capitate, on peduncles 2–6 cm. long and much shorter than petioles; pedicels 1–3 mm. long; fruit suborbicular, 2–3 mm. long, the ribs obscure.

Shallow pools and muddy shores, lower altitudes: South Coast Ranges, southern California; east to Atlantic.

Fig. 285. *Hydrocotyle verticillata* var. *triradiata: a,* fruit, strongly flattened laterally, × 12; *b,* bud, × 12; *c,* inflorescence, the stylopodia conspicuous at the summit of each fruit, × 8; *d,* carpel, × 12; *e,* habit, × 1½.

Fig. 286. *Hydrocotyle ranunculoides: a,* habit, × ⅔; *b,* umbel in flower, × 8; *c,* flower, × 8; *d,* umbel in fruit, × 8; *e,* carpel, × 32; *f,* fruit, × 12.

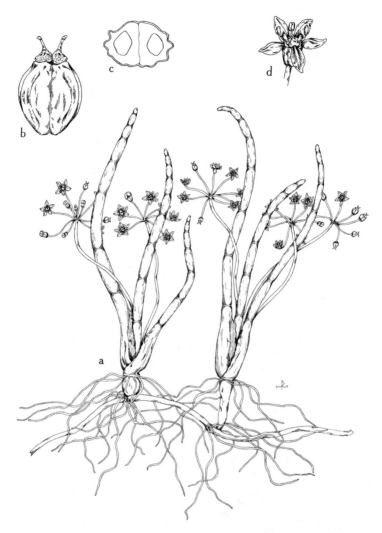

Fig. 287. *Lilaeopsis occidentalis: a,* habit, × 1½; *b,* fruit, × 8, *c,* fruit (cross section), × 8; *d,* flower, × 4.

LILAEOPSIS

1. Lilaeopsis occidentalis Coult. & Rose, Bot. Gaz. 24:48. 1897. Fig. 287.

Lilaeopsis lineata var. *occidentalis* Jepson, Man. Fl. Pl. Calif. 714. 1925.

Small, glabrous perennial; stems fistulous, rhizomatous, rooting in mud; leaves erect, reduced to hollow, septate, terete phyllodes, 2.5–15 (–25) cm. long, 1–4 mm. wide; umbels simple, axillary, 5- to 12-flowered, erect on slender, unbranched peduncles, 0.5–4 cm. long, involucre of a few minute bracts, the fruiting pedicels 2–8 (–10) mm. long; sepals small; petals white; stylopodium lacking; fruit subglobose, 1.5–2 mm. long; carpels with dorsal ribs obscure and filiform, the lateral ones broad, corky and thickened next to commissure; oil tubes solitary or sometimes 2 in the intervals, 2–6 on the commissure.

Salt marshes or brackish mud flats, at elevations of 1–100 feet: north along coast from San Francisco Bay; north to British Columbia and Alaska.

On the basis of vegetative characters there appear to be two forms of this species: (1) the coastal form, extending southward to Marin County, California, from British Columbia, which has somewhat broad, often flattened, and conspicuously septate phyllodes; and (2) the San Francisco Bay and river-mouth form, which has very fine, terete, and only obscurely septate phyllodes. Additional collections and further study may show that these merit taxonomic recognition.

OENANTHE. WATER PARSLEY

1. Oenanthe sarmentosa Presl in DC., Prodr. 4:138. 1830. Fig. 288.

Aquatic, glabrous herb; stems decumbent or prostrate, succulent, 6–15 dm. long from heavy rootstocks, often rooting from nodes; leaves pinnately compound, the upper leaves bipinnate, the lower ones simply pinnate or bipinnate toward apex, 3–6 dm. long; leaflets ovate, serrate or coarsely toothed, 2–6 cm. long; umbels terminal, compound, involucre present or absent, involucel bractlets numerous, linear; rays 2–3 cm. long; pedicels 2–6 mm. long; sepals prominent, persistent; petals white; styles slender, elongating after anthesis to become nearly as long as fruit; fruit glabrous, oblong-cylindric, somewhat flattened laterally, 2.5–3.5 mm. long; ribs very corky, broad, obtuse; oil tubes solitary in the intervals, 2 on commissure.

In ponds or slow streams, sometimes growing in massive colonies, at elevations of 5–3,000 feet: coastal southern California, Coast Ranges, occasionally in the northern Sierra Nevada; north to British Columbia.

Fig. 288. *Oenanthe sarmentosa: a,* fruit, × 8; *b,* carpel, × 8; *c,* umbel, × ⅔; *d,* umbellet, × 2; *e,* flower, × 8; *f,* habit, basal part of plant, × ⅓; *g,* habit, upper part of plant, × ⅖; *h,* heavy rootstock, × ⅖.

OXYPOLIS

1. Oxypolis occidentalis Coult. & Rose, U. S. Nat. Herb. 7:196. 1900.

Glabrous, erect, aquatic herb; roots fleshy, tuberous, fascicled on a short rootstock; stem simple or little-branched, 4–15 dm. tall; basal leaves simply pinnate, long-petioled, 12–30 cm. long; leaflets 5–13, orbicular or broadly ovate to linear-acuminate, serrate or crenate to dentate, entire at base, 3–7 cm. long; cauline leaves few with leaflets few, the upper leaves often reduced to phyllodes; umbel compound, long-pedunculate; involucre of 1 or 2 bracts, the involucel of a few linear bractlets, both bracts and bractlets sometimes absent; rays 12–24, 2–8 cm. long; pedicels slender, 6–10 mm. long; sepals evident; petals white; stylopodium short-conical; fruit oblong to obovate, 5–6 mm. long, strongly flattened dorsally; carpels with dorsal and intermediate ribs filiform, lateral ribs broadly winged, the wings strongly nerved at inner margin, thus giving to carpel the appearance of having 5 dorsal ribs; oil tubes solitary in the intervals, 2–6 on the commissure.

In shallow water of marshy meadows, at elevations of 4,500–8,500 feet: Sierra Nevada; north to southern Oregon.

SANICULA. SNAKEROOT

1. Sanicula maritima Kell. in Wats., Bot. Calif. 2:451. 1880.

Glabrous perennial; stems from a thickened taproot, 1.5–3.5 dm. tall, branching above; basal leaves numerous, long-petioled, the blades orbicular to cordate, entire or somewhat serrate, 2–5 cm. long; cauline leaves few, 3-parted, the divisions roundish, subentire or coarsely toothed; peduncles few, elongate; umbels compound with rays 1–4, the involucre with bracts foliaceous, the involucel of numerous, small, lanceolate bractlets; umbellets capitate, perfect and staminate flowers in same head, the staminate ones short-pediceled; sepals somewhat foliaceous, persistent; petals yellow; fruit subglobose, 4–5 mm. long; carpels lacking ribs, densely covered with hooked bristles, except at base; oil tubes many, irregularly distributed.

Coastal adobe flats bordering salt marshes: only known in San Luis Obispo County; originally collected near San Francisco Bay, where it is now apparently extinct.

SIUM

1. Sium suave Walt., Fl. Carol. 115. 1788. Hemlock water parsnip.

Sium cicutaefolium Schrank, Baier. Fl. 1:558. 1789.

Aquatic or subaquatic, glabrous, perennial herb; stem stout, erect, usually simple, 8–15 dm. tall, from a cluster of fibrous, fleshy roots; leaves simply pinnate or pinnately compound with lower leaves long-petiolate and 1.5–4 dm. long including petiole, the upper leaves subsessile; leaflets 6–18, 6–12 cm. long, those of submersed leaves dissected, others linear to lanceolate, serrate; umbels compound; flowers white; involucre and involucel with many lanceolate-ovate bracts and bractlets; rays 10–20, 20–50 mm. long; sepals minute; styles short and stylopodium depressed; fruit oval or oblong, 2.5–3 mm. long, compressed laterally, glabrous; carpels with prominent, corky ribs, the intervals broad, red-brown; oil tubes 1–3 in the intervals.

Swamps, ponds, and bogs, at elevations of 3,000–6,500 feet: northern Sierra Nevada and north to Modoc and Siskiyou counties; across the continent.

SPHENOSCIADIUM

1. Sphenosciadium capitellatum Gray, Proc. Am. Acad. Arts & Sci. 6:537. 1866. Button parsley.

Selinum capitellatum (Gray) Benth. & Hook. ex Wats. in King, Geol. Expl. 40th Par. 5:126. 1871.

Perennial; roots thick; stems stout, 8–15 dm. tall, little-branched, glabrous below inflorescence; leaves grayish green, 1–4 dm. long, large, simply or compoundly pinnate, glabrate or scabrous, the cauline leaves with petioles bladdery and dilated, the leaflets or segments 4–12 cm. long, linear-lanceolate to ovate, entire to serrate or coarsely dentate; umbel compound, the rays 4–18, tomentose, 2–10 cm. long; umbellets globose, capitate, 7–12 mm. across, the pubescent flowers sessile and densely crowded together; involucre none; involucel of numerous linear-setaceous, deciduous bractlets; sepals none; petals scarious, white or purplish; stylopodium small, conical; fruit cuneate-obovate, strongly flattened dorsally, 5–8 mm. long, subglabrous or tomentose; carpels prominently ribbed, winged above; oil tubes solitary in intervals, 2 on commissure.

Wet places at springs or along streams or in swampy flats, at elevations of 3,000–8,500 feet: southern California, Sierra Nevada, higher North Coast Ranges; north and east to Oregon, Idaho, and Nevada.

CORNACEAE

Trees, shrubs, or suffrutescent plants. Leaves usually opposite, simple, entire. Flowers small, perfect, regular, epigynous, in cymes or heads. Sepals 4 and minute, or none. Petals white or greenish, valvate, inserted with sepals at apex of hypanthium. Stamens 4, alternate with petals. Ovary inferior, 2-celled, with filiform style and terminal, simple stigma. Fruit a drupe with 2-celled, 2-seeded stone.

CORNUS. DOGWOOD

Flowers in sessile, involucrate umbels, appearing with or before leaves; mature drupe
  black .... 1. *C. sessilis*
Flowers in compound, pedunculate, noninvolucrate cymes, appearing after leaves; mature
  drupe bluish or white.
  Leaf blades lighter and pubescent beneath, veins prominent; branches reddish or pur-
    plish .... 2. *C. stolonifera* var. *californica*
  Leaf blades nearly alike on both surfaces, veins not prominent; branches mostly gray
    to brown .... 3. *C. glabrata*

1. Cornus sessilis Torr. ex Durand, Jour. Acad. Nat. Sci. Phila. II. 3:89. 1855.

Spreading shrubs or small trees, 1–3 m. or more tall, the branches green; leaves deciduous, ovate to broadly elliptic, 5–9 cm. long, on short petioles; flowers few to several in small sessile umbels, yellowish, appearing with or before leaves, the umbels subtended by 4 small, caducous bracts; pedicels 2–4 mm. long, densely pubescent; drupe oblong or ovoid, greenish white to red when immature, shiny black when fully ripe, 10–12 mm. long, on pedicels as long.

Stream banks in foothills, usually forming thickets: Sierra Nevada foothills from Amador County north, Siskiyou Mountains, North Coast Ranges.

2. Cornus stolonifera var. **californica** (C. A. Mey.) McMinn, Illus. Man. Calif. Shrubs 377. 1939. Creek dogwood.

Cornus californica C. A. Mey., Bull. Acad. St. Pétersb. Phys. Math. 3:372. 1845.

Shrub 2–5 m. tall, erect, branches spreading, reddish or purplish; leaves deciduous, ovate or ovate-lanceolate, shortly acute, 3–10 cm. long, the upper surface dark green, glabrate, the lower surface lighter green, with prominent veins, and more or less tomentose with usually numerous, long, curly or straight, simple or 2-branched hairs, and a few short, straight, 2-branched ones; cymes 2–5 cm. across, their branches and peduncles sparsely to densely pubescent; flowers whitish, appearing after leaves; style glabrous; mature drupe subglobose, white to bluish, the stone 4–5 mm. wide, smooth or furrowed.

Stream banks and moist areas in canyons and along waterways in valleys, from near sea level to an elevation of 2,500 feet or somewhat higher: Sierra Nevada, delta area of Sacramento and San Joaquin rivers, Coast Ranges, coastal southern California.

The major morphological distinction between this variety and the more widespread *Cornus stolonifera* proper consists in the pubescence of the leaves: in the latter the leaves are microscopically puberulent beneath with appressed, straight, 2-branched hairs, as compared with the more abundant tomentum described above for *C. stolonifera* var. *californica*. However, typical *C. stolonifera* is found across North America, usually at altitudes above 5,000 feet on mountain or canyon slopes away from the immediate vicinity of water, whereas the variety *californica* is limited to wet areas and watercourses in California, usually at lower altitudes.

3. Cornus glabrata Benth., Bot. Voy. Sulphur 18. 1844. Brown dogwood.

Erect shrub 2–5 m. tall, with spreading or trailing, usually glabrous, smooth, gray or brownish twigs; leaves deciduous, narrowly ovate to elliptic, acute at each end, 3–6 cm. long, both surfaces green and obscurely pubescent with short, appressed hairs, the veins not prominent; petioles slender, 3–5 mm. long; flowers many, white, appearing after leaves, in small, open, flat-topped cymes, their branches and peduncles slightly pubescent to glabrous; style pubescent; drupe globose, the flesh bluish to white; stone globose, smooth or only obscurely furrowed.

Stream banks, moist flats, or borders of swamps, often forming dense thickets, from near sea level to an elevation of 4,000 feet: Sierra Nevada foothills, Siskiyou Mountains, delta region of the Sacramento and San Joaquin rivers, Coast Ranges, coastal southern California.

ERICACEAE

Perennial herbs, shrubs, or trees. Leaves simple, usually alternate, mostly evergreen and coriaceous. Flowers usually regular, 4- or 5-whorled. Calyx usually synsepalous, or sepals nearly distinct. Corolla usually sympetalous, or petals free. Stamens attached to receptacle, usually twice as many as lobes of corolla; anthers 2-celled, usually opening by a terminal pore, often with awned appendages. Ovary superior to inferior, 4- to 10-celled, placentation axile, the ovules numerous; style 1; stigma 1, entire or lobed.

Ovary superior; fruit a capsule (surrounded by the fleshy calyx in *Gaultheria*).
    Corolla with petals distinct (choripetalous) .... *Ledum*
    Corolla with petals united (sympetalous).
        Stamens 5 .... *Rhododendron*
        Stamens 10.
            Leaves opposite .... *Kalmia*
            Leaves alternate.
                Calyx of 5 sepals, not enveloping fruit .... *Leucothoe*
                Calyx tubular, becoming fleshy and enveloping fruit .... *Gaultheria*
Ovary inferior; fruit a berry .... *Vaccinium*

GAULTHERIA. WINTERGREEN

Erect to prostrate, evergreen shrubs or subshrubs. Leaves alternate, shining above, broad. Flowers solitary and axillary or in terminal panicles. Calyx 5-cleft. Corolla 5-lobed, urceolate or globose. Stamens 10. Ovary 5-lobed and

5-celled. Fruit a capsule invested in the fleshy calyx which forms a globose, berry-like fruit.

Calyx glabrous; leaf blades frequently oval, 10–15 mm. long; branches rooting
1. *G. humifusa*
Calyx hairy; leaf blades ovate, 15–30 mm. long; branches not rooting .. 2. *G. ovatifolia*

1. Gaultheria humifusa (Gray) Rydb., Mem. N. Y. Bot. Gard. 1:300. 1900.

Depressed shrub with creeping, prostrate, rooting stems 2–20 cm. long, forming mats; leaves broadly ovate to elliptic or oval, rounded at both ends or shortly acute at apex, 10–15 mm. long, subentire or finely serrate, short-petioled; flowers solitary, axillary; calyx 2.5–3 mm. long, glabrous; corolla white, slightly longer than the calyx, subcampanulate; anthers not awned; fruiting calyx berry-like, red, 5–6 mm. wide.

Wet meadows and slopes, at elevations of 8,000–9,000 feet: Sierra Nevada locally in Fresno, El Dorado, Tuolumne, and Mono counties; north to British Columbia.

2. Gaultheria ovatifolia Gray, Proc. Am. Acad. Arts & Sci. 19:85. 1883.

Low, slender shrub with prostrate or ascending branches, 1–4 dm. long from long rootstocks; stems, petioles, and calyx with loose, spreading hairs; leaf blades ovate, rounded or subcordate at base, apically acute, glabrous or glabrate, minutely denticulate, short-petioled, 15–30 mm. long; flowers solitary, axillary; calyx thickly rusty-tomentose, 2 mm. long, the lobes longer than the tube; corolla 3–4 mm. long, white, broadly campanulate; fruiting calyx berry-like, red, 6–7 mm. wide.

Wet slopes, bogs, or along streams, at elevations of 3,000–4,000 feet: North Coast Ranges from Humboldt County north, Sierra Nevada in El Dorado County; north to British Columbia.

KALMIA

Erect shrubs, ours low or diminutive. Leaves evergreen, coriaceous, entire, alternate, opposite, or verticillate. Inflorescence of terminal or lateral corymbose or umbellate flower clusters, the pedicels viscid. Calyx 5-parted, persistent. Corolla rotately spreading, 5-lobed, white or pink, with 10 sacs on the short tube and keels running from these to lobes and sinuses. Stamens 10, included, the filaments subulate or filiform, often pubescent, the anthers opening by pores. Ovary 5-celled, the style slender. Capsule globose or subglobose, 5-valved, within the persistent calyx.

The ten sacs on the corolla tube constitute a pollen-discharging mechanism distinctive to the genus. The ten sacs pocket the ten anthers; disturbances of the flower release the anthers, which are flung up, discharging the pollen.

Shrubs tall (mostly 2–6 dm. tall); leaves 12–25 (–30) mm. long, the margins somewhat
revolute; corolla 12–18 mm. broad .... 1. *K. polifolia*
Shrubs low (mostly 1–2 dm. tall); leaves 10–15 mm. long, the margins usually strongly
revolute; corolla 8–12 mm. broad .... 1a. *K. polifolia* var. *microphylla*

1. Kalmia polifolia Wang., Beob. Gesell. Naturf. Freunde Berlin 2²:130. 1899.
Pale laurel.

Erect, to 6 dm. tall; herbage without bristles or glandular hairs; leaves normally opposite, oblong-elliptic, narrowed at both ends, glossy green above,
glaucous and simple-pubescent or glabrous beneath, often somewhat revolute-
margined, 12–25 mm. long, nearly sessile; corymbs axillary to normal foliage
leaves; corolla pink to rose-purple, 12–18 mm. broad; capsule globose, 5–6
mm. broad, glabrous.

Wet meadows and swamps, at elevations of 3,500–8,000 feet: Sierra Nevada
from El Dorado County north, northern North Coast Ranges and Siskiyou
Mountains; north to Alaska and east to Atlantic.

1a. Kalmia polifolia var. **microphylla** (Hook.) Rehd. in Bailey, Cycloped. Am.
Hort. 2:854. 1900. Alpine laurel.

Often only 1 dm. tall, sometimes to 2 dm.; leaves usually oval, 1–2 cm.
long, the margins strongly revolute; corolla 8–12 mm. broad. Otherwise like
the species.

Wet meadows or swamps, at elevations of 7,000–12,000 feet: Sierra Nevada
from Tulare County north to Shasta County; north to British Columbia and
east to Colorado.

LEDUM

1. Ledum glandulosum Nutt., Trans. Am. Philos. Soc. II. 8:270. 1843. Labrador tea.

Shrub to 1.5 m. tall; herbage fragrant; leaves alternate, numerous, the blades
plane or nearly so, oblong, acute at both ends, apically mucronate, 2–7 cm.
long, green, glabrous and glaucous beneath except for scattered glands; flowers
small, in terminal, condensed corymbs on slender pedicels; sepals 5, almost
distinct, minute; petals 5, white, distinct, ovate to obovate, spreading, 4–6 mm.
long; stamens 10, the filaments hairy at base; ovary impressed at summit; fruit
a 5-celled, many-seeded capsule borne on the recurved pedicel.

Margins of wet meadows and marshy places: Sierra Nevada at elevations
above 4,000 feet, Coast Ranges to an elevation of 500 feet from Santa Cruz
Mountains to Del Norte County; north to Oregon, east to Rocky Mountains.

LEUCOTHOE

1. Leucothoe Davisiae Torr. in Gray, Proc. Am. Acad. Arts & Sci. 7:400. 1868. Sierra laurel.

Erect shrub 6–20 dm. tall; leaves coriaceous, alternate, sometimes evergreen, glabrous, the blades oblong, obscurely serrulate or entire, 2.5–7 cm. long; inflorescence of terminal, scaly-bracted racemes 5–10 cm. long, the flowers pendulous; calyx of 5 distinct sepals; corolla about 6 mm. long, white, ovate, the opening narrow; stamens 10, the anthers mucronate, opening by terminal pores; capsule depressed-globose, somewhat 5-lobed, 5-chambered.

Bogs and wet ground, at elevations of 6,000–8,000 feet: Sierra Nevada from Madera County to Butte County, Siskiyou Mountains, North Coast Ranges in Trinity County; southern Oregon.

RHODODENDRON

1. Rhododendron occidentale Gray, Bot. Calif. 1:458. 1876. Western azalea.

Widely branching shrub 1–7 m. tall; leaves alternate, entire, thin, deciduous, short-petioled, the blades obovate, basally cuneate, apically rounded to acute, mucronulate, 2.5–10 cm. long, nearly glabrous; inflorescence of terminal corymbs or umbels; calyx small, lobes ovate, 2–4 mm. long; corolla yellowish white or somewhat pinkish, 4–5 (–6) cm. long, somewhat irregular, tube funnelform, lobes comparatively short, ovate, the upper lobe yellow-splotched; stamens 5, long-exserted; ovary glandular-pubescent; fruit a 5-valved capsule.

Stream margins and moist mountain meadows at elevations of 3,500–6,000 feet: mountains of southern California (absent from some areas), Sierra Nevada, outer Coast Ranges; north to southern Oregon.

VACCINIUM. HUCKLEBERRY, BILBERRY, BLUEBERRY

Shrubs, sometimes diminutive. Leaves alternate, deciduous in our species. Flowers small, axillary, solitary or in small clusters, pediceled. Calyx epigynous, limb 5-parted or entire. Corolla globose or urn-shaped, 4- or 5-toothed. Stamens twice as many as corolla lobes, the anthers (in ours) with 2 upwardly curved awns, and each cell with an apical prolongation opening by a pore at tip. Ovary 4- or 5-celled, each cell several- to many-seeded. Fruit a berry, in ours black.

Leaves entire or nearly so; calyx 5-lobed.
Leaves thick, prominently reticulate, apically obtuse or rounded .... 1. *V. uliginosum*
Leaves thin, obscurely reticulate, acute or acutish .... 2. *V. occidentale*

Leaves serrulate; calyx truncate or undulate-margined.
    Stems 0.5–2 m. tall; leaf blades 2–4 cm. long .... 3. *V. membranaceum*
    Stems less than 0.5 m. tall; leaf blades 1–2 cm. long .... 4. *V. caespitosum*

1. Vaccinium uliginosum L., Sp. Pl. 350. 1753. Bog huckleberry.

Low, branching shrub to 6 (or to 10) dm. tall; branchlets terete; leaf blades 10–30 mm. long, oval or obovate, rounded at apex or obtuse, somewhat obscurely mucronulate, crowded on branches, thick and entire, firm, glabrous, dark green above, paler beneath and reticulate-veined, short-petioled; flowers solitary or 2–4 together, pendent on short pedicels; calyx lobes to 1.5 mm. long; corolla pink, ovoid or globose, 5–7 mm. long, shallowly lobed; berry bluish black, glaucous, globose, 6–7 mm. wide, pendent.

Sphagnum bogs at low altitudes: Humboldt County north in the North Coast Ranges; circumboreal.

2. Vaccinium occidentale Gray, Bot. Calif. 1:451. 1876. Western huckleberry.

Low, much-branched shrub 2–4 or –6 dm. tall; branchlets terete, herbage glabrous; leaf blades firm but thin, obovate or oblanceolate, 1–2.5 cm. long, entire, paler beneath, usually acutish apically, acute to rounded basally, obscurely reticulate; flowers pendent, solitary or in 2's; calyx lobed, the deltoid lobes to 1 mm. long; corolla white or pink, oblong-ovoid, 4–5 mm. long, the lobes very short; berry blue-black, glaucous, globose, 4–5 or –6 mm. wide, pendent or erect on a pedicel.

Bogs and swampy places, at elevations of 5,000–10,000 feet: Sierra Nevada from Tulare County north, Humboldt County north in North Coast Ranges; north to Washington and east to Utah.

3. Vaccinium membranaceum Dougl. ex Hook., Fl. Bor. Am. 2:32. 1834.
    Mountain huckleberry, thin-leaved huckleberry.

Slender-branching shrub to 2 m. tall; branches angled; herbage glabrous; leaves ovate to obovate or oval, 2–6 cm. long, acute to obtuse at apex, membranous, nearly alike on upper and lower surfaces, sharply serrulate; flowers solitary in axils on long, recurved pedicels which become erect in fruit; calyx truncate or undulate-margined; corolla globose, yellowish white or greenish white, 4–5 mm. wide; berry dark red, usually becoming purplish black, not glaucous, globose, 7–10 mm. wide.

Bogs, at elevations of 5,000–7,000 feet: northern California in Humboldt, Del Norte, Siskiyou, and Modoc counties; north to British Columbia.

4. Vaccinium caespitosum Michx., Fl. Bor. Am. 1:234. 1803. Dwarf huckleberry.

Dwarf, rigid shrub to 1 (or to 3) dm. tall; branchlets terete; herbage usually glabrous; leaf blades obovate, obtuse or acute, sharply but very finely serrulate, 1–2 (–3) cm. long, basally cuneate, glossy green on upper surface; flowers solitary in axils, nodding on recurved pedicels; calyx truncate or undulate; corolla ovoid, pink or white, 5–6 mm. long; berry on a curved pedicel, blue-black, glaucous, globose, 3–6 mm. wide.

Wet mountain meadows or on rocky slopes that are either coastal or in the mountains at elevations of 6,500–11,500 feet: Sierra Nevada from Tulare County north, North Coast Ranges from Humboldt County north; north to Alaska, east across North America.

PRIMULACEAE

Annual or perennial herbs. Leaves simple, exstipulate, alternate, opposite or basal. Flowers perfect, regular, usually 5-merous (sometimes 3- to 9-merous), solitary in axils or in terminal or axillary racemes, umbels, or spikes. Calyx herbaceous, synsepalous, usually persistent. Corolla sympetalous, deeply to shallowly lobed, the lobes spreading or reflexed. Stamens epipetalous, the same number as and opposite the corolla lobes. Ovary superior (semi-inferior in *Samolus*), 1-celled, the ovules on a free central placenta; style 1; stigma capitate. Fruit a 2- to 6-valved capsule.

Ovary superior.
  Stems elongate, leaves opposite; inflorescence axillary, flowers solitary or in spikelike racemes.
    Corolla present, yellow .... *Lysimachia*
    Corolla none; calyx corolla-like, purplish or white .... *Glaux*
  Stems short, leaves basal; inflorescence a terminal umbel on an elongate, naked scape
                                           Dodecatheon
Ovary half-inferior, adherent at base to the calyx; inflorescence of axillary racemes;
  flowers long-pediceled .... *Samolus*

DODECATHEON. SHOOTING STAR

Herbaceous, glabrous or glandular-puberulent perennials. Roots fleshy, fibrous, from a short rootstock. Leaves basal, forming a rosette, linear to oblanceolate to broadly obovate, entire. Scapes terminating in an involucrate umbel of few to many flowers. Calyx 4- or 5-lobed, the lobes strongly reflexed at anthesis, erect in fruit. Corolla 4- or 5-lobed, the tube short, the throat thickened, the lobes long, narrow, and strongly reflexed at anthesis. Stamens inserted on corolla throat, the filaments short, free or united, the anthers basally attached and connivent around style, the connective prominent and rugose or smooth, the pollen sacs yellow or deep maroon. Ovary 1-celled; style single, filiform; stigma capitate. Capsule 5-valved, splitting by the valves or circumscissile just below the style and then splitting by the valves. Seeds numerous, reticulate.

Thompson, Henry J., The biosystematics of *Dodecatheon*. Contr. Dudley Herb. 4:73–154. 1953.

Fig. 289. *Glaux maritima: a,* habit, unbranched plant, × ⅖; *b,* flowers at anthesis, × 4; *c,* habit, branched plant, × ⅖; *d,* young fruit, × 6; *e–g,* seeds, × 16.

The three species which are considered below comprise the section *Capitulum* Thompson. They represent a well-distinguished unit in the genus, particularly in respect to habitat (wet or boggy areas) and altitude (montane), but they are sometimes difficult to delimit from one another because of intermediates.

At least a few capsules circumscissile by a small apical operculum and valvate only below
the operculum . . . . 1. *D. Jeffreyi*
Capsules entirely valvate, all lacking an operculum.
Corolla 4-merous; herbage glabrous; base of anthers exserted from corolla tube
2. *D. alpinum*
Corolla 5-merous; herbage glandular-pubescent; base of anthers included in corolla
tube . . . . 3. *D. redolens*

1. Dodecatheon Jeffreyi Van Houtte, Fl. Serres 16:99. 1865.

Roots lacking bulblets; herbage glandular-puberulent to glabrous; leaves 8–50 cm. long including petiole, 1–3 cm. wide, oblong-oblanceolate, narrowed below to a short-winged petiole; scapes 25–60 cm. tall, the umbels 4- to 15-flowered, glandular-pubescent to glabrous, the pedicels 4–10 cm. long; corolla 4- or 5-merous, maroon at base, with a yellow ring above, the lobes narrow, deep lavender to white, 15–25 mm. long, closely reflexed exposing base of anthers; filaments very short, deep maroon or black, the anthers 8–10 mm. long, linear, obtuse, the pollen sacs maroon to yellow, the connective rugose, deep maroon or black; stigma enlarged; capsule 7–12 mm. long, ovoid, dehiscence circumscissile by a small operculum, then splitting by valves, or sometimes valvate for the entire length.

Montane, in wet meadows, at elevations of 7,000–9,000 feet: Sierra Nevada, North Coast Ranges.

2. Dodecatheon alpinum (Gray) Greene, Erythea 3:39. 1895.

Roots lacking bulblets; herbage completely glabrous; leaves 2–15 cm. long including petiole, linear to linear-oblanceolate, obtuse, entire, gradually tapering into petiole; scapes 10–30 cm. tall, the umbels 1- to 10-flowered, the pedicels 1–3 cm. long; corolla 4-merous, maroon at base, yellow above, the lobes magenta to lavender, the tube reflexed so as to expose base of anthers; filaments short, free, black, the anthers 5–7 mm. long, linear, obtuse, the pollen sacs dark, the connective rugose, deep maroon to black; stigma enlarged; capsule 7–8 mm. long, ovoid, dehiscing entirely by valves.

In wet, montane meadows. According to Thompson (*op. cit.,* p. 110), this occurs as two subspecies: *Dodecatheon alpinum* subsp. *majus* occupies intermediate altitudes in southern California, the Sierra Nevada, North Coast Ranges, and Siskiyou Mountains; whereas *D. alpinum* subsp. *alpinum* is confined, in California, to high altitudes of the Sierra Nevada, and also occurs in Oregon, Nevada, and Arizona.

3. Dodecatheon redolens (Hall) H. J. Thompson, Contr. Dudley Herb. 4:143. 1953.

Dodecatheon Jeffreyi var. *redolens* Hall, Bot. Gaz. 31:392. 1901.

Roots lacking bulblets; entire plant strongly glandular-pubescent; herbage with strong odor; leaves 20–40 cm. long, oblanceolate, obtuse, tapering gradually into petiole; scapes 20–60 cm. tall, the umbels 5- to 10-flowered, the pedicels 2–5 cm. long; corolla 5-merous, the tube yellow, lacking maroon band, not closely reflexed and thus covering anther bases, the lobes magenta to lavender, 15–25 mm. long; filaments free, short, black, the anthers 7–10 mm. long, lanceolate, obtuse, the pollen sacs dark, the connective rugose, deep maroon to black; stigma enlarged; capsule 8–15 mm. long, ovoid, dehiscing entirely by valves.

Wet mountain meadows at high altitudes (above 8,000 feet): southern Sierra Nevada, San Jacinto Mountains, San Bernardino Mountains; Nevada and Utah.

GLAUX

1. Glaux maritima L., Sp. Pl. 207. 1753. Sea milkwort. Fig. 289.

Glabrous and glaucous perennial herb; rootstocks horizontal, slender; stems slender, branched, ascending, 5–20 cm. long; leaves sessile, oval to linear-oblong, 4–10 mm. long, fleshy, obtuse to acute at apex, the lower ones opposite, the upper ones usually alternate; flowers sessile or subsessile in leaf axils about midway along stem; calyx petaloid, campanulate, 3–4 mm. long, 5-cleft, whitish; corolla absent; stamens 5, inserted at base of calyx tube and alternate with its lobes, the filaments as long as the calyx lobes, the anthers hairy; ovary superior, ovoid; style filiform; stigma capitate; capsule beaked by persistent style.

Coastal salt marshes: central California; north to Washington.

LYSIMACHIA

Perennial herbs. Leaves opposite (in ours), usually gland-dotted. Calyx 5- to 7-lobed. Corolla rotate, 5- to 7-lobed, the lobes convolute. Filaments more or less united at base or free. Ovary globose; style filiform.

Flowers solitary, axillary; stems creeping .... 1. *L. Nummularia*
Flowers in spikelike, axillary racemes; stems erect .... 2. *L. thyrsiflora*

1. Lysimachia Nummularia L., Sp. Pl. 148. 1753. Creeping loosestrife. Fig. 290.

Glabrous perennial; stems slender, creeping, rooting at nodes, 2–10 dm. long; leaves more or less orbicular, obtuse, 1–2.5 cm. long, short-petioled,

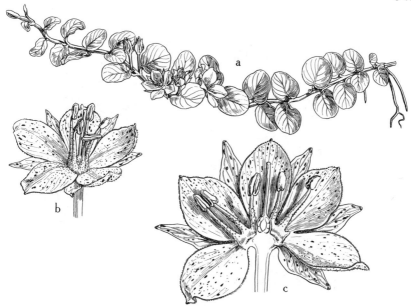

Fig. 290. *Lysimachia Nummularia: a,* habit, × ⅔; *b,* flower, showing glands and red spots, × 2½; *c,* flower (longitudinal section), × 3.

glandular-punctate; flowers solitary (or occasionally 2) in leaf axils, nodding on peduncles 1–3.5 cm. long; calyx lobes ovate, acute, 6–8 mm. long; corolla yellow, the lobes 7–10 mm. long, broadly ovate, black-dotted, rotate; filaments slightly united below, glandular; capsule shorter than sepals.

Occasional in bogs: northern Sierra Nevada; naturalized from Europe.

2. Lysimachia thyrsiflora L., Sp. Pl. 147. 1753. Fig. 291.

Naumburgia thyrsiflora Duby in A. DC., Prodr. 8:60. 1844.

Glabrous or sparsely pubescent perennial; stems erect, simple, 3–8 dm. tall; leaves lanceolate to linear-lanceolate, acute, sessile, 5–10 cm. long, dotted with dark glands; inflorescences solitary in the axils, the peduncles 1.5–3 cm. long, the racemes dense, to 3 cm. long, spikelike with pedicels very short; calyx lobes linear, 1–2 mm. long; corolla yellow, the lobes distinct nearly to base, 3–4 mm. long; filaments distinct; stamens and style exserted; capsule slightly longer than sepals, glandular-dotted.

Borders of lakes and in bogs and marshy places: occasional in the northern Sierra Nevada; north to Washington.

Fig. 291. *Lysimachia thyrsiflora: a,* capsule, × 8; *b* and *c,* habit, × ⅖; *d,* flower, show-ing glands on calyx and ovary, × 4; *e,* seed, abaxial view, × 20; *f,* flower, spread open, × 6; *g,* open capsule, showing seeds around the central placenta, × 8; *h,* seed, adaxial view, × 20.

SAMOLUS

1. Samolus floribundus H.B.K., Nov. Gen. & Sp. 2:224. 1817. Water pimpernel. Fig. 292.

Glabrous perennial herb; stems erect, simple or branched, 10–40 cm. tall; basal leaves more or less rosette-like, obovate to oblong-spatulate, 2–7 cm. long, obtuse at apex, narrowed at base into a broad, short petiole; cauline leaves alternate, reduced upward and often sessile; flowers in elongate, open, terminal racemes, on slender pedicels 8–20 mm. long, these with minute bractlets about midway along their length; calyx tube basally adnate to ovary, the limb 5-lobed, persistent; corolla campanulate, perigynous, 5-lobed, about 1.5 mm. broad, white; staminodia in sinuses of corolla lobes and alternating with the stamens; filaments short, the anthers cordate; capsule 5-valvate, 2.5 mm. wide.

Marshes and streams, uncommon: San Francisco Bay area to southern California; north and east across North America.

PLUMBAGINACEAE

Perennial herbs with leaves in a basal cluster and with scapose stems. Flowers regular, 5-merous, perfect. Calyx tubular or funnelform, plicate, usually 10-ribbed. Corolla with petals united only at base. Stamens hypogynous, opposite the petals; anthers versatile, 2-celled. Ovary superior, 1-celled, 1-ovuled. Styles 5, distinct. Fruit a utricle, enclosed by the persistent calyx.

LIMONIUM

1. Limonium commune S. F. Gray var. **californicum** (A. Gray) Greene, Man. Bot. Bay Reg. 235. 1894. Marsh rosemary.

Limonium californicum Heller, Cat. N. Am. Pl. 6. 1898.

Root woody, red; stems 20–60 cm. tall; leaves fleshy, the blades obovate to oblong-spatulate, obtuse, 4–8 cm. long, tapering to a petiole often as long as the blade; inflorescence secund, paniculate; calyx obconic, the ribs hairy at base, the lobes membranous above; corolla united only at base, the lobes long-clawed, violet-purple; stamens epipetalous.

Salt marshes and beaches along the coast: Los Angeles and Orange counties north to Humboldt County.

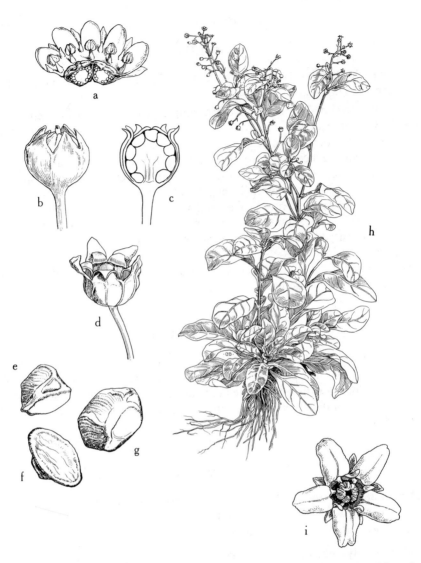

Fig. 292. *Samolus floribundus: a,* flower, spread open, × 8; *b,* fruit, enclosed by calyx, × 8; *c,* fruit (longitudinal section), × 8; *d,* capsule, after dehiscence, × 8; *e–g,* seeds, × 40; *h,* habit, × ⅖; *i,* flower, × 8.

OLEACEAE

Shrubs or trees. Leaves opposite, pinnately compound or simple. Flowers small, regular, unisexual or perfect, usually in compact panicles or clusters. Calyx 4-cleft or lacking. Corolla (in ours) choripetalous or none. Stamens 1–4. Ovary superior, 2-celled. Fruit a samara, drupe, or capsule.

Leaves simple; fruit a 1-seeded drupe .... *Forestiera*
Leaves pinnately compound; fruit a 1-seeded samara with conspicuous terminal wing
Fraxinus

FORESTIERA

1. Forestiera neo-mexicana Gray, Proc. Am. Acad. Arts & Sci. 12:63. 1876. Desert olive.

Glabrous shrub 1.5–3 m. tall, with spiny branchlets and simple, opposite, or fascicled leaves; leaves spatulate-oblong or obovate, acute, entire or serrulate, 1–4 cm. long, usually more than 6 mm. wide; flowers inconspicuous, appearing in small panicles before leaves, polygamous or dioecious; calyx minute or lacking; corolla usually lacking; stamens 2–4; ovary 2-celled, 2 ovules per cell; fruit a 1-seeded, black, ovoid drupe.

Valley flats, along streams, or on canyon slopes, to an elevation of 6,000 feet: South Coast Ranges, mountains of southern California, bordering ranges of Mojave Desert; east to Colorado and Texas.

FRAXINUS. ASH

Trees or shrubs with pinnately compound, petioled, deciduous leaves. Flowers perfect, dioecious, or polygamous, appearing before leaves. Calyx truncate with 4-cleft or toothed margin. Petals 2 or none. Stamens 2. Fruit a samara with 1 seed and a terminal wing.

Leaflets oblong or oval, lateral ones commonly sessile; fruit 2.5–3.5 cm. long; cismontane
    California .... 1. *F. latifolia*
Leaflets lanceolate, petiolulate; fruit 1.5–2.5 cm. long; desert ranges ..... 2. *F. velutina*

1. Fraxinus latifolia Benth., Bot. Voy. Sulphur 33. 1844. Oregon ash.

Fraxinus oregona Nutt., Sylva 3:59. 1849.

Tree 8–12 m. tall, with terete branchlets; leaves 15–30 cm. long; leaflets 5–7, oblong to oval, abruptly acute, sessile (except the terminal one), subentire, 3–8 cm. long; flowers dioecious; corolla none; style conspicuously 2-lobed; samara 2.5–3.5 cm. long including wing, 4–5 mm. wide, oblong-lanceolate, margins of wing decurrent on ¼–½ of the body.

Streamsides or wet bottomlands at elevations below 5,000 feet: localized in its preferred habitats in North Coast Ranges, Siskiyou Mountains, delta region of Central Valley, possibly also in South Coast Ranges and in cismontane southern California; north to British Columbia.

2. Fraxinus velutina Torr. in Emory, Notes Mil. Reconn. 149. 1848. Arizona ash.

Tree 5–10 m. tall, with terete, puberulent, or tomentose branchlets; leaves 10–12 cm. long; leaflets 3–7, lanceolate to narrowly ovate, acute to attenuate, 2–8 cm. long, all petioled; flowers dioecious; corolla none; style conspicuously 2-lobed; samara 1.5–2.5 cm. long including wing, 4–5 mm. wide, the wing only narrowly decurrent ¼ or less the length of the body, or strictly terminal.

Borders of streams, lakes, or springs: mountain ranges bordering California desert areas; east to Texas.

This is a highly variable species, for which a number of varieties have been described. The extreme forms could well be taken for distinct species were it not for the presence of many intermediate forms. *Fraxinus velutina* var. *coriacea* (Wats.) Rehder (Proc. Am. Acad. Arts & Sci. 53:206, 1917), a thick-leaved form with coarsely serrate leaflets and less dense pubescence, occurs in Inyo County and along the eastern edge of the San Bernardino Mountains, but according to Rehder (*op. cit.*, p. 207) most California specimens distributed to herbaria as *F. coriacea* are a glabrous or glabrescent form of *F. latifolia*.

GENTIANACEAE

Glabrous herbs. Leaves simple, entire, the basal ones petiolate, those of the stem opposite and sessile. Flowers perfect, regular, terminal or axillary, and often in a cymose inflorescence. Calyx persistent, hypogynous, 4- or 5-lobed. Corolla sympetalous, 4- or 5-lobed, usually convolute in bud, usually withering-persistent. Stamens borne on the petals and alternate with them; anthers 2-celled, longitudinally dehiscent. Ovary superior, 1-celled, with 2 parietal placentae; style 1; stigma entire or 2-lobed. Fruit a 1-celled capsule, dehiscent apically by 2 valves, the incurved edges of the latter bearing the numerous, minute seeds.

Corolla red or pink (occasionally white or yellow); anthers spirally twisted after anthesis
Centaurium
Corolla blue or white; anthers not spirally twisted after anthesis.
    Style filiform, mostly deciduous in fruit; stamens inserted on corolla throat .. *Eustoma*
    Style stout, short, persistent, or none; stamens inserted on corolla tube ..... *Gentiana*

CENTAURIUM. CANCHALAGUA

Annual herbs. Inflorescences cymose. Calyx deeply parted, the tube cylindric, the lobes linear or linear-lanceolate, often keeled. Corolla salverform, red or pink to white, sometimes yellow. Stamens inserted on corolla throat, the lobes usually shorter than the tube; filaments slender; anthers usually exserted, spirally twisted after shedding pollen. Style filiform, deciduous; stigma 2-lobed, oblong to fan-shaped. Capsule oblong-ovate to fusiform.

Flowers all on slender pedicels, at least some of the pedicels becoming long (to 20 cm. or more) .... 1. C. exaltatum
Flowers in the forks of the cymes sessile or subsessile, the lateral ones sessile, subsessile, or short-pediceled .... 2. C. floribundum

1. Centaurium exaltatum (Griseb.) Wight in Piper, Contr. U. S. Nat. Herb. 11:449. 1906.

Stems slender, usually strict, 6–35 cm. tall; leaves lanceolate to linear, acute, 1.5–2 cm. long; flowers usually numerous, those in the main forks on pedicels 15–25 mm. long, the others on shorter pedicels; calyx lobes subulate, 8–9 mm. long; corolla pink, 12–14 mm. long, the lobes 3–4 mm. long; anthers about 1 mm. long; stigma lobes fan-shaped; capsule cylindrical.

Moist places along watercourses or around springs, at elevations of 3,000–4,500 feet: cismontane southern California, north along the east side of the Sierra Nevada to Modoc County; north to Washington, east to Utah.

2. Centaurium floribundum (Benth.) Robins., Proc. Am. Acad. Arts & Sci. 45:396. 1910.

Stems usually branching freely, 10–50 cm. tall; leaves obovate to oblong or ovate-lanceolate, 1.5–2 cm. long; flowers sessile or subsessile in the forks, others pediceled; bract subtending each flower bearing a rudimentary flower in its axil; corolla pink, 10–14 mm. long, the lobes oblong to oval, 4–5 mm. long; anthers 1.5–1.7 mm. long; stigma lobes appressed; capsule cylindrical.

Moist places and waterways, to an elevation of 1,500 feet: Central Valley and adjoining cismontane areas.

Other species of *Centaurium* might also qualify for treatment here on the basis of their occasional preference for wet habitats: *C. Muhlenbergii* (Griseb.) Wight, *C. Davyi* (Jepson) Abrams, and *C. trichanthum* (Griseb.) Robins. Species delimitation in the genus is not always clear-cut, and a monographic study is needed.

EUSTOMA

1. Eustoma exaltatum (L.) Griseb. in A. DC., Prodr. 9:51. 1845.

Eustoma silenifolium Salisb., Parad. London. pl. 34. 1806.

Annual or short-lived perennial with erect, glaucous stems from a taproot, these solitary or 2 or 3, 2–7 dm. tall, branched above; basal leaves obovate to spatulate, narrowed to a broad petiole; cauline leaves clasping, somewhat connate, broadly oblong, 2–5 cm. long, reduced in inflorescence; inflorescence an open, cymose panicle, the flowers on long peduncles; calyx deeply cleft, the lobes 5 or 6, lanceolate, acuminate, keeled, 10–15 mm. long; corolla blue or white, broadly campanulate, the tube 1 cm. long, the 5 or 6 oblong or obovate lobes 1.5–2 cm. long and erose-denticulate; stamens inserted on corolla throat, the anthers not coiling; style 4–5 mm. long; stigma 2-lobed; capsule oblong, 8–12 mm. long, obtuse.

Streams or wet meadows, at elevations of 500–1,500 feet: cismontane southern California, western edge of Colorado Desert; east to Florida, south into Mexico.

GENTIANA. GENTIAN

Annual or perennial herbs. Flowers showy, terminal and solitary or clustered, or axillary. Calyx 4- or 5-cleft. Corolla campanulate or funnelform, showy, blue or white, often with plaited folds in the sinuses of the lobes. Stamens inserted on corolla tube; anthers versatile, straight or recurved after anthesis. Style short or none, persistent; stigmas 2. Capsule oblong, enclosed in withering-persistent corolla.

Annual or biennial; calyx tube without an inner membrane.
    Corolla not more than 12 mm. long.
        Flowers numerous; leaves not scarious-margined; capsule sessile ... 1. *G. Amarella*
        Flowers solitary; leaves scarious-margined; capsule long-stipitate .. 2. *G. Fremontii*
    Corolla larger.
        Stem solitary, simple; calyx lobes lacking black midrib; seeds smooth, cylindric
                    3. *G. simplex*
        Stems usually several, branched; calyx lobes often with black midrib; seeds oval, scaly
                    4. *G. holopetala*
Perennial; calyx tube with an inner membrane projecting somewhat above base of lobes.
    Plaits in corolla sinuses not lacerate or appendaged .... 5. *G. sceptrum*
    Plaits in corolla sinuses lacerate or appendaged.
        Uppermost pair of leaves united at base at least on one side; plaits of the sinuses setae-like, nearly as long as lobes .... 6. *G. setigera*
        Uppermost pair of leaves little or not at all united; plaits of the sinuses triangular-subulate, ½ as long as lobes .... 7. *G. calycosa*

This key includes only the more strictly bog-dwelling species; the list could be amplified to include all species of this genus which occur in California.

1. Gentiana Amarella L. var. **acuta** (Michx.) Herd., Acta Hort. Petrop. 1:428. 1872.

Annual; stems erect, strict, abundantly branching, 2–4 dm. tall; basal leaves obovate or oblanceolate, obtuse, 2–5 cm. long, the cauline leaves oblong at base, lanceolate and acute above; flowers numerous, in small, axillary clusters on pedicels 6–12 mm. long; calyx lobes linear or oblong, about ½ as long as corolla tube; corolla 10–12 mm. long, the tube greenish, the lobes pale blue to dark blue, lanceolate, acute, a row of slender, yellow fimbriae within near base of each lobe; capsule sessile, fusiform-cylindric, dehiscent across summit.

Boggy meadows, at elevations of 4,500–9,000 feet: San Bernardino Mountains, Sierra Nevada, Trinity Mountains; north to Alaska and east to Labrador.

2. Gentiana Fremontii Torr. in Frem., Rep. Expl. Exped. 94. 1845.

Gentiana humilis Steven, Acta Mosq. 3:258. 1812. Not *G. humilis* Salisb. 1796.

Annual or biennial; stems simple or branched near base, 3–10 cm. tall; basal leaves scarious-margined, orbicular or obovate and 5–6 mm. long, the cauline leaves oblong to linear; flowers terminal, solitary; calyx narrowly funnelform, the lobes acute, scarious-margined; corolla tubular, greenish white, 5–7 mm. long, the tube shorter than the calyx, the lobes with minutely toothed plaits in the sinuses; capsule at maturity exserted on an elongate stipe, 2-valved, dehiscent from summit, the lobes spreading.

Boggy meadows, at elevations of 7,500–8,500 feet: San Bernardino Mountains; east to Rocky Mountains, Asia.

3. Gentiana simplex Gray, Pacif. R. R. Rep. 6:87. 1857.

Annual; stems erect, simple, 8–20 cm. tall, with 3–5 pairs of linear-oblong to lanceolate leaves 1–2 cm. long; flower solitary on a bractless peduncle, 2–6 cm. long; calyx 15–20 mm. long, the lanceolate lobes about as long as the tube, never with a black midrib; corolla 20–35 (–40) mm. long, blue, the lobes obovate, subentire or finely toothed, the plaits in the sinuses lacking; capsule on a stipe, the stipe about as long as the capsule; seeds cylindric, striate but otherwise smooth, or roughened only at ends.

Boggy meadows, at elevations of 4,000–9,000 feet: San Bernardino Mountains, Sierra Nevada, North Coast Ranges; western Nevada and southern Oregon.

4. Gentiana holopetala (Gray) Holm, Ottawa Nat. 15:110. 1901.

Annual; stem simple to much-branched at base, 8–25 (–40) cm. tall; basal leaves obovate, obtuse, 1–2 cm. long, those of the lower stem oblong; upper part of stem naked, terminating in a bractless peduncle bearing a single flower; calyx about 2 cm. long, the lanceolate lobes about as long as the tube, the lobes often with a black midrib; corolla 2.5–3 cm. long, blue, the lobes broad and short, entire or strongly erose, lacking plaits in the sinuses; capsule on a stipe, the stipe about as long as the capsule; seeds oval, scaly.

Wet meadows, at elevations of 6,000–11,000 feet: San Bernardino Mountains, Sierra Nevada; Oregon, Nevada.

5. Gentiana sceptrum Griseb. in Hook., Fl. Bor. Am. 2:57. 1838.

Perennial; stems stout, erect, several from the stout base, 5–10 dm. tall, often branched above, leafy to the top; leaves oblong to linear-oblong, obtuse, 4–8 cm. long, the lowest ones merely connate bracts; flowers 1 to several at ends of branches or sometimes also axillary, on short or elongate stout peduncles; calyx tube with an inner, truncate membrane projecting above base of the lanceolate, unequal lobes; corolla 3–3.5 (–4) cm. long, deep blue with greenish dots, the lobes short, broad, with truncate, usually entire plaits in the sinuses; capsule on a stout stipe about 10 mm. long; seeds winged on ends.

Swamps or bogs, at elevations of 25–4,000 feet: North Coast Ranges; north to British Columbia.

6. Gentiana setigera Gray, Proc. Am. Acad. Arts & Sci. 11:84. 1875.

Perennial; stems stout, several from a short caudex, ascending, simple, 1–3 dm. tall; leaves crowded toward base, oblong to obovate, obtuse, thick, 2.5–4 cm. long, narrower above; flowers terminal, solitary or 2 or 3 in a cluster; uppermost pair of leaves united at base and partly concealing the flowers; calyx lobes broad, about as long as the tube, with an inner membrane projecting above lobes; corolla greenish-tinged outside, deep blue within and mottled with green, 1.5–2 cm. long, the lobes 7–12 mm. long, ovate, acute to apiculate; plaits of the sinuses deeply lacerate and setae-like; seeds winged evenly all around.

Wet meadows at elevations of 4,000–6,500 feet: North Coast Ranges.

7. Gentiana calycosa Griseb. in Hook., Fl. Bor. Am. 2:58. 1838.

Perennial; stems several from a short caudex, usually simple, 1–3 dm. tall; leaves thick, ovate, obtuse to acute, 2–3 cm. long; flowers terminal, solitary or a few crowded into a cluster, occasionally with 1 to several axillary flowers below on long pedicels; calyx lobes ovate, about as long as the tube, its inner membrane truncate; corolla deep blue, 3–3.5 cm. long, with yellow dots within; lobes 7–10 mm. long, ovate, acute or obtuse, the plaits in the sinuses triangular-subulate, ½ as long as the lobes; seeds wingless.

Wet meadows, at elevations of 4,500–10,500 feet: Sierra Nevada to Siskiyou Mountains, Trinity Mountains; north to British Columbia, Wyoming and Montana.

MENYANTHACEAE

Aquatic perennials with horizontal rhizomes; herbage glabrous. Leaves alternate, blade trifoliate. Flowers perfect, regular. Petals 5, margins rolled inward in bud, otherwise valvate. Stamens 5, inserted on corolla. Styles long or short (heterostylous). Fruit a 2-valved capsule. Included in Gentianaceae in many western manuals.

MENYANTHES

1. Menyanthes trifoliata L., Sp. Pl. 145. 1753. Fig. 293.

Glabrous perennial; rootstocks thick, creeping, clad with sheathing bases of old petioles; leaves all basal, alternate, the blades trifoliate, the leaflets oblong to obovate, narrowed to base, 2–6 cm. long, entire, pinnate-veined; petioles 3–30 cm. long, stout, with sheathing base; inflorescence a 5- to 20-flowered raceme on an elongate, naked peduncle 2–4 dm. tall, the pedicels stout, bracteate, 5–25 mm. long; flowers perfect, regular; calyx persistent, 5-lobed, the lobes 3–5 mm. long; corolla white or purplish, short-funnelform, the tube as long as the calyx, the spreading lobes 5 mm. long and bearded within with white hairs; stamens 5, epipetalous, alternate with corolla lobes; filaments short; anthers sagittate, longitudinally dehiscent, scarcely exserted (if style is long) or well exserted (if style is short); disc hypogynous, of 5 glands; ovary superior, 1-celled, the placentae 2, parietal; style subulate; stigma 2-lobed; capsule 2-valved, irregularly and tardily dehiscent, ovoid, 8 mm. long; seeds few, subglobose, shining.

Bogs and lake shores: Sierra Nevada from Tulare County north; north to Alaska and circumpolar.

CONVOLVULACEAE

Mainly herbs, often twining or trailing. Leaves alternate, exstipulate. Flowers 5-merous, regular, perfect, sympetalous. Inflorescence axillary, cymose, or terminal and solitary. Calyx usually with shallow lobes, the lobes imbricated, persistent. Corolla campanulate or funnelform, the lobes usually shallow, in bud the lobes folded longitudinally and twisted. Stamens epipetalous, included, alternate with corolla lobes; anthers 2-celled, longitudinally dehiscent. Pistil 1, ovary superior, 2-celled (to 4-celled), 2 ovules per cell; styles 1 or 2. Fruit usually a capsule with 2–4 valves.

Style 1, entire or cleft only at apex; twining herbs; corolla lobes plicate and twisted in bud
Convolvulus
Styles 2, distinct; corolla lobes imbricate in bud.
  Low, canescent perennial herbs .... *Cressa*
  Twining, leafless parasites; stems yellow or orange .... *Cuscuta*

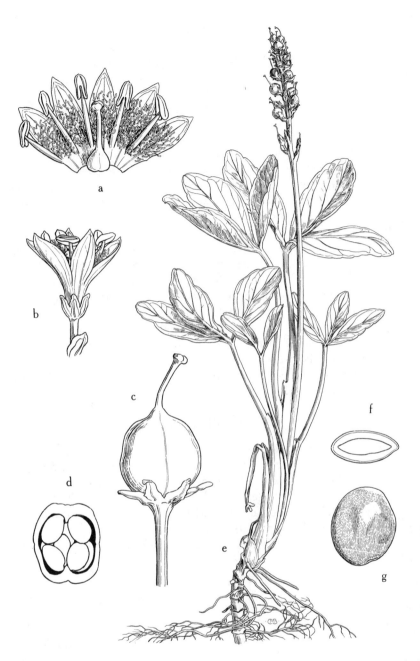

Fig. 293. *Menyanthes trifoliata: a,* flower, spread open, × 2; *b,* flower, × 2; *c,* fruit, × 3; *d,* fruit (cross section), × 3; *e,* habit, × ⅓; *f,* seed (cross section), × 12; *g,* seed, light brown, shiny, × 12.

CONVOLVULUS. MORNING-GLORY

1. Convolvulus sepium var. **repens** (L.) Gray, Man. 348. 1848.

Rootstock slender, horizontal; stems slender, twining, 5–10 dm. long; herbage glabrate to pubescent; leaf blades lanceolate or ovate-lanceolate, acuminate apically, sagittate at base, 4–8 cm. long, petioles slender, shorter than blade; peduncles as long as or longer than leaves, 1-flowered; bracts ovate, acuminate, cordate at base, longer than calyx and enclosing it; corolla pink to white, 4–5 cm. long.

Saline marshes: San Francisco Bay area, coastal southern California; Atlantic coast, Europe.

Convolvulus sepium var. *Binghamiae* (Greene) Jepson (Fl. Calif. 3:118, 1939) occurs very locally in coastal southern California from Santa Barbara south to Orange County. It has finely puberulent herbage, small leaves, and bracts which only half conceal the calyx. It has been considered as a distinct species (*C. Binghamiae* Greene, Bull. Calif. Acad. Sci. 2:417, 1887), and is so treated by Abrams (Illus. Fl. Pacif. States 3:383, 1951).

CRESSA

1. Cressa truxillensis H.B.K., Nov. Gen. & Sp. 3:119. 1818. Alkali weed.
  Cressa cretica of western American authors, not L.

Perennial, low, tufted herb; stems 6–25 cm. tall; herbage gray, strigose; leaves oblong-ovate, 4–10 mm. long, entire, nearly sessile; flowers solitary in the axils, short-pediceled; sepals oblong-ovate, acute, 4 mm. long; corolla white, 6 mm. long, the lobes oblong-ovate, 2 mm. long, imbricate in the bud, pubescent without; ovary 2-celled, 4-ovuled (usually only 1 ovule develops), long-hirsute at summit; styles 2, distinct; stigmas capitate; seeds flattened, irregular in outline, reddish brown, dull, microscopically pitted.

Lowland alkaline areas: throughout California; southeastern Oregon, east to Utah and Arizona, Mexico, South America.

CUSCUTA. DODDER

1. Cuscuta salina Engelm. in Gray, Bot. Calif. 1:536. 1876. Salt-marsh dodder.
  Fig. 294.

Parasitic herb lacking chlorophyll; stem twining on host, delicate, slender, yellow or orange; leaves reduced to minute scales, these alternate; flowers in cymose clusters, 2–3 (–4.5) mm. long, campanulate, numerous, 5-merous; calyx 2–3 mm. long, lobes ovate-lanceolate, acute; corolla about as long as calyx, the lobes as long as the tube, ovate, acute; scales of corolla tube narrow,

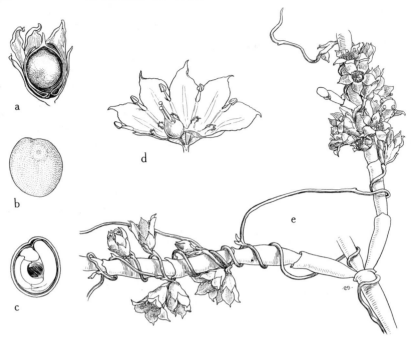

Fig. 294. *Cuscuta salina: a,* fruit (longitudinal section), × 6; *b,* seed, × 8; *c,* seed (longitudinal section), × 8; *d,* flower, spread open, × 4; *e,* habit, plant growing on *Salicornia,* × 2.

oblong, fringed, closely attached to tube below stamens; anthers oval, about as long as the short, subulate filaments; ovary ovoid, 2-celled, with 4 ovules; styles 2, distinct, persistent, slender, shorter than or about as long as ovary; capsule indehiscent, mostly 1-seeded, surrounded by withered corolla.

Parasitic on plants growing in saline areas (*Cressa, Suaeda, Salicornia,* and others) in coastal or sometimes inland salt marshes: Humboldt County to southern California; north to British Columbia, east to Utah and Arizona.

Yuncker, T. G., The genus *Cuscuta.* Mem. Torrey Club 18:168–169. 1932.

Plants with flowers larger and more broadly campanulate can be separated as *Cuscuta salina* var. *major* Yuncker. This variety is found in coastal marshes mainly on *Salicornia;* it intergrades with the typical form of the species.

POLEMONIACEAE

Ours annual or perennial herbs. Leaves alternate or opposite, entire, lobed, pinnately or palmately divided, or pinnately compound. Inflorescence usually a paniculate, glomerate, or flat-topped cyme, or the flowers congested in densely bracteate heads, rarely flowers solitary. Flowers regular. Calyx syn-sepalous, herbaceous to variously membranous, 5-lobed. Corolla sympetalous, usually 5-lobed (sometimes 4-lobed). Stamens 5 (sometimes only 4), inserted on corolla, alternate with its lobes. Pistil superior; ovary usually 3-celled (sometimes 1- or 2-celled); style with 3 (occasionally 1–4) stigma lobes. Capsule usually 3-valved, usually regularly dehiscent, sometimes irregularly so or indehiscent.

Annuals; leaves pinnate or pinnatifid to entire, the segments rigid, subulate, spinose; calyx membranous or scarious between the lobes .... *Navarretia*
Ours a perennial herb; leaves pinnately compound, the leaflets not rigid and subulate; calyx entirely herbaceous .... *Polemonium*

NAVARRETIA

Erect, spreading, or prostrate annuals, 1–30 cm. high, with rigid stems, simple or divaricately branched, occasionally with several branches proliferating from beneath terminal heads. Herbage glandular-puberulent to villous. Leaves alternate, entire to pinnately or palmately toothed, cleft, or lobed, becoming bracteate above, acerose or spinose-tipped; lobes often proliferating, or supplemented by extra lobes arising at base or from back of rachis. Bracts usually with broadened, coriaceous rachis, with or without proliferating lobes, the inner bracts reduced. Flowers sessile or subsessile in dense, spiny-bracted heads. Calyx cleft to base into unequal, entire or toothed, usually acerose lobes, united in sinuses in lower ¼–¾ by a scarious membrane to form a pseudotube. Corolla funnelform or salverform, white, yellow, blue, violet, or pink, 5-lobed or rarely 4-lobed, 4–19 mm. long, with tube 1–10 mm. long. Stamens equally or unequally inserted in throat or in sinuses of corolla lobes, included or exserted; filaments glabrous at base. Stigma entire, or 2- or 3-lobed, included or exserted. Capsule ovoid or obovoid, 1- to 3-celled, 3- to 8-valved, with regular or irregular dehiscence, usually with circumscissile dehiscence about the base and thence splitting into valves from the base upward, or, less commonly, breaking irregularly when wetted. Seeds brown, minutely pitted, ovoid, or irregularly angled, 1 to many in each cell.

Capsule at length thin-membranous, the walls disintegrating on dehiscence, not splitting into discrete valves but sometimes irregularly circumscissile.
  Stamens inserted about midway on the corolla throat; corolla white or cream-colored; plants erect, caespitose, or decumbent.
    Bracts highly dissected, usually some lobes proliferating from the back; corolla 7–10 mm. long; plants erect or corymbosely spreading, rarely becoming prostrate .... 1. *N. leucocephala*
    Bracts pinnately lobed, the lobes all marginal or submarginal, the terminal lobe often very much longer than lateral lobes; corolla 4–6 mm. long; plants caespitose or decumbently spreading .... 2. *N. minima*

Stamens inserted in or immediately below the sinuses of the corolla lobes; corolla blue to purple or white; plants prostrate or decumbent, rarely erect.

Bracts within the heads pinnately lobed or dissected, occasionally with only a few short lobes proliferating from the base; foliaceous bracts 1–2 times as long as the head; corolla 5–6 mm. long.

Stems 0.2–0.5 mm. in diameter; heads 2- to 10-flowered; spongy cortex of the hypocotyl 2–4 times as thick as the vascular cylinder . . . . . . 3. *N. pauciflora*

Stems 0.9–1.5 mm. in diameter; heads 20- to 50-flowered; spongy cortex of the hypocotyl as thick as or thinner than the vascular cylinder.

Plant with prostrate habit; flowers blue . . . . 4. *N. plieantha*

Plant not prostrate, but erect or spreading; flowers white . . . . 5. *N. Bakeri*

Bracts within the heads pinnately toothed, rarely somewhat lobed; foliaceous bracts 2–5 times as long as the head; branches prostrate, proliferating radiately from below the terminal, usually acaulescent head; corolla 7.5–8.5 mm. long

6. *N. prostrata*

Capsule walls thicker, regularly dehiscent from the base upward.

Corolla typically 4-merous, yellow; stamens inserted in the sinuses of the petals

7. *N. cotulaefolia*

Corolla 5-merous, tube white, lobes blue with purple spot at base of each; stamens inserted on upper ½ of throat . . . . 8. *N. Jepsonii*

1. Navarretia leucocephala Benth., Pl. Hartw. 324. 1849.

Erect to rarely prostrate annual, 3–17 cm. high; stems white or reddish-streaked, simple or racemosely branched, rarely proliferating from below the heads, glabrous to sparsely pubescent below, becoming densely pubescent above, with white, retrorse, crisped hairs; lower leaves linear, pinnately lobed or entire, 1–8.5 cm. long, glabrous, the lobes blunt to cuspidate, 3–10 mm. long; upper leaves pinnate to bipinnate, 0.5–5 cm. long, the lobes cuspidate to acerose, glabrous to sparsely pubescent below; flowers sessile or subsessile in heads; bracts 0.4–1.5 cm. long, pinnate to bipinnate, proliferating on abaxial side between lobes, often ciliate-membranous-winged on margins and between lobes, outer bracts foliaceous; calyx 5–7 mm. long, cleft to base, the sepals acerose, entire, pubescent about midway on abaxial side, sinuses about ¾ filled with membrane which is truncate and ciliate across top; corolla funnel-form, 7–9.5 mm. long, white, the tube 3.5–5.5 mm. long, the throat 1.5–2 mm. long, the lobes 1.5–2 mm. long; stamens inserted equally at about middle of throat, 2–5 mm. long, exserted; style exserted; stigma less than 0.25 mm. long, 2-lobed; capsule ovoid, 2.5 mm. long, indehiscent, the locules 4- to 6-seeded.

Vernal pools: northern California and south in Inner Coast Ranges to San Benito County, Central Valley, Sierra Nevada foothills from Butte County to El Dorado County; southern Oregon.

2. Navarretia minima Nutt., Jour. Acad. Nat. Sci. Phila. II. 1:160. 1848.

Prostrate to suberect annual, 2.5–10 cm. high, with a spread of 1–11 cm.; stems white or sometimes reddish-tinged, simple, caespitose or divaricately branched, pubescent with white, crisped, retrorse hairs; lower leaves linear, entire, or pinnately dissected, cuspidate, 1–1.5 cm. long, glabrous, the upper ones 1–2.5 cm. long, linear, entire or pinnate with 1–3 pairs of lobes, acerose; inflorescence capitate; bracts 0.7–1.7 cm. long, pinnate with 1 to several pairs of acerose lobes, the base of rachis membranous-winged, pubescent inside at base; calyx 4–5 mm. long, cleft to base, the sepals usually entire, unequal in size, acerose, sparsely pubescent on abaxial side below, the sinuses ½ or more filled by membrane; corolla funnelform, 4–6 mm. long, white, the tube 2–2.5 mm. long, the throat 1–1.5 mm. long, the lobes 1–2 mm. long; stamens equally inserted in middle of throat, 2.5–3 mm. long, exserted; style exserted; stigma less than 0.25 mm. long, 2-lobed; capsule ovoid, 2 mm. long, indehiscent.

Boggy or dry places: northeastern California from Sierra County to Modoc County; north to eastern Washington.

3. Navarretia pauciflora Mason, Madroño 8:200. 1946.

Prostrate annual, 1–4 cm. high and 2–8 cm. in diameter; hypocotyl with a thick, spongy cortex; stems slender, filiform, 0.2–0.5 mm. thick, white with streaks of purple, densely clothed with short, white, retrorse, crisped hairs or almost glabrous; leaves 1–2.5 cm. long, linear and entire or pinnately parted into 1 or 2 pairs of linear, cuspidate lobes each about 2 mm. long, glabrous; outer bracts foliaceous, few, 1½–3 times as long as the head, with several pairs of lobes below the middle, membranous-winged below, the inner bracts scarcely surpassing the calyx, the membranes ciliate-margined, the lobes acerose to cuspidate; flowers sessile or subsessile in 2- to 10-flowered heads, the heads 4–10 mm. broad; calyx cylindric, 4–5 mm. long, membranous except for the lobes and the narrow band of tissue below them, this often reduced to a single vascular strand, the membrane in the sinus truncate across the top and ciliate on the upper margin, the lobes pubescent within; corolla funnelform, 5–6 mm. long, blue or white, fading blue, the tube 3 mm. long, the throat 1.5 mm. long, the lobes 1.5 mm. long; stamens inserted in the sinuses of corolla lobes, equal in length, somewhat longer than the petals and well exserted from throat; stigma exserted, 2-lobed, the lobes minute; capsule irregularly dehiscent, the somewhat thickened top falling away irregularly from the membranous sidewalls; seeds 1 to several, minutely pitted, reddish brown.

Known only from the type locality: vernal pools in volcanic rubble, five miles north of Lower Lake, Lake County.

4. Navarretia plieantha Mason, Madroño 8:199. 1946.

Prostrate annual forming a mat 5–20 cm. broad with several stout branches, but not proliferating from below a central head, the main axis often with crisped, retrorse hairs, the lateral stems glabrate, the epidermis often exfoliating as a white, membrane-like tissue; leaves 3–4 cm. long, linear and entire or pinnate with a few remote filiform lobes; outer bracts foliaceous, 3 or 4 to each head, 1–2 times as long as the head, pinnate, the lobes often 2–4 times proliferated or the bract once-pinnate, the rachis flanked by a ciliate membrane

Fig. 295. *Navarretia Bakeri: a,* flower, showing hairs at margins of calyx membrane, × 8; *b,* habit, × ⅘; *c,* capsule, 1- to 3-seeded, × 10; *d* and *e,* seeds, × 20; *f,* flower, the aristate calyx lobes unequal in length, × 8; *g,* calyx in fruit, × 6; *h,* floral bract, inner side, × 8.

below; bracts within the inflorescence with from 1 to several pairs of lobes below the middle, entire above or with a pair of acerose teeth; flowers in heads 1.5–2 cm. broad, the heads 20- to 50-flowered; calyx somewhat constricted above, 4–5 mm. long, membranous throughout except for the herbaceous lobes and a line of herbaceous tissue immediately below the lobes, glabrous, or with a few weak hairs, except for the ciliate margin of the truncated membrane in the sinus of calyx lobes; corolla 5–6 mm. long, blue, funnelform, the tube 3–3.5 mm. long, included in calyx tube, the throat 0.5 mm. long, the lobes 2 mm. long; stamens inserted in the sinuses of the corolla lobes, 2.5 mm. long; stigma exserted, 2-cleft to 2-lobed or entire; capsule not regularly dehiscent, the somewhat thickened top breaking away irregularly from the membranous walls when wetted, the seeds working out of the constricted orifice of the calyx and resting on top; seeds about 3 to each capsule, reddish brown and minutely pitted.

Known only from the type locality: Boggs Lake, northwest slope of Mount Hannah, Lake County.

5. Navarretia Bakeri Mason, Madroño 8:198. 1946. Fig. 295.

Erect spreading annual 2–5 cm. tall; stems racemosely branched, 0.5–1.5 mm. thick, densely clothed with retrorse, crisped hairs; lower leaves linear, entire to few-toothed or pinnatifid, the upper leaves dissected, the lobes often proliferating, glabrate below, pilose with short, crisped hairs above; outer bracts foliaceous, pinnatifid with highly dissected proliferations; bracts within head pinnate with 1 or 2 pairs of teeth in upper ⅓ and 1 pair of lobes below middle, with proliferating lobes from the base of the bract or from the abaxial surface of the rachis; flowers in heads; calyx lobes unequal in length, the longest lobes 5.5 mm. long, slender, aristate, with a few weak hairs, membranous to base in sinuses, the free margin of membrane ciliate; corolla white, 5–7 mm. long, the tube 4 mm. long, the throat 0.5–1 mm. long, the lobes 1–1.5 mm. long; stamens inserted in the sinuses of corolla lobes, 2.5 mm. long, exserted from throat; style exserted; stigma minutely 2-lobed; capsule about 2 mm. long, the somewhat thickened top breaking away with irregularly circumscissile dehiscence from the membranous base; seeds few, minutely pitted, reddish brown.

Vernal pools in meadows: inner North Coast Ranges from Lake County to Trinity County.

6. Navarretia prostrata (Gray) Greene, Pittonia 1:130. 1887. Fig. 296.

Gilia prostrata Gray, Proc. Am. Acad. Arts & Sci. 17:223. 1881.

Prostrate annual with spread of 8–15 cm.; branches proliferating from beneath a terminal acaulescent head; stems densely pubescent with white, retrorse hairs below heads, becoming almost glabrous at base; leaves 3–7 cm. long, pinnately to bipinnately lobed, with linear rachis, the longest lobes above,

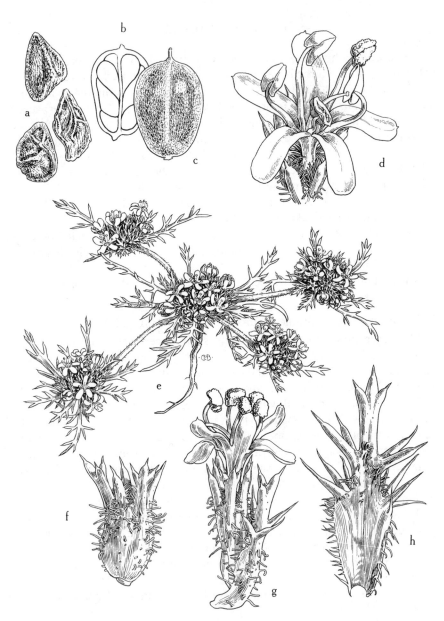

Fig. 296. *Navarretia prostrata: a,* seeds, × 20; *b,* capsule, diagram (longitudinal section), showing arrangement of seeds, × 10; *c,* capsule, × 10; *d,* flower, × 8; *e,* habit, × 1; *f,* fruiting calyx, × 6; *g,* flower and subtending bract, the calyx with some of its teeth trifid, × 6; *h,* floral bract, × 6.

glabrous; inflorescence capitate; outer bracts foliaceous, once-pinnate, 1–4 cm. long, glabrous or sparingly pubescent ventrally, the inner bracts pinnately toothed above, often trifid, at base coriaceous, ciliate, membranous-margined and pubescent; flowers sessile; calyx 4–5 mm. long, cleft to base, the sepals entire or some trifid at apex, pubescent on abaxial surface mostly below middle, united by a sinus membrane for about ⅔ of their length to form a pseudotube, the membrane ciliate at top; corolla broadly funnelform, white to violet, 7.5–8.5 mm. long, the tube 5–5.5 mm. long, the throat 1–1.5 mm. long, the lobes 1.5–2 mm. long; stamens equally inserted in sinuses, 2.5 mm. long, exserted; style exserted; stigma minute, slightly 2-lobed; capsule 2-celled, indehiscent, many-seeded.

Vernal pools and low places in valleys and foothills: western Merced County and southern Monterey County, through coastal southern California to San Diego County.

7. Navarretia cotulaefolia (Benth.) Hook. & Arn., Bot. Beechey Voy. 368. 1840.

Erect annual, 4–25 cm. tall; stems reddish, simple or divaricately branched, puberulent to glabrate; lower leaves 2–3.5 cm. long, pinnate or bipinnate with about 5 pairs of minutely cuspidate lobes, but often entire below the middle, the rachis and lobes linear, puberulent; upper leaves 1.5–4.5 cm. long, becoming bracteate, with the rachis somewhat broader than in the lower leaves; inflorescence capitate; bracts to 1 cm. long, broad and coriaceous, with 2–7 pairs of acerose lobes, coarsely villous with broad white hairs; flowers sessile; calyx 5.5–7 mm. long, cleft to base into 4 lobes which are unequal in length (2 long and 2 short), the longer lobes usually with a pair of acerose teeth near apex, puberulent, but with a few long coarse hairs about midway, the sinus membrane forming a pseudotube 3–4 mm. long; corolla funnelform, 4-lobed, yellow to cream-colored, 9–10.5 mm. long, the tube 5–6 mm. long, the throat 2–3 mm. long, the lobes 1–1.5 mm. long; stamens 4, inserted equally in or below the sinuses, 3 mm. long, exserted; styles exserted; stigma 2-lobed, to 0.5 mm. long; capsule 4-valved, 1-celled, dehiscent from base, obovoid, 2 mm. long; seeds obovoid, light brown, 1 in each capsule.

Low places in valleys: Inner Coast Ranges from Mendocino County to San Benito County, inland to western borders of Sacramento Valley.

8. Navarretia Jepsonii V. Bailey ex Jepson, Fl. Calif. 3:154. 1943.

Erect, ascending, or more often broadly spreading annual; herbage often with a reddish or purple pigment; leaves finely dissected into short, linear, acerose lobes which are forked and often proliferated from the base; inflorescence capitate; bracts with a broad, oblong rachis, finely dissected into linear, acerose lobes, coarsely pilose with papery multicellular hairs, the outer bracts red or red-tipped; calyx deeply cleft into linear acerose lobes, the longer lobes sometimes with a few lateral teeth near the tip, the sinuses with a membrane in the lower part which ascends the margins of the lobes, the lobes coarsely pilose, the tube puberulent; corolla funnelform, 10 mm. long, the tube 5 mm. long, white, the throat 2.5 mm. long, ample and rounded, white, the lobes 2.5 mm. long, blue with a purple spot at base of each; stamens inserted at base of throat, filaments 8 mm. long, anthers 1 mm. long, white; style long-exserted; stigma 2-lobed; capsule obovoid, thin-walled, almost membranous, circumscissile about the base and splitting upward into 8 valves.

Boggy or dry places: Coast Ranges from Colusa County to Napa County, and eastern borders of Sacramento Valley.

POLEMONIUM

1. Polemonium occidentale Greene, Pittonia 2:75. 1890.

Perennial herb 3–9 dm. tall; stems simple, solitary, erect from a decumbent base or horizontal rootstock, glabrous to finely pubescent; leaves alternate, pinnate, the leaflets in 7–13 pairs, these sessile, ovate-lanceolate, apically acute, rounded and oblique at base, the upper 3 leaflets often fused, 1–3 cm. long, 2–8 mm. broad, glabrous; inflorescence cymose; bracts entire to pinnatifid, 2–7 mm. long, glandular-pubescent; pedicels short, subequal to the calyx in length; calyx herbaceous throughout, 4–10 mm. long, the lobes subequal to the tube in length or commonly slightly longer, glandular-pubescent; corolla rotate-campanulate, 1.5–2 cm. broad, blue, the broadly ovate lobes slightly longer than the tube; stamens inserted near middle of corolla tube and not exserted beyond the corolla lobes, the filaments pubescent at base; ovary and style conspicuously longer than corolla.

Wet places, at elevations of 2,000–10,000 feet: San Bernardino Mountains, Sierra Nevada; north to British Columbia and east to Rocky Mountains.

BORAGINACEAE. BORAGE FAMILY

Ours herbs, usually bristly. Leaves simple, alternate, rarely opposite or whorled, commonly entire and pubescent, hispid or setose. Flowers perfect and regular, in one-sided spikes or racemes, coiled spirally and uncoiling as flowering proceeds. Calyx commonly 5-lobed or 5-parted, persistent and often accrescent in fruit. Corolla 5-lobed, the throat often crested. Stamens 5, inserted on the tube or throat of corolla and alternate with its lobes and usually included; filaments short. Ovary superior, 4-celled, shallowly or deeply 4-lobed, the style simple, terminal or inserted between the lobes. Fruit mostly of 4 1-seeded nutlets.

Ovary undivided or shallowly lobed, the style borne on its summit, the stigma capitate
Heliotropium
Ovary 4-lobed, the style borne on the gynobase and arising between the lobes of the ovary, the stigma obscurely lobed.
Receptacle flat or merely convex; nutlets attached by their bases .... *Myosotis*
Receptacle more or less elongated and produced upward into a slender structure to which the nutlets are usually attached laterally .... *Allocarya*

ALLOCARYA

Low, spreading annuals (except *Allocarya mollis*), commonly branching from near base. Leaves linear, entire, at least the lowest ones opposite. Flowers small, in bractless or bracteate terminal spikes or racemes. Pedicels persistent, thickened at the summit. Calyx persistent, 5-parted, the segments narrow, indurated, somewhat elongate in fruit. Corolla white, salverform, the tube short, the throat yellow with crests usually present. Stamens included. Ovary deeply 4-lobed, the style short, inserted between the lobes. Nutlets ovoid to ovoid-lanceolate, smooth, variously roughened, or some with prickles, the attachment scar on ventral side of the nutlets basal or above the base, the scar concave or sometimes raised.

Several species not described in this flora sometimes occur in marshy habitats in California: *Allocarya hispidula* Greene, *A. lithocarya* Greene, *A. cognata* Greene, *A. Greenei* (Gray) Greene, *A. leptoclada* Greene, *A. Chorisiana* Greene, *A. trachycarpa* (Gray) Greene.

Johnston, Ivan M., Studies in the Boraginaceae—IX, Contr. Arnold Arb. 3:5–82. 1932.

Plants perennial, covered with dense, long, soft hairs .... 1. *A. mollis*
Plants annual, slender.
Stems floriferous to near the base, prostrate; lower pedicels stout and recurved; calyx lobes with hardened midrib, irregularly spreading or recurved in age; nutlets lanceolate-ovoid, dull, granulate and tuberculate, scar extending along 1/5 of the length of the nutlet .... 2. *A. humistrata*
Stems not floriferous to base, or, if so, the pedicels not stout and recurved.
Scar large, lateral, deeply excavated; nutlets with or without prickles.
Nutlets broad (about 2/3 as broad as long), dorsal (abaxial) keel and transverse ridges with long and slender prickles, glochidiate at apex
3. *A. acanthocarpa*
Nutlets slender (about 1/2 as broad as long), very angular, dorsal (abaxial) surface between keel and margins transversely rugose, without prickles
4. *A. glyptocarpa*
Scar small, slightly if at all excavated (sometimes becoming concave because of upturned margins).
Nutlet attachment exactly or almost basal, often substipitate; plant somewhat succulent .... 5. *A. stipitata*

Nutlet attachment lateral or rarely obliquely basal.
   Scar linear or nearly so, or rarely cuneate and sessile, lying in the basal, ventral
      (adaxial) longitudinal groove .... 6. *A. undulata*
   Scar broad, not linear.
      Scar in a cavity which is broader than long; sharp edges of the scar sometimes
         incurved, nutlets glossy, not granulate .... 7. *A. Cusickii*
      Scar in a cavity which is longer than broad; nutlets granulate.
         Corolla 3–7 mm. broad; scar of nutlet lateral, lying in plane of ventral
            keel .... 8. *A. tenera*
         Corolla 1–3 mm. broad; scar of nutlet and ridge surrounding it more or
            less oblique to the plane of ventral keel, almost basal
                                      9. *A. bracteata*

1. Allocarya mollis (Gray) Greene, Pittonia 1:20. 1887.

Plagiobothrys mollis Johnst., Contr. Gray Herb. 68:74. 1923.

Perennial with a fleshy taproot, covered throughout with long, soft, mostly spreading hairs, 1–5 dm. tall; stems several from base, ascending or trailing; leaves numerous, opposite, linear to linear-spatulate, obtuse or rounded; racemes mostly solitary at the ends of the branches, sometimes bracteate below; mature calyx 4–5 mm. long, the lobes lanceolate; corolla 5–10 mm. broad; nutlets ovoid, about 1.5 mm. long, gray, the dorsal side with median keel distinct only near apex but occasionally extending to the middle, the transverse ridges irregular, merging at their ends to form an indefinite lateral ridge, or sometimes replaced by tubercles toward the base, the ventral side with conspicuous ovate or triangular scar.

Moist alkaline flats and borders of ponds: east side of the Sierra Nevada from Sierra County to Modoc County; adjacent western Nevada, northward to northeastern Oregon.

2. Allocarya humistrata Greene, Pittonia 1:16. 1887.

Plagiobothrys humistratus Johnst., Contr. Gray Herb. 68:77. 1923.
Allocarya humistrata var. *similis* Jepson, Man. Fl. Pl. Calif. 853. 1925.

Stems several from the base, mostly prostrate, rather stout, glabrous or sparsely puberulent, somewhat succulent; branches floriferous to base; leaves linear, 1–2 cm. long; spikes 5–10 cm. long, rather remotely flowered in age; fruiting pedicels stout, erect or sometimes spreading or slightly deflexed; calyx elongate in fruit, 6–10 mm. long, erect, sometimes all turned to one side and thus forming a single row; corolla inconspicuous, 1–2 mm. broad; nutlets ovoid, the dorsal side sparsely short-bristly and tuberculate, the ventral side keeled and sparingly rugulose; scar ovate-deltoid, subbasal.

Low places, hog wallows: Livermore Valley, Central Valley.

3. Allocarya acanthocarpa Piper, Contr. U. S. Nat. Herb. 22:87. 1920.

Plagiobothrys acanthocarpus Johnst., Contr. Arnold Arb. 3:33. 1932.

Stems slender, usually branched below, spreading or erect, 1–4 dm. long, strigose; lower leaves linear to spatulate-linear, 2–6 cm. long, the upper leaves and bracts linear to narrowly oblong; racemes bracteate, becoming loose and elongate; mature calyx 3–6 mm. long, in age often stellately spreading; corolla scarcely longer than the calyx, 1–2.5 mm. broad; nutlets ovoid, mostly 1.5–2 mm. long, contracted toward the apex, the dorsal side keeled, reticulate with thin ridges, the keel and ridges armed with minutely barbed, subulate spines, the interspaces tuberculate, the ventral side distinctly keeled down to the scar, the sides bearing transverse ridges, the keel and ridges unarmed, the scar deeply excavated, ovate or deltoid.

Vernal pools and adobe flats: lower Sacramento Valley, San Joaquin Valley and southern South Coast Ranges, south to San Diego County; northern Baja California, Mexico.

4. Allocarya glyptocarpa Piper, Contr. U. S. Nat. Herb. 22:80. 1920.

Plagiobothrys glyptocarpus Johnst., Contr. Arnold Arb. 3:37. 1932.

Stems slender, branching at or near the base, ascending, 1–5 dm. tall; herbage strigose; leaf blades linear to narrowly spatulate, 4–8 cm. long; racemes simple, loosely flowered, elongate, bracteate near the base; calyx lobes becoming 3–5 mm. long; corolla 5–9 mm. broad; nutlets narrowly ovoid, beaked or abruptly acute, about 2 mm. long, the dorsal side strongly keeled, the transverse ridges sharp and irregular, angular, the interspaces finely tuberculate, the ventral side with a high, thin keel down to the pitlike scar, the sides with diagonal ridges, the scar deeply excavated, narrowly triangular, nearly half as long as nutlet.

Moist ground along streams: Lake and Butte counties; north to Jackson County, Oregon.

5. Allocarya stipitata Greene, Pittonia 1:19. 1887.

Stems few or usually many, branching at or above the base, somewhat succulent, spreading, commonly 1–5 dm. tall; herbage greenish yellow and shining, appearing glabrous but really very finely strigulose; leaf blades linear to linear-oblanceolate, 3–8 cm. long, attenuate into a broad petiole-like base; racemes at length elongate, mostly one-sided, leafy-bracted below; pedicels slightly stout; calyx lobes at length brownish and often spreading, strongly elongate in age, becoming 5–8 mm. long; corolla 5–12 mm. broad, white with yellow eye; nutlets narrowly ovoid, 1.5–2.5 mm. long, the dorsal side somewhat flattened, keeled above the middle, transversely rugose and finely tuberculate, the ventral side strongly keeled down to the base, the apex of nutlet beaked, the scar basal, small, sessile, and obscurely stipitate.

Hog wallows and damp depressions of plains and valley flats: Coast Range valleys from San Benito County to Napa County, lower San Joaquin Valley, Sacramento Valley; north to Oregon.

Allocarya stipitata var. *micranthus* Piper, a common plant of the Central Valley, differs only in its smaller corolla (2.5 mm. broad) and occurs throughout California and into southeastern Oregon.

A closely related species, *A. glabra* (Gray) Macbride, of fleshy, stout, fistulose habit, is found in salt marshes and alkaline flats in Alameda and Santa Clara counties and also in San Benito County.

6. Allocarya undulata Piper, Contr. U. S. Nat. Herb. 22:104. 1920.

Plagiobothrys undulatus Johnst., Contr. Arnold Arb. 3:46. 1932.

Stems 1 or several from the base, 1–3 dm. tall, erect at first but sprawling in age; herbage thinly strigose or ascending-hirsute; lower leaves oblong to linear, 3–6 cm. long; racemes loose, slender, with some flowers bracted below the middle; calyx densely hirsute, somewhat elongate; corolla 1.5–2 mm. broad; nutlets ovoid, the dorsal side somewhat flattened, keeled toward the apex and transversely rugose with crowded, wavy ridges above, these reduced to tubercles at base, finely granulate, the ventral side puncticulate or rugulose, keeled from the apex to just below the middle, the scar linear, rarely cuneate-linear, about ⅕ the length of the nutlet, usually set in an elongate depression.

Moist places and beds of vernal pools in valleys and mesas near the coast: San Diego County to Marin County.

7. Allocarya Cusickii Greene, Pittonia 1:17. 1887.

Plagiobothrys Cusickii (Greene) Johnst., Contr. Arnold Arb. 3:63. 1932.

Stems branched from the base, slender, prostrate or erect, 5–20 cm. long; herbage sparsely strigose; leaves linear, 3–10 cm. long, the lower surface pustulate-hairy, the upper surface nearly or quite glabrous, the hairs often more abundant and more spreading on the margins; racemes loosely or sometimes densely flowered, bracteate, at least below the middle; pedicels slender, 1 mm. long or less; calyx finely appressed-hispidulous, only slightly elongated in age, the lobes 1.5–4 mm. long; corolla 1–1.5 mm. broad; nutlets lanceolate to oblong-ovoid, 1–2 mm. long, usually abruptly angled at base, the dorsal side glossy, not granulate, keeled near the apex, with irregular (oblique) ridges, tuberculate in the interspaces and toward the base, the ventral side keeled to well below the middle, bearing a deep, small, lateral scar, the scar ovate or deltoid on the axial nutlet, the scars of the other nutlets cuneate to oblong to linear with sharp, erect, or infolding scar margins.

Saline marshes and along margins of moist meadows and irrigation ditches: northeastern California; adjacent Idaho and Nevada, northward to east side of the Cascades in Oregon and Washington.

8. Allocarya tenera Greene, Pittonia 3:109. 1896.

Plagiobothrys tener (Greene) Johnst., Contr. Arnold Arb. 3:66. 1932.

Stems branching from the base, spreading or ascending, very slender, 1–3 dm. long, sparsely and inconspicuously strigose; basal leaves narrowly linear-oblanceolate, 2–6 cm. long, the stem leaves few; racemes slender, loosely flowered, if bracted, only bracteate at base; pedicels, at least the lower ones, slender, 1–2 mm. long; corolla 3–7 mm. broad; fruiting calyx tube distinctly 4-ribbed, the lobes erect, narrowly linear-lanceolate; nutlets ovoid, about 1 mm. long, the dorsal side keeled toward apex, reticulate-rugulose, the ridges thin and tuberculate-roughened on the margins, the interspaces finely tuberculate and sometimes with short bristles, the ventral side diagonally rugulose, keeled above the scar, the scar lanceolate-ovate, extending from the base almost to the middle of the nutlet.

Moist meadows or stream borders: Lake County, north in the Coast Ranges and east to Modoc and Shasta counties.

9. Allocarya bracteata Howell, Fl. NW. Am. 481. 1901.

Allocarya Cusickii var. *vallicola* Jepson, Fl. Calif. 3:364. 1943.
Plagiobothrys bracteatus Johnst., Contr. Arnold Arb. 3:68. 1932.

Stems branched from the base, decumbent or ascending, rarely erect, 1–4 dm. high, usually rather thinly strigose; leaves linear, the lower ones 4–10 mm. long; racemes slender and elongate in age, bracteate below; pedicels about 1 mm. long or the lowermost longer; calyx more or less elongate in age, the lobes lanceolate, 2–4 mm. long; corolla 1–3 mm. broad; nutlets about 2 mm. long, ovoid, the dorsal surface somewhat keeled above the middle, granulate, the sides with oblique transverse wrinkles, those below the middle becoming obscure and often replaced by small tubercles, the interspaces narrow and often sparsely tuberculate, the ventral side keeled to well below the middle, the scar small, obliquely situated or almost basal.

Chiefly in wet places: southern California north through the valleys to southwestern Oregon, northern Baja California, Mexico.

HELIOTROPIUM

1. Heliotropium curassavicum L., Sp. Pl. 130. 1753. Seaside heliotrope. Fig. 297.

Annual or short-lived perennial herb, fleshy, glaucous and glabrous throughout; stems diffusely branching, 1–6 dm. high; leaves succulent, entire, linear to obovate to spatulate, 1–4 cm. long, obtuse, narrowed to a thick petiole; scorpioid spikes usually 2 (sometimes 3–5); calyx 5-lobed, linear; corolla funnelform, 3–5 mm. long, white, the throat open and bearing a violet-purple

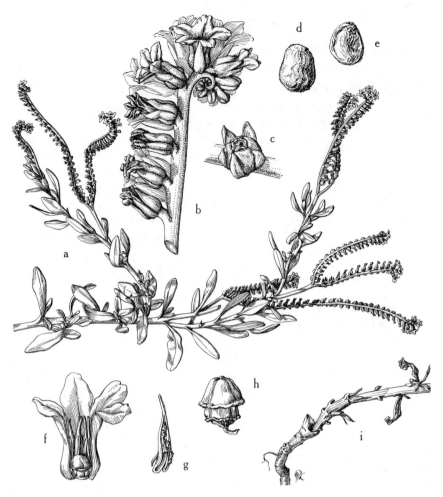

Fig. 297. *Heliotropium curassavicum: a,* branch, showing semiprostrate habit, × ⅔; *b,* apex of inflorescence, × 3; *c,* ovary, × 3; *d,* nutlet, abaxial view, × 12; *e,* nutlet, adaxial view, × 12; *f,* flower, with 1 lobe and 1 anther removed, × 6½; *g,* anther, × 12; *h,* pistil, × 12; *i,* basal part of stem and stout root, × ⅔.

eye, the lobes 5, imbricated and bent inward at the tip; stamens included, the anthers acuminate, nearly sessile; ovary 4-celled; stigma capitate, glabrous; fruit depressed-ovoid, at length separating into 4 nutlets.

Marshes, stream beds, and alkaline flats or plains: southern California and north through the state; north to Washington, east to Atlantic, and south into Mexico.

MYOSOTIS

Slender annual or perennial herbs with alternate leaves. Flowers small, blue, pink, or white, in many-flowered, one-sided racemes, bractless or sometimes leafy-bracted at base. Calyx 5-cleft, the narrow lobes spreading or erect in fruit. Corolla short-salverform, 5-lobed, the lobes rounded, the throat crested. Stamens 5, included, inserted on the corolla tube. Ovary 4-divided; style filiform. Nutlets ovoid, glabrous or pubescent, attached by their bases to the gynobase, the scar small, flat.

Myosotis virginica (L.) B.S.P. and *M. versicolor* (Pers.) Smith are occasionally found at the margins of moist meadows.

Stems coarse, angled, often stoloniferous at base; corolla 6–9 mm. broad
1. M. scorpioides
Stems slender, terete, branches at base, without stolons; corolla 3–6 mm. broad
2. M. laxa

1. Myosotis scorpioides L., Sp. Pl. 131. 1753. True forget-me-not.

Myosotis palustris With. of Jepson, Man. Fl. Pl. Calif. 842. 1925.

Perennial with slender rootstock or stolons; herbage thinly strigulose with straight, pointed hairs; stems 1.5–4 dm. long, decumbent or ascending, rooting at lower nodes; leaves oblanceolate to spatulate, 2.5–8 cm. long, the upper stem leaves sessile, the lower ones narrowed to a winged petiole; racemes slender, loosely flowered, in fruit to 2 dm. long, bractless; fruiting pedicels as long as or longer than the calyx; calyx with straight, appressed hairs, its lobes equal in size, shorter than the tube, more or less spreading in fruit; corolla blue with a yellow eye, the tube longer than the calyx, 6–9 mm. broad; style longer than the nutlets.

Escaped from cultivation and established in wet meadows and margins of streams; native of Europe and Asia.

2. Myosotis laxa Lehm., Asperif. 1:83. 1818.

Perennial; herbage with appressed, pointed hairs; stems slender, erect or decumbent, 1.5–5 dm. long, rooting at the nodes; leaves linear-oblong to spatulate, sessile or short-petioled; racemes loosely flowered; pedicels much longer than calyx, widely spreading; calyx with straight, appressed hairs, the

Fig. 298. *Lippia nodiflora: a,* flower, × 12; *b,* pistil, × 12; *c,* nutlets, × 12; *d,* habit, decumbent plant, the spikes in fruit, × ⅔; *e,* habit, prostrate plant, the spikes in flower, × ⅔; *f,* bract of the inflorescence, × 12; *g,* flower, split longitudinally on adaxial side, × 12.

lobes equal in size, as long as the tube; corolla blue, about 4 mm. broad, the tube about as long as the calyx; style shorter than the nutlets.

Marshes and along streams: Del Norte County; north to British Columbia, east to Newfoundland and northeastern United States.

VERBENACEAE. VERBENA FAMILY

Ours herbs with opposite or whorled simple leaves. Flowers perfect. Calyx persistent. Corolla sympetalous, more or less irregular. Stamens 4, in 2 pairs. Ovary superior, undivided, 2- to 4-celled, separating at maturity into as many 1-celled, 1-seeded achene-like nutlets; style single, entire; stigmas 2 or 1.

Calyx 2-cleft; nutlets 2; creeping herbs .... *Lippia*
Calyx 5-toothed; nutlets 4; diffuse or erect herbs .... *Verbena*

LIPPIA

Ours creeping, often densely matted, perennial herbs from woody roots, rooting along the stem. Leaves opposite. Flowers sessile in short heads or spikes on axillary peduncles. Bracts closely imbricated. Calyx 2-cleft, the lobes lateral. Corolla bilabiate, the upper lip emarginate, the lower lip 3-lobed or 3-cleft, white to pink or pale blue. Stamens 4. Fruit composed of 2 nutlets which do not separate readily.

Leaves broadest below middle (elliptic-ovate), tapering both at base and apex, thin, 10–40 mm. broad, green .... 1. *L. lanceolata*
Leaves broadest above the middle (cuneate-obovate or spatulate), rounded at apex, thick, 5–20 mm. broad, minutely canescent .... 2. *L. nodiflora*

1. Lippia lanceolata Michx., Fl. Bor. Am. 2:15. 1803.

Phyla lanceolata Greene, Pittonia 4:17. 1899.

Stems creeping or decumbent, sometimes the tips ascending; leaves 3–6 cm. long, thin, lanceolate-elliptic, short-petiolate, tapering at each end, green, toothed to below the middle; peduncles slender, from the leaf axils, 2–3 times as long as the subtending leaf; spike short; corollas pink to pale blue or white.

Wet bottomlands: Central Valley to southern California; east to Atlantic. Less frequent than *Lippia nodiflora*.

The California material is considered by Fernald to be included in the variety *Lippia lanceolata* var. *recognita* Fern. & Grisc.

2. Lippia nodiflora (L.) Michx., Fl. Bor. Am. 2:15. 1803. Fig. 298.

Phyla nodiflora Greene, Pittonia 4:46. 1899.

Stems creeping, 3–15 dm. long, rooting at the nodes; leaves 1–2 cm. long, thick, spatulate, oblanceolate, cuneate, sharply serrate above the middle, sessile, minutely canescent; peduncles in the axils of leaves, 3–4 times as long as the subtending leaf; spike cylindric in fruit, 8–25 mm. iong; bracts attenuate; calyx teeth triangular; corolla white, 2–3 mm. broad.

Wet banks of lowlands: Central Valley; east to Atlantic.

VERBENA

Erect or decumbent herbs with opposite leaves and square stems. Flowers in elongate, bracteate spikes, 1 or more terminal on the branches. Calyx 5-angled and 5-lobed. Corolla salverform, obscurely 2-lipped, the limb rotate. Stamens 4. Ovary 4-celled, ultimately divided into 4 nutlets, the commissural face often ornamented with waxy granules or rough or striated processes.

Stems square with sharp-ridged angles; lower leaves sessile or narrowed to a margined or winged petiole.
 Base of lower leaves subcordate and clasping the stem . . . . 1. *V. bonariensis*
 Base of lower leaves not subcordate or clasping the stem . . . . 2. *V. litoralis*
Stems square, the angles not ridged; lower leaves petioled, those of the inflorescence petioled or sometimes sessile.
 Bracts shorter to slightly longer than calyx; plants erect or ascending, 5–15 dm. tall.
  Fruits on the elongated spike remote, not imbricate.
   Spikes thick, 6–8 mm. broad; herbage coarsely pilose or hirsute, glandular
                3. *V. lasiostachys*
   Spikes very slender, 2–4 mm. broad; herbage hispid, scabrous . . . . . . 4. *V. scabra*
  Fruits on the elongated spike closely imbricated, not interrupted; spikes many in a dense panicle . . . . 5. *V. hastata*
 Bracts distinctly longer than the calyx; plants decumbent-spreading . . . 6. *V. bracteata*

1. Verbena bonariensis L., Sp. Pl. 20. 1753.

Perennial, 1–1.5 m. tall; stems erect, square, sharply angled, the angles conspicuously ridged, the nodes slightly constricted, scabrous-pubescent; leaves lanceolate, coarsely and unequally serrate-toothed except at tip and base, 6–15 cm. long, sessile, at least the lower leaves subcordate and clasping the stem, rugose and hirtellous above, spreading-pubescent beneath; spikes several to many in terminal panicles, pediceled, elongating in fruit, 3–10 cm. long, 4–6 mm. thick; bracts subequal to the calyx in length or slightly longer, not glandular; corolla tube slightly exserted from calyx; commissure of nutlet covered with white waxy granules.

Ditch banks and wet ground: Yuba, Stanislaus, Merced, Solano, and Marin counties; native of South America.

2. Verbena litoralis H.B.K., Nov. Gen. & Sp. 2:276. 1817.

Closely resembling *Verbena bonariensis* in habit and size, differing most obviously in the leaves, which in *V. litoralis* taper to the base and are either subsessile or short-petioled, but not subcordate or clasping.

Along sloughs and ditches, infrequent: Clinton in Amador County, Bouldin Island in San Joaquin County; native of Central and South America.

3. Verbena lasiostachys Link, Enum. Hort. Berol. 2:122. 1822.

Verbena prostrata of Jepson, Man. Fl. Pl. Calif. 858. 1925.

Erect, spreading, or decumbent perennial, 3–15 dm. high; stems square, the angles rounded; herbage densely soft-pubescent; leaves broadly ovate or oblong-ovate, sometimes the lower leaves laciniately lobed, 3-parted or 3-divided, coarsely serrate, tapering to a margined petiole 5–10 cm. long; spikes often 10–30 cm. long, solitary, or 3–5 aggregated at the ends of a branch; bracts subequal to the calyx in length; calyx not glandular; corolla purple; fruits remote on the elongated spike; nutlets muriculate on the commissure.

Very common in moist places: throughout California; north to Oregon, south to Mexico.

4. Verbena scabra Vahl, Eclog. Am. 2:2. 1798.

Verbena urticifolia of Jepson, Man. Fl. Pl. Calif. 858. 1925. Not L.

Erect perennial, 5–15 dm. tall, paniculately branched or simple; stem hispidulose, scabrous; leaves ovate to oblong-lanceolate, acuminate, serrate-dentate, on short petioles, the blades scabrous above; spikes very slender, interrupted; bracts ovate-acuminate, shorter than calyx; fruiting calyx 2.5–3 mm. long, hispidulous, the lobes connivent over fruit, forming a short beak; corolla white, pink, or bluish; nutlet clothed with white waxy granules on the commissure.

Marshy ground: southern California; northern Baja California, Mexico.

5. Verbena hastata L., Sp. Pl. 20. 1753.

Rhizomatous perennial; stems erect, 5–15 dm. tall, square, the corners angled but not with a conspicuous rib; herbage short-hispid to somewhat silky or sometimes scabrous; leaves ovate-lanceolate, simple or occasionally some hastately lobed at base, 7–15 cm. long, on short petioles, the margin serrate; spikes several to many, in terminal erect panicles, 2–4 mm. thick, 5–10 cm. long, not leafy; flowers and fruits closely imbricate; bracts subulate, shorter than calyx; calyx pubescent, the lobes incurved, overtopping the fruit; corolla violet-blue; nutlets nearly smooth on the commissure.

Floodlands and ditch banks: Central Valley, Sierra Nevada foothills to eastern Shasta County; east to Atlantic. Often forming large, conspicuous colonies on wet land.

6. Verbena bracteata Lag. & Rodr., Anal. Cienc. Nat. [Madrid] 4:260. 1801.

Verbena bracteosa Michx., Fl. Bor. Am. 2:13. 1803.

Plant decumbent or ascending; stems 5–20 cm. long, hirsute to hispidulous throughout; leaves obovate or oblong, coarsely serrate or incised or lobed, 1–6 cm. long, narrowed to a winged petiole; spikes sessile, terminal, elongating in fruit, thick, the flowers dense; bracts linear, divergent, much longer than calyx, recurving in age; calyx 3–4 mm. long, hirsute, the lobes connivent over the

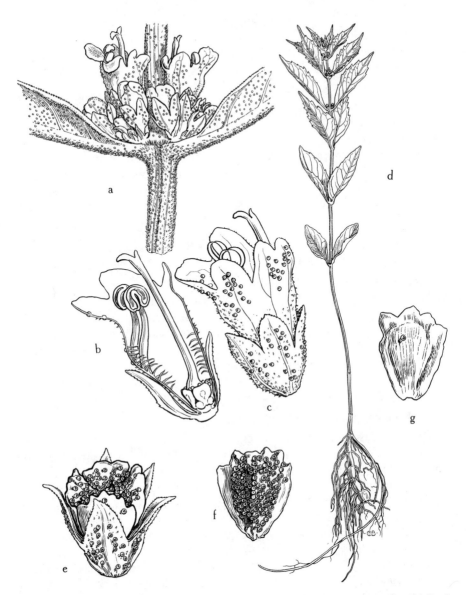

Fig. 299. *Lycopus uniflorus: a,* flower clusters in leaf axils, × 8; *b,* flower (longitudinal section), × 1⅕; *c,* flower, the calyx and corolla gland-dotted, × 12; *d,* habit, the rootstock tuber-like, × ⅖; *e,* calyx containing mature nutlets, × 16; *f,* nutlet, gland-dotted, adaxial view, × 20; *g,* nutlet, abaxial view, × 20.

fruit; corolla purple, the limb 2.5–3 mm. broad; nutlets muriculate-scabrous on the commissure.

Common near water in marshes and floodlands of valleys and foothills: throughout California; north to British Columbia, east and south to Florida and Mexico.

LABIATAE. MINT FAMILY

Annual or perennial herbs or low shrubs with aromatic herbage, square stems, and opposite, simple leaves. Flowers perfect, solitary or in small clusters or whorls, axillary or with bracts reduced and the whorls crowded at apex, thus simulating terminal spikes. Calyx synsepalous, 2-lipped to 5-toothed and regular. Corolla sympetalous, usually bilabiate, with 3 lobes in the lower lip and 2 in the upper one, or the corolla 5-lobed and nearly regular. Stamens 4, inserted on the corolla tube, in 2 pairs, or the upper pair wanting. Ovary superior, deeply 4-lobed; style single, apically 2-cleft, often attached in depression between the 4 ovary lobes. Fruit splitting into 4 small, 1-seeded, smooth nutlets.

Calyx distinctly 2-lipped with the lips entire .... *Scutellaria*
Calyx regular, or 2-lipped with the lips toothed.
  Corolla markedly 2-lipped.
    Corolla tube with no hairy ring inside; calyx unequally 5-cleft, ciliate or hirsute
                                                                    Pogogyne
    Corolla tube with hairy ring inside; calyx with 5 equal or subequal, spine-tipped teeth
                                                                     Stachys
  Corolla nearly regular, the lobes subequal.
    Fertile stamens 2 .... *Lycopus*
    Fertile stamens 4 .... *Mentha*

LYCOPUS

Perennial rhizomatous or tuber-bearing herbs. Stems erect, simple or branched. Leaves opposite, sessile or short-petioled. Flowers clustered in the leaf axils, white to purple. Calyx campanulate, 4- or 5-toothed, nearly regular. Corolla funnelform to campanulate, the upper lip entire to emarginate, the lower lip 3-lobed. Stamens 2, sometimes with 2 additional rudimentary filaments, the anther sacs parallel. Ovary deeply 4-parted; style 2-cleft. Nutlets smooth, truncate or toothed above, narrowed below, gland-dotted on front side.

Calyx lobes obtuse or rounded at summit, as long as or shorter than nutlets
                                                        1. *L. uniflorus*
Calyx lobes lanceolate to attenuate or subulate, longer than nutlets.
  Leaves unequally serrate or incised to pinnatifid .... 2. *L. americanus*
  Leaves equally serrate .... 3. *L. lucidus*

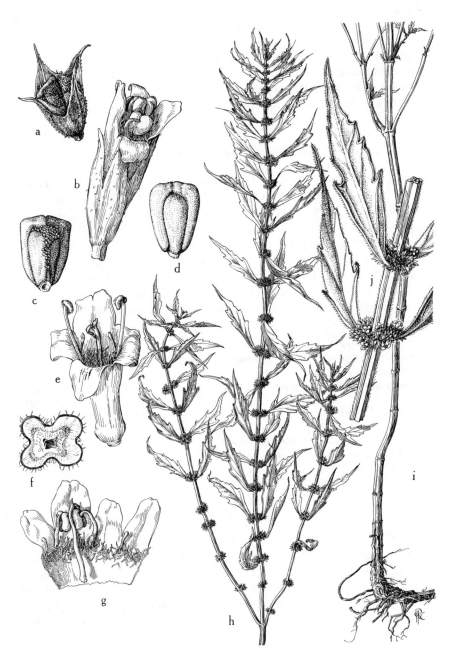

Fig. 300. *Lycopus americanus: a,* calyx containing mature nutlets, × 12; *b,* flower, × 12; *c,* nutlet, adaxial view, × 20; *d,* nutlet, abaxial view, × 20; *e,* corolla, showing rudimentary and fertile stamens and the pubescent tube, × 12; *f,* stem (cross section), × 12; *g,* corolla, spread open to show rudimentary stamens and the pubescence in the tube, × 12; *h,* habit, tip of flowering plant, × ⅖; *i,* habit, base of plant, showing stout rhizome, × ⅖; *j,* flower clusters in leaf axils, × 1½.

1. Lycopus uniflorus Michx., Fl. Bor. Am. 1:14. 1803. Fig. 299.

Plant from thick, often tuber-like rootstock and often stoloniferous from lower nodes; stems erect, slender, simple or branched, 1–6 dm. tall; herbage green or purplish; leaves sessile or subsessile, lanceolate, acute to ovate, unequally serrate to dentate-serrate; calyx lobes oblong-ovate to triangular, obtuse; corolla 2–3 mm. long; stamens 2, rudimentary stamens absent or much reduced; nutlets slightly shorter than the sepals to longer, somewhat irregular across the truncate tip.

Wet places along the coast: Humboldt and Del Norte counties, Swamp Lake in Tuolumne County; north to Alaska, east to Atlantic.

2. Lycopus americanus Muhl. ex Barton, Fl. Phila. Prodr. 15. 1815. Fig. 300.

Rhizomatous to stoloniferous perennial; stems erect, simple or branched, 3–10 dm. tall; herbage glabrous to sparingly pubescent; leaves lanceolate to lanceolate-ovate, narrowed to a sessile or subsessile petiole-like base, deeply and irregularly serrate to incised or laciniate-pinnatifid; calyx lobes lanceolate, acuminate to cuspidate, minutely serrulate or occasionally more conspicuously so; upper corolla lobes fused, entire or notched, the tube pubescent within; rudimentary stamens conspicuous, club-shaped; nutlets shorter than calyx, truncate above, smooth and rounded.

Marshes, ditch banks: widely distributed at low and middle altitudes; across the continent.

3. Lycopus lucidus Turcz. ex Benth. in A. DC., Prodr. 12:178. 1848. Fig. 301.

Rhizomatous or tuber-bearing herb; stems erect, simple or branched; herbage pubescent, especially stems and midveins of leaves, rarely glabrate; leaves lanceolate to oblong-lanceolate, sessile to subsessile, conspicuously but evenly serrate; calyx lobes lanceolate-attenuate or acuminate, serrate, longer than calyx tube; corolla barely surpassing the calyx lobes, the lower lip conspicuously 3-lobed; rudimentary stamens slender, club-shaped; nutlets much shorter than calyx, somewhat irregular across the truncate tip.

Wet ground and marshes: northern California; to British Columbia, Asia.

MENTHA

Perennial rhizomatous herbs. Stems erect or diffuse. Leaves sessile or petioled. Flowers in whorls or interrupted in congested terminal spikes. Calyx cylindric to campanulate, 10-nerved, regular to slightly bilabiate, 5-toothed. Corolla conspicuously bilabiate, the tube included in calyx, the upper lip entire or emarginate, the lower lip 3-lobed. Stamens 4, equal in length; filaments glabrous. Ovary 4-parted; style 2-cleft. Nutlets ovoid.

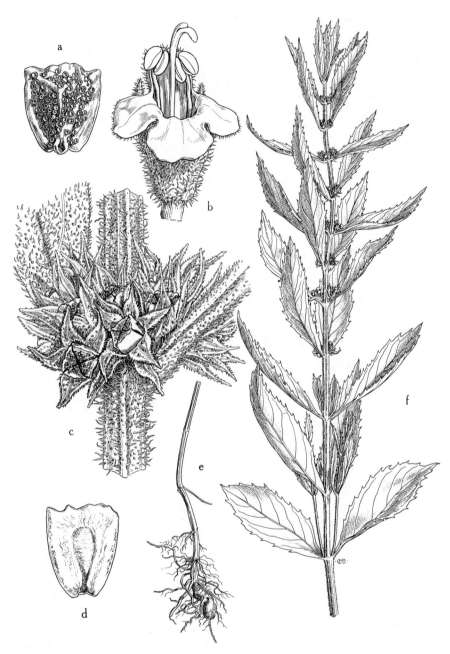

Fig. 301. *Lycopus lucidus:* *a,* nutlet, adaxial view, × 12; *b,* flower, showing pubescent calyx, × 8; *c,* flower cluster in leaf axils, × 4; *d,* nutlet, abaxial view, × 12; *e,* habit, basal part of plant, showing tuberous rootstock, × ⅖; *f,* habit, tip of flowering plant, × ⅖.

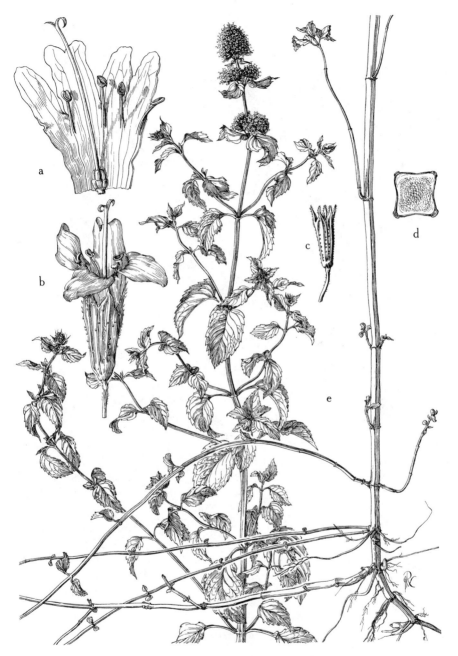

Fig. 302. *Mentha citrata: a,* corolla, spread open, × 8; *b,* flower, × 8; *c,* calyx, × 4; *d,* stem (cross section), × 3; *e,* habit, × ⅖.

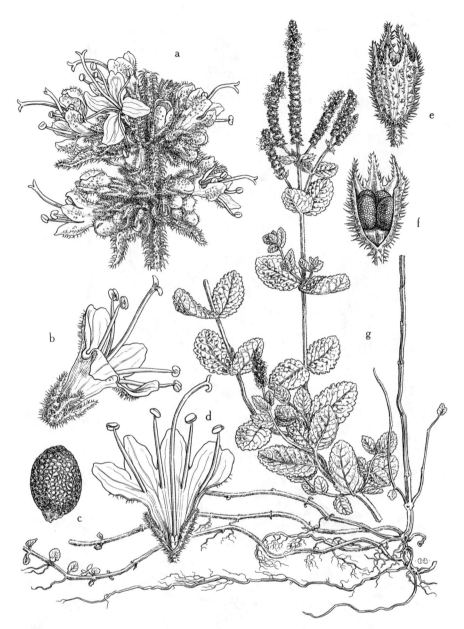

Fig. 303. *Mentha rotundifolia: a,* flower clusters at nodes, × 4; *b,* flower, × 8; *c,* nutlet, × 28; *d,* flower, spread open, × 8; *e,* calyx, × 12; *f,* calyx, spread open to show the 4 mature nutlets, × 12; *g,* habit, × ⅖.

Flowers in terminal spikes, sometimes the lower part interrupted and leafy-bracted.
  Stamens included; plants glabrous or nearly so.
    Leaves sessile or nearly so; spikes slender .... 1. *M. spicata*
    Leaves short-petioled; spikes broad.
      Leaves mostly lanceolate or ovate-lanceolate, acute; sepals ciliate .. 2. *M. piperita*
      Leaves mostly ovate or elliptic, obtuse; sepals not ciliate .... 3. *M. citrata*
  Stamens long-exserted; plants tomentose or villous.............. 4. *M. rotundifolia*
Flowers in the axils of normal or reduced leaves; stamens long-exserted.
  Subtending leaves reduced and reflexed; herbage canescent .... 5. *M. Pulegium*
  Subtending leaves not much reduced; herbage variously pubescent but not canescent
                                                                    6. *M. arvensis*

1. Mentha spicata L., Sp. Pl. 576. 1753. Spearmint.

Perennial herb, glabrous or pubescent only at nodes; stems 3–12 dm. tall, often purple; leaves sessile or subsessile, ovate to oblong-lanceolate, acute or acuminate, 3–6 cm. long, serrate; flowers in interrupted, leafless, terminal spikes 6–8 cm. long; bracts lanceolate to subulate, as long as or longer than calyx, ciliate; calyx lobes subulate, as long as the tube, ciliate; corolla lavender; stamens included; style 2-cleft; nutlets somewhat rough, not reticulate.

Moist meadows at low and middle elevations; weed naturalized from Europe, but not common.

2. Mentha piperita L., Sp. Pl. 576. 1753. Peppermint.

Rhizomatous or stoloniferous perennial; stems erect or decumbent, simple or branched, 3–8 dm. long; herbage glabrous, green or reddish; leaves lanceolate to ovate, serrate, short-petioled; flowers in dense, interrupted, terminal spikes 2–12 cm. long; calyx teeth hirsute, the tube glabrous; corolla glabrous, pink, purple, or white; stamens included.

Wet places at low and middle altitudes throughout the state; native of Europe.

3. Mentha citrata Ehrh., Beitr. Nat. 7:150. 1792. Bergamot mint. Fig. 302.

Stoloniferous herb; stems decumbent or ascending; herbage green to purple, glabrous; leaves petioled, ovate to nearly orbicular, round or subcordate at base, 2–5 cm. long, shallowly serrate; flowers in thick, rounded, interrupted, terminal spikes and in upper leaf axils; calyx lobes subulate, glabrous, shorter than calyx tube; corolla pink; stamens included in or barely exserted from throat.

Wet seepage areas; sparingly naturalized from Europe.

Our specimens seem to vary in calyx pubescence toward *Mentha piperita.* *Mentha citrata* is treated by some authors as a subspecies of *M. piperita.* Both are plagued with tremendous horticultural variation, and their dissemination from their Mediterranean home has been mostly through horticultural sources.

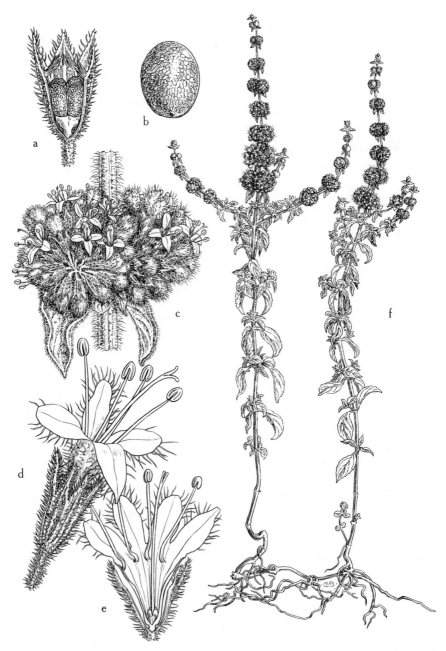

Fig. 304. *Mentha Pulegium: a,* calyx, spread open to show the 4 mature nutlets, $\times$ 8; *b,* nutlet, $\times$ 28; *c,* flower cluster at node, $\times$ 2; *d,* flower, $\times$ 8; *e,* corolla, spread open, $\times$ 6; *f,* habit, $\times$ $\frac{2}{5}$.

4. Mentha rotundifolia (L.) Huds., Fl. Angl. 221. 1762. Apple mint. Fig. 303.

Stoloniferous perennial; stems erect or decumbent, 5–15 dm. long, simple or branched; herbage sparsely tomentose, viscid; leaves oblong-ovate, sessile or short-petioled, 2–3 cm. long, serrate to serrate-dentate or serrate-crenate, rugose, reticulate; flowers in dense, narrow, terminal spikes 5–10 cm. long, the lower ones somewhat interrupted; bracts acuminate, shorter than flowers; calyx campanulate, the lobes attenuate, about as long as the tube, densely pubescent; corolla white, campanulate, the lobes pubescent on abaxial surface, the tube glabrous; stamens 4, long-exserted; nutlets finely reticulate.

Sparingly naturalized in wet seepage places: Lone Pine in Inyo County; native of Europe.

5. Mentha Pulegium L., Sp. Pl. 577. 1753. Pennyroyal. Fig. 304.

Rhizomatous perennial; stems erect or decumbent, 3–6 dm. long; herbage canescent, short-villous; leaves elliptic to elliptic-ovate, tapering to a short petiole or subsessile, serrate to entire; flowers in a narrow interrupted spike subtended by reduced leaves; calyx hirsute on nerves and teeth, the lobes unequal in size, white-villous within; corolla lavender, twice as long as calyx, the petals villous outside; stamens exserted; nutlets finely reticulate.

Common in vernally wet ground and around marshes at low altitudes; native of Europe.

6. Mentha arvensis L., Sp. Pl. 577. 1753. Field mint. Fig. 305.

Stoloniferous perennial; stems simple or freely branched, erect or decumbent, 1–8 dm. long; herbage retrorse-pubescent to woolly-lanate; leaves oblong-ovate, petioled, serrate, short-villous, 2–5 cm. long; flowers all axillary; calyx pubescent, the lobes triangular-subulate, as long as the tube; corolla pink, violet, or white, essentially glabrous; stamens long-exserted; nutlets smooth.

An exceedingly variable species in leaf shape and pubescence.

Common in wet places: widespread in the state at low and middle altitudes; eastern North America, Eurasia.

POGOGYNE

Annual herbs with erect stems, simple or branching from base. Leaves obovate to spatulate. Flowers in dense, bracteate spikes or verticillate at base of inflorescence. Calyx 5-cleft, the lobes unequal in length, the tube 15-nerved, glabrous within. Corolla tubular-funnelform, blue or purple, the tube exserted beyond the calyx, the lower lip 3-lobed, spreading, the upper lip erect, entire. Stamens 4, all anther-bearing or the upper ones sterile; filaments pubescent. Style bearded.

Fig. 305. *Mentha arvensis: a,* habit, × ⅖; *b,* corolla, spread open, × 8; *c,* flower, × 8; *d,* stem (cross section), × 3; *e,* calyx, spread open to show the 4 nutlets, × 8; *f,* nutlet, adaxial side, × 12; *g,* nutlet, abaxial side, × 12; *h,* inflorescence, × 1½.

Stamens all with fertile anthers.
Bracts and sepals bristly-ciliate and hirsute .... 1. *P. Douglasii*
Bracts and sepals glabrous .... 2. *P. nudiuscula*
Only the lower pair of stamens with fertile anthers, the upper pair with rudimentary or
no anthers .... 3. *P. zizyphoroides*

1. Pogogyne Douglasii Benth., Lab. Gen. & Sp. 414. 1834.

Stems erect, simple or branching, 5–50 cm. tall; herbage glabrous or puberulent, except in inflorescence; leaves oblanceolate to elliptic, attenuate below, entire or remotely toothed, 1–2 cm. long; inflorescence a dense spike or the lower verticels remote; bracts linear, pungent, bristly-ciliate; calyx lobes unequal, ciliate-margined, the 2 lower lobes 1–3 times as long as the tube, the upper ones short, the tube glabrous to puberulent; corolla 1–2 cm. long, lavender to purple, the lower lip often mottled with yellow; stamens all fertile, the filaments pubescent above; style pubescent below stigmas; nutlets brown.

Chiefly in vernal pools: central California from Lake and Butte counties to Kern and San Luis Obispo counties.

2. Pogogyne nudiuscula Gray, Bot. Calif. 1:597. 1876.

Pogogyne Abramsii J. T. Howell, Proc. Calif. Acad. Sci. IV. 20:119. 1931.

Stems simple or branched, erect or spreading, 1–3 dm. long; herbage glabrous, except the sparsely pubescent inflorescence; leaves oblanceolate, narrowed below to a petiole, entire to subentire, glabrous, 1–2 cm. long; inflorescence a dense capitate spike with 1 or 2 verticels below; bracts oblanceolate to linear, glabrous to sparsely ciliate; calyx lobes unequal, sparsely ciliate; corolla 10–15 mm. long, sparsely pubescent, lavender, sometimes mottled; stamens all fertile, the lower pair about 3 times as long as the upper pair, the filaments pubescent; style pubescent below stigmas.

Beds of vernal pools: on coastal mesas, San Diego County.

Pogogyne Abramsii appears to be a striking color variant of *P. nudiuscula*.

3. Pogogyne zizyphoroides Benth., Pl. Hartw. 330. 1849.

Stems erect or spreading, 2–20 cm. long; herbage glabrous below to bristly-ciliate above; leaves ovate-elliptic, narrowed to slender petiole, 15–20 mm. long, the upper leaves bristly-ciliate along petioles; inflorescence spicate above, the lower verticels separate; bracts spatulate, ciliate; calyx lobes unequal, bristly on veins, the tube glabrous; corolla lavender, 4–8 mm. long; lower stamens fertile, the upper ones sterile; nutlets obovoid, brown.

Beds of vernal pools: Central Valley and North Coast Ranges; to southern Oregon.

SCUTELLARIA

Ours perennial herbs with rhizomes, stolons, or tubers. Flowers solitary or in axillary racemes. Calyx campanulate, gibbous, 2-lipped, the lips entire, often with a horizontal crest across base of upper lip. Corolla exserted, bilabiate, often abruptly dilated at throat, the upper lip arched, entire or emarginate, the lower one 3-lobed. Stamens 4, included, the upper pair 2-celled, the lower pair 1-celled. Style 2-cleft. Nutlets subglobose.

Flowers solitary or few in the leaf axils; nutlets rugose.
    Corolla blue, slightly enlarged at throat . . . . 1. *S. galericulata*
    Corolla white, abruptly enlarged at throat . . . . 2. *S. Bolanderi*
Flowers on slender axillary racemes; nutlets smooth . . . . 3. *S. lateriflora*

1. Scutellaria galericulata L., Sp. Pl. 599. 1753.

Scutellaria epilobifolia Ham., Monogr. Gen. Scut. 32. 1832.

Stoloniferous perennial without tubers; stems simple or few-branched, 2–5 dm. tall; herbage glabrous to thinly puberulent; leaves lanceolate to ovate-lanceolate, cordate to subtruncate at base, remotely crenate-serrate, short-petioled; flowers solitary in axils of upper leaves, short-pediceled; calyx 4 mm. long; corolla 15–20 mm. long, blue; nutlets rugose.

Swamps: northern Sierra Nevada to Modoc and Shasta counties; north to Alaska, east to Atlantic.

2. Scutellaria Bolanderi Gray, Proc. Am. Acad. Arts & Sci. 7:387. 1868.

Rhizomatous perennial; stems erect, simple or occasionally much-branched, often very leafy; leaves ovate, truncate or rounded at base, crenate, serrate to entire, 1–3 cm. long, sessile to subsessile, sparsely pilose; flowers few, borne in leaf axils; sepals 3–4 mm. long; corolla white, 16–18 mm. long, abruptly expanded at throat, the upper lip smaller than lower one, the lower lip shallowly 3-lobed, purple-dotted; nutlets rugose.

Moist meadows or around water holes and along streams: Sierra Nevada at low and middle altitudes to mountains of southern California.

Epling segregates material from southern California, which lacks purple dots on the corolla, as *Scutellaria Bolanderi* var. *austromontana* (Madroño 5:158, 1939).

3. Scutellaria lateriflora L., Sp. Pl. 598. 1753.

Stoloniferous perennial; herbage glabrous to puberulent; stems erect or ascending, simple or branched, 2–15 dm. tall; leaves ovate to oblong or ovate-lanceolate, obtuse or subcordate below, coarsely serrate-dentate, 3–7 cm. long, reduced above, slender-petioled; flowers secund, in slender, axillary, many-flowered racemes; corolla blue to white, 6–10 mm. long; nutlets smooth.

Marshy ground: Bouldin Island in San Joaquin County, also Inyo County; Oregon to British Columbia and east to Atlantic.

STACHYS

Annual or perennial herbs; herbage variously pubescent. Flowers verticil-late in interrupted terminal spikes and in the upper leaf axils. Calyx campanu-late, 5- to 10-nerved, nearly regular. Corolla 2-lipped, the upper lip concave, entire or emarginate, the lower lip 3-lobed, spreading, rarely 2-lobed, the tube not dilated. Stamens 4, the anther sacs divergent. Ovary 4-lobed; style 2-cleft. Nutlets ovoid or oblong.

Corolla red or purple; constriction at the ring of hairs not prominent externally, the
    corolla tube not gibbous, the ring of hairs horizontal or nearly so.
    Corolla purplish throughout, 15–25 mm. long .... 1. *S. Chamissonis*
    Lower lip of corolla mottled with white, 10–15 mm. long .... 2. *S. Emersonii*
Corolla white, or sometimes pink to rose with purple veins; constriction at ring of hairs
    prominent externally by a gibbous or saccate swelling, the ring of hairs oblique.
    Lower leaves sessile, the middle ones petioled .... 3. *S. palustris*
    Lower leaves petioled.
        Upper lip of corolla 2–3 mm. long .... 4. *S. stricta*
        Upper lip of corolla 3–5 mm. long.
            Plants, at least when young, variously tomentose, usually with straight hairs.
            Leaves oval to ovate, round to subcordate at base, not parallel-sided.
                Spike interrupted, elongate; flowers pink .... 5. *S. rigida*
                Spike short, typically not interrupted; flowers whitish; herbage strong-scented
                                        6. *S. pycnantha*
            Leaves oblong, narrowed at base, at least the upper ones often parallel-sided
                                          7. *S. ajugoides*
            Plants, at least when young, densely white-villous to floccose-tomentose
                                          8. *S. albens*

1. Stachys Chamissonis Benth., Linnaea 6:80. 1831.

Rhizomatous perennial, 1–3 m. high; herbage soft, pubescent; stems re-trorsely scabrous; leaves petioled, often viscid, ovate, 5–15 cm. long, ovate-rounded to cordate below; inflorescence of 1–3 spikes or racemes at the ends of the branches; flowers whorled; calyx clavate, tubular, 10–18 mm. long; corolla rose-red, tubular, 2–3 cm. long, much exserted, the upper lip entire, the lower one 3-lobed, the tube with a horizontal hairy ring just above the base.

Swamps or wet wooded areas near the coast: San Luis Obispo County; north to Oregon.

2. Stachys Emersonii Piper, Erythea 6:31. 1898.

Rhizomatous perennial herb; stems branched, 4–8 dm. high; herbage hir-sute, the hairs on angles of stems pustulate at base; leaves 2–4 cm. long, on slender petioles, or the upper ones nearly sessile, ovate-lanceolate, cordate to subcordate at base, coarsely crenate-serrate; flowers short-pediceled, in an interrupted spike and in the axils of the upper leaves; calyx 5–7 mm. long, the lobes subulate; corolla reddish purple, 10–15 mm. long, the tube with a horizontal band of hairs within well above the base, the lower lip 6–8 mm. long, spotted with white.

Wet places: northwestern California from Mendocino County to Del Norte County; north to Alaska, east to Great Lakes.

3. Stachys palustris subsp. **pilosa** (Nutt.) Epling, Rep. Sp. Nov. 80:63. 1934.

This entity differs from *Stachys stricta* chiefly in its sessile and subsessile basal leaves.

It has been reported in northeastern California, whence it extends north to Alaska and east to the Great Lakes.

4. Stachys stricta Greene, Erythea 2:122. 1894.

Stachys ajugoides var. *stricta* Jepson, Fl. West. Middle Calif. 457. 1901.

Rhizomatous perennial; stems erect or decumbent, 6–12 dm. high; herbage villous-hirsute throughout and resinous-glandular; leaves petioled below, sessile or subsessile above, oblong-lanceolate, acute or rounded at apex, subcordate below, crenate-serrate, 5–15 cm. long; spikes at length interrupted, the verticels globose, 8- to 12-flowered; calyx 5–6 mm. long, the teeth as long as the tube, deltoid; corolla white, the tube barely exserted, 6 mm. long, the inner hairy ring situated well below the middle and slightly oblique, a slight spur on lower side of the tube, the upper lip about ½ as long as the lower one, not hooded, the lower lip with lateral lobes reduced to recurved teeth; filaments included, pubescent below.

Bogs and wet meadows: North Coast Ranges to Mendocino County, northern Sierra Nevada foothills south to Merced County.

5. Stachys rigida Nutt. ex Benth. in A. DC., Prodr. 12:472. 1848.

Stachys rivularis Heller, Muhlenbergia 1:33. 1904.
Stachys quercetorum Heller, Muhlenbergia 2:318. 1907.
Stachys rigida subsp. *lanata* Epling, Madroño 4:270. 1938.
Stachys ajugoides var. *rigida* Jepson & Hoover in Jepson, Fl. Calif. 3:426. 1943.

Rhizomatous perennial; stems erect or decumbent, simple or branched, 3–12 dm. high; herbage densely villous-hirsute; leaves long-petioled, oblong-ovate to oblong-lanceolate, round or subcordate below, acute or obtuse above, crenate-serrate; spikes 10–20 cm. long, becoming interrupted with age; calyx campanulate, the lobes equal to the tube in length; corolla 12–16 mm. long, rose-purple, the tube longer than calyx, saccate below, the hairy ring oblique, the upper lip 3–4 mm. long, the lower lip 5–6 mm. long; stamens included, the filaments densely hairy.

Common in wet or in shady places: central and northern California; north to Washington.

This very common species perhaps constitutes a series of hybrid complexes. Because of intergradation with *Stachys ajugoides*, Jepson included the entire complex in that species. This, I believe, is unnecessary, and it may be that a thorough field analysis of the problem might justify further segregation centering around material represented by the names *S. rivularis* Heller, *S. quercetorum* Heller, and *S. rigida* subsp. *lanata* Epling, all of which both Epling the Abrams treat as subspecies under *S. rigida*. *Stachys rivularis* has some leaf characters in common with *S. ajugoides,* and perhaps this is why Jepson put the complex in with *S. ajugoides* rather than *S. rigida*.

6. Stachys pycnantha Benth., Pl. Hartw. 331. 1849.

Rhizomatous perennial; stems simple or branched from base, erect, 3–10 dm. tall; herbage soft-villous to glandular-puberulent; leaves ovate to oblong-lanceolate, subcordate or rounded at base, crenate-serrate, long-petioled to subsessile; spikes short to subcapitate, 4–5 cm. long, occasionally somewhat interrupted; calyx campanulate, 6–7 mm. long, the teeth cuspidate, shorter than tube; corolla white, the upper lip 4 mm. long, the lower lip 6 mm. long, the tube included in calyx, the spur indistinct, the hairy ring oblique; filaments glabrous.

Wet soil: central California, Coast Ranges from Marin County to San Benito and San Luis Obispo counties.

7. Stachys ajugoides Benth., Linnaea 6:80. 1831.

Rhizomatous perennial; stems erect, or decumbent at base, 10–60 cm. long; herbage villous to hirsute and somewhat glandular; leaves oblong to oblanceolate, narrower at base, crenate to crenate-serrate, long-petioled to subsessile; spikes 8–20 cm. long, dense or interrupted, verticels 6-flowered; calyx 6–8 mm. long, the teeth lanceolate to deltoid, cuspidate; corolla white to rose, 10–15 mm. long, the upper lip 4–6 mm. long, the lower lip 5–7 mm. long, the tube 7–9 mm. long, saccate near the base, the hairy ring oblique; filaments pubescent.

Wet ground at low altitudes: Central Valley and bordering foothills, Coast Ranges south to Los Angeles County.

8. Stachys albens Gray, Proc. Am. Acad. Arts & Sci. 7:387. 1868. Fig. 306.
Stachys albens var. *juliensis* Jepson, Man. Fl. Pl. Calif. 877. 1925.

Rhizomatous perennial; stems erect, 3–20 dm. tall; herbage densely white-woolly, from silky-villous to lanate-floccose; leaves 3–12 cm. long, oblong to ovate or lanceolate-ovate, cordate to subcordate or truncate at base, silky-villous to tomentose or lanate; spikes 8–20 cm. long, becoming interrupted; calyx 7 mm. long, the teeth triangular-lanceolate, cuspidate; corolla white to rose, purple-veined, bilabiate, the upper lip 6 mm. long, the lower lip 6–8 mm. long, the tube subequal to the calyx in length, saccate below middle, the hairy ring oblique; filaments hairy.

Swamps, bogs, and moist stream banks at low and middle altitudes: central California to southern California.

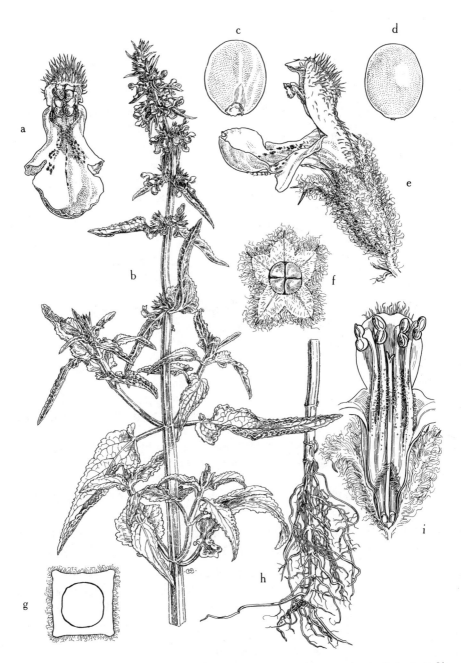

Fig. 306. *Stachys albens: a*, corolla, face view, × 4; *b*, habit, upper part of plant, × ⅖;
c, nutlet, adaxial view, × 4; *d*, nutlet, abaxial view, × 4; *e*, flower, side view, × 4; *f*, calyx,
top view, showing nutlets, × 4; *g*, stem (cross section), × 3; *h*, basal part of stem and
roots, × ⅖; *i*, flower, split longitudinally, × 4.

SOLANACEAE. NIGHTSHADE FAMILY

Herbs or shrubs, usually with alternate leaves. Flowers perfect and regular, solitary or axillary in umbels or cymes or panicles. Calyx 5-lobed. Corolla 5-lobed, rotate to tubular, the lobes valvate or mostly plicate in bud. Stamens 5, alternate with the lobes and inserted on the tube. Ovary entire, superior, mostly 2-celled; style 1; stigma terminal. Fruit a berry or capsule. Seeds numerous.

Fruit a berry; corolla rotate or campanulate.
    Anthers converging around the style; fruiting calyx not conspicuously enlarged
                                                      Solanum
    Anthers distinct, not converging around the style; fruiting calyx large and bladdery
                                                      Physalis
Fruit a capsule; corolla tubular or funnelform.
    Flowers small, less than 1 cm. long, solitary in the axils . . . . *Petunia*
    Flowers large, more than 1 cm. long, in racemes or panicles . . . . *Nicotiana*

NICOTIANA

Heavy-scented, usually viscid-puberulent annual or perennial herbs or shrubs. Leaves alternate, usually entire, petioled or sessile. Inflorescence few- to many-flowered, in terminal racemes or panicles. Calyx tubular-campanulate or ovoid, 5-toothed, persistent. Corolla funnelform, salverform, or tubular, the shallowly 5-lobed limb usually spreading. Stamens 5, the filaments filiform, inserted near the base of the corolla tube. Style slender; stigma capitate. Ovary 2-celled. Capsule ovoid, smooth, acute, 2- to 4-valved at the summit. Seeds many, small, ovoid to reniform, minutely reticulate-punctate.

Stamens inserted about equally at bottom of throat or at apex of corolla tube proper; corolla limb 4–6 mm. broad; cauline leaves petioled . . . . 1. *N. attenuata*
Stamens inserted unequally and high in throat of corolla; corolla limb 1.5–3 cm. broad; cauline leaves sessile . . . . 2. *N. Bigelovii*

Both of these species are reported to have been used as smoking tobacco by the Indians, who cultivated them to some extent.

1. Nicotiana attenuata Torr. ex Wats. in King, Geol. Expl. 40th Par. 5:276, pl. 27. 1871. Coyote tobacco.

Erect, branched annual, 3–10 dm. tall; herbage glandular-pubescent throughout, ill-smelling; lower leaves broadly ovate, 8–15 cm. long, the upper ones narrowly lanceolate, all petioled; inflorescence racemose or paniculate; flowers vespertine; calyx ovoid-campanulate, 7–10 mm. high at anthesis, pockmarked when mature, the teeth deltoid, shorter than the tube; corolla narrowly funnelform, white, 2–2.5 cm. long, essentially glabrous except near apex, the tube proper about 5 mm. long, the limb 4–6 mm. broad, the corolla lobes obtuse; capsule 8–12 mm. long, 4-valved at the apex; seeds 0.5–0.7 mm. long, dull gray-brown, the surface fluted-reticulate.

At low elevations in moist to dried stream bottoms and river beds, and as a weed in waste places: Siskiyou and Humboldt counties, south to San Diego County; Washington and Idaho south to Arizona and Texas.

2. Nicotiana Bigelovii (Torr.) Wats. in King, Geol. Expl. 40th Par. 5:276, pl. 27, figs. 3 and 4. 1871. Indian tobacco.

Branched annual, 3–20 dm. tall; herbage glandular-pubescent, ill-smelling; basal leaves oblong-ovate to narrowly lanceolate, sometimes petioled, the upper cauline leaves lanceolate, sessile or tapering to a winged petiole; flowers in racemes along the branches; calyx 1.5–2 cm. long, the narrowly lanceolate teeth unequal in length, the longer ones about as long as the tube; corolla 2.5–5 cm. long, the limb 1.5–3 cm. broad, white, faintly tinged with green; stamens unequally inserted in the upper part of throat; capsule ovoid, about 1.5 cm. long, shorter than the calyx lobes; seeds about 0.9 mm. long, dull brown, the surface fluted-reticulate.

Along streams of foothills and valley floors: southern California to southern Oregon.

PETUNIA

1. Petunia parviflora Juss., Am. Mus. Hist. Nat. Paris 2:216, pl. 47, fig. 1. 1803.

Annual herb, prostrate, diffusely branched, glandular-puberulent; leaves entire, the upper leaves opposite, oblong-linear to spatulate, 4–12 mm. long, rather fleshy; calyx 5-parted, the lobes linear, 2–3 mm. long in flower, elongating to 5–6 mm. in fruit; corolla funnelform, purplish with whitish tube, 5–7 mm. long, the short lobes somewhat unequal in size; stamens 5, unequally inserted near the base of corolla tube; capsule ovoid, 3–4 mm. long; seeds about 0.6 mm. long, pale brown, reticulated.

Damp sandy soil of stream beds and lake margins: Sacramento Valley, central Coast Ranges, and southern California; south and east to Arizona, Texas, Florida, Mexico, and South America.

PHYSALIS. GROUND CHERRY

Annual or perennial herbs with solitary, axillary flowers. Leaves entire to sinuate-dentate. Calyx 5-toothed, campanulate to tubular-campanulate, enlarged and inflated in fruit and enclosing the berry, 5- to 10-angled, 10-ribbed, reticulate-veined. Corolla obscurely 5-lobed, campanulate to campanulate-rotate, yellow, whitish, or purplish, the center often of a different or deeper shade. Stamens 5, inserted near the base of corolla tube, the anthers oval or oblong, dehiscing longitudinally, not connivent. Style slender, the stigma faintly bilobed. Fruit a many-seeded berry. Seeds numerous, flattened, reniform, finely pitted.

Corolla large, rotate, 10–15 mm. broad; pedicels shorter than the fruiting calyx
.... 1. *P. ixocarpa*
Corolla small, narrowly campanulate, 5–6 mm. broad; pedicels as long as or longer than the fruiting calyx .... 2. *P. lanceifolia*

1. Physalis ixocarpa Brot. ex Hornem., Hort. Hafn. Suppl. 26. 1819. Tomatillo.

Much-branched annual 3–10 dm. high, the stems angular; herbage glabrous or the young leaves and calyces sparsely puberulent; leaves ovate to elliptic, toothed or a few subentire, 2–6 cm. long, the base cuneate and somewhat asymmetrical, the apex acute to short-acuminate; petioles 1.5–3.5 cm. long; pedicels very short in flower, elongating (to about 5–10 mm.) in fruit, but shorter than the fruiting calyx; calyx campanulate, sparsely puberulent or glabrous, the lobes short-triangular; corolla 10–15 mm. in diameter, bright yellow with brownish purple spots at the base; anthers tinged with green or purple; fruiting calyx ovoid, 1.5–2 cm. long, obscurely 10-angled, sparsely villous or glabrous; berry purple.

An escape from cultivation, rice-field levees: Sacramento Valley, Marin County southward; east to Atlantic.

2. Physalis lanceifolia Nees, Linnaea 6:473. 1831.

Erect annual 5–8 dm. tall; stems ascending, angled, glabrous, the foliage subglabrous; leaves narrowly lanceolate to lanceolate, 3.5–7 cm. long, attenuate at both ends, entire to shallowly sinuate-toothed, dark green, subglabrous or with a few stiff, short hairs on the veins; pedicels filiform, 1.5–3 cm. long; calyx puberulent, tubular-campanulate, 3–4 mm. high, the lobes deltoid; corolla yellow, narrowly campanulate, 5–6 mm. in diameter; anthers ovoid, 1.5–2 mm. long, greenish or purplish; fruiting calyx broadly ovoid, 2–2.5 cm. long, on pedicels as long as or longer than the calyx, the lobes broadly deltoid.

Rice-field levees and canal borders in the Sacramento Valley, moist fields and along rivers in the Imperial Valley; east to Texas and south to Mexico.

SOLANUM

Herbs or shrubs; herbage glabrous, pubescent or tomentose. Leaves alternate, simple, entire or lobed. Flowers in umbels or cymes, white, blue, purple, or yellow. Calyx campanulate or rotate, 5-toothed or 5-cleft. Corolla rotate, 5-lobed. Stamens 5; filaments short; anthers converging around the style to form a cylinder; cells dehiscing by a terminal pore. Ovary 2-celled; stigma

Fig. 307. *Solanum nodiflorum: a,* habit, × ⅖; *b,* inflorescence, × 4; *c,* seeds, × 6½; *d,* fruit, × 1½.

small, capitate or obscurely bilobed. Fruit usually a globose berry with numerous more or less flattened seeds.

In addition to the following, *Solanum triflorum* and *S. elaeagnifolium* may occur on floodlands.

Stebbins, G. L., and Elton F. Paddock, The *Solanum nigrum* complex in Pacific North America. Madroño 10:70–81. 1949.

Ripe berries reddish; seeds 1.8–2 mm. long; leaves hastate and lobed; anthers 5 mm. long
1. S. Dulcamara
Ripe berries black; seeds 1.2–1.8 mm. long; leaves not hastate or lobed; anthers 1.5–3.4 mm. long.
  Flowers large; corolla lobes 6–11 mm. long; style exserted about 2 mm. beyond the anthers; anthers 2.6–4 mm. long . . . . 2. *S. Douglasii*
  Flowers small; corolla lobes about 3 mm. long; style barely exserted beyond the anthers; anthers 1.5–2.4 mm. long . . . . 3. *S. nodiflorum*

1. Solanum Dulcamara L., Sp. Pl. 185. 1753.

Perennial straggling or climbing vine, woody below with branches to 3 m. long, sparsely puberulent or glabrate; leaves ovate to hastate, 5–12 cm. long, acute or acuminate at apex, rarely entire, often with 1 lobe on one side near the base or deeply 3-lobed, the basal lobes smaller than the terminal one; flowers in lateral compound cymes, drooping; calyx 3–4 mm. deep; corolla deeply 5-cleft, 12–16 mm. across, blue, the lobes triangular-lanceolate; anthers about 5 mm. long; berry oval to globose, bright red, 8–12 mm. long; seeds light-colored, minutely tessellated with low, rounded tubercles.

An escape from cultivation in marshy stream and lake borders, moist shady places, and waste places; northeastern California; north to Washington and from Minnesota to Nova Scotia. Native of Europe.

2. Solanum Douglasii Dunal in A. DC., Prodr. 13¹:48. 1852.

Solanum nigrum var. *Douglasii* Gray, Syn. Fl. N. Am., 2d ed., 2¹:228. 1886.

Bushy perennial, 0.6–2 m. tall, with angled stems; herbage puberulent to subglabrate, the simple, antrorse hairs with heavy, conical bases; leaves ovate, 2–10 cm. long, coarsely sinuate-dentate, acute to short-acuminate, cuneate to subtruncate at base, sparsely puberulent; inflorescence several-flowered, the peduncle 1–3 cm. long, remaining erect at maturity, the pedicels slender, 5–12 mm. long; calyx 2–3 mm. long at anthesis, the lobes lanceolate-oblong; corolla white with greenish basal spots, the lobes lanceolate-oblong, 6–11 mm. long; anthers 2.6–4 mm. long; style well exserted beyond anthers; seeds light yellow, minutely reticulate-pitted.

Streamsides, swales in coastal dunes, drying floodlands, dry slopes, and waste places: coastal North Coast Ranges in Mendocino County, South Coast Ranges from San Mateo County south to southern California; Baja California, Mexico, east and south through Arizona and New Mexico to northern Mexico.

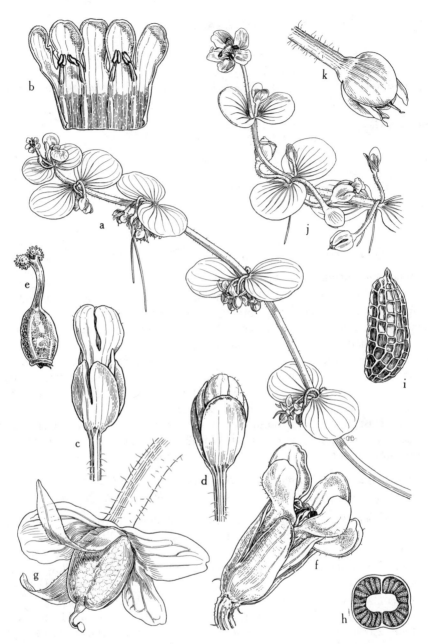

Fig. 308. *Bacopa. a–i, B. Nobsiana: a,* habit, × ⅖; *b,* corolla, spread open, × 4; *c* and *d,* calyx, two views, showing zygomorphic character, × 4; *e,* pistil, × 8; *f,* flower, × 8; *g,* mature fruit, showing spreading calyx lobes, × 6; *h,* ovary (cross section), × 6; *i,* seed, showing reticulately sculptured surface, × 40. *j* and *k, B. Eisenii: j,* habit, × ⅘; *k,* calyx lobes appressed over mature fruit, × 2.

3. Solanum nodiflorum Jacq., Icon. Pl. Rar. 2:288, pl. 326. 1786. Fig. 307.

Solanum nigrum of western American authors, not L.

Straggling perennial with branches to 6 dm. long; stems glabrous or sparsely pubescent, angular or rounded; leaves ovate to elliptic, to 12 cm. long, entire or sinuate-dentate, cuneate to truncate at base; inflorescence umbelliform; calyx lobes small, thick and reflexed at maturity; corolla white or faintly tinged with purple, 4–6 mm. in diameter, the lobes lanceolate-ovate; anthers 1.5–2.4 mm. long, on hairy filaments; berry globose, 5–6 mm. in diameter, black or purple-black, the granules at base 3–6 or wanting; seeds about 40–60 in each berry, pale cream-colored, minutely granulose, with low, round tubercles.

A common weed in marshes and on ditch banks, on drying floodlands and cultivated land, and in waste places: coastal California and the Central Valley to southern California; north to Washington. Widely distributed as a weed in the tropics of both Eastern and Western hemispheres; probably a native of the Western Hemisphere.

SCROPHULARIACEAE

Herbs, shrubs, or trees, chiefly with opposite leaves, but sometimes leaves alternate. Flowers hypogynous, irregular. Calyx synsepalous, 4- or 5-lobed. Corolla of 5 petals, these often fused to simulate 4 petals, usually bilabiate. Stamens 2 or 4, rarely 5, usually fewer than the petals, sometimes a rudimentary fifth stamen represented by a sterile filament. Ovary 2-celled with a single style and a 2-lobed stigma. Fruit a capsule.

Upper lip of the corolla not produced into a beak; leaves opposite or the upper ones
    alternate.
    Anther-bearing stamens 4, the fifth stamen present as a gland or scale; corolla reddish
        purple; tall plant with many-flowered panicles .... *Scrophularia*
    Anther-bearing stamens 2 or 4, the fifth stamen wholly absent.
        Corolla 2-lipped, 2 lobes in upper lip and 3 in lower one, or the corolla nearly
            regular and 5-lobed.
        Anther-bearing stamens 4.
            Calyx tubular, 5-toothed, or if lobed the lobes of equal size.
                Plants erect.
                    Calyx 5-angled, folded on the angles .... *Mimulus*
                    Calyx not 5-angled .... *Mimetanthe*
                Plants caespitose and stoloniferous .... *Limosella*
            Calyx 5-parted, the lobes unequal in size; creeping or floating plants; leaves
              orbicular-cuneate, sessile .... *Bacopa*
        Anther-bearing stamens 2.
            Corolla strongly 2-lipped, pale blue to blue-violet.
                Upper leaves much reduced, scale-like; herbage sparsely stipitate-glandular;
                  sterile filament minute .... *Dopatrium*
                Upper leaves ovate, not much reduced; herbage glabrous; sterile filaments
                  forked .... *Lindernia*

Corolla lobes weakly 2-lipped, white or yellow .... *Gratiola*
Corolla 4-lobed; stamens 2; capsule usually obcordate .... *Veronica*
Upper lip of the corolla produced into a beak or galea; leaves alternate.
Calyx 2- or 4-cleft, but not to base.
Lower lip of corolla 3-toothed, upper lip or beak much longer than the lower one;
bracts bright red .... *Castilleja*
Lower lip of corolla 3-saccate, upper lip not much longer than the lower one; bracts
white, yellow, or pink .... *Orthocarpus*
Calyx cleft to base in front, tonguelike behind .... *Cordylanthus*

BACOPA

Floating or prostrate annuals or perennials, rooting at some nodes; leaves opposite, sessile, orbicular-cuneate to cuneate, often somewhat clasping, palmately nerved; flowers in axillary clusters of 1–4; sepals 4 or 5, growing with fruit, often appearing as though an inner and an outer series, the inner series often membranous; corolla campanulate, the lobes nearly equal in size, weakly disposed toward a grouping of 2 and 3 petals; stamens 3 or 4, inserted on throat of corolla, the anthers versatile; ovary 2-celled, asymmetrical; capsule ovoid, 4 mm. long; seeds many, the testa bladdery, reticulately sculptured.

Pedicels in anthesis as long as or longer than the subtending leaves; mature capsule
globose to oblong, enclosed by the appressed sepals .... 1. *B. Eisenii*
Pedicels in anthesis about ½ as long as the subtending leaves; mature capsule ovoid,
usually exposed by the spreading sepals .... 2. *B. Nobsiana*

1. Bacopa Eisenii (Kell.) Pennell, Proc. Acad. Nat. Sci. Phila. 98:96. 1946.
Fig. 308.

Bacopa rotundifolia of western American authors, not Wettstein.

Aquatic annual, creeping and rooting at some of the nodes or at length floating on the surface of shallow pools; stems fistulose, sometimes slightly constricted at the nodes, pilose-hirsute, becoming glabrate in age, 1–5 dm. long; leaves opposite, 10–20 mm. long, almost orbicular to orbicular-cuneate, sessile by a broad base, with 10–12 palmate veins; flowers solitary or in pairs in the leaf axils, on slender pedicels longer than the leaves, 1–3 cm. long, erect and emersed in anthesis, becoming submersed and reflexed in fruit; calyx somewhat bilabiate, flattened in bud, the upper sepal broad, orbicular-cuneate, entire, the lower 2 sepals joined as 1 but 2-lobed or 2-cleft or 2-parted, the lateral 2 sepals lanceolate to linear, ciliate-margined or entire, connivent over fruit; corolla 8–12 mm. long, white, the throat and tube yellow, the lobes imbricate in bud, campanulate, the 5 petals nearly regular; stamens 3 or 4, the anthers versatile; style branched at summit, each branch tipped with a capitate stigma; fruit a capsule splitting lengthwise into 2 or 4 valves, the valves disarticulating completely; seeds many, with a bladdery testa marked with 10–12 rows of large areolae.

Shallow ponds and marshes: rice fields of the San Joaquin Valley, and Lone Pine in Inyo County.

Jepson reports this species, under the name *Bacopa rotundifolia,* as perennial. We have seen no evidence of its perennial nature.

2. Bacopa Nobsiana Mason, Madroño 11:206. 1952. Fig. 308.

Floating or prostrate and creeping herb, rooting at some of the nodes; stems and pedicels pilose to hirsute when young, becoming glabrate in age; leaves opposite, sessile, 1–2 or –3 cm. long, nearly as wide as long, orbicular-cuneate, clasping the stem, conspicuously palmately 5- to 7-nerved, glabrous, succulent; flowers 1–4 in each leaf axil; pedicel plus flower shorter than the subtending leaf, pedicel sometimes elongating in fruit; calyx growing with fruit, the sepals 4 or 5, usually spreading in age, appearing as though in an inner and an outer series, the outer 2 sepals each deeply cleft or parted, almost orbicular, foliaceous, 5- to 7-nerved, the inner 2 sepals oblong, 2 mm. long, 1 mm. wide, minutely ciliate-margined below or sometimes glabrous, at first membranous, becoming firmer in age; corolla campanulate, white with yellow throat, the lobes subequal in size; stamens 4, inserted on throat of corolla, the anthers versatile; capsule broadly ovoid, asymmetrical, 4 mm. long, conspicuously grooved up the sides and across the top along the septum, thus often appearing as though 2-lobed; seeds many, ellipsoid to oblong, the testa bladdery, reticulately sculptured with 7 or 8 transverse areolae.

Shallow ponds and rice fields: San Joaquin and Sacramento valleys, less common in the San Joaquin Valley.

CASTILLEJA

Annual or perennial herbs or subshrubs with alternate, simple to pinnatifid leaves. Flowers in spikes or spikelike racemes subtended by foliaceous or colored bracts. Sepals variously cleft, growing with the capsule. Corolla sympetalous, irregular, the upper pair of petals fused and elongated to form a galea, the lower 3 petals much reduced, forming the lower lip. Stamens 4, the anther cells unequally placed. Ovary asymmetrical. Style elongate; stigma bilobed. Fruit an ovoid capsule, dehiscent by 2 basally persistent valves. Seeds with a loose, membranous, reticulate coat.

A genus of about two hundred genetically complicated species, found chiefly in the New World and northern Asia.

A species not included in the treatment below, but originally found in a bog, is *Castilleja Leschkeana,* which was described by J. T. Howell from a single plant on Point Reyes.

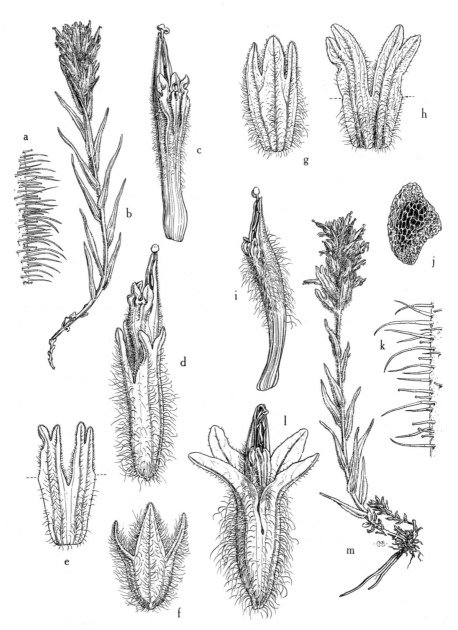

Fig. 309. *Castilleja. a–f, C. Lemmonii: a,* stem hairs, × 16; *b,* habit, × ⅔; *c,* corolla, front view, showing the 3-saccate lower lip, × 3; *d,* flower, × 3; *e,* calyx, spread open, the cut tube extending from the base to the dotted lines, × 2; *f,* bract, × 2. *g–m, C. Culbertsonii: g,* bract, × 2; *h,* calyx, spread open, the cut tube extending from the base to the dotted lines, × 2; *i,* corolla, side view, the lower lip obscurely saccate, × 3; *j,* seed, × 12; *k,* stem hairs, × 16; *l,* flower, × 3; *m,* habit, × ⅔.

Plants perennial.

Corolla tube distended into 3 pouchlike lobes or folds below lower lip, the lower lip of corolla usually exserted from calyx.

Lower lip of corolla with conspicuous, free lobes; calyx not deeply cleft laterally .... 1. *C. Lemmonii*

Lower lip of corolla with rudimentary or inconspicuous, free lobes; calyx deeply cleft laterally .... 2. *C. Culbertsonii*

Corolla tube not pouchlike below lower lip, the lower lip of corolla included in or barely exserted from calyx.

Leaves oblong to oblong-ovate from a broad, sessile base; flowers in an elongate spike .... 3. *C. uliginosa*

Leaves linear-lanceolate to broadly lanceolate, usually narrowed at the base; flowers in a branched inflorescence of short spikes.

Spikes lax, the bracts entire to toothed or shallowly lobed .... 4. *C. elata*

Spikes dense, the bracts mostly deeply lobed .... 5. *C. miniata*

Plants annual.

Corolla 2–3 cm. long .... 6. *C. stenantha*

Corolla 1–2 cm. long.

Corolla well exserted from calyx, pubescence not coarse, spike slender .. 7. *C. minor*

Corolla included within calyx or barely exserted from it, pubescence coarse, spike stout .... 8. *C. exilis*

1. Castilleja Lemmonii Gray, Syn. Fl. N. Am. 2[1]:297. 1878. Fig. 309.

Perennial herb from a taproot; the lateral branches often hypogeous and radiating as short rhizomes 2–8 cm. long; stems then erect, 1–2 dm. tall; herbage puberulent to short-villous, the stems becoming glabrous below; leaves linear to linear-lanceolate, 1–3 cm. long; bracts with 1 or 2 pairs of linear lobes, purple-tipped; calyx medianly cleft, the lobes again short-cleft or notched at apex; corolla 18–22 mm. long, the galea attenuate, ½ as long as tube, glandular-hairy to glabrate on back, the lower lip somewhat distended or 3-saccate below, the free lobes abruptly horizontal or ascending to erect, each with a palate-like swelling at base; capsule 10–12 mm. long.

Moist meadows and bogs: Sierra Nevada from Shasta County to Inyo County.

This species is very closely related to *Castilleja Culbertsonii* but differs in the more highly developed lower lip of the corolla and in the less conspicuous calyx lobes. The saccate lower lip resembles that of some species of *Orthocarpus*.

2. Castilleja Culbertsonii Greene, Leafl. Bot. Obs. & Crit. 1:78. 1904. Fig. 309.

Perennial herb from a taproot, branches few to several, often radiating as short rhizomes; stems then erect, 5–20 cm. tall, simple or sparingly branched; herbage glandular-villous, the stems becoming glabrous below; leaves linear to linear-lanceolate, entire or occasionally with a pair of linear lobes near the summit; bracts with 1 or 2 pairs of lobes, the tip rounded at summit, purple; calyx 16–18 mm. long, medianly cleft about halfway, the lobes again deeply cleft and often petaloid; corolla 20–25 mm. long, pubescent on back, the galea

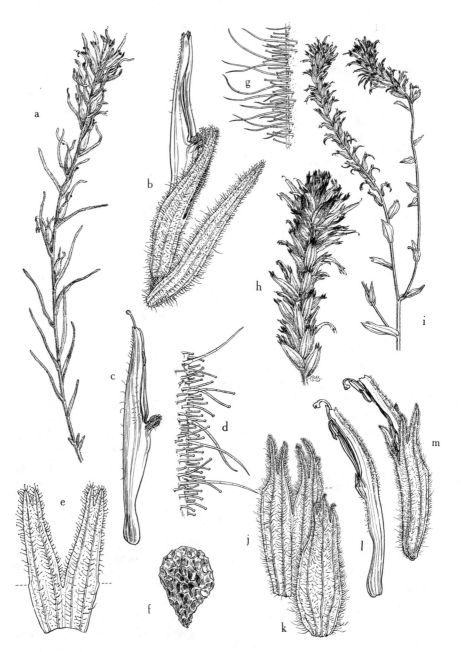

Fig. 310. *Castilleja. a–f, C. stenantha: a,* habit, × ⅖; *b,* flower and subtending bract, × 2; *c,* corolla, side view, showing the divergent teeth of the lower lip, × 2; *d,* stem hairs, × 6; *e,* calyx, spread open, the cut tube extending from the base to the dotted lines, × 2; *f,* seed, × 12. *g–m, C. uliginosa: g,* stem hairs, × 8; *h,* upper part of inflorescence, × ⅔; *i,* habit, × ⅕; *j,* calyx, spread open, the cut tube extending from the base to the dotted lines, × 2; *k,* bract, × 2; *l,* corolla, showing the outer segments of the protuberant lower lip longer than the inner segment, × 2; *m,* flower, × 2.

attenuate, about ½ as long as the tube, the lower lip 3–4 mm. long, glabrous, green, obscurely saccate or folded below, often with rudimentary lobes or teeth above; capsule 10–12 mm. long.

High mountain meadows and bogs: southern Sierra Nevada north to Tuolumne County.

3. Castilleja uliginosa Eastw., Leafl. West. Bot. 3:117. 1942. Fig. 310.

Weak, straggling, procumbent, perennial herb, openly branched, the branches ascending, 3–5 dm. long; herbage sparsely villous, the hairs very unequal in length; leaves oblong-lanceolate to ovate-lanceolate, from a broad, sessile base, entire, 3–5 cm. long, 1–2 cm. broad, much reduced upward; spikes 1–3 dm. long, 3–4 cm. broad, the lower bracts leaflike and entire, the upper ones 3-toothed or 3-lobed, yellowish green to reddish; calyx about 20–25 mm. long, medianly cleft about halfway, the lobes again cleft into linear to linear-spatulate or lanceolate segments; corolla 22–30 mm. long, the galea long-attenuate, equal or subequal to the tube in length, finely pubescent along the back to below base and thence around the top of tube, the lower lip protuberant, the outer segments longer than inner ones, the segments lanceolate, acuminate, sparsely pubescent.

Known only from Pitkin Marsh, Sonoma County.

4. Castilleja elata Piper, Smithson. Misc. Coll. 50:201. 1907. Fig. 311.

Perennial herb 4–8 dm. tall; herbage glabrous below, glandular-pubescent to hirsute in inflorescence; leaves linear-lanceolate, entire; spikes often many and short; bracts leaflike, dull red to orange in anthesis, a lateral tooth or lobe on each side near the tip, the lowermost bracts occasionally entire; calyx 9–17 mm. long, medianly cleft to below middle, the lobes again cleft for 2–3 mm.; corolla 15–25 mm. long, the galea attenuate, about as long as the tube, not toothed at summit, glandular-puberulent, the lower lip incurved, 1–2 mm. long; capsule 6–8 mm. long.

Bogs and wet serpentine meadows: northwestern California; adjacent Oregon.

This species is closely related to *Castilleja miniata*.

5. Castilleja miniata Dougl. ex Hook., Fl. Bor. Am. 2:106. 1838. Fig. 311.

Perennial herb 2–8 dm. tall; herbage glabrous or pubescent, becoming villous and often glandular in the inflorescence; leaves entire or the upper ones 3-toothed or 3-lobed; bracts usually with a pair of slender lobes below tip, often bright red or paler; calyx 17–25 mm. long, cleft medianly to below the middle, the lateral lobes again cleft for 3–7 mm.; corolla 15–30 mm. long, the galea 8–20 mm. long, with a pair of slender teeth at summit, glandular-puberulent, the lower lip 1–2 mm. long, incurved, usually included in calyx; capsule 10–12 mm. long.

Moist meadows and stream banks, occasionally in thickets in bogs: middle

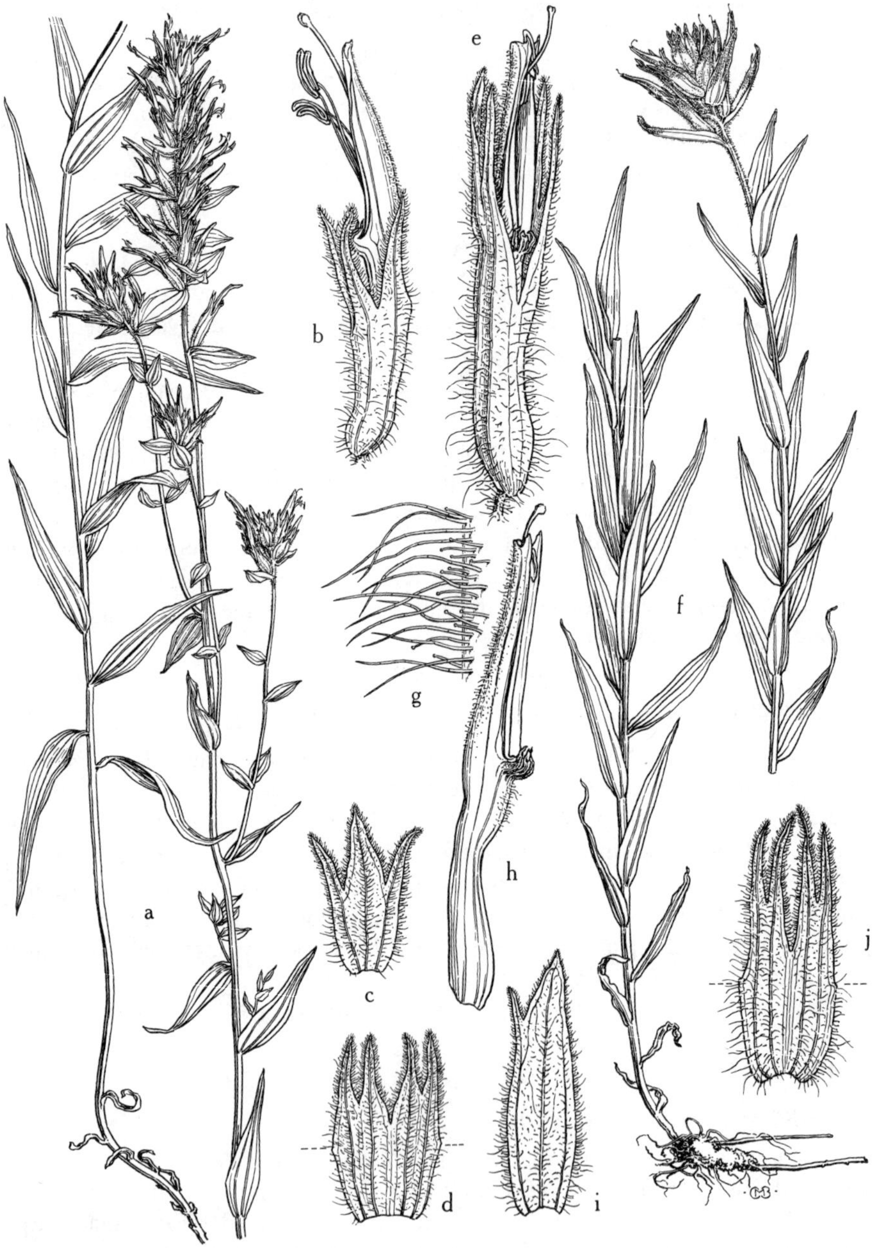

Fig. 311. *Castilleja. a–d, C. elata: a,* habit, × ⅖; *b,* flower, × 2; *c,* bract, × 1⅕; *d,* calyx, spread open, the cut tube extending from the base to the dotted lines, × 1⅕. *e–j, C. miniata: e,* flower, × 2; *f,* habit, × ⅖; *g,* stem hairs in inflorescence, × 4; *h,* corolla, × 2; *i,* bract with only 1 lateral lobe (atypical), × 1⅕; *j,* calyx, spread open, the cut tube extending from the base to the dotted lines, × 1⅕.

altitudes throughout the mountains; north to British Columbia, east to Rocky Mountains.

6. Castilleja stenantha Gray, Syn. Fl. N. Am. 2¹:295. 1878. Fig. 310.

Castilleja spiralis Jepson, Fl. West. Middle Calif. 412. 1901.

Annual herb; stems solitary to few, simple or branched from base, or occasionally a few lateral branches above, 3–10 dm. tall; herbage pubescent, some hairs gland-tipped; leaves lanceolate-attenuate, simple, entire, often coiling at tip; bracts linear, entire, scarlet-tipped, becoming green in age, longer than the flowers; flowers in an elongate spikelike raceme, the lower flowers somewhat remote, short-pediceled; calyx cleft medianly into ovate-attenuate lobes, these notched or cleft at tip; corolla 25–35 mm. long, the galea 15–20 mm. long, yellow, well exserted, about as long as or longer than the tube, the lower lip 2–3 mm. long, conspicuously divergent, yellow, green, purple, or red; capsule 10–12 mm. long.

Marshes, bogs, or stream banks: central Coast Ranges and Sierra Nevada from Colusa and Tuolumne counties to San Diego County.

Northern California material at one time was segregated by Jepson as *Castilleja spiralis,* on the basis of the dark-colored lower lip of the corolla. He later abandoned *C. spiralis* in favor of *C. stenantha* Gray. The differences hold, but it is questionable that two species are involved.

7. Castilleja minor (Gray) Gray, Bot. Calif. 1:573. 1876. Fig. 312.

Slender annual; stems erect, simple or branched from near base, 2–10 dm. tall; herbage variously pubescent but with some gland-tipped hairs, occasionally glabrate below, usually somewhat viscid or "clammy"; leaves linear to lanceolate-attenuate, 4–10 cm. long, minutely but densely pilose, interspersed with gland-tipped hairs; bracts entire, linear to lanceolate-attenuate, the lower bracts green, the upper ones red-tipped; flowers in a spikelike raceme, pediceled, the lower flowers remote; calyx cleft medianly into 2 ovate-attenuate lobes, these cleft or notched at apex; corolla short, but the galea well exserted, the lower lip included in calyx or barely exserted, the galea about ½ to nearly as long as corolla tube; capsule included in growing calyx, 10–12 mm. long.

Subsaline to alkaline marshes and bogs east of the Sierra Nevada crest; north to Canada, south to Mexico, east to Rocky Mountains. An extremely variable species.

8. Castilleja exilis Nels., Proc. Biol. Soc. Wash. 17:100. 1904. Fig. 312.

Slender to coarse annual; stems simple, erect, or with few branches from base, 3–10 dm. tall; pubescence various, often of coarsely hispid-pilose hairs intermixed with fine hairs or sometimes with gland-tipped hairs; leaves lanceolate-attenuate to linear, entire; bracts foliaceous, erect and scarlet-tipped when young, soon becoming green, entire; flowers in a stout, spikelike raceme, the flowers becoming more remote after anthesis; calyx green, 15–18 mm. long,

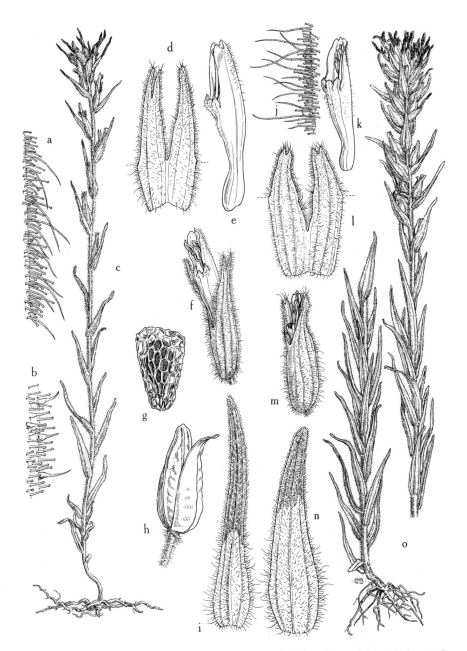

Fig. 312. *Castilleja. a–i, C. minor: a,* stem hairs, × 6; *b,* calyx and bract hairs, × 8; *c,* habit, × ⅖; *d,* calyx, spread open, the cut tube extending from the base to the dotted lines, × 2; *e,* corolla, side view, × 2; *f,* flower, × 1½; *g,* seed, × 12; *h,* capsule, × 2; *i,* bract, × 2. *j–o, C. exilis: j,* stem hairs, × 4; *k,* corolla, × 2; *l,* calyx, spread open, the cut tube extending from the base to the dotted lines, × 2; *m,* flower, × 1½; *n,* bract, × 2; *o,* habit, × ⅖.

cleft medianly for about ⅔ of its length into ovate-attenuate lobes, these entire, toothed, or emarginate at apex, rarely cleft; corolla 14–20 mm. long, included within or barely exserted from the calyx, the galea about ½ as long as tube, the lower lip included; capsule 8–12 mm. long.

Saline or alkaline bogs or marshes of the northern Great Basin area. In California *Castilleja exilis* is known only from Surprise Valley in Modoc County, but the species has also been found on the west side of Carson Valley, near Mottsville, Nevada.

In herbaria *C. exilis* is often confused with *C. minor,* from which it may be distinguished by its stouter inflorescence, its shorter, usually included corollas, and its coarser pubescence. Our material had a pungent, mintlike fragrance when fresh.

CORDYLANTHUS

Annual herbs, erect. Leaves alternate, entire or pinnatifid, glabrous to densely hispid, hirsute, or canescent. Flowers in a bracteate spike or scattered along the branches. Calyx spathelike, cleft to the base down the back. Corolla tubular, 2-lipped, the lips subequal in length, the upper lip curved or hooked at tip, the lower one crenately 3-lobed. Stamens 4 or 2, the anthers 2-celled or 1-celled, sometimes both types in the same flower, the cells pubescent at base. Fruit a flattened capsule, 2-celled. Seeds with a membranous, loose coat.

There are approximately thirty species of *Cordylanthus* in California, some of which, on occasion, occur on floodlands or near marshes. Only the following do so consistently.

Stamens 4, one pair 2-celled, the other 1-celled; bracts entire, or with a pair of short lobes near the tip.
    Bracts with a pair of short lobes near the tip; corolla shorter than the calyx; coastal
        species .... 1. *C. maritimus*
    Bracts entire; corolla longer than the calyx; east of the Sierra Nevada .. 2. *C. canescens*
Stamens 2; bracts deeply lobed, chiefly from near the base; Central Valley to San Fran-
    cisco Bay .... 3. *C. mollis*

1. Cordylanthus maritimus Nutt. ex Benth. in A. DC., Prodr. 10:598. 1846.

Annual herb 1–3 dm. tall; herbage glaucous to white-pubescent; leaves lanceolate-oblong, 20–25 mm. long, entire; bracts entire or with a pair of shallow lobes near the tip, not gland-tipped; calyx entire or notched at apex; corolla about as long as the bracts, white to purplish, pubescent, 2-lipped, the lips about equal in length, the lower lip shallowly 3-lobed, the upper one curved at apex; stamens 4, both anthers of the upper pair 2-celled, their filaments glabrous, both anthers of the lower pair 1-celled, their filaments hairy; style hooked at apex, exserted; fruit a capsule; seeds favose-pitted.

Salt marshes along the coast from southern Oregon to northern Baja California, Mexico.

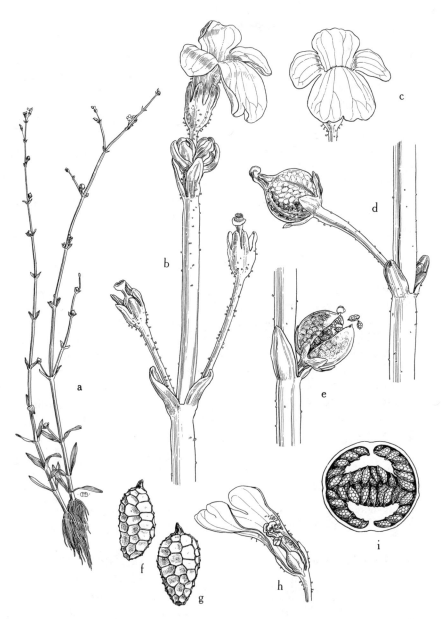

Fig. 313. *Dopatrium junceum: a,* habit, × ⅖; *b,* inflorescence, × 6; *c,* flower, front view, × 6; *d,* capsule, × 6; *e,* capsule, showing dehiscence, × 6; *f* and *g,* seeds, × 40; *h,* flower, spread open, side view, × 6; *i,* ovary (cross section), × 12.

2. Cordylanthus canescens Gray, Proc. Am. Acad. Arts & Sci. 7:383. 1868.

Cordylanthus maritimus var. *canescens* Jepson, Man. Fl. Pl. Calif. 947. 1925.
Cordylanthus maritimus var. *Parryi* Jepson, *loc. cit.*

Annual herb 1–3 dm. tall; herbage pubescent, glandular, the exudate becoming powdery; leaves lanceolate and glabrescent, glaucous green, sometimes becoming purplish; bracts ovate-lanceolate, foliaceous, canescent-pubescent; calyx lanceolate, enclosing the corolla tube, 13 mm. long, cleft at apex; corolla exserted beyond calyx, 15–17 mm. long, pale yellow to whitish, pubescent near the apex, the lower lip inflated; stamens 4, the upper pair with rudimentary anthers, the anthers of the lower pair fused; capsule about 1 cm. long.

Saline and alkaline flats around water holes: east of the Sierra Nevada crest from Modoc County south to Mono County; east to Utah.

3. Cordylanthus mollis Gray, Proc. Am. Acad. Arts & Sci. 7:384. 1868.

Bushy annual herb, erect or ascending, 1–3 dm. high; herbage pubescent to densely villous, often glandular-viscid; leaves 8–15 mm. long, linear to ovate-lanceolate, the lower leaves entire, the upper ones pinnatifid from the lower part of the blade; bracts pinnately lobed; rachis broad, green to purplish; flowers in bracteate spikes; calyx entire or cleft at the tip; corolla 2-lipped, white to purple, 15–20 mm. long, the tube slender, slightly gibbous at base, the lower lip inflated, with 3 short lobes at tip, the upper lip curved, both pubescent; stamens 2, the anthers 2-celled, hairy at base, the filaments glabrous; style hooked or bent at apex, exserted; seeds reniform, reticulate-favose.

Salt marshes and saline and alkaline floodplains: borders of San Francisco Bay, Central Valley.

Pennell, in his treatment of this complex (in Abrams, Illus. Fl. Pacif. States, 3:850, 1951), divides it into *Cordylanthus mollis* Gray, *C. palmatus* (Ferris) Macbride, *C. hispidus* Pennell, and *C. carnulosus* Pennell. He overlooked *C. mollis* var. *viridis* Jepson, from near Bakersfield. The characters used by both Pennell and Jepson are the cleft versus the entire calyx, the disposition of the lobing of the bracts, the color of the herbage, and the color of the corolla. In my field experience with these plants, acquired in an attempt to resolve the problems in the group, I found that these characters are very inconclusive. The calyx often varies appreciably on an individual plant, from entire to deeply cleft adaxially. The color of the herbage seems to be associated with hemiparasitism, as evidenced by the fact that plants having colored bracts have their roots attached to the roots of other plants; however, these relationships will vary within an individual colony. When the herbage is colored, the corolla is also sometimes colored. Finally, the lobing of the bracts, whether spreading or ascending, is so variable on an individual plant as to be inconclusive. I have been unable to find any significant concomitance of characters sufficient to warrant use in the differentiation of taxa. Sometimes a given population

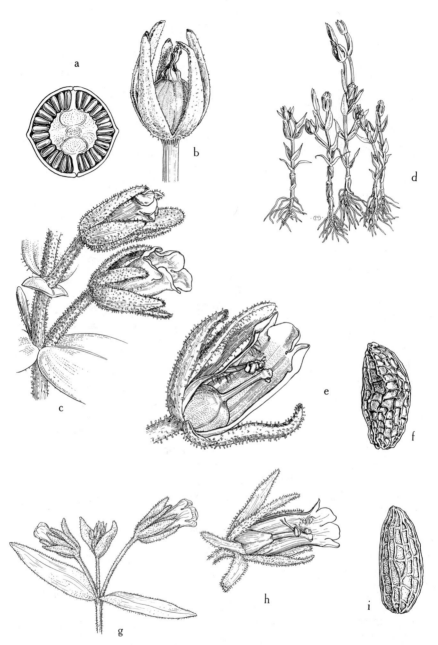

Fig. 314. *Gratiola. a–f, G. ebracteata: a,* ovary (cross section), × 6; *b,* mature capsule, shorter than the calyx lobes, × 4; *c,* part of inflorescence, showing a single flower borne in the leaf axil, × 2½; *d,* habit, erect form, × ⅕; *e,* flower, × 4; *f,* seed, × 40. *g–i, G. neglecta: g,* inflorescence, × 2½; *h,* flower, spread open to show stamen insertion, × 4; *i,* seed, × 40.

may be marked off by characters of the pubescence. This by itself, however, seems inadequate to the needs of taxonomy. It is perhaps best to regard this exceedingly variable complex as being in an active state of speciation, with minor habitat races becoming evident. Of these, *C. palmatus* represents one extreme and *C. mollis* another.

DOPATRIUM

1. Dopatrium junceum Ham. in Benth., Scroph. Indicae 31. 1835. Fig. 313.

Small, slender annual, fleshy and sparingly gland-dotted; leaves few, opposite, the upper leaves remote, about 5 mm. long or shorter, minute, the lower stem leaves to 2 cm. long; calyx 5-lobed, the tube about as long as the obtuse lobes; corolla 2-lipped, pale blue with slender tube slightly longer than the calyx, the upper lip erect, slightly 2-lobed, the lower lip 3-lobed, the middle lobe much longer than the lateral lobes; anther-bearing stamens 2, the rudimentary stamens 2, minute, all included, the filaments short; pistil with short style; style 1; stigma 2-lobed; capsule globose; seeds reticulate.

Rice fields: Butte County; native of Eurasia.

GRATIOLA

Annuals with fistulose stems and succulent leaves; herbage glabrous to puberulent or glandular-pubescent. Leaves opposite, sessile, entire to dentate. Flowers solitary in the leaf axils, bracteate or ebracteate. Sepals unequal in size, free or partially united. Corolla tubular, 2-lipped, tube yellow, limb white. Anther-bearing stamens 2, sometimes 4, sterile ones also present. Stigma dilated. Capsule 4-valved.

Calyx with a pair of bracts at base; corolla 2–3 times as long as the calyx .. 1. *G. neglecta*
Calyx without a pair of bracts at base; corolla as long as or slightly longer than the calyx.
    Sepals lanceolate, attenuate, separate to base (occasionally slightly joined at base);
        corolla white .... 2. *G. ebracteata*
    Sepals oblong, obtuse and emarginate, the upper 3 joined for ⅓ of their length or
        more; corolla yellow except for the 2 white lower lobes ...... 3. *G. heterosepala*

1. Gratiola neglecta Torr., Cat. Pl. N. Y. 89. 1819. Fig. 314.

Gratiola virginiana of western American authors, not L.

Erect annual; stems often much-branched, 5–30 cm. tall; herbage puberulent to viscid; leaves ovate to oblong-lanceolate, acute, crenate-serrate to entire, 2–4 cm. long; sepals unequal in size, free, lanceolate, with a pair of long bracts at base; corolla much longer than the calyx, the yellow tube and throat usually conspicuously exserted, the limb white, 2-lipped; capsule ovoid, about as long as the calyx; seeds reticulate with prominent longitudinal ribs.

Wet ground, stream banks and along irrigation ditches: northern Sierra

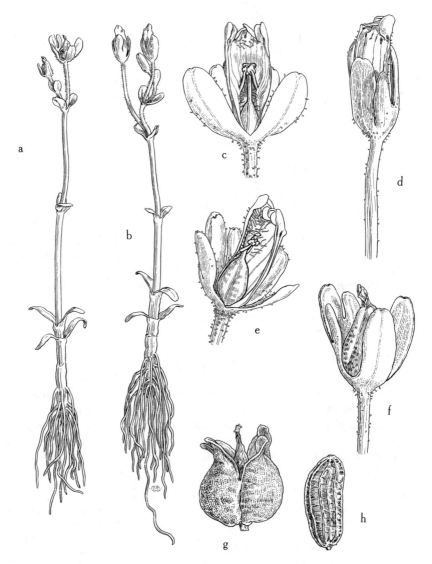

Fig. 315. *Gratiola heterosepala: a* and *b,* habit, × 1⅕; *c,* flower, front view, the lower lip removed, × 4; *d,* flower, side view, × 4; *e,* flower, side view, spread open, × 5; *f,* capsule and sepals, × 4; *g,* capsule after dehiscence, × 6; *h,* seed, × 40.

Nevada from Mariposa County north, and into Lassen, Modoc, and Shasta counties; east to Atlantic.

2. Gratiola ebracteata Benth. in A. DC., Prodr. 10:595. 1846. Fig. 314.

Erect or spreading annual, simple or few-branched; stems fistulose, 2–10 cm. high, somewhat glandular-pubescent; leaves lanceolate, entire, 6–15 mm. long; sepals unequal in size, lanceolate, free or sometimes slightly fused at base, surpassing the mature capsule in length; corolla equal or subequal to the calyx, often campanulate, somewhat 2-lipped, the upper lip bifid, with a band of glandular hairs below the sinus, the lower lip of short, broad, notched lobes; capsule globose, somewhat 4-angled, as long as the calyx lobes; seed reticulate with prominent longitudinal ribs.

In shallow water or wet mud on the margins of ponds or vernal pools, at low and middle altitudes: throughout northern California; north to Washington.

3. Gratiola heterosepala Mason & Bacigalupi, Madroño 12:150. 1954. Fig. 315.

Erect annual; stems stout and fistulose, 2–10 cm. tall, sparsely glandular-pubescent in the inflorescence; leaves entire, the lowermost linear-lanceolate, 1–2 cm. long, the upper ones much shorter, oblong, obtuse; calyx unequally cleft, the 3 upper sepals joined for ⅓ of their length, the middle lobe longer than the other 2, obtuse, the 2 lower sepals distinct, emarginate, 4–6 mm. long; corolla yellow, 6–8 mm. long, slightly longer than the calyx, the lobes 1–2 mm. long, the upper pair joined nearly to the tip; capsule pyriform, about as long as the calyx; seed with distinct longitudinal ribs and indistinct horizontal ribs.

In shallow water of lake margin: known only from Boggs Lake in Lake County.

LIMOSELLA. MUDWORT

Small rosulate plants of aquatic or wet habitats, usually acaulescent. Leaves basal, erect, rarely cauline. Flowers solitary on naked, 1-flowered scapes, white to pink or pale blue. Calyx campanulate. Corolla campanulate, nearly regular, the upper surface of petals minutely to sparsely papillate. Stamens 4. Style terminal or subterminal. Capsule globose to ellipsoid, 2-celled by a very thin partition, many-seeded.

Leaves terete to subulate, without blade; capsule spheroid; style longer than ovary
.... 1. *L. subulata*
Leaves flat, linear to linear-spatulate or with an oval blade and long petiole; capsule ellipsoid; style equal to or shorter than ovary.
    Leaves linear to linear-spatulate .... 2. *L. acaulis*
    Leaves with an elliptic-oblong blade and long petiole .... 3. *L. aquatica*

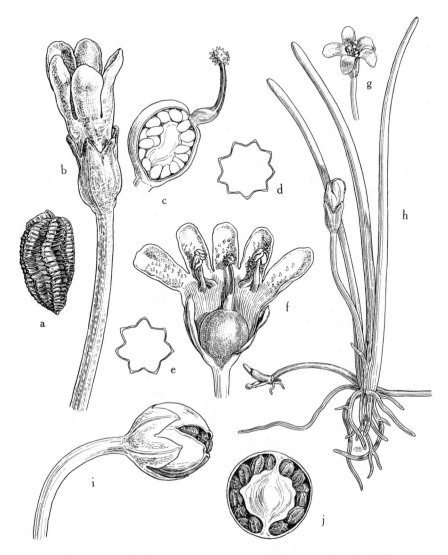

Fig. 316. *Limosella subulata: a,* mature seed, × 40; *b,* young flower, × 10; *c,* ovary (longitudinal section), × 16; *d* and *e,* seeds (cross sections), the longitudinal ridges 6–9, × 40; *f,* corolla, spread open, showing minute papillae on the lobes, × 10; *g,* flower, × 4; *h,* habit, showing stolons, leaves, and bud, × 4; *i,* mature capsule, dehiscing, × 10; *j,* capsule (cross section), × 10.

This genus is much in need of cultural study to determine the nature of the characters customarily relied upon to differentiate species. Material from northeastern California is strikingly like specimens of *Limosella americana* Glück. It will be necessary to resolve some of the problems of variation in the group before positive identification can be made.

1. Limosella subulata Ives, Trans. Phys. Med. Soc. N. Y. 1:440. 1817. Fig. 316.

Tufted stoloniferous plant; leaves erect, terete to subulate, not differentiated into blade and petiole; stipules with tapered shoulders; flowers solitary on slender scapes; calyx campanulate; corolla campanulate, nearly regular, the lobes oblong, minutely papillate on inner face; stamens 4; style about 1 mm. long, longer than ovary; seeds many, ridged and reticulate.

Sandy shores in the tidal zone of the lower San Joaquin River near Antioch Bridge; Atlantic coast.

2. Limosella acaulis Sessé & Mociño, Fl. Mex., 2d ed., 143. 1894. Fig. 317.
  Limosella aquatica var. *tenuifolia* of Jepson, Man. Fl. Pl. Calif. 930. 1925.

Caespitose, stoloniferous plants, often forming mats; leaves flat, linear to linear-spatulate, 1–6 cm. long, the stipules somewhat auriculate; flowers solitary on erect scapes; calyx campanulate; corolla nearly regular, white; petals oblong, sparsely papillate on inner face; stamens 4; style 0.2–0.7 mm. long, equal to or shorter than ovary; capsule ovoid, 2-celled by a thin partition; seeds many, ridged and reticulate.

Margins of ponds, lakes, and streams: widely scattered in California in the valleys and mountains; south to Mexico. Some intergradation with the next taxon makes identification insecure; however, there is usually a clear-cut population identity.

3. Limosella aquatica L., Sp. Pl. 631. 1753. Fig. 318.
  Limosella aquatica var. *americana* Glück, Notizblatt 12:75. 1934. Not *L. americana* Glück.

Tufted annual 5–12 cm. tall, from threadlike rhizomes or stolons, rooting at nodes; leaves on long, slender petioles, 3–10 cm. long, the blades from linear-spatulate to broadly oblong-elliptic, 1–3 cm. long, 3–12 mm. wide, with broad sheathing base and conspicuous hyaline stipules; peduncles shorter than leaves, 1-flowered; calyx campanulate, regular; corolla scarcely longer than calyx, the petals oblong, acute, sparsely papillate within, white or pink, nearly regular; stamens 4; style short, 0.2–0.4 mm. long; stigma obscurely 2-lobed; capsule ovoid, 2-celled, many-seeded, ridged, reticulate.

Along margins of pools, on stream banks, and in irrigation ditches: widely scattered throughout California; east to Atlantic, Europe.

Material from northeastern California tends to intergrade with that of

Fig. 317. *Limosella acaulis: a,* mature seed, × 40; *b,* seed (cross section), × 40; *c,* habit, showing stolon, leaves, and capsules, × 2; *d,* mature capsule, × 6; *e,* corolla, spread open, × 12; *f,* base of petiole, showing auriculate stipules, × 8; *g,* capsule (cross section), × 8.

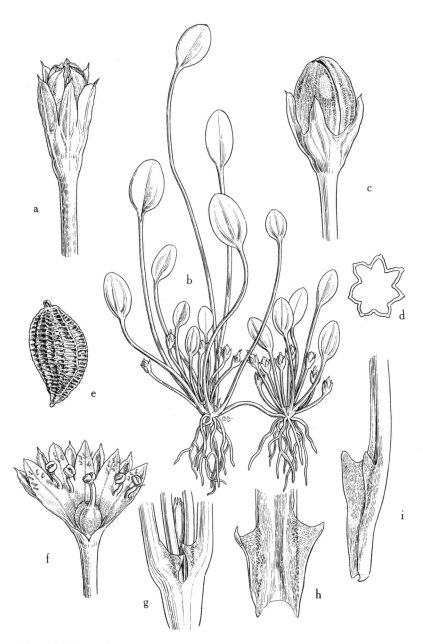

Fig. 318. *Limosella aquatica: a,* young flower, × 10; *b,* habit, showing the slender rhizome, the long petioles and broad leaf blades, and the short peduncles, × ⅘; *c,* mature capsule, × 6; *d,* seed (cross section), × 40; *e,* mature seed, × 40; *f,* corolla, spread open, the lobes sparsely papillate, × 10; *g–i,* bases of petioles, showing hyaline stipules, × 8.

Fig. 319. *Lindernia anagallidea: a,* flower (longitudinal section), × 6; *b,* capsule, × 6; *c,* capsule, after dehiscence, × 6; *d,* capsule (cross section), × 8; *e,* habit, × ⅔; *f,* seed, × 60.

Limosella americana Glück, which has thus far been reported in the United States only from the Mogollon Mountains of New Mexico.

LINDERNIA

1. Lindernia anagallidea (Michx.) Pennell, Monogr. Acad. Nat. Sci. Phila. 1:152. 1935. Fig. 319.

Ilysanthes dubia of western American authors, not Barnhart.
Lindernia dubia as applied to western American plants.

Annual 5–12 cm. tall; stems slender, diffusely branched; herbage glabrous; leaves opposite, sessile and clasping or narrowed at base, often both types on same plant, the blades ovate to oblong, 8–15 mm. long, usually reduced upward, denticulate to entire; flowers on slender axillary pedicels or the upper leaves reduced and bracteate, rendering the inflorescence a bracteate raceme; calyx 2 mm. long, the sepals nearly distinct; corolla tubular, 6–8 mm. long, bluish to white, the upper lip short, erect, 2-cleft, the lower lip spreading, 3-cleft; fertile stamens 2; stigma 2-lobed; capsule many-seeded; seeds smooth or reticulate, 1⅓–3 times as long as broad.

Margins of ponds or streams: Central Valley to middle altitudes in the mountains; east to Atlantic.

Both *Lindernia dubia* subsp. *major* and *L. anagallidea* have been reported as occurring in California, but the characters whereby they are usually distinguished are too inconclusive to serve as a basis for separation.

MIMETANTHE

1. Mimetanthe pilosa (Benth.) Greene, Bull. Calif. Acad. Sci. 1:181. 1885.

Mimulus pilosus Wats. in King, Geol. Expl. 40th Par. 5:225. 1871.

Low, erect, simple or much-branched annual, 10–25 cm. tall; herbage white-villous, glandular-viscid; leaves lanceolate to oblong-ovate, entire, sessile; flowers on slender pedicels in the leaf axils; calyx tubular to short-campanulate, 5-cleft, the tube 5-sulcate, the upper tooth longest; corolla yellow, 6–8 mm. long, slightly surpassing the calyx, obscurely 2-lipped, the lower lobe with brown spots; capsule oblong-ovate, attenuate.

Beds of vernal streams: valleys and foothills; north to Washington.

MIMULUS

Annual or perennial herbs. Flowers solitary in the leaf axils, regular. Calyx prismatic, folded on the angles, 5-toothed. Corolla tubular to funnelform,

2-lipped or nearly so, palate conspicuous or inconspicuous, sometimes reduced to a pair of bearded or naked ridges in the throat on the lower lip. Stamens 4. Stigma of 2 orbicular, sensitives lobes, closing when touched. Capsule variously dehiscent or indehiscent. Seeds many, very small, reticulate to nearly smooth.

There are several other species of *Mimulus* to be expected in boggy or wet places. The following are the more common and most conspicuous in such habitats.

Flowers yellow.
　Plants from slender, tuber-bearing rhizomes; corolla slightly 2-lipped.
　　Flowers on slender pedicels, spreading from the erect, leafy stem; capsule not splitting at apex .... 1. *M. moschatus*
　　Flowers on erect, scapelike pedicels; leaves basal; capsule splitting through the placental tissue at apex .... 2. *M. primuloides*
　Plants from stout rhizomes; corolla strongly 2-lipped, the throat closed by a prominent palate .... 3. *M. guttatus*
Flowers bright red or pink to nearly white.
　Flowers bright red; leaves ovate to ovate-lanceolate; low altitudes .. 4. *M. cardinalis*
　Flowers pink to almost white; leaves lanceolate to oblong; high altitudes
　　　　　　　　　　　　　　　　　　　　　　　　　　　　　5. *M. Lewisii*

1. Mimulus moschatus Dougl. in Lindl., Bot. Reg. 13: pl. 1118. 1828. Muskflower.

Plant from tuber-bearing, perennial rhizome; stems erect or ascending, weak, 1–2 dm. long; herbage slimy-viscid; lower leaves woolly with long white hairs, the leaves oblong-ovate, denticulate to remotely serrate, 2–5 cm. long, petioled; flowers on slender pedicels in the leaf axils; calyx narrowly campanulate, folded in the angles, the teeth acute, unequal; corolla yellow, 20–25 mm. long, 2–3 times as long as the calyx, obscurely 2-lipped, the lobes nearly equal, broad and rounded; capsule oblong-ovate, sessile on the receptacle, abruptly tapering to the persistent style, shorter than calyx.

Bogs, springs, and along streams: throughout California; north to British Columbia.

A form with sessile leaves, *Mimulus moschatus* var. *sessilifolius* Gray (Syn. Fl. N. Am., 2d ed., 2¹:447, 1886; *M. inodorus* Greene, Bull. Calif. Acad. Sci. 1:119, 1885), occurs in the Coast Ranges from Santa Cruz County to Humboldt County.

2. Mimulus primuloides Benth., Scroph. Indicae 29. 1835.

Stoloniferous herb, bearing tubers; stems 2–10 cm. tall; herbage sparsely soft-hairy, occasionally somewhat clammy-viscid; leaves elliptic, oblong to oblanceolate, 1–4 cm. long, serrate; flowers on long, slender, erect, scapelike pedicels; calyx somewhat urceolate, the angles folded, pubescent on teeth and ribs, the lobes short, triangular-subulate; corolla regular to slightly 2-lipped, the limb nearly rotate, yellow; anthers bristly-ciliate; capsule shorter than calyx, splitting through the placental tissue and style base.

Bogs, marshes, and wet meadows: throughout the mountains at elevations above 4,000 feet; western states.

3. Mimulus guttatus Fischer ex DC., Cat. Pl. Hort. Monsp. 127. 1813. Fig. 320.

Annual or perennial herb; stems simple or branched, erect or declined, 1–6 dm. long; herbage glabrous or somewhat pubescent; leaves elliptical, irregularly serrate to dentate or the blades lobed at base, the lower leaves short-petioled, the upper ones sessile; flowers in terminal bracteate racemes on slender pedicels; calyx campanulate, folded on the angles, the teeth connivent in age; corolla yellow, with reddish brown spots, sometimes conspicuously spotted, 2–4 cm. long, the upper lip of 2 erect lobes, the lower lip of 3 reflexed lobes, with a conspicuous palate closing the throat; capsule ovate, flattened, 2-celled, incompletely partitioned.

The most common and most variable species of *Mimulus,* occurring almost everywhere where water stands on rich soil, principally in bogs, swamps, marshes, and stream banks: throughout the state; north to Alaska.

Many subspecies and varieties have been segregated within the group, but there is little that can be definitely said about them until the problem is approached from an ecogenetic point of view and some idea is formed concerning the relative stability of characters and the ecological and geographic pattern of variation. For additional synonymy, see Pennell (in Abrams, Illus. Fl. Pacif. States 3:712, 1951).

4. Mimulus cardinalis Dougl. ex Benth., Scroph. Indicae 28. 1835. Scarlet monkey flower.

Coarse perennial from a thick rhizome; stems erect or ascending, 3–10 dm. tall, rooting where they touch the ground, sometimes much-branched; herbage viscid-pubescent; leaves broadly lanceolate to linear-lanceolate, dentate to entire; flowers on slender pedicels in the leaf axils, the pedicels 2–8 cm. long; calyx tubular, campanulate, the lobes nearly equal; corolla 3–5 cm. long, bright scarlet, the upper lip erect, 2-lobed, the lower lip with 3 reflexed lobes, the throat closed; anthers 4, clothed with scale-like hairs; capsule ovate, acuminate.

Along streams and in bogs at low and middle altitudes: mountains throughout California; north to Oregon, east and south to Utah, Arizona, and Baja California, Mexico.

5. Mimulus Lewisii Pursh, Fl. Am. Sept. 2:427, pl. 20. 1814.

Rhizomatous perennial herb; stems usually several to many, simple, erect, 3–6 dm. tall; herbage viscid-pubescent; leaves lanceolate to ovate-lanceolate, sessile, 2–5 cm. long, remotely repand-dentate; flowers on slender, elongate

Fig. 320. *Mimulus guttatus: a,* calyx enclosing mature fruit, × 1⅕; *b,* capsule, dehiscent by lateral sutures, × 2; *c,* seed, × 40; *d,* habit, basal part of plant, × ⅖; *e,* habit, inflorescence, × ⅔; *f,* pistil, showing the fimbriolate stigma lobes, × 2; *g,* capsule, showing dehiscence, × 4.

pedicels in the leaf axils; calyx tubular-campanulate, folded on the angles, the teeth lanceolate, acuminate; corolla deep rose to almost white, the limb obliquely rotate; anthers with scale-like hairs.

Bogs, streams, and springs, at high altitudes: Sierra Nevada; north to British Columbia, Rocky Mountains.

ORTHOCARPUS

1. Orthocarpus castillejoides Benth., Scroph. Indicae 13. 1835.

Annual; stems branched from base, 15–30 cm. tall; herbage hirsute; leaves 2–8 mm. wide, alternate, sessile, entire or with lacerate tips; spikes short, dense, the bracts white or yellowish at tips; sepals linear; corolla 12–20 mm. long, white with purple spots, or sacs yellow, the galea slightly longer than teeth; capsule oblong, 10–12 mm. long.

Marshes near the coast: central California; north to British Columbia.

SCROPHULARIA

1. Scrophularia californica Cham. & Schl., Linnaea 2:585. 1827.

Perennial herb 5–20 dm. tall; herbage glabrous or finely glandular-pubescent in inflorescence; leaves opposite, on short petioles, the blades ovate, cordate at base, incised-serrate, 6–15 cm. long; flowers in a cymose, terminal panicle; calyx 5-parted; corolla 6–10 mm. long, dull purple, the tube swollen, the upper lip erect, 2-lobed, the 2 lateral petals smaller and erect, the lower lip small and reflexed; fertile stamens 4, a fifth stamen sterile; capsule ovoid-conic, septicidal; seeds numerous, oblong, longitudinally furrowed, rugose, and sometimes finely reticulate.

Marshes in Sacramento Valley and Suisun Bay; also on drier soil from southern California to British Columbia.

VERONICA

Annual or perennial herbs, erect or reclining. Leaves opposite or the upper ones sometimes alternate, linear to lanceolate or oval to oblong or orbicular, sessile to short-petioled. Flowers in axillary racemes or solitary in the leaf axils. Sepals 4 or 5, free. Corolla rotately spreading, deeply 4-lobed, the upper lobes broader and representing 2 fused petals. Stamens 2. Ovary 2-celled, flattened, usually obcordate. Capsule loculicidal.

Fig. 321. *Veronica. a–d, V. peregrina: a,* habit, × ⅘; *b,* flower in leaf axil, × 6; *c,* seed, × 32; *d,* capsule, × 6. *e–g, V. americana: e,* capsule, × 4; *f,* habit, × ⅖; *g,* corolla, × 4.

Flowers solitary in the leaf axils .... 1. *V. peregrina*
Flowers in axillary racemes.
   Leaves, at least some, on short petioles.
      Leaf blade lanceolate-ovate, much longer than wide .... 2. *V. americana*
      Leaf blade orbicular, as wide as or wider than long .... 3. *V. Beccabunga*
   Leaves all sessile.
      Leaves lanceolate to oblong-lanceolate; flowers on slender but not filiform pedicels.
         Capsule globose, emarginate .... 4. *V. Anagallis-aquatica*
         Capsule obcordate, deeply notched .... 5. *V. connata*
      Leaves linear to linear-lanceolate; flowers on filiform, spreading pedicels
                                                 6. *V. scutellata*

1. Veronica peregrina L., Sp. Pl. 14. 1753. Fig. 321.

Annual; stems erect, 1–3 dm. tall, simple or branched from base; herbage glabrous to finely puberulent; leaves alternate or the lowest ones opposite, sessile or subsessile, oblong to linear, entire or dentate, 1–3 cm. long; flowers solitary in the axils, on short peduncles; corolla white; stamens included; capsule globose, 3 mm. long.

Common in wet places on the margins of ditches and ponds, or in rice fields: Central Valley; native of eastern North America.

2. Veronica americana (Raf.) Schwein. ex Benth. in A. DC., Prodr. 10:468. 1846. Fig. 321.

Veronica Beccabunga var. *americana* Raf., Med. Fl. 109. 1830.

Plant perennial; stems erect or decumbent, branching, 1–8 dm. long, stoloniferous; herbage glabrous; leaves ovate to oblong or oblong-lanceolate, short-petioled, serrate, 3–8 cm. long, truncate, rounded or subcordate at base; racemes peduncled, axillary, the pedicels filiform, longer than subtending bracts; calyx 2–3 mm. long; corolla 4 mm. broad, bright blue with white center; capsule orbicular, biconvex, slightly emarginate, 3 mm. long.

Common in fresh-water marshes, along streams, and near springs: almost throughout California; east to Atlantic.

3. Veronica Beccabunga L., Sp. Pl. 12. 1753.

Prostrate to ascending or trailing perennial; herbage glabrous, succulent; leaves short-petioled, oval to short-oblong or orbicular, rounded at tip, crenate; racemes axillary in the middle leaf axils, the peduncles 7–10 cm. long, the pedicels slender, 1–2 times as long as the capsule; sepals 4, broadly lanceolate, subequaling the fruit; corolla blue to purple; capsule globose, the locules plump.

European weed, introduced near Bridgeport, Mono County. This is the only known locality in California.

Fig. 322. *Veronica Anagallis-aquatica: a,* habit, × ⅖; *b,* flower, × 6; *c,* corolla, showing stamen insertion, × 6; *d,* capsule, × 6; *e* and *f,* seeds, × 32.

4. Veronica Anagallis-aquatica L., Sp. Pl. 12. 1753. Fig. 322.

Veronica Anagallis L. of western American authors.

Perennial herb from rhizomes or stolons; stems 1–5 dm. long; herbage glabrous below, often glandular-pubescent in the inflorescence; leaves sessile, lanceolate, serrate, 2–5 cm. long; racemes axillary, peduncled, few-flowered, loose and elongate, 2 to a node; pedicels slender, 4–8 mm. long; calyx lobes 2 mm. long; corolla blue, marked with purple; capsule orbicular, emarginate, many-seeded; seeds plano-convex.

Swamps and stream banks: widely distributed but not common; introduced from Europe.

5. Veronica connata subsp. **glaberrima** Pennell, Monogr. Acad. Nat. Sci. Phila. 1:368. 1935.

Plant aquatic or in very wet soil, glabrous; stems submersed or distally emersed; leaves sessile, clasping, oblong-lanceolate, acute, crenate-serrate; flowers in small-bracted, axillary racemes; sepals 4; petals white with purple veins; capsule obcordate, prominently notched.

Scattered in marshes and bogs throughout California, but not common; western North America.

Veronica connata differs from *V. Anagallis-aquatica* chiefly in the obcordate capsule, which is deeply notched at the apex.

6. Veronica scutellata L., Sp. Pl. 12. 1753. Fig. 323.

Plant perennial, stoloniferous; stems slender, simple or branched, 1–6 dm. tall, decumbent or ascending, rooting at lower nodes; herbage glabrous or somewhat hairy; leaves linear to linear-lanceolate, sessile and somewhat clasping, denticulate or entire, 2–8 cm. long; racemes axillary, exceeding the leaves in height, the pedicels filiform, 6–17 mm. long, longer than the subtending bract; corolla blue, 4 mm. long; capsule broader than long, flat, deeply emarginate at apex.

Marshes and wet places: central and northern California, fairly common; cosmopolitan in Northern Hemisphere.

LENTIBULARIACEAE

Aquatic or terrestrial, annual or perennial plants, often "insectivorous." Leaves cauline or in a basal rosette, simple or dissected. Flowers 1 to several on an erect scape. Calyx 2-lobed. Corolla deeply 2-lipped, the lower lip 3-lobed, spurred at base, and with a conspicuous palate. Stamens 2. Ovary superior, 1-celled, the placentation free-central. Capsule ovoid to globose, 2-valved. Seed minute, with a poorly differentiated embryo and no endosperm.

Fig. 323. *Veronica scutellata: a,* habit, × ⅖; *b,* lower part of stem, showing roots developing at the nodes, × ⅖; *c,* flower, × 3; *d,* capsule, × 3.

Plants submersed; aquatic with filiform-dissected, bladder-bearing leaves; flowers in ours
   yellow .... *Utricularia*
Plants not submersed; leaves entire, in a basal rosette, not bladder-bearing; flowers in
   ours violet .... *Pinguicula*

PINGUICULA

1. Pinguicula vulgaris L., Sp. Pl. 17. 1753.

Caespitose herb; leaves in a basal rosette, spatulate, elliptic, 2–6 cm. long,
fleshy, greasy to the touch; scapes 1-flowered, 5–15 cm. tall; corolla violet,
2-lipped, the lips unequal, with a conspicuous, straight spur, the palate hairy
and spotted; capsule ovoid to globose, longer than calyx.

Bogs: Del Norte County; northward and circumpolar.

UTRICULARIA

Perennial, aquatic herbs; stems submersed, creeping, or floating. Leaves
dissected and usually bearing bladder-like traps with a valvelike action for
trapping micro-organisms. Flowers on slender scapes, ours yellow, 2-lipped,
the lower lip 3-lobed and with a conspicuous palate closing the throat, the
upper lip erect, usually entire. Stamens 2. Ovary superior, 1-celled, with free-
central placenta. Capsule 2-valved.

Corollas 10–18 mm. broad.
   Bladders borne on all leaves; stems 1–3 m. long .... 1. *U. vulgaris*
   Bladders borne only on special leafless branches; stems 1–10 dm. long
                                                        2. *U. intermedia*
Corollas 4–6 mm. broad.
   Leaf segments flat, 2–4 times forked; lower lip of corolla much longer than upper lip;
      scapes 5–10 cm. tall; seeds wingless, reticulate .... 3. *U. minor*
   Leaf segments filiform, often rootlike, usually with 2 or 3 divisions; lower lip of corolla
      shorter than or about as long as upper lip; seeds winged.
      Bladders on leaves only; corolla spur much shorter than lower lip; body of the seed
         smooth .... 4. *U. gibba*
      Bladders on both stems and leaves; corolla spur slightly longer than the lower lip;
         body of the seed rough-tuberculate .... 5. *U. fibrosa*

1. Utricularia vulgaris L., Sp. Pl. 18. 1753. Common bladderwort. Fig. 324.

Perennial, submersed aquatic herb, floating just beneath the surface; stems
1–2 or –3 m. long, the terminal bud ultimately becoming the winter bud;
leaves elliptic to elliptic-ovate, 14–20 mm. long, much dissected, with capillary
segments bearing many bladder-like traps; scapes erect, 1–6 dm. tall, emersed,
stout, bearing 5–20 flowers; corolla 12–18 mm. broad, the lips closed, the
lower lip a little longer than the upper one and more or less 3-lobed, the palate

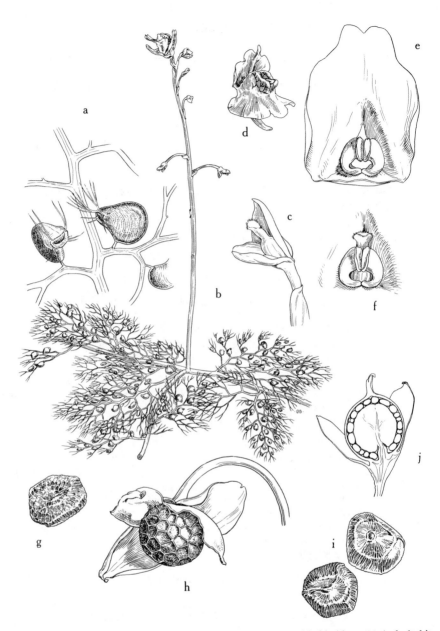

Fig. 324. *Utricularia vulgaris: a,* capillary leaf segments with bladders, × 4; *b,* habit, showing the erect flowering scape and the floating stems, leaves, and bladders, × ⅔; *c,* calyx and pistil, × 4; *d,* corolla, × 1⅕; *e,* upper lip of corolla, showing stamens in front of throat, × 4; *f,* throat of corolla, showing position of stamens, style, and stigma, × 4; *g,* mature seed, × 20; *h,* mature capsule at dehiscence, the seeds in a tight mass, × 6; *i,* mature seeds, × 20; *j,* capsule (longitudinal section), × 4.

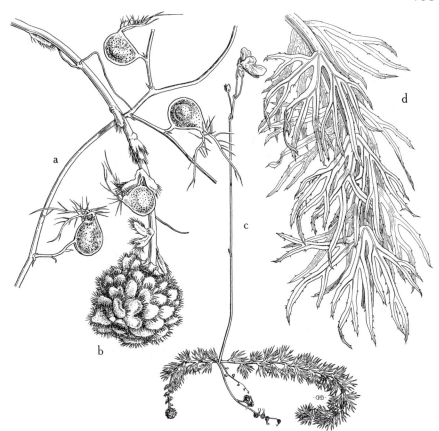

Fig. 325. *Utricularia intermedia: a,* bladder-bearing branch, × 4; *b,* winter bud, produced at tip of leafy branch, × 4; *c,* habit, showing leafy branches, bladder-bearing branches, and the slender flowering scape, × ⅔; *d,* part of a leafy branch, × 4.

conspicuous, the spur conical, the upper lip nearly entire, the fruiting pedicels reflexed; seeds brown, shiny, striate-reticulate.

Common in fresh-water ponds, ditches, and slow-moving streams: almost throughout the state and the continent. Often forming dense masses on the surface of the water.

2. Utricularia intermedia Hayne in Schrad., Jour. für die Bot. 1:18. 1800. Fig. 325.

Perennial herb of shallow water; stems creeping along the bottom, the leafy branches plumelike, producing few or no bladders; leaves 2-ranked, crowded on stem, 4–8 mm. long, dissected into flat linear segments; terminal bud developing into a winter bud; bladders produced on special rootlike, leafless branches; scapes erect, emersed, leafless, 1- to 5-flowered; corolla yellow, lower

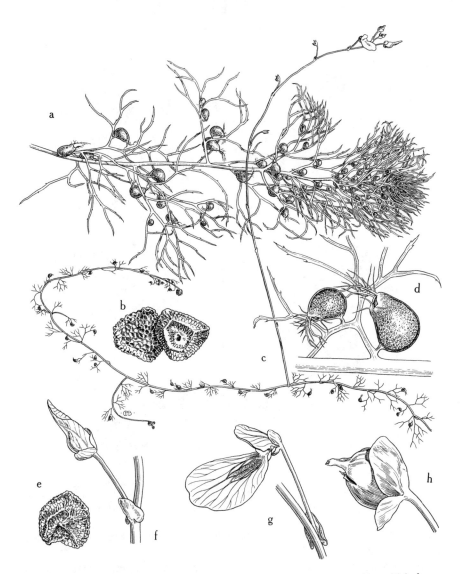

Fig. 326. *Utricularia minor: a*, stem, showing abundant leaves and bladders, × 2; *b*, mature seeds, × 20; *c*, habit, showing threadlike stem with few leaves and bladders and a long flowering scape, × ⅔; *d*, stem with bladder-bearing leaf, × 8; *e*, mature seed, × 20; *f*, bud arising in axil of auricled bract, × 3; *g*, flower, × 3; *h*, mature capsule, × 6.

lip about twice as long as the upper one, ovate, 1–1.5 cm. broad, the spur closely appressed, nearly as long as the lower lip, the upper lip deltoid; fruiting pedicels erect.

Bogs and quagmires: Fresno County, Plumas County to Modoc County; east to New Jersey, north to Canada.

3. Utricularia minor L., Sp. Pl. 18. 1753. Fig. 326.

Delicate, submersed, threadlike stems, creeping or floating; leaves 2–4 times forked, 2–12 mm. wide, the segments flat, usually abundantly laden with bladders or sometimes with few or no bladders; scapes slender, 5–18 cm. tall, with auricled bracts, 2- to 9-flowered; corolla yellow, the upper lip short, equaling or slightly surpassing the depressed palate, the lower lip 4–6 mm. broad, twice as long as the upper lip or longer, the spur short and saccate or wanting; fruiting pedicels recurving, 4–8 mm. long; capsule exserted, globose, 2 mm. in diameter; seeds reddish brown, the center nearly black, reticulate.

Bogs and ponds: northern Sierra Nevada, rarely collected in flower, sterile specimens much more common; Northern Hemisphere.

4. Utricularia gibba L., Sp. Pl. 18. 1753. Fig. 327.

Plant from a thick mat of closely interlaced, delicate, threadlike stems in shallow water or floating; stems radiating from base of erect scapes and often penetrating the surface debris, colorless, bearing chlorophyll when exposed to light; leaves usually 2-forked, bearing occasional scattered bladders; scapes erect, 2–6 cm. tall, 1- to 3-flowered; upper and lower lips of corolla about equal in length, the lower lip 6–8 mm. broad with a conspicuous palate, the spur oblong-conic, subequaling to much shorter than the lower lip; fruiting pedicels ascending, the lower ones 3–12 mm. long; capsule 2–3 mm. thick, barely longer than the calyx; mature seeds broadly winged, smooth on body of seed.

Marshes and bogs, often on masses of floating debris in shallow water or along the shore: Sacramento Valley delta region, North Coast Ranges and the northern Sierra Nevada; east to Atlantic, Central America.

5. Utricularia fibrosa Walt., Fl. Carol. 64. 1788. Fig. 328.

Plant very similar to *Utricularia gibba* in habit and size; stems radiating from base of scape; leaves usually 3-forked at base or dichotomous; bladders borne mainly on the stems and occasionally on the leaves; flowers yellow, the upper and lower lips subequal in length, the palate conspicuous, the spur equaling to slightly longer than lower lip; capsule globose; mature seeds broadly winged, the body rough-tuberculate.

Shallow fresh water or the margins of ponds, often on compacted floating debris or floating and stems elongate: Sacramento Valley, Caspar in Mendocino County; southern and eastern states.

Superficially *Utricularia fibrosa* closely resembles *U. gibba*. It may be dis-

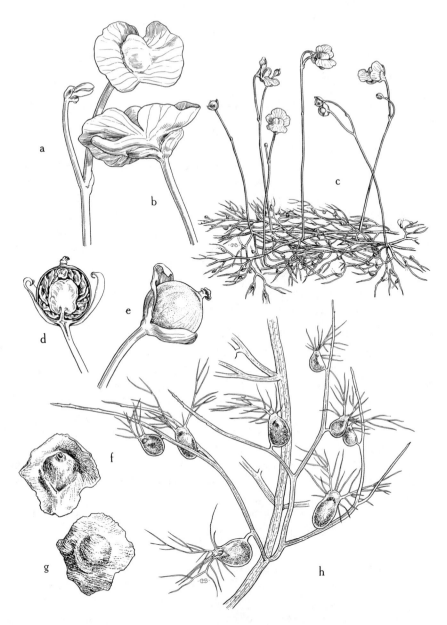

Fig. 327. *Utricularia gibba: a* and *b,* flowers, showing lips more or less equal in size, rounded palate, and short spur, × 3; *c,* habit, showing the erect flowering and fruiting scapes and the floating leaves and bladders, × ⅖; *d,* capsule (longitudinal section), × 4; *e,* mature capsule, × 4; *f* and *g,* mature seeds, the wings broad, × 24; *h,* part of stem, showing bladders borne on the leaf segments, × 6.

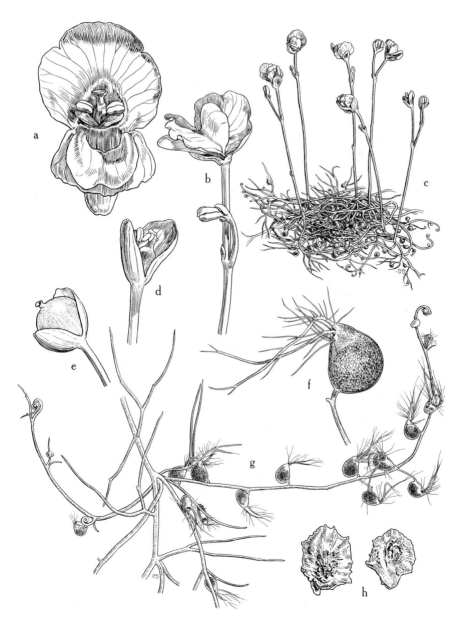

Fig. 328. *Utricularia fibrosa: a,* flower, the palate bent down to show stamens and pistil, × 6; *b,* flowering scape, showing young bud and flower, the lips subequal in size and the palate conspicuous, × 4; *c,* habit, showing dense mass of stems, leaves, and bladders, and the erect scapes, × 1⅕; *d,* calyx, × 6; *e,* capsule, × 4; *f,* bladder, × 12; *g,* bladders on stems and leaves, × 3; *h,* mature seeds, × 12 .

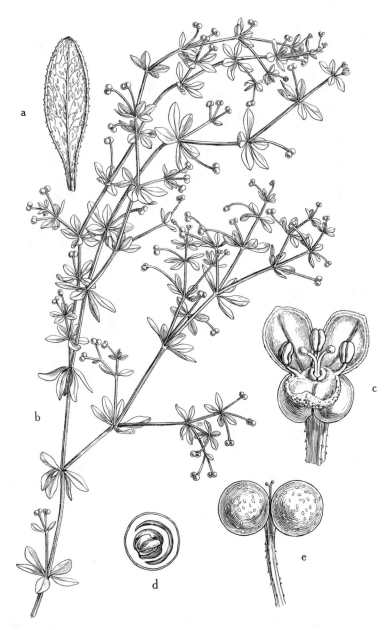

Fig. 329. *Galium trifidum: a,* leaf, × 4; *b,* habit, × ⅖; *c,* flower, × 20; *d,* fruit (longitudinal section), × 8; *e,* fruit, × 8.

tinguished, however, by the fact that the bladders are borne chiefly on the stems and only occasional bladders are on the leaves, whereas in *U. gibba* they are confined to the leaves. Also, the body of the winged seed is rough-tuberculate in *U. fibrosa* and smooth in *U. gibba*.

RUBIACEAE

Herbs or shrubs. Leaves simple, entire, opposite or whorled. Stipules present. Flowers perfect, regular. Calyx epigynous, the limb 4- or 5-lobed or lacking. Corolla sympetalous, 4- or 5-lobed. Stamens alternate with corolla lobes and inserted on corolla tube. Ovary inferior, 2-celled, often lobed, ovules 1 to many; fruit a capsule, berry, drupe, or of separating nutlets.

Herbaceous perennial; leaves whorled; stipules foliaceous .... *Galium*
Large shrub; leaves opposite or sometimes ternate; stipules inconspicuous .. *Cephalanthus*

CEPHALANTHUS

1. Cephalanthus occidentalis L., Sp. Pl. 95. 1753.

Shrub or tree to 9 m. tall; leaves opposite or sometimes ternate, oblong or elliptic, apically somewhat pointed, truncate or obtuse at base, entire, 5–15 cm. long, on petioles about 4 mm. long, the stipules 2–3 mm. long; flowers densely clustered in globose heads 2–3 cm. wide, on peduncles 2.5–8 cm. long; calyx limb 4-lobed, greenish; corolla funnelform, 4-lobed, white, 5–8 mm. long; stamens with filaments short and anthers included; style filiform, much exserted; stigma capitate; fruit dry, hard, 4-angled, splitting into 2–4 1-seeded nutlets.

Along running water at elevations below 1,000 feet: common in Central Valley and the Sierra Nevada foothills; north to Canada and east across North America.

GALIUM

1. Galium trifidum L., Sp. Pl. 105. 1753. Fig. 329.

Rootstocks very slender; stems slender, weak, ascending, many-branched and much-tangled, to 30 cm. long, 4-angled, nearly smooth or retrorsely scabrous; branches 2 at a node; leaves 4 or 5 in a whorl, oblong to oblanceolate, obtuse, 4–15 mm. long, thin, marginally somewhat scabrous; flowers minute, terminal or axillary, solitary on elongate, capillary pedicels; calyx limb lacking; corolla rotate, white; ovary 2-lobed; styles 2; fruit glabrous, dry, separating into 2 indehiscent, 1-seeded segments.

Marshes at elevations below 5,000 feet: almost throughout cismontane California; cosmopolitan.

CAPRIFOLIACEAE

Erect shrubs, sometimes treelike. Leaves opposite, simple or compound. Flowers perfect and complete. Calyx tube adnate to the ovary, the lobes small, usually inconspicuous. Corolla tubular or the limb rotate, regular or irregular, the lobes 4 or 5. Stamens as many as the lobes and inserted on the corolla tube. Ovary 2- to 5-celled, inferior. Fruit a berry or sometimes drupelike.

Leaves simple; corolla tubular, irregular; flowers in cymose pairs in the leaf axils
Lonicera

Leaves compound; corolla rotate (saucer-shaped), regular; flowers in flat-topped or paniculate cymes .... *Sambucus*

LONICERA

Ours erect shrubs. Leaves simple and entire, sometimes those subtending the inflorescence connate-perfoliate. Flowers in axillary pairs on a slender peduncle. Calyx limb, when present, 5-toothed or truncate, sometimes absent. Corolla 5-lobed, from nearly regular to strongly 2-lipped, sometimes the tube gibbous at base. Stamens 5, inserted on corolla tube. Ovary ovoid to globose, sometimes those of adjacent flowers fused. Fruit a few- to several-seeded berry.

Fruits of the pair of flowers distinct; limb of corolla regular, the tube somewhat gibbous, corolla orange .... 1. *L. involucrata*
Fruits of the pair of flowers connate; limb of corolla 2-lipped, the upper lip 4-lobed, the lower lip a single lobe, corolla purple .... 2. *L. conjugialis*

1. Lonicera involucrata (Richards.) Banks in Spreng., Syst. Veg. 1:759. 1825. Twinberry.

Low shrub to 1 m. tall; leaves 5–10 cm. long, short-petioled, the blade ovate, more or less abruptly attenuate, pilose beneath; peduncles 2–3 cm. long, borne in leaf axils and bearing at the summit a conspicuous pair of connate (often purple) bracts which subtend the pair of sessile flowers; corolla orange-yellow, 1–2 cm. long, tubular, the limb spreading, its lobes subequal, the tube gibbous below; ovary of each flower in the pair distinct; fruit fleshy, black; seeds flattened, irregularly ovate, finely pitted.

High montane stream banks and marshy meadows: Sierra Nevada from Tulare County north; north to British Columbia.

A coastal form, *Lonicera involucrata* var. *Ledebourii* (Esch.) Jepson, is differentiated by its greater size, since it is 1–2.5 m. tall, and the more saccate nature of the base of the corolla tube. It occurs on canyon bottoms and streamsides in the Coast Ranges from Santa Barbara County north to Humboldt County.

2. Lonicera conjugialis Kell., Proc. Calif. Acad. Sci. 2:67. 1863. Siamese twin-berry.

Shrub rarely more than 1 m. tall; herbage finely pilose-hispid; leaves ovate to oblong-ovate, 2–6 cm. long, on short petioles; peduncles 1–2.5 cm. long, with inconspicuous terminal bracts; calyx limb scarcely evident; corolla black-purple, 4–6 mm. long, campanulate in form but strongly 2-lipped, the upper lip of 4 shallow lobes, the lower lip linear-oblong and deflexed on the corolla tube, the corolla throat filled with soft white hairs; stamens 5, the filaments hairy; style hairy, the ovaries of the paired flowers fused toward base; fruits fused and thus appearing double; seeds ovate, flat, thin, very finely pitted.

Frequent along streams in the high mountains at elevations above 6,000 feet: Sierra Nevada and North Coast Ranges, Modoc County; east to Nevada, north to Washington.

SAMBUCUS. ELDERBERRY

Shrubs or small trees. Leaves pinnate with a terminal leaflet, deciduous. Flowers white to yellowish, in flat-topped or paniculate cymes, the pedicels jointed at top. Calyx 5-toothed. Corolla regular, rotately spreading, 5-lobed. Stamens 5, inserted on corolla tube and alternate with its lobes. Ovary 3- to 5-celled; stigmas 3–5 on a short style. Fruit a small, berry-like drupe.

Berries blue; cyme flat-topped .... 1. *S. coerulea*
Berries red; cyme pyramidal or elongate .... 2. *S. racemosa*

1. Sambucus coerulea Raf., Alsog. Am. 48. 1838. Blue elderberry.
Sambucus glauca of western American authors.

Shrub, sometimes treelike, to 8 m. tall; leaves pinnate, the leaflets 5–7, ovate to oblong-lanceolate, serrate except at the entire apex, spreading from rachis; cyme flat-topped, 5–25 cm. broad; flowers white to yellowish; berries blue with a whitish, waxy bloom; seeds ovoid to triangular-ovoid, transversely rugose.

Along streams in the lowlands, but occurring away from streams in cooler uplands to an elevation of 5,000 feet: throughout California west of the Sierra Nevada; north to British Columbia, east to Utah.

The southern California populations are regarded as *Sambucus coerulea* var. *arizonica* Sarg., on the basis of having fewer leaf segments and being puberulent. It occurs along stream banks from Kern County southward.

2. Sambucus racemosa L., Sp. Pl. 270. 1753. Red elderberry.

Low shrub, rarely more than 1 m. high, often thickly branched and spreading; leaves thin and glabrous, the leaflets 5, ascending on the rachis, ovate to ovate-lanceolate or acuminate, the margin serrate, the apex entire; flowers in

Fig. 330. *Campanula linnaeifolia: a,* capsule, $\times$ 3; *b,* seed, $\times$ 32; *c,* habit, $\times$ ⅖; *d,* flower, the corolla removed to show the stamens, style, and stigmas, $\times$ 4; *e,* leaf, $\times$ 2.

a conical or pyramidal, paniculate cyme, white; berries red; seeds ovoid, finely rugose.

Stream banks and springs at high altitudes: Sierra Nevada; north to British Columbia.

The coastal form is segregated as *Sambucus racemosa* var. *callicarpa* (Greene) Jepson on the basis of its larger size and its leaflets, which are serrate to the tip. It occurs from San Mateo County to Humboldt County and north to British Columbia.

CAMPANULACEAE

Slender, small, perennial herbs with milky juice. Leaves simple, alternate. Flowers regular, complete, solitary to racemose. Calyx tube adnate to ovary, the limb commonly 5-parted, persistent. Corolla campanulate to rotate, 5-lobed, inserted with the 5 stamens where calyx becomes free from the ovary. Anthers generally free. Ovary inferior, 2- to 5-celled; style 1; stigmas 2–5. Fruit a many-seeded capsule.

CAMPANULA

1. Campanula linnaeifolia Gray, Proc. Am. Acad. Arts & Sci. 7:366. 1868. Fig. 330.

Perennial herb from slender rhizome; stems slender, weak, and often reclining, simple or sparingly branched above, retrorsely scabrous on the angles; leaves ovate to oblong-ovate, 1–3 cm. long, sessile or subsessile, crenulate above, retrorsely scabrous on margins; flowers few, on slender pedicels (peduncles); sepals 5, lanceolate; corolla pale blue, campanulate, regular, 1–2 cm. long; stamens 5, on slender, basally dilated filaments; ovary inferior, 5-celled; style subequal to the corolla in length; capsule globular, opening by perforations near the middle or the base of capsule.

Bogs or swamps: Point Reyes in Marin County, Mendocino County.

LOBELIACEAE

Ours annual or perennial herbs with alternate or opposite, simple leaves. Flowers solitary in the axils of the leaves, irregular and usually bilabiate, the subtending leaves often reduced and bractlike. Sepals 4 or 5, equal or unequal in size. Petals 5, rarely 4, united or sometimes the tube split on upper side to the base, the lower lip 3-lobed, usually spreading or reflexed, the upper lip usually erect, sometimes recurved, the lobes parallel or divergent. Stamens united into a tubelike column; filaments wholly united or free only at base; anthers united, rarely only connivent; orifice of anther tube usually subapical.

Fig. 331. *Downingia elegans: a,* habit, × ⅘; *b,* flower, front view, × 3; *c,* flower, side view, × 3; *d,* seed, × 30; *e,* ovary (cross section), × 5; *f,* capsule, × 1½; *g,* stamen column, showing abrupt bend at base of anther tube, the anthers with a hooked bristle at the apex, and the exserted stigma lobes, × 4.

Ovary wholly inferior or the tip free, 1-, 2-, or 3-celled; placentae parietal or axile. Fruit a capsule.

Ovary and capsule linear, elongating in fruit, dehiscent along longitudinal sutures, sessile in the axils of leaves or leafy bracts .... *Downingia*
Ovary and capsule obconic to subglobose, on slender pedicels.
  Ovary obconical.
    Corolla tube shorter than or about as long as the lobes.
      Corolla tube not split on upper side to base, the lobes very unequal in size; sepals at least ½ as long as capsule; flowers showy .... *Porterella*
      Corolla split to base on upper side, the lobes nearly equal in size; sepals less than ½ as long as capsule; flowers inconspicuous .... *Legenere*
    Corolla tube much longer than the lobes, white, the lobes blue, entire .. *Palmerella*
  Ovary subglobose; corolla tube split to base on upper side, the flowers deep red
                                                 Lobelia

DOWNINGIA

Annual herbs of vernal pools or vernally wet soil; erect or ascending. Leaves often dimorphic, alternate (except in *Downingia humilis*), the submersed leaves linear and often filiform, the emersed ones lanceolate to oblanceolate. Flowers solitary and sessile in the leaf axils, forming upon emergence. Sepals linear, subequal in size, erect to rotately spreading. Corolla often variable in size, strongly bilabiate, rarely the lobes approximate, the lower lip spreading or reflexed, usually with a white patch and yellow or purple spots or ridges, the upper lip of 2 erect, spreading, or recurved lobes, the lateral sinuses usually deeper than the sinus between the 2 upper lobes of the corolla. Stamens united into a tubular column, the filament tube longer than the anther tube. Ovary linear, much elongating in fruit. Capsule sessile, splitting along longitudinal sutures, 1-celled or 2-celled by a delicate longitudinal septum; seeds smooth, plane, longitudinally or spirally striate, somewhat twisted in a few species.

McVaugh, R., Campanulales: Campanulaceae, Lobelioideae. N. Am. Fl. 32A¹:1–134. 1943.
Hoover, R. F., New or imperfectly known Californian species of *Downingia*. Leafl. West. Bot. 2:1–6. 1937.

Stamen column always sharply bent at junction of filament tube and anther tube.
  Ovary 1-celled; anthers not granular-roughened on backs .... 1. *D. elegans*
  Ovary 2-celled; anthers granular-roughened throughout .... 2. *D. insignis*
Stamen column straight or curved (if bent at junction of filament tube and anther tube the sinus of the lower corolla lip cut below the level of the platform).
  Corolla 8–20 mm. long.
    Seeds not spirally striate; lateral sinuses of corolla sometimes deeper than upper sinus
      Anther tube tapered at the tip .... 3. *D. pulchella*
    Anther tube blunt or rounded at tip.
      Sinus between the upper corolla lobes distended backward into a lip or hornlik process .... 4. *D. ornatissima*

Fig. 332. *Downingia insignis: a,* habit, × ⅖; *b,* flower, front view, × 3; *c,* flower, side view, × 3; *d,* stamen column, showing sharp bend at base of anther column, × 6; *e,* seed, × 30; *f,* capsule, × 1½.

Sinus between the upper corolla lobes not distended.
    Bristles at apex of anthers tightly twisted together .... 5. *D. bicornuta*
    Bristles of anthers not twisted together.
        Ovary 2-celled.
            Upper corolla lobes ciliate-margined .... 6. *D. concolor*
            Upper corolla lobes not ciliate-margined .... 7. *D. bella*
        Ovary 1-celled.
            Anthers glabrous to sparsely short-hispid near tip .... 8. *D. yina*
            Anthers pilose-bearded at tip .... 9. *D. montana*
Seeds spirally striate; lateral sinuses of corolla shallower than upper sinus
                                                          10. *D. cuspidata*
Corolla 2–7 mm. long.
    Seeds spirally striate and twisted; corolla 2–4 mm. long .... 11. *D. pusilla*
    Seeds not spirally striate or twisted; corolla 4–7 mm. long .... 12. *D. laeta*

1. Downingia elegans (Dougl.) Torr., U. S. Expl. Exped. (Wilkes) 17:375. 1874. Fig. 331.

Erect, simple or divaricately branched annuals; plants glabrous except for the puberulent or scabrous ovary and sepals; stems slender to stout, straight or somewhat zigzag, 1–5 dm. tall; leaves 5–25 mm. long, lanceolate; sepals linear to linear-elliptic, obtuse to subacute, 3–10 mm. long; corolla blue to purple, pink or occasionally white, the lower lip concave, not reflexed, with a central white spot and 2 white-margined, yellow ridges near base, the lobes of upper lip lanceolate, erect or ascending or recurved, somewhat divergent, the tube short, broadly campanulate, often with 3 purple blotches at base below lower lip, the lateral sinuses much deeper than the upper; stamen column abruptly bent at anther tube, the filament tube slender, glabrous, often well exserted, the anther tube slender, 4–5 mm. long, smooth, glabrous to sparsely ciliate, the orifice subapical, the tip of each of the lower anthers white-tufted and with a hornlike scale; ovary eventually 15–55 mm. long, often very stout at base, 1-celled; seeds not twisted.

Vernal pools and wet meadows, valleys and mountains at low and middle altitudes: Humboldt and Plumas counties; north to eastern Washington, east to Nevada and Idaho.

McVaugh (*op. cit.,* p. 25) established two varieties in *Downingia elegans,* based upon the length of the filament tube and the size of the corolla: *D. elegans* var. *elegans* has the filament tube 4.5–10.5 mm. long and the corolla 8–13 mm. long, while *D. elegans* var. *brachypetala* has the filament tube less than 4.5 mm. long and the corolla 5–9 mm. long.

2. Downingia insignis Greene, Pittonia 2:80. 1890. Fig. 332.

Erect, slender plants, often with zigzag stems, 10–30 cm. tall; leaves 5–20 mm. long, elliptic to ovate, obtuse, rounded to subacute; sepals elliptic-obtuse or rounded, 3–12 mm. long, ascending; corolla 9–15 mm. long, glabrous, blue with darker veins, the tube 3–5 mm. long, broadly campanulate, the lateral sinuses deeper than the upper sinus, the lower lip concave, not reflexed, shorter

Fig. 333. *Downingia pulchella: a,* habit, × ⅖; *b,* flower, front view, × 2; *c,* flower, side view, × 2; *d,* stamen column, × 6; *e,* capsule, × 2; *f,* seed, × 30.

Fig. 334. *Downingia ornatissima: a,* habit, × ⅘; *b,* flower, front view, × 3; *c,* flower, side view, × 3; *d,* seed, × 30; *e,* capsule, × 1½; *f,* anther tube and exserted stigma lobes, × 8; *g,* stamen column, × 6.

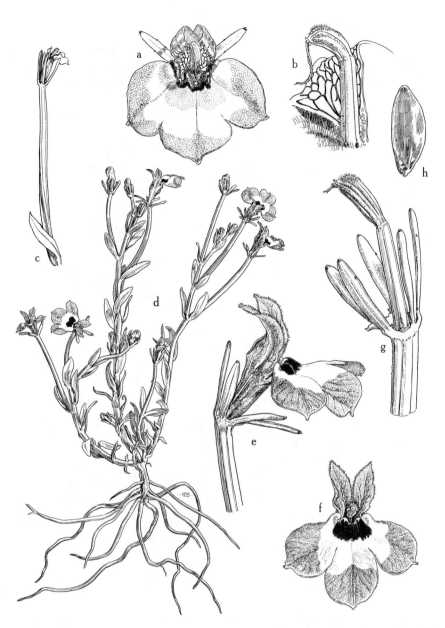

Fig. 335. *Downingia. a* and *b, D. bicornuta* var. *picta: a,* flower, front view, × 3; *b,* stamen column, showing twisted anther bristles, × 6. *c–h, D. concolor: c,* capsule, × 1½; *d,* habit, × 1; *e,* flower, side view, × 3; *f,* flower, front view, × 3; *g,* stamen column, × 8; *h,* seed, × 30.

than upper lip, with 2 yellow folds near base in a patch of dark purple, the upper lobes parallel, ascending, elliptic-acute, 6–10 mm. long; stamen column sharply bent at junction of filament tube and anther tube, the filament tube 7–12 mm. long, exserted, the anther tube 2.5–3.5 mm. long, minutely granular-roughened, the orifice subapical, the shorter anthers white-tufted at tip and with a scale-like bristle; ovary 3–8 cm. long, 2-celled; seeds not twisted.

Wet clay soil and margins of vernal pools: Central Valley from Stanislaus County north to Modoc and Lassen counties; into Washoe County, Nevada.

3. Downingia pulchella (Lindl.) Torr., Pacif. R. R. Rep. 4:116. 1857. Fig. 333.

Erect annual herb; stems to 4 dm. tall; herbage glabrous; leaves 4–25 mm. long, linear to elliptic, lanceolate or ovate, sometimes irregularly crenulate-serrulate; sepals lanceolate or elliptic, obtuse or rounded, usually rotately spreading, subequal in size, 3–10 mm. long; corolla 8–13 mm. long, glabrous, deep blue, the tube short, 2–3 mm. long, broadly campanulate, the lateral sinuses slightly deeper than the upper sinus, the lower lip 4–6 mm. long, with a broad white spot enclosing 2 egg-yellow spots along a pair of folds alternating with 3 purple spots, the lobes of upper lip elliptic to oblanceolate, widely divergent; stamen column elongated and slightly curved, the filament tube exserted, glabrous, the anther tube 2.5–3.5 mm. long, attenuate or pointed at tip, the orifice subapical, the shorter anthers with slender, scale-like bristles; ovary 3–8 cm. long, sometimes scabrous; seeds not twisted.

Vernal pools and wet ground at low elevations: central California from Monterey and Merced counties north to Colusa County.

4. Downingia ornatissima Greene, Pittonia 2:80. 1890. Fig. 334.

Erect or ascending annual herb; stems simple or branched, to 3 dm. tall; herbage glabrous; leaves lanceolate to elliptic-ovate, 6–15 mm. long, obtuse or acute; sepals linear to broadly elliptic, subequal in length or the lower ones shorter, 2–9 mm. long, erect to rotately spreading; corolla 8–13 mm. long, the tube narrowly campanulate, sparsely pilose within on lower side, 2–4 mm. long, the lateral and upper sinuses cleft to equal depth, the edge of the upper sinus projected backward into a lip or beaklike process, the lower lip plane or concave, the folds nipple-like, bright dark blue to pale lilac or white, with 2 yellow or yellow-green spots which appear to enter throat as palate-like folds, the upper lobes lanceolate to triangular, 2–6 mm. long, rolled backward into a ring or horizontally spreading, somewhat divergent; stamen column 4–9 mm. long, straight or sometimes bent abruptly at junction of filament tube and anther tube, the filament tube 3–6 mm. long, pubescent near base, the anther tube 2–3 mm. long, wholly exserted, the orifice subapical, the anthers all white-tufted at tips, the 2 shorter anthers each with scale-like bristles; ovary 25–65 mm. long, 2-celled, usually scabrous; seeds not twisted.

Wet mud flats and margins of vernal pools: Central Valley and lowermost foothills.

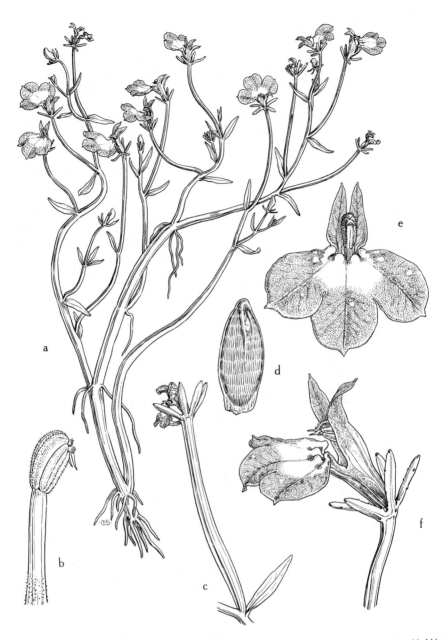

Fig. 336. *Downingia bella: a,* habit, × ⅔; *b,* stamen column, × 6; *c,* capsule, × 1½; *d,* seed, × 30; *e,* flower, front view, × 3; *f,* flower, side view, × 3.

McVaugh (*op. cit.*, p. 18) recognizes the typical form of *Downingia ornatissima,* with glabrous upper corolla lobes, as occurring from the Sacramento Valley southward to Merced County, and *D. ornatissima* var. *eximia,* with pubescent upper lobes, as occurring in the lower foothills and plains bordering the San Joaquin Valley and reaching as far north as Contra Costa and Calaveras counties.

5. Downingia bicornuta Gray, Syn. Fl. N. Am., 2d ed., 2^1(suppl.):395. 1886.

Erect, glabrous, annual herb, 1–4 dm. tall; leaves 3–30 mm. long, the submersed leaves linear, the emersed ones linear to broadly lanceolate; sepals linear to linear-elliptic, 3–13 mm. long, 3 long and 2 short; corolla blue, the tube broadly campanulate, white-bearded within on lower side, the lateral sinuses deeply cleft, the lower lip folded or reflexed, purplish blue, the central area white, yellow, or greenish, with 2 yellow or green spots and 2–4 horns or nipples at the folds, the upper lip darker than the lower one; stamen column 4–7 mm. long, straight to slightly curved, the orifice subapical, the shorter anthers each with an elongate scale-like bristle, these tightly twisted together; ovary 2-celled; seeds neither spirally striate nor twisted.

Vernal pools or wet grasslands: Sacramento Valley through the mountains to northeastern California; southeastern Oregon and adjacent Idaho and Nevada. Rare in San Joaquin Valley.

Hoover (*op. cit.*) has recognized *Downingia bicornuta* var. *picta* as having the bristles as long as or longer than the anther tube, while the corolla is somewhat smaller than is typical of the species. It is distributed in the Central Valley from Shasta County to Fresno County. (See fig. 335.)

6. Downingia concolor Greene, Bull. Calif. Acad. Sci. 2:153. 1886. Fig. 335.

Erect or decumbent annual, the stems to 3 dm. long; herbage glabrous; leaves 5–20 mm. long, linear to elliptic or ovate; sepals subequal in size, elliptic to linear or oblanceolate, obtuse to subacute, rotate or ascending; corolla 7–13 mm. long, blue, the lower lip with a quadrate or 2-lobed purple spot at base of central white area, with 2 ridges of nipple-like processes, glabrous, the upper lip of 2 lanceolate, ciliate lobes, the tube 3–5 mm. long, narrow-campanulate or turbinate, the lateral sinuses cut below plane of abruptly reflexed lower lip; stamen column straight to somewhat curved, the filament tube 2–4 mm. long, glabrous, the anther tube included or tip exserted, 2–3 mm. long, glabrous or pubescent on back, the orifice subapical, the lower pair of anthers each bearing short, scale-like bristles; ovary occasionally scabrous, becoming 3–6 cm. long in fruit, 2-celled; seeds not twisted.

Vernal pools: central California, Coast Range valleys at low elevations from Monterey County to Lake County, Cuyamaca Lake in San Diego County.

McVaugh (*op. cit.*, p. 16) accepts two varieties as distinguished by the character of the capsule: *Downingia concolor* var. *tricolor* Jepson, with the capsule

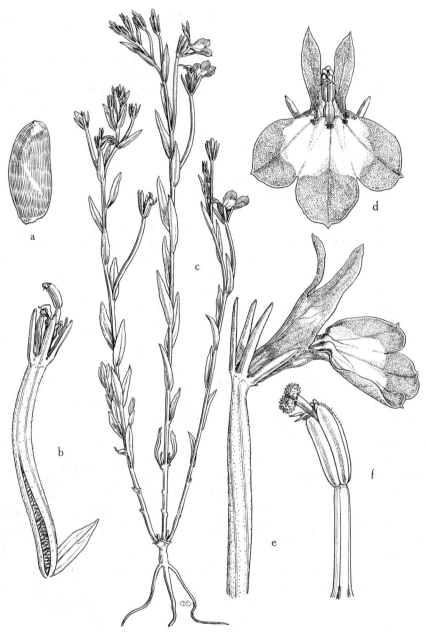

Fig. 337. *Downingia yina: a,* seed, × 30; *b,* capsule, × 2; *c,* habit, × ⅘; *d,* flower, front view, × 3; *e,* flower, side view, × 3; *f,* stamen column, × 6.

3–5 cm. long, is the central California plant; *D. concolor* var. *brevior* Mc-Vaugh, with the capsule 12–25 mm. long, is known only from Cuyamaca Lake in San Diego County.

7. Downingia bella Hoover, Leafl. West. Bot. 2:2. 1937. Fig. 336.

Erect or decumbent annual to 3 dm. long; stems sometimes thick-fistulose at base; leaves 5–20 mm. long, lanceolate to elliptic-ovate; sepals subequal in size, elliptic to linear or oblanceolate, obtuse to subacute, rotate or ascending; corolla 7–13 mm. long, light blue to purple, the tube campanulate, the lateral sinuses more deeply cleft than the upper sinus, the lower lip sharply reflexed, the central white area surrounding a yellow spot and with 2 yellow ridges extending into throat, the purple area, if present, reduced to 3 small spots, the lobes of upper lip not ciliate-margined; stamen column curved, barely exserted, the anthers sparsely pubescent, the orifice subapical, the lower pair of anthers each with a scale-like bristle; ovary 2-celled; seeds neither striate nor twisted.

Margins of vernal pools: San Joaquin Valley from San Joaquin County to Merced County, Boggs Lake in Lake County.

Similar in many respects to *Downingia concolor,* but differing chiefly in the absence of ciliation of the upper corolla lobes, the usual absence of purple coloration on lower lip of corolla or its reduction to small spots, and the frequently fistulose stems.

8. Downingia yina Applegate, Contr. Dudley Herb. 1:97. 1929. Fig. 337.

Erect or ascending annual, simple or often branched from base; stems 10–35 cm. long; leaves 1–3 cm. long, linear to broadly lanceolate-elliptic or ovate, entire to toothed, often the plant appearing very leafy; flowers solitary in the leaf axils; sepals linear-subulate to linear-lanceolate, equal in size or somewhat subequal, ascending; corolla 7–15 mm. long, the tube longer than calyx, narrowly campanulate, the lower lip not sharply reflexed, with a central white patch enclosing a yellow area over 2 low ridges which enter the throat, the upper lip erect or recurved, the lobes slightly spreading; stamen column slightly curved, barely exserted from the corolla tube, the anthers glabrous to sparsely short-hispid on the back, the orifice subapical; ovary 1-celled, the placentae parietal.

Bogs, lakes, and ponds: Lake County, Humboldt County, Siskiyou County; north to eastern Washington.

Apparently not common in California. Differs from *Downingia montana* in its usually broader leaves, the sepals more nearly equal in size, and the absence of bearding on the anthers. The specimen *Schreiber 2339,* from Sheep Camp, Snow Mountain, Lake County, is referred to *D. yina,* even though it is more slender in most details than typical members of this species.

McVaugh (*op. cit.,* p. 24) recognized *D. yina* var. *major,* based upon the type of *D. willamettensis* Peck, which is larger in most details than typical *D. yina.*

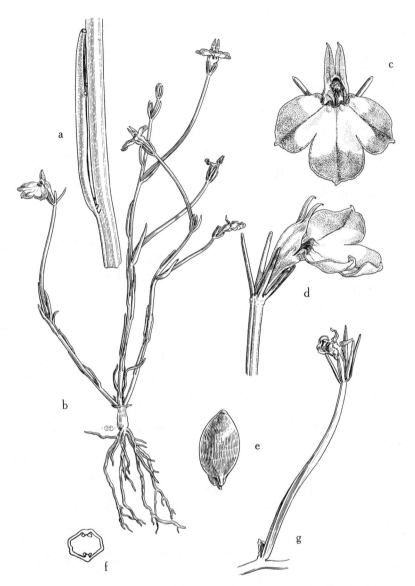

Fig. 338. *Downingia montana: a,* part of stem, showing closely appressed leaf, × 6; *b,* habit, × 1; *c,* flower, front view, × 3; *d,* flower, side view, × 3; *e,* seed, × 30; *f,* ovary (cross section), showing the single cell, × 8; *g,* capsule, × 2.

9. Downningia montana Greene, Pittonia 2:104. 1890. Fig. 338.

Downningia bicornuta var. *montana* Jepson, Madroño 1:102. 1922.

Erect or ascending annual herb; stems to 15 cm. long, glabrous throughout or sometimes scabrous above; leaves linear-subulate to lanceolate or elliptic, 8–16 mm. long, sometimes minutely toothed; sepals linear-subulate, unequal in size, 3 long and 2 short, ascending or appressed in anthesis; corolla blue to violet, 9–12 mm. long, glabrous, the tube 3–5 mm. long, narrowly campanulate to narrowly funnelform, often the upper side somewhat gibbous, the lateral sinuses cut more deeply than the upper sinus, the lower lip with 2 greenish folds near base, alternating with purple grooves, the lobes of upper lip erect, lanceolate, almost parallel; stamen column short, slightly curved, almost concealed by upper lip of corolla, the filament tube 3–4 mm. long, glabrous or minutely pubescent below, the anther tube 1.5–2 mm. long, the orifice subapical, all anthers prominently bearded, the shorter ones with scale-like bristles; ovary 1-celled, the placentae parietal; seeds not twisted, the striations longitudinal.

Wet, grassy meadows: middle altitudes of the northern Sierra Nevada, north to Shasta County.

10. Downningia cuspidata (Greene) Greene ex Jepson, Fl. West. Middle Calif., 2d ed., 403. 1911.

Downningia pulchella var. *arcana* Jepson, Madroño 1:100. 1922.
Downningia immaculata Munz & Johnst., Bull. Torrey Club 51:300. 1924.

Erect or ascending herb; stems to 3 dm. high; herbage glabrous; leaves 3–15 mm. long, linear to broadly elliptic or lanceolate; sepals elliptic to oblanceolate, subequal in size, 3–10 mm. long, ascending; corolla blue to lavender or occasionally white, the tube 3–5 mm. long, narrowly campanulate to almost cylindrical, the lateral sinuses only slightly deeper than the upper sinus, barely below plane of lower lip, the lower lip plane or nearly so, with inconspicuous yellow ridges at base, and with a broad white area having a central or basal yellow patch; stamen column curved but not conspicuously bent, the filament tube 2–4 mm. long, glabrous, the anther tube included at base, 1–2.5 mm. long, the anthers sparsely pubescent on backs, all white-tufted, the shorter ones bearing also scale-like bristles, the orifice subapical; ovary 15–55 mm. long, 2-celled; seeds twisted, with thin, spiral striations.

Vernal pools and wet soil: Coast Ranges and central Sierra Nevada foothills, mountains of southern California.

This is a very variable species, and its variations, as at present understood, cannot be correlated either with geography or with ecology.

Fig. 339. *Downingia pusilla: a,* ovary and flower, × 8; *b,* stamen column, × 12; *c,* habit, × 1½; *d,* seed, × 30; *e,* seed, × 30; *f,* stamen column, × 12; *g,* flower, × 8; *h,* ovary and flower, × 1½; *i,* habit, × ⅕.

11. Downingia pusilla (G. Don) Torr., U. S. Expl. Exped. (Wilkes) 17:375. 1874. Fig. 339.

Downingia humilis Greene, Leafl. Bot. Obs. & Crit. 2:45. 1910.

Erect or ascending annual, simple or branched from base; stems 3–15 cm. long, thick, somewhat flaccid below, glabrous, the internodes unequal in length; leaves linear, 5–10 mm. long, alternate, sessile, 3-nerved, 2 of the veins marginal; sepals unequal in size, with or without 2 or 3 marginal callous glands, 3–6 mm. long, usually longer than the petals; corolla blue or whitish, 4–5 mm. long, distinctly bilabiate, the lower lip distinctly divergent but not reflexed, with a yellow spot on the palate, the lobes about as long as the narrowly campanulate tube, the upper lobes 2–2.5 mm. long, lanceolate; stamen column 1.5–2 mm. long, the anther tube in a parallel plane and continuous with filament tube, 1 mm. long, the anthers unequal in length, the upper ones curved over the end of the tube, the orifice thus subapical, the longest anther bearing a pair of bristles; ovary rapidly elongating, becoming 2–3 cm. long, linear-ellipsoid to linear-oblong; seed brown, the coat spirally striate.

About vernal pools: Central Valley from Merced and San Joaquin counties to Sacramento County.

Under the name *Downingia pusilla* we include a variable lot of material which further study may reveal to represent two distinct taxa. Greene described *D. humilis* from plants from the Sonoma Valley. Material from this locality and other localities immediately north of San Francisco Bay has chiefly white flowers and a corolla tube longer than the lobes. The lobes are chiefly approximate. In our collections in the Central Valley, the corolla tube is generally shorter than the lobes, and the lower lip is conspicuously spreading and often has a distinct palate. The corolla is, furthermore, blue. However, dried material at hand does not lend itself to conclusive decisions on this question, and hence the problem must await detailed field study. Should it be concluded that two taxa are represented, it would raise the further problem of determining to which form the name *D. pusilla* should be attached, since that name is based upon plants from Chile. In figure 339, parts *a–d* represent one of these entities, and *e–i* the other.

12. Downingia laeta (Greene) Greene, Leafl. Bot. Obs. & Crit. 2:45. 1910. Fig. 340.

Erect or decumbent annual; stems 1–3 dm. long, often fistulose below; herbage glabrous; leaves of submersed stems linear-filiform, 1–2 cm. long, about 0.5 mm. wide, those of emersed stems lanceolate to lanceolate-elliptic, sessile, 10–12 mm. long, about 2 mm. wide; sepals linear to linear-elliptic, subequal in size, rounded to acute at apex, 3–10 mm. long, 1–2 mm. wide, longer than the corolla; corolla lobes longer than the tube, 4–8 mm. long, glabrous, blue to purplish, the lower lip with white or yellow area and a trans-

Fig. 340. *Downningia laeta: a,* flower, front view, × 6; *b,* flower, side view, × 6; *c,* habit, × ⅔; *d,* seed, × 30; *e,* capsule, × 2.

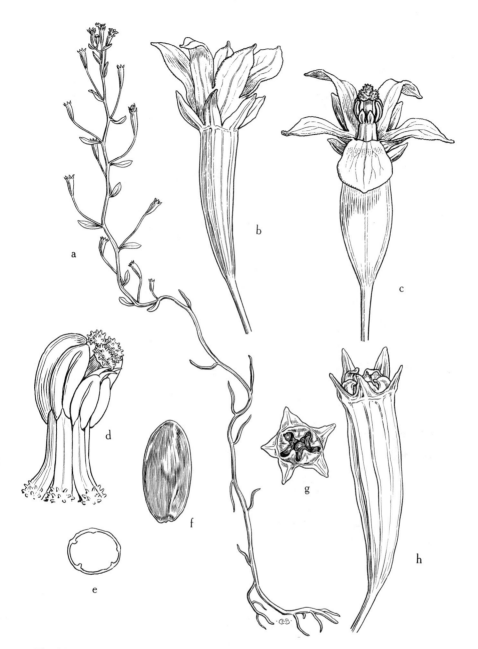

Fig. 341. *Legenere limosa: a,* habit, × ⅓; *b,* flower, side view, × 8; *c,* flower, front view, × 8; *d,* stamen column, × 24; *e,* capsule (cross section), × 6; *f,* seed, × 30; *g,* apex of capsule, showing method of dehiscence, × 6; *h,* mature capsule, × 6.

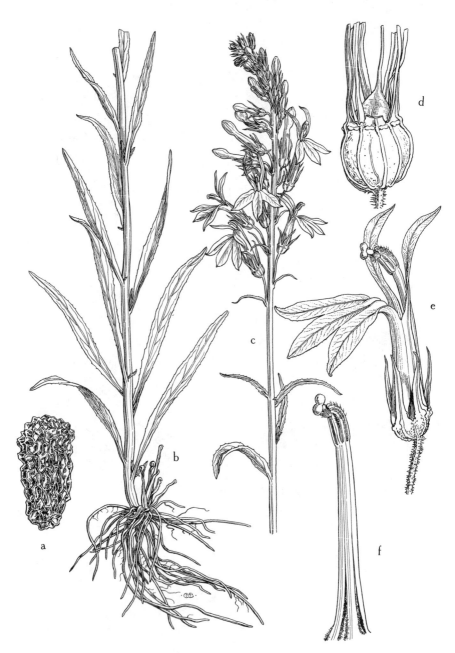

Fig. 342. *Lobelia splendens: a,* seed, × 30; *b* and *c,* habit, × ⅖;
d, capsule, × 3; *e,* flower, × 1½; *f,* stamen column, × 2½.

verse purple band or row of purple spots at base; stamen column straight to slightly curved, the filament tube stout, 1.5–3 mm. long, glabrous, the anther tube 1–2 mm. long, the orifice subapical, the anthers glabrous or ciliate on back, the 2 below the orifice white-tufted and with scale-like bristles; ovary becoming 2–4 cm. long in fruit, 2-celled; seeds straight or only slightly twisted.

Marshy ponds: northeastern California in Lassen and Modoc counties; eastern Oregon, northeast to Wyoming and Saskatchewan.

LEGENERE

1. Legenere limosa (Greene) McVaugh, N. Am. Fl. 32A[1]:13. 1943. Fig. 341.

Howellia limosa Greene, Pittonia 2:81. 1890.

Procumbent annual; submersed leaves linear, 1–3 cm. long, the emersed leaves oblong-lanceolate to oblanceolate, 1–2 cm. long, glabrous; flowers solitary in the axils of the upper leaves on slender pedicels, these elongating in fruit often to 2 or 3 times as long as the subtending leaf; sepals minute, narrowly triangular, equal in size or somewhat subequal; corolla present or absent, 2–4 mm. long, the tube split on upper side to near base, the lobes subequal in size, the middle lobe of lower lip broadest; stamen column erect, about 2 mm. long, the anthers connate or united into a tube, subequal in length to the filament tube, the orifice subapical, minutely appendaged; ovary obconic; stigma 2- or 3-lobed; fruit a 1-celled capsule 6–10 mm. long, the placentation parietal; seeds smooth, shiny, brown.

Beds of vernal pools: lower Sacramento and San Joaquin valleys, also at Snow Lake in Lake County.

LOBELIA

1. Lobelia splendens Willd., Hort. Berol. pl. 86. 1806. Fig. 342.

Lobelia cardinalis L. of some western American authors.

Perennial herb, 2–7 dm. tall; stems simple or sometimes branched; leaves linear-lanceolate or oblanceolate, acute or more rarely obtuse, entire to serrulate or denticulate or distinctly and coarsely serrate or dentate, sessile or the lower leaves narrowed to a petiole-like base, the upper ones sessile and becoming much reduced and bractiform in the inflorescence; flowers in a terminal, erect, racemose inflorescence 1–3 dm. long, flowering from below upward, the pedicels pilose; calyx tube short, its limb 5-cleft, the lobes linear-subulate, 7–12 mm. long; corolla strongly 2-lipped with the long cylindrical tube split to the base between the upper 2 lobes, red, yellow, or orange suffused with red, the lower lip 3-lobed, deep red, the lobes 1–2 cm. long, 3–4 mm. broad, oblong-elliptic to oblong-obovate, short-acuminate to mucronate or apiculate,

Fig. 343. *Palmerella debilis: a,* habit, × ⅖; *b,* seed, × 40; *c,* flower, × 3; *d,* flower (longitudinal section), showing stamen insertion, × 3; *e,* capsule, × 4.

the lobes of upper lip erect or recurved, linear-lanceolate, 8–12 mm. long, deep red; stamens 5, united into a tube except at the very base of the filaments, the filament tube slightly exserted, the anther tube erect, the orifice subapical, the tip of shorter pair of anthers with a dense tuft of hairs, the other 3 anthers sparsely ciliate; ovary inferior; style elongate; stigma of 2 unequally sized lobes; fruit a turbinate capsule as wide as or wider than long, with a low conical apex, loculicidally dehiscent at the conical apex; seeds conspicuously ridged and pitted.

Wet stream banks or seepage bogs: mountains of southern California; east to Nevada and Arizona, south to Mexico.

McVaugh (*op. cit.*, p. 81) treats these plants as subspecific to *Lobelia cardinalis*. There is little question but that the two forms are related, and there may be clear justification for McVaugh's course. Since, however, he neither discusses the problem nor documents his conclusions with an adequate series of cited specimens, he has failed to make his case clear. Furthermore, the geographic pattern of his subspecies, for the most part, fails to make taxonomic sense, and hence suggests that the taxa are artificially drawn. Although, for these reasons, it seems necessary here to return to former local usage in nomenclature, it is nevertheless clear that ultimately the name *L. splendens* will have to give way to another.

PALMERELLA

1. Palmerella debilis Gray, Proc. Am. Acad. Arts & Sci. 11:80. 1876. Fig. 343.

Laurentia debilis var. *serrata* McVaugh, Bull. Torrey Club 67:144. 1940.

Perennial herb from slender rhizomes; stems slender, erect or decumbent, 1–8 dm. high, simple or racemosely branched, glabrous below the inflorescence; leaves thin, almost membranous, the lower leaves glabrous, the upper ones sometimes pubescent, narrowly lanceolate to broadly obovate, often very variable in the same individual, ours sharply to bluntly serrate to serrate-dentate, acute to obtuse, sessile or the lower leaves narrowed to a petiole, reduced and bractiform in the inflorescence; flowers in a subcapitate or elongate raceme on slender pedicels; calyx tube obconic to campanulate, the lobes linear-attenuate, 8–11 mm. long; corolla bicolored, the tube white, the limb deep blue, the upper sinus slightly deeper than the lateral ones, the tube 8–15 mm. long, closely pubescent within, the limb 2-lipped, the lower lip of 3 obovate-elliptic lobes 5–10 mm. long, the upper lip of 2 erect, recurved or spreading linear lobes 5–7 mm. long; filaments 9–14 mm. long, unequally inserted on corolla tube, united into a tube above, the anther tube 2–3 mm. long, the orifice subapical, the 2 shorter anthers white-tufted at tip, the others pilose on back; ovary inferior or nearly so, bristly-pubescent to glabrous; capsule campanulate to obconic, 6–12 mm. long; seeds smooth.

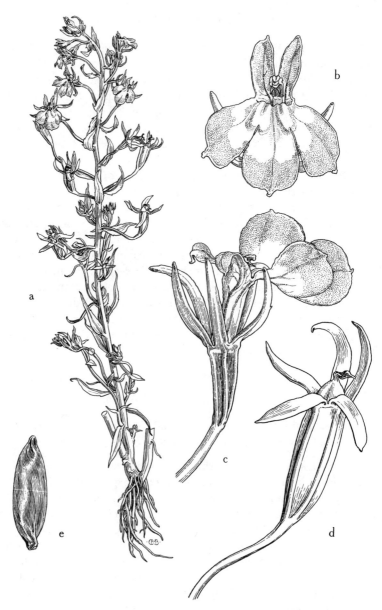

Fig. 344. *Porterella carnulosa: a,* habit, × ⅘; *b,* flower, front view, × 4;
c, flower, side view, × 4; *d,* capsule, × 3; *e,* seed, × 30.

Margins of streams: mountains of southern California, north along the coast to Big Sur River and Arroyo Seco in Monterey County.

PORTERELLA

1. **Porterella carnosula** (Hook. & Arn.) Torr. in Hayden, Rep. Geol. Surv. Mont. 488. 1872. Fig. 344.

Laurentia carnosula (Hook. & Arn.) Benth. & Hook. ex Gray, Bot. Calif. 1:444. 1876.

Erect, somewhat succulent, annual herb; stems branched, slender to stout, 6–30 cm. tall; herbage glabrous; leaves linear to lanceolate, sessile, 1–3 cm. long, the tip acute or acuminate; flowers solitary in the leaf axils on slender, spreading or ascending pedicels; sepals linear-lanceolate, 3–10 mm. long; corolla blue to white, bilabiate, the tube short-cylindrical to somewhat enlarged above, 4–6 mm. long, the lower lip 3-lobed, rotately spreading, the lobes broadly obovate to nearly orbicular, with a broad white area and a central greenish yellow spot conforming to 2 low ridges near the base of the lip, the upper lobes oblong-lanceolate, united at base, somewhat divergent; stamen column united throughout, the filament tube included, 3–7 mm. long, the anther tube 1.5–3 mm. long, the orifice subapical, the anthers all tufted at tip, the 2 short ones each with a scale-like bristle; ovary inferior, obconic to turbinate, 5-angled, bilocular, 8–14 mm. long.

Vernal pools and wet meadows: eastern Siskiyou County and Modoc County, southward through the Sierra Nevada to Tulare County; Oregon, Wyoming, Utah, and Arizona.

COMPOSITAE. SUNFLOWER FAMILY

Herbs or shrubs. Flowers in a close head (anthodium) simulating a single flower, each head on the enlarged summit of the common peduncle (receptacle) and surrounded by 1 to several series of common bracts (phyllaries) forming an involucre. Receptacle either with scale-like or bristle-like bracts among the flowers (bracteate) or without bracts among the flowers (naked). Calyx tube fused with ovary, the lobes often continued above ovary as bristles or scales (pappus). Heads with corollas either all tubular and (in ours) 5-toothed at summit (heads discoid), or corollas all strap-shaped (ligulate) and toothed at apex, or both kinds in the same head (radiate), the flowers with strap-shaped corollas then marginal (ray flowers) and the flowers with tubular corollas central (disc flowers). Stamens 5, on the corolla tube, united by their elongated anthers or these rarely free; filaments usually free. Style 2-cleft at apex. Ovary inferior, 1-celled, 1-ovuled. Fruit (achene) 1-seeded, with or without pappus of scales (paleae), awns, or bristles at apex.

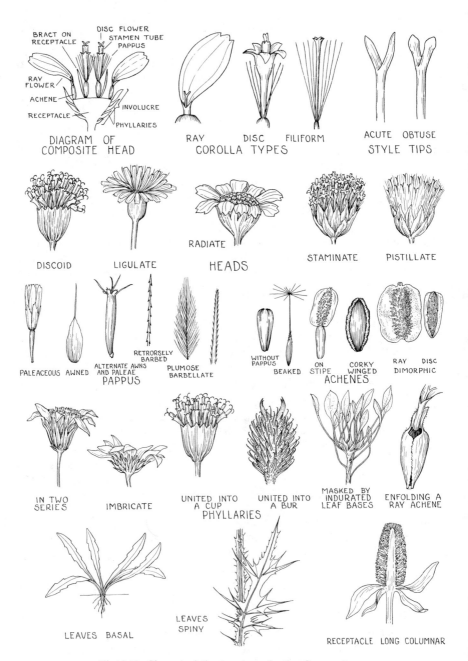

Fig. 345. Characteristic structures in the Compositae.

KEY TO THE TRIBES

Heads composed wholly of flowers with ligulate corollas; herbs with milky juice
CICHORIEAE

(*Agoseris, Apargidium, Lactuca, Leontodon, Phalacroseris, Taraxacum*)

Heads composed of ray and disc flowers or of disc flowers only; flowers of the disc essentially regular, sometimes those on the margins of the disc enlarged on one side; ray flowers ligulate.

Heads discoid, never radiate; anther sacs caudate at the base.

Herbs mostly thistles or thistle-like; phyllaries prickly or with fringed margins; flowers all perfect, the corollas deeply cleft into linear lobes; anthers with an elongate appendage at the tip; receptacle covered with bristles ..... CYNAREAE
(*Cirsium*)

Herbs or shrubs not thistle-like, mostly white-woolly; phyllaries membranous and entire, never fringed; flowers unisexual, the pistillate corollas filiform; anthers not appendaged at the tip; receptacle naked or with chaffy bracts .... INULEAE
(*Evax, Gnaphalium, Pluchea, Psilocarphus*)

Heads radiate or discoid; anther sacs not caudate at the base.

Style branches stigmatic only on the lower half, thickened upward; flowers perfect, white or pink, never yellow; rays none .... EUPATORIEAE
(*Brickellia, Trichocoronis*)

Style branches stigmatic to the tip; disc flowers yellow to purple, occasionally white. Receptacle naked.

Margins of phyllaries papery or scarious; herbage usually strong-scented (except *Cotula*); leaves often (but not always) finely dissected ... ANTHEMIDEAE
(*Achillea, Anthemis, Artemisia, Cotula*)

Margins of phyllaries neither papery nor scarious.

Phyllaries usually well imbricated and in several series, never differentiated into an inner and an outer series; style branches of the perfect flowers with appendages, these hairy outside, glabrous inside, never with a ring of longer hairs .... ASTEREAE
(*Aster, Baccharis, Conyza, Erigeron, Grindelia, Haplopappus, Solidago*)

Phyllaries in few series, little imbricated.

Pappus awnlike, bristly, paleaceous, or none; if plumose, the phyllaries penicillate; receptacle conical or flat; rays yellow or none; disc yellow, sometimes white; style branches of perfect flowers usually without appendages .... HELENIEAE
(*Amblyopappus, Baeria, Helenium, Jaumea, Lasthenia*)

Pappus of soft capillary or plumose bristles; disc corollas and commonly the rays yellow; receptacle flat; phyllaries usually in 1 series, if numerous then usually differentiated into an inner and an outer series, those of the outer series often very short .... SENECIONEAE
(*Arnica, Senecio*)

Receptacle with chaffy bracts.

Rays present (if absent, the heads bisexual); anthers usually united.

Phyllaries in 1 series, each embracing or enfolding a ray achene; bracts of the receptacle often in a single series between the ray and the disc flowers, or a bract sometimes subtending each flower of the disc ..... MADIEAE
(*Centromadia*)

Phyllaries in more than 1 series, plane.
  Phyllaries wholly herbaceous; rays commonly yellow, occasionally white
                                                   HELIANTHEAE
        (*Bidens, Eclipta, Helianthus, Rudbeckia, Verbesina*)
  Phyllaries scarious-margined; herbage very strongly aromatic; leaves finely
    dissected; rays white, rarely pink . . . . ANTHEMIDEAE
        (*Achillea, Anthemis, Artemisia, Cotula*)
Rays absent; heads mostly unisexual; pistillate corollas absent or rudimentary;
  anthers distinct or nearly so; fruit sometimes with a burlike or a horizontal,
  winglike process enclosing the achene . . . . AMBROSIEAE
        (*Ambrosia, Iva, Xanthium*)

KEY TO THE GENERA

Herbs with milky juice; heads with all corollas strap-shaped.
  Achenes of two kinds; the outer achenes with pappus of short paleae, the inner ones
    with pappus of plumose bristles . . . . *Leontodon*
  Achenes all alike; pappus never plumose.
    Achenes without pappus or with a short crown . . . . *Phalacroseris*
    Achenes with pappus of awns or bristles.
      Leaves all basal; heads solitary on long, naked peduncles.
        Mature achenes beaked.
          Phyllaries in 2 distinct series, those of the outer series very short; achenes
            4- or 5-ribbed, roughened or spinose above . . . . *Taraxacum*
          Phyllaries not in 2 distinct series, imbricated, graduated; achenes 10-ribbed,
            smooth . . . . *Agoseris*
        Mature achenes not beaked . . . . *Apargidium*
      Leaves all along the tall stems; heads in a paniculate inflorescence . . . . . . *Lactuca*
Herbs or shrubs without milky juice; heads discoid (all corollas tubular) or radiate
  (marginal flowers rays).
  Heads discoid, that is, without ray flowers.
    Heads unisexual, containing either all staminate flowers or all pistillate flowers, the
      two kinds usually quite unlike.
      Plants woody, at least at base, or shrubby; herbage more or less glutinous; sta-
        mens and pistils occurring on separate plants (plants dioecious); phyllaries
        distinct, never becoming burlike . . . . *Baccharis*
      Plants annual or perennial herbs; herbage not glutinous; stamens and pistils borne
        in separate heads on the same plant (plants monoecious), the staminate heads
        in terminal clusters, the pistillate heads axillary and usually solitary, their
        1–4 flowers enclosed in a burlike involucre.
      Involucre of staminate heads with phyllaries united, cuplike; fruiting involucre
        of pistillate heads a tuberculate bur 1–1.5 mm. long . . . . *Ambrosia*
      Involucre of staminate heads with phyllaries distinct; fruiting involucre of pis-
        tillate heads a hardened bur 1–2 cm. long, covered with hooked prickles
                                         Xanthium
    Heads all bisexual and alike, a single head containing all perfect flowers, or perfect
      and pistillate flowers, or perfect and staminate flowers, or pistillate and stami-
      nate flowers.
      Outer pistillate flowers lacking corollas, their achenes on stipes as long as pyhl-
        laries; disc corollas bright yellow . . . . *Cotula*

All flowers with corollas, their achenes sessile.

Leaves spiny-toothed; pappus bristles strongly plumose (thistles) .... *Cirsium*

Leaves not spiny-toothed; pappus none or its bristles not plumose (sometimes hairy at base in *Gnaphalium*).

Herbage usually very white-woolly.

Heads with outer flowers pistillate and central flowers staminate; pistillate flowers enfolded by or enclosed in phyllaries or in bracts of receptacle; pappus absent.

Leaves alternate; heads in a dense terminal cluster; phyllaries in fruit indurated and merely enfolding achenes of pistillate flowers .. *Evax*

Leaves opposite; heads scattered, axillary, or terminal; true phyllaries lacking; receptacle bracts saclike, apically hooded, laterally beaked, each completely enclosing a pistillate flower .... *Psilocarphus*

Heads with outer flowers pistillate and central flowers perfect, all fertile; phyllaries scarious, plane, not enclosing pistillate flowers; receptacle bracts absent; pappus present, deciduous .... *Gnaphalium*

Herbage not at all white-woolly in any part.

Pappus absent.

Heads in leaf axils, nodding; leaves entire, never white-tomentose below; phyllaries herbaceous, 5–10, united into a cup or free .... *Iva*

Heads in terminal panicles, usually erect; leaves (in ours) usually pinnately lobed to bipinnately divided, or if some entire, these white-tomentose below; phyllaries many, imbricated, scarious .. *Artemisia*

Pappus present.

Pappus paleaceous, at least in part, or of 2 to several stiff awns.

Leaves mostly opposite.

Flowers flesh-colored; achenes 5-angled; pappus of minute paleae alternating with barbellate bristles .... *Trichocoronis*

Flowers yellow; achenes flattened or 4-angled; pappus of retrorsely barbed, stiff awns .... *Bidens*

Leaves alternate.

Disc flowers dark purple, rays yellow; receptacle long-columnar; pappus a short, lacerate or denticulate, paleaceous crown

Rudbeckia

Disc and ray flowers both yellow; receptacle low-conical; pappus of 8–12 short, rounded paleae .... *Amblyopappus*

Pappus of numerous capillary bristles, these scabrous, barbellate, or smooth.

Corollas yellow; pappus bristles pure white, soft, copious, hairlike

Senecio

Corollas purplish to white; pappus bristles often scabrous or barbellate, somewhat rigid, not very copious or hairlike, usually sordid white.

Flowers of two kinds, marginal flowers pistillate with filiform corollas, central perfect flowers with tubular corollas; phyllaries not strongly graduated, not striate.

Phyllaries frequently lavender-tinged; corollas lavender to reddish purple .... *Pluchea*

Phyllaries green with scarious, whitish margins; corollas whitish

Conyza

Flowers all alike, perfect, whitish; phyllaries strongly graduated and striately nerved .... *Brickellia*

Heads radiate, that is, with both ray and disc flowers.
  Rays blue, purple, pink, or white.
    Pappus none; rays white.
      Leaves simple, serrate, not aromatic; receptacle bracts of filiform bristles
        Eclipta
      Leaves finely dissected, aromatic; receptacle naked or with stiff, chaffy, or acuminate bracts.
        Annual; heads solitary on long, terminal peduncles .... *Anthemis*
        Perennial; heads small, closely clustered in a terminal compound corymb
          Achillea
    Pappus of numerous capillary bristles; rays blue, purple, pink, or white.
      Phyllaries usually strongly graduated, sometimes subequal; ray corollas usually fairly broad; style tips of disc flowers elongate, acute .... *Aster*
      Phyllaries always subequal, scarcely graduated; ray corollas usually narrow; style tips of disc flowers short, broadly triangular or obtuse ..... *Erigeron*
  Rays yellow.
    Pappus of numerous bristles.
      Phyllaries unequal and imbricated in 2 or more series; pappus bristles more or less rigid, at least not very soft or hairlike.
        Pappus dull white; heads many, small, in compound inflorescences .. *Solidago*
        Pappus sordid or yellow; heads solitary in axils of reduced upper leaves, hence racemose .... *Haplopappus*
      Phyllaries equal, in 1 or 2 series (or a few minute outer bracts at base); pappus of many soft capillary hairs.
        Leaves opposite .... *Arnica*
        Leaves alternate .... *Senecio*
    Pappus of 2–12 awns or paleae or none.
      Phyllaries in a single series, each enfolding a ray achene .... *Centromadia*
      Phyllaries in 1 or more series, not enfolding ray achenes.
        Phyllaries united into a cup .... *Lasthenia*
        Phyllaries distinct.
          Receptacle naked.
            Leaves linear, all opposite.
              Leaves fleshy; pappus none .... *Jaumea*
              Leaves not fleshy; pappus usually present .... *Baeria*
            Leaves broader, all alternate or only the lower ones opposite.
              Phyllaries in several series, imbricated, gummy .... *Grindelia*
              Phyllaries in a single series, not gummy .... *Helenium*
          Receptacle with conspicuous, chaffy bracts.
            Receptacle flat or low-convex.
              Phyllaries imbricated in 2–4 series; achenes thick, quadrangular; pappus of 2 paleaceous awns sometimes with intervening squamellae .... *Helianthus*
              Phyllaries differentiated into an outer and an inner series; achenes usually flattened (4-angled in *Bidens cernua*); pappus of 2–5 bristle-like, retrorsely barbed awns .... *Bidens*
            Receptacle conical to columnar.
              Receptacle conical; achenes strongly flattened, with corky wings
                Verbesina
              Receptacle high-columnar; achenes quadrate, not winged .. *Rudbeckia*

ACHILLEA

1. **Achillea borealis** subsp. **californica** (Pollard) Keck, Carnegie Inst. Wash. Publ. 520:299. 1940. Yarrow, milfoil. Fig. 346.

Achillea californica Pollard, Bull. Torrey Club 26:369. 1899.
Achillea millefolium L. var. *californica* Jepson, Man. Fl. Pl. Calif. 1137. 1925.
Achillea millefolium var. *lanulosa* of authors, not Piper.

Robust perennial herb 6–10 dm. tall; stems erect from horizontal rootstocks, simple below, somewhat branched above; herbage sparsely pubescent with long, appressed hairs; leaves alternate, numerous, linear-lanceolate, 6–12 cm. long, glabrate, bipinnatifid with fine, distinct, linear segments, the tips of segments spinulose; heads 5–6 mm. high, numerous in a dense, terminal, compound, flat-topped or convex corymb 2–10 cm. across; involucre oblong, the phyllaries imbricated, graduated, stramineous, papery- or scarious-margined, the keels greenish; receptacle flat, somewhat chaffy; ray flowers 4–6, conspicuous, fertile, the ligules 3–5 mm. long, white; disc flowers white, fertile; achene obcompressed, linear, narrowly callous-margined, without pappus.

Dry slopes or in moist bottomlands, at low altitudes: Coast Ranges and lowest foothills of the Sierra Nevada; north to Oregon and Washington.

Although by no means always a marsh inhabitant, this plant occurs in great abundance in some marshlands, notably the Suisun marshes in Solano County.

AGOSERIS

Ours perennial, scapose herbs with milky juice and large taproots. Leaves basal, elongate, entire to lobed. Heads single, large, terminal on long, naked peduncles. Phyllaries imbricated, graduated, the outer ones often ovate, the inner ones gradually narrower. Receptacle naked. Corollas all ligulate, yellow or reddish. Achenes terete, elongate, 10-ribbed, beaked; pappus of soft, white, capillary bristles, usually longer than achenes.

The taxonomy of this genus is greatly in need of basic and thorough revision, and a recent comprehensive study of the group has not yet been published. Since the species dealt with below are only marginal to this study, I have attempted no re-evaluation of their status.

Beak of achene much shorter than body, thick, ribbed; heads conspicuous, ligules much
 longer than involucre .... 1. *A. glauca* var. *laciniata*
Beak of achene as long as or longer than body, slender, not ribbed; heads inconspicuous,
 ligules about as long as or slightly longer than involucre.
 Beak about as long as body or a little shorter .... 2. *A. elata*
 Beak much longer than body, often twice as long .... 3. *A. gracilens*

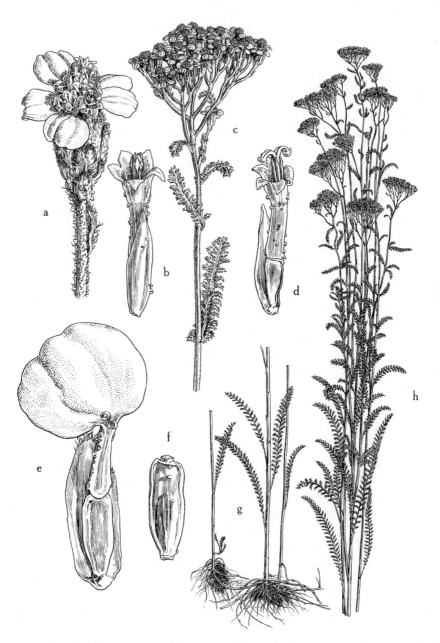

Fig. 346. *Achillea borealis* subsp. *californica: a*, flowering head, × 4; *b*, disc flower at early stage in anthesis, × 8; *c*, inflorescence, × ⅘; *d*, disc flower at later stage in anthesis (compare with *b*), × 8; *e*, ray flower and subtending bract, × 12; *f*, achene, × 12; *g* and *h*, habit, × ⅛.

1. Agoseris glauca (Pursh) D. Dietr. var. **laciniata** (D. C. Eat.) Smiley, Univ. Calif. Publ. Bot. 9:404. 1921.

Scapes 0.5–3 dm. tall; herbage glabrate or puberulent; leaves lanceolate, entire or variably toothed or lobed; heads usually showy, the ligules pale yellow, longer than the involucre; involucre 1.5–2 cm. high, usually glabrous; achenes gradually tapering into a stout, ribbed beak, this usually much shorter than the body.

Montane meadows, often in wet spots, at elevations of 4,000–7,500 feet: Sierra Nevada and Siskiyou Mountains; north to Oregon, east to Rocky Mountains.

2. Agoseris elata (Nutt.) Greene, Pittonia 2:177. 1891.

Scapes stout, 30–60 dm. tall; herbage glabrate or puberulent, but peduncles tomentose, at least just beneath heads; leaves oblanceolate, usually entire; heads not showy, the ligules pale yellow, about as long as or slightly longer than the involucre; involucre 1–2.5 cm. high; achenes tapering into a narrow, nonribbed beak about as long as or a little shorter than the body.

Montane meadows, at elevations of 6,500–7,000 feet: Sierra Nevada; north to Oregon.

3. Agoseris gracilens (Gray) Kuntze, Rev. Gen. Pl. 1:304. 1891.

Scapes stout, 30 dm. tall; herbage mostly glabrate; upper end of peduncles and the involucres frequently glandular-puberulent; leaves subentire to remotely toothed or pinnatifid; heads small, few-flowered, the ligules dull orange or reddish, slightly longer than the involucre; involucre 1.5–2.5 cm. high; achenes with slender beak often twice as long as body.

Grassy areas or in moist spots, at elevations of 100–4,000 feet: North Coast Ranges and the Sierra Nevada; north to Washington.

AMBYLOPAPPUS

1. Ambylopappus pusillus Hook. & Arn. in Hook., Jour. Bot. 3:321. 1841.

Paniculately branched annual 1–4 dm. tall; herbage sweet-scented; leaves alternate, pinnately dissected into 3–5 linear segments or the upper leaves linear and entire; heads small, yellow, discoid; involucre campanulate, 2–3 mm. high, the phyllaries 5, in 1 series, obovate and wholly herbaceous; receptacle conical; flowers 10–25 per head; achenes short-clavate, 4-sided; pappus paleaceous, the paleae 8–12, oblong or rounded.

Salt marshes: southern California north to San Luis Obispo County; Baja California in Mexico, South America.

AMBROSIA. RAGWEED

1. Ambrosia psilostachya DC., Prodr. 5:526. 1836. Western ragweed.

Coarse, aromatic, perennial herb; stems 3–6 dm. tall, erect from long, slender, horizontal rootstocks; herbage harshly pubescent with stiff gray hairs; leaves opposite below, frequently alternate above, once- or sometimes twice-pinnately incised; heads small, discoid, unisexual, greenish; staminate heads several- to many-flowered, nodding in erect, spikelike, terminal racemes, the phyllaries united into a broadly hemispherical cup, at least the outer flowers with receptacular bracts, the corollas 5-lobed; pistillate heads usually solitary in axils of upper leaves at base of staminate racemes, the involucres closed, turbinate, containing a single flower with no corolla and no pappus, enlarging in fruit to form a pear-shaped bur 2.5–3 mm. long, with usually a single row of tubercles or protuberances at the truncate apex, and a central apical point.

Uncultivated waste lands at low altitudes, sometimes in marshy flats: throughout cismontane California; east through middle-western states, north to Canada.

ANTHEMIS. CAMOMILE

1. Anthemis Cotula L., Sp. Pl. 894. 1753. Mayweed, dog fennel.

Many-branched annual herb, 3–6 dm. tall; herbage with pungent odor, thinly pilose or nearly glabrous; leaves alternate, pinnately 2 or 3 times dissected into many linear divisions with cuspidate tips; heads solitary on naked, long peduncles arising at ends of branches, 1.5–2.5 cm. across; involucre hemispheric, the phyllaries scarious, imbricated, narrow, acute; receptacle conical, naked below, with slender, stiff, acuminate bracts toward summit; ray flowers white, pistillate, reflexed in age; disc flowers perfect, yellow; achenes rugose, 10-ribbed; pappus none.

Weed introduced from Europe, frequent in fields and waste areas, sometimes in marshy lowlands: throughout most of the United States.

APARGIDIUM

1. Apargidium boreale Torr. & Gray, Fl. N. Am. 2:474. 1843.

Perennial herb with milky juice, glabrous; leaves linear or lanceolate, 1–2 dm. long, in a basal rosette, often from a branched crown, entire or remotely denticulate, tapering to a broad petiole; heads solitary on slender scapes 2–4 dm. high, the buds nodding; involucre campanulate, 10–14 mm. high, the phyllaries in several series, lanceolate-acuminate; corollas yellow, all ligulate, well exserted; achenes strongly ribbed, linear, 4–5 mm. long; pappus of 30–45 persistent, minutely barbellate bristles.

Wet meadows: mountains of Humboldt County; north to Alaska.

ARNICA

1. Arnica Chamissonis Less., Linnaea 6:238. 1831.

Perennial herb; rhizomes pale, naked except for nodal scales, runner-like, mostly unbranched; stems single, usually unbranched to inflorescence, 2–5 dm. tall, coarse; herbage variously pubescent or villous, often glandular; leaves opposite, numerous, lanceolate to lanceolate-elliptic to oblanceolate, acute, the lowermost leaves petioled and connate at base into a membranous sheath surrounding stem; heads 3–15 in a corymbose panicle, 12–18 mm. high, hemispheric-campanulate; phyllaries equal in size, in 2 series, thin, lanceolate, pilose at tip; receptacle flat, naked; ray flowers pale yellow, elliptic-oblong, 15–20 mm. long, the ligule elliptic-oblong; disc flowers many, yellow; achenes 4–6 mm. long, slender, pubescent to glabrate; pappus a single row of straw-colored to tawny, soft, barbellate to subplumose bristles.

Maguire, Bassett, A monograph of the genus *Arnica*. Brittonia 4:386–510. 1943.

Within *Arnica Chamissonis* local variation is rampant; according to Maguire, this species can be divided into three rather natural subspecies, one of which is definitely within our sphere of interest (*A. Chamissonis* subsp. *incana*), and a second of which is marginally so (subspecies *foliosa*).

Herbage not silvery-tomentose; leaves irregularly denticulate
　　　　　　　　　　　　　　　　　　　　1a. *A. Chamissonis* subsp. *foliosa*
Herbage silvery-tomentose; leaves entire . . . . 1b. *A. Chamissonis* subsp. *incana*

1a. Arnica Chamissonis subsp. **foliosa** (Nutt.) Maguire, Rhodora 41:508. 1939.
　Arnica foliosa var. *bernardina* Jepson, Man. Fl. Pl. Calif. 1157. 1925.
　Arnica foliosa var. *Sonnei* (Greene) Jepson, *loc. cit.*

Herbage variously pubescent, but not silvery-tomentose or pilose; stems usually not succulent, and without elongated internodes (unless plants occur in standing water); cauline leaves lanceolate to oblanceolate, more or less spreading, usually irregularly denticulate, to 2.5 cm. wide and to 12 cm. long; pappus stramineous and barbellate, sometimes becoming tawny and subplumose.

Meadows and moist places: south through California to San Bernardino Mountains; western Canada and western United States.

A highly polymorphic unit, intergradation occurring between all the local populations and to a greater or lesser extent with the other subspecies.

1b. Arnica Chamissonis subsp. **incana** (Gray) Maguire, Brittonia 4:466. 1943.
　Arnica foliosa Nutt. var. *incana* Gray, Am. Nat. 8:213. 1874.

Herbage densely silvery-tomentose or pilose; stems succulent, the internodes long; cauline leaves narrow, strict, usually entire, to 2 cm. wide and to 9 cm. long; pappus stramineous, barbellate.

Common in shallow standing water in wet meadows and swales: from San

Bernardino Mountains north through the Sierra Nevada; to Oregon, Washington, and British Columbia.

Typically, *Arnica Chamissonis* subsp. *incana* occurs only in wet habitats, usually in standing water. It is easily recognized by the narrow, mostly entire, erect leaves, the dense, silvery tomentum, the long internodes, and the naked, long stem bases which are left after the lower leaves have been shed. Although very distinctive when within the ecological limits of its specialized habitat, it is not geographically isolated, and in drier habitats it appears to merge wth the other subspecies. Suspicion is raised (and indicated by Maguire, *op. cit.*, p. 467) that this unit has more of an environmental than a genetical basis. However, two collections made during the marsh plant survey would tend to refute this suspicion. These specimens were from Grass Lake, Siskiyou County (*McMillan & Nobs 797* and *798*), and were growing in twelve inches of water. Number *797* is *A. Chamissonis* subsp. *foliosa;* number *798* is subspecies *incana*. They are strikingly different, and according to the collectors' notes they are "intermixed with only very slightly intermediate forms." The lower stems are naked in both collections, however, a fact indicating that this condition is environmentally induced in either subspecies when it grows in standing water.

ARTEMISIA. SAGEBRUSH

Annual, biennial, or perennial herbs or evergreen shrubs, usually aromatic and bitter. Leaves alternate, entire, or more usually 1–3 times pinnately dissected. Heads small, inconspicuous, greenish, numerous, discoid, nodding or erect, spicate or racemose, or more frequently panicled. Phyllaries imbricated, dry, scarious. Receptacle flat to hemispheric, naked or rarely chaffy. Ray flowers absent. Outer flowers pistillate and with an irregularly 2- or 3-toothed corolla, or all flowers perfect and with 5-toothed corolla. Achenes short, thick, 2- to 5-ribbed or 2- to 5-angled, glabrous or sometimes pubescent, without pappus.

Biennial from woody taproot; herbage without odor, glabrous; leaves bipinnately divided into numerous divisions .... 1. *A. biennis*
Perennial from horizontal rootstocks; herbage aromatic, gray-tomentose; leaves entire to coarsely few-lobed, the lobes entire .... 2. *A. Douglasiana*

1. Artemisia biennis Willd., Phytogr. 11. 1794.

Introduced biennial weed from woody taproot; stems erect, simple, 3–30 dm. tall, nonwoody; herbage glabrous and without aromatic odor; leaves 4–15 cm. long, bipinnately divided into very numerous, serrulate, acute divisions, or the upper leaves merely incised or pinnatifid; inflorescence of numerous heads crowded on short, ascending, lateral branchlets to form a long, dense, more or less leafy, spikelike panicle; heads erect, sessile, with marginal flowers pistillate, these and central perfect ones all fertile; style divided; achenes glabrous.

European weed of roadsides and waste places, often occurring in river bottoms and borders of marshes: at many isolated localities throughout California; widely distributed in North America.

2. Artemisia Douglasiana Bess. in Hook., Fl. Bor. Am. 1:323. 1834. Wormwood.

Artemisia heterophylla Nutt., Trans. Am. Philos. Soc. II. 7:400. 1841. In part. Not Bess.

Artemisia vulgaris var. *heterophylla* Jepson, Man. Fl. Pl. Calif. 1142. 1925.

Perennial from horizontal rootstocks; stems tall (5–20 dm.), stout, herbaceous to base, erect, usually simple, occasionally branched; herbage gray-tomentose; leaves variable, entire or coarsely few-lobed or -toothed, 7–15 cm. long, 1–3 cm. wide (exclusive of entire lobes), white-tomentose beneath, glabrate green above; inflorescence a leafy, elongate panicle 3–15 cm. broad, with heads mostly erect in dense spikes; involucre oblong, 3–4 mm. high, tomentose, the phyllaries 8–14; marginal flowers pistillate, the central flowers perfect, all fertile.

Common along stream banks and flats, low to middle altitudes: throughout California; from Baja California in Mexico north to Washington, east into Nevada.

ASTER

Herbs or sometimes shrubs. Leaves simple, toothed or entire. Heads usually in simple or compound inflorescences, occasionally single and scapose. Phyllaries usually unequal in length and imbricate, occasionally almost equal, herbaceous or in part coriaceous and sometimes partially scarious, usually at least green-tipped. Receptacle flat, pitted. Ray flowers pistillate, the rays white, blue, purple, violet, or pink. Disc flowers perfect, yellow, sometimes suffused with purple or brown; style branches flattened, elongate or at least acute. Achenes usually flattened; pappus of copious bristles.

Aster may be distinguished from *Erigeron* by its imbricate phyllaries, which are unequal in length, by its acute or elongate style tips, and by the characteristically broader ray corollas.

The genus *Aster* presents a complicated taxonomic problem. For our treatment we are indebted to Roxana S. Ferris of Dudley Herbarium, Stanford University, for valuable assistance. In the asters of marshy areas the problem centers chiefly around the species related to *A. foliaceus* Lindl. and *A. chilensis* Nees, and involves what have been described as *A. Douglasii, A. Oregonus, A. adscendens, A. Menziesii, A. coerulescens, A. occidentalis,* as well as many other entities the names of which are freqeuently found in synonymy. The group has been treated by Cronquist (Am. Midl. Nat. 29:429–468, 1943), who ascribes the taxonomic difficulties to hybridization between geographically close individuals. A study of the variation from both the point of view of geography and that of ecology suggests that in addition to hybridization some ecotypic selection has been involved, and the numerous compact geographic variants suggest that some aspect of apomixis might also have been operative. Apomixis would tend to reproduce and maintain the variants indefinitely and thus account for the great complexity of the group. These facts will have to be established before a sound taxonomy will be possible. The treatment here is a modification of that by Cronquist. Owing to great variability of material, the user may experience considerable frustration in reaching satisfactory identifications. Most of the material from the habitats here being treated fall into two groups, centering around *A. chilensis* and *A. occidentalis*. For most purposes of nomenclature these names will be adequate.

I. Plants Perennial

Stems with a single head, subscapose from a caespitose caudex.
  Stem and involucres more or less woolly; plants 2–30 cm. tall
                         1. *A. alpigenus* subsp. *Andersonii*
  Stem and involucres completely glabrous; plants 30–70 cm. tall . . . . 2. *A. elatus*
Stems branched, with few to many heads.
  Plant woody, at least toward base.
    Plants not glaucous, woody only toward base, often spiny in or above leaf axil; heads radiate, white; achenes glabrous . . . . 3. *A. spinosus*
    Plants glaucous, shrubby, stems not spiny; heads discoid, yellow; achenes pubescent
                                       4. *A. intricatus*
  Plant wholly herbaceous.
    Lower leaves not grasslike, but toothed and sessile or petioled.
      Inflorescence a leafy panicle, heads many (few in *A. subspicatus*).
        Involucre of phyllaries subequal in length, the phyllaries not strongly graduated from short to long; larger leaves in inflorescence ½–⅓ as long as basal leaves.
          Leaves usually linear to linear-lanceolate, entire or only the lowermost rarely serrate.
            Outer involucral bracts lanceolate-acute to lanceolate-attenuate, with hyaline margins and herbaceous tips . . . . 5. *A. hesperius*
            Outer involucral bracts oblong to obovate, obtuse, herbaceous throughout or only the base chartaceous . . . . 6. *A. Eatonii*
          Leaves chiefly oblanceolate to lanceolate, the lower and middle leaves serrate or serrulate . . . . 7. *A. subspicatus*
        Involucre with phyllaries strongly graduated, the outer phyllaries short, the inner ones progressively longer; leaves in inflorescence ¹⁄₁₀–¹⁄₂₀ as long as the lower leaves.
          Inflorescence an open, often widely divergent, panicle; involucre 7–10 mm. high; herbage chiefly dark green, rarely gray . . . . 8. *A. chilensis*
          Inflorescence a close panicle or raceme; involucre 5–6 mm. high; herbage cinereous gray; southern California . . . . 9. *A. bernardinus*

Inflorescence essentially a naked cyme or cymose panicle, heads few (to many in
   A. adscendens).
  Heads, if solitary, not on widely divergent branches.
    Larger leaves broadly lanceolate or linear, 0.5–3 cm. wide.
      Involucre with phyllaries definitely graduated, in 3 or 4 series
                                                10. *A. adscendens*
      Involucre with phyllaries scarcely graduated, in at most 2 series
                                                11. *A. occidentalis*
    Larger leaves oblanceolate, narrowed to a petiole, with conspicuously clasp-
      ing bases . . . . 12. *A. foliaceus*
  Heads solitary on widely divergent, wirelike branches . . . . 13. *A. paludicola*
Lower leaves grasslike, sessile; upper leaves often much reduced and closely ap-
   pressed to stem; plant glabrous except for glandular inflorescence
                                                14. *A. pauciflorus*

II. PLANTS ANNUAL OR BIENNIAL

Heads small.
  Phyllaries subequal in length or lightly graduated, oblanceolate to oblong, the outer
   ones not scarious-margined; rays inconspicuous but longer than the involucre
                                              15. *A. frondosus*
  Phyllaries distinctly graduated, all scarious-margined; rays small but conspicuous
                                             16. *A. exilis*
Heads large; plant biennial . . . . 17. *A. tephrodes*

1. Aster alpigenus subsp. Andersonii (Gray) Onno, Bibl. Bot. 26:15. 1932.

Erigeron Andersonii Gray, Proc. Am. Acad. Arts & Sci. 6:540. 1865.

Caespitose (or essentially so) perennial from a thick caudex; leaves chiefly
basal, erect, either narrowly linear or in some plants linear-lanceolate, 5–15
cm. long, glabrous, those of the scapes reduced and scale-like; heads solitary
on scapelike stems 1–5 dm. high, conspicuous, 2–4 cm. broad; phyllaries nearly
equal in length, oblong, acute, scarious-margined, occasionally somewhat
woolly at base, the pubescence extending down peduncle; rays purple to blue,
8–12 mm. long; style branches filiform.

Narrow-leaved and broad-leaved forms of *Aster alpigenus* subsp. *Andersonii* occur frequently. No evidence of intermediate leaf types has been found.

Common in wet meadows at high altitudes: throughout the North Coast
Ranges and the Sierra Nevada, south to southern California.

2. Aster elatus (Greene) Cronquist, Leafl. West. Bot. 5:80. 1948.

Oreastrum elatus Greene, Pittonia 3:147. 1896.

Caespitose, fibrous-rooted, glabrous perennial; basal and lower leaves linear-
elliptic, 8–25 cm. long, 0.5–1 cm. wide at broadest part, the upper leaves
much reduced; flowering stems 30–70 dm. high, single-flowered; involucre
hemispheric, 11–14 mm. high, the phyllaries imbricate, glabrous, firm, broadly
linear, the apex triangular-subulate, green, pale-chartaceous below; ray flowers about 25, violet or lavender, 7–12 mm. long; achenes several-nerved, glabrous or sparsely hairy at apex; pappus whitish, of about 40 bristles.

Wet meadows: Plumas and Lassen counties.

3. Aster spinosus Benth., Pl. Hartw. 20. 1839. Mexican devil-weed.

Leucosyris spinosa Greene, Pittonia 2:244. 1897.

Herbaceous perennial from a woody base; stems 0.6–3 m. tall, reedy, much-branched, light green, glabrous, striate, often bearing subterete or flattened axillary or supra-axillary spines 1.5 cm. long or shorter, essentially leafless above, the branches erect, broomlike; lower leaves linear or linear-spatulate, usually 1–3 cm. long, 1–3 mm. wide, acute, sessile, entire, rough-margined, 1-nerved, the upper leaves mostly reduced to minute scales; heads white, radiate, 0.8–1.5 cm. wide, solitary at tips of short or long branchlets, racemosely or paniculately disposed; involucre campanulate-hemispheric, 4- or 5-seriate, 3.5–6 mm. high, strongly graduated, the phyllaries appressed, lanceolate or lanceolate-ovate, usually acuminate, with greenish center and subscarious, sometimes ciliolate margins; rays white, turning brown, about 22, 3–4 mm. long; style appendages triangular, much shorter than the stigmatic region; achenes glabrous, 4-nerved.

Stream banks, irrigation ditches, and moist places: along Colorado River from Needles to Imperial County, west to San Diego County; south to Baja California in Mexico, east and south to Texas, Louisiana, Mexico, Central America.

4. Aster intricatus (Gray) Blake, Jour. Wash. Acad. Sci. 27:398. 1937. Shrubby alkali aster.

Aster carnosus Gray ex Hemsl., Biol. Centr.-Am. Bot. 2:120. 1881.
Bigelovia intricata Gray, Proc. Am. Acad. Arts & Sci. 17:208. 1882.

Shrubby perennial; stems 0.6–0.8 m. tall, rigidly and intricately branched, glaucescent and essentially glabrous throughout; leaves linear, 1–2 cm. long, 1–2 mm. wide, entire, mucronulate, fleshy, those of inflorescence reduced, mostly appressed, scale-like, 1–4 mm. long; heads solitary at tips of branches, discoid, 5–8 mm. wide, 8–10 mm. high; involucre turbinate-campanulate, 6–7 mm. high, about 5-seriate, the phyllaries strongly graduated, appressed, linear or lanceolate-linear, acute or acuminate, the outer phyllaries cuspidulate, glabrous or obscurely ciliolate, chartaceous, whitish, with green midline; corollas usually yellowish; achenes terete, many-ribbed, appressed-pilose; style appendages lanceolate-subulate, longer than the stigmatic region.

Alkaline meadows: Mono County and south through the Mojave Desert to Los Angeles and San Bernardino counties, San Joaquin Valley from Madera County to Kern County; east to Nevada and western and southern Arizona.

5. Aster hesperius Gray, Syn. Fl. N. Am. 1²:192. 1884. Marsh aster.

Aster coerulescens of authors, not DC.
Aster foliaceus var. *hesperidus* Jepson, Man. Fl. Pl. Calif. 1047. 1923.

Herbaceous perennial; stem 1–2 m. tall, with numerous ascending branches, pubescent in lines above, very leafy; leaves lanceolate-linear, thick, with midvein prominent on abaxial side, 6–13 cm. long, 3–15 mm. wide, attenuate, clasping, entire or the lower leaves serrate, rough-margined; heads about 2 cm. wide, usually very numerous, in narrow or spreading panicles; involucre graduated, 6–7 mm. high, sometimes subtended by a few hispid-ciliate, chiefly herbaceous, linear bracts, the phyllaries from not graduated to somewhat graduated, linear, acuminate, ciliolate, erect or loose, the hyaline margin of the inner phyllaries extending to the tip or nearly so; rays about 25, white or purple, about 8 mm. long.

Marshy meadows and margins of streams and irrigation ditches: Sierra Nevada from Mono County south through southern California; east and north to North Dakota and Alberta, Canada.

6. Aster Eatonii (Gray) Howell, Fl. NW. Am. 310. 1900.

Aster foliaceus var. *Eatonii* Gray, Syn. Fl. N. Am. 1²:194. 1884.
Aster oreganus of authors, not Nutt.

Herbaceous perennial; stem 1 m. tall or less, pubescent in lines below inflorescence, with numerous, erect or ascending branches; leaves chiefly lanceolate-linear or linear, rather thin, with midvein prominent beneath, 5–13 cm. long, 0.5–2 cm. wide, acuminate, sessile with truncate base, usually entire, rough-margined, often scabrous above; heads very numerous, in an oblong, terminal, subspicate or racemose panicle, usually crowded toward tips of branches, 1.5–3 cm. wide; involucre 5–9 mm. high, occasionally subtended by herbaceous bracts; phyllaries subequal to slightly graduated, thin, oblanceolate or spatulate to linear, acute, ciliolate, glabrous on back, the outer phyllaries with spreading, herbaceous tips; rays about 20–35, lavender or violet, about 7 mm. long.

Moist ground along streams: Inyo County to Modoc, Siskiyou, and Trinity counties; north to British Columbia and Alberta, Canada.

7. Aster subspicatus Nees, Gen. & Sp. Aster 74. 1832.

Aster Dougiasii Lindl. in DC., Prodr. 5:239. 1836.

Slender, herbaceous perennial with erect or ascending branches; stems about 1 m. or less tall, pubescent in lines; leaves practically glabrous, serrate, rough-margined, the lowest leaves oblanceolate, narrowed to a winged petiolar base, the middle leaves lanceolate, 7–13 cm. long, 1–2 cm. wide, slightly or not clasping, usually serrate above the middle; inflorescence a cymose panicle, the heads many to few, 2–3 cm. wide; involucre 5–6 mm. high, the phyllaries subequal to graduated or loosely imbricate, linear, acute or the outer ones often spatulate, with green, obtuse or acute, herbaceous tips; rays 20–30, violet, about 1 cm. long.

Moist places along coastal stream banks: Monterey County to Del Norte County; north to Alaska, east to Montana.

8. Aster chilensis Nees, Gen. & Sp. Aster 123. 1832. Common California aster.

Aster Menziesii Lindl. in DC., Prodr. 5:243. 1836.

Herbaceous perennial; stems erect, 0.5–1 m. tall, pubescent throughout to nearly glabrous; lowest leaves often obovate, narrowed to a somewhat clasping base, 12 cm. long, 3.5 cm. wide, the middle leaves usually lanceolate, 4–9 cm. long, 0.5–2 cm. wide, entire or serrulate, rough-margined, often rough above, essentially glabrous beneath; inflorescence racemose-paniculate, or a simple raceme; heads usually numerous, 2–2.5 cm. wide, borne on leafy branches; involucre 5–7 mm. high, 4- or 5-seriate, the phyllaries strongly and closely graduated, linear-oblong to oblong, with whitish, 1-nerved base and shorter, somewhat rhombic, obtuse to acute, mucronulate, pale-margined, appressed, green tips, ciliolate, glabrous on back; rays about 20–30, violet, bluish, or white, 6–12 mm. long.

Type collection erroneously reported to have come from Chile. A tremendously variable complex, the extremes of which often differ more widely from one another than other species differ from *A. chilensis*. They are connected, however, by intergrading forms that make it impossible to separate them. The varieties treated below seem to be clearly recognizable.

Plants glabrous or nearly so.
  Inflorescence conspicuously leafy-bracted; heads large .... 8a. *A. chilensis* var. *lentus*
  Inflorescence with smaller bracts; heads few and solitary.
    Plants somewhat glaucous; marshes about San Francisco Bay
                                              8b. *A. chilensis* var. *sonomensis*
    Plants not glaucous; watercourses in lower foothills around the Central Valley
                                              8c. *A. chilensis* var. *media*
Plants scabrous to short-hirsute-pubescent; inflorescence many-flowered
                                              8d. *A. chilensis* var. *invenustus*

8a. Aster chilensis var. **lentus** (Greene) Jepson, Fl. West. Middle Calif. 566. 1901.

Plants 1 m. or more tall, glabrous or nearly so; leaves linear-lanceolate; inflorescence ample, widely branching, the branches conspicuously leafy-bracted, the flowers large.

Suisun marshes in Solano County. Also, specimens approaching this form have been collected in salt marshes along the coast in Humboldt and Mendocino counties.

8b. Aster chilensis var. **sonomensis** (Greene) Jepson, Fl. West. Middle Calif. 567. 1901.

Plants 3–3.5 dm. tall, glabrous and rather glaucous; basal leaves oblong, the stem leaves reduced, lanceolate; inflorescence a cymose, leafy-bracted panicle, the heads few and solitary at the ends of the branches.

Marshes of San Francisco Bay, in Sonoma, Napa, and Santa Clara counties.

8c. Aster chilensis var. **media** Jepson, Fl. West. Middle Calif. 566. 1901.

Plants 1 m. or less tall, glabrous or nearly so, very leafy; leaves small, lanceolate or oblong-lanceolate, mostly spreading; inflorescence with widely divaricate branches, the leafy bracts somewhat smaller than the stem leaves, the heads few.

Along watercourses: lower foothills of Amador and Tuolumne counties and in Sacramento and San Joaquin counties.

8d. Aster chilensis var. **invenustus** (Greene) Jepson, Fl. West. Middle Calif. 566. 1901.

Plants stout, to 6 dm. tall; herbage scabrous to short-hirsute; leaves lanceolate, the lower stem leaves lanceolate-spatulate, 4.5–7 cm. long; inflorescence many-flowered, in an ample cymose panicle.

Northern California from Sonoma County to Siskiyou County, and from Tulare County northward in the Sierra Nevada foothills.

9. Aster bernardinus Hall, Univ. Calif. Publ. Bot. 3:79. 1907.

Herbaceous perennial from a woody root; stems erect, 1 to several, 3–10 dm. tall, densely cinereous-pubescent, rarely somewhat glabrate, densely leafy; leaves 3–5 cm. long, 3–6 mm. wide, sessile, linear to linear-lanceolate, early-deciduous; heads in simple, short-branched, leafy-bracted racemes or an elongated narrow panicle, the uppermost heads congested on the branchlets; involucre 5–6 mm. high, the phyllaries strongly graduated, pubescent to nearly glabrate, green-tipped, the margins white-chartaceous and ciliate, the outermost phyllaries obtuse or rounded, the inner ones acute; rays 30–35, purple, 6–10 mm. long; achenes canescent, the pappus sordid.

Meadows and drainage ditches: Los Angeles and San Bernardino counties south to San Diego County, east to western edge of the Mojave Desert in Los Angeles and San Bernardino counties.

Aster defoliatus Parish (Bot. Gaz. 38:461, 1904) is known only from the original collection (*Parish 5336,* from damp meadows, at an elevation of about 900 feet; San Bernardino Valley, San Bernardino County). Because of the possibility that it might be re-collected, it is included here. Slender, rigid perennial about 1 m. tall, short-pubescent; leafless below by flowering time; branchlets of the inflorescence widely divaricate, 6–15 cm. long, clothed with linear, acute leaflets about 1 cm. long and longer than the internodes; heads solitary, or sometimes 2 at the tips of the branchlets; phyllaries strongly graduate, broadly scarious-margined below; pappus copious.

10. Aster adscendens Lindl. in Hook., Fl. Bor. Am. 2:8. 1834.

Herbaceous perennial; stems slender, 0.3–0.6 m. tall, usually erect, pubescent all around below or only in lines above; lower leaves narrowly oblanceolate, tapering to petiole-like base, the middle and upper leaves lanceolate to linear, usually entire, somewhat clasping, rough-margined, glabrous to more or less pubescent; heads few to many, in a nearly naked, closed or open cyme or cymose panicle about 2.5 cm. wide; involucre 4–7 mm. high, the phyllaries usually strongly graduated, erect, linear or linear-oblong, the outermost phyllaries usually spatulate, obtuse to acute, the inner ones acuminate, ciliolate, glabrous or pubescent on back; rays 22–35, violet or purple, about 8 mm. long.

A widespread complex group breaking up into ill-defined forms. All have strongly graduated involucres and a cymose or cymose-paniculate, mostly naked inflorescence with ascending branches, but are highly variable in respect to pubescence, leaf texture and habit.

Wet or dry areas: mountains of southern California and north through the Sierra Nevada; north to British Columbia.

11. Aster occidentalis (Nutt.) Torr. & Gray, Fl. N. Am. 2:164. 1841.

Aster adscendens var. *Fremontii* Torr. & Gray, Fl. N. Am. 2:503. 1843.

Herbaceous perennial from creeping rhizomes; stems usually 0.5 m. or less tall, slender, glabrous below, usually pubescent in lines below or throughout above; lower leaves usually persisting, oblanceolate or linear-oblanceolate, tapering to a ciliate petiole, entire or serrulate, rough-margined, thickish, the middle and upper leaves linear-lanceolate, 3–10 mm. wide, entire, scarcely clasping; heads 1 to few, in a nearly naked cyme or cymose panicle about 2.5 cm. wide; involucre about 6 mm. high, the phyllaries subequal to slightly graduated, chiefly linear, acute or somewhat obtuse, appressed, ciliolate; rays about 30, violet or purple, 6–9 mm. long.

Mountain meadows and river thickets: Sierra Nevada from Tulare County north; north to British Columbia, east to Colorado and Idaho.

Leaves not clasping, linear, acute . . . . 11a. *A. occidentalis* var. *yosemitanus*
Leaves clasping or somewhat clasping, lanceolate to oblanceolate.
　　Lower leaves oblanceolate; involucres 5–8 mm. long
　　　　　　　　　　　　　　　　　　　11b. *A. occidentalis* var. *Parishii*
　　Lower leaves lanceolate; involucres 8–10 mm. long
　　　　　　　　　　　　　　　　　　　11c. *A. occidentalis* var. *delectabilis*

11a. Aster occidentalis var. **yosemitanus** (Gray) Cronquist, Am. Midl. Nat. 29:467. 1943. Yosemite aster.

Aster adscendens var. *yosemitanus* Gray, Syn. Fl. N. Am. 1²:191. 1884.

Stems slender, leafy to the inflorescence; leaves gradually reduced upward, thin, 2–3 cm. long, linear, acute; heads solitary or several, 1.5–2 cm. broad, the phyllaries linear-lanceolate, loose.

Wet meadows: Sierra Nevada north from Tulare County, north into Siskiyou County; southern Oregon.

11b. Aster occidentalis var. **Parishii** (Gray) Ferris, comb. nov.

Plants about 3 dm. or more tall; stems leafy; leaves somewhat clasping, linear-lanceolate and reduced upward, the lower leaves oblanceolate; heads in a short cymose panicle; involucres 5–8 mm. high, the phyllaries usually linear, loosely imbricate, acute.

Mountain meadows: southern California; south to San Pedro Martir in Baja California, Mexico.

11c. Aster occidentalis var. **delectabilis** (Hall) Ferris, comb. nov.

Plants 1.5–4 dm. tall; stems leafy; leaves narrowly lanceolate, 12–20 cm. long, the bases definitely clasping; heads solitary or few; involucres large, 8–10 mm. high, the phyllaries linear, acute, loosely imbricated.

Mountain meadows: southern Sierra Nevada in Fresno and Tulare counties, San Jacinto and San Bernardino mountains; south to San Pedro Martir in Baja California, Mexico.

12. Aster foliaceus Lindl. ex DC., Prodr. 5:228. 1836.

Herbaceous perennial from a creeping rootstock, 20–50 dm. tall; stems (in ours) reddish, glabrous below; lower leaves oblanceolate, 12–20 cm. long, 16–24 mm. wide, narrowed to a conspicuously clasping petiolar base, the margins entire, ciliate-appressed, the stem leaves sessile; inflorescence mostly 1-headed, or subcymose with 4–6 heads; heads to 3.5 cm. wide, 9–12 mm. high, usually pubescent below the involucre; phyllaries green, subequal in size, additional large foliaceous bracts sometimes present, glabrous on back, ciliate-margined, oblong, obtuse, or broadly acute at the apex; rays 10–17 mm. long, purple; achenes glabrous or sparsely pubescent.

Damp places: Alaska through British Columbia and Alberta, south to Olympic and Cascade mountains in Washington, east to Montana.

Aster foliaceus in the typical form does not reach California from the north. In our region it is represented by the following variants:

Inflorescence subcymose or heads solitary . . . . 12a. *A. foliaceus* var. *apricus*
Inflorescence corymbose; heads few to several . . . . 12b. *A. foliaceus* var. *Parryi*

12a. Aster foliaceus var. **apricus** Gray, Syn. Fl. N. Am. 1²:193. 1884.

Plant 1.5–2.5 dm. tall, caespitose, the rootstocks branched; stem pubescent above; heads usually solitary or sometimes subcymose.

Wet alpine meadows: northern California; north to British Columbia, east to Montana and Colorado.

12b. Aster foliaceus var. **Parryi** Gray, Syn. Fl. N. Am. 1²:193. 1884.

Plant about 5 dm. tall, with large, usually persistent lower leaves and few stem leaves; inflorescence corymbose, of few to several heads.

Damp places: central Sierra Nevada; north to Washington and Montana, and east to Colorado and New Mexico.

Fig. 347. *Aster exilis: a* and *b*, habit, × ⅖; *c*, part of inflorescence, showing the phyllaries imbricate in several series, × 4; *d*, ray flower, × 8; *e*, disc flower, × 8; *f*, mature achene with pappus, × 8.

13. Aster paludicola Piper, Contr. U. S. Nat. Herb. 16:210. 1913. Western bog aster.

Herbaceous perennial from slender rootstocks; stems very slender, 1.5–8 dm. tall or less, usually pubescent in lines, with few or many wiry branches above; leaves narrowly linear, 3–15 cm. long, 2–6 mm. wide, sessile, subclasping, entire, rough-margined, reduced above; heads few to many, about 1–1.5 cm. wide, cymose, solitary at tips of widely divaricate, wiry branches clothed with appressed, lanceolate, leafy bracts; involucre about 5–7 mm. high, the phyllaries definitely graduated, erect, linear, acute or acuminate or the outer phyllaries sometimes linear-oblong and somewhat obtuse, the base chartaceous and whitish, the thin green tips lanceolate or rhombic-lanceolate and usually purple-margined and -tipped; rays about 22–38, white, rose-colored, violet, or purple, about 8 mm. long; achenes glabrous or pubescent.

Bogs: Siskiyou Mountains; southern Oregon.

14. Aster pauciflorus Nutt., Gen. 2:154. 1818. Marsh alkali aster.

Perennial herb from creeping rootstocks; stems erect, simple or branched from the base, 2–9 dm. tall, glabrous except for glandular inflorescence; leaves somewhat fleshy, with midvein not prominent, linear or lanceolate-linear, entire, sessile, acuminate at apex, 6–12 cm. long, 3–6 mm. wide, reduced and bractlike on the branches of the corymbiform inflorescence; heads 6–12 mm. wide, 6–8 mm. long; involucre densely glandular-puberulent, the phyllaries linear-lanceolate, herbaceous except for a narrow hyaline margin, rather loose, of 2 or 3 different lengths but scarcely graduate; ligules pale purple or whitish, 6–10 mm. long; achenes appressed-pubescent.

In alkaline soil about springs and streams: Death Valley region in Inyo County; north to Saskatchewan, east to Texas, and south to Mexico.

15. Aster frondosus (Nutt.) Torr. & Gray, Fl. N. Am. 2:165. 1841.

Low stout annual; stems 1 to several from base, 1–3 dm. tall, erect or diffuse, simple or branched; herbage nearly glabrous; leaves linear, entire, somewhat hispid-ciliate, 1–5 cm. long; inflorescence a narrow raceme of heads or of 1 to several compact clusters of heads, the heads 6–8 mm. high, hemispherical; phyllaries linear to spatulate, obtuse, the outer phyllaries foliaceous; rays pink to purple, barely longer than the disc flowers; achenes strigose-pubescent; pappus copious, becoming longer than the ray flowers.

Readily distinguished by its copious pappus, which is longer than the rays.

Marshes and water holes, at elevations of 2,500–5,000 feet: chiefly east of the Sierra Nevada from Inyo County to Modoc and Siskiyou counties; north to Idaho, east to New Mexico.

16. Aster exilis Ell., Bot. S. C. & Ga. 2:344. 1824. Fig. 347.

Annual herb; stems slender, erect, 3–15 dm. tall; herbage glabrous; leaves linear to occasionally lanceolate or oblanceolate, sessile, 5–10 cm. long, 2–4

Fig. 348. *Baccharis Douglasii: a,* staminate flower with conspicuous corolla, the ovary aborted but the style present, × 8; *b,* staminate head, × 4; *c,* pistillate flower with inconspicuous, filiform, tubular corolla, × 8; *d,* pistillate head, flowers removed to show the conical receptacle, × 4; *e,* mature achene, × 8; *f,* young staminate head, showing phyllaries, × 4; *g* and *h,* habit, staminate plant, × ⅖.

mm. wide, entire or less commonly serrate, much reduced in inflorescence; heads 4–6 mm. high on bracteate peduncles to nearly sessile, often tending to be glomerate; phyllaries linear, acute, scarious-margined; rays pink to purplish, 4 mm. long, inconspicuous; achene somewhat ribbed on angles, sparsely pilose-hispid; pappus very fine.

Common weed of saline or alkaline soil, marshes, and rice fields, abundant on floodlands: Central Valley to coastal southern California; east to Atlantic.

17. Aster tephrodes (Gray) Blake in Tidestr., Contr. U. S. Nat. Herb. 25:563. 1925.

Aster canescens var. *tephrodes* Gray, Syn. Fl. N. Am. 1²:206. 1884.

Erect biennial; stems simple below, often paniculately branched above, 2–8 dm. tall, stipitate-glandular and cinereous-puberulent or cinereous-pilosulous with mostly incurved hairs; larger leaves oblanceolate or lanceolate, 3–10 cm. long, 4–13 mm. wide, acute, the lower leaves tapering to a petiole-like base, others sessile and often slightly clasping, shallowly toothed with spinescent-mucronate teeth, pubescent chiefly along margin and often stipitate-glandular, more or less distinctly triplinerved, the upper leaves gradually reduced to small, entire bracts; heads solitary at tips of cymosely or paniculately arranged branches, 2.5–4 cm. wide; involucre hemispheric, 8–10 mm. high, the phyllaries 6- or 7-seriate, graduated, narrowly linear-lanceolate, with whitish, chartaceous base, the tip herbaceous, subulate-attenuate, mucronate, spreading or reflexed, cinereous-pilosulous; rays about 23–40, violet or purple, 1–1.2 cm. long; achenes striate, finely pubescent; pappus stiffish, scarcely graduate.

River bottomlands: eastern San Bernardino County, south and west to Imperial, Riverside, and San Diego counties; southern Nevada, Arizona, New Mexico, and Baja California, Mexico.

BACCHARIS

Perennial herbs, shrubs, or straggly subshrubs. Herbage often glutinous or resinous-glandular, essentially glabrous in ours. Stems striate-angled. Leaves alternate, simple, usually elongate, margin entire or toothed or sometimes lobed. Heads dioecious, discoid, usually aggregated into corymbs or corymbose panicles. Phyllaries imbricated in several series, ovate to lanceolate, the outer phyllaries shorter or sometimes nearly equal in length, the margin somewhat scarious and erose to fimbriate or ciliate. Receptacle usually naked, flat or conical. Flowers whitish or yellowish, often the staminate flowers and pistillate ones appearing very unlike. Staminate flowers with well-developed, tubular-funnelform corolla, aborted ovary, and scant, crinkly pappus, the head drying after anthesis and the pappus becoming inconspicuous; pistillate flowers with slender, filiform corolla usually much shorter than the long-exserted style,

Fig. 349. *Baeria Fremontii: a* and *b,* disc flowers, showing stages in anthesis, × 8; *c,* ray flower, × 8; *d,* achene, showing pappus composed of long awns and short scales, × 12; *e,* habit, × ⅖; *f,* flowering head, the phyllaries in more than 1 series (overlapping at base), × 2; *g,* flowering head (longitudinal section), showing dome-shaped receptacle and arrangement of ray and disc flowers, the outer disc flowers opening first, × 3; *h,* leaf, × 2.

the pappus very copious and conspicuous in fruit, smooth to somewhat scabrous. Achenes longitudinally ribbed with 4, 5, or 10 nerves.

In some species the staminate and pistillate flowers are so unlike in aspect that they are frequently mistaken for different species. Among the Compositae this situation is nearly unique.

Low subshrubs, woody only at base, herbaceous above; receptacle conical
1. *B. Douglasii*
Tall, willow-like shrubs with erect, whiplike branches; receptacle flat or low-convex.
Herbage not or scarcely resinous; leaves dull green, entire or subentire; inflorescences both terminal and distributed along stem on short, lateral branches
2. *B. viminea*
Herbage usually distinctly resinous and sticky to the touch; leaves lustrous green, usually prominently toothed; inflorescences mainly terminal and subterminal
3. *B. glutinosa*

1. Baccharis Douglasii DC., Prodr. 5:400. 1836. Fig. 348.

Subshrub, woody only at base, 12–15 dm. tall; herbage glutinous; stems erect, simple; leaves ovate-lanceolate to lanceolate, acute, 7–10 cm. long, 1–2 cm. wide, serrulate to entire; heads numerous in an often large, terminal, compound corymb; phyllaries linear, greenish when young or occasionally pinkish; receptacle high-conical.

Salt marshes and moist spots: about San Francisco Bay and south along coast to southern California.

2. Baccharis viminea DC., Prodr. 5:400. 1836. Mule fat.

Large, straggly shrub with many whiplike, loosely branching, woody stems 10–40 dm. tall; herbage dull, not or only slightly resinous and sticky to the touch; leaves willow-like, lanceolate to oblong, acute, 2–10 cm. long, 0.5–1 cm. wide, usually entire; heads in corymbose panicles terminating the main stems and also on short, lateral branches distributed along the stems; phyllaries ovate to lanceolate; receptacle low-convex or flat.

Washes and floodplains of streams, often very abundant, especially in the northern part of its range: Sierra Nevada foothills, Central Valley and the valleys of the Coast Ranges from central California south to coastal southern California, and occasionally eastward in desert areas; east to Arizona.

3. Baccharis glutinosa Pers., Syn. Pl. 2:425. 1807. Water wally, seep willow.

Straggling shrub with several slender, erect stems 10–30 dm. tall, woody below and more or less herbaceous above; herbage usually obviously resinous, sticky to the touch, and lustrous; leaves lanceolate, 2–12 cm. long, 0.5–2.5 cm. wide, usually obviously toothed, the teeth serrate to denticulate, acute; heads in large, compound, corymbose panicles arising in upper axils of main branches; phyllaries ovate to lanceolate, pinkish when young; receptacle low-conical or flat.

Desert washes and stream courses: Mojave and Colorado deserts and north

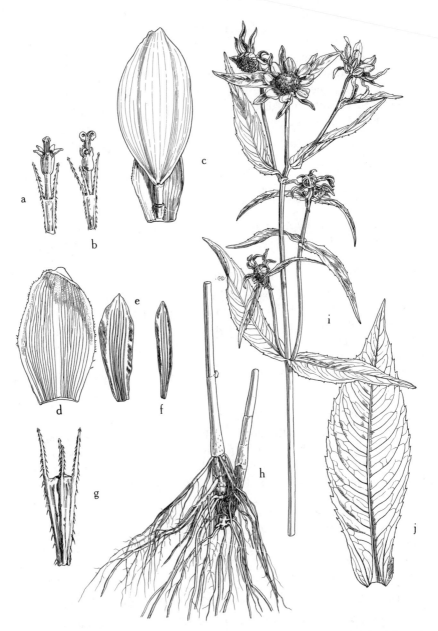

Fig. 350. *Bidens cernua: a* and *b,* disc flowers, showing stages in anthesis of stamens and style, × 3; *c,* ray flower and base of adjacent phyllary, × 3; *d,* inner phyllary, × 4; *e* and *f,* receptacle bracts, × 4; *g,* mature achene, showing the callous-thickened margins, the 4 pappus awns, and retrorse pubescence, × 4; *h,* base of plant, showing the fascicled roots, × ⅖; *i,* upper part of plant, showing leaves and inflorescence, × ⅖; *j,* leaf, × ⅖.

to Owens Valley; east to western Texas and south to Sonora in Mexico, Chile.

This species and the preceding one are closely related, and are not always easy to distinguish from each other.

BAERIA. GOLD-FIELDS

Mostly spring annuals. Leaves opposite, sessile. Heads pedunculate, many-flowered. Involucre of 3–15 distinct or slightly united phyllaries. Receptacle naked, conic to subglobose, muricate. Ray flowers yellow, sometimes minute. Disc flowers yellow, with narrow tube, campanulate throat, and 5-lobed limb. Achenes somewhat 4-angled or flattened, linear or linear-clavate. Pappus of awns or scales or both, sometimes wanting.

Phyllaries in a single series (not overlapping basally), 3-nerved, reticulate-veined; pappus
    paleae all alike (ovate at base and attenuate into an awn) .... 1. *B. platycarpha*
Phyllaries in more than 1 series (at least strongly overlapping basally), 1-nerved; pappus
    paleae of two kinds (long, slender awns alternating with short scales)
                                           2. *B. Fremontii*

1. Baeria platycarpha (Gray) Gray, Proc. Am. Acad. Arts & Sci. 9:196. 1874.

Low, slender annual; herbage sparsely villous to glabrous; stems wiry, purplish, usually several from base, the branching strict, 1–1.5 dm. tall; leaves narrowly linear or laciniately cleft, 1–4 cm. long; involucre 6–7 mm. high, of 6 or 7 phyllaries in a single series (not overlapping basally), the phyllaries 3-nerved, reticulate-veined, the middle nerve thickening at base in age; ray flowers 6–7, about 8 mm. long; disc flowers many; receptacle sharply conical, with prominent papillae to which achenes are attached; achenes linear to linear-clavate, 4-angled or somewhat flattened, glabrous to pubescent with upwardly appressed hairs; pappus paleae 4–7, sordid to white, ovate, abruptly tapered into a slender awn, the whole as long as or slightly longer than the corolla tube.

Salt marshes or alkaline places at low altitudes: Central Valley, from Merced County north along the Sacramento River.

The closely related *Baeria chrysostoma* Fischer & C. A. Mey. may be distinguished by its nonreticulate, 1-nerved phyllaries, which are not strictly in 1 series, and by its hirsute pubescence.

2. Baeria Fremontii (Torr.) Gray, Proc. Am. Acad. Arts & Sci. 9:196. 1874. Fig. 349.

Slender annual, simple to many-branched, erect, 1–4 dm. tall, glabrous below to hirsute-villous in the inflorescence; leaves linear or laciniately lobed, 2–6 cm. long; involucre of 10–12 1-nerved phyllaries overlapping in 2 or 3 series, 4–6 mm. high; receptacle subglobose or dome-shaped, with obtuse papillae to which the achenes are attached; ray flowers 10–13, the ligules

Fig. 351. *Bidens laevis: a,* habit, upper part of plant, × ⅖; *b* and *c,* disc flowers, showing stages in anthesis of style and stamens, × 2½; *d,* ray flower and base of adjacent phyllary, × 2½; *e,* mature achene, showing pappus awns and retrorse pubescence, × 4; *f,* flowering head, showing the outer and the inner phyllaries (rays removed), × ⅘.

elliptic, entire to toothed at apex; disc flowers many; achenes linear, angled, with pubescence of short, upwardly appressed hairs; pappus sordid, of 4 or 5 slender awns as long as or slightly longer than the achene and alternating with short, toothed scales.

Common in vernal pools: Central Valley from Kern County north to Butte County, North Coast Range valleys.

BIDENS. BUR MARIGOLD, STICKTIGHT, BEGGAR-TICKS

Annual or perennial herbs. Roots fascicled at lower stem nodes. Stems erect, often decumbent at base, or sometimes floating. Leaves opposite or rarely the upper leaves verticillate or alternate, simple or pinnately incised to pinnate. Heads few to many, solitary to paniculate. Phyllaries usually differentiated into an inner and an outer series, those of the outer series often foliaceous. Receptacle flat, with a deciduous bract subtending each flower in the head. Ray flowers usually present, usually sterile, ours yellow. Achenes flat (dorsiventrally compressed) or sometimes quadrate in cross section; pappus of 2–5 stiff, bristle-like, retrorsely barbed awns.

In the following key, *Bidens laevis* and *B. cernua* present a problem in clear differentiation, yet I agree with previous workers in regarding them as distinct. In general, *B. laevis* is large-flowered, whereas the flowers of *B. cernua* are small or of medium size, but all characters used to differentiate the two tend to break down if enough material is investigated.

Leaves simple, sessile or short-petioled, basally somewhat connate; margins of achenes retrorsely hispid.
  Ray flowers when present 1–2 cm. long (1–3 times as long as the disc flowers); mature achenes usually with both margins and median keels prominent and strongly callous-thickened, thus 4-angled and more or less quadrate in cross section
  .... 1. *B. cernua*
  Ray flowers 2–4 cm. long (3–4 times as long as the disc flowers); mature achenes strongly flattened, margins acute and seldom callous-thickened, median keels scarcely or not developed .... 2. *B. laevis*
Leaves pinnate or pinnately lobed, petioled, not basally connate; margins of achenes antrorsely hispid.
  Outer phyllaries usually much longer than the inner ones; mature achenes flat, the body about as long as the inner phyllaries; awns 2 .... 3. *B. frondosa*
  Outer phyllaries as long as or shorter than the inner ones; mature achenes linear-fusiform, conspicuously longer than the involucre; awns 2–5 .... 4. *B. pilosa*

1. Bidens cernua L., Sp. Pl. 832. 1753. Fig. 350.

Annual; stems erect or ascending, 1–10 dm. tall, occasionally somewhat succulent, glabrous to scabrous-hispid; leaves normally opposite, sometimes verticillate, normally linear-lanceolate or lanceolate, unequally serrate, narrowed to sessile or subsessile, with connate base, 4–17 cm. long; heads at first

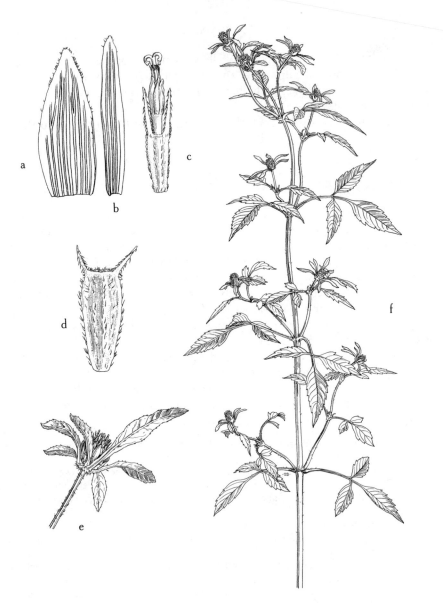

Fig. 352. *Bidens frondosa:* *a,* inner phyllary, × 6; *b,* receptacle bract, × 6; *c,* disc flower, × 6; *d,* mature achene, showing divergent pappus awns with retrorse barbs, and the margins of achene with ascending hairs, × 4; *e,* flowering head, showing foliaceous outer phyllaries, × ⅘; *f,* habit, upper part of plant, × ⅖.

erect, becoming nodding, radiate or discoid, in anthesis 1–5 cm. wide, 0.6–1.2 cm. high; outer phyllaries 5–8, foliaceous, linear-lanceolate, sparsely acicular-ciliate or glabrous, unequal in length, usually longer than the disc flowers (to 4 cm. long), the inner phyllaries membranous, striate, orange-brown, ovate-lanceolate or obovate-lanceolate, scarcely longer than the disc flowers; receptacle bracts similar but narrower; ray flowers when present 6–8, ovate-lanceolate, 1–1.5 (–2) cm. long, about 1–2 times as long as the disc; achenes often purplish, normally cuneate and apically thickened, marginal and median angles mostly prominent and usually strongly callous-thickened, the body thus more or less 4-angled and quadrate in cross section, at least subapically, 5–6.5 mm. long, retrorsely hispid, especially on angles; awns 4 (sometimes only 2 or 3), usually 1 at each angle of achene apex, 3–4 mm. long, retrorsely barbed.

Marshy ground from sea level to an elevation of 6,000 feet: San Francisco and Marin counties, north through Lake, Trinity, and Siskiyou counties, and south in the Sierra Nevada to Tuolumne and Mono counties; north to Canada, east to Atlantic, South America, Europe, Asia.

2. Bidens laevis (L.) B.S.P., Prelim. Cat. N. Y. 29. 1888. Fig. 351.

Annual or short-lived perennial; stems erect or decumbent at base, 2–20 dm. tall, sometimes tending to be succulent, glabrous; leaves lanceolate, usually regularly serrate-dentate, tapering to the sessile or subsessile and connate base, 8–20 cm. long; heads erect or nodding, 4–7 cm. wide, 0.8–1.5 cm. high, radiate; outer phyllaries 6–8, linear-lanceolate, sometimes longer than the inner phyllaries, rarely longer than the disc flowers, often serrulate, the inner phyllaries membranous, striate, about as long as the disc flowers, ovate to obovate, orange-yellow; bracts of receptacle similar but oblanceolate; ray flowers 5–10, conspicuous, 3–4 times as long as the disc flowers, 2–4 cm. long; achenes sometimes purplish, narrowly cuneate, strongly flattened or only slightly thickened at apex, the median keels weakly or not at all developed, the margins acute and slightly or not at all callous-thickened, the body glabrous or scantily retrorsely hispid, especially on margins, 5–8 mm. long; awns usually 2 (these arising marginally) or sometimes 3 or 4 (the additional 2 smaller and arising medially), retrorsely barbed, 2–3 (–4) mm. long.

Marshes and irrigation ditches in warm valleys and foothills: southern California to Central Valley; east to Florida, thence north to Indiana and New Hampshire.

3. Bidens frondosa L., Sp. Pl. 832. 1753. Fig. 352.

Annual; stems erect, 3–13 dm. tall, glabrous or sparsely pilose; leaves pinnately compound or only cleft, petioled, with 3–5 leaflets, the terminal leaflet with a distinct stalk (petiolule) or sometimes nearly sessile, the segments irregularly lanceolate, coarsely and irregularly toothed or lobed; outer phyllaries rotately spreading, usually 1½ to 2 or 3 times as long as the disc flowers, from linear-spatulate and entire to distinctly foliaceous and then toothed or some-times lobed, narrowed to the base, the margin more or less ciliate, the inner phyllaries erect, ovate-lanceolate, distinctly striate, approximately as long as the disc flowers, the margin hyaline; bracts of receptacle deciduous, narrowly oblong; ray flowers 1–5, much reduced or sometimes absent, the achenes incompletely developed; disc flowers 2.5–3 mm. high; mature achenes flat, obovate, the body about 5–7 mm. long, about as long as the inner phyllaries, the margins and body with few to many ascending (antrorse) hairs; awns 2, divergent, with usually retrorse barbs.

In moist and wet habitats at low and middle altitudes: almost throughout California; across United States and southern Canada.

4. Bidens pilosa L., Sp. Pl. 832. 1753.

Annual; stems erect, 3–18 dm. tall, quadrate, glabrous or sparsely pilose, often purplish or straw-colored; leaves petioled, ovate, pinnately compound, the leaflets thin, serrate, the terminal leaflet tapering at base to a short stalk; heads numerous, discoid, 5–7 mm. high; outer phyllaries as long as or shorter than the disc flowers and the inner phyllaries, linear to linear-spatulate, the margin somewhat ciliate, the inner phyllaries lanceolate, striate, hyaline-margined; ray flowers usually absent or imperfectly developed, if present, yellow or white; mature achenes linear-fusiform, the body to 16 mm. long, conspicuously elongate and much longer than the involucre, the margins and body usually with short, stiff, callous-based, ascending (antrorse) hairs; awns 2–5, divergent, retrorsely barbed, yellow.

Along irrigation ditches and in moist habitats: southern California north along coast to San Mateo County; widely distributed in the warmer parts of the earth.

BRICKELLIA

1. Brickellia californica (Torr. & Gray) Gray, Mem. Am. Acad. Arts & Sci. II. 4:64. 1849.

Low, rounded shrub with many weak, brittle stems, 6–10 dm. tall; leaves alternate, round to triangular-ovate, irregularly serrate, 1–6 cm. long, with a rough, sordid puberulence, petioled; heads in glomerules, racemosely disposed or more densely compacted into spikelike inflorescences, 8–12 mm. high; phyllaries much imbricated, striately nerved, the tips thin, straight; receptacle naked; flowers all perfect, discoid; corolla slender, 5-toothed, whitish; achenes 2–3 mm. long; pappus of numerous, sordid, slightly scabrid, capillary bristles.

Stream courses in the foothills: Siskiyou County to coastal southern California.

CENTROMADIA

1. Centromadia pungens Greene, Man. Bot. Bay Reg. 196. 1894.

Hemizonia pungens Torr. & Gray, Fl. N. Am. 2:399. 1843.

Annual; stems rigid, freely branching, 2–6 dm. or more tall; herbage rough-hispid with both long and short spreading hairs, not or only minutely glandular, heavy-scented; leaves alternate, spinescent, the lower leaves pinnatifid, 3–4 cm. long, the upper leaves lanceolate-subulate, 1–2 cm. long; heads terminal on short, lateral stems which are axillary along the main branches, these flowering stems with condensed internodes and crowded leaves, the latter appearing whorled or involucrate beneath the shorter heads; phyllaries in a single series and about as long as the disc flowers, spinescent, each enclosing a ray achene at base; receptacle flat or convex, chaffy, the bracts spinulose, all free, persistent; ray flowers bifid, pale yellow, 25–40, the disc flowers yellow; ray achenes fertile, triangular, glabrous, smooth or minutely roughened, apiculate, without pappus, the disc achenes sterile, the pappus none.

Frequent in proximity of alkaline bogs and of marshy areas: upper San Joaquin Valley, also widely distributed in drier areas of California.

CIRSIUM. THISTLE

Coarse, weedy, biennial or perennial herbs, usually with numerous stout prickles. Leaves alternate, sinuately lobed to pinnatifid, the lobes, teeth, and margins usually beset with prickles. Receptacle thickly beset with hairs or bristles. Flowers usually perfect; ray flowers lacking. Corollas tubular, the throat slightly dilated, the lobes linear. Achenes smooth and glabrous, not ribbed; pappus united in a ring at base, plumose or barbellate above, at length deciduous.

In addition to the species here described, the common weed *Cirsium vulgare* (*C. lanceolatum*) often occurs on floodlands and ditch banks. It is readily recognized by the decurrent leaf bases, which form a prickly wing down the stems. *Cirsium Vaseyi, C. quercetorum, C. campylon, C. Andersonii, C. crassicaulis,* and *C. mohavensis* may also be occasionally encountered in wet habitats.

Herbage at length green, the arachnoid tomentum thin and deciduous; Suisun marshes
1. *C. hydrophyllum*
Herbage white-lanate .... 2. *C. Breweri*

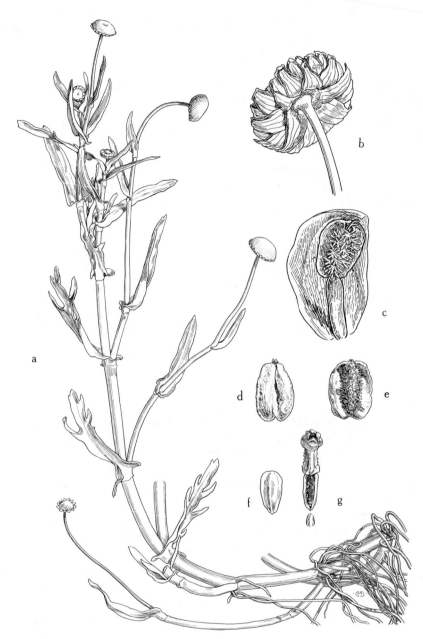

Fig. 353. *Cotula coronopifolia: a,* habit, showing decumbent stems rooting at the nodes, × ⅔; *b,* flowering head, showing phyllaries, × 3; *c,* outer pistillate flower and subtending phyllary, showing the long stipe supporting the ovary, × 16; *d* and *e,* mature achenes of outer pistillate flower, showing the inflated, membranous coat, the smooth abaxial surface, and the papillate adaxial surface, × 8; *f,* mature achene of disc flower, showing smooth abaxial surface, × 8; *g,* disc flower, showing achene with narrow, membranous margin and papillate adaxial surface, × 8.

1. Cirsium hydrophyllum (Greene) Petrak in Jepson, Fl. West. Middle Calif. 507. 1901.

Cirsium Vaseyi Jepson var. *hydrophyllum* (Greene) Jepson, Man. Fl. Pl. Calif. 1165. 1925.

Erect biennial, 4–10 dm. tall; stems sparsely arachnoid to glabrate; basal leaves lanceolate, 1–3 dm. long, remotely sinuately lobed, the lobes toothed, short-spine-tipped, the margins spiny-serrulate, glabrate above, closely arachnoid to glabrate below, the cauline leaves much reduced and more conspicuously spiny; heads solitary or a few in a cluster, 2–5 cm. high, distinctly peduncled; phyllaries broadly lanceolate-attenuate, sparsely arachnoid to glabrate, with a prominent, often glandular midrib; flowers purple to rose, the corolla lobes shorter than the throat; pappus plumose in basal half.

Suisun marshes in Solano County. Often confused with *Cirsium Vaseyi* from serpentine soil on Mount Tamalpais.

2. Cirsium Breweri (Gray) Jepson, Fl. West. Middle Calif. 507. 1901.

Cnicus Breweri Gray, Proc. Am. Acad. Arts & Sci. 10:43. 1874.

Erect, simple or branched biennial, 5–25 dm. tall; herbage white-tomentose, rarely glabrate, beset with spines; leaves narrowly oblong, 6–50 cm. long, sinuately pinnatifid, spinose-toothed, the spines short, slender; heads several to numerous, peduncled, thinly arachnoid to glabrate; phyllaries broadly lanceolate, slenderly spine-tipped; flowers pink to white or pale purple; pappus awns plumose almost to tip.

Swamps and stream banks: North Coast Ranges, northeastern California; north to Oregon.

CONYZA

1. Conyza Coulteri Gray, Proc. Am. Acad. Arts & Sci. 7:355. 1868.

Annual herb; stems erect, simple below and paniculately branched above, 0.3–1 m. tall; herbage rough-hairy and somewhat glandular; leaves alternate, obovate to oblanceolate, toothed or coarsely pinnatifid, sessile, somewhat clasping; inflorescence a large panicle; heads small, numerous, rayless, of many pistillate flowers and a few central perfect ones; phyllaries imbricated in 2 or more series, linear, acuminate, the outer phyllaries glandular-hairy with narrow scarious margins, the inner ones more completely scarious; receptacle naked; marginal flowers pistillate, fertile, the corolla tubular-filiform, sordid white, apically truncate, the style much exserted; central flowers perfect, about 5, mostly fertile, the corolla tubular-funnelform, 5-lobed, the style but little exserted; achenes small (less than 1 mm. long), hispidulous, compressed; pappus a single series of copious, dull white capillary bristles, about 3 mm. long and longer than the involucre.

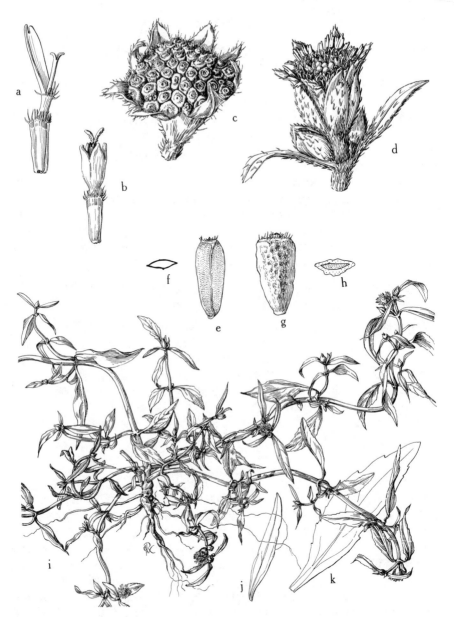

Fig. 354. *Eclipta alba: a,* ray flower, $\times$ 10; *b,* disc flower, $\times$ 10; *c,* flowering head, showing mature achenes, $\times$ 4; *d,* flowering head, the central flowers in bud, the marginal ones with styles exserted, $\times$ 4; *e* and *f,* achene of the thin-walled, flattened type, side view and cross section, $\times$ 10; *g* and *h,* achene of the thick-walled, tuberculate type, side view and cross section, $\times$ 10; *i,* habit, $\times$ ⅔; *j* and *k,* leaves, $\times$ 1½.

Moist areas, often in alkaline soil, at low altitudes: occasional in San Joaquin Valley and southern California; southeast to Colorado and Mexico.

Superficially, *Conyza Coulteri* bears a very close resemblance to some species of *Erigeron,* but it differs in being rayless, all the flowers having tubular corollas.

COTULA

1. Cotula coronopifolia L., Sp. Pl. 892. 1753. Brass buttons. Fig. 353.

Low, diffuse, strong-scented, glabrous, perennial herb; stems many, decumbent, to 3 dm. long; leaves alternate, 2–6 cm. long, entire to toothed or pinnatifid, linear to oblong, sheathing stem at base; heads on slender, naked peduncles, discoid, bright clear yellow, depressed, to 1 cm. broad; phyllaries thin, scarious, greenish, in 1 or 2 ranks; receptacle essentially flat, naked; ray flowers absent, their places taken by 1 outer row of pistillate flowers which lack corollas and are on stipes as long as the involucre; central (disc) flowers with 4-toothed corollas, on shorter stipes; achenes pedicellate, compressed, papillate on inner face, smooth on outer face, those of the outer pistillate flowers with a loose membranous coat which is expanded on the margins and extends over the back, those of the central flowers with a narrow membranous margin; pappus wanting.

In saline and marshy situations at low altitudes, often almost aquatic: frequent along the coast, particularly in salt marshes; weed introduced from South Africa.

ECLIPTA

1. Eclipta alba (L.) Hassk., Pl. Jav. Rar. 528. 1848. Fig. 354.

Low, weak, annual herb; herbage rough-hairy, the hairs appressed; stems decumbent or ascending, 2–6 dm. long, rooting from lowest nodes; leaves opposite, lanceolate, serrate, sessile or subsessile; heads pedunculate, arising singly in upper leaf axils, many-flowered, 4–8 mm. wide; phyllaries foliaceous, ovate-lanceolate, indistinctly 2-rowed; receptacle flat, with a bristle-like, barbed bract subtending each flower; corollas white, those of the disc flowers tubular, 4-toothed at apex, those of the ray flowers bilobed at apex; achenes dimorphic, but alike in each head, most heads composed entirely of straw-colored achenes, quadrangular or triangular in cross section, with the sides from irregularly weakly to strongly tuberculate, 2–2.5 (usually 2.25) mm. long, 1–1.5 mm. wide, other heads composed entirely of dark gray to black achenes, flattened and biconvex in cross section, with sides smooth (microscopically transversely corrugated), 2–2.5 (usually 2) mm. long, 0.8–1 mm. wide, sometimes with a narrow wing along each margin and on the middle of each side, both types of achene with few to many hyaline hairs at apex; pappus of inconspicuous teeth, or lacking.

Fig. 355. *Erigeron philadelphicus: a,* habit, basal part of plant, × ⅓; *b,* flower head, showing the little-imbricate phyllaries and the numerous, linear-filiform rays, × 2; *c,* achene and pappus, the bristles of one series short and those of the other series long, × 12; *d,* habit, upper part of plant, × ⅖; *e,* ray flower, the corolla linear-filiform, the stigma lobes linear and obtuse, × 8; *f,* disc flower, the corolla tubular, the stigma lobes broad and triangular-tipped, × 8.

Occasional along streams and ditches and in moist waste places: southern California north to Central Valley, where it occurs along drainage courses of the San Joaquin and Sacramento rivers; widely distributed in warmer parts of the world.

ERIGERON

Annual, biennial, or perennial herbs with simple or occasionally dissected leaves. Heads in panicles or corymbs or sometimes on scapes, either radiate or discoid. Involucres campanulate to hemispherical or cylindrical; phyllaries little imbricated, nearly equal in length. Receptacle convex. Rays usually numerous, often very numerous, linear to narrowly lanceolate or oblanceolate, in some species absent or inconspicuous, all pistillate, white to pink or purple, rarely yellow. Disc flowers perfect, yellow. Stigma lobes of the disc flowers broadly triangular or obtuse, those of the ray flowers linear. Achenes flattened, usually pubescent and nerved; pappus often of an inner series of long, and an outer series of very short, capillary bristles, usually fragile.

Often confused with *Aster,* but distinguishable from it by the short, angular or blunt stigma lobes of the disc flowers and by the nonimbricated phyllaries, all of equal length.

Heads 1–4 cm. broad; plants perennial or rarely annual or biennial.
    Heads several on short, corymbosely arranged peduncles; rays filiform; plants chiefly
        of low altitudes . . . . 1. *E. philadelphicus*
    Heads solitary on slender, erect peduncles or branches; rays linear to oblanceolate,
        never filiform.
        Rays white (rarely purplish), narrowly linear . . . . 2. *E. Coulteri*
        Rays purple to violet, lanceolate to oblanceolate . . . . 3. *E. peregrinus*
Heads 2–3 mm. broad; plants annual . . . . 4. *E. canadensis*

1. Erigeron philadelphicus L., Sp. Pl. 863. 1753. Fig. 355.

Erect biennial or short-lived perennial herb; stems 4–10 dm. tall, often simple below and branched above, leafy throughout; herbage hispid; basal leaves oblanceolate, spatulate or obovate, 1–3 dm. long, entire to coarsely and irregularly serrate, the cauline leaves much reduced, sessile and auriculate-clasping; heads in terminal corymbs on peduncles of various lengths; rays numerous, linear-filiform, white to pink, 8–10 mm. long.

Widely distributed but occasional, in bogs and marshes at low and middle altitudes: coastal southern California through North Coast Ranges, Sierra Nevada; north to British Columbia, east to Atlantic.

Often said to be annual, but ours is usually a short-lived perennial.

2. Erigeron Coulteri Porter in Porter & Coult., Fl. Colo. 61. 1874.

Perennial herb from slender rhizomes; stems 1 to several, simple, erect, 3–6 dm. tall, single-headed or occasionally the stems branched from below and giving rise to several single-headed stems; herbage variously hirsute, pubescent to glabrous; lower leaves oblanceolate to obovate, serrulate to entire, narrowed to a winged petiole, 5–15 cm. long, the upper leaves shorter, oblong-lanceolate, sessile, clasping; heads usually solitary, 2–5 cm. broad; rays 50–80, narrow-linear but not filiform, usually white, rarely purplish, 1–2 cm. long.

Wet meadows, bogs, or swamps at middle and higher altitudes: Sierra Nevada; east to Rocky Mountains.

3. Erigeron peregrinus (Pursh) Greene, Pittonia 3:166. 1897.

Aster peregrinus Pursh, Fl. Am. Sept. 2:556. 1814.
Erigeron salsuginosus (Richards.) Gray, Proc. Am. Acad. Arts & Sci. 16:93. 1880.

Rhizomatous perennial herb; stems solitary or few, simple, erect or ascending, 1–6 dm. tall; herbage glabrous or sometimes pubescent above; lower leaves oblanceolate to oblong-spatulate, erect, 5–15 cm. long, minutely ciliate, the cauline leaves reduced, lanceolate to oblanceolate or ovate-oblong, sessile, becoming bractlike above; heads solitary or few, 3–4 cm. broad; phyllaries linear-attenuate, somewhat spreading, viscidulous; rays 40–70, 10–20 mm. long, purple to violet, narrowly elliptic.

Wet meadows, bogs, and swamps, at middle and higher altitudes: Sierra Nevada; north to Washington, east to New Mexico.

4. Erigeron canadensis L., Sp. Pl. 863. 1753.

Erect annual; stems simple below, paniculately branched above, 2–20 dm. tall; herbage hispid with scattered hairs or glabrous, usually dark or bright green; leaves linear to lanceolate, sessile or the lower leaves narrowed to a petiole, entire or toothed, 5–10 cm. long; heads 3–4 mm. high in a dense panicle; phyllaries scarious-margined, nearly glabrous; rays inconspicuous, white, toothed.

Common weed in waste places, often abundant on floodlands and along streams. Throughout California at low altitudes; to eastern United States. It is not common in wet habitats, but as a wasteland weed it is to be expected on floodlands.

Erigeron linifolius Willd., a plant of similar stature and form, is often confused with *E. canadensis*. It differs in having gray-hairy herbage and in having only a few and somewhat larger heads.

EVAX

1. Evax caulescens (Benth.) Gray in Torr., Pacif. R. R. Rep. 4:101. 1857.

Rigid, very white-woolly annual; stems erect or caespitose, simple or branched from base, 2–15 cm. tall; leaves alternate, simple, entire, the blade spatulate to obovate or rhomboid, 10–25 mm. long, narrowed to a petiole as long as or longer than the blade, the petiole again broadening at base, the whorl of many leaves subtending the head, the expanded petiole bases at length becoming strongly indurated around the mature head; heads discoid, terminal, hemispherical, containing both pistillate and staminate flowers; phyllaries closely imbricated, becoming indurated and persistent, subtending the filiform pistillate flowers and in fruit enfolding their achenes; receptacle columnar, elongate and bearing a cuplike, at length deciduous whorl of 5 bracts surrounding 2–4 staminate flowers; pappus none.

In dried beds of vernal pools and on vernally wet ground, at low altitudes: Central Valley and Coast Range valleys of central California.

A dwarf form has been segregated as *Evax caulescens* var. *humilis* Jepson. It was found near Antioch in Contra Costa County.

Evax sparsiflora Jepson, an erect form with heads scattered in the leaf axils, may occasionally occur in vernally wet habitats.

GNAPHALIUM. CUDWEED

White-woolly herbs. Leaves alternate, entire, sessile or decurrent. Inflorescence of dense clusters of small heads aggregated into panicles, corymbs, or spikes. Heads discoid, containing both pistillate and perfect flowers with tubular corollas, all flowers fertile. Phyllaries scarious or hyaline, white, clear, or colored, numerous, imbricated. Receptacle flat, naked. Pistillate flowers in several series, the corollas very slender. Perfect flowers central, the corollas tubular and 5-lobed. Achenes terete or somewhat flattened; pappus a single series of minutely scabrous, capillary bristles, these all distinct and falling separately, or lightly cohering and falling in groups, or united at thickened base and falling as a ring.

Heads mostly terminal in a spikelike, often elongate and interrupted, bracteate cylinder; pappus bristles united into a thickened ring at base and falling as a unit
1. *G. purpureum*
Heads axillary or terminal, aggregated in a dense, globose cluster or forming corymbose panicles; pappus bristles either free or only slightly coalesced at base, falling separately or in groups.
  Heads in small, leafy-bracted, axillary and terminal clusters; phyllaries narrowly linear with white, scarious tips; pappus bristles not hairy at base .... 2. *G. palustre*
  Heads in larger, mostly terminal clusters, often forming corymbose panicles, not leafy-bracted; phyllaries ovate to lanceolate, the upper part hyaline; pappus bristles hairy at base.
    Heads 4–6 mm. high, yellowish or straw-colored; corollas yellowish; achenes smooth, glabrous; pappus bristles usually not cohering at base, falling separately
3. *G. chilense*
    Heads 3–3.5 mm. high, greenish to light brownish; corollas purplish red; achenes scabrous-pubescent or papillate; pappus bristles tending to cohere somewhat at base and to fall in groups .... 4. *G. luteo-album*

1. Gnaphalium purpureum L., Sp. Pl. 854. 1753. Purple cudweed.

Annual or biennial; stems several, 1–4 dm. tall, usually simple and erect; herbage densely white-woolly; leaves 2–6 cm. long, spatulate, obtuse, tapering below to a short, winged petiole, the cauline leaves sessile and somewhat auriculate, all leaves glabrate early and becoming green on upper surface; inflorescence mainly terminal, forming a short or elongate, spikelike, continuous or interrupted cylinder of crowded heads, these clustered in the axils of bracts, which are strongly reduced upward; outer phyllaries densely woolly only at base, obovate, acute, brownish, the inner phyllaries oblanceolate, acute, green with an expanded brown tip; corollas purplish; achenes sparsely scabrous; pappus bristles united at base into a ring and falling as a unit.

On dry, open ground or in vernally wet areas, at elevations of 50–1,500 feet: cismontane California; widely distributed in the United States.

2. Gnaphalium palustre Nutt., Trans. Am. Philos. Soc. II. 7:403. 1841. Lowland cudweed.

Annual; stems 5–20 cm. tall, many and branching from often decumbent base, erect or ascending; herbage loosely floccose-woolly, the wool long and eventually more or less deciduous; leaves spatulate, acute, 1–3 cm. long, tapering to a sessile or subsessile base, the upper leaves oblong or lanceolate and little reduced; inflorescences scattered, the densely aggregated heads in globose, terminal and axillary clusters, the latter at the tips of short or reduced branches, all clusters leafy-bracted, the encircling leaves longer than the heads; phyllaries loosely woolly, linear, the tips whitish, scarious; achenes smooth or scabrous; pappus bristles falling separately, not hairy at base.

Stream beds or low, moist areas: throughout cismontane California; north and east to British Columbia, Wyoming, and New Mexico.

3. Gnaphalium chilense Spreng., Syst. Veg. 3:480. 1826. Cotton-batting plant.

Annual or biennial; stems usually several, 2–6 dm. tall, erect from usually decumbent base, often very leafy; herbage silky-hairy with a greenish tomentum; leaves 2–5 cm. long, narrowly spatulate, auriculate to strongly decurrent at base, the upper leaves linear and reduced; inflorescence mainly terminal, of dense clusters of heads which often form large, corymbose panicles; heads 4–6 mm. high; involucre yellowish or straw-colored, woolly only at base, the phyllaries ovate to oblanceolate or lanceolate, usually hyaline except for herbaceous lower part, which is often thickly sprinkled with yellow glands, the tips obtuse or erose; achenes smooth, glabrous; pappus bristles hairy below but usually falling separately.

Open, often moist ground in valleys and low hills, frequently in waste places: throughout cismontane California; north to Washington, east to Texas.

Gnaphalium chilense var. *confertifolium* Greene, occurring with the species, is not well marked.

Plants with more than one head in a leaf axil have been segregated as *Haplopappus racemosus* var. *glomerellus* Gray. For a discussion of the very complicated synonymy of thirty-three members of the genus, see Hall (Carn. Inst. Wash. Publ. 389:136–139, 1928).

HELENIUM. SNEEZEWEED

Erect perennial herbs. Leaves alternate, simple, the lower leaves usually petioled, the upper ones sessile and often extensively decurrent, thereby forming wings along the stems. Heads solitary or occasionally the plant much-branched. Involucre with the phyllaries in a single series, the phyllaries linear or lanceolate, becoming reflexed in anthesis. Receptacle hemispherical to nearly globose, without bracts. Ray flowers often numerous and conspicuous (in one of ours inconspicuous), yellow, often drooping or reflexed. Achenes turbinate; pappus of 5–7 paleae.

Ray flowers shorter than the disc flowers, closely reflexed and often inconspicuous
1. *H. puberulum*
Ray flowers longer than the disc flowers, drooping or spreading, usually very conspicuous.
  Flowers 6–8 cm. broad; peduncles stout and felty-tomentose at apex . . 2. *H. Bolanderi*
  Flowers 2–6 cm. broad; peduncles slender, glabrous or puberulent at top
3. *H. Bigelovii*

1. Helenium puberulum DC., Prodr. 5:667. 1836.

Erect perennial herb, 4–15 dm. tall; stems branched, winged with decurrent leaf bases; herbage puberulent; leaves sessile, decurrent, alternate, lanceolate to linear; heads solitary on slender peduncles, globose; ray flowers reflexed and concealing the phyllaries, often inconspicuous; disc flowers reddish brown; achenes obovoid, somewhat flattened; pappus ovate, tipped with a short awn.

Along stream courses and about springs at low altitudes: Coast Ranges and northern Sierra Nevada foothills; south to Baja California in Mexico.

2. Helenium Bolanderi Gray, Proc. Am. Acad. Arts & Sci. 7:358. 1868.

Helenium Bigelovii var. *festivum* Jepson, Man. Fl. Pl. Calif. 1132. 1925.

Perennial herb with stout, often simple stems 4–6 dm. tall, often somewhat flocculent-pilose; leaves oblanceolate to ovate-lanceolate, acute or rounded at summit, 8–20 cm. long, sessile or petioled, often the petioles enlarged at base and clasping the stem or the upper leaves somewhat decurrent; peduncles solitary or 2 or 3, stout, expanding below head and here densely tomentose; phyllaries in a single series, oblong-ovate, densely felty-tomentose; rays 15–25 mm. long; disc flowers brown or brownish yellow, forming a hemispherical dome 2–3 cm. broad; achenes rufous-hairy; pappus scales lanceolate-attenuate with a few lacerate teeth.

Coastal marshes: Point Arena in Mendocino County, and north to Coos Bay in southern Oregon.

Fig. 356. *Helenium Bigelovii: a,* habit, $\times$ ⅒; *b,* disc flower, $\times$ 6; *c,* flower head, showing the involucre of phyllaries in a single series, $\times$ ⅘; *d,* mature achene, showing the tapered pappus paleae, $\times$ 6; *e,* flower head, $\times$ 1; *f,* flower head (longitudinal section), showing the nearly globose receptacle, $\times$ ⅘; *g,* stem, showing the decurrent leaf bases, $\times$ ⅘; *h,* ray flower, $\times$ 4.

3. Helenium Bigelovii Gray in Torr., Pacif. R. R. Rep. 4:107. 1857. Fig. 356.

Perennial herb; stems erect, several, 5–12 dm. tall, simple or occasionally much-branched, angled and sometimes winged by the decurrent bases of the upper leaves; leaves oblong-lanceolate to linear, sessile or the lower leaves petioled, the upper ones sometimes decurrent on stem, glabrous or puberulent; heads solitary on slender peduncles, rarely more than 1 to a peduncle; phyllaries lanceolate; rays conspicuous, numerous, 1–2 cm. long, lobed at apex; disc flowers brown to yellowish, forming a subhemisphere; achenes hairy; pappus paleae 5–8, tapering to a slender awn and often with a few slender, lacerate lobes.

Widespread in marshes and bogs from near sea level to high in the mountains: Coast Ranges from Monterey County northward, San Bernardino, San Gabriel, and San Jacinto mountains, Sierra Nevada, northern California; southern Oregon.

HELIANTHUS. SUNFLOWER

Coarse, stout, erect, annual or perennial herbs. Lowest leaves opposite, the upper leaves usually alternate, the blades simple, rough, 3-nerved (sometimes obscurely so), usually more or less decurrent on petioles. Heads bisexual, radiate, terminal, solitary or usually in cymose panicles. Phyllaries imbricated in 2–4 series, herbaceous. Receptacle flat or low-convex, a bract subtending each flower at maturity, the bracts conspicuous, chaffy, persistent, embracing the achenes. Ray flowers usually conspicuous, yellow, sterile; disc flowers perfect, the corolla yellow to reddish brown or purplish brown, bulbous at base. Achenes thick, hairy or glabrous, quadrangular, obovate, compressed and somewhat elliptic in cross section; pappus of 2 elongated paleaceous awns, sometimes with short, intervening squamellae, all deciduous.

Watson, E. E., Contributions to a monograph of the genus *Helianthus*. Papers Mich. Acad. Sci. 9:305–475. 1929.

Heiser, C. B., Study in the evolution of the sunflower species *Helianthus annuus* and *H. Bolanderi*. Univ. Calif. Publ. Bot. 23:157–208. 1949.

————, Notes on western North American sunflowers (*Helianthus* spp.). Contr. Dudley Herb. 4:315–317. 1955.

Helianthus is a genus (mainly North American) of many species of bewildering variability. Most species have some propensity for starting their growth in at least vernally wet situations, a habit which has rendered difficult the selection of species for inclusion here. This problem is not aided by the conspicuous lack of habitat notes on most herbarium specimens.

The polymorphy of the group is such that, in the construction of a key, diagnostic characters seem to vanish, and the liberal use of such qualifying terms as "frequently," "usually," "more or less," "seldom," and so on, is demanded. In our key this is evident, not only between the two perennial species, but also between the annual and perennial groups, wherein difficulty was encountered in finding reliable characters to bolster the not always clearcut duration of life span. It is felt that the present treatment, however, will least disturb the "status quo" of nomenclature, and is preferable to making changes on the basis of a study of selected species apart from the whole genus and on the basis of morphological criteria only. Such a procedure is necessary in a work which must be completed within a reasonable length of time. However, Heiser's cytogenetical approach to the study of this genus, now in progress, and his critical assessment of this manuscript, have been of valuable aid to me.

Annuals; stem more or less scabrous throughout; inflorescence usually of relatively few heads, inflorescence branches and peduncles usually long; receptacle bracts obviously 3-cuspidate at apex, middle cusp elongate.
  Middle cusp of receptacle bracts lanceolate, not longer than the disc flowers at maturity, hispid to tip; phyllaries usually not or only slightly longer than the disc flowers
  .... 1. *H. annuus*
  Middle cusp of receptacle bracts attenuate into a pointed awn, longer than the disc flowers at maturity, glabrous except at base; disc 2–2.5 cm. broad at maturity; phyllaries lanceolate, gradually attenuate to a subulate tip, the outer phyllaries frequently much longer than the disc flowers .... 2. *H. Bolanderi*
Perennials; stems usually quite glabrous and frequently glaucous; heads often numerous, inflorescence branches and peduncles usually short and the heads clustered; receptacle bracts obtuse to acute, apex entire or only obscurely 3-cuspidate.
  Inflorescence frequently densely flowered, the heads often crowded on short peduncles; phyllaries often 30–35 mm. long, gradually tapering to a long, acuminate, reflexed tip .... 3. *H. californicus*
  Inflorescence usually fewer-flowered, the heads often on longer peduncles; phyllaries seldom more than 20–25 mm. long, lanceolate-subulate, usually not reflexed
  .... 4. *H. Nuttallii*

1. Helianthus annuus L., Sp. Pl. 904. 1753. Common sunflower.

Stout annual; stem very hispid and rough, simple or profusely branching, 3–30 dm. tall; leaf blades 7–30 cm. long, usually broadly ovate, serrate, truncate to subcordate, rough-scabrous, green, the petiole often as long as or longer than the blade; heads single and terminal on long, stout peduncles, or the inflorescence forming an open, cymose, long-branched, long-peduncled, leafy-bracted panicle; phyllaries 1.5–2 cm. long, 5–10 mm. wide, broadly ovate, as long as or sometimes longer than the disc flowers, often conspicuously ciliate and densely hispid to hirsute, the apex abruptly narrowed and produced into a long, tail-like acumination; receptacle bracts apically 3-cleft, all 3 cusps acute, the lateral ones somewhat lacerate, the longer, lanceolate middle cusp hispid almost to its purple, acuminate tip, about as long as the disc flowers and appressed until the latter open; rays 2–4 cm. long; disc 2.5–4 cm. broad at maturity, disc flowers reddish or purplish (in wild forms); achenes 4–7 mm. long, glabrous or sparingly pubescent, variously colored and striped or speckled; pappus of 2 lanceolate paleae without intervening squamellae.

Often in moist areas on valley lands, plains, or prairies. An introduced species widely distributed in the United States, but occurring only in areas disturbed by man. Domesticated forms occur in cultivation almost throughout the world.

In the Mojave Desert of California and in the adjacent part of Nevada, a race with narrower leaves, smaller heads, and narrower phyllaries which lack the long, tail-like acumination has been recognized by Heiser as *Helianthus annuus* subsp. *Jaegeri* (Heiser) Heiser (*op. cit.,* 1955).

2. Helianthus Bolanderi Gray, Proc. Am. Acad. Arts & Sci. 6:544. 1865.

Helianthus exilis Gray, Proc. Am. Acad. Arts & Sci. 6:545. 1865.

Slender to fairly stout annual; stem erect, hirsute to rough-scabrous, usually branched, 3–13 (usually 7) dm. tall; leaf blades 3–9 cm. long, ovate to ovate-lanceolate, green with variable pubescence (scabrous to hirsute), entire to serrate, cuneate at base to rarely truncate, somewhat decurrent, the usually distinct petiole mostly about ⅓ as long as the blade; heads solitary at apices of the long, slender peduncles, the inflorescence often an open, long-branched, leafy-bracted, cymose panicle; phyllaries 1.5–2.5 (–3) cm. long, 3–4.5 mm. broad, oblong to lanceolate, with gradually acuminate, subulate tips, the outer phyllaries often longer and narrower, surpassing the disc, spreading and sometimes more or less reflexed, green, often densely hirsute, especially toward the base, with long, shaggy, white hairs; receptacle bracts apically 3-cleft, the lateral cusps lacerate, the middle cusp much longer (surpassing the disc flowers) and attenuate into a long, subulate, purple awn which is slightly hispid at base only and erect as soon as the phyllaries open; receptacle 2–2.5 cm. broad; rays 10–17, 1–2 cm. long; disc flowers yellow to red-purple; achenes 3–4.5 mm. long, gray to reddish brown, usually speckled, glabrous to villous; pappus of 2 lanceolate paleae without intervening squamellae.

Dry or wet valley lands: northern and central California, occasionally south as far as Tulare and San Luis Obispo counties; southern Oregon.

What has been regarded in California botanical literature as *Helianthus exilis* is considered as syonymous with *H. Bolanderi* by Heiser (*op. cit.,* 1949). He points out, however, that these taxa represent two races of *H. Bolanderi:* the one, *H. exilis,* occurs on dry serpentine outcrops of the foothills area; the other, *H. Bolanderi* (*sensu stricto*), the unit in which we are interested here, is a weedy and more widespread plant often occurring in moist habitats, and is characterized by larger leaves, discs, and achenes, more numerous rays, and the usually hirsute pubescence tending to be merely hispid or scabrous.

3. Helianthus californicus DC., Prodr. 5:589. 1836. California sunflower.

Helianthus californicus var. *mariposianus* Gray, Syn. Fl. N. Am. 1²:277. 1884.

Erect perennial from somewhat tuber-like roots; stems 10–35 dm. tall, glabrous and smooth, usually glaucous; leaf blades 7–15 cm. long, 1–2.5 mm.

wide, oblong to lanceolate or linear-lanceolate (sometimes the lower leaves ovate), acuminate, usually entire, tapering at base and gradually decurrent below confluence of the usually very weak lateral veins onto the obscured petiole, scabrous, green; heads usually numerous on usually short peduncles in a terminal, cymose, leafy-bracted panicle, the peduncles scabrous below heads; phyllaries 15–30 (or –35) mm. long, 3–4 mm. wide, rough-puberulent or hispidulous above, subglabrous below, the lower margins obscurely ciliate, lanceolate, gradually tapering to a long, acuminate tip, very loose and reflexed; receptacle bracts linear, obviously shorter than mature disc flowers, straw-colored, glabrous except for darker, hispid keel and tip, the tip of mature bract abruptly acute or almost truncate and mucronulate, entire or sometimes with 2 very small lateral teeth; rays 15–20, about 2 cm. long; achenes glabrous; pappus of 2 (infrequently 3) lanceolate paleae, sometimes with intervening squamellae.

Along streams or near springs, at elevations below 6,000 feet: Sacramento and San Joaquin valleys, cismontane southern California, Coast Ranges, Sierra Nevada; south to northern Baja California, Mexico.

In this highly plastic species, *Helianthus californicus* var. *mariposianus* represents too tenuous a unit, both geographically and morphologically, to retain.

4. Helianthus Nuttallii Torr. & Gray, Fl. N. Am. 2:324. 1842.

Perennial from short, tuber-like, fascicled roots; stem 3–10 (or –20) dm. tall, simple or branched, smooth and glabrous, usually glaucous; leaf blades 5–15 cm. long, about 1–2 cm. wide, narrowly linear-lanceolate, acute to acuminate, entire to somewhat serrulate, tapering at base and gradually decurrent below confluence of lateral veins onto a winged petiole about ¼ as long as the blade, scabrous or hispid, paler green beneath; heads few to several, in a sometimes many-branched, leafy-bracted, cymose panicle, the peduncles long or short, scabrous below heads; phyllaries 12–20 mm. long, lanceolate-subulate, gradually tapered from base to apex, seldom hispid but often clothed with whitish hairs, somewhat hirsute-ciliate on margins, the outer phyllaries often extending considerably beyond the disc and loose but not reflexed; receptacle bracts linear, straw-colored, glabrous below, brown and pubescent on the back, especially toward the tip, entire or with 2 very obscure lateral teeth, the acute apex sometimes produced as a short mucro or awn from the distally somewhat keeled midrib, shorter than or of almost the same length as the mature disc flowers; rays 8–24, 2–2.5 cm. long; disc yellow, 1.5–2 cm. across; achenes glabrous; pappus of 2 linear-lanceolate paleae, rarely with some intermediate squamellae.

In springy places, sloughs, or on dry ground in valleys and plains: northeastern California, Inyo County, San Bernardino County; Alberta and Saskatchewan south to Nevada, New Mexico, and northern Arizona.

Helianthus Nuttallii, a well-known, perennial species of the middle-western plains and prairies, is not reported for California in any of the regional manuals. It is recorded from Nevada and southeastern Oregon; and herbarium material, as well as collections made during the marsh plant survey, indicate its presence in northeastern California. It is also reported from Inyo County and northern San Bernardino County, according to Heiser (correspondence). On the basis of its occasional predilection for habitats which are wet during the early growth period, it is included here.

Most previous workers have considered *Helianthus Nuttallii* and *H. californicus* as distinct species, and because the two have not been treated concurrently in any of the regional manuals, the question of their interrelationship has not ordinarily arisen. When specimens of the two species are assessed simultaneously, however, reliable diagnostic characters seem to vanish, and one is confronted by an excellent example of the dilemma outlined in our introductory statement to this genus, wherein every claim to distinction must be hedged by qualifications. Nevertheless, their almost complete geographic separation and their continued recognition by botanists indicate the advisability of preserving the status quo until it may be modified by more intense future study.

4a. Helianthus Nuttallii var. **Parishii** (Gray) Jepson, Man. Fl. Pl. Calif. 1077. 1925.

Helianthus Parishii Gray, Proc. Am. Acad. Arts & Sci. 19:7. 1883.
Helianthus Oliveri Gray, Proc. Am. Acad. Arts & Sci. 20:299. 1885.
Helianthus californicus var. *Oliveri* Blake in Munz, Man. So. Calif. Bot. 549. 1935.

Leaf blades linear-lanceolate, gray-tomentose, especially beneath, with appressed but not tangled hairs, the upper leaflike bracts of inflorescence extending beyond the heads; involucre canescent at base, the phyllaries usually shorter than in the species.

Swamps and stream margins: San Bernardino County to Orange and Los Angeles counties.

Of limited occurrence and of comparatively easy recognition. The more densely pubescent form of this entity has been treated as *Helianthus Oliveri* Gray; I know of no recent collections of this form, and according to Heiser (correspondence) it may now be extinct.

IVA

Ours coarse perennial herbs. Leaves alternate above, often opposite below, simple and entire (in ours), rather coriaceous. Inflorescence spicate, the heads axillary, small, greenish, nodding, discoid. Involucre hemispherical, the phyllaries united into an irregularly lobed or toothed cup or phyllaries distinct and 5–10. Receptacle chaffy. Marginal flowers pistillate, fertile, 1–5, the corollas

Fig. 357. *Iva axillaris: a,* habit, × ⅖; *b,* central, sterile flower, × 10; *c,* inflorescence, × 1½; *d,* marginal pistillate flower, × 10; *e,* flower head, the toothed involucral cup enclosing flowers, × 6; *f,* involucral cup with distinct lobes, × 4; *g* and *h,* achenes, the outer and inner sides with surface resin-dotted, × 10; *i* and *j,* chafflike receptacle bracts, × 10; *k,* achene (cross section), × 10.

rudimentary; central flowers perfect, with 5-lobed, funnelform corolla, not producing mature fruits. Achenes obovate, thickened but slightly compressed laterally, golden brown, more or less resinous, without pappus.

Phyllaries united into a lobed cup; heads solitary in upper leaf axils ...... 1. *I. axillaris*
Phyllaries distinct; heads clustered on short peduncles, the inflorescence forming a narrow, spikelike, bracteate panicle .... 2. *I. Hayesiana*

1. Iva axillaris Pursh, Fl. Am. Sept. 2:743. 1814. Poverty weed. Fig. 357.

Herbage glabrate to rough-pubescent, more or less resinous-dotted; stems many, very leafy, straw-colored, fascicled on creeping, woody rootstocks, suffrutescent and decumbent to prostrate at base, ascending and herbaceous above, 10–20 dm. long; leaves elliptic to obovate, 1–3 cm. long, obtuse, narrowed at base, sessile; heads solitary in leaf axils on very short peduncles or sessile; involucral cup with 5–8 lobes or merely toothed.

Waste lands and cultivated fields, often on alkaline plains or bordering salt marshes: throughout California at low altitudes; British Columbia to Mexico in western half of North America.

2. Iva Hayesiana Gray, Proc. Am. Acad. Arts & Sci. 11:78. 1876.

Herbage glabrate to rough-pubescent, more or less resinous-dotted; stems woody below, not densely leafy, openly branched, the branches ascending from base, to 1 m. tall; leaves spatulate to linear, 3–6 cm. long, narrowed to a short petiole, usually obtuse; heads clustered on short peduncles in a narrow, spikelike, bracteate panicle, the bracts leaflike but reduced; phyllaries distinct, oval to orbicular, 5–10, deciduous in age.

Alkaline flats or brackish areas: San Diego County; south to Baja California, Mexico.

JAUMEA

1. Jaumea carnosa (Less.) Gray in Torr., U. S. Expl. Exped. (Wilkes) 17:360. 1874. Fig. 358.

Glabrous, somewhat succulent, perennial herb; stems simple, erect, 1–2 dm. tall, numerous from woody root crown or from prostrate stems of previous year; leaves fleshy, linear, entire, 2–3 cm. long, opposite, connate at base; heads solitary, terminal or occasionally axillary, radiate, many-flowered, 1–1.5 cm. wide; involucre narrowly campanulate, 1–1.5 cm. high, the phyllaries imbricated, broadly ovate, obtuse, the outer phyllaries shorter and fleshy; receptacle conical, naked; ray flowers 9–12, fertile, yellow, inconspicuous, 2–3 mm. long; disc flowers yellow, fertile, the style branches apically thickened; achenes linear, 10-ribbed; pappus none.

Salt marshes and mud flats along the coast; north to British Columbia.

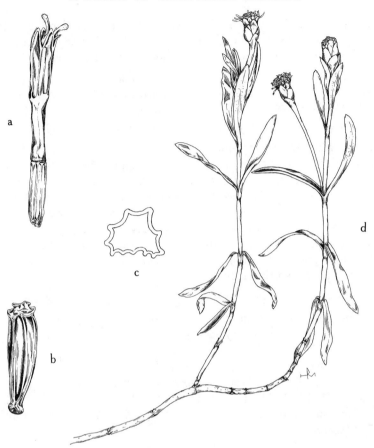

Fig. 358. *Jaumea carnosa: a,* disc flower, showing the apically thickened style branches, × 4; *b,* achene, × 10; *c,* achene (cross section), × 20; *d,* habit, × ⅘.

LACTUCA

Annuals, biennials, or perennials with milky juice. Stems tall, leafy. Leaves alternate, entire to deeply lobed or pinnatifid. Inflorescence paniculate, the usually small heads with yellow, cream, purple, or blue ligulate flowers only. Involucre cylindrical or campanulate, the phyllaries imbricated in 2 or more series and of unequal length, the outer phyllaries usually short. Receptacle naked. Corollas all ligulate. Achenes flattened, ribbed, beaked, the apex expanded; pappus of soft, copious, capillary bristles which fall separately.

Plants perennial; rays bright blue or purplish, much longer than the involucre; leaves not
    spinulose .... 1. *L. pulchella*
Plants annual or biennial; rays yellow, little longer than the involucre; leaves spinulose
    on margin and midrib .... 2. *L. scariola*

1. Lactuca pulchella DC., Prodr. 7:134. 1838. Blue lettuce.

Perennial with deep rootstock; stems simple, erect, very leafy, 6–12 dm. tall; herbage glaucous and glabrous; leaves numerous, variable, linear to lanceolate or oblong, entire or incised-dentate or runcinately lobed or pinnatifid, sessile or with winged petiole, 10–15 cm. long; panicle of few to many heads, scaly-bracted below heads; involucre about 10–15 mm. high; rays well exserted, linear, blue or purplish; achenes with short beak; pappus whitish.

Moist ground of montane ravines or meadows: mostly on eastern slope of the Sierra Nevada at elevations of 3,000–6,000 feet; north to British Columbia, east to Michigan.

2. Lactuca scariola L., Sp. Pl., 2d ed., 1119. 1763. Prickly lettuce.

Annual or biennial from heavy taproot; stems erect, simple or branched, 5–15 dm. tall, glabrous or often prickly or hirsute below; leaves oblong or oblanceolate, sessile and clasping at base, pinnatifid or lobed, the margin and lower side of midrib spinulose, the lower surface of leaf blade often hirsute; heads numerous in an open panicle; involucre about 10–12 mm. high, cylindrical; rays pale yellow, extending slightly beyond the involucre; beak about as long as body of achene; pappus white.

Throughout valleys at low or middle altitudes, sometimes bordering wet places; weed introduced from Europe, found locally in waste places and fields almost throughout the United States.

LASTHENIA

Annuals, essentially glabrous and somewhat succulent. Stems erect or ascending, simple or branched, from slender taproot. Leaves opposite, usually entire, linear, sessile, more or less connate at base. Heads medium-sized, radiate, terminating the branches on slender peduncles arising from upper leaf axils. Involucre a single series of phyllaries united into a hemispheric or campanulate, toothed cup. Receptacle conical, bractless, with slender projections to which achenes are attached. Ray flowers yellow, fertile, 5–15. Disc flowers yellow, perfect. Achenes linear, flattened, with paleaceous pappus or none.

Rays conspicuous; pappus none .... 1. *L. glabrata*
Rays very inconspicuous; pappus of 5–10 paleae .... 2. *L. glaberrima*

1. Lasthenia glabrata Lindl., Bot. Reg. 21: pl. 1780. 1836. Fig. 359.

Stems erect, 1–4 dm. tall; leaves conspicuously connate-sheathing at base, succulent, the upper leaves sometimes toothed; peduncles long, erect, pubescent below heads; involucral cup broadly hemispherical, the teeth acuminate, ciliate; ray flowers 5–10 mm. long; disc flowers with corolla lobes papillate outside; achenes minutely roughened; pappus none.

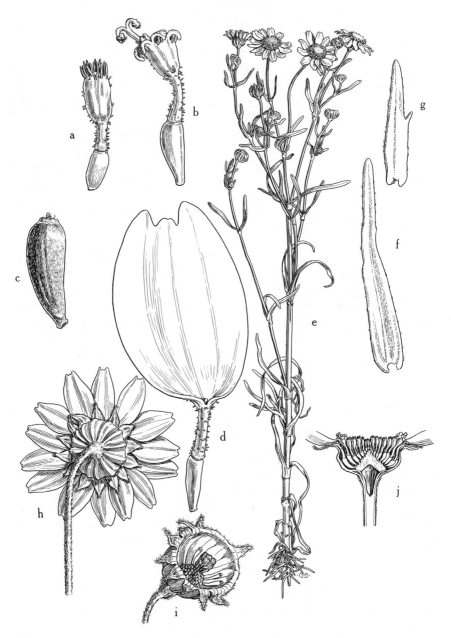

Fig. 359. *Lasthenia glabrata:* *a* and *b,* disc flowers at different stages in anthesis, × 8; *c,* achene, × 12; *d,* ray flower, × 6; *e,* habit, × ⅖; *f* and *g,* leaves, showing variation in shape, the pubescence sparse, *f,* × 1⅕, *g,* × 2; *h,* flowering head, showing toothed involucral cup and rays, × 2; *i,* involucral cup, after mature flowers have fallen, × 2; *j,* flowering head (longitudinal section), showing conical receptacle and arrangement of ray and disc flowers, the outer disc flowers opening first, × 1½.

Borders of saline or alkaline meadows or marshes at elevations below 5,000 feet: Coast Ranges, Central Valley, southern California.

Variations in leaves and achenes are indicated by the trinomials *Lasthenia glabrata* var. *californica* Jepson and *L. glabrata* var. *Coulteri* Gray.

2. Lasthenia glaberrima DC., Prodr. 5:664. 1836.

Stems weak, ascending, 1–4 dm. long; peduncles usually short; involucre broadly campanulate, slightly villous, the teeth triangular and acute; ray flowers minute, shorter than achenes; achenes with appressed, ascending hairs; pappus paleae brown, rigid, lacerate or erose, ½ to ⅓ as long as achenes.

Semiaquatic in swampy ground, at elevations of 10–3,000 feet: central Coast Ranges, North Coast Ranges, and Central Valley; north to Oregon.

LEONTODON. HAWKBIT

1. Leontodon nudicaulis (L.) Banks ex Lowe, Trans. Cambridge Philos. Soc. 4:28. 1831.

Low, acaulescent perennial with milky juice; scapes 6–25 cm. tall, numerous from woody taproot, glabrous or somewhat villous; leaves in basal cluster, narrowly oblanceolate, 5–15 cm. long, sinuate-pinnatifid to subentire, rough-hispid; heads solitary, with corollas all ligulate; involucre 8–10 mm. high, with 1 or 2 series of longer, lanceolate phyllaries and a few shorter (calyculate) ones at base; receptacle naked; corollas all ligulate, yellow, 10–15 mm. long; achenes linear, about 5 mm. long, dimorphic, the outer achenes smooth, the pappus a crown of short scales more or less united at base, the inner achenes papillose-roughened, the pappus of 10 paleaceous-based, long-plumose bristles in addition to an outer ring of short scales.

Isolated localities along California coast, sometimes in marshy areas; European weed occasional in the United States and Canada.

PHALACROSERIS

1. Phalacroseris Bolanderi Gray, Proc. Am. Acad. Arts & Sci. 7:364. 1868.

Perennial, caespitose herb with milky juice; scapes 2–6 dm. tall, 1 to several from a basal rosette of leaves; leaves linear to oblong-lanceolate, entire, narrowed to the petiole, 1–3 dm. long, somewhat succulent; heads 2–3 cm. broad, solitary on naked scapes; involucre campanulate, 8–10 mm. high; receptacle naked; corollas well exserted, all ligulate, yellow; achenes plump, oblong, short; pappus none or represented by a short crown.

Wet meadows: central and southern Sierra Nevada at elevations of 7,000–8,000 feet.

Fig. 360. *Pluchea sericea: a,* habit, upper branch, × ⅖; *b,* flower heads, × 2;
c, outer pistillate flower, × 8; *d,* inner perfect flower, × 8.

PLUCHEA

Leafy shrubs or herbs. Leaves alternate, entire. Inflorescence of numerous heads arranged in terminal, corymbose cymes. Phyllaries imbricated, often strongly lavender-tinged, not scarious. Receptacles flat, naked. Marginal flowers abundant, pistillate, fertile, with filiform, tubular, truncate or 2- to 4-toothed, lavender to reddish purple corolla and 2-cleft, exserted style. Central flowers few, perfect, with tubular, 5-toothed, reddish purple corolla and entire or 2-cleft style, not producing mature fruits. Achenes grooved; pappus a single series of capillary bristles.

Annual herb; leaves glandular, broadly ovate to lanceolate, serrate, to 10 cm. long, usually petioled .... 1. *P. camphorata*
Shrub; leaves silvery-hairy, linear to lanceolate, entire, to 4 cm. long, sessile
2. *P. sericea*

1. Pluchea camphorata (L.) DC., Prodr. 5:452. 1836. Salt-marsh fleabane.

Annual herb; stems erect, branching above, 3–8 dm. tall; herbage glandular-puberulent, green; leaves broadly ovate to lanceolate, glandular-dentate, 6–10 cm. long, 2–3 cm. wide, petioled below; inflorescences large, the heads about 5 mm. high, numerous in compound, corymbose cymes borne in upper leaf axils; phyllaries ovate-lanceolate, chartaceous, to 5 mm. long; achenes pubescent; pappus alike in all flowers, the bristle tips not dilated.

Salt marshes and alkaline flats at low elevations: region of San Francisco Bay, Central Valley, and coastal southern California; east to Texas and Atlantic coast.

2. Pluchea sericea Cov., Contr. U. S. Nat. Herb. 4:128. 1893. Arrow-weed. Fig. 360.

Willow-like, erect shrub 1–4 m. tall; herbage silvery-silky; leaves linear to lanceolate, entire, acute, sessile, 1–4 cm. long, 3–6 mm. wide, often densely clothing branches; inflorescences of small, compact, corymbose clusters of heads on terminal and on short, lateral branches; heads 10–15 mm. high; phyllaries to 4 mm. high, the outer phyllaries ovate, coriaceous, more or less woolly and glandular, the inner ones narrower, thinner, glabrous; achenes glabrous; pappus bristles of central flowers crinkly and bent at the slightly dilated tip, those of marginal flowers neither crinkly nor dilated.

Forming dense thickets in river bottoms and wet places: cismontane southern California, to Colorado and Mojave deserts and Inyo County; east to Texas.

PSILOCARPHUS

Inconspicuous annual herbs. Stems prostrate, white-woolly. Leaves opposite, the uppermost often crowded like an involucre around the heads. Heads small, sessile, globose, solitary, terminal, or appearing clustered in the forks of the

Fig. 361. (For explanation, see facing page.)

branches, rayless and without true involucre. Receptacle globose. Bracts of the receptacle woolly, ontogenetically enlarging into a saclike hood bearing a hyaline appendage at the front in the upper half, open from the appendage to the base of the bract, at maturity each bract loosely enclosing an achene. Staminate flowers few, central, borne at apex of receptacle. Pistillate flowers with filiform corolla. Achenes straight or curved; pappus none.

In addition to the species treated herein, *Psilocarphus tenellus* occurs commonly in California, in drier habitats. *Psilocarphus elatior* Gray (*P. oreganus* var. *elatior* Gray), described on the basis of its apical style, is considered by Cronquist (Res. Stud. State Coll. Wash. 18:80–83, 1950) as being confined to the Northwest, but some of our material approaches it in achene characters.

Leaves broadest at the apiculate tip, gradually tapering to the base; herbage with a close tomentum, never loosely floccose; heads not at all concealed by subtending leaves
1. *P. oregonus*
Leaves linear, or broadest at base and tapering to obtuse or acute tip; herbage loosely floccose, at least in inflorescence; heads at least partly concealed by the subtending leaves . . . . 2. *P. brevissimus*

1. Psilocarphus oregonus Nutt., Trans. Am. Philos. Soc. II. 7:341. 1841.

Stems slender, diffusely branched from base, 5–15 cm. long, erect or spreading; herbage closely tomentose but not loose-woolly; leaves spatulate, gradually narrowed to the base, apiculate, 1–3 cm. long; heads 4–5 mm. wide, solitary or in clusters in leaf axils and much shorter than the leaves.

Dried beds of vernal pools: chiefly in the Central Valley and Coast Range valleys; north to Oregon.

2. Psilocarphus brevissimus Nutt., Trans. Am. Philos. Soc. II. 7:340. 1841. Fig. 361.

Psilocarphus globiferus Nutt., *loc. cit.*
Psilocarphus oreganus var. *brevissimus* Jepson, Fl. West. Middle Calif. 549. 1901.
Psilocarphus brevissimus var. *brevissimus* (Jepson) Cronquist, Res. Stud. State Coll. Wash. 18:78. 1950.

Stems 1 to many, erect, prostrate, or ascending, 2–10 cm. long; herbage loose-woolly, especially in inflorescence; leaves linear-oblong to lanceolate-

Fig. 361. *Psilocarphus brevissimus: a,* saclike receptacular bract containing achene, × 10; *b,* achene, × 10; *c,* habit, × 1⅕; *d,* young pistillate flower, showing receptacular bract with conspicuous hyaline scale, × 20; *e,* young flowering head, showing the centrally placed staminate flowers, × 2; *f,* receptacular bract containing the mature achene (the hyaline tooth has remained about the same size as it was in the young flower— compare with *d*), × 10; *g,* achene, × 10; *h,* young flowering head (longitudinal section), the pistillate flowers below the central staminate ones, the hyaline tooth of the receptacular bract appearing apical in the young pistillate flowers, × 12; *i,* mature flowering head (longitudinal section), showing the change in position and relative proportion of the hyaline tooth in the mature receptacular bracts (compare with *d* and *h*), × 4; *j,* mature flowering head, the staminate flowers completely obscured by the enlarged receptacular bracts, × 2.

oblong or sometimes triangular, 5–25 mm. long, often crowded beneath heads and extending beyond them; heads in the forks of the branches and terminal, often concealed or partly concealed by the leaves and a dense, arachnoid tomentum.

Dried beds of vernal pools: Central Valley and adjacent foothills to an elevation of 3,000 feet, coastal southern California, north-central California; north to Washington. Often conspicuous in the beds of vernal pools because of its white-tomentose, ball-like inflorescences.

RUDBECKIA. CONEFLOWER

Erect, coarse, biennial or perennial, North American herbs. Stems simple or somewhat branched. Leaves alternate, simple or pinnately divided, especially below, ovate to lanceolate, 2–12 cm. long. Heads large, usually long-peduncled, terminal; phyllaries in 2 or 3 rows, foliaceous, spreading. Receptacle prominently conical to columnar, its bracts (each of which subtends a flower) more or less carinate and embracing the achenes, usually longer than mature achenes. Ray flowers, when present, yellow, showy, sterile. Disc flowers purple to brown. Achenes quadrangular, apically truncate, tapering at base; pappus a short, persistent, lacerate or denticulate crown, or wanting.

Rays present.
  Herbage essentially glabrous or with inconspicuous appressed pubescence; pappus present . . . . 1. *R. californica*
  Herbage conspicuously rough-hairy, the hairs usually spreading; pappus none
                                                                    2. *R. hirta*
Rays none; pappus present . . . . 3. *R. occidentalis*

1. Rudbeckia californica Gray, Proc. Am. Acad. Arts & Sci. 7:357. 1868. California coneflower.

Perennial; roots from a short, woody rhizome; stems simple, arising singly, 6–18 dm. tall, glabrous, strongly ribbed; leaf blades ovate or more usually broadly lanceolate, tapering gradually into petiole at base, glabrous to hispidulous to appressed-pubescent, the lower blades long-petioled, often irregularly toothed to deeply incised, the upper ones sessile, entire, reduced; heads solitary on long terminal peduncles, the peduncles apically minutely pubescent; phyllaries linear, appressed-pubescent; receptacle to 5 cm. high at maturity, the bracts apically canescent and acute; rays 10–15, 3–5 cm. long; disc 3–4 cm. high, extending much beyond the involucre; achenes somewhat flattened; pappus crown 4- to 5-toothed.

Moist or meadow areas, at elevations of 5,000–7,000 feet: Siskiyou Mountains, North Coast Ranges south to Humboldt County, Sierra Nevada from Kern County to Mariposa County; southwestern Oregon.

2. Rudbeckia hirta L., Sp. Pl. 907. 1753. Black-eyed Susan.

Biennial; roots fibrous, from short root crown; stems 0.3–1 m. tall, 1 to several from base, simple or branched above; herbage rough-hairy throughout; leaf blades lanceolate, simple, entire or sparingly toothed, the lower ones sub-petiolate, the upper ones sessile and amplexicaul; heads terminal on branches, long-peduncled; outer phyllaries linear, acute, very hairy, reflexed, strongly foliaceous, the inner ones shorter, erect, obtuse; receptacle bracts with acute to attenuate, bristly-ciliate or merely hispid apex; ray flowers orange-yellow, often darker-colored at base, 2–4 cm. long; disc 1–2 cm. high, dark brown; achenes without pappus.

Introduced in meadows of the Sierra Nevada from Mariposa County to Amador County, and also along irrigation ditches in Stanislaus County. A native of the eastern United States with a wide distribution as a weed.

3. Rudbeckia occidentalis Nutt., Trans. Am. Philos. Soc. II. 7:355. 1841. Western coneflower.

Perennial; roots from a short, woody rhizome; stems arising singly, simple or branched above, 10–14 dm. tall; herbage nearly glabrous to hirsutulous or scabrous; leaves simple, broadly ovate to lanceolate, usually subcordate to truncate at base, irregularly serrate, 8–15 cm. long, the lower leaves long-petioled, the upper ones sessile; heads 1–3 on long peduncles; phyllaries broadly lanceolate, acute, essentially in 1 series, reflexed; receptacle bracts apically canescent and acute; ray flowers none; disc dark purple, 1.8–4 cm. high, at maturity extending much beyond the involucre; achenes flattened; pappus a low, denticulate crown.

Moist areas at middle altitudes: Sierra Nevada from Placer County to Butte County, Siskiyou Mountains, and North Coast Ranges in Glenn County; eastern Nevada, Oregon, Washington, Montana.

Rudbeckia occidentalis and *R. californica* are closely related. The most striking difference between them is that ray flowers are present in *R. califor-nica* and absent from *R. occidentalis*. Combined with this are almost equally obvious differences in the leaves—frequently deeply laciniate and with trun-cate to subcordate base in *R. californica*, entire or shallowly serrate and with tapering base in *R. occidentalis*. In other aspects, these two species are very similar.

SENECIO. GROUNDSEL

Annual or perennial herbs, or shrubs. Leaves alternate. Heads solitary or in terminal cymes, many-flowered, radiate or discoid. Flowers yellow, the ray flowers pistillate when present, sometimes lacking, the disc flowers perfect, fertile. Involucre cylindrical to campanulate, the phyllaries in 1 or 2 series and of equal length, often with some short, inconspicuous bracts at base.

Receptacle usually flat and naked. Achenes terete, 5- to 10-ribbed; pappus abundant, of soft, white, capillary bristles.

Stems abundantly leafy to the inflorescence; upper leaves gradually reduced; leaf blade
  usually toothed, cordate or truncate at base, abruptly petioled .... 1. *S. triangularis*
Stems sparsely leafy above; upper leaves much reduced; leaf blade entire, narrowed at
  base into a margined petiole ....·.................. 2. *S. hydrophilus*

1. Senecio triangularis Hook., Fl. Bor. Am. 1:332. 1834.

Stems erect, 1 to a few from a heavy horizontal rootstock, 5–20 dm. tall; herbage tomentose when young, usually becoming glabrous; leaves many, gradually reduced upward, lanceolate to triangular-ovate, 5–20 cm. long, cordate, hastate, or merely truncate at base, sharply serrate, dentate, denticulate, or nearly entire, the lower leaves on petioles as long as the blade; heads usually many in an open or dense corymbose terminal cyme; involucre 6–8 mm. high; rays 6–12, 8–10 mm. long.

Bogs, marshy areas, and stream margins in mountains, at elevations of 4,000–9,000 feet: mountains of southern California, the Sierra Nevada, and North Coast Ranges; north to British Columbia and east to Colorado.

2. Senecio hydrophilus Nutt., Trans. Am. Philos. Soc. II. 7:411. 1841. Swamp
  senecio.

Stems solitary from short rootstock, simple, erect, 5–15 dm. tall, stout, fistulose; herbage somewhat succulent, glabrous and somewhat glaucous; leaves usually few, the basal and lower ones linear-lanceolate to oblong, usually entire, 10–30 cm. long, gradually attenuate at base into a long, margined petiole, the upper leaves erect, reduced and clasping; inflorescence a large, spreading, corymbose cyme or the cyme condensed and almost capitate; involucre campanulate, 5–8 mm. high; rays none or few and short.

Marshes and swamps, at elevations of 10–7,000 feet: North Coast Ranges and the northern Sierra Nevada; north to British Columbia and east to Colorado.

Several other species of *Senecio* might reasonably be included here if their propensity for sometimes occurring in moist habitats were emphasized: in particular, *S. Clarkianus* Gray, *S. lugens* Richards., and *S. Clevelandii* Greene. The first two are montane species of the Sierra Nevada, the third a localized North Coast Range species of Napa and Lake counties.

SOLIDAGO. GOLDENROD

Perennial herbs, often rhizomatous. Leaves alternate, simple, entire or toothed. Heads numerous, in open or in dense thyrsoid or corymbose panicles. Phyllaries imbricate in 1 or 2 series, narrow, chartaceous. Receptacle naked.

Flowers yellow. Achenes terete, 5- to 10-nerved; pappus of equal, scabrous, capillary bristles, dull white.

Stems much-branched; heads scattered in corymbs across top of plant; ray flowers more
  numerous than disc flowers .... 1. *S. occidentalis*
Stems simple; heads in dense panicles; ray flowers fewer than disc flowers.
  Leaves mostly entire, the upper ones much shorter than the lower ones; South Coast
    Ranges, southern California .... 2. *S. confinis*
  Leaves serrate, at least toward tip, the upper ones not much reduced; North Coast
    Ranges, Sierra Nevada .... 3. *S. elongata*

Solidago corymbosa Nutt. of the high mountains is sometimes found in wet habitats. *Solidago guiradonis* Gray, an endemic of the San Carlos Range, is in most respects very close to *S. confinis,* and may be but a few-headed variant of that species.

1. Solidago occidentalis (Nutt.) Torr. & Gray, Fl. N. Am. 2:226. 1842. Fig. 362.

Rhizomatous perennial; stems simple, erect, 1–2 m. tall, paniculately branched in inflorescence; herbage glabrous; leaves numerous, linear, entire, dark-punctate, 2–5 cm. long; heads small, numerous, corymbosely disposed throughout inflorescence; phyllaries linear-lanceolate, chartaceous; ray flowers 16–20, the disc flowers fewer; achenes turbinate.

Very common in marshes and along irrigation ditches: throughout the valleys and foothills and south to coastal southern California; north to British Columbia, east to Rocky Mountains.

2. Solidago confinis Gray, Proc. Am. Acad. Arts & Sci. 17:191. 1882.

Stems 5–15 dm. tall, simple; herbage pale green; leaves oblong to narrowly lanceolate, acuminate, the lower leaves 5–15 cm. long and petioled, the upper ones much shorter, sessile, entire or sometimes serrate; panicle dense, either broad or narrow; heads about 4 mm. high; ray flowers slightly longer than disc; achenes canescent.

Moist and swampy ground: South Coast Ranges to coastal southern California.

Jepson reports a more luxuriant variety about hot springs in the San Bernardino Mountains and in Ventura County.

3. Solidago elongata Nutt., Trans. Am. Philos. Soc. II. 7:327. 1841.

Rhizomatous perennial; stems simple, erect, to 1 m. tall; leaves bright green, oblong to broadly lanceolate or oblanceolate, usually acute, tapered to a petiole-like base, coarsely serrate to sometimes entire, 5–10 cm. long, glabrous or sometimes scabrous-margined; heads 3–5 mm. high, in a broad, dense panicle; phyllaries linear; ray flowers 10–16.

Moist soil: coastal northern California, North Coast Ranges, and the Sierra Nevada; north to British Columbia.

Fig. 362. *Solidago occidentalis: a,* disc flower, × 8; *b,* ray flower, × 8; *c,* rhizome and base of stem, × ⅖; *d,* habit, upper part of plant, × ⅖; *e–h,* phyllaries, × 8; *i,* flower head, showing involucre, and ray and disc flowers, × 5; *j,* achene and pappus of scabrous, capillary bristles, × 8; *k,* flowering head with mature achenes ready for dispersal, × 5.

TARAXACUM

1. Taraxacum officinale Web. in Wiggers, Primit. Fl. Holsat. 56. 1780. Common dandelion.

Taraxacum vulgare (Lam.) Schrank, Baier. Reise 11. 1786.

Perennial herb with milky juice and simple or branched crown; leaves in a basal rosette, oblong or spatulate, sinuate-pinnatifid; scapes several to many, hollow, each bearing a single flowering head; involucre with the outer phyllaries short and becoming reflexed and the inner phyllaries long and erect, linear-lanceolate; receptacle naked; flowers yellow; corollas all ligulate, longer than phyllaries; achenes 4- or 5-ribbed, the ribs roughened or minutely denticulate to spinose above; pappus of numerous, soft, capillary bristles, borne at the apex of the elongated beak of the achene.

Common in wet meadows in the mountains, also as a weed in gardens and lawns; introduced from Europe.

TRICHOCORONIS

1. Trichocoronis Wrightii Gray, Mem. Am. Acad. Arts & Sci. II. 4:65. 1849.

Straggling annual herb; stems creeping and ascending, or sometimes erect, rooting at nodes; leaves opposite, sessile, auriculate-clasping, oblong to lanceolate, obtuse or acute; heads on slender peduncles in a branched inflorescence; involucre saucer-shaped, the phyllaries numerous, thin, herbaceous, 1-nerved, oblong-lanceolate, 3–4 mm. long; flowers all discoid; corolla funnelform, the tube thin below and abruptly dilated above, flesh-colored; achenes sharply 5-angled, minutely hispidulous near summit, about 1 mm. long; pappus of several minute paleae and barbellate bristles.

On muddy banks, infrequent: Central Valley and southern California.

VERBESINA

1. Verbesina encelioides (Cav.) Benth. & Hook. var. **exauriculata** Robins. & Greenm., Proc. Am. Acad. Arts & Sci. 34:544. 1899. Crownbeard.

Coarse annual; stem erect, freely branching, to 1 m. tall, striate; herbage more or less canescent; leaves 4–8 cm. long, alternate, deltoid-ovate to deltoid-lanceolate, sinuate-dentate, usually white-canescent beneath, less so and green above, the petioles without wings or auricles; inflorescence of numerous scattered heads on naked peduncles; heads 2–4 cm. wide, radiate; involucre hemispheric, herbaceous, the phyllaries in 2 or 3 series, linear to ovate; receptacle conical, the bracts membranous, becoming chaffy in age, concave and one enfolding each disc flower; ray flowers many, orange-yellow, 1–1.5 cm. long;

achenes strongly flattened, more or less silky-pubescent, with 2 thin, wide, corky wings, obovate or oblong, notched at apex between wings; pappus of 2 attenuate, caducous awns arising at inner angle of each wing.

Weedy plant of waste ground, frequently found on moist summer floodplains of winter streams: possibly only adventive in California; across the United States, mainly in more southern areas and more common in the east.

XANTHIUM

Coarse annual weeds. Stems stout, widely branching. Leaves alternate, toothed or lobed, petioled. Inflorescence of unisexual, discoid, greenish heads, the staminate flowers in terminal clusters, the pistillate ones axillary. Staminate involucre a single series of distinct, narrow phyllaries; receptacle cylindrical, bracteate; flowers many. Pistillate involucre coriaceous, closed, 2-celled, each cell with 1 flower, forming in fruit a hardened, bilocular bur covered with hooked prickles, and with 1 or 2 beaks at apex.

Leaves lanceolate, narrowed at both ends, white-pubescent beneath, with yellow, 3-parted, stipular spines at base .... 1. *X. spinosum*
Leaves deltoid-ovate, truncate or cordate at base, without stipular spines
2. *X. pensylvanicum*

1. Xanthium spinosum L., Sp. Pl. 987. 1753. Spiny cocklebur.

Stems erect or ascending, branching, 2–10 dm. tall, puberulent; leaves lanceolate, 4–8 cm. long, with a pair of long, narrow lobes on lower half of blade, sometimes with a few small lobes above middle, green above, densely white-pubescent on lower surface, shortly petioled, each with a pair of long, yellow, 3- or 4-parted, stipular spines at base; fruiting bur weakly spiny, tomentose, about 1 cm. long, the beaks inconspicuous.

Abundant in waste fields, sometimes along dikes and edges of marshy areas; introduced European weed occasional throughout the United States.

2. Xanthium pensylvanicum Wallr., Beitr. Bot. 1:236. 1842. Cocklebur.

Xanthium canadense of California references, not Mill.

Stems erect, usually branched, 2–9 dm. tall; leaves thick, harsh, deltoid-ovate, cordate at base or subtruncate, irregularly serrate to somewhat 3-lobed, green on both sides, on petioles as long as blades; fruiting bur 1–2 cm. long, cylindric, densely set with hooked yellowish prickles 3–7 mm. long, these often glandular and sparsely pubescent at base, the 2 beaks strongly developed, hooked at tip.

Very abundant in low, marshy lands: a weed throughout the United States, native to Atlantic coast.

Glossary

This glossary defines the botanical terms as they are used in the body of the text. References to text figures which serve to illustrate the terms are also given.

Abaxial.—Pertaining to the side of an organ away from the axis, such as the lower surface of a leaf (fig. 97, *a, i*). Compare *Adaxial.*

Acaulescent.—Seemingly without a stem; term applied to a plant which is apparently stemless, the stem being very short or subterranean (fig. 367).

Accrescent.—Becoming enlarged, as do certain parts of a flower after anthesis.

Accumbent.—Lying against anything; applied to cotyledons having edges against the radicle.

Acerose.—With a sharp, slender, needle-like point (fig. 295, *g*).

Achene.—A hard, dry, indehiscent, one-seeded fruit with a single cavity (fig. 367).

Acicular.—Needle-shaped.

Acuminate.—Tapering gradually to a sharp point at the end (fig. 365).

Acute.—Ending in a point which is less than a right angle, but not so tapering as "acuminate" (fig. 365).

Adaxial.—Pertaining to the side of an organ toward the axis, such as the upper surface of a leaf (fig. 97, *b, h*). Compare *Abaxial.*

Adnate.—United to an organ of a different kind, as are the stamens in flowers of the Scrophulariaceae (epipetalous), or stipules in certain members of *Potamogeton* (fig. 13, *a*).

Adventitious.—Occurring out of regular order in either time or place; term applied, for example, to a bud developing on a tree trunk.

Aestivation.—The arrangement of the perianth in the bud.

Alate.—Winged.

Alliaceous.—Having the odor of onions.

Alternate.—Said of leaves occurring one at a node, those of successive nodes forming a definite sequence around the stem; said also of members of adjacent whorls in the flower when any member of one whorl is in front of or behind the junction of two adjacent members of the succeeding whorl.

839

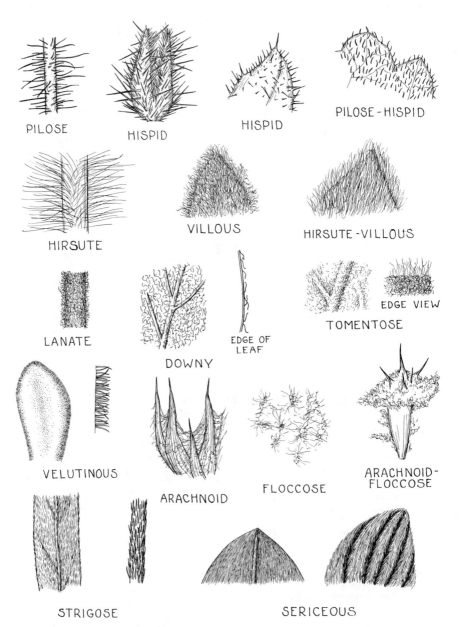

PILOSE

HISPID

HISPID

PILOSE-HISPID

HIRSUTE

VILLOUS

HIRSUTE-VILLOUS

LANATE

DOWNY

EDGE OF LEAF

TOMENTOSE

EDGE VIEW

VELUTINOUS

ARACHNOID

FLOCCOSE

ARACHNOID-FLOCCOSE

STRIGOSE

SERICEOUS

Fig. 363. Types of pubescence.

Alveolate.—Pitted, honeycombed, as are the seeds in certain Scrophulariaceae.

Ament.—A lax, spikelike inflorescence, as in the willows or alders (fig. 366).

Amplexicaul.—Clasping the stem (fig. 18, *g*).

Androecium.—A collective term applied to all structures in the stamen whorl or whorls.

Annual.—A plant which completes its life history within a year.

Annular.—In the form of a ring.

Anterior.—Pertaining to the side away from axis.

Anther.—The pollen-bearing part of a stamen.

Antheriferous.—Anther-bearing.

Anthesis.—The period during which a flower is open; the act of opening of a flower.

Anthodium.—A head which resembles a flower, as in the sunflower (figs. 345, 366).

Antrorse.—Directed forward and upward.

Apetalous.—Without petals (fig. 9, *d*).

Apical.—At the apex or tip.

Apiculate.—Ending abruptly in a minute point.

Apomixis.—Any form of asexual reproduction.

Appendage.—A lateral organ on a stem, usually at a node.

Appressed.—Closely pressed against.

Arachnoid.—Covered with long hairs so entangled as to give a cobwebby appearance (fig. 363).

Arcuate.—Curved as a bow (fig. 27, *b*).

Areolate.—Bearing areoles, divided into distinct spaces.

Areole.—A small, angular pit on a surface, as may occur between the veins of a leaf or on a seed coat.

Aristate.—Awned; provided with a bristle, usually at the end (fig. 295, *f*).

Aristulate.—Bearing a short awn.

Articulate.—Jointed (fig. 211, *g*).

Articulation.—A joint; the area in a stem or in a leaf petiole where separation occurs naturally.

Ascending.—Directed or rising upward obliquely.

Asexual.—Characterized by reproduction which does not involve the fusion of a sperm and an egg.

Attenuate.—Gradually narrowed to a point at apex or base.

Auricle.—An earlike appendage (fig. 9, *t*).

Auriculate.—Eared (fig. 365).

Awl-shaped.—Tapering gradually from the base to a slender tip, as does a needle.

Awn.—A stiff, bristle-like appendage, usually at the end of an organ (fig. 58, *b*).

Awned.—Provided with an awn (fig. 345).

Axil.—The upper angle between an organ and the axis which bears it, such as the angle between the leaf and the stem bearing the leaf.

Axile placentation.—Placentation in fruits the seeds of which are borne attached to the placenta situated in the angles of the cross walls along the axis (fig. 367).

Axillary.—Growing in an axil.

Axis.—The main or central line of development of a plant, structure, or organ, such as the main stem.

Balanced hair.—A hair seemingly attached at the middle (fig. 364).

Banner.—The upper, broad, more or less erect petal of a papilionaceous flower; standard (fig. 253, *d*).

Barbed.—Furnished with reflexed projections (figs. 85, *a*; 364).

Barbellate.—Finely barbed (fig. 345).

Barbulate.—Having fine beards.

Basal placentation.—The attachment of the ovule at the base of the ovary.

Base.—Basal or lower part of a plant or organ; through growth this basal part may eventually become uppermost.

Basifixed.—Attached by the base.

Beak.—A long, substantial point, which may be terete or angular (fig. 345).

Beaked.—Ending in a beak (fig. 345).

Bearded.—Furnished with long, stiff hairs or bristles.

Berry. — A fleshy fruit, few- to many-seeded.

Bidentate.—Two-toothed (fig. 123, *g*).

Biennial.—A plant requiring two years in which to complete its life history; the first year the vegetative growth occurs, and the second year it flowers, seeds, and dies.

Bifid.—Split into two parts; two-cleft.

Bifurcate.—Forked; said of Y-shaped hairs, for example.

Bilabiate.—Divided into two separate parts or lips (fig. 306, *e*).

Bipinnate.—Said of leaves wherein both the primary and the secondary divisions are pinnate.

Bisulcate.- –Having two grooves or furrows.

Biternate.—Said of leaves wherein the three main divisions are themselves divided into three parts.

Bladdery.—Appearing as though inflated.

Blade.—The lamina, or expanded part of a leaf (fig. 365).

Bloom.—The white powder or dust covering stems, leaves, fruits, or flowers.

Brackish.—Said of water with a high concentration of dissolved substances, usually somewhat salty.

Bract.—A reduced or modified leaf, particularly the scale-like leaves in a flower cluster. Also said of any bract-like emergence.

Bractiform.—Having the form of a bract.

Bractlet.—A small bract, or sometimes applied to bracts in secondary positions.

Bristle.—A stiff, sharp hair (fig. 364).

Bulb.—A much shortened axis bearing fleshy leaf blades.

Bulbiferous.—Bulb-bearing.

Bulblet.—A small bulb, especially one borne in a leaf axil.

Bur.—A fruit or fruiting involucre bearing prominent spines or hooks (figs. 46, *b; 345*).

Caducous.—Falling off early or prematurely, as do the sepals of California poppy.

Caespitose. — Tufted; having several to many stems in a close basal tuft (figs. 95, *g;* 104, *l*).

Callus.—A thick, leathery, or hardened protuberance, or part of an organ; new tissue covering a wound (fig. 210, *b* and *c*).

Callous grain.—Callus on perianth segments of *Rumex.*

Calycine.—Resembling a calyx; said of involucres or involucels.

Calyculate.—Having bracts around an involucre or calyx, these bracts resembling an outer involucre or calyx, as in the common dandelion.

Calyx.—The outermost whorl of the floral envelopes, composed of separate or united sepals; it may sometimes be petaloid.

Campanulate.—Bell-shaped.

Cancellate.—Latticed, or resembling lattice construction.

Canescent.—Gray-pubescent.

Capillary.—Hairlike, threadlike, very slender.

Capitate.—Aggregated into a dense, compact cluster or head.

Capsule.—A dry, dehiscent fruit originating from two or more carpels.

Carinate.—Provided with a longitudinal ridge on the lower, or abaxial, surface; keeled.

Carpel.—One of the foliar units of which a pistil is composed. If one carpel forms the pistil the latter is simple; if more than one, the pistil is compound.

Cartilaginous.—Hard and tough.

Caryopsis.—An achene in which the pericarp is united with the seed; developed from a superior, one-carpeled ovary (fig. 55, *f* and *g*).

Catkin.—A deciduous, erect or lax spike, consisting of unisexual, apetalous flowers (fig. 366).

Caudate.—Having a long, soft, terminal, tail-like appendage.

Caudex.—The trunk or stem of a plant; a term applied particularly to the persistent stem of an herbaceous perennial.

Caudicle.—Sterile stalk of the pollen mass of certain orchids (fig. 193, *g* and *h*).

Caulescent.—Having an evident stem above the ground level (fig. 367).

Cauline.—Pertaining to the stem (fig. 193, *b*).

Chaff.—Thin, dry scales or bracts.

Chaffy.—Having the texture of chaff.

Chartaceous.—Thin but stiff; having the texture of thin paper.

Choripetalous.—Term applied to a corolla having its petals distinct from one another.

Ciliate.—Having marginal hairs that form a fringe (fig. 364).

Cinereous.—Ash gray.

Circinate.—Coiled downward and inward, like the scroll of a fiddle (fig. 297, *b*). See *Scorpioid.*

Circumscissile.—Dehiscent by a horizontal line cutting through the middle, the top part falling away as a lid (fig. 367).

Clavate.—Club-shaped; gradually thickened upward (fig. 23, *a*).

Claw.—The narrowed, petiole-like base of some petals or sepals.

Cleft.—Cut about halfway to the midvein (fig. 365).

Cleistogamous.—Said of self-fertilized flowers that never open.

Coalescence.—Union of similar parts or organs, as of petals to form a corolla.

Coherent.—Having like parts united; said of two or more organs of the same kind which are united in the same whorl by ontogenetic fusion.

Cohesion.—The state of cohering.

Column.—Structure formed by the union of filaments of stamens, or by the union of stamens and pistils; term also applied to the receptacle structure around which the carpels are situated in Malvaceae and related groups (fig. 262, *c*).

Coma.—The tuft of hairs which is to be found at the end of some seeds (fig. 277, *e*).

Commissure.—The line of meeting of the margins of carpels; the plane or face along which two carpels adhere.

Comose.—Said of organs, such as seeds, which have a tuft of hairs at one end.

Compound. — Formed of several parts united in one common whole, as is a compound pistil; or of leaves composed of two or more distinct leaflets (fig. 365).

Compressed.—Flattened laterally.

Connate.—United; a term especially applied to such similar structures as the bases of two opposite leaves joined through toral growth.

Connective.—The tissue connecting the two "cells" of an anther.

Connivent.—Approximate but not organically united.

Convoluted.—Said of flower parts when rolled in the bud with the edge of one part overlapping the adjacent part.

Cordate.—Heart-shaped, such as the base of a leaf (fig. 365).

Cordate-clasping.—Said of sessile appendages the basal lobes of which surround the stem (fig. 18, *g*).

Coriaceous.—Leathery.

Corm.—A solid, bulblike stem, usually found underground.

Corniculate.—Furnished with horns or hornlike processes.

Corolla.—The second whorl of the floral envelope, the units of which are petals; frequently the showy part of a flower.

Corrugated.—Crumpled or folded irregularly.

Corymb.—A racemose type of inflorescence in which the lower pedicels are successively elongated, thus forming a flat-topped inflorescence in which the outer flowers open before the inner ones do (fig. 366).

Corymbose.—Said of flowers arranged in corymbs.

Cotyledon.—One of the embryo leaves to be found in a seed.

Crenate.—Having a margin with low, rounded lobes (fig. 365).

Crested.—Having a ridgelike process across the top of a structure.

Cristate.—Crested; having a crest.

Cruciferous.—Having a crosslike form, as in members of the mustard family.

Crustaceous.—Having a surface with a crustlike texture.

Culm.—The aboveground stem of grasses or grasslike plants (fig. 80, *b*).

Cuneate.—Wedge-shaped; tapering toward the point of attachment (fig. 365).

Cuneiform.—The same as *Cuneate.*

Cuspidate.—Terminating in a short, abrupt point (fig. 365).

Cyme.—A form of inflorescence in which the main axis terminates in a single flower which opens before the lateral flowers arising beneath (fig. 366).

Cymose.—Bearing cymes.

Deciduous.—Losing leaves seasonally.

Decompound.—Said of compound leaves having divisions that are again dissected (fig. 365).

Decumbent.—Reclining on the ground, with ascending apex (fig. 55, *c*).

Decurrent.—Said of an appendage joined to the stem in a manner to appear as though the margin extends downward and is attached along the stem.

Decussate.—With successive pairs of organs arranged at right angles to one another, causing them to appear 4-ranked.

Deflexed.—Turned back from point of attachment.

Dehiscent.—Opening and shedding contents; said of fruits and stamens (fig. 367).

Deltoid.—Triangular (fig. 365).

Dendritic hairs.—Hairs that branch like a tree (fig. 364).

Dentate.—Having marginal teeth pointing outward and not forward (fig. 365). Compare *Serrate.*

Denticulate.—Bearing minute teeth directed outward.

Depauperate.—Much reduced and imperfect in structure and development.

Di-.—Prefix meaning two or twice.

Diadelphous.—Said of stamens which are united by their filaments into two groups (sometimes with only one filament in one group) (fig. 253). Compare *Monadelphous.*

Dichotomous.—Branching into two forks, the successive segments repeatedly twice forking (fig. 366).

Dicotyledon.—A plant the seeds of which bear two cotyledons or seed leaves.

Digitate.—Having finger-like divisions.

Dimorphic.—Said of a given structure when it occurs in two forms on the same plant or on different plants of the same species; e.g., ray achenes differing from disc achenes in heads of some of the Compositae (fig. 345).

Dioecious.—Having stamens and pistils borne on different plants.

Disc flower.—A flower of the family Compositae in which the corolla is tubular, in contradistinction to the ligulate ray flowers (fig. 345).

Discoid.—Having the form of a discus; or, in members of the Compositae, having all flowers with tubular corollas and none with ligulate corollas (fig. 345).

Disposed.—Referring to the ultimate arrangement, irrespective of point of origin; thus, spirally arranged leaves may be disposed in two ranks so as to appear as though coming from opposite sides of the stem.

Dissected.—Divided into several to many separate parts; said, for example, of the blade of a leaf (figs. 243, *a*; 365).

Distal.—Pertaining to the end opposite that of attachment; toward the end opposite that of attachment.

Divaricate.—Diverging widely; said of branching when the branches separate at a wide angle (fig. 131, *a*).

Divided.—Referring to the blade of an appendage when it is cut into distinct divisions to, or almost to, the midvein, as though cut with scissors (fig. 365).

Dorsal.—Pertaining to the upper (or back) side of a structure. See *Abaxial.*

Downy.—Covered with very short, weak hairs (fig. 363).

Drupe.—A fleshy, one-seeded fruit, such as the plum or cherry (fig. 367).

Drupelet.—A small drupe, an aggregation of which comprises a compound fruit such as the raspberry or blackberry.

Elater.—An appendage within the sporangium which aids in dispersal of spores; in *Equisetum,* the clubbed hygroscopic bands attached to the spores.

Elliptic.—Said of an organ which is essentially oblong, with broadly rounded ends and sides (fig. 365).

Emarginate.—Said of leaves, sepals, or petals, and other structures that are notched at the apex (figs. 153, *g;* 365).

Embryo.—Rudimentary plant within the seed.

Endemic.—Restricted to a particular area or condition.

Ensiform.—Sword-shaped.

Entire.—Having a margin devoid of any indentations, lobes, or teeth; said of the margin of appendages such as leaves, bracts, stipules, sepals, and petals (fig. 365).

Ephemeral.—Referring to an organ living a very short time, usually a day or less.

Epigynous.—Said of a flower having sepals, petals, and stamens that are borne on a structure at the top of the ovary, the ovary thus being inferior.

Epipetalous.—Said of stamens when they are inserted on the corolla (fig. 315, *c*).

Equitant.—Said of leaves disposed in a plane parallel to the radius of the axis and with their bases enfolding or clasping the stem, such as the leaf of *Iris* and some species of *Juncus* (fig. 183, *e, m*).

Erose.—Uneven; said of margins that give the appearance of having been torn, or of margins with very small teeth of irregular shape and size.

Evergreen. — Retaining leaves throughout the year.

Exserted.—Extending beyond [some enclosing part of the plant]; said of any structure in respect to its position relative to another structure, such as stamens that extend beyond the corolla.

Falcate.—Curved like a sickle; said of appendages.

Farinaceous.—Containing starch or starch-like substance; term applied to a surface with a mealy or scurfy coating. See *Scurfy.*

Fascicle.—Borne as though in bundles tied at base, or as though branching from a common base (fig. 8, *h*).

Favose.—Pitted in a manner to give the appearance of a honeycomb.

Fertile.—Said of seed-bearing fruit or flowers capable of producing seeds, or of pollen-bearing stamens; also applied, incorrectly, to female flowers.

Filament.—The stalk bearing the anther, or any threadlike structure.

Filamentose.—Having the character of a filament (figs. 88, *a;* 119, *a*).

Filiform.—Filament-like, long and very slender (fig. 37, *d*).

Fimbriate.—Lacerate into regular segments so as to appear fringed (figs. 150, *a;* 364).

Fistulous (fistulose). — Hollow; said of some stems or petioles, or of leaves such as those of the onion.

Flexuous.—More or less zigzag or wavy (fig. 121, *g*).

Floccose.—Said of pubescence which gives the impression of irregular tufts of cotton or wool, the hairs usually loosely tangled (fig. 363).

Floret.—One of the flowers in a close inflorescence of small flowers, such as in the spikelet of a grass or in the head of a member of the Compositae.

Floristic.—Having to do with the composition and organization of a flora.

Floristics.—That aspect of phytogeography that deals with taxonomic composition and the geographic and quantitative relations of floras.

Flower.—An axis bearing either functional stamens or pistils or both, these either naked or subtended by a perianth.

Foliaceous.—Leaflike.

Follicle.—A fruit, usually developing from a simple pistil and dehiscing along one margin (fig. 231, *c*).

Foveolate.—Deeply pitted.

Free.—Neither attached to a member of the same whorl nor to a member of another whorl.

Free-central.—Said of placentation when the seeds are attached to a column which arises from the base and is not otherwise attached to the ovary wall (fig. 367).

Fruit.—The matured pistil or pistils and their accessory structures, bearing the ripened seeds.

Fruticose.—Shrubby; shrublike.

Fugacious.—Falling soon after maturing, as do flowers or flower parts.

Funiculus.—The stalk which connects the ovule to the placenta.

Funnelform.—Having the shape of a funnel; said of corolla or calyx.

Fusiform.—Tapering at both ends; term applied to any structure.

Galea.—That part of an irregular sympetalous corolla (usually the upper lip) that is extended as a spur or helmet (fig. 310, *c*).

Geminate hair.—Pair of hairs from a common base (fig. 364).

Gibbous.—Said of a calyx or corolla tube or segment which has a distended, rounded swelling on one side (fig. 42, *b*).

Glabrous.—Without pubescence of any kind.

Gland.—Any special secreting organ; (as commonly employed) any regularly occurring, anomalous, small protuberance anywhere on the plant (fig. 185, *a*).

Glandular.—Bearing glands or having any glandular secretion.

Glaucous.—Having a frosted or whitish waxy appearance from a waxy bloom or powdery coating.

Globose.—Shaped like a globe or sphere.

Glochidia.—A barbed hair or process.

Glochidiate.—Having barbs.

Glomerate.—Gathered in compact groups; said of flowers occurring in small clusters.

Glomerule.—A small cluster of flowers consisting usually of a compacted cyme.

Glume.—A member of a pair of bracts (often chaffy) subtending the spikelet of the grasses (fig. 79, *b*).

Glutinous.—Sticky.

Graduated.—Said of phyllaries when the outer ones comprising the involucre are successively shorter than the inner ones.

Grain.—The seed or seedlike fruit of a member of the grass family; a small, hard, often superficial structure having the appearance of a grain, such as the callous grain in *Rumex* (fig. 208, *c*).

Gynobase.—An enlargement of the torus or receptacle to form a platform or disc upon which the ovary rests.

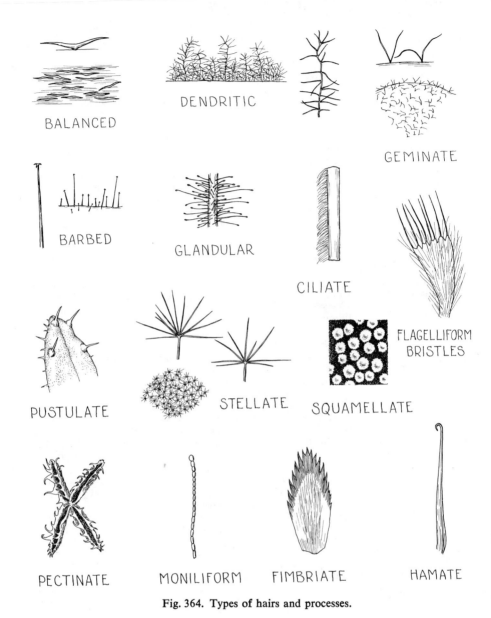

BALANCED

DENDRITIC

GEMINATE

BARBED

GLANDULAR

CILIATE

FLAGELLIFORM
BRISTLES

PUSTULATE

STELLATE

SQUAMELLATE

PECTINATE

MONILIFORM

FIMBRIATE

HAMATE

Fig. 364. Types of hairs and processes.

Gynoecium.—That part of the flower in which fertilization takes place and in which the seeds develop; the total of structures, including carpels and accessory parts, occurring on the axis of the flower morphologically above the stamens or androecium.

Habit.—The growth form of the plant.

Habitat.—The precise set of environmental conditions in which the plant occurs.

Hamate.—Said of a spine which is hooked at the tip (fig. 364).

Hastate.—Said of arrow-shaped leaves with basal lobes that spread or extend downward and outward (figs. 215, *b;* 365).

Head.—Compact group of flowers that occur together and form a spheroid aggregate at the end of a stem or branch (figs. 10, *e, g;* 366).

Hemi-.—Prefix meaning half.

Herb.—A plant, either annual, perennial, or biennial, of which the parts aboveground are not woody.

Herbaceous.—Having the structure or texture of an herb, not woody.

Herbage.—All of the aboveground, nonwoody part of a plant.

Heterostylic.—Having long styles in some flowers and short styles in others on the same plant, or in different plants of the same species.

Hilum.—The scar of the point of attachment of a seed.

Hirsute.—Clothed with long, shaggy hairs, often rough to the touch (fig. 363).

Hirsutulous.—Minutely hirsute.

Hirtellous.—Minutely hirsute.

Hispid.—Clothed with stiffish hairs that are sometimes spinelike (fig. 363).

Hispidulous. — Having fine, short, stiff hairs.

Hyaline.—Of thin, membranous, transparent or translucent texture.

Hypogeous.—Occurring below the surface of the ground.

Hypogynous.—Said of flowers in which the perianth and stamens are inserted upon the receptacle below the gynoecium.

Imbricated.—Said of organs that are so placed as to overlap like the shingles on a roof (fig. 345).

Incised.—Having the margin deeply cleft.

Included.—Not projecting beyond [an enclosing part of the plant]; said of an inner structure of a flower relative to an outer structure when the inner structure does not extend beyond the outer one or beyond some specified part of the outer one; for example, of stamens included in the corolla tube (figs. 297, *f;* 313, *h*).

Indehiscent.—Said of fruits that remain closed and do not shed their seeds.

Indeterminate.—Said of inflorescences in which the terminal flowers open last; also said of conditions in which growth and differentiation are not arrested.

Indurate.—Hardened and thereby often persistent.

Indusium.—The scale-like covering that invests the sorus in ferns (fig. 6, *a*).

Inferior.—Occurring below; said of the ovary when it occurs in such a manner as to appear to be below the other flower parts (fig. 3).

Inflated.—Hollow and swollen in a manner to appear as if distended with air.

Inflorescence.—An aggregation of flowers occurring clustered together in a particular manner which is usually characteristic of a given kind of plant (fig. 366).

Insertion.—The place of attachment of one structure on another.

Internode.—The part of a stem between any two adjacent nodes.

Involucel.—A secondary involucre, such as one subtending an umbellet in a compound umbel (fig. 366).

Involucre.—A group of closely placed, free or united bracts that subtend or enclose an inflorescence (figs. 127, *d;* 366).

Involute.—Said of margins that are rolled inward (toward the adaxial side), as in a petal or a leaf.

Irregular.—Said of the members of a given kind of structure when they are unlike in shape or size, such as the unequal lobes of a corolla (fig. 3).

Jointed.—Having one or more constrictions marking a point of articulation (figs. 213, *b, d*).

Keel.—The folded edge or ridge of any structure, alluding to its resemblance to the keel of a boat (fig. 124, *a*); in papilionaceous flowers, the two united front petals (fig. 253, *d*).

Lacerate.—Torn irregularly.

Laciniate.—Deeply incised into irregular, pointed lobes.

Lanate.—Woolly, with long, intertwined, curled hairs (fig. 363).

Lanceolate.—Much longer than broad; from a broad base tapering to the apex; lance-shaped (figs. 203, *g;* 365).

Lax.—Loose and often scattered; often said of flowers in an inflorescence.

Leaflet. — A discrete segment of a compound leaf.

Legume.—A one-celled fruit dehiscent on two sutures; specifically, the fruit of a member of the pea family.

Lemma.—The outer (abaxial) bract subtending a floret in the flower of a grass, often chaffy (figs. 2; 55, *d*).

Lenticular.—Shaped like a lens, having opposite sides convex (fig. 200, *a* and *b*).

Ligulate.—Shaped like a ligule (fig. 345).

Ligule.—An elongate, flattened structure; specifically, in monocotyledons, especially grasses, the bractlike emergence from the top of the leaf sheath at the base of the blade (fig. 55, *b*); in the Compositae, a strap-shaped corolla (fig. 345). (All corollas are ligules in the tribe Cichoreae; in many other Compositae only the marginal ray corollas are ligules.)

Limb.—The spreading part of a sympetalous corolla or synsepalous calyx; usually referring only to the corolla lobes (fig. 3).

Linear.—Long and slender, with more or less parallel sides (fig. 365).

Lip.—The upper or lower part of a bilabiate corolla or calyx.

Lobe.—An outward projection from the margin of an organ, usually with the margin indented on either side of the projection, as in leaves.

Lobed.—Characterized by having lobes.

Locule (loculus).—A compartment or cell, such as that of an ovary or anther.

Loculicidal.—Said of capsules that are dehiscent along the loculus or back of the carpels (fig. 367). Compare *Septicidal.*

Lodicule.—One of two or three scales at the bottom of the ovary, as in many grasses (fig. 2).

Marcescent.—Withering, but remaining attached (fig. 259, *c*).

Maritime.—Occurring in an area near the sea that is strongly influenced by environmental conditions imposed by the sea.

Massulae.—The group of cohering pollen grains, as in orchids.

Membranous.—Having a thin, soft, pliable texture.

Meristem.—An area of actively dividing and growing cells, as at stem and root tips.

-merous.—A suffix indicating the number of members in any given structure or whorl, such as a whorl of flower parts; for example, 5-merous or few-merous.

Midrib.—The conspicuous central vein in the vascular system of an appendage.

Monadelphous.—Having the stamens united into a single structure. Compare *Diadelphous.*

Moniliform.—Constricted so as to simulate a string of beads (figs. 191, *c;* 364).

Monocotyledon.—A plant the seeds of which bear only one cotyledon.

Monoecious.—Having the stamens and pistils in different flowers on the same plant.

Mucro.—A sharp, abrupt point or spiny tip.

Mucronate.—Said of appendages that come to an abrupt point (fig. 365).

Mucronation.—The abrupt point of an appendage.

Mucronulate.—Coming to a small, abrupt point.

Muricate.—Having a rough surface texture owing to many small, sharp projections (fig. 241, *d*).

Naked.—Without vestiture of any kind.

Nectariferous pit.—A depression or cavity bearing nectar, which may occur on a sepal, petal, or stamen (fig. 238, *b*).

Nerve.—A vein.

Net-veined.—Having the veins intricately branched and anastomosing.

Node.—The region on the stem where a leaf or leaves occur.

Nodose.—With knobs or knots.

Nut.—An indehiscent, one-seeded fruit from more than one carpel and having a woody coat.

Obcompressed.—Flattened at right angles to the radius of the axis.

Obcordate.—Heart-shaped, with the notched part away from the point of attachment (fig. 365).

Oblanceolate.—Pointed at the apex, broadest above the middle, and tapering to the base (fig. 365).

Oblique.—Said of a leaf having one side of the blade lower on the petiole than the other (fig. 365).

Oblong.—Longer than broad, the sides nearly parallel for most of their length (fig. 365).

Obovate.—Ovate in shape, but with the broadest part near the distal end (fig. 365).

Obtuse.—Having a blunt or rounded terminal part (fig. 365).

Ochrea (ocrea).—A nodal sheath formed by the fusion of stipules, as in the Polygonaceae (fig. 201, *i*).

Odd-pinnate.—Said of a pinnately compound leaf having a terminal leaflet, thus having an odd, rather than an even, number of pinnae (fig. 365).

Operculate.—Having a lid.

Operculum.—A lid.

Opposite.—As said of leaves: occurring two at a node on opposite sides of the stem. As said of flower parts: when one part occurs in front of another.

Orbicular.—Circular in outline.

Orthotropous.—Said of seeds that are erect and having their micropyle at the apex.

Ovary.—The part of the pistil bearing the ovules and maturing to form at least part of the fruit which bears the seeds.

Ovate.—Said of a plane structure having the shape of the outline of an egg (fig. 365).

Ovoid.—Egg-shaped.

Ovule.—An unfertilized egg.

Palea(e).—A hyaline scale; specifically, in the grasses, the upper bract of two sterile bracts subtending a floret; in the Compositae, said of the scale-like pappus (fig. 345).

Paleaceous.—Scale-like.

Palmate.—Having several lobes radiating from a common base like the fingers from the palm of the hand (fig. 365).

Palmately compound.—Said of a leaf divided into discrete segments to a common basal area at the top of the petiole (fig. 365).

Palmately lobed. — Said of appendages when the lobes are so disposed as to appear to radiate from a common basal point (fig. 365).

Palmately veined.—Said of veins when they radiate from a common basal point.

Palustrine.—Occurring in marshy places.

Paludose.—Occurring in marshy places.

Panicle.—A compound inflorescence, that is, one in which the axis is branched one or more times (figs. 55, *h;* 366).

Papilionaceous.—Butterfly-like; said of the flowers of leguminous plants having a corolla composed of an upright banner and two lateral wings, each representing a single petal, and a keel comprised of two petals variously united (fig. 253, *d*).

Papilla(ae).—A short protuberance.

Papillate, papillose.—Bearing papillae (fig. 48, *a*).

Pappus.—The chaffy, scaly, bristle-like, or plumose structure at the junction of the achene and the corolla in the Compositae (fig. 345).

Papule.—A nipple-like projection.

Parallel-veined.—Said of an organ in which the veins are so placed relative to one another that they approximate parallel lines.

Parietal placentation.—Said of ovaries in which the seeds are borne on structures on the ovary wall, or on structures raised from the ovary wall (fig. 367).

Parted.—Cleft to below the middle.

Pectinate.—Said of an organ which is cleft into divisions in such a way as to resemble a comb (fig. 364).

Pedicel.—Stalk or stem of a flower in a flower cluster (fig. 366).

Peduncle.—The stem of a solitary flower or the main stem of a flower cluster (fig. 366).

Pellucid-punctate. — Having translucent dots.

Peltate.—Said of a plane structure that is attached at a point on its surface rather than on the margin, such as the leaf of the garden nasturtium, *Tropaeolum* (fig. 229, *h*).

Pendulous.—Hanging.

Penicillate.—Like a brush, such as the tuft of hairs on the style in certain vetches, or at the tip of the phyllaries in certain Compositae.

Perennial.—Living three or more seasons.

Perfect.—Said of flowers that have both stamens and pistils.

Perfoliate.—Said of opposite or whorled bracts or leaves that are united into a collar-like structure around the stem that bears them (fig. 365).

Perianth.—The nonessential appendages of the flower situated outside the stamen whorl, and including both sepals and petals or other segments homologous with them.

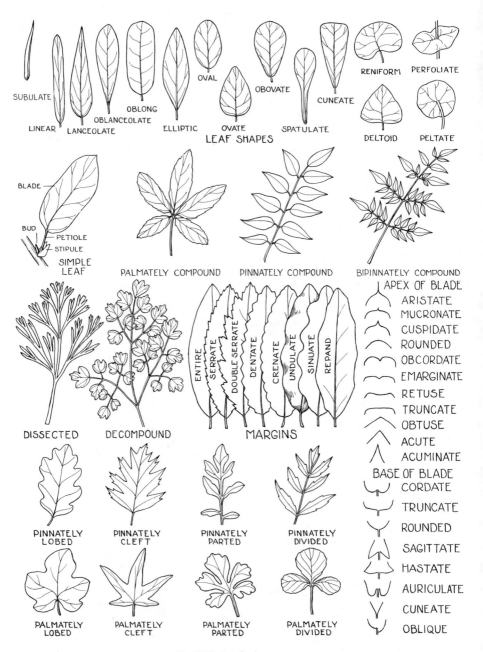

Fig. 365. Leaf characters.

Pericarp.—The ovary wall as it matures in the fruit.

Perigynium.—The sac or sheath enveloping the fruit in *Carex* and often appearing as though it were the ovary wall (fig. 86, *d*).

Perigynous.—Said of flowers in which the perianth and stamens are inserted on the receptacle around the gynoecium, or of flowers in which the ovary is partly embedded in the receptacle, e.g., the flowers of cherries and plums.

Persistent.—Said of an organ that remains attached after ceasing to perform its usual biological function.

Petal.—A unit segment of the corolla presumed to be homologous with a leaf.

Petaloid.—Having the form and structure and sometimes also the arrangement of petals (fig. 195, *a*).

Petiole.—The stem or stalk of a leaf (fig. 365).

Petiolule.—The stalk of a discrete segment of a compound leaf.

Phyllary.—A bract of the involucre in the Compositae (fig. 345).

Pilose.—Having a vestiture or pubescence of scattered, long, slender, but not harsh, hairs (fig. 363).

Pilosulous.—Finely pilose.

Pinna(ae).—A leaflet or primary segment of a pinnately compound leaf.

Pinnate. — Having a common elongate rachis or axis, with segments arranged either oppositely or alternately along either side (fig. 252, *c*).

Pinnately compound. — Said of structures the lateral segments of which are discrete and arranged along a common axis (fig. 365).

Pinnatifid.—Cleft in a pinnate manner.

Pinnule.—A secondary pinnately disposed part of a twice or more pinnately compound leaf.

Pistil.—One of the essential organs of a flower, consisting usually of stigma, style, and ovary, the ovary containing the ovule or ovules.

Pistillate.—Bearing pistils only.

Placenta.—The structure or tissue in the ovary bearing the ovules.

Placentation.—Disposition of the ovules on placentae within the ovary.

Plaited.—With more or less equal lengthwise folds; plicate.

Plane.—With a flat surface; projected in the manner of a plane from some designated point or level.

Plicate.—Repeatedly folded, usually lengthwise, though not necessarily so; plaited.

Plumose.—With hairlike branches as in a feather; said of pappus segments and sometimes of branched hairs (fig. 345).

Pollen.—The powdery grains which bear the sperm nuclei and which are contained in the anther.

Pollinium (pollinia).—A mass of coherent pollen grains characteristic of orchids and milkweeds (fig. 193, *g*).

Polygamous. — Having perfect, pistillate, and staminate flowers on an individual plant.

Polymorphous.—Occurring in more than one form.

Procumbent.—Trailing or lying flat, but not rooting.

Proliferous.—Bearing supplementary structures such as buds or flowers, either in an abnormal manner, or in a manner that is normal but from adventitious tissues.

Prostrate.—Prone, said of stems or leaves that lie on the ground.

Pruinose.—Having a waxy secretion forming a powdery covering.

Puberulent. — Covered with a pubescence of very fine, short hairs, not densely spaced.

Pubescent.—Hairy; the general term for hairiness (fig. 363).

Punctate.—With depressed dots scattered over the surface.

Puncticulate. — With very fine depressed dots (fig. 135, *f*).

Pungent.—Having a long, sharp point; or penetrating, as said of an odor.

Pustulate hair.—Hair with a bulbous base (fig. 364).

Quadrate.—Four-angled in cross section.

Raceme.—An inflorescence with a single axis, the flowers arranged along it on pedicels (fig. 366).

Racemose. — Characterized by having its parts disposed as in a raceme.

Rachilla.—The axis of a spikelet, bearing the florets in the Gramineae (fig. 66, *e*).

Rachis.—The prolongation of the peduncle through a flower cluster, or of a petiole through a compound leaf (figs. 72, *a;* 127, *f*).

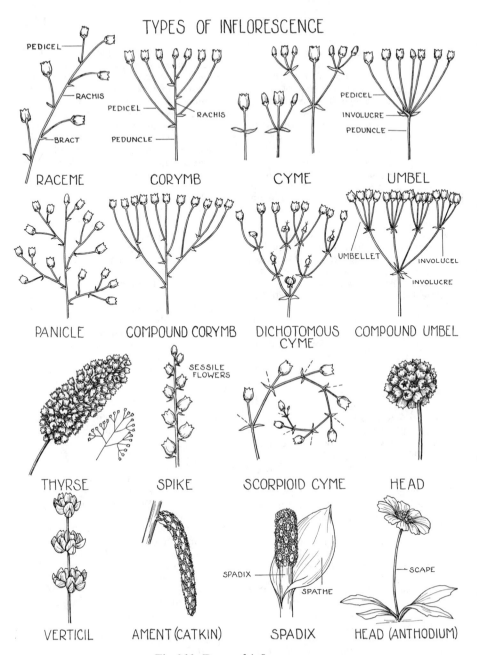

Fig. 366. Types of inflorescences.

Radiate. — Disposed in a plane from a common center like the spokes of a wheel; also said of inflorescences in which the marginal corollas spread in this manner (figs. 228, *c;* 345).

Ray.—Any spreading segment of a radiate structure; often used specifically for the marginal, ligulate corollas of the Compositae (fig. 345).

Ray flower.—One of the marginal flowers in Compositae when the flower head bears ligulate corollas, in contrast to the tubular disc corollas of the central flowers (fig. 345).

Receptacle.—The axis of a flower and its toral proliferation that bears the flower parts; term also applied to the discoid platform which bears a group of flowers in the headlike inflorescence of members of the Compositae (figs. 195, *c;* 345).

Recurved.—Turned backward in a curving manner (fig. 47, *a*).

Reflexed.—Turned backward from the point of attachment.

Regular.—Said of a flower or of parts of a flower when the members within each whorl are alike.

Remote.—Separated from one another.

Reniform.—Kidney-shaped (fig. 365).

Repand. — Exhibiting a slightly uneven margin when viewed at right angles to the plane (fig. 365).

Reticulate.—Netted (fig. 255, *e*).

Retrorse.—Having hairs or other processes turned or pressed toward the base (fig. 345).

Retuse.—With a shallow, rounded notch at the apex (fig. 365).

Revolute.—Said of margins that are rolled backward (toward the abaxial side), as in some leaves.

Rhizome.—A horizontal underground stem.

Rib.—A thickened structure, usually surrounding a primary vein on a leaf; also, one of more or less parallel ridges on fruits, seeds, or stems.

Rootstock. — A horizontal underground stem bearing both roots and aerial stems along its axis or from its tip.

Rosette. — A group of organs, such as leaves, clustered and crowded around a common point of attachment.

Rosulate.—Having the form of a rosette.

Rotate.—Radiately spreading in one plane.

Rudimentary. — Said of organs in which development has been arrested.

Rugose.—With a wrinkled surface (figs. 154, *a;* 243, *g*).

Runcinate.—Deeply incised, with the segments directed toward the base.

Runner.—A horizontal stem with long internodes that trails along the surface of the ground (fig. 367). See *Stolon*.

Saccate.—Having a saclike swelling; said of petals or sepals and sometimes of stamens and leaves (fig. 309, *i*).

Sagittate.—Shaped like an arrowhead; said of the basal margins of a leaf which are drawn into points on either side of the petiole (fig. 52).

Salverform.—Said of a corolla in which the tube is essentially cylindrical and the lobes are rotately spreading.

Samara.—A dry, indehiscent, one-seeded fruit bearing a wing, or two wings in a double samara (fig. 367).

Scaberulous.—Finely scabrous.

Scabrellate.—Rough as a result of minute surface protuberances.

Scabrid.—Slightly rough.

Scabrous.—Rough; said of a surface that is rough and harsh to the touch.

Scale.—A small, thin, platelike lamina arising from the surface of a stem or other organ.

Scale leaf.—A leaf having the size and form of a scale.

Scandent.—Climbing without aid of tendrils.

Scape.—An erect, naked peduncle of an acaulescent plant arising at the surface or from below the surface of the ground (fig. 366).

Scapose.—Bearing a scape.

Scarious.—Thin and membranous, usually dry.

Scorpioid.—Said of structures that grow as though uncoiling (fig. 297, *b*).

Scurfy.—Having flakes or scales adhering to the surface.

Secund.—Disposed on one side of a stem.

Segment.—An ultimate natural division of an organ or whorl of organs.

Sepal.—One of the segments of the calyx.

Septate.—Partitioned by walls (fig. 180, *f*).

Septicidal.—Said of carpels dehiscing at their junction (fig. 367). Compare *Loculicidal*.

Septum.—The wall which separates cavities, such as those of the ovary.

Sericeous.—Covered with soft, silky hairs, which usually point in one direction

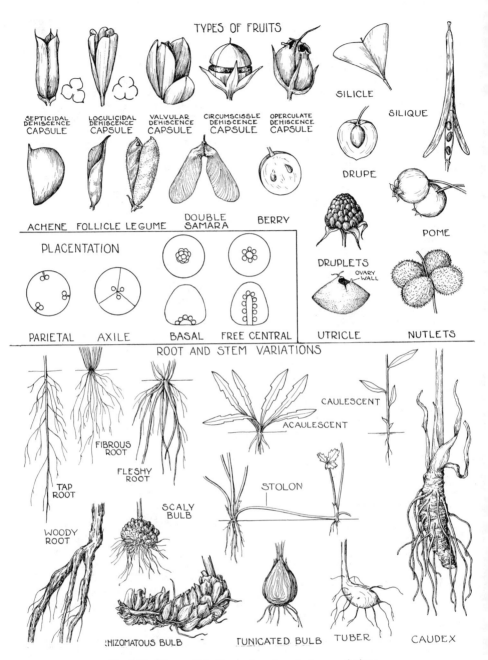

Fig. 367. Types of fruits, and root and stem variations.

Sericeous (cont.)
and thus give the appearance of silk (fig. 363).

Serrate.—Having marginal teeth pointing forward (fig. 365).

Serrulate. — Having very small marginal teeth; minutely serrate.

Sessile.—Joined directly by the base without a stalk, pedicel, or petiole (fig. 10, *f*).

Seta.—A bristle (fig. 364).

Setaceous.—Bearing bristles.

Setose.—Bearing bristles.

Sheath.—The basal part of a lateral organ that closely surrounds or invests the stem (fig. 55, *b*).

Silicle.—A short, two-celled fruit, usually broader than long, composed of two valves which separate from a central partition; a shortened silique (fig. 367).

Silique.—An elongate, many-seeded, two-celled fruit with two parietal placentae, usually with two valves that separate from the partition on dehiscence; occasionally indehiscent (fig. 367).

Silky.—Having the appearance or texture of silk.

Simple.—Neither branched nor otherwise compound.

Sinuate.—Having a wavy margin in the plane of the blade (fig. 365).

Sinus.—The angle between lobes, as between two sepals of a synsepalous calyx, or petals of a sympetalous corolla.

Sori (*sorus*).—The clusters of sporangia appearing as dots on the back of the leaf of a fern.

Spadix. — The spikelike inflorescence enclosed in a spathe (figs. 158, *b*; 366).

Spathe.—A sheathing lateral organ or pair of organs usually open on one side and enclosing an inflorescence (figs. 38, *d*; 158; 366).

Spatulate.—Shaped like a spatula, that is, gradually widening above and rounded at the tip (figs. 257, *d*; 365).

Spheroidal.—Shaped like a sphere or ball.

Spicate.—Arranged in such a way as to resemble a spike.

Spike.—A type of inflorescence in which the axis is somewhat elongated and the flowers are numerous and sessile (figs. 7, *l–n*; 366).

Spikelet.—The segment of the inflorescence

of grasses enclosed by a pair of glumes.

Spine. — A rigid, sharp-pointed structure usually modified from a stem.

Spinescent.—Bearing spines.

Spinose.—Bearing spines.

Spinule.—A diminutive spine.

Spur.—Any hollow, elongate, pointed or blunt outgrowth of the corolla or calyx, as in *Delphinium* (fig. 193, *d*).

Squamellate.—Having a vesture of scales (fig. 364).

Squamellae.—Small scales.

Squarrose.—Having thickly crowded but spreading and rigid leaves, bracts, or other processes.

Stalk.—A short or elongate structure bearing or supporting another structure.

Stamen.—The pollen-bearing organ; usually consisting of the stalk or filament and the anther containing the pollen.

Staminate.—Said of plants or structures bearing stamens and not bearing pistils.

Staminodium.—A sterile organ in the stamen whorl, presumed to be of staminal origin.

Standard. — The broad, usually upright, petal in a papilionaceous flower such as that of the pea (fig. 254, *d*).

Stellate. — Star-shaped; said of hairs or scales that branch in such a manner as to radiate from a central point of attachment (figs. 261, *d*; 364).

Stem.—The part of the plant bearing the foliar and floral organs and composed of nodes and internodes, or the latter much reduced.

Sterile.—Not fertile; said of flowers which for any reason are not bearing fruit. Sometimes, but not correctly, applied to staminate flowers.

Stigma.—That part of the pistil that receives the pollen and in which pollination is effected (figs. 2 and 3).

Stigmatic.—Pertaining to the stigma.

Stipe.—The stalk of an ovary, or, in the Compositae, of an achene (fig. 345).

Stipitate.—Said of glands or of ovaries when they are borne on stalks.

Stipule.—An appendage frequently occurring at the base of a leaf.

Stolon.—A stem with elongate internodes that trails along the surface of the ground, often rooting at the nodes. See *Runner*.

Stoloniferous.—Bearing stolons, as in the strawberry.

Stramineous.—Straw-colored.

Striate.—Marked with fine parallel lines (figs. 162, *j* and *k*; 168, *b*).

Strict.—Erect and straight.

Strigose.—Characterized by stiff, often appressed hairs, these usually pointing in one direction (figs. 202, *f; 363*).

Strophiolate. — Said of seeds having a strophiole.

Strophiole.—An appendage at the hilum of some seeds.

Style.—A short or long, simple or branched stalk arising from the ovary and bearing the stigma or stigmas; the part of the pistil which connects ovary and stigma (figs. 2 and 3).

Stylopodium.—An outgrowth from the base of the style and covering the top of the ovary, as in the Umbelliferae (fig. 285, *c*).

Subacute.—Between acute and obtuse.

Subcoriaceous.—Somewhat leathery.

Subulate.—Shaped like an awl or prong (fig. 365).

Succulent. — Fleshy; composed of soft, watery tissue.

Suffrutescent.—Woody, at least at the base; said of a subshrub.

Suffruticose.—Woody at base and definitely herbaceous above.

Sulcate.—Bearing grooves.

Superior.—Wholly above and not adnate to other organs, as a superior ovary (fig. 2).

Suture.—A groove marking the line along which a structure opens; any lengthwise groove that forms a junction between two parts.

Sympetalous.—A corolla in which the petals are united.

Syncarpous.—Having the carpels fused to form a compound pistil.

Synsepalous.—Having the sepals united.

Taproot. — A single, main primary root bearing small lateral roots (fig. 367).

Tawny.—Of the color of natural leather, light brown with a dull reddish or yellowish hue.

Taxon.—A concept of a class of organisms at any categorical level that is predicated on the similarity of the constituents to one another with respect to a set of properties taken as the defining type of the class.

Tendril.—A modified stem, leaf, or part of a leaf, in a climbing plant, that twines around an object and thus supports the plant.

Terete.—Cylindrical or tapering, and circular in cross section, as said of the stem of a plant.

Ternate. — Thrice-forked, -branched, or -divided.

Tessellate.—Having a checkerboard appearance or pattern.

Throat.—The usually expanded part of the corolla tube immediately below the lobes.

Thyrse.—A compact, compound panicle with an indeterminate axis (fig 366).

Tomentose.—Densely covered with short, matted hairs (fig. 363).

Tomentum.—Pubescence of densely matted, short hairs.

Toothed.—Bearing teeth.

Torus. — Specifically, the structure of a flower surrounding the gynoecium and bearing the other flower parts on its margin; may be used for any tubular outgrowth that originates as a ring of growth from the meristem.

Trichome.—A hair or hairlike structure.

Tricostate.—Having three ribs.

Trifid.—Cleft into three parts.

Trifoliate.—Bearing three leaves, as does *Trillium.*

Trifoliolate.—Having a leaf comprised of three leaflets, as does clover.

Trigonous.—Said of an achene or other structure which is three-sided or triangular in cross section (fig. 112, *c*).

Triquetrous.—Having three projecting angles.

Truncate.—Cut squarely across, either at the base or at the apex (fig. 365).

Tuber.—An enlarged, fleshy, underground stem, such as the potato (fig. 131, *e*).

Tubercle.—A nodule on the surface, or a thickened, solid or spongy crown or cap, as on an achene (fig. 136, *c*).

Tuberculate.—Bearing tubercles on the surface.

Tubular.—Cylindrical and hollow.

Tufted.—With a dense cluster of elongate structures spreading from what appears as a common point of attachment.

Tumid.—Swollen.

Tunicated.—Said of a bulb having its scales arranged in concentric layers as in the bulb of an onion (fig. 367).

Turbinate. — Inversely conical, or top-shaped.

Umbel.—An inflorescence of few to many flowers on stalks of approximately equal length arising from the top of a scape or peduncle (fig. 366).

Umbellate.—Arranged like an umbel (fig. 128, *a*).

Umbellet.—The umbel-like segment of a compound umbel (fig. 366).

Undulate.—Having a wavy margin; said of leaves, petals, or sepals (fig. 365).

Unisexual.—In flowering plants, said of a plant or flower that either bears only stamens or only pistils, but not both.

Urceolate.—Said of a corolla with united petals and a tube that is expanded below the middle and narrowed at the top.

Utricle.—A usually one-seeded, indehiscent fruit with a thin, bladdery, persistent ovary wall (fig. 367).

Valvate.—Said of appendages (leaves, sepals, petals, or carpels) when arranged margin to margin with adjacent structures in the bud, or in the fruit.

Valve.—A rigid or semirigid segment joined to adjacent structures by its margin.

Vascular bundles. — The elements of the conducting or vascular system of a plant.

Veins.—The ultimate branches or divisions of the vascular system, as in leaves or petals.

Velum.—A veil; a membranous indusium, as in *Isoetes* (fig. 5, *d*).

Venation.—The pattern of the veins in an organ.

Velutinous.—Having pubescence which is velvety in texture (fig. 363).

Venous.—Characterized by having more or less conspicuous veins.

Ventral. — Pertaining to the under (or front) side of a structure. See *Adaxial.*

Ventricose.—Swelling unequally or inflated on one side.

Vernal.—Occurring in springtime; said of pools that contain water in spring and are dry in summer.

Vernation.—The arrangement of leaves in the bud.

Verrucose.—Having wartlike nodules on the surface.

Versatile.—Turning freely on its support, as an anther attached crosswise at its mid-point to the apex of the filament.

Verticil.—A ring of organs or flowers at a node; a whorl.

Verticillate.—Occurring in verticils (fig. 366).

Villous.—Densely beset with shaggy hairs.

Virgate.—Long, slender, and straight, like a wand.

Viscid.—Sticky.

Whorl.—A ring of leaves, flower parts, or flowers occurring at a single node (fig. 366). See *Verticil.*

Whorled.—Occurring in a whorl (fig. 43, *a*).

Wing.—A thin, flat extension from an angle or margin (fig. 100, *a*).

Woolly.—Densely beset with wavy, curly, or twisted hairs (fig. 361, *a*).

Zygomorphic. — Having the members of any whorl unlike; irregular (fig. 308, *c* and *d*).

Index

Synonyms are in italic type. If a name is both a legitimate name and a synonym, its use as a synonym is indicated by page numbers in italic.

Abama californicum, 383
Acer, 568
  circinatum, 569
  glabrum, 569
  macrophyllum, 569
  Negundo var. californicum, 569
Aceraceae, 568
Achillea, 775
  borealis subsp. californica, 775, fig. 346
  californica, 775
  millefolium
    var. *californica,* 775
    var. *lanulosa,* 775
Aconitum columbianum, 495
Acorus calamus, 324
Agoseris, 775
  elata, 777
  glauca var. laciniata, 777
  gracilens, 777
Agrostis, 127
  alba, 129, fig. 55
  exarata, 129, fig. 54

  longiligula, 127
  scabra, 129
  semiverticillata, 131, fig. 56
  stolonifera, 129, 131
  verticillata, 131
Aira
  caespitosa, 143
  danthonioides, 143
  elongata, 144
Aizoaceae, 475
Alder, 409
  Mountain, 410
  Red, 410
  Sitka, 410
  White, 410
Alisma, 103
  brevipes, 105
  Geyeri, 105, fig. 44
  gramineum var. *Geyeri,* 105
  Plantago-aquatica subsp. brevipes, 105, fig. 43
Alismaceae, 103

Alkali weed, 657
Allenrolfea occidentalis, 451, fig. 213
Allium validum, 377
Allocarya, 667
    acanthocarpa, 669
    bracteata, 671
    Cusickii, 670
        var. *vallicola*, 671
    glabra, 670
    glyptocarpa, 669
    humistrata, 668
        var. *similis*, 668
    mollis, 668
    stipitata, 669
        var. micrantha, 670
    tenera, 671
    undulata, 670
Alnus, 409
    oregona, 410
    rhombifolia, 410
    rubra, 410
    sinuata, 410
    sitchensis, 410
    tenuifolia, 410
    viridis var. *sinuata*, 410
Alopecurus, 131
    aequalis, 135, fig. 57
    californicus, 136
    geniculatus, 135, fig. 58
    saccatus, 136
Amaranthaceae, 472
Amaranthus, 473
    californicus, 473
    carneus, 473
Ambrosia psilostachya, 778
Ambylopappus. pusillus, 777
Ammannia, 595
    auriculata, 597, fig. 273
    coccinea, 595, fig. 273
Andropogon, 136
    glomeratus, 136
    macrourum, 136
Anemopsis californica, 397, fig. 195
Angelica, 620
    genuflexa, 620
    Kneeling, 620
Anthemis Cotula, 778
Anthochloa colusana, 176
Anthoxanthum odoratum, 137
Apargidium boreale, 778
Apium graveolens, 621
Aquilegia, 495
    eximia, 495
    Tracyi, 495
Araceae, 324
Aralia californica, 619

Araliaceae, 617
Arenaria, 483
    paludicola, 483, fig. 226
    serpyllifolia, 483
Arnica, 779
    Chamissonis, 779
        subsp. foliosa, 779
        subsp. incana, 779
    foliosa
        var. *bernardina*, 779
        var. *incana*, 779
        var. *Sonnei*, 779
Arrow-weed, 829
    Family, 95
Artemisia, 780
    biennis, 780
    Douglasiana, 781
    heterophylla, 781
    vulgaris var. *heterophylla*, 781
Arthrocnemum subterminale, 451, fig. 214
Arum family, 324
Aruncus, 543
    sylvester, 543
    vulgaris, 543
Arundo donax, 137
Ash, 649
    Arizona, 650
    Oregon, 649
Aspen, 398
    Quaking, 399
Aster, 781
    adscendens, 787
        var. *Fremontii*, 788
        var. *yosemitanus*, 788
    Alkali
        Marsh, 791
        Shrubby, 784
    alpigenus subsp. Andersonii, 783
    barnardinus, 787
    canescens var. *tephrodes*, 793
    carnosus, 784
    chilensis, 786
        var. invenustus, 787
        var. lentus, 786
        var. media, 787
        var. sonomensis, 786
    coerulescens, 784
    Common California, 786
    defoliatus, 787
    Douglasii, 785
    Eatonii, 785
    elatus, 783
    exilis, 791, fig. 347
    foliaceus, 789
        var. apricus, 789
        var. *Eatonii*, 785

var. *hesperidus,* 784
var. Parryi, 789
frondosus, 791
hesperius, 784
intricatus, 784
Marsh, 784
Menziesii, 786
occidentalis, 788
var. delectabilis, 789
var. Parishii, 789
var. yosemitanus, 788
oreganus, 785
paludicola, 791
pauciflorus, 791
peregrinus, 810
spinosus, 784
subspicatus, 785
tephrodes, 793
Western bog, 791
Yosemite, 788
Athyrium filix-femina, 27
Atriplex, 452
argentea, 456
bracteosa, 455
cordulata, 455
var. *tularensis,* 456
coronata, 457
Coulteri, 457
fruticulosa, 458
microcarpa, 457
pacifica, 457
Parishii, 457
patula, 453
Phyllostegia, 453
pusilla, 456
rosea, 456
semibaccata, 458
Serenana, 455
truncata, 455
tularensis, 456
vallicola, 455
Avens, Big-leaf, 543
Awlwort, Water, 531
Azalea, Western, 639
Azolla, 31
filiculoides, 31, fig. 6
mexicana, 33, fig. 6

Baccharis, 793
Douglasii, 795, fig. 348
glutinosa, 795
viminea, 795
Bacopa, 702
Eisenii, 702, fig. 308
Nobsiana, 703, fig. 308
rotundifolia, 702

Baeria, 797
chrysostoma, 797
Fremontii, 797, fig. 349
platycarpha, 797
Barbarea, 521
americana, 521
vulgaris, 521
Barley, 165
Foxtail, 167
Meadow, 167
Bassia hyssopifolia, 461
Batidaceae, 473
Batis maritima, 473
Bayberry family, 408
Beckmannia Syzigachne, 137, fig. 59
Beet, Common, 458
Beggar-ticks, 799
Bergia texana, 579, fig. 264
Berula erecta, 621
Beta vulgaris, 458
Betula, 411
fontinalis, 411
glandulosa, 411
occidentalis f. *inopina,* 411
Betulaceae, 409
Bidens, 799
cernua, 799, fig. 350
frondosa, 801, fig. 352
laevis, 801, fig. 351
pilosa, 802
Bigelovia intricata, 784
Bilberry, 639
Birch, 411
Family, 409
Water, 411
Black-eyed Susan, 833
Bladderwort, Common, 733
Blueberry, 639
Bluegrass, 192
Bluestem, 136
Bog asphodel, 383
Boisduvalia, 603
cleistogama, 604
densiflora, 603
glabella, 604
macrantha, 603
stricta, 604
Bolandra californica, 536
Borage family, 666
Boraginaceae, 666
Box elder, 569
Boykinia, 536
elata, 537
major, 536
rotundifolia, 537
Brasenia Schreberi, 491, fig. 229

Brass buttons, 807
Brickellia californica, 802
Brodiaea, 377
  hyacinthina, 379
  peduncularis, 379
Buckwheat family, 413
Bulrush, 303
  California, 319
  Great, 323
  River, 307
Bur-reed
  Broad-fruited, 45
  Family, 43
  Simple-stemmed, 47
Buttercup
  Aquatic, 517
  Creeping, 511
  Cursed, 509
  Desert, 517
  Dwarf, 513
  Spiny, 513

Cabombaceae, 491
Calamagrostis, 139
  Bolanderi, 139
  crassiglumis, 139
  nutkaensis, 141
Calla, 327
  aethiopica, 327
  palustris, 327
Callitrichaceae, 555
Callitriche, 555
  autumnalis, 557
    var. *bicarpellaris,* 557
  hermaphroditica, 557, fig. 256
    var. bicarpellaris, 557
  longipedunculata, 557, fig. 257
  marginata, 561, fig. 258
    var. *longipedunculata,* 557
  palustris, 557, fig. 255
    var. *Bolanderi,* 557
    var. *stenocarpa,* 557
Calochortus uniflorus, 379, fig. 185
Caltha biflora subsp. Howellii, 495, fig. 231
Calycanthaceae, 519
Calycanthus occidentalis, 519, fig. 245
Camass
  Common, 381
  Death, 386
Camassia, 379
  Leichtlinii subsp. Suksdorfii, 381
  quamash, 381, fig. 186
Camomile, 778
Campanula linnaeifolia, 745, fig. 330
Campanulaceae, 745
Canchalagua, 651

Caprifoliaceae, 742
Cardamine, 521
  bellidifolia, 522
  Breweri, 522
  Gambelii, 522
  Lyallii, 522
  oligosperma, 523
  pennsylvanica, 523
Carex, 204
  ablata, 247
  amplifolia, 245
  arcta, 235, fig. 105
  atherodes, 255, fig. 117
  athrostachya, 231, fig. 101
  Barbarae, 213, fig. 90
  breviligulata, 235
  Buxbaumii, 253, fig. 116
  californica, 245, fig. 111
  comosa, 257, fig. 120
  Cusickii, 237, fig. 106
  debiliformis, 253
  densa, 233
  diandra, 237
  Douglasii, 221, fig. 96
  Dudleyi, 235
  exsiccata, 263, fig. 123
  feta, 229, fig. 100
  gynodynama, 249
  Harfordii, 231, fig. 103
  Hassei, 209, fig. 86
  Hindsii, 221, fig. 95
  hystricina, 257
  lanuginosa, 241, fig. 109
  Lemmonii, 247
  leptalea, 241, fig. 108
  livida, 245, fig. 110
  luzulina, 247, fig. 112
  Lyngbyei, 213, fig. 89
  mendocinensis, 253, fig. 115
  nebraskensis, 217, fig. 94
  obnupta, 209, fig. 88
  ormantha, 241
  pansa, 229, fig. 99
  phyllomanica, 241, fig. 107
  praegracilis, 223, fig. 98
  rostrata, 261, fig. 122
  salinaeformis, 209, fig. 87
  Schottii, 213, fig. 91
  senta, 217, fig. 92
  Sheldonii, 255, fig. 118
  simulata, 223, fig. 97
  sitchensis, 217, fig. 93
  sonomensis, 247, fig. 113
  spissa, 255, fig. 119
  stipata, 233, fig. 104
  sub-bracteata, 231, fig. 102

Tracyi, 231
vesicaria, 261, fig. 121
vicaria, 235
viridula, 249, fig. 114
Caryophyllaceae, 479
Castilleja, 703
   Culbertsonii, 705, fig. 309
   elata, 707, fig. 311
   exilis, 709, fig. 312
   Lemmonii, 705, fig. 309
   miniata, 707, fig. 311
   minor, 709, fig. 312
   spiralis, 709
   stenantha, 709, fig. 310
   uliginosa, 707, fig. 310
Cat-tail
   Common, 39
   Family, 36
   Hybrid, 41
   Narrow-leaved, 41
Celery, 621
Centaurium, 651
   exaltatum, 651
   floribundum, 651
Centromadia pungens, 803
Cephalanthus occidentalis, 741
Ceratophyllaceae, 493
Ceratophyllum demersum, 493, fig. 230
Chaetochloa
   geniculata, 201
   glauca, 199
   lutescens, 199
   verticillata, 201
Chenopodiaceae, 449
Chenopodium, 458
   album, 459
   ambrosioides, 459
   rubrum, 459
Chickweed family, 479
Chrysosplenium glechomaefolium, 537
Cicuta, 621
   Bolanderi, 622
   californica, 622
   Douglasii, 622
Cirsium, 803
   Breweri, 805
   hydrophyllum, 805
   Vaseyi var. hydrophyllum, 805
Cladium, 263
   californicum, 263
   mariscus var. californicum, 263
Clover, 553
   Cow, 554
   Elk, 619
   Sweet, 552
      White, 552
      Yellow, 552

White-tip, 554
Club rush, 303
   Low, 319
   Nevada, 315
Cnicus Breweri, 805
Cocklebur, 838
   Spiny, 838
Compositae, 769
Coneflower, 832
   California, 832
   Western, 833
Conioselinum, 622
   chinense, 622
   pacificum, 622
Conium maculatum, 623
Convolvulaceae, 655
Convolvulus, 657
   sepium
      var. Binghamiae, 657
      var. repens, 657
Conyza Coulteri, 805
Cordylanthus, 711
   canescens, 713
   maritimus, 711
      var. canescens, 713
      var. Parryi, 713
   mollis, 713
Cornaceae, 634
Cornus, 635
   californica, 635
   glabrata, 636
   sessilis, 635
   stolonifera var. californica, 635
Cotton-batting plant, 812
Cottonwood, Smooth-barked, 398
Cotula coronopifolia, 807, fig. 353
Crassulaceae, 535
Cress
   American winter, 521
   Swamp, 522
   Water, 527
Cressa, 657
   cretica, 657
   truxillensis, 657
Crownbeard, 837
Cruciferae, 521
Crypsis, 141
   aculeata, 141
   niliaca, 141, fig. 60
Cudweed, 811
   Lowland, 812
   Purple, 812
Cuscuta salina, 657, fig. 294
Cut-grass, Rice, 169
Cynodon dactylon, 141
Cynosurus cristatus, 142
Cyperaceae, 203

Cyperus, 263
  acuminatus, 273, fig. 129
  aristatus, 273, fig. 128
  Coarse, 279
  difformis, 273, fig. 130
  Eragrostis, 269, fig. 127
  erythrorhizos, 277, fig. 133
  esculentus, 273, fig. 131
  ferax, 279, fig. 135
  inflexus, 273
  laevigatus, 269, fig. 126
  lateriflorus, 273
  melanostachys, 265
  niger var. capitatus, 265, fig. 124
  Parishii, 277
  rivularis, 269, fig. 125
  rotundus, 277, fig. 132
  Shining, 269
  sphacelatus, 277
  strigosus, 279, fig. 134
  vegetus, 269
  virens, 269
Cypripedium californicum, 389, fig. 191
Cypselea humifusa, 475

Dactylis glomerata, 142
Damasonium californicum, 107, fig. 45
Dandelion, Common, 837
Darlingtonia californica, 531, fig. 247
Datisca glomerata, 593
Datiscaceae, 593
Delphinium, 497
  scopulorum var. glaucum, 497
  uliginosum, 497
Dentaria, 523
  californica, 524
    var. *integrifolia,* 524
  gemmata, 523
  integrifolia, 524
    var. *californica,* 524
  pachystigma, 523
Deschampsia, 142
  caespitosa, 143
  danthonioides, 143
    var. *gracilis,* 143
  elongata, 144
Desert olive, 649
Devil-weed, Mexican, 784
Digitaria sanguinalis, 145
Distichlis, 145
  spicata, 147, fig. 61
  texana, 145
Ditch grass, 81
  Family, 81
Dock
  Bitter, 445

Curly-leaved, 443
Fiddle, 443
Golden, 447
Western, 442
Willow, 439
Dodder, 657
  Salt-marsh, 657
Dodecatheon, 641
  alpinum, 643
  Jeffreyi, 643
    var. *redolens,* 644
  redolens, 644
Dog fennel, 778
Dogtail, Crested, 142
Dogwood, 635
  Brown, 636
  Creek, 635
Dopatrium junceum, 715, fig. 313
Downingia, 747
  bella, 757, fig. 336
  bicornuta, 755
    var. *montana,* 759
    var. picta, 755
  concolor, 755, fig. 335
    var. brevior, 757
    var. tricolor, 755
  cuspidata, 759
  elegans, 749, fig. 331
  humilis, 761
  immaculata, 759
  insignis, 749, fig. 332
  laeta, 761, fig. 340
  montana, 759, fig. 338
  ornatissima, 753, fig. 334
  pulchella, 753, fig. 333
    var. *arcana,* 759
  pusilla, 761, fig. 339
  yina, 757, fig. 337
Drosera, 533
  longifolia, 533, fig. 249
  rotundifolia, 533, fig. 248
Droseraceae, 533
Duckweed
  Family, 327
  Inflated, 331
  Stan, 329
Dulichium arundinaceum, 279

Echinochloa, 149
  colonum, 149, fig. 62
  crusgalli, 153, fig. 63
    var. frumentacea, 149
  pungens, 149
Echinodorus cordifolius, 109, fig. 46
Echinopsilon hyssopifolium, 461, fig. 216
Eclipta alba, 807, fig. 354

Eichornia crassipes, 345, fig. 167
Elatinaceae, 577
Elatine, 579
  ambigua, 583, fig. 266
  brachysperma, 587, fig. 269
  californica, 583, fig. 265
  chilensis, 585, fig. 268
  gracilis, 583, fig. 267
  heterandra, 587, fig. 271
  obovata, 587, fig. 270
  rubella, 585, fig. 266
  triandra, 585
    var. *brachysperma, 587*
    var. *genuina, 585*
    var. *obovata, 587*
Elderberry, 743
  Blue, 743
  Red, 743
Eleocharis, 281
  acicularis, 283, fig. 136; *282*
    var. *bella, 282*
    var. *occidentalis, 283*
  arenicola, 291
  atropurpurea, 291, fig. 140
  bella, 282
  Bolanderi, 289
  capitata, 293
  caribaea, 293
  coloradoensis, 285, fig. 136
  disciformis, 289
  Engelmannii, 293
  flaccida, 291
  flavescens, 291, fig. 139
  geniculata, 293, fig. 141
  leptos, 285
  Lindheimeri, 283
  macrostachya, 297, fig. 143
  montana, 291
  montevidensis, 291
  monticola, 293
  obtusa, 293, fig. 142
  pachycarpa, 287
  palustris, 297
  Parishii, 289, fig. 138
  parvula, 283, fig. 136
  pauciflora, 285
    var. bernardina, 287
    var. Suksdorfiana, 287
  quadrangulata, 297, fig. 144
  radicans, 283
  rostellata, 289, fig. 137
  Suksdorfiana, 287
  uniglumis, 297
Elodea, 121
  canadensis, 121, fig. 53
  densa, 123, fig. 53

Planchonii, 121
Elymus triticoides, 153, fig. 64
Epilobium, 604
  angustifolium, 605
  brevistylum, 607
  californicum, 607, fig. 277
  glaberrimum, 609
  latifolium, 605
  oregonense, 605
  Watsonii, 607
Epipactis gigantea, 391, fig. 192
Equisetaceae, 35
Equisetum, 35
  hiemale var. californicum, 36
  maximum, 36
  palustre, 36
  telemateia, 36
    var. *brauni, 36*
Eragrostis hypnoides, 154, fig. 65
Erianthus ravennae, 154
Ericaceae, 636
Erigeron, 809
  Andersonii, 783
  canadensis, 810
  Coulteri, 809
  linifolius, 810
  peregrinus, 810
  philadelphicus, 809, fig. 355
  salsuginosus, 810
Eriochloa, 155
  aristata, 156
  contracta, 155
  gracilis, 155
Eriophorum gracile, 299
Eryngium, 623
  alismaefolium, 624
  aristulatum, 624
  articulatum, 624
    var. *Bakeri, 624*
  castrense, 625
  globosum, 625
  Parishii, 624
  Vaseyi, 625
    var. castrense, 625
    var. globosum, 625
Eustoma, 652
  exaltatum, 652
  silenifolium, 652
Evax caulescens, 810
Evening primrose, 611
  Family, 602

Fern
  Family, 27
  Giant chain, 27
  Lady, 27

Fescue, 156
  Red, 156
Festuca rubra, 156
Fimbristylis, 299
  capillaris, 300
  thermalis, 300
  Vahlii, 300
Fireweed, 605
Fleabane, Salt-marsh, 829
Forestiera neo-mexicana, 649
Forget-me-not, True, 673
Foxtail, 131
  Pacific, 136
  Short-awn, 135
  Water, 135
Frankenia, 589
  Family, 589
  grandifolia, 589, fig. 272
  Palmeri, 589
Frankeniaceae, 589
Fraxinus, 649
  latifolia, 649
  oregona, 649
  velutina, 650
    var. coriacea, 650
Frogbit family, 119

Galium trifidum, 741, fig. 329
Gastridium ventricosum, 156
Gaultheria, 636
  humifusa, 637
  ovatifolia, 637
Gentian, 652
Gentiana, 652
  Amarella var. acuta, 653
  calycosa, 654
  Fremontii, 653
  holopetala, 653
  humilis, 653
  sceptrum, 654
  setigera, 654
  simplex, 653
Gentianaceae, 650
Geum macrophyllum, 543
Gilia prostrata, 663
Glaux maritima, 644, fig. 289
Glinus lotoides, 475, fig. 222
Glyceria, 157
  borealis, 161, fig. 68
  elata, 159
  grandis, 161, fig. 67
  leptostachya, 161, fig. 69
  occidentalis, 165, fig. 70
  pauciflora, 157, fig. 66
Gnaphalium, 811
  chilense, 812

  luteo-album, 813
  palustre, 812
  purpureum, 812
Goat's beard, 543
Goldenrod, 834
Gold-fields, 797
Goosefoot, 458
  Family, 449
  Red, 459
Gramineae, 123
Grape
  California wild, 571
  Family, 571
Grass
  Alkali, 197
    Weeping, 197
  Arrow, Slender, 99
  Barnyard, 153
  Bent, 127
    Creeping, 129
    Long-tongue, 127
    Spike, 129
    Spreading, 129
    Water, 131
  Bermuda, 141
  Bristle, 199
    Knotroot, 201
    Yellow, 199
  Bushy beard, 136
  Canary, 185
    California, 187
    Lemmon, 185
    Reed, 187
  Colusa, 176
  Cord, 202
    California, 202
  Cotton, Slender, 299
  Crab, 145
    Hairy, 145
  Cup, 155
    Bearded, 156
    Prairie, 155
    Southwestern, 155
  Cut, Rice, 169
  Dallis, 183
  Ditch, 81
    Family, 81
  Eel, 93
  Family, 123
  Finger, 145
  Golden-eyed, 389
    Sierra, 389
  Hair, 142
    Annual, 143
    Slender, 144
    Tufted, 143

Johnson, 201
Love, 154
    Creeping, 154
Manna, 157
    American, 161
    Northern, 161
    Northwestern, 165
    Slim-head, 161
    Tall, 159
    Weak, 157
Nit, 156
Nut, 277
    Yellow, 273
Orchard, 142
Plume, 154
Prickle, 141
Ravenna, 154
Reed, 139
    Pacific, 141
    Redwood, 139
Salt, 145, 147
Saw, 263
Semaphore, 191
    Nodding, 192
Shore, 173
Slough, American, 137
Star, Water, 345
Surf, 91
Teal, 154
Tickle, 129
Vernal, 137
    Sweet, 137
Water, 149
Wire, 415
Witch, 179
    Common, 179
Grass-of-Parnassus family, 541
Grass wrack, 89
Gratiola, 715
    ebracteata, 717, fig. 314
    heterosepala, 717, fig. 315
    neglecta, 715, fig. 314
    virginiana, 715
Greasewood, Black, 469
Grindelia, 813
    camporum, 814
    cuneifolia, 814
    robusta, 813
Ground cherry, 696
Groundsel, 833
Gum plant, 813

Habenaria, 391
    leucostachys, 391
    sparsiflora, 395, fig. 193
Haloragaceae, 613
Haplopappus racemosus, 814

Hawkbit, 827
Helenium, 815
    Bigelovii, 817, fig. 356
        var. *festivum,* 815
    Bolanderi, 815
    puberulum, 815
Heleochloa schoenoides, 165, fig. 60
Helianthus, 817
    annuus, 818
    Bolanderi, 819
    californicus, 819
        var. *mariposianus,* 819
        var. *Oliveri,* 821
    exilis, 819
    Nuttallii, 820
        var. Parishii, 821
    Oliveri, 821
    Parishii, 821
Heliotrope, Seaside, 671
Heliotropium curassavicum, 671, fig. 297
Hemicarpha, 300
    micrantha var. minor, 301
    occidentalis, 301
Hemizonia pungens, 803
Hemlock, Poison, 623
Hemp, Colorado River, 553
Hesperoscordum hyacinthinum, 379
Heteranthera dubia, 345, fig. 168
Hibiscus californicus, 571, fig. 261
Hippuris vulgaris, 613
Hordeum, 165
    boreale, 167
    brachyantherum, 167
    jubatum, 167
Hornwort, 493
Horsetail
    Family, 35
    Giant, 36
Hosackia
    gracilis, 549
    oblongifolia, 551
Howellia limosa, 765
Huckleberry, 639
    Bog, 640
    Dwarf, 640
    Mountain, 640
    Thin-leaved, 640
    Western, 640
Hydastylus
    californicus, 389
    Elmeri, 389
Hydrocharitaceae, 119
Hydrocotyle, 627
    ranunculoides, 627, fig. 286
    umbellata, 627
    verticillata, 627
Hypericaceae, 577

Hypericum, 577
  anagalloides, 577, fig. 263
  mutilum, 577

Ibidium
  porrifolium, 397
  Romanzoffianum, 395
Ilysanthes dubia, 723
Iodine bush, 451
Iridaceae, 387
Iris, 387
  Family, 387
  longipetala, 387
  missouriensis, 387
  pseudacorus, 387
  Western, 387
Isoetaceae, 33
Isoetes, 33
  Bolanderi, 34, fig. 5
  Braunii, 35
  Howellii, 34
  Nuttallii, 34
  occidentalis, 35
  Orcuttii, 34
Isolepis koilolepes, 319
Iva, 821
  axillaris, 823, fig. 357
  Hayesiana, 823

Jaumea carnosa, 823, fig. 358
Juncaceae, 347
Juncaginaceae, 95
Juncus, 347
  acuminatus, 361, fig. 177
  acutus var. sphaerocarpus, 349, fig. 169
  articulatus, 365, fig. 178
  balticus, 351, fig. 171
  Bolanderi, 361, fig. 176
  bufonius, 357, fig. 174
  Cooperi, 349
  Covillei, 375
  effusus
    var. brunneus, 355, fig. 172
    var. pacificus, 355, fig. 172
  ensifolius, 369, fig. 180
  falcatus, 373
  latifolius, 375
  Leseurii, 349, fig. 170
    var. *elatus,* 351
  longistylis, 375
  mexicanus, 353
  nevadensis, 365, fig. 179
  orthophyllus, 375, fig. 184
  oxymeris, 371, fig. 182
  patens, 357, fig. 173
  phaeocephalus, 369, fig. 181
    var. *glomeratus,* 369

  var. paniculatus, 371
  rugulosus, 359
  sphaerocarpus, 359
  supiniformis, 359
  textilis, 351
  Torreyi, 361, fig. 175
  xiphioides, 373, fig. 183
Jussiaea, 609
  californica, 609, fig. 278
  repens var. *peploides,* 609

Kalmia, 637
  polifolia, 638
    var. microphylla, 638
Knotgrass, 183
Knotweed, Floating, 419
Kochia, 461
  americana, 461
    var. *californica,* 463
  californica, 463

Labiatae, 679
Labrador tea, 638
Lactuca, 824
  pulchella, 825
  scariola, 825
Lady's slipper, California, 389
Lady's thumb, 427
Lamb's quarters, 459
Lasthenia, 825
  glaberrima, 827
  glabrata, 825, fig. 359
Lathyrus, 547
  Jepsonii, 549, fig. 253
  palustris, 549
Lauraceae, 519
Laurel
  Alpine, 638
  California, 519
  Family, 519
  Pale, 638
  Sierra, 639
Laurentia
  carnosula, 769
  debilis var. *serrata,* 767
Leather-root, 553
Ledum glandulosum, 638
Leersia oryzoides, 169, fig. 71
Legenere limosa, 765, fig. 341
Leguminosae, 547
Lemna, 329
  cyclostasa, 333
  gibba, 331, fig. 160
  minima, 334, fig. 162
  minor, 331, fig. 161
    var. *cyclostasa,* 333
  oligorrhiza, 335

paucicostata, 333
perpusilla, 333, fig. 161
trisulca, 329, fig. 159
valdiviana, 333
Lemnaceae, 327
Lentibulariaceae, 731
Leontodon nudicaulis, 827
Lepidium, 524
  acutidens, 525
  dictyotum, 525
    var. *acutidens,* 525
  latipes, 525
  oxycarpum, 525
    var. *acutidens,* 525
Leptochloa, 169
  fascicularis, 171, fig. 73
  filiformis, 169, fig. 72
  imbricata, 171
  uninervia, 171, fig. 74
Lepturus cylindricus, 173
Lettuce
  Blue, 825
  Prickly, 825
Leucosyris spinosa, 784
Leucothoe Davisiae, 639
Lilaea, 101
  scilloides, 101, fig. 42
  subulata, 101
Lilaeaceae, 101
Lilaeopsis, 631
  lineata var. *occidentalis,* 631
  occidentalis, 631, fig. 287
Liliaceae, 375
Lilium, 381
  pardalinum, 383
  parvum, 383
Lily
  Family, 375
  Leopard, 383
  Tiger, Small, 383
Limnanthaceae, 561
Limnanthes, 561
  alba, 567, fig. 260
    var. alba, 567
    var. *detonsa,* 567
    var. versicolor, 568
  Bakeri, 565
  Douglasii, 562
    var. Douglasii, 562
    var. nivea, 563, fig. 259
    var. rosea, 563
    var. sulphurea, 563
  floccosa, 568
  gracilis var. Parishii, 565
  Howelliana, 562
  montana, 567

rosea, 563
  var. *candida,* 562
  striata, 565
  versicolor, 568
    var. *Parishii,* 565
Limnorchis
  leucostachys, 391
  sparsiflora, 395
Limonium, 647
  californicum, 647
  commune var. californicum, 647
Limosella, 719
  acaulis, 719, fig. 317
  aquatica, 719, fig. 318
    var. *americana,* 719
    var *tenuifolia,* 719
  subulata, 719, fig. 316
Lindernia, 723
  anagallidea, 723, fig. 319
  dubia, 723
Lippia, 675
  lanceolata, 675
  nodiflora, 675, fig. 298
Lizard's-tail family, 397
Lobelia, 765
  cardinalis, 765
  splendens, 765, fig. 342
Lobeliaceae, 745
Lonicera, 742
  conjugialis, 743
  involucrata, 742
    var. Ledebourii, 742
Loosestrife
  Common, 599
  Creeping, 644
Lophotocarpus
  californicus, 111
  calycinus, 111
Lotus, 549
  formosissimus, 549, fig. 254
  oblongifolius, 551
  Stream, 551
Ludwigia, 609
  natans, 611
  palustris, 611, fig. 279
Lupine, 551
Lupinus polyphyllus, 551
Lycopus, 679
  americanus, 681, fig. 300
  lucidus, 681, fig. 301
  uniflorus, 681, fig. 299
Lysichitum, 325
  americanum, 325, fig. 158
  camtschatcense, 325
Lysimachia, 644
  Nummularia, 644, fig. 290

Lysimachia (*cont.*)
  thyrsiflora, 645, fig. 291
Lythraceae, 595
Lythrum, 597
  adsurgens, 599
  californicum, 599
  hyssopifolia, 599, fig. 274
  tribracteatum, 597

Machaerocarpus californicus, 107
Mallow
  Alkali, 573
  Family, 571
Malvaceae, 571
Maple
  Big-leaf, 569
  Sierra, 569
  Vine, 569
Mare's tail, 613
Marigold, Bur, 799
Mariscus californicus, 263
Marsilea, 29
  Family, 29
  mucronata, 29, fig. 4
  vestita, 29
Marsiliaceae, 29
Mayweed, 778
Meadow foam
  Family, 561
  Mountain, 567
Melilotus, 552
  alba, 552
  indica, 552
Mentha, 681
  arvensis, 687, fig. 305
  citrata, 685, fig. 302
  piperita, 685
  Pulegium, 687, fig. 304
  rotundifolia, 687, fig. 303
  spicata, 685
Menyanthaceae, 654
Menyanthes trifoliata, 655, fig. 293
Mesembryanthemum family, 475
Mexican tea, 459
Milfoil, 775
  American, 615
  Western, 615
Milkwort, Sea, 644
Millet, 199
Mimetanthe pilosa, 723
Mimulus, 723
  cardinalis, 725
  guttatus, 725, fig. 320
  inodoratus, 724
  Lewisii, 725
  moschatus, 724
    var. sessilifolius, 724

pilosus, 723
  primuloides, 724
Mint
  Apple, 687
  Bergamot, 685
  Family, 679
  Field, 687
Mitella
  ovalis, 539
  pentandra, 539
Mollugo verticillata, 476
Monanthochloe littoralis, 173
Monerma cylindrica, 173, fig. 75
Monkey flower, Scarlet, 725
Monochoria vaginalis, 347
Monolepis Nuttalliana, 463
Montia, 477
  Chamissoi, 477, fig. 224
  fontana, 477, fig. 223
  sibirica, 479, fig. 225
Morning-glory, 657
Mudwort, 719
Muhlenbergia, 173
  asperifolia, 175
  californica, 175
  filiformis, 176
Muhly, 173
  Alkali, 175
  California, 175
  Pull-up, 176
Mule fat, 795
Musk-flower, 724
Myosotis, 673
  laxa, 673
  palustris, 673
  scorpioides, 673
Myosurus, 497
  alopecuroides, 505
  aristatus, 501, fig. 234
    subsp. montanus, 503, fig. 234
  cupulatus, 501, fig. 233
  lepturus var. *filiformis,* 501
  major, 501
  minimum, 499, fig. 232
    var. apus, 501
    var. filiformis, 501, fig. 233
    subsp. *montanus,* 503
    var. *sessiliflorus,* 503
  sessilis, 503, fig. 235
    subsp. alopecuroides, 505, fig. 235
Myrica, 408
  californica, 409
  Hartwegii, 409
Myricaceae, 408
Myriophyllum, 613
  brasiliense, 615, fig. 280
  exalbescens, 615, fig. 281

hippurioides, 615, fig. 282
spicatum var. *exalbescens,* 615
verticillatum, 617, fig. 283

Naias
flexilis var. *guadalupensis,* 87
graminea var. *Delilei,* 85
microdon var. *guadalupensis,* 87
Najadaceae, 83
Najas, 85
flexilis, 87, fig. 35
graminea, 85, fig. 34
guadalupensis, 87, fig. 36
marina, 85, fig. 33
Narthecium californicum, 383, fig. 187
Naumburgia thyrsiflora, 645
Navarretia, 659
Bakeri, 663, fig. 295
cotulaefolia, 665
Jepsonii, 665
leucocephala, 660
minima, 660
pauciflora, 661
plieantha, 661
prostrata, 663, fig. 296
Neostapfia colusana, 176
Nettle, 412
Nicotiana, 695
attenuata, 695
Bigelovii, 696
Nightshade family, 695
Ninebark, 543
Nitrophila occidentalis, 463
Nuphar polysepalum, 489, fig. 228
Nymphaea polysepala, 489
Nymphaeaceae, 489

Oenanthe sarmentosa, 631, fig. 288
Oenothera, 611
heteranthera, 611
Hookeri, 611
subacaulis, 611
Oleaceae, 649
Onagraceae, 602
Onion, Swamp, 377
Orchid
Bog
Green, 395
White, 391
Family, 389
Spiral, 395
Hooded, 395
Western, 397
Orchidaceae, 389
Orchis, Stream, 391

Orcuttia, 177
Beardless, 177
californica, 177
var. inaequalis, 178
var. viscida, 178
Greenei, 177
pilosa, 178
tenuis, 178
Oreastrum elatus, 783
Oregon myrtle, 519
Orthocarpus castillejoides, 727
Oryza sativa, 179
Oxypolis occidentalis, 633

Palmerella debilis, 767, fig. 343
Panicum, 179
barbipulvinatum, 179
capillare, 179
var. *occidentale,* 179
crusgalli, 153
glaucum, 199
occidentale, 181
Parapholis incurva, 181, fig. 76
Parnassia, 542
californica, 542
palustris var. *californica,* 542
Parnassiaceae, 541
Parsley
Button, 634
Water, 631
Paspalum, 183
dilatatum, 183, fig. 78
distichum, 183, fig. 77
Patata, 463
Pectiantia, 537
ovalis, 539
pentandra, 539
Peltiphyllum peltatum, 539, fig. 251
Pennyroyal, 687
Pennywort, Marsh, 627
Peppermint, 685
Petunia parviflora, 696
Phalacroseris Bolanderi, 827
Phalaris, 185, fig. 79
arundinacea, 187, fig. 79
brachystachys, 185
californica, 187
Lemmonii, 185
minor, 185
paradoxa, 185
tuberosa var. stenoptera, 185
Philotria Planchonii, 121
Phleum, 187
alpinum, 189
pratense, 189
Pholiurus incurvatus, 181

Phragmites communis, 189, figs. 80, 81
Phyla
 lanceolata, 675
 nodiflora, 675
Phyllospadix, 91
 Scouleri, 93, fig. 38
 Torreyi, 91, fig. 38
Physalis, 696
 ixocarpa, 697
 lanceifolia, 697
Physocarpus capitatus, 543
Piaropus crassipes, 345
Pickerel-weed family, 343
Pickleweed, 463
 Common, 467
Pigweed, 458
Pilularia americana, 31, fig. 5
Pimpernel, Water, 647
Pinguicula vulgaris, 733
Pink family, 479
Pinkweed, 429
Pitcher plant, 531
Plagiobothrys
 acanthocarpus, 669
 bracteatus, 671
 Cusickii, 670
  var. vallicola, 671
 glyptocarpus, 669
 humistratus, 668
 mollis, 668
 tener, 671
 undulatus, 670
Plane tree, California, 542
Platanaceae, 542
Platanus racemosa, 542
Pleuropogon, 191
 californicus, 191
 Davyi, 191
 refractus, 192
Pluchea, 829
 camphorata, 829
 sericea, 829, fig. 360
Plumbaginaceae, 647
Poa trivialis, 192
Pogogyne, 687
 Abramsii, 689
 Douglasii, 689
 nudiuscula, 689
 zizyphoroides, 689
Polemoniaceae, 658
Polemonium occidentale, 666
Polygonaceae, 413
Polygonum, 413
 acre, 423
 amphibium, 419
 argyrocoleon, 417, fig. 196

aviculare, 415
 bistortoides, 417, fig. 197
 coccineum, 419, fig. 198
 esotericum, 417
 Hydropiper, 425, fig. 201
 hydropiperoides, 425, fig. 202
 lapathifolium, 429, fig. 205
 mexicanum, 427, fig. 204
 Muhlenbergii, 419
 natans, 419, fig. 199
 pennsylvanicum, 429
 Persicaria, 427, fig. 203
 punctatum, 423, fig. 200
Polypodiaceae, 27
Polypogon, 193
 Ditch, 195
 elongatus, 195, fig. 82
 interruptus, 195
 lutosus, 195
 maritimus, 193
 monspeliensis, 193
 Rabbit-foot, 193
Pond lily, Yellow, 489
Pondweed
 Broad-leaved, 77
 Common American, 75
 Family, 49
 Sego, 53
 Western, 55
Pontederiaceae, 343
Poplar, 398
Populus, 398
 acuminata, 398
 angustifolia, 399
 Fremontii, 399
  var. Macdougallii, 399
 Macdougalii, 399
 tremuloides, 399
 trichocarpa, 398
Porterella carnosula, 769, fig. 344
Portulaca oleracea, 479
Portulacaceae, 477
Potamogeton, 49
 alpinus, 71, fig. 26
 americanus, 75
 amplifolius, 75, fig. 27
 Berchtoldii, 71, fig. 25
 compressus, 63
 crispus, 63, fig. 21
 dimorphus, 57
 diversifolius, 57, fig. 17
 epihydrus, 77, fig. 30
 fibrillosus, 67
 filiformis, 53, fig. 13
 foliosus, 67, fig. 23
  var. californicus, 67

gramineus, 81, fig. 31
heterophyllus, 81
illinoensis, 63, fig. 20
interior, 53
latifolius, 55, fig. 15
lucens, 63
natans, 77, fig. 29
nodosus, 75, fig. 28
panormitanus, 69
pectinatus, 53, fig. 14
  var. *latifolius,* 55
perfoliatus var. *Richardsonii,* 61
praelongus, 61, fig. 19
pusillus, 69, fig. 24; *71*
Richardsonii, 61, fig. 18
Robbinsii, 57, fig. 16
vaginatus, 53
zosteriformis, 63, fig. 22
Potamogetonaceae, 49
Potentilla, 545
  Anserina, 546, fig. 252
    var. *grandis,* 545
  Marsh, 545
  millegrana, 546
  pacifica, 545
  palustris, 545
Poverty weed, 823
Primulaceae, 641
Psilocarphus, 829
  brevissimus, 831, fig. 361
    var. *brevissimus,* 831
  globiferus, 831
  oreganus var. *brevissimus,* 831
  oregonus, 831
Psoralea, 552
  macrostachya, 553
  orbicularis, 552
Puccinellia, 197
  distans, 197
  grandis, 197
  Nuttalliana, 197
Purslane
  Family, 477
  Lowland, 476

Quaking aspen, 399
Quillwort
  Family, 33
  Flowering, 101

Radicula
  curvisiliqua, 529
  Nasturtium-aquaticum, 527
  palustris, 529
  sinuata, 527
Ragweed, 778
  Western, 778

Ranunculaceae, 493
Ranunculus, 505
  alismaefolius, 511, fig. 239
  alveolatus, 513
  aquatilis var. capillaceus, 517, fig. 244
  Bloomeri, 509, fig. 237
  Cymbalaria var. saximontanus, 517,
    fig. 242
  delphinifolius, 509
  flabellaris, 509
  flammula var. ovalis, 511, fig. 240
  Gormanii, 513
  hydrocharoides, 511
  Lobbii, 517, fig. 243
  muricatus, 513, fig. 241
  pusillus, 513
  repens, 507, fig. 236
  sceleratus, 509, fig. 238
Redtop, 129
Reed, 189
  Common, 189
  Giant, 137
Rhododendron occidentale, 639
Rice, 179
  Wild, 202
    Northern, 202
Rorippa, 525
  Columbiae, 529
  curvisiliqua, 529
  hispida, 529
  Nasturtium-aquaticum, 527, fig. 246
  obtusa, 531
  palustris, 529
  sinuata, 527
  subumbellata, 529
Rosaceae, 542
Rosemary, Marsh, 647
Rotala, 601
  dentifera, 601, fig. 275
  indica, 602, fig. 276
  ramosior, 602, fig. 276
Rubiaceae, 741
Rubus spectabilis, 546
Rudbeckia, 832
  californica, 832
  hirta, 833
  occidentalis, 833
Rumex, 431
  Acetosella, 433
  californicus, 435
  conglomeratus, 442
  crassus, 437
  crispus, 443, fig. 211
  fenestratus, 442
  fueginus, 447, fig. 212
  lacustris, 439, fig. 210
  obtusifolius, 445

Rumex (*cont.*)
  occidentalis, 442, *442*
  paucifolius, 435, fig. 206
  persicarioides, 447, *447*
  pulcher, 443
  salicifolius, 439, fig. 208; *435, 437, 439, 441*
    var. *denticulatus,* 439, 441
  transitorius, 439, fig. 209
  triangulivalvis, 441
  utahensis, 437
  venosus, 435, fig. 207
  violascens, 445
Ruppia maritima, 81, fig. 32
Ruppiaceae, 81
Rush
  Basket, 351
  Beaked, 301
  Bog, 355
  Common, 355
  Family, 347
  Jointed, 365
  Pointed, 371
  Salt, 349
  Soft, 355
  Spiny, 349
  Three-stamened, 369
  Toad
    Common, 357
    Round-fruited, 359
Rye
  Wild, 153
    Creeping, 153
Rynchospora, 301
  alba, 302
  californica, 302
  capitellata, 302
  globularis, 303
    var. recognita, 303

Sagebrush, 780
Sagina, 484
  apetala, 484
  crassicaulis, 485
  Linnaei, 485
  occidentalis, 484
  procumbens, 484
  saginoides, 485
Sagittaria, 109
  arifolia, 117
    var. *stricta,* 117
  calycina, 111, fig. 47
  cuneata, 117, fig. 50
  Greggii, 115, fig. 49
  longiloba, 115
  latifolia, 117, figs. 51, 52
    var. latifolia, 119

    var. obtusa, 119, fig. 52
  montevidensis subsp. *calycina,* 111
  obtusa, 119
  Sanfordii, 113, fig. 48
  sinensis, 119
Salicaceae, 397
Salicornia, 463
  Bigelovii, 465, fig. 217
  depressa, 465, fig. 219
  europea, 465
  mucronata, 465
  pacifica, 467, fig. 220
  rubra, 465, fig. 218
  subterminalis, 451
Salix, 399
  argophylla, 402
  Breweri, 407
    var. *delnortensis,* 407
  californica, 407
  caudata, 401
  commutata, 405, *407*
  cordata, 405
    var. *Mackenziana,* 405
  Coulteri, 408
  delnortensis, 407
  Eastwoodiae, 407
  exigua, 402
    var. *Parishiana,* 402
  Geyeriana, 403
  glauca
    var. *orestera,* 407
    var. *villosa,* 407
  Gooddingii, 402
  Hindsiana, 402
  Hookeriana, 405
  Jepsonii, 406
  laevigata, 401
  lasiandra, 401
  lasiolepis, 404
  Lemmonii, 406
  lutea, 404
  Mackenziana, 405
  melanopsis, 403
    var. *Bolanderiana,* 403
  monica, 406
  nigra, 402
    var. *vallicola,* 402
  orestera, 407
  Parksiana, 403
  phycifolia var. *monica,* 406
  Piperi, 404
  pseudocordata, 405
  Scouleriana, 408
  sessilifolia var. *Hindsiana,* 402
  sitchensis var. *Coulteri,* 408
  subcoerulea, 406
  Tracyi, 404

Salmonberry, 546
Salsola Kali var. tenuifolia, 467
Salt cedar, 173
Salt tree, 591
Saltbush, 452
  Silver, 456
Salviniaceae, 31
Sambucus, 743
  coerulea, 743
    var. arizonica, 743
  glauca, 743
  racemosa, 743
    var. callicarpa, 745
Samolus floribundus, 647, fig. 292
Sanguisorba microcephala, 546
Sanicula maritima, 633
Saponaria officinalis, 483
Sarcobatus vermiculatus, 469
Sarraceniaceae, 531
Saururaceae, 397
Saxifraga, 539
  arguta, 540
  integrifolia var. *sierrae,* 541
  Marshallii, 540
  Mertensiana, 540
  oregana, 541
Saxifragaceae, 535
Saxifrage
  Bog, 541
  Brook, 540
  Spotted, 540
Scheuchzeria palustris, 95
Scheuchzeriaceae, 95
Schoenolirion, 383
  album, 385, fig. 188
  bracteosum, 385
Schoenus, 303
  capitellatus, 302
  nigricans, 303
Scirpus, 303
  acutus, 323, fig. 157
  americanus, 315, fig. 152
    var. polyphyllus, 315
  bernardinus, 287
  californicus, 319, fig. 156
  carinatus, 319
  cernuus var. californicus, 319, fig.155
  Clementis, 324
  Congdonii, 307, fig. 147
  criniger, 305, fig. 145
  fluviatilis, 307, fig. 148
  heterochaetus, 323
  koilolepis, 319, fig. 155
  mamillata, 297
  microcarpus, 307, fig. 146
  mucronatus, 317, fig. 154
  nanus, 283

  nevadensis, 315, fig. 150
  Olneyi, 317, fig. 153
  paludosus, 309
  pauciflorus, 285
  robustus, 309, fig. 149
  rubiginosus, 323
  setaceus, 315, fig. 151
  tuberosus, 309
  validus, 323
  yosemitanus, 324
Scrophularia californica, 727
Scrophulariaceae, 701
Scutellaria, 690
  Bolanderi, 690
  epilobifolia, 690
  galericulata, 690
  lateriflora, 690
Sedge
  Awned, 255
  Beaked, 261
  Black, 303
  Bristly, 257
  California, 245
  Deceiving, 209
  Dense, 233
  Family, 203
  Field, Clustered, 223
  Green, 249
  Inflated, 261
  Porcupine, 257
  Rough, 217
  San Diego, 255
  Sand-dune, 229
  Short-beaked, 223
  Slough, 209
  Walking, 289
  Woolly, 241
Seep willow, 795
Selinum capitellatum, 634
Senecio, 833
  hydrophilus, 834
  Swamp, 834
  triangularis, 834
Sesbania macrocarpa, 553
Sesuvium sessile, 476
Setaria, 199
  gracilis, 201
  geniculata, 201, fig. 84
  lutescens, 199, fig. 83
  verticillata, 201, fig. 85
Shooting star, 641
Sida hederacea, 573
Sidalcea, 573
  calycosa, 575
  neo-mexicana var. parviflora, 575
  parviflora, 575
  rhizomata, 575, fig. 262

Silverweed, 546
  Pacific, 545
Sisyrinchium, 387
  californicum, 389, fig. 190
  Elmeri, 389
Sium, 633
  cicutaefolium, 633
  suave, 633
Skunk cabbage, Yellow, 325
Smartweed, 413
  Annual, 425
  Giant, 427
  Perennial, 423
  Water, 425
  Willow, 429
Snakeroot, 633
  Button, 623
Snakeweed, 417
Sneezeweed, 815
Solanaceae, 695
Solanum, 697
  Douglasii, 699
  Dulcamara, 699
    nigrum, 701
      var. Douglasii, 699
    nodiflorum, 701, fig. 307
Solidago, 834
  confinis, 835
  elongata, 835
  occidentalis, 835, fig. 362
Sorghum
  halepense, 201
  sudanensis, 202
  vulgare, 202
Sparganiaceae, 43
Sparganium, 43
  affine, 47
  angustifolium, 47
  californicum, 45
  eurycarpum, 45, fig. 10
    var. Greenei, 47, fig. 11
  Greenei, 47
  hyperboreum, 49
  minimum, 49
  simplex, 47, fig. 12
Spartina
  foliosa, 202
  gracilis, 202
Spearmint, 685
Spergularia, 485
  macrotheca, 485, fig. 227
  marina, 487
  rubra, 485
  salina, 487
Sphenosciadium capitellatum, 634
Spicebush, 519

Spike rush, 281
  Common, 297
  Creeping, 297
  Dwarf, 285
Spiranthes, 395
  porrifolia, 397
  Romanzoffiana, 395, fig. 194
Spirodela, 335
  oligorrhiza, 335, fig. 163
  polyrrhiza, 335, fig. 164
Sporobolus asperifolius, 175
Sprangle-top, 169
  Bearded, 171
  Mexican, 171
  Red, 169
Stachys, 691
  ajugoides, 693
    var. rigida, 692
    var. stricta, 692
  albens, 693, fig. 306
    var. juliensis, 693
  Chamissonis, 691
  Emersonii, 691
  palustris subsp. pilosa, 692
  pycnantha, 693
  quercetorum, 692
  rigida, 692
    subsp. lanata, 692
  rivularis, 692
  stricta, 692
Star tulip, Large-flowered, 379
Stellaria, 487
  borealis, 487
  calycantha, 487
  littoralis, 489
  longipes, 489
Stenophyllus capillaris, 300
Sticktight, 799
Suaeda, 469
  californica, 472
    var. pubescens, 472
  depressa, 471
  fruticosa, 472, fig. 221
  minutiflora, 469
  Moquinii, 472
  occidentalis. 471
  ramosissima, 472
  taxifolia, 472
  Torreyana, 47 i
    var. ramosissima, 471
Subularia aquatica, 531
Sundew
  Oblong-leaved, 533
  Round-leaved, 533
Sunflower, 817
  California, 819

Common, 818
Family, 769
Sweet bush, 519
Sweet flag, 324
Sweet-shrub family, 519
Sycamore, California, 542
Syntherisma sanguinale, 145

Tamaricaceae, 591
Tamarisk, 591
  Desert, 592
  Four-petaled, 591
Tamarix, 591
  aphylla, 591
  articulata, 591
  parviflora, 591
  pentandra, 592
  tetrandra, 591
Taraxacum, 837
  officinale, 837
  vulgare, 837
Thistle, 803
  Russian, 467
Three-square, 315
Tillaea, 535
  aquatica, 535, fig. 250
  aquaticum, 535
Timothy, 187, 189
  Alpine, 189
  Swamp, 165
Tinker's penny, 577
Tobacco
  Coyote, 695
  Indian, 696
Tofieldia occidentalis, 385, fig. 189
Tolmiea Menziesii, 541
Tomatillo, 697
Toothwort, 523
Trianthema Portulacastrum, 476
Trichocoronis Wrightii, 837
Trifolium, 553
  Bolanderi, 554
  fimbriatum, 554
  involucratum, 554
  spinulosum, 554
  variegatum, 554
  Wormskjoldii, 554
Triglochin, 95
  concinna, 99, fig. 40
  elata, 97
  maritima, 97, fig. 39
  maritimum var. *debilis,* 99
  striata, 99, fig. 41
Triteleia peduncularis, 379
Tule, 323
Tumbleweed, 467

Twinberry, 742
  Siamese, 743
Typha, 37
  angustifolia, 41, fig. 9
    var. *elongata,* 41
    × *latifolia,* 41
    var. *longispicata,* 41
  domingensis, 39, fig. 7
  elongata, 41
    × glauca, 41
  latifolia, 39, fig. 8
    × *angustifolia,* 41
    var. *elongata,* 41
Typhaceae, 36

Umbelliferae, 619
Umbellularia californica, 519
Urtica, 412
  Breweri, 412
  californica, 412
  gracilis var. *holosericea,* 412
  holosericea, 412
Urticaceae, 411
Utricularia, 733
  fibrosa, 737, fig. 328
  gibba, 737, fig. 327
  intermedia, 735, fig. 325
  minor, 737, fig. 326
  vulgaris, 733, fig. 324

Vaccinium, 639
  caespitosum, 640
  membranaceum, 640
  occidentale, 640
  uliginosum, 640
Veratrum, 385
  californicum, 386
  fimbriatum, 386
Verbena, 676
  bonariensis, 676
  bracteata, 677
  bracteosa, 677
  Family, 675
  hastata, 677
  lasiostachys, 677
  litoralis, 676
  prostrata, 677
  scabra, 677
  urticifolia, 677
Verbenaceae, 675
Verbesina encelioides var. exauriculata, 837
Veronica, 727
  americana, 729, fig. 321
  Anagallis, 731

Veronica (*cont.*)
  Anagallis-aquatica, 731, fig. 322
  Beccabunga, 729
    var. *americana,* 729
  connata subsp. glaberrima, 731
  peregrina, 729, fig. 321
  scutellata, 731, fig. 323
Vine maple, 569
Viola, 592
  blanda var. *Macloskeyi,* 593
  Macloskeyi, 593
  occidentalis, 593
  palustris, 592
Violaceae, 592
Vitaceae, 571
Vitis californica, 571

Wapato, 117
Water carpet, Pacific, 537
Water hemlock, 621
Water hyacinth, 345
Water lentil, 331
Water-lily family, 489
Water nymph
  Common, 87
  Family, 83
  Holly-leaved, 85
  Rice-field, 85
  Slender, 87
Water parsnip
  Cut-leaved, 621
  Hemlock, 633
Water plantain, 103
  Family, 103
  Star, 107
Water wally, 795
Waterweed, 121
Wax myrtle, 408

Willow, 399
  Arroyo, 404
  Dusky, 403
  Red, 401
  Sand-bar, 402
  Seep, 795
  Yellow, 404
  Yellow tree, 401
Willow herb, 604
Wind bags, 331
Wintergreen, 636
Wireweed, Persian, 417
Wolffia, 339
  arrhiza, 341, fig. 165
  columbiana, 341, fig. 165
  cylindracea, 341, fig. 165
Wolffiella lingulata, 343, fig. 166
Woodwardia Chamissoi, 27
Wormwood, 781
Wrack, 93

Xanthium, 838
  canadense, 838
  pensylvanicum, 838
  spinosum, 838

Yarrow, 775
Yerba mansa, 397

Zannichellia palustris, 89, fig. 37
Zannichelliaceae, 89
Zantedeschia aethiopica, 327
Zizania aquatica var. angustifolia, 202
Zostera, 93
  marina, 93
    var. *latifolia,* 93
Zosteraceae, 91
Zygadene, Deadly, 386
Zygadenus veneosus, 386